A Dictionary of

Ecology

THIRD EDITION

Edited by
MICHAEL ALLABY

OXFORD
UNIVERSITY PRESS

OXFORD
UNIVERSITY PRESS

Great Clarendon Street, Oxford OX2 6DP

Oxford University Press is a department of the University of Oxford.
It furthers the University's objective of excellence in research, scholarship,
and education by publishing worldwide in

Oxford New York

Auckland Cape Town Dar es Salaam Hong Kong Karachi
Kuala Lumpur Madrid Melbourne Mexico City Nairobi
New Delhi Shanghai Taipei Toronto

With offices in

Argentina Austria Brazil Chile Czech Republic France Greece
Guatemala Hungary Italy Japan Poland Portugal Singapore
South Korea Switzerland Thailand Turkey Ukraine Vietnam

Oxford is a registered trade mark of Oxford University Press
in the UK and in certain other countries

Published in the United States
by Oxford University Press Inc., New York

First published 1994 as an Oxford University Press paperback
Second edition 1998
Reissued with new covers 2004
Third edition 2005

British Library Cataloguing in Publication Data

Data available

Library of Congress Cataloging in Publication Data

Typeset by SNP Best-set Typesetter Ltd., Hong Kong
Printed in Great Britain by Clays Ltd., St Ives plc

ISBN 0-19-860905-1
ISBN 978-0-19-860905-6

1

Contents

From the Preface to the First Edition

Ecology is a relatively new scientific discipline. The study of plants and animals must have begun with the realization that living organisms are of different kinds. In the modern sense, however, the life sciences can trace their origin to ideas that were current in Greece around 500 BC and which derived in part from the much older study of medicine. Aristotle (384–322 BC) emphasized the careful observation and description of organisms that is usually accepted as laying the foundation of biology, although the word 'biology' was not introduced until 1802, by the German G. R. Treviranus (or Trevirons) (1776–1837), in the title of his *Biologie oder die Philosophie der lebenden Natur*. J. B. de Lamarck may have introduced the word independently at about the same time.

The study of the relationships among organisms and between them and their abiotic environment shares some of this history. Aristotle was an accomplished naturalist and ecology is descended partly from natural history. 'Ecology' was also coined first in German (as *ökologie*), in 1866 by Ernst Haeckel, but interest in the new subject was overwhelmed by the upsurge in professional and popular fascination with evolutionary theory that followed the publication of Darwin's *Origin of Species*—oddly enough, itself essentially an ecological theory. It was not until the end of the century that attention returned to the relationship between organisms and their environment and what we understand today as the independent discipline of ecology is barely a century old.

In addition to students who plan to become ecologists or to enter some other branch of the life sciences, ecology is now a component of courses taken by geographers, architects, planners, land managers, industrial chemists, civil engineers, and probably many more. It is for them, and also for those with a less formal interest in the subject, that this Dictionary has been compiled.

The general subject embraces concepts, and therefore words and expressions, garnered from a wide range of disciplines. In addition to terms pertaining to environmental pollution and conservation and those derived from ecology itself, together with relevant terms from biogeography, animal behaviour, evolutionary theory, and taxonomy, the dictionary also contains terms from plant and animal physiology, climatology and meteorology, oceanography, hydrology, pedology, glaciology, and geomorphology. It also includes brief biographical notes on a number of individuals who have made important contributions to the discipline.

This Dictionary aims to explain concepts and describe processes in sufficient detail to be helpful to the user. This results in a wide variation in the length of entries and, in some cases, the definition of more than one term within an entry (see, for example, the entry for moraine). Such embedded definitions appear in their alphabetical positions as cross-references. Where an entry uses a term defined in its own right, this is indicated by an asterisk before the term (e.g. *till in the entry for moraine). In some cases, words are cross-referenced to entries defined as different parts of speech (an adjective might be cross-referenced to its noun, for example) if the two are so similar as to allow the appropriate entry to be found easily. Other cross-references use the conventional *see*, *see also*, and *compare*. Cross-references are included within entries only where they might be helpful, in order to control what might otherwise have developed as a proliferation of cross-references that made entries more difficult to read. The dictionary defines a total of about 5000 terms.

MICHAEL ALLABY
Wadebridge, Cornwall

Preface to the Second Edition

It is now four years since the first edition of this Dictionary appeared. Like all young disciplines, ecology is developing very rapidly, and during these years new terms have been introduced and the meaning of others modified. This has made a new edition desirable and I have taken the opportunity to make many revisions. In doing so I have been greatly helped by Robin Allaby, who has scrutinized my revisions and additions and contributed many of his own.

In particular, the Dictionary now includes a small number of entries pertaining to molecular ecology, a subdiscipline which barely existed at all when the first edition was compiled. Its existence and growing importance mark the widening influence molecular biology exerts throughout the life sciences, based on its power to illuminate evolutionary processes and taxonomic relationships at the most fundamental level. Its necessary inclusion here may also amount to a comment on the modern tendency to group the life sciences under two headings: molecules and cells; and organisms and communities. That ecologists are now required to possess some understanding of genetics demonstrates that, in practice, the two groups are inseparable.

Other entries appearing here for the first time reflect the large amount that has been learned in recent years about ecological relationships and, most especially, about conservation and the management of habitats. Not all these terms are new. Some were omitted from the first edition by oversight or because they did not seem to be in such wide use then as they do now.

All the entries in the first edition have been examined afresh and many have been modified. The modifications are often small, but they bring the definitions up to date, in some cases by providing a little more detail.

Despite the changes, the character of this Dictionary was formed with the help of those who contributed to the compilation of the first edition: Robin Allaby, Kenneth Gregory, Tim Whitmore, and Kenneth Mellanby. That, and their contribution to it, remain unaltered.

It is with great sadness that I must record the death of Kenneth Mellanby, just before Christmas, 1993. A leading world authority on environmental pollution as well as ecology, Kenneth was a wise and thoughtful adviser. He was also a dear friend.

MICHAEL ALLABY
Tignabruaich, Argyll
www.michaelallaby.com

Preface to the Third Edition

Six years have elapsed since the second edition of this Dictionary appeared and although it has been reprinted from time to time, affording opportunities to make very minor alterations, the preparation of a new edition has allowed me to make much more extensive revisions. Every entry has been re-examined and amended wherever that seemed necessary or helpful. Many new entries have been added and some earlier entries have been rewritten to bring them up to date and, in certain cases, to clarify them. I have also added more illustrations where these contribute materially to the explanations contained in the entries.

In preparing the new edition I have been greatly helped by Dr Peter D. Moore, of King's College, University of London. He has scrutinized all the entries, reading the Dictionary from cover to cover, and has made very many helpful suggestions for additions and clarifications. I am most grateful to him.

Despite the many changes, the underlying character of the Dictionary was formed with the help of those who contributed to the first and second editions: Robin Allaby, Kenneth Gregory, Tim Whitmore, and Kenneth Mellanby. That remains unchanged and my debt to them remains undiminished.

MICHAEL ALLABY
Tighnabruaich, Argyll
www.michaelallaby.com

Michael Allaby has written many books on environmental science and especially on climatology and meteorology. These include *Encyclopedia of Weather and Climate*; *The Facts On File Weather and Climate Handbook*; and the *DK Guide to Weather*. He is the General Editor of the Oxford Dictionaries of Ecology, Zoology, and Plant Sciences and co-author of the Dictionary of Earth Sciences.

a The usual abbreviation for *year (from the Latin *annus*, meaning year).

AAC (Antarctic convergence) *See* ANTARCTIC POLAR FRONT.

Aapa mires *Fens, sometimes called 'string bogs', of circumpolar distribution, found in Fennoscandia (*see* BALTIC SHIELD) and *boreal Canada. They are *soligenous mires with ridges arranged roughly normal to the slope, along the contours of the terrain. Water occupies the linear depressions between the ridges.

abaptation The process by which an organism is fitted to its environment as a consequence of the characters it inherits, which have been filtered by *natural selection in previous environments. Because present environments seldom differ greatly from recent past environments, abaptive fitness can resemble *adaptation. In this sense, however, adaptation appears to imply advance planning, or design, which is misleading.

abaxial Of a plant organ, facing away from the axis or main stem (e.g. the lower surface of leaves). *Compare* ADAXIAL.

abiogenesis The development of living organisms from non-living matter, as in the supposed origin of life on Earth, or in the concept of spontaneous generation which was once held to account for the origin of life but which modern understanding of evolutionary processes (*see* EVOLUTION) has made outdated.

abiotic Non-living; devoid of life. *Compare* BIOTIC.

ablation **1.** Removal of snow and ice by melting and by direct alteration from the solid to the gaseous phase (sublimation). The rate of loss is controlled chiefly by air temperature, wind velocity, *humidity, rainfall, and solar radiation. Ablation on snowfields is also influenced by aspect, depth of snow, and the nature of the underlying surface. Ablation *till is the glacial deposit that may be released. The ablation zone of a glacier is that area in which losses, including *calving, exceed additions. **2.** The removal of rock material, especially by wind action.

ablation till *See* ABLATION (1) and TILL.

abrasion (corrasion) The erosive (*see* EROSION) action that occurs when rock particles of varying sizes are dragged over or hurled against a surface. Some common agents of abrasion are the *bed load of streams, rock debris embedded in the bases of *glaciers, and *sand and *shingle transported by wind or waves.

abscission The rejection of plant organs (e.g. of leaves in autumn). This occurs at an abscission zone, where hydrolytic enzymes reduce cell adhesion. The process can be promoted by abscisic acid and inhibited by respiratory poisons, and is controlled in nature by the proportion and gradients of auxin and ethylene. Other hormones may be involved.

absent rings *See* TREE RING.

absolute age (true age) The age of a geologic phenomenon measured in present Earth years, rather than its age relative to other geologic phenomena (*compare* RELATIVE AGE). The term 'absolute age' has been considered rather misleading, as the means for measuring ages (*radiometric dating, *dendrochronology, *varve analysis) are subject to experimental error and the dates obtained are not precise. The alternative term *'apparent age' has been suggested, but nowadays it is usual to link a statement of age with the technique used to determine it, to distinguish that age from an age in calendar or solar years (e.g. to state an age in 'radiocarbon years'). *See also* DATING METHODS.

absolute dating *See* GEOCHRONOLOGY.

absolute humidity *See* HUMIDITY.

absolute pollen frequency (APF) The expression of *pollen data from sedi-

a

ments, expressed in terms of the absolute numbers (for each species, genus, or family) per unit of area of surface and, where deposition rates are known, per unit of time. In certain circumstances this approach gives clearer information than does the traditional way of expressing pollen data as *relative pollen frequencies (RPF). APFs are particularly useful in site comparisons in which one or more high pollen producers vary. For example, when trees first appear in the regional pollen rain their prolific pollen may, in an RPF method, give the impression of declining herbaceous species, whereas examination by an APF method will show constant values for herb species.

absolute porosity *See* POROSITY.

absorption The uptake of substances, usually nutrients, water, or light, by cells or tissues.

absorption spectrum A graph that shows the percentage of each wavelength of light absorbed by a pigment (e.g. chlorophyll, which absorbs mainly in the red and blue parts of the spectrum).

abstraction (extraction) The artificial removal of water from a well, reservoir, or river.

ABW *See* ANTARCTIC BOTTOM WATER.

abyssal Applied to the deepest part of the ocean, at depths between about 2 000 m and 6 000 m (*see* OCEAN DIVISIONS). The abyssal zone lies seaward of, and deeper than, the *bathyal zone, and covers approximately 75 per cent of the ocean floor, making it the most extensive environment on Earth. Conditions there are cold (about 4°C), dark, and the pressure ranges from approximately 20 MPa to 100 MPa. Currents are slow-moving (less than a few centimetres per second). Organisms inhabiting the abyssal zone are typically black or grey, delicately structured, and not streamlined. *Compare* NERITIC ZONE.

abyssal fish Fish that live in the deepest part of the ocean, below about 2 000 m. Many abyssal species have prominent snouts; a tapering, rat-tailed body consisting of flabby, watery tissue; and a lightweight skeleton.

abyssalpelagic zone The part of the *pelagic zone of the ocean that overlies the *abyssal plain. It extends from a depth of between 2 000 m and 4 000 m to about 6 000 m. The temperature is about 4°C and the pressure 20–60 MPa. *See* OCEAN DIVISIONS.

acceleration **1.** *Evolution that occurs by increasing the rate of ontogenetic (*see* ONTOGENY) development, so that further stages can be added before growth is completed. This form of *heterochrony was proposed by E. H. *Haeckel as one of the principal modes of evolution. **2.** Change of speed or direction of a moving body.

accelerator mass spectrometry (AMS) A technique for *radiocarbon dating in which a cyclotron is used to measure directly the number of ^{14}C atoms or the proportion of ^{14}C to ^{12}C and ^{13}C atoms in the sample. Results are obtained much more quickly than by alternative methods and can be used on much smaller samples of material.

accessory pigments In *photosynthesis, pigments that can absorb light energy and pass the electrons they emit to primary pigments.

accidental species One of five classes of fidelity used by the *Braun-Blanquet school of phytosociology in the description and classification of plant communities. Accidentals are rare species in the community, present either as chance invaders from another community or as *relicts from a previous community. *Compare* EXCLUSIVE SPECIES; INDIFFERENT SPECIES; PREFERENTIAL SPECIES; and SELECTIVE SPECIES.

acclimation **1.** A response by an animal or plant that enables it to tolerate a change in a single factor (e.g. temperature) in its environment. The term is applied most commonly to animals used in laboratory experiments and implies a change in only one factor. *Compare* ACCLIMATIZATION (1). **2.** *See* ACCLIMATIZATION (2).

acclimatization **1.** A reversible, adaptive response that enables animals or plants to tolerate environmental change (e.g. seasonal climatic change) involving several factors (e.g. temperature and availability of food). The response is physiological, but may affect behaviour (e.g. when an animal responds physiologically to falling temperature in ways that make *hibernation possible, and behaviourally by seeking a nesting site, nesting materials, and food). *Compare* ACCLIMATION (1). **2. (acclimation, hardening)** The changes involving the synthesis of proteins, membranes, and metabolites that occur in a plant in response to chilling or freezing temperatures, which protect tissues or confer tolerance of the cold. The term may also be applied to a range of physiological adjustments which occur in a plant when it is subjected to unusual environmental conditions.

accommodation (fatigue, synaptic accommodation) The exhaustion of a neurotransmitter at the synapse when a stimulus is repeated frequently. This may result in a decrease in behavioural responsiveness.

accretion **1.** The process by which an inorganic body grows in size by the addition of new particles to its exterior. **2.** The accumulation of sediments from any cause, representing an excess of deposition over *erosion. **3.** The addition of material to the edge of a continent, thus enlarging it.

accumulated temperature The sum, counted in degrees, by which the actual air temperature rises above or falls below a *datum level over a prolonged period. The datum level is usually set at a value relevant to an ecological study or to crop production (e.g. the critical temperature for sustained plant growth is 6°C). If the mean temperature on a particular day is $m°$ above or below ($-m°$) the datum level and remains there for n hours ($= n/24$ days), then the accumulated temperature for that day is $mn/24$ degree-days. Degree-days can be added to give the accumulated temperature over a week, month, season, or year.

accumulation zone That part of a *glacier where the mean annual gain of ice, *firn, and snow is greater than the mean annual loss. The zone consists of stratified firn and snow together with ice from frozen melt water. Its lower boundary is the *equilibrium line.

accumulator **1.** In plant *succession studies, a *pioneer plant species whose activities are claimed to enrich the *abiotic environment with nutrients. **2.** A plant that accumulates certain elements, such as selenium.

-aceae A standardized suffix used to indicate a family of plants in the recognized codes of classification (e.g. Rosaceae, the rose family).

acephalous Lacking a distinct head.

acervulus An asexual structure, bearing conidia (i.e. thin-walled, asexual spores), that is formed by certain fungi parasitic (*see* PARASITISM) in plants. It consists of a mat of fungal tissue which bears a layer of conidiophores (i.e. specialized hypha that bear conidia); initially formed within the plant tissues, it later breaks through to the surface to release conidia.

achene A small, usually single-seeded, dry, *indehiscent *fruit.

acicular Pointed or needle-shaped.

acid According to the Brønsted–Lowry theory (proposed in 1923 by both J. N. Brønsted in Copenhagen and T. M. Lowry in Cambridge, who were working independently of one another), a substance that in solution liberates hydrogen ions or protons. The Lewis theory (proposed by G. N. Lewis, also in 1923) states that it is a substance that acts as an electron-pair acceptor. An acid reacts with a base to give a salt and water (neutralization), and has a *pH of less than 7.

acidic dye A dye which consists of an organic *anion that combines with and stains positively charged macromolecules. It is used particularly for staining cytoplasm. *Compare* BASIC DYE.

acidic grassland A grassland that occurs on *acid soil: it is usually derived from former woodland as a consequence of centuries of grazing and, to a lesser extent, burning. In Britain and much of north-western Europe the dominant grasses are species of *Agrostis* (bent) and *Festuca* (fescue). This type of vegetation is most

a

extensive in upland areas, but the associated plant species tend to be different, and the name 'grass heath' is considered more appropriate. In North America broomsedge (*Andropogon virginicus*), Elliott's broomsedge (*A. elliotti*), supina bluegrass (*Poa supina*), and Canada bluegrass (*P. compressa*) are typical grasses of acidic parts of the prairie.

acidophile An *extremophile (*see* ARCHAEA) that thrives in environments where the pH is below 5.0.

acidophilic **1.** Refers to the propensity of a cell, its components, or its products to become stained by an *acidic dye. **2.** *See* ACIDOPHILOUS.

acidophilous **(acidophilic)** Applied to 'acid-loving' organisms (i.e. organisms which grow best in acidic habitats).

acid precipitation *Precipitation that has a *pH lower than about 5.0, which is the value produced when naturally occurring carbon dioxide, sulphate, and nitrogen oxides dissolve into water droplets in clouds. The increased acidity may be caused naturally (e.g. by gases and *aerosols ejected by a volcanic eruption) or by *anthropogenic emissions (e.g. from the burning of fuel). The effects of acid precipitation on vegetation, soils, and surface waters are complex, their severity depending on the form of deposition (e.g. acid rain washes rapidly from plant surfaces, but may affect soil; acid mist tends to coat leaves, making it more harmful than rain), and the pH and natural buffering of the soil and water into which it falls. *See* BUFFER.

acid rain *See* ACID PRECIPITATION.

acid rock *Igneous rock that contains more than about 60 per cent silica (SiO_2) by weight, most of the silica being in the form of silicate minerals, but with the excess of about 10 per cent as free quartz. Typical acid rocks are granites, granodiorites, and rhyolites. *See also* ALKALINE ROCK; *compare* BASIC ROCK and INTERMEDIATE ROCK.

acid soil *Soil that has a *pH less than 7.0. Degrees of soil acidity are recognized. Soil is regarded as 'very acid' when the reaction is less than pH 5.0. The *USDA lists five standard ranges of soil acidity (less than pH 4.5, extremely acid; 4.5–5.0, very strongly acid; 5.1–5.5, strongly acid; 5.6–6.0, medium acid; and 6.1–6.5 slightly acid). Surface *soil horizons of acid *brown earths have a reaction of pH 5.0 or less.

acme zone **(peak zone, flood zone, epibole)** Informally, a body of strata that contains the maximum abundance of a particular fossil taxon occurring within the stratigraphic range of that taxon, and after which the zone is named.

acquired characteristics Characteristics that are acquired by an organism during its lifetime, according to early evolutionary theorists (e.g. *Lamarck and *Buffon). Lamarck further suggested a kind of directional, non-random evolution, in that *traits acquired in one generation as a result of behavioural changes in response to environmental stimuli would be transmitted to the *gametes and inherited by the next generation. Thus, over several generations, a particular type of organism would become better adapted to its environment. The kinds of acquisition envisaged by Lamarck and their heritability are now discredited, although *organic selection simulates them.

acre A unit of area, the British statute acre being equal to 0.4047 ha (4 840 square yards or 10 square *chains). Originally, unenclosed land and later enclosed, cultivated land; as a measure, probably the area that could be ploughed in a day by a yoke of oxen (and therefore varying according to the type of soil). In Britain, historically a woodland acre varied from forest to forest, but was always larger than the statute measure. *Assarts were surveyed at the statute measure.

Acrisols *Acid soils that have an *argic horizon with a *cation-exchange capacity lower than 24 $cmol_c$/kg. Acrisols are a reference soil group in the *FAO *soil classification.

acrocarpous moss A type of moss in which the stems are erect and in which the archegonia (i.e. female sex organs), and hence the *capsules, are borne at the tips of stems or branches. Acrocarpous mosses may branch extensively; once they have fruited, branches take over the erect growth. *Compare* PLEUROCARPOUS MOSS.

acropetal Growing or developing upwards from the base or point of attachment, so that the oldest parts are at the base and the youngest are at the tip. *Compare* BASIPETAL.

acropetal movement The movement of substances within the plant toward its root and shoot apices. *Compare* BASIPETAL MOVEMENT.

acrotelm The upper layer of a peat *bog, in which organic matter decomposes *aerobically and much more rapidly than in the underlying, *anaerobic catotelm. As litter accumulates at the surface the size of the catotelm increases, because the thickness of the acrotelm is limited to depth at which aerobic respiration can occur. Although the rate of decomposition per unit volume of material is much greater in the acrotelm than in the catotelm, a point is reached at which the difference in volume between the two layers is such that the total rate of decomposition in the catotelm is equal to that in the acrotelm. This limits the thickness to which the bog can grow. Should there be a climate change (e.g. an increase in precipitation) growth can resume. Bogs therefore preserve a record of climatic conditions. Most water movement occurs within this layer, because the hydraulic conductivity of the acrotelm is about 100 times that of the catotelm.

actinium series *See* DECAY SERIES.

Actinobacteria A phylum of *Bacteria comprising organisms, many of which form mycelia (*see* MYCELIUM). Most live in soil and some fix nitrogen (*see* NITROGEN FIXATION), forming *root nodules on certain non-leguminous plants. Because they form mycelia they were formerly mistaken for fungi and known as actinomycetes.

actino- A prefix that means 'radiating', derived from the Greek *aktis, -inos*, meaning 'ray'.

actinomorphic Radially symmetrical, as is a daisy flower.

actinomycete *See* ACTINOBACTERIA.

actinomycosis Any disease of humans or other animals in which the causal agent is an *actinomycete.

actinorrhiza *See* ROOT NODULE.

action spectrum A graph of the efficiency of different wavelengths of light in promoting a given photoresponse, as in *photosynthesis or heliotropism.

activation energy **(energy of activation)** The energy that must be delivered to a system in order to increase the incidence within it of reactive molecules, thus initiating a reaction. It is an important feature of *enzymes that they greatly lower the activation energy of many metabolic reactions.

activator A metal ion that functions in conjunction with either an *enzyme or its *substrate in order to bring about a reaction.

active chamaephyte *See* CHAMAEPHYTE.

active dispersal *See* DISPERSAL.

active immunity Resistance to a disease that is acquired by an animal as the result of the production of *antibodies in response to *antigens produced by the disease organism while inside the host animal.

active layer A seasonally thawed surface layer, between a few centimetres and about 3 m thick, that lies above the permanently frozen ground in a *periglacial environment. It may be subject to considerable expansion on freezing because of the water contained, and this is especially the case if silt-sized particles dominate, with important engineering implications. *See also* MOLLISOLS and PERMAFROST.

active pool That part of a *biogeochemical cycle in which the nutrient element under consideration is exchanged rapidly between the *biotic and *abiotic components. Usually the active pool is smaller than the *reservoir pool, and it is sometimes referred to as the 'exchange' or 'cycling' pool.

activity Broadly, the rate or extent of a change associated with some substance or system. For example, it may be the tendency of a metal high in the electromotive series to replace another metal lower in the series (e.g. magnesium displacing copper from most of its compounds). It may also be used to describe the rate of decay of atoms by radioactivity.

actual evapotranspiration **(*AE*)** The amount of water that evaporates from the surface and is transpired by plants if the total amount of water is limited. *Compare* POTENTIAL EVAPOTRANSPIRATION.

actualism The theory that present-day processes provide a sufficient explanation for past geomorphological phenomena, although the rate of activity of these processes may have varied. The theory was first clearly expressed in 1749 by G. L. L. de *Buffon, who proposed an age for the Earth of about 75 000 years, and was developed by James Hutton (*see* UNIFORMITARIANISM). As 'actualism', it came to be generally accepted as a result of the much more convincing arguments advanced by Sir Charles Lyell (1797–1875) in his *Principles of Geology* (published 1830).

aculeate Prickly, pointed. The term is applied to organisms that are armed with a sting (e.g. insects of the hymenopteran division Aculeata, which have stings). The word is derived from the Latin *aculeatus*, meaning 'stinging', from *acus*, needle.

acuminate Tapering to a point.

acute **(of disease)** Applied to a disease that develops rapidly and is of short duration; symptoms tend to be severe.

adaptation **1.** Generally, the adjustments that occur in animals in respect of their environments. The adjustments may occur by *natural selection, as individuals with favourable genetic traits breed more prolifically than those lacking these traits (genotypic adaptation), or they may involve non-genetic changes in individuals, such as physiological modification (e.g. *acclimatization) or behavioural changes (phenotypic adaptation). *Compare* ABAPTATION. **2. (evol.)** That which fits an organism both generally and specifically to exploit a given *adaptive zone (e.g. wings allow birds to fly, whereas the hooked beak and sharp talons of birds of prey are more specialized adaptations well suited to a predatory way of life). The word also implies that the feature has survived because it assists its possessor in its existing *niche. *Compare* EXAPTATION. **3.** A decrease over time of the frequency of the impulses leaving a sensory receptor when a stimulus is repeated frequently. *See* ACCOMMODATION and HABITUATION.

adaptedness The condition of being adapted, as distinct from the process of *adaptation.

adaptive breakthrough Evolutionary change by the acquisition of a distinctive *adaptation that permits a population or taxon to move from one *adaptive zone to another. At the most extreme such moves might be from water to land or from land to air.

adaptive pathway A series of small adaptive steps, rather than a single large one, which leads from one *adaptive zone across an environmental and *adaptive threshold into another adaptive zone. In effect, small changes accumulate so that the organism is virtually pre-adapted (*see* PRE-ADAPTATION) to enter the new zone.

adaptive peaks and valleys Features on a symbolic contour map that shows the *adaptive value of genotypic combinations. Such a map will usually display adaptive peaks and valleys occurring at points where the adaptive value is relatively strong or weak. The population of a given *species will therefore be distributed more densely at the adaptive peaks and more sparsely at the valleys.

adaptive radiation **1.** A burst of evolution, with rapid divergence from a single ancestral form, that results from the exploitation of an array of *habitats. The term is applied at many taxonomic levels (e.g. the radiation of the mammals at the base of the *Cenozoic is of ordinal status, whereas the radiation of *'Darwin's finches' in the *Galápagos Islands resulted in a proliferation of species). **2.** A term used synonymously with *cladogenesis by some authors. (Such a use is misleading, because cladogenesis involves the division of one species into two and adaptive radiation imposes no such constraint.)

adaptive threshold The limit imposed on the *adaptation of a taxon within a particular *adaptive zone (e.g. the fins of a flying fish are adapted for gliding flight, but the fish can remain airborne for only a limited time, because it is incapable of powered flight).

adaptive type A population or *taxon that has distinctive adaptive attributes, expressed as a particular morphological theme, characteristic of a particular *habitat or mode of life. In evolutionary terms, the appearance of a new adaptive type is frequently followed by radiations that yield variants; these partition the environment and exploit it more effectively.

adaptive value (Darwinian fitness, fitness, selective value) The balance of genetic advantages and disadvantages that determines the ability of an individual organism (or *genotype) to survive and reproduce in a given environment. The environment, and the competition or struggle for survival within it, determine which individuals are fittest to achieve this, the 'fittest' being the individual (or genotype) that produces the largest number of offspring that later reach reproductive maturity. Such *natural selection has been described as 'the survival of the fittest'. *See also* SELECTION.

adaptive zone The abstract area that a taxon occupies by virtue of its *adaptations, when it is considered together with its associated environmental regime(s), *habitat, or *niche. The adaptive specialization that fits the taxon to its environment, and hence the adaptive zone, may be narrow (as with the giant panda, which eats only certain types of bamboo shoots) or broad (as with the brown bear, which is omnivorous).

adaxial Of a plant organ, facing towards the axis or main stem (e.g. the upper surface of leaves). *Compare* ABAXIAL.

additive genetic variance *See* HERITABILITY.

additive tree A phylogenetic (*see* PHYLOGENY) tree in which the genetic distance between any two *nodes is proportional to the *branch lengths connecting them.

adenine A *purine base found in *nucleic acids. *See also* CYTOSINE, GUANINE, and THYMINE.

adiabatic Applied to the changes in temperature, pressure, and volume in a *parcel of air that occur as a consequence of the vertical movement of the air, and without any exchange of energy with the surrounding air. *See also* DRY ADIABATIC LAPSE RATE and SATURATED ADIABATIC LAPSE RATE.

adnate Applied to two organs that are attached to one another by most or all of their width (e.g. the gill and stipe of an agaric fungus or the leaf and stem of a plant). *Compare* ADNEXED.

adnexed Applied to two organs that are attached to one another by only part of their width. *Compare* ADNATE.

adobe A silty *clay, often calcareous, that is found in dry, desert-lake basins. This fine-grained sediment is usually deposited by desert floods which have eroded wind-blown *loess deposits. It is widely used locally as a building material. The term is of Spanish origin.

adpressed Pressed close (e.g. as the leaves of heather (*Calluna vulgaris*) are pressed close to the stem of the plant).

adret Applied to the south-facing slopes of Alpine valleys, which receive a high degree of insolation as a result of their aspect. Cultivation and settlement patterns often favour such slopes. *Compare* UBAC.

adsorption The physical binding of a particle of a particular substance to the surface of another by adhesion or penetration. In soils it is the attachment of an ion, molecule, or compound to the charged surface of a particle, usually of *clay or *humus, where it may be subsequently replaced or exchanged. Ions carrying positive charges (e.g. those of calcium, magnesium, sodium, and potassium) become attached to, or adsorbed by, negatively charged surfaces (e.g. those of clay or humus).

adsorption complex Various materials of the soil, mainly *clay and *humus and to a lesser degree other particles, that

a

are capable of adsorbing ions and molecules.

advection The horizontal transfer of heat by means of a moving gas (usually air).

advection fog *Fog that forms when relatively warm, moist air moves horizontally across cold ground at 3–9 m/s^{-1}.

adventitious Growing from an unusual position (e.g. roots from a leaf or stem).

advertisement A form of *display in which an individual makes itself as conspicuous as possible. It is used most commonly by male animals holding a *territory, in order to ward off rivals and to attract females. The petals of a flower can also be regarded as a form of advertisement.

AE *See* ACTUAL EVAPOTRANSPIRATION.

-ae A standardized suffix used to indicate a class of plants in the recognized codes of classification.

aeolian **(eolian)** Wind-borne; applied to the processes of erosion, transport, and deposition of material due to the action of the wind at or near the Earth's surface. Aeolian processes are at their most effective when the vegetation cover is discontinuous or absent.

aeolianite Generally, the sedimentary products of wind (aeolian) deposition, which are cemented to make a rock.

aerial mycelium The part of a *mycelium that is held aloft from the *substrate.

aerobe *See* AEROBIC (2).

aerobic **1.** Of an environment: one in which oxygen is present. **2.** Of an organism: one requiring the presence of oxygen for its existence (i.e. an aerobe). **3.** Of a process: one that occurs only in the presence of oxygen.

aerobic fermentation Cellular respiration that requires oxygen. The process involves glycolysis, the citric-acid cycle, and the respiratory chain.

aerochory *See* ANEMOCHORY.

aerodynamic method A method for monitoring rates of *primary productivity; it is particularly useful in studies of woodland or forest communities. The carbon-dioxide concentration from ground level to above canopy height is measured at regular intervals or continuously by sensors attached to a mast. Differences in carbon-dioxide concentration reflect differences in rates of *photosynthesis and respiration and hence in productivity. The chief advantages of this approach are that it is non-destructive and causes minimal disturbance to the environment of the vegetation being studied. Other non-destructive methods usually enclose the vegetation (e.g. in polythene or plexiglass chambers), which might make results unrepresentative of normal field conditions. *Compare* CHLOROPHYLL METHOD; GAS-EXCHANGE METHOD; and HARVEST METHOD. *See* FLUX STUDY.

aerodynamic roughness An uneven flow of air caused by irregularities in the surface (which may be that of a solid or of air of different density) over which the flow takes place.

aerogenic Applied to *Bacteria that can produce gas during the metabolism of certain types of substrate.

aerological diagram A diagram that demonstrates variations with height of the physical characteristics of the atmosphere, particularly its temperature, pressure, and humidity.

aeropalynology The study of *pollen grains and *spores in the atmosphere. This is important for allergy studies (e.g. of hay fever), and for understanding the spread of diseases in humans, other animals, and agricultural plants; and recently for *palaeoecologists attempting to improve the interpretation of microfossil data.

aerosol A colloidal substance, either natural or man-made, that is suspended in the air because the small size (0.01–10 µm) of its particles makes them fall slowly. Aerosols in the *troposphere are usually removed by precipitation and their *residence time is measured in days or weeks. Aerosols that are carried into the *stratosphere usually remain there much longer. Tropospheric aerosols may act as *Aitken nuclei but the general effect of aerosols is to absorb, reflect, or scatter radiation. Stratospheric aerosols, mainly sulphate particles resulting from volcanic eruptions, may reduce *insolation significantly. About 30 per cent of tropospheric dust particles are the result of human activities. *See* ATMOSPHERIC STRUCTURE; MIE SCATTERING; RAYLEIGH SCATTERING; and VOLCANIC DUST.

aerotaxis A change in direction of locomotion, in a *motile organism or cell, made in response to a change in the concentration of oxygen in its immediate environment.

aerotolerant Applied to organisms (usually bacteria) that, although normally *anaerobic, are not killed by the presence of air.

aesthetic injury level *See* ECONOMIC INJURY LEVEL.

aestival In the early summer. The term is used with reference to the six-part division of the year used by ecologists, especially in relation to the study of terrestrial and freshwater communities. The full series comprises: prevernal (in early spring); vernal (in late spring); aestival; serotinal (in late summer); autumnal (in the autumn); and hibernal (in the winter).

aestivation (estivation) 1. Dormancy or sluggishness that occurs in some animals (e.g. snails and hagfish) during a period when conditions are hot and dry. Aestivation is analogous to *hibernation in cold environments and normally lasts the length of the dry period or season. **2.** The arrangement of sepals and petals in a flower bud before opening.

aethalium In certain acellular slime moulds (Myxomycetes), a large *fruit body consisting of masses of *spores contained within a surface crust; it may be stalked or *sessile.

affinity index A measure of the relative similarity in composition of two samples. For example, $A = c/\sqrt{(a + b)}$, where A is the affinity index, a and b are the numbers of species in one sample but not in the other and vice versa respectively, and c is the number of species common to both. The reciprocal $\sqrt{(a + b)}/c$, indicates the ecological and phytosociological distance (D) between samples.

afforestation 1. The establishment of forest by natural *succession or by the planting of trees on land where they did not grow formerly. **2.** Historically, the act of defining an area of land that henceforth would become subject to *forest laws.

African–Indian desert floral region Part of the African subkingdom of R. Good's (1974, *The Geography of the Flowering Plants*) palaeotropical kingdom. The flora is relatively poor and specialized, the number of endemic (*see* ENDEMISM) genera probably just exceeding 50. *Phoenix dactylifera* (the date palm) belongs to this region, but, not surprisingly perhaps, there are no garden plants. *See also* FLORAL PROVINCE and FLORISTIC REGION.

African rain forest Rain forest (*tropical rain forest) occurs in perhumid climates in tropical lowlands where rainfall is heavy and where there is little or no dry season. Floristically, three main regions of the rain forest are recognized: the African, *American, and *Indo-Malesian. These regions share few species and genera in common.

Afro-alpine vegetation On the highest mountains of Africa, above the dwarf or *elfin woodlands, are found shrublands and grasslands. These are alpine (*alpine zone) in character (hence the term 'Afro-alpine') and, besides containing plants found at lower altitudes in middle and high latitudes, they also include endemics (*see* ENDEMISM), e.g. the giant lobelias and groundsels on Ruwenzori and Kilimanjaro in Kenya.

Afro-Tethyan mammal region A zoogeographic region which appears to reflect mammal distributions more satisfactorily than the traditional zoogeographic regions. It includes the *Ethiopian and *Oriental faunal regions, as well as the Mediterranean subregion.

Aftonian The earliest (1.3–0.9 Ma) of four *interglacial stages in North America, following the Nebraskan glacial

a

episode, and approximately equivalent to the *Donau/Günz Interglacial of Alpine terminology. Climatically it was marked by mild summers and winters warmer than those in present-day North America.

agamic generation (asexual generation) In the life cycle of some species (e.g. many members of the insect family Cynipidae) a generation that comprises parthenogenetically reproducing females that are genetically *diploid. The unfertilized eggs laid may be male (having undergone *meiosis) or female (not having undergone meiosis). In Cynipidae, agamic females usually emerge from stout galls designed to overwinter or to survive a period when resources are scarce.

agamospermy The formation of seeds without fertilization. *See also* APOMIXIS.

agar A complex polysaccharide obtained from certain types of seaweed. When heated with water and subsequently cooled to about 45°C, agar readily forms a gel (jelly); an agar gel supplemented with nutrients is used widely as a medium for the culture of bacteria and other micro-organisms.

agaric **1.** Any fungus that forms a mushroom-like fruit body (i.e. one which is umbrella-shaped and bears gills). **2.** Any fungus belonging to the order Agaricales or, more specifically, to the family Agaricaceae.

agaricide A natural or synthetic chemical substance that is toxic to *agarics.

agarophyte A seaweed from which *agar can be obtained.

Agassiz, Jean Louis Rodolphe (1807–73) A Swiss geologist who worked initially on fossil fish, Agassiz is better known for his *glacial theory (1837). He met the geologist William Buckland (1784–1856) in 1840, and persuaded him that *drift deposits in Britain were evidence of a glacial epoch. In 1846 he moved to the USA to become professor of zoology and geology at Harvard, where he founded the Museum of Comparative Zoology (1859).

age **1.** The interval of geologic time equivalent to the chronostratigraphic unit 'stage'. Ages are subdivisions of epochs and may themselves be subdivided into chrons. An age takes its name from the corresponding stage, so like the stage name it carries the suffix '-ian' (or sometimes '-an'); the term 'age' is often capitalized when used in this formal sense, e.g. 'of Oxfordian Age'. **2.** An informal term to denote a time span marked by some specific feature (e.g. 'Villefranchian mammalian age').

age-and-area hypothesis The idea (suggested by J. C. Willis in 1922) that, all other things being equal, the area occupied by a taxon is directly proportional to the age of that taxon. Thus in a polytypic genus, the species with the smallest area of distribution would be the youngest in the genus. However, other things rarely are equal, and the idea has never gained acceptance as a law or rule.

ageostrophic wind A wind above the *boundary layer that blows at a different speed from that of the *geostrophic wind predicted by the *pressure-gradient force.

age polyethism *See* POLYETHISM.

age structure Within a population, the number or proportion of individuals in each age group (e.g., in a human population, the number of individuals aged 0–4, 5–14, 15–29, 30–44, 45–59, 60–69, over 70). Knowledge of the age structure of a population is used (with other factors) to calculate future changes in age structure and total population size.

agglomerative method A system of hierarchical classification that proceeds by grouping together the most similar individuals, and subsequently groups into progressively larger and more heterogeneous units. At each stage the groups or individuals linked are those giving the minimum increase in group heterogeneity.

aggradation The accumulation of unconsolidated sediments on a surface, thereby raising its level.

aggregate **1.** In the building and construction industry, a mixture of mineral substances (bulk minerals), e.g. sand, gravel, crushed rock, stone, slag, and other

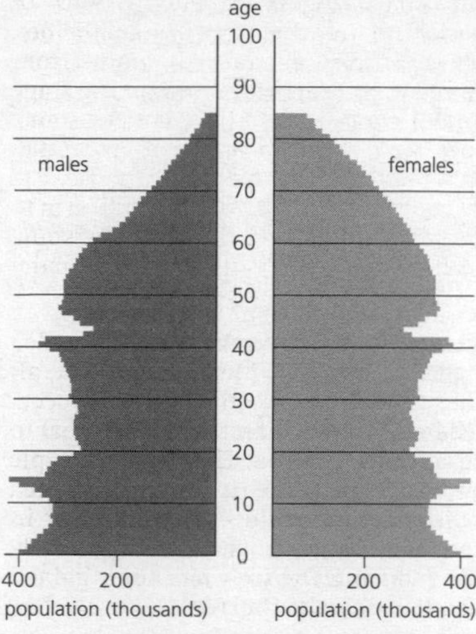

England & Wales 2002 (*Source*: Population Estimates Unit, Office of National Statistics)

Age structure

materials (e.g. colliery spoil, pulverized fuel ash) which, when cemented, forms concrete, mastic, mortar, plaster, etc. Uncemented, it can be used as road-making material, railway ballasts, filter beds, and in some manufacturing processes as flux. In road-making, aggregate mixed with bitumen is called 'coated stone', and different physical characteristics are required for the different layers comprising the road pavement. **2.** A group of soil particles adhering together in a cluster; the smallest structural unit, or ped, of soil. Aggregates join together to make up the major structural soil units.

aggregation **1.** The group of animals that forms when individuals are attracted to an environmental resource to which each responds independently. The term does not imply any social organization. **2.** The process in which soil particles coalesce and adhere to form soil aggregates. Aggregation is encouraged by the presence of bonding agents such as organic substances, clay, iron oxides, and ions (e.g. calcium and magnesium). **3.** The progressive attachment of particles (e.g. ice or snow) or droplets around a nucleus, thereby causing its growth.

aggression Behaviour in an animal that serves to intimidate or injure another animal, but that is not connected with predation.

aggregative response The preference for consumers to spend most of their feeding time in patches containing the highest density of prey. *See also* PARTIAL REFUGE.

agonistic Applied to behaviour between two rival individuals of the same species that may involve *aggression, *threat, *appeasement, or *avoidance, or that may be ritualized (e.g. as birdsong (*see* SONG), or as certain forms of *display). Agonistic behaviour often arises from a conflict between aggression and fear.

agouti In *mammals, the hair colour produced when each hair consists of alternate *eumelanic and *phaeomelanic bands. According to the theory of *metachromism, agouti is the evolutionarily primitive mammalian hair colour. The pattern is named after the agouti (*Dasyprocta aguti*), a South American rodent with hair banded in this way.

agric horizon A mineral-soil diagnostic *horizon, formed from an accumulation of *clay, *silt, and *humus, which has moved down from an overlying, cultivated soil layer. It is a *soil horizon created by agricultural management, and is identified by its near-surface position, and by *colloids accumulated in the pores of the soil. The name is from the Latin *ager*, meaning field.

agroclimatology The scientific study of the ways climate affects agriculture.

agro-ecosystem An agricultural ecosystem, e.g. a cereal crop.

agroforestry A system of mixed arable farming and forestry that has been practised for millenniums in many parts of the tropics (e.g. in home gardens in Java), especially in rain-forest environments. Tree crops are interspersed with food-crop patches on a continuous basis, sometimes as hedges (so-called 'alley-cropping' or 'corridor farming'). The system ensures a continuous food supply, some continuous economic return, and the avoidance of soil degradation by exposure to tropical sunlight and rainfall during prolonged clearance; and it is suitable for small-scale family farmers. The trees act as 'nutrient pumps' and bring plant mineral nutrients from the depths up to the surface soil layers that are occupied by the roots of the crop plants.

agrometeorology The study of the relationship between conditions in the lowest layer of the atmosphere and those in the surface of the Earth, as this affects agriculture.

agronomy A branch of agriculture concerned with the theoretical and practical production of crops, and with the management of soils.

Agulhas current Part of the large-scale circulation of the southern *Indian Ocean. It is a surface-water current that flows off the east coast of southern Africa between latitudes 25°S and 40°S in a south-westerly direction. Flow velocity varies seasonally between 0.2 and 0.6 m/s.

ahermatypic Applied to corals that lack *zooxanthellae and that are not reef-forming.

air frost The condition in which the air temperature is below freezing. Unlike *ground frost, air frost can damage plants.

air mass (airmass) A large body of air (sometimes of oceanic or continental proportions) that is identified primarily by an approximately constant wet-bulb-potential temperature (i.e. the lowest temperature to which the air can be cooled by the evaporation of water into it). The temperature and *humidity characteristics of an air mass, which are roughly the same within the one air mass at a particular latitude and height, are modified by and modify the atmospheric environment through which the air mass passes.

Aitken nuclei counter A device invented by the Scottish physicist John Aitken (1839–1919) that is used to estimate the concentration of particles with radii of more than 0.001 μm in a sample of air. Air is made to expand in a chamber. This causes it to cool and water vapour in it condenses on to particles, forming a mist whose opacity allows estimation of the number of particles present. *See also* AITKEN NUCLEUS.

Aitken nucleus A suspended, solid, atmospheric particle with a radius of less than 0.2 μm; most Aitken nuclei are about 0.5 μm. On average, their concentration varies from less than 1 000/cm³ over oceans to 150 000/cm³ in urban areas. *See* AITKEN NUCLEI COUNTER.

AIW *See* ANTARCTIC INTERMEDIATE WATER.

aklé A French name for a network of sand *dunes that is found especially in the western Sahara. The basic unit of the network is a sinuous ridge, at right angles to the wind, made up of crescent-shaped sections which alternately face the wind (linguoid) and back the wind (barchanoid). Aklé patterns require winds from one direction and a large quantity of sand.

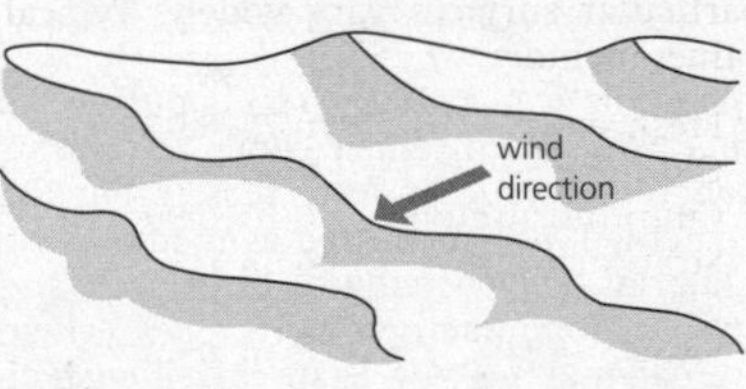

Aklé dune

alarm response Signals emitted by an animal that serve to warn others of a danger the individual has perceived. The signals may be visual (e.g. the white tail of a rabbit, displayed when running), aural (e.g. the call of a member of a flock of ground-feeding birds), or olfactory (substances emitted by some fish and invertebrates). *See* PRONKING and STOTTING.

alarm substance A substance released from the skin of injured fish that causes an immediate alarm reaction among fish nearby, leading to the dispersal of a school formed by members of the same species. Originally discovered in minnows and other Cyprinidae, the alarm substance is probably present in several fish families.

alas A *thermokarst depression with relatively steep sides and a flat floor, which may be occupied by a lake. Alases are well developed in Siberia (the word 'alas' is of Yakutian origin) where they can occupy 40–50 per cent of the land surface.

Alaska Current An oceanic water boundary current produced by the deflection of the *North Pacific Current by the North American continent. It flows in a north-westward direction along the south-eastern margin of Alaska. The Alaska Current is also called the *Aleutian Current in some texts.

alate Winged, or having appendages resembling wings.

albedo The measure of the proportion of sunlight that is reflected from cloud tops, the atmosphere, and surfaces. It is expressed as a percentage of incoming sunlight, written as a percentage, whole number omitting the percentage sign, or a decimal fraction. The average albedo for the Earth is 0.30 (30 per cent). Values for particular surfaces vary widely. Typical values include:

Fresh snow	0.75–0.95
Old snow	0.40–0.70
Cumuliform cloud	0.70–0.90
Stratiform cloud	0.59–0.84
Cirrostratus	0.44–0.50
Sea ice	0.30–0.40
Dry sand	0.35–0.45
Wet sand	0.20–0.30
Desert	0.25–0.30
Meadow	0.10–0.20
Field crops	0.15–0.25
Deciduous forest	0.10–0.20
Coniferous forest	0.05–0.15
Concrete	0.17–0.27
Black road	0.05–0.10

Radiation that is reflected cannot be absorbed and consequently light-coloured surfaces absorb less radiation than dark surfaces. Altering a surface often changes the albedo (e.g. by clearing coniferous forest, 0.05–0.15, to grow field crops, 0.15–0.25).

Albeluvisols Soils that have an *argic horizon with an irregular upper boundary. Albeluvisols are a reference soil group in the *FAO *soil classification.

Alberta low Storms that are common in Alberta, Canada, and associated with heavy rain and snow. The storms form as a result of cyclone regeneration after passage over the Canadian Rockies: as they move eastwards they bring very cold conditions, with blizzards.

albic Applied to an almost white soil in which there is little *clay or oxides coating the sandy or silty particles. The albic *horizon lies at or below the surface.

albinism In animals, the heritable condition observed as the inability to form the dark pigment melanin in the hair, skin, or vascular coat of the eyes. It is due to a deficiency in the enzyme tyrosinase, and is usually inherited as an autosomal *recessive gene (i.e. a recessive gene on a *chromosome other than a sex chromosome in the cell nucleus). In plants, a deficiency of chromoplasts (i.e. organelles containing carotenoid pigments that colour ripe fruits and flowers).

alcohol A hydrocarbon in which a hydroxyl (OH) group is substituted for a hydrogen atom. An alcohol is designated as primary, secondary, or tertiary, according to whether the carbon to which the hydroxyl group is attached is bound to one, two, or three other carbons. *See* ETHANOL.

alcoholic fermentation A process that occurs in plants which do not require oxygen for respiration. Glucose is broken down to *ethanol and carbon dioxide, with the release of energy that is used for ATP production.

alcrete *See* DURICRUST.

aldrin An *organochlorine insecticide that was formerly used as a seed dressing. In the 1950s seed-eating birds were exposed to lethal doses and predatory birds accumulated doses ingested from their prey, which caused them to lay eggs with shells too thin to protect the embryo adequately. Restrictions on the use of aldrin, introduced in the 1960s, were followed by a recovery in populations of birds of prey.

-ales A standardized suffix used to indicate an order of plants in the recognized codes of classification (e.g. Filicales, the ferns).

a

aletophyte A plant that grows on roadside verges or in a *mesic environment.

Aleutian Current (Subarctic Current) The oceanic current that flows westwards south of the Aleutian Islands and parallel to, but north of, the *North Pacific current. The water mass is a mixture of water from the *Kuroshio and *Oyashio Currents. *See also* ALASKA CURRENT.

Aleutian low A region of the North Pacific, near the Aleutian Islands, where the average value of atmospheric pressure is low, owing to the frequency of low-pressure systems (cyclones) moving into and occupying the region. Any one of these systems, when present on an individual day, may be called 'an Aleutian low'. Some of them are intense, others much less so. The term is the Pacific equivalent of *'Iceland low', used in the Atlantic.

Alfisols (grey-brown podzolics) Mineral soils that are grouped as an order. These soils have *clay-enriched *argillic horizons, are alkaline to intermediate in reaction, with the *base saturation in the B horizon more than 35 per cent, are usually derived from base-rich parent materials, and are drier than –15 bars moisture potential for at least three months when plants could grow.

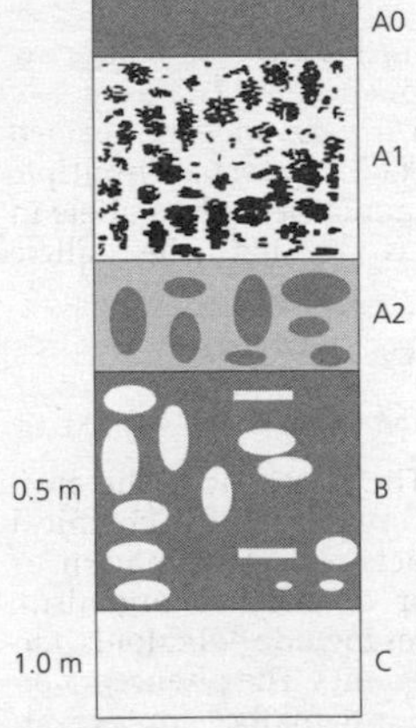

Alfisols

alga (pl. algae) The common (non-taxonomic) name for a *protist resembling a relatively simple plant that is never differentiated into root, stem, and leaves. It contains chlorophyll *a* as the primary photosynthetic pigment, has no true vascular (water-conducting) system, and there is no sterile layer of cells surrounding the reproductive organs. There are three different types of algae: red (Rhodophyceae); brown (Phaeophyta); and several groups of green algae. Algae range in form from single-celled *eukaryotes to plant-like organisms several metres long. Seaweeds are algae. Green plants have evolved from green algae. Algae can be found in most habitats on Earth, although the majority occur in freshwater or marine environments.

algal bloom A sudden growth of algae in an aquatic *ecosystem. It can occur naturally in spring or early summer when primary production exceeds consumption by aquatic herbivores (*see* PRIMARY PRODUCTIVITY). Algal blooms may also be induced by nutrient enrichment of waters due to pollution. They are a characteristic symptom of *eutrophication.

algal mat A sheet-like accumulation of cyanobacteria ('blue-green bacteria') that develops in shallow marine *subtidal to *supratidal environments, as well as in lakes and swamps. The algae cover the sediment surface and themselves trap sediment to produce a laminated alternation of dark, organic-rich algal layers and organic-poor sediment layers. *See also* STROMATOLITE.

alginite *See* COAL MACERAL.

algology (phycology) The study of algae.

alignment The comparison of homologous sequences of DNA or *amino acids in order to identify sites of *homology and the location of insertions, deletions, and *substitutions.

Alisols Soils that have an *argic horizon with a *cation-exchange capacity of more than 24 mol_c/kg clay, and a *base saturation of less than 50 per cent within 100 cm of the surface. These are soils with high concentrations of aluminium.

Alisols are a reference soil group in the *FAO *soil classification.

alkali-aggregate reaction A chemical reaction that can lead to damage in concrete structures. Free lime (CaO) in *cement reacts with CO_2 in the atmosphere to precipitate $CaCO_3$ around the cement grains. This protects them from weathering and also gives an alkalinity level (*pH higher than 7) which helps to protect steel from corrosion. If the aggregate contains soluble silica, however, new minerals may be precipitated by reaction between the aggregate and the cement. These may absorb water, causing the concrete to swell and eventually crack. Water entering these cracks may cause rusting of reinforcement bars and repeated wetting and drying may eventually destroy a structure.

alkalic *See* ALKALINE.

alkaline 1. (alkalic) Having a *pH greater than 7. **2.** *See* ALKALINE ROCK.

alkaline rock An *igneous rock that contains a relatively high concentration of the alkali (lithium, sodium, potassium, rubidium, caesium, and francium) and alkaline earth metals (magnesium, calcium, strontium, barium, and radium). Both silica-saturated and silica-under-saturated varieties exist, expressed in the presence of alkali feldspars and feldspathoids respectively. Alkali ferromagnesian minerals are usually present, and their identity depends on the composition of the rock. *Compare* ACID ROCK.

alkaline soil Soil with a *pH greater than 7.0. Degrees of soil alkalinity are recognized. The *USDA lists soils with pH 7.4–7.8 as mildly alkaline; 7.9–8.4 as moderately alkaline; 8.5–9.0 as strongly alkaline; and more than 9.0 as very strongly alkaline. Soil is not regarded as highly alkaline unless the reaction is between 8.0 and 10.0. The full range of the pH scale (0–14) is not used in soils, as the reaction of most soils is between pH 3.5 and pH 10.0. A *base saturation of 100 per cent indicates a pH of about 7.0 or higher. Plants characteristic of such soils are called calcicole.

alkaliphile An *extremophile (*Archaea) that thrives in environments where the pH is above 9.0.

alkaloid One of a group of more than 1 000 basic, nitrogenous, normally heterocyclic compounds of a complex nature that exist in combination with organic acids in plants belonging to approximately 1 200 species. Alkaloids are *secondary plant compounds that in some species may confer a degree of protection against attacks by herbivores. Many alkaloids are very poisonous to humans and in small amounts are the active ingredients in many therapeutic drugs. Pharmacologically powerful alkaloids derived from plants include caffeine, cocaine, morphine, nicotine, and strychnine.

Allee effect The social dysfunction and failure to mate successfully that occurs in some species when their population density falls below a certain threshold. This is believed to have caused the extinction of the American passenger pigeon (*Ectopistes migratorius*), probably because habitat destruction and hunting reduced the population density to a level at which individuals lacked the social stimulation necessary to prompt mating. The effect was first described in 1949 by W. C. Allee, O. Park, A. E. Emerson, T. Park, and K. P. Schmidt.

allele Common shortening of the term 'allelomorph'. One of two or more forms of a gene that arise by *mutation and occupy the same *locus on homologous chromosomes. When in the same cell, alleles may undergo pairing during *meiosis. They may be distinguished by their differing effects on the *phenotype. The existence of two forms of a gene may be termed 'diallelism', and of many forms, 'multiple allelism'. The commonness of an allele in a population is termed the 'allele frequency'.

allele frequency *See* ALLELE.

allelomorph *See* ALLELE.

allelopathy The release into the environment by an organism of a chemical substance that acts as a germination or growth inhibitor to another organism. Typical substances include *alkaloids, terpenoids, and phenolics. The phenomenon was described originally for heath and scrub communities, notably the

a

Californian *chaparral, but is now thought to be a widespread anti-competition mechanism in plants (e.g. barley inhibits competing weeds by means of root secretions). It is, however, extremely difficult to demonstrate in natural ecosystems. Allelopathy is also found in other organisms (e.g. antibiotics may be produced by fungi to inhibit competing bacteria, when the term 'antibiosis' may be used). It is a form of interference competition, substituting space for a resource.

Allen's rule A corollary to *Bergmann's rule and *Gloger's rule, holding that a race of warm-blooded species in a cold climate typically has shorter protruding body parts (nose, ears, tail, and legs) relative to body size than another race of the same species in a warm climate. This is because long protruding parts emit more body heat, and so are disadvantageous in a cool environment but advantageous in a warm environment. The idea is disputed, critics pointing to many other adaptations for heat conservation which are probably more important, notably fat layers, feathers, fur, and behavioural adaptations to avoid extreme temperatures. The rule was proposed by the American zoologist J. A. Allen in 1876.

Allerød A period during the last glacial that marks a prolonged warmer oscillation or *interstadial during the general phase of ice retreat in north-western Europe. *Radiocarbon dating suggests it lasted from about 12 000 BP to 10 800 BP. Pollen records for the north-west European area indicate a cool temperate flora with birch (*Betula* species) widespread, in marked contrast to the preceding and following, colder, *Dryas phases. In Britain the Allerød is not usually distinguished from the *Bølling.

alley-cropping *See* AGROFORESTRY.

alliance In *phytosociology, a grouping of closely related associations. *See also* PLANT ASSOCIATION.

allochthon A body of rock that has been transported to its present position, usually over a considerable distance. *See* ALLOCHTHONOUS.

allochthonous Not indigenous; acquired. Applied to material which did not originate in its present position (e.g. plant material in a deposit, such as lake sediment, which did not grow at that location but was introduced by some process). *Compare* AUTOCHTHONOUS.

allodapic Applied to materials that have been deposited by turbidity (*see* TURBIDITY CURRENT) or *mass flow. The term is particularly used in relation to limestones deposited by mass flow.

allogenic Applied to successional change caused by a change in *abiotic environmental conditions. *Compare* AUTOGENIC; *see also* SUCCESSION.

allogenic stream A stream that originates outside a particular area and whose continuation is inconsistent with its new surroundings. Type examples are the Nile and the Indus, whose discharges are sufficient to carry them through arid regions, and the Neretva in Bosnia Hercegovina, which is large enough to pass over permeable limestone.

allogrooming *Grooming performed by one animal upon another animal of the same species. *Compare* AUTOGROOMING.

allometric macroecology The branch of *ecology that seeks to explain differences among species in terms of patterns related to their body sizes. For example, there are six orders of magnitude in body size separating the smallest terrestrial mammals (shrews) from the largest (elephants) and *metabolic rate in mammals is proportional to the $^3/_4$ power of body mass.

allometry A differential rate of growth, such that the size of one part (or more) of the body changes in proportion to another part, or to the whole body, but at a constant exponential rate. Strictly speaking, 'allometry' is an umbrella term describing three distinct processes. Ontogenetic allometry refers to the differential growth rates of different body parts (i.e. juveniles are not merely diminutive adults; for example, the extinct Irish elk (*Megaceros giganteus*) was the largest of all cervids (Cervidae), but its antlers were 2.5 times larger than would be predicted from its body size, and reached an adult span of up to 3.5 m in the largest individuals). Static allometry refers to shifts in proportion among a series of related taxa of different size. Evolutionary allometry refers to gradual shifts in proportions as size changes in an evolutionary line (e.g. in the evolution of the horse the face became relatively longer and longer). In other cases allometry may be negative, leading to comparatively smaller parts.

allopatric Applied to species or populations that occupy habitats which are geographically separated, and that do not occur together in nature. *Compare* SYMPATRIC.

allopatric speciation The formation of new *species from ancestral species as a result of the geographical separation or fragmentation of the breeding population. Separation may be due to climatic change, causing the gradual fragmentation of the population in a few surviving favourable areas (e.g. during glaciation or developing aridity), or it may arise from the chance migration of individuals across a major dispersal barrier. Genetic divergence in the newly isolated daughter populations ultimately leads to new species; divergence may be gradual or, according to punctuationist models (*see* PUNCTUATED EQUILIBRIUM), very rapid. The populations must evolve some sort of sexual or genetic isolating mechanism that prevents them from interbreeding should they come into contact again later. *Compare* SYMPATRIC SPECIATION.

allopatry The occurrence of species in different geographical regions. The differences between closely related species usually decrease (i.e. the characteristics converge) when species are separated, in a process called character displacement, which may be morphological or ecological. *Compare* SYMPATRY.

allopolyploid A polyploid (*see* POLYPLOIDY) that is formed from the union of genetically distinct chromosome sets, usually from different species (e.g. bread wheat (*Triticum aestivum*) derived from three *genomes).

allothetic Applied to information concerning an animal's orientation in an environment that is obtained by the animal from external spatial clues. *Compare* IDIOTHETIC.

allozyme An allelic (*see* ALLELE) form of an *enzyme.

alluvial Applied to the environments, processes, and products of rivers or streams. Alluvial deposits (alluvium) are clastic, detrital materials transported by a stream or river and deposited as the river flood-plain. The term is also applied to surface flow, as in *alluvial fans, *bajadas, etc.

alluvial cone *See* ALLUVIAL FAN.

alluvial fan (alluvial cone) A mass of sediment deposited at some point along a stream course at which there is a sharp decrease in gradient (e.g. between a mountain range and a plain). Essentially, a fan is the terrestrial equivalent of a river-delta formation.

alluvium An *alluvial deposit.

alpha decay The decay of a radionuclide (radioactive *nuclide) by the spontaneous emission from its nuclei of alpha particles. An alpha particle is composed of two protons and two neutrons and has a charge of +2. It also has an appreciable mass and its ejection from the nuclide creates a certain amount of recoil energy in the nucleus. The total energy (E_x) created by alpha decay is, therefore, the sum of the kinetic energy of the particle, the recoil energy given to the new nucleus, and the total energy of any emitted gamma rays.

alpha-mesohaline water *See* BRACKISH and HALINITY.

alpine zone A region that occurs above the *tree line and below the *snow line on temperate and tropical mountains. The vegetation is characterized by an absence of trees and varies greatly with aspect, the greatest contrasts being between the wet side and the dry, leeward side of the mountains concerned. The elevation of the lower limit of the zone increases from about 1 000 m above sea level in Scotland to over 2 000 m in the Swiss Alps, and to 3 700 m in the western Himalaya.

alternation of generations The alternate development of two types of

a

individual in the life cycle of an organism. Usually one type reproduces asexually and the other sexually (the term is often restricted to organisms which have a *haploid generation alternating with a *diploid generation). No instances of such alternation are known among multicellular animals; where two types of individual occur, both capable of reproduction, they are invariably diploid and the process is better described as *metagenesis. In many parasitic protozoa (e.g. the malaria parasite *Plasmodium*) the sexual phase inhabits one host and the asexual phase another. In mosses, vascular plants, many algae, and some fungi, a haploid phase, during which *gametes are produced by *mitosis (gametophyte phase), alternates with a spore-producing diploid (sporophyte) phase. In an isomorphic alternation of generations (found in some algae, for example) the sporophyte and gametophyte are morphologically similar or identical; in a heteromorphic alternation of generations they are dissimilar (e.g. in mosses the gametophyte is the dominant and conspicuous generation, whereas in higher plants it is the sporophyte that forms the conspicuous plant).

altiplanation The process of relief reduction or planation (i.e. the smoothing of the surface) under *periglacial conditions. Two mechanisms are involved: the destruction of upstanding relief features by *frost wedging or *nivation, and the accumulation of debris in depressions or as terraces. In many areas only partial altiplanation has been achieved, with the emergence of altiplanation terraces, such as those of Cox Tor on Dartmoor, England.

altitudinal vegetation zones With increased altitude conditions usually become cooler and damper, so that the vegetation of mountains of considerable elevation shows a corresponding zonation. In the tropics the zonation may extend from rain forest on the lower slopes to alpine communities at heights above about 3 500 m. On the other hand, in progressively higher latitudes the elevation of the *tree line gradually descends, eventually to sea level.

altocumulus From the Latin *altum*, 'height' and *cumulus*, 'heap', the name of a genus of clouds composed largely of water droplets and consisting of grey-white sheets, or banded layers and rolls, which may also be broken up into cells. Sometimes it has a banded appearance, occasionally giving a *mackerel-sky effect; this is probably associated with strong vertical wind shear in middle altitudes. *See also* CLOUD CLASSIFICATION.

altostratus From the Latin *altum*, 'height' and *stratus*, 'spread out', the name of a genus of clouds that consists of greyish sheets or layers; the cloud may be striated, fibrous, or uniform. It may be composed of ice crystals as well as water droplets. *See also* CLOUD CLASSIFICATION.

altricial Applied to young mammals (e.g. rats, mice, cats, dogs) that are helpless at birth. Their eyes and ears are sealed, and they cannot walk, maintain their body temperature, or excrete without assistance. *Compare* PRECOCIAL.

altruism Sacrifice, if necessary of life itself, so that others, commonly offspring or other genetically close younger relatives, may survive or otherwise benefit. On the face of it this seems to reduce the *adaptive value of the altruist. However, by saving offspring at the cost of its own life, the altruist may 'save' more of its own genes than if the situation were reversed, particularly if the reproductive potential of the altruist is exhausted or nearly so.

amatoxins Poisonous substances present in the *fruit bodies of certain species of fungi belonging to the genus *Amanita*. When eaten by humans, amatoxins cause severe gastrointestinal symptoms and degeneration of liver and kidneys. They may be lethal even in small quantities.

Amazon floral region Part of R. Good's (1974, *Geography of the Flowering Plants*) *Neotropical floral kingdom, corresponding to one of the blocks of the rain forest. Because of extensive flooding by the Amazon river, the vegetation can be divided into that which is above the flood-level, called 'terra firme', and that which is below it, called 'igapo' or 'varzea'. The flora of this vast region is one of the richest in the world. *See also* FLORAL PROVINCE and FLORISTIC REGION.

amber Fossil conifer resin, which is brittle and hard, translucent to transparent, and yellow to brown in colour. It is found in sediments or on the shore and takes a fine polish.

ambient pressure Atmospheric pressure in the surrounding air.

ambient temperature The dry-bulb temperature prevailing in the surrounding air.

ambivalent Applied to apparently confused behaviour by an animal, arising most commonly from a conflict produced by impulses to behave in contradictory ways (e.g. to fight or to flee). Ambivalent behaviour may be alternate (i.e. repeatedly switching between the two responses) or simultaneous, causing the animal to freeze.

ameba *See* AMOEBA.

amensalism An interaction of species populations, in which one population is inhibited while the other (the amensal) is unaffected. It is the opposite of *commensalism. *See also* SYMBIOSIS. *Compare* COMPETITION; MUTUALISM; NEUTRALISM; PARASITISM; PREDATION; and PROTOCOOPERATION.

American rain forest Like the other three regions of the rain forest, the American region is distinctive not by virtue of its physiognomic characteristics, but because of its floristic composition. For instance, of the non-tree plants, bromeliads (Bromeliaceae) and cacti (Cactaceae) are important, as well as the orchids and ferns more generally associated with the rain forest. *See also* AFRICAN RAIN FOREST and INDO-MALESIAN RAIN FOREST.

Amersfoort An *interstadial during the last *glaciation of the Netherlands (somewhere between 60 000 and 70 000 years ago). The July temperature (based on floral evidence) was perhaps 15–20 °C.

amictic Applied to a lake in which no thermal stratification of the water occurs and there is no overturn. Such lakes occur in the Arctic.

amine A compound formed by the replacement of one or more of the hydrogen atoms in a molecule of ammonia (NH_3) by an organic group. Amines are classified as primary, secondary, or tertiary according to the number of hydrogen atoms replaced. *See also* NITROSAMINE.

amino acid An organic compound containing an acidic carboxyl (COOH) group and a basic amino (NH_2) group. The general formula for naturally occurring amino acids is $R{-}CH(NH_2){-}COOH$, where R is a variable grouping of atoms (fundamentally a carbon chain or ring). Amino acids constitute the basic building blocks of *peptides and *proteins and are classified as: (*a*) neutral, basic, or acidic; or (*b*) non-polar, polar, or charged (depending on the net electrical charge resulting from the configuration of the molecule).

ammonium fixation The adsorption of ammonium ions (NH^+_4) into inter-layer sites of the *clay minerals, similar to locations of potassium in hydrous mica, which renders them unavailable to plants.

ammonotelic Applied to organisms (e.g. many aquatic invertebrates and teleost fish) that excrete nitrogenous waste derived from amino-acid catabolism in the form of ammonia.

amoeba **(ameba)** A single-celled *eukaryote that is naked and that changes shape and moves by the irregular extension and retraction of *pseudopodia. Amoebae are *heterotrophs, feeding mainly on *bacteria, although some are parasites (*see* PARASITISM) or *symbionts of mammals. They are *protozoa, classified as members of the *Protoctista.

amoebiasis Any disease in which the causal agent is an *amoeba. In humans the most common form of amoebiasis is amoebic dysentery, caused by *Entamoeba histolytica*. This disease varies in severity from mild diarrhoea to severe or fatal dysentery.

amoebic dysentery *See* AMOEBIASIS.

amorphous Applied to materials that have no regular atomic structure, or in which no extensive portions have a regular structure; but small units, irregularly aligned or stacked, may have a regular structure.

a

amorphous cloud A continuous cover of low, featureless cloud (e.g. *nimbostratus), that often produces rain.

amosite *See* ASBESTOS.

amphi- The Greek *amphi*, meaning 'both' or 'on both sides', used as a prefix.

amphi-Atlantic species The plants that are found on both sides of the northern Atlantic, along the seaboards of eastern North America and Europe. This disjunct distribution pattern is thought to be largely inherited from a single circum-North-Atlantic coastal-plain flora. Typical representatives are pipewort (*Eriocaulon septangulare*), a slender aquatic herb, and Scots lovage (*Ligusticum scoticum*), an umbelliferous plant.

Amphibia (amphibians) A chordate class represented today by just three groups, of which the salamanders (Urodela) and frogs and toads (Anura) are the best known. They are *poikilothermic vertebrates. The majority are terrestrial but develop by a larval phase (tadpoles) in water. The skin is soft, naked (non-scaly), rich in mucous and poison glands, and important in cutaneous respiration. There are about 3 000 extant species. Most are found in damp environments and they occur on all continents except Antarctica. The class was formerly much more varied and in the *Triassic some forms (e.g. *Mastodontosaurus*) grew up to 6 m long. Amphibians appeared first in the *Devonian, having evolved from the Rhipidistia (lobe-finned fish), flourished in the *Carboniferous and *Permian, but declined thereafter. The first modern types were established during the Triassic.

amphibiotic **1.** Applied to an organism that can live with a host organism either parasitically or mutualistically. *See* MUTUALISM and PARASITISM. **2.** Living in water (larval form) and later on land (adult form).

amphidiploid An *allopolyploid in which the genetic behaviour of the constitutive *genomes is *diploid, such that *bivalents between chromosomes originating from different genomes do not form during *meiosis.

amphidromic point The centre of a tidal system; it is a no-tide or nodal point around which the crest of a standing wave or a high-water level rotates once in each tidal cycle. The tidal range increases progressively with increasing distance from the central point. The high water rotates anticlockwise around the central point in the northern hemisphere and clockwise in the southern hemisphere. The word 'amphidromic' is derived from an Attic festival, the Amphidromia, at which friends carried a young child around the hearth and then named it.

amphidromous Applied to the migratory behaviour of fish moving from fresh water to the sea, and vice versa. Such migration is not for breeding purposes, but occurs regularly at some stage of the life cycle (feeding, overwintering, etc.). *Compare* DIADROMOUS. *See also* ANADROMOUS; CATADROMOUS; and POTAMODROMOUS.

amphimixis Sexual reproduction. *Compare* APOMIXIS.

amphitheatre A flat-topped, steep-walled, depression that is shaped like a horseshoe and resembles an ancient Greek theatre. It may be the result of glacial erosion, forming a *cirque, or the collapse of a volcano. In the case of Mount St Helens, Washington, the catastrophic collapse of the summit and northern slopes of the cone and the subsequent pyroclastic flow eruption on 18 May 1980 created a north-facing amphitheatre enclosing a small active vent, later occupied by a small dacite dome.

amphitrophic Applied to an organism that can carry out photosynthesis in the presence of light and that can also grow chemotrophically in the dark.

amphitropical species Species that have disjunct distribution patterns, one part of the range being to the north of the Equator, the other to the south, the different parts being geographically quite separate. These disjunctions probably arose in the *Pleistocene when the climatic belts were telescoped, and migration across the Equator would have been easier.

ampulliform Flask-shaped or bottle-shaped.

AMS *See* ACCELERATOR MASS SPECTROMETRY.

amygdale (adj. amygdaloidal) A spheroidal, ellipsoidal, or (literally) 'almond-shaped' cavity or vesicle within a lava that is filled with secondary minerals (e.g. *calcite, quartz, or *zeolites).

amygdaliform Almond-shaped.

amygdaloidal *See* AMYGDALE.

amylolytic Capable of digesting starch.

anabatic wind A wind that blows up a slope, often gently, and usually when the sloping ground surface has been warmed by the Sun. *Compare* KATABATIC WIND.

anabranching channel A type of *distributary channel that separates from its trunk stream and may flow parallel to it for several kilometres before rejoining it. The anabranching channel remains undivided, and so differs from an anastomosing channel which has major distributaries that branch and then rejoin it.

anachoresis The habit of living in holes or crevices as a means of avoiding predators.

anadromous Applied to the migratory behaviour of fish that spend most of their lives in sea, but then migrate to fresh water to spawn (e.g. salmon and lamprey). *See also* AMPHIDROMOUS and DIADROMOUS. *Compare* CATADROMOUS and POTAMODROMOUS.

anaerobe *See* ANAEROBIC (2).

anaerobic **1.** Of an environment: one in which oxygen is absent. **2.** Of an organism: one able to exist only in the absence of oxygen, i.e. an anaerobe. Organisms may be facultative anaerobes (e.g. yeasts) or obligatory anaerobes (e.g. sulphur bacteria). **3.** Of a process: one that can occur only in the absence of oxygen.

anaerobiosis Living or growing in an environment lacking molecular oxygen.

anafront A *front (warm or cold) at which there is upward movement of the warm-sector air, commonly producing clouds and precipitation. *Compare* KATAFRONT.

anagenesis In the original sense, evolutionary advance; the term is now often applied more widely to virtually all sorts of evolutionary change along a single, unbranching lineage.

analog data (analogue data) Data which are recorded continuously, as opposed to digital data, which are recorded by discrete (digital) sampling.

analogous variation The occurrence of features with similar functions which have developed independently in unrelated taxonomic groups, in response to a similar way of life, or a similar method of locomotion, or a similar food source, etc. Thus the wings of birds and insects are analogous and phyllodes (flattened petioles performing leaf functions) found on *Acacia* species are analagous to leaves. *Compare* CONVERGENT EVOLUTION.

analysis of variance A statistical procedure that is used to compare more than two means.

anamniote Applied to a type of development typical of lower vertebrates (fish and amphibians). The egg lacks a shell and protective embryonic membranes; consequently it must be laid in water or in a suitably damp environment.

anastomosing channel *See* ANABRANCHING CHANNEL.

anastomosis In woody plants, the linking of branches.

ancient countryside In Britain, an area in which most of the fields, woods, roads, etc. were present before 1700.

ancient woodland Both *primary and *secondary woodland that originated prior to a particular date, which in Britain is often set at around 1600, as secondary woodland was rarely created by planting before then. *Compare* OLD-GROWTH FOREST.

Andean floral region Part of R. Good's (1974, *The Geography of the Flowering Plants*) *Neotropical floral kingdom. The region is elongated and extends from Colombia to southern Chile, spanning both tropical and temperate climates; it also includes the Galápagos Islands, included as, although they have numerous

a

endemic (*see* ENDEMISM) species, they have only two important endemic genera: *Leiocarpus* and *Scalesia*. The mainland flora is rich in endemic genera, however, and has supplied many economic and garden plants. *See also* FLORAL PROVINCE and FLORISTIC REGION.

andhis A local term, used in the north-west of the Indian subcontinent, for dust storms accompanying violent *squalls, caused by strong convection.

andic horizon A *soil horizon formed by the moderate *weathering of *pyroclastic materials. The name is from the Japanese *an* meaning dark, and *do* meaning soil.

Andosols Soils that have developed from volcanic material and either have a *vitric horizon more than 30 cm below the surface, or have weathered (*see* WEATHERING) *pyroclastic material within 25 cm of the surface. Andosols are a reference soil group in the *FAO *soil classification.

androchory (anthropochory, brotochory) Dispersal of spores or seeds by humans.

androdioecious Applied to plants in which male and hermaphrodite flowers are borne on separate plants. *Compare* GYNODIOECIOUS.

androecious Applied to a plant that possesses only male flowers.

andromonoecious Applied to plants in which male and hermaphrodite flowers are borne on the same plant. *Compare* GYNOMONOECIOUS.

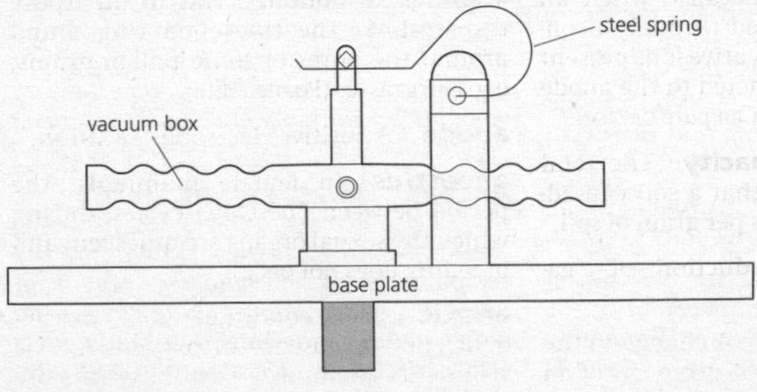

Aneroid barometer

anemo- A prefix derived from the Greek *anemos*, meaning 'wind', used to associate the remainder of the word with the concept of wind.

anemochory (aerochory) Dispersal of spores or seeds by wind.

anemometer An instrument for measuring wind speed, usually either from the rotation of a wind-driven turbine ('rotating-cups anemometer') or from the wind pressure through a tube aligned by a vane to point into the wind ('pressure-tube anemometer').

anemophilous Pollinated by wind.

anemophily Pollination of plants by wind.

aneroid barometer An instrument for measuring atmospheric pressure as represented by the height (or thickness) of a sealed, flexible, metal, concertina-shaped drum from which air has been partially evacuated. Variations in the height of the drum are transmitted to a pressure scale either mechanically, through a series of levers and a pointer, or electronically.

aneuploid Applied to a cell or organism whose nuclei possess a *chromosome number that is greater by a small number than the normal chromosome number for that species. Instead of having an exact multiple of the *haploid number of chromosomes, one or more chromosomes are represented more times than the rest.

Angara A continental mass of Asia, China, and the Far East that existed during the *Palaeozoic. In the latest Palaeozoic it became joined to *Euramerica along the line of the present Ural Mountains.

Angaraland The name given by the Austrian geologist Eduard Suess (1831–1914) to a small shield in north-central Siberia, where *Precambrian rocks are exposed, and which was considered to be the nucleus for subsequently developed structural features in Asia.

angiosperm A flowering plant, distinguished by producing seeds that are fully enclosed by fruits. All angiosperms are included in the plant division Anthophyta. *Compare* GYMNOSPERM.

Anglian **1.** A glacial stage of the middle *Pleistocene in Britain. **2.** A middle Pleistocene, cold-climate series of deposits. There were a number of glacial advances in East Anglia (England), which are difficult to understand. Near Lowestoft there are two *till sheets separated by stratified sands. The lower, North Sea Drift, contains Scandinavian *erratics, and the upper, thicker till contains erratics of *Jurassic or *Cretaceous material. The highest terraces of the Thames may belong to this period and perhaps could also be correlated with the *Elsterian deposits of Europe.

animal behaviour *See* ETHOLOGY.

Animalia The taxonomic kingdom that includes all multicellular animals.

anion A negative ion, i.e. an atom, or complex of atoms, that has gained one or more electrons and thereby carries a negative electric charge (e.g. Cl^-, OH^-, and SO_4^{2-}). It is so called because when an electric current is passed through a conducting solution the negative ions present in the solution are attracted to the anode (the positive electrode). *Compare* CATION.

anion-exchange capacity The total exchangeable *anions that a soil can adsorb, measured as moles per gram of soil.

anisogamy The production of *gametes of different sizes.

anisometric growth A change in the ratio between the sizes of two parts of an organism during its *ontogeny, such that if an organism grows anisometrically its shape will change.

annual Applied to a plant that completes its *life cycle (from germination to flowering to seed production and the death of vegetative parts) within a single *growing season. *See also* THEROPHYTE. *Compare* BIENNIAL; EPHEMERAL; and PERENNIAL.

annual fish Certain small fish (e.g. killifish and toothcarps of the family Cyprinodontidae) whose adult life is confined to a single year. Spawning takes place before the onset of the dry season; the adults then die, but the hardy eggs survive, hatching with the arrival of the rains.

annual ring *See* TREE RING.

annual snowline *See* FIRN LINE.

annulate Having ring-shaped markings.

annulus **1.** One of a series of concentric rings or bands of varying width and opacity which are formed in the scales of bony fish. Winter rings are often narrower and denser than summer rings. The number of annuli is indicative of the age of a fish. **2.** In the fruit bodies of certain *agarics, a remnant of the partial veil that adheres as a ring of tissue around the stipe. **3.** A ring of cells around the sporangium of some ferns. The walls of these cells become progressively thicker around the circumference, so inducing tension as they dehydrate, eventually causing the rupture of the sporangium (the point of breaking is called the stomium). The lid of the sporangium then curls back, releasing the tension induced by the. annulus in a spring-like motion serving to aid *spore dispersal. **4.** The thickened ring found around the pores of some pollen grains, e.g. the grasses (Poaceae).

anode A positive electrode. *See* ANION.

anoestrus In female mammals, the period between *oestrous cycles, during which the sexual organs are quiescent and breeding does not occur.

anoxic The condition of oxygen deficiency or absence of oxygen. Anoxic sediments and anoxic bottom waters are commonly produced where there is a

a

deficiency of oxygen owing to very high organic productivity, and a lack of oxygen replenishment to the water or sediment, as in the case of stagnation or stratification of the body of water.

antagonistic resources Two or more resources that can substitute for one another, but when taken together some partially offset the effects of others. The consumer requires more of the resources when they are taken together than when they are taken separately. *Compare* COMPLEMENTARY RESOURCES.

Antarctic air A very cold, continental polar *air mass, which originates over the frozen surface of Antarctica and the surrounding pack ice, and over the coldest waters of the *Antarctic Ocean. As this air moves northward over warmer waters it becomes convectively unstable.

Antarctic bottom water (ABW) A dense bottom-water mass, formed in the Weddell and Ross Seas, which moves in an easterly direction around Antarctica under the influence of the deep-reaching, wind-driven, surface *Antarctic Circumpolar Current (West Wind Drift). It is typified by a *salinity of 34.66 parts per thousand and a low temperature (–2°C to –0.4°C). The high salinity and dense nature is caused by the removal of pure water as sea ice.

Antarctic Circumpolar Current (West Wind Drift) The largest and most important ocean current in the southern hemisphere. It flows in an eastward direction around Antarctica, and occupies a wide tract of water in the South *Pacific, South *Atlantic, and *Indian Oceans. There is very little separation between surface- and bottom-water circulation within this area. The current is remarkably constant and is characterized by low *salinity (less than 34.7 parts per thousand), and cold waters (–1 to 5°C). It is the only current which flows right around the world.

Antarctic convergence (AAC) *See* ANTARCTIC POLAR FRONT.

Antarctic front The frontal boundary between the cold, dry, continental arctic air that covers the Antarctic continent throughout the year and the milder, moister, polar air covering the Antarctic (Southern) Ocean. The front is almost continuous around Antarctica and almost permanent.

Antarctic intermediate water (AIW) A water mass formed at the surface near to the *Antarctic convergence, at about 50°S. It is typified by a low *salinity (33.8 parts per thousand) and low temperature (2.2°C). As it spreads northwards it sinks to depths of 900 m and can be traced even in the North Atlantic at 25°N.

Antarctic Ocean (Southern Ocean) The oceanic waters surrounding Antarctica. It extends northwards to about 40°S latitude, the limit of the northward drift of ice from the Antarctic region, where there is a marked change in water temperature and *salinity. Water temperatures are low (–1.8°C to 10°C).

Antarctic polar current A surface current which flows in a westward direction around Antarctica under the influence of easterly winds blowing off the ice cap.

Antarctic polar front A convergence line in the seas that circle Antarctica between latitudes 50°S and 60°S. It is where the cold waters from the Antarctic region meet and sink beneath the warm waters from the middle latitudes, so forming the *Antarctic intermediate water. This convergence line was formerly known as the Antarctic convergence (AAC).

ante- The Latin word *ante*, meaning 'before', used as a prefix meaning 'preceding' or 'previous'.

antennate In insects, to communicate information by touching antennae. *See* DANCE LANGUAGE.

anteriad Pointing forward.

anthesis **1.** The time of flowering in a plant. This appears to be a response to a combination of factors including day length, temperature, and rainfall, but may also be initiated by the addition of gibberellins, one of a group of growth-promoting substances. **2.** The opening of a flower bud.

Anthophyta *See* ANGIOSPERM.

anthracnose A general term for any of several plant diseases in which symptoms include the formation of dark, often sunken spots on leaves, fruit, etc. An example is bean anthracnose, a disease of dwarf and runner beans caused by *Colletotrichum lindemuthianum*. Grape anthracnose, caused by *Elsinoe ampelina*, is a serious disease in parts of northern Europe; leaves develop irregular greyish spots, and portions of dead leaf tissue may drop out.

anthropic horizon The surface *horizon of a mineral soil which is produced by very long periods of cultivation and fertilization by humans. It is dark brown in colour, contains at least 1 per cent organic carbon, is relatively deep (more than 50 cm), has a base saturation of more than 50 per cent, and has more than 250 ppm P_2O_5 soluble in 1 per cent citric acid. The name is from the Greek *anthropos*, meaning human.

anthropochory *See* ANDROCHORY.

Anthropogene The name that is sometimes given informally to the *Holocene, since this is the time during which human activity has significantly altered the natural environment.

anthropogenic Strictly, pertaining to anthropogeny, which is the study of human origins (anthropogenesis, first used in 1839, from the Greek *anthropos*, meaning 'human being', and *gen-*, 'be produced') More recently the term has acquired a wider, secondary meaning, applied to substances, processes, etc. of human origin or resulting from human activity.

anthropogenic indicator An organism, usually present as a fossil, that indicates the past involvement of human activity. An example is the pollen of the herbs *Plantago lanceolata* (ribwort) or *Ambrosia* species (ragweeds), indicating human disturbance of vegetation in the *Holocene history of vegetation in Europe and North America respectively.

anthropogeomorphology The study of those land-forms and processes that are a direct result of human activity, including accelerated *erosion, channelized river channels (i.e. rivers made to flow along fixed, sometimes concrete-lined channels), the melting of *permafrost, and ground subsidence caused by the extraction of water or minerals. Particular examples include the Norfolk Broads, England, which are essentially flooded peat quarries, and the Zuider Zee, whose damming has had a major impact on the coastal morphology of the Netherlands.

anthropomorphism The attribution of human characteristics to non-human animals, most commonly by supposing non-human behaviour to be motivated by a human emotion that might motivate superficially similar human behaviour.

Anthrosols Soils that have been strongly affected by human activities such as ploughing, irrigation, or the addition of manure. Anthrosols are a reference soil group in the *FAO *soil classification.

anti- The Greek word *anti*, meaning 'against', used as a prefix meaning 'against' (in the sense of opposed to), 'opposite', or 'preventing'.

antibiosis *See* ALLELOPATHY and ANTIBIOTIC.

antibiotic An antimetabolite obtained from or produced by a living bacterium, fungus, or plant, which, in very small amounts, is toxic or lethal to other organisms (usually other bacteria or fungi). The term may also refer to chemical derivatives of naturally occurring antibiotics or to synthetic substances with similar properties. Under natural conditions the ability to produce an antibiotic presumably confers a competitive advantage on the organism. Some antibiotics are important in the treatment of animal diseases caused by micro-organisms (e.g., in humans chloramphenicol (developed in 1947) is active against many bacterial infections, including those that cause typhoid and rickettsial fevers).

antibody A complex protein that is produced in response to the introduction of a specific *antigen (which is normally foreign to it) into an animal. Antibodies are usually highly specific, combining only with antigens of a particular kind. Antibodies belong to a class of proteins

a

called immunoglobins, which are formed by plasma cells in the blood as a defence mechanism against invasion by parasites, notably bacteria and viruses, either by killing them or rendering them harmless. The specificity of their binding reaction with a particular antigen is owing to the configuration of a particular small area, known as the active site, on the surface of the antigens. Thus when a parasite (or its poisonous products) enters the tissues of its host, the antigens deriving from the parasite each produce a particular response according to the specific antibody that binds to that antigen. This 'recognition' by the host of the species or strain of parasite which has entered is sometimes applicable to other parasites should they share the same antigen. For example, the vaccinia and smallpox viruses share the same antigen, so that immunity to one confers immunity to the other. Antibodies may persist in the body long after the disappearance of the antigen, so conferring immunity to any new infection by the same strain or species of parasite. Vaccination or inoculation provides immunity to an organism by the injection of particular foreign proteins (not necessarily from a parasite) which then stimulate the production of antibodies.

anticline An arch shaped fold in rocks in which the oldest rocks are in the core.

anticoincidence circuit A device to minimize errors that may occur when measurements are made to date radiocarbon samples. These measurements must be extremely accurate because of the very low level of activity (*see* RADIOCARBON DATING). The error quoted on a radiocarbon age determination is solely an error in counting statistics. Such errors may rise from spurious counts generated by contamination of the sample, cosmic activity detected by the counter, or radioactive contaminants in the equipment being used. Initially the counter was shielded by surrounding it with large amounts of iron, lead, distilled mercury, or paraffin wax mixed with boric acid. An anticoincidence circuit is an alternative to material absorbers, and consists of a series of tangentially placed Geiger tubes operated in anticoincidence (i.e. they do not require input signals to arrive within specified intervals in order to be activated). These are positioned within an iron shield and around the central counting chamber. Radiation from outside, or from within the shield, is detected by this ring of Geiger tubes and can be discounted. Special counters have now been developed in which the anticoincidence counters are built into the same tube as the main counter, so that the same gas is used in the whole system. The wall of such a counter usually consists of a polystyrene foil covered on both sides with aluminium. This is then surrounded by a ring of wires forming the *anode for the anticoincidence circuit.

anticyclogenesis The process whereby an *anticyclone or a ridge of high pressure is formed and developed.

anticyclolysis The process whereby an *anticyclone or ridge of high pressure is dissipated or weakened.

anticyclone An area or system of high atmospheric pressure that has a characteristic pattern of air circulation, with subsiding air and horizontal divergence of the air near the surface in its central region. Continental-scale anticyclones form *source regions for *air masses. Winds are generally light. They flow clockwise around the centre in the northern hemisphere and anticlockwise in the southern hemisphere. A *temperature inversion is common at the base of the air subsidence, and this restricts the vertical development of cloud. The systems are often slow-moving or stationary. Weather conditions are generally settled. Cold anticyclones form over continents in winter and over polar areas at any time, accompanied by strong inversions; in the clear air, pronounced frosts and very cold surface conditions result. Warm anticyclones (so called because of the warm, subsided air aloft) over land areas typically bring spells of settled and often warm weather. *See also* ANTICYCLONIC GLOOM.

anticyclonic gloom A condition of low visibility associated with *anticyclones and accompanied, in the colder months of the year, by well-developed *temperature inversions that can trap dust and other pollutants and often have *radiation fog in the lower layers. The stability of the high-pressure system can make the resultant reduced visibility very persistent and may establish *smog conditions.

antigen A molecule, normally of a protein although sometimes of a polysaccharide, usually found on the surface of a cell in an animal, whose shape causes the production in the invaded organism of *antibodies that will bind specifically to the antigen. The reaction is so highly specific that it can be used in the identification of viruses and to distinguish between proteins extracted from different plant species.

antiserum An immune serum containing specific *antibodies that is prepared from the blood of a human or of another animal that has been immunized against *antigens from bacteria, viruses, or other parasites.

antitoxin An *antibody that neutralizes or inactivates a specific *toxin (forming an *antigen).

antitrade An upper wind in low latitudes that flows counter (i.e. poleward) to the *trade wind below.

anvil **1.** A flat-topped stone against which a bird smashes the shell of a mollusc in order to eat the soft parts of its prey. **2.** *See* INCUS.

apatite *See* PHOSPHATE ROCK.

APF *See* ABSOLUTE POLLEN FREQUENCY.

aphicide A natural or synthetic chemical substance that is toxic to aphids (Aphididae).

aphotic zone **(disphotic zone)** A deep-water area of marine ecosystems below the depth of effective light penetration (the *compensation level), analogous to the *profundal zone in freshwater ecosystems. *See* OCEAN DIVISIONS.

Apicomplexa *See* PROTOZOA.

aplanate Lying in a plane; leaves may be displayed on the twigs of a plant to give aplanate foliage.

aplasia The failure of an organ to develop.

apo- The Greek *apo*, meaning 'away' or 'from', used as a prefix imparting a sense of 'away from' or 'separate'.

apodous Without legs.

apogeotropism The type of *tropism in which plant organs (e.g. shoots) grow against the force of gravity.

apomixis In plants, a form of asexual reproduction that gives the appearance of sexual reproduction. *Parthenogenesis is an example of apomixis, as is the seed production of dandelions (*Taraxacum officinale*). Since apomixis does not involve fertilization or *meiosis, the progeny are genetically identical to their parents.

apomorph **(adj. apomorphic)** An evolutionarily advanced ('derived') character state (the opposite of plesiomorph) that is possessed by a group of biological organisms and distinguishes those organisms from others descended from the same ancestor. The long neck of the giraffe is apomorphic; the short neck of its ancestor is plesiomorphic. Apomorphic features may be *autapomorphic (uniquely derived) or *synapomorphic (derived and present in more than one group of organisms). The term means 'new-featured' and refers to 'derived' characters which have appeared during the course of evolution.

apomorphic *See* APOMORPH.

aposematic coloration **(aposematism)** Warning coloration in which conspicuous markings on an animal serve to discourage potential predators. Usually the animal is poisonous (e.g. has a venomous bite or sting) or unpalatable. Examples include the conspicuous bands or stripes on the back and sides of venomous snakes (e.g. coral snakes), and contrasting markings on the wings of distasteful monarch butterflies.

aposematism *See* APOSEMATIC COLORATION.

apostatic selection A type of selection which operates on a polymorphic species. Classically, the term is used in relation to prey species that have several different morphological forms. It occurs when, in proportion to their frequency in the population, rare forms of a species are

a

preyed on less than common forms, thus conferring a selective advantage on them. Such selection may produce a stable genetic *polymorphism.

apparent age **1.** An alternative name that has been proposed for *absolute age. **2.** An idea current in the nineteenth century among certain naturalists who opposed evolutionary theory (e.g. Philip Henry Gosse, 1810–88), that God had placed *fossils on the Earth to give it an apparently great age as a test of their faith.

apparent cohesion The cohesion of grains caused by surface tension in the surrounding pore water.

appeasement Behaviour by an animal that serves to reduce *aggression shown towards it by another member of the same species, but that does not involve *avoidance or escape.

appetite A complex phenomenon, not to be confused with hunger, that in humans is the comparatively pleasant, though at times compelling, anticipation of certain foods. In other animals appetite is studied by observing feeding behaviour and its relation to specific feeding stimuli, with no assumption of subjective experiences.

appetitive A general and rather imprecise adjective that is applied to the behaviour exhibited by an animal that is exploring its environment or seeking a goal. *See* CONSUMMATORY.

apple scab A very common disease of apple trees (*Malus*) in which the most obvious symptom is the appearance of superficial, dark, corky scabs on the fruit. In general the disease is not serious in itself, although more serious secondary infections can occur. The causal agent of apple scab is a fungus, *Venturia inaequalis*.

aptation A character that suits its possessor to its *environment; it may be an *abaptation, *adaptation, or *exaptation.

apterous Literally, without wings (from the Greek *a*, not and *pteron*, wing), and applied to insect species that are primitively wingless, or secondarily so due to a parasitic life style.

apterygote Applied to members of the Apterygota, the smaller of the two subclasses of insects, which are primitively wingless. There are about 580 species, distributed worldwide. Most feed on fungi, lichens, pollen, etc. *Compare* PTERYGOTE.

aquiclude **(aquifuge)** A rock with a very low value of hydraulic conductivity (*see* PERMEABILITY). Although it may be saturated with *groundwater, it is almost impermeable with respect to groundwater flow and may act as a boundary to an *aquifer.

aquic moisture regime The moisture balance of humid climates and soils, where annual precipitation exceeds the combined actual evaporation and transpiration, and where the soil moisture status is normally above *field capacity.

aquifer A body of permeable rock (e.g. unconsolidated gravel or a *sand stratum) that is capable of storing significant quantities of water, that is underlain by impermeable material, and through which *groundwater moves. An unconfined aquifer is one in which the *water-table defines the upper water limit. A confined

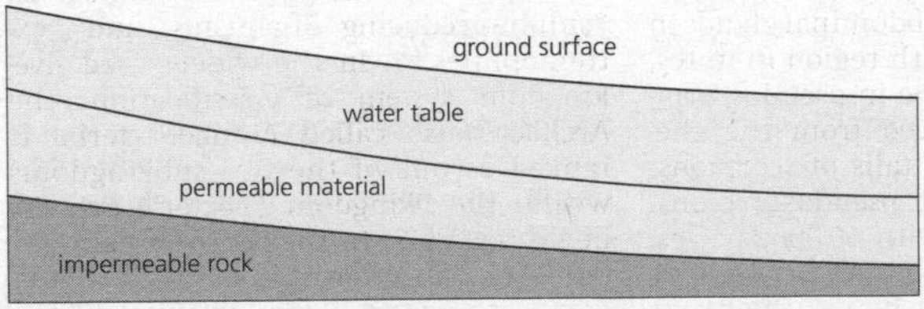

Aquifer

aquifer is sealed above and below by impermeable material. A perched aquifer is an unconfined groundwater body supported by a small impermeable or slowly permeable unit. *See* ARTESIAN WATER.

aquifuge *See* AQUICLUDE.

Arachnida (harvestmen, mites, palpigrades, pseudoscorpions, scorpions, spiders, whip scorpions, etc.) A class of *Arthropoda (subclass Chelicerata) which have book lungs or tracheae derived from gills, indicating their aquatic derivation. They have invaded most terrestrial habitats and have secondarily invaded aquatic habitats, although to a very much smaller extent (there is only one species of aquatic spider, and only one species of mite in ten is aquatic). Except for many plant and animal parasites found among the mites, and some scavenging harvestmen (order Opiliones or Phalangida), most arachnids are predatory. Scorpions have been recorded from the *Silurian Period and a Silurian scorpion, *Palaeophonus nuncius*, was perhaps the first terrestrial animal. The first fossil spiders are known from the *Devonian. The class is extremely diverse, but apart from the mites the body is in two portions. The number of eyes varies up to 12 in some scorpions, but generally vision is poor, many species being nocturnal and equipped with sensory hairs to detect prey. Pedipalps function as hands, and the chelicerae as jaws or teeth. In all arachnids the mouth is small, and food is generally predigested by enzymes from the mid-gut. Reproductive organs are placed on the ventral surface of the abdomen, and courtship may be complex and prolonged, with parental care of the young common to all. All arachnids are *dioecious. The production of silk and poison is characteristic of some orders, but the methods of production and their origins are varied. Silk is produced from abdominal glands in spiders, from the mouth region in mites, and from the chelicerae in pseudoscorpions. Poison is produced from the chelicerae of spiders, the tails of scorpions, and the pedipalps of pseudoscorpions. There are 11 orders, with 60 000 species. Members of 5 orders occur in northern Europe, the remainder being tropical in distribution.

arachnoid Resembling a spider's web (spiders belong to the class *Arachnida).

aragonite A colourless, white, grey, or yellowish mineral ($CaCO_3$) that occurs in hot springs and in association with gypsum, also in veins and cavities, and in the oxidized zone of ore deposits with other secondary minerals. Aragonite is a polymorph of *calcite, from which it is distinguished by its lack of cleavage and its higher specific gravity (2.9). Calcite is the more stable form of $CaCO_3$ and many fossil shells that were made originally of aragonite have either been converted to calcite or undergone replacement by some other mineral. Present-day mollusc shells are formed of aragonite crystals. The name is derived from the Aragon province of Spain. Aragonite usually occurs as prismatic or *acicular crystals but is sometimes fibrous and stalactitic.

aragonite mud A fine carbonate mud, with particles less than 4 μm, that is composed mainly of *aragonite needles. The aragonite is generally believed to have been deposited from the break-up of calcareous green algae (Chlorophyta).

Araucaria *See* DISJUNCT DISTRIBUTION.

arboriculture The cultivation and management of individual specimens of ornamental trees. *Compare* SILVICULTURE.

arbovirus A (non-taxonomic) term applied to a virus that can replicate in both vertebrates and arthropod vectors.

archae- (arche-) A prefix from the Greek *arkhaios*, 'ancient', itself derived from *arkhe*, 'beginning'. It adds the meaning 'ancient', with the implication 'first', to words to which it is attached.

Archaea Single-celled organisms including the *phenotypes: *methanogens; *sulphur-reducing organisms; and *extremophiles. **1.** In the widely used five-kingdom system of classification, the Archaea (also called Archaebacteria) is ranked as one of the two subkingdoms within the *kingdom *Bacteria (*see also* EUBACTERIA). **2.** In the three-domain system of classification, Archaea is one of the *domains, comprising organisms formerly known as the archaebacteria and placed in a kingdom of that name. The former kingdom Archaebacteria has been split into two kingdoms: *Crenarchaeota and *Euryarchaeota.

Archaean An eon of geologic time following the *Hadean and preceding the *Proterozoic eons, which together comprise what was formerly known as the *Precambrian. The Archaean eon lasted from 3 800–2 500 Ma and is divided into the Neoarchaean (2 800–2 500 Ma), Mesoarchaean (3 200–2 800 Ma), Palaeoarchaean (3 600–3 200 Ma), and Eoarchaean (3 800–3 600 Ma) eras. The Archaean was formerly known as the Archaeozoic or Azoic.

Archaebacteria *See* ARCHAEA.

archaeomagnetism The non-destructive study of the magnetic properties of objects and materials from archaeological sites. Such studies include magnetic dating, reconstruction of objects and structures, sourcing artefacts, determining past firing temperatures, etc. Artefacts often cause a local slight distortion of the Earth's magnetic field, which can be detected by a magnetometer. Materials that have been raised to a high temperature (fired pottery, kilns, etc.) can be dated by their thermoremanent magnetism (TRM). When substances containing iron and nickel compounds are heated strongly, the metals become randomly oriented with respect to the Earth's magnetic field, but as they cool below their Curie temperature (760°C for iron, 365°C for nickel) they become oriented with the magnetic field. Dating is achieved by comparing their magnetic orientation with the Earth's present magnetic field and relating this to a master sequence of changes caused by the wandering of the magnetic North Pole.

Archaeozoic *See* ARCHAEAN.

arche- *See* ARCHAE-.

archipelago A group of islands; a sea containing many scattered small islands.

Arctic air A very cold *air mass, generally formed north of the Arctic Circle. As air from this source moves southwards, it cools the regions in which it arrives; but being itself heated in the process it becomes convectively unstable. Polar lows sometimes form and the accompanying wintry precipitation is often heavy.

Arctic-alpine species A species which is found both in the Arctic and on the higher mountains in the temperate zone. Examples include *Salix herbacea* (least willow) and *Saxifraga oppositifolia* (opposite-leaved saxifrage). Although characteristic of the Arctic region and the Alps, Rockies, and Himalayas, they also occur in upland areas between the mountain ranges and the Arctic, and represent relicts of the late *Pleistocene *tundra.

Arctic and subarctic floristic region Part of R. Good's (1974, *The Geography of the Flowering Plants*) boreal kingdom, corresponding to the treeless wastes north of the *boreal conifer zone. There are virtually no endemic (*see* ENDEMISM) genera, which suggests that this is possibly the youngest flora in the world. The flora has two components: one Arctic, which includes the endemic species; the other *Arctic-alpine. *See also* FLORAL PROVINCE and FLORISTIC REGION.

Arctic front **1.** The frontal boundary between cold, *Arctic air and warmer *air masses, usually lying to the south of it. Many depressions originate on it. In north-western Canada in winter, for example, the frontal zone incorporates cold, dry, continental polar air and modified maritime Arctic air from the Gulf of Alaska to the north of continental tropical air. **2.** A front that forms in winter over snow and sea ice when the wind is weak or blows parallel to the edge of the ice. The front is shallow, but can trigger the development of a small but intense polar low.

Arctic heath Heath that is found in the low and middle Arctic belts of the *tundra. Normally it is dominated by members of the heath family (Ericaceae) or by heath-like plants, e.g. *Vaccinium vitis-idaea* (crowberry). Arctic heaths tend to be restricted to relatively well-drained sites that are sheltered and snow-covered in winter.

Arctic Ocean The smallest and shallowest of the major oceanic areas, the shallowness being caused by the surrounding wide continental shelves (up to 1 700 km wide). For much of the year the surface is covered by floating pack ice.

Arctic scrub A scrub composed of plants that average about 60 cm high, and including dwarf willows (*Salix*), birches (*Betula*), or both. It occurs in damp hollows and along the edge of water in the low Arctic belt of the *tundra. Northwards it becomes more stunted and limited in area, and more or less disappears beyond the centre of the middle Arctic belt.

Arctic sea smoke (frost smoke) *Fog that appears in very cold air from the Arctic ice or frozen-land regions, when it comes over the warmer water of open parts of the *Arctic Ocean. The rapid heating induces convection currents which rise in the air: these carry moisture upwards from the water surface, and this becomes visible as the moisture quickly condenses again in the very cold surrounding air. Thus a fog of rising columns of condensing water vapour is formed. The fog is usually fairly shallow, wispy, and smoke-like. This, and its common occurrence in coastal seas around cold land masses (e.g. Labrador, Greenland, and Norway), gave rise to the name. Similar steam fogs may be seen in winter over the open water of rivers when the air is 10 °C or more colder than the water.

Arctogea A traditional *zoogeographical region that comprises the northern continents, Africa, and Indo-China. It is usually subdivided into the *Palaearctic and *Nearctic (together Holarctica), *Ethiopian, and *Oriental faunal regions. *See* FAUNAL REGION and FAUNAL ZOOGEOGRAPHIC KINGDOM. *Compare* NEOGEA and NOTOGEA.

arcuate Curved or arched.

arcus The Latin *arcus*, meaning 'arch', used to describe a cloud feature that has a rolled appearance, with fragmented edges on the leading surface of *cumulonimbus and occasionally *cumulus. When well developed the feature has a prominent arch-like form. *See also* CLOUD CLASSIFICATION.

area cladistics A technique that employs relationships between organisms to reconstruct past distributions and the positions of continents, independently of geological data. Genetic and morphological characteristics reveal patterns of rela-

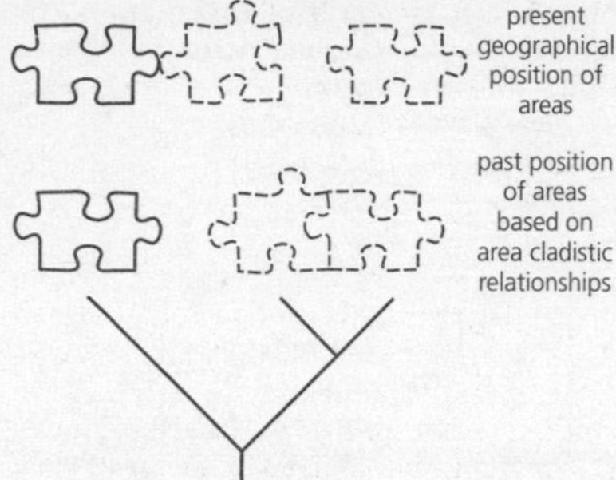

Area cladistics

tionships from which former geographic distributions can be inferred. Plotting past distribution patterns identifies the location of geographic barriers responsible for the isolation of organisms and the consequent divergence of species. This allows the geographic areas to be arranged hierarchically by adding the geographic information to *cladograms, producing diagrams called *areagrams.

area-effect speciation Speciation by the increased differentiation of two *subspecies with incompatible gene complexes, so that selection is strongly against *hybrids. The phenomenon is observed among slow-moving or sedentary organisms (e.g. snails and plants) occupying areas of habitat that remain unchanged for long periods (sometimes thousands of years) and within which stable local populations of particular polymorphs occur. As a result of strong selection for certain loci (*see* LOCUS), particular *alleles become more frequent, together with those genes that are compatible with them, thus forming gene complexes and consequent heterogeneity among the polymorphic populations. The fecundity of hybrids is low and a steep *cline develops, sometimes over a distance of metres.

areagram A diagram derived from a *cladogram but with the addition of data referring to the former geographic distribution of organisms. The areagram resembles a cladogram, but instead of showing the relationships among groups of organisms it shows geographic relationships

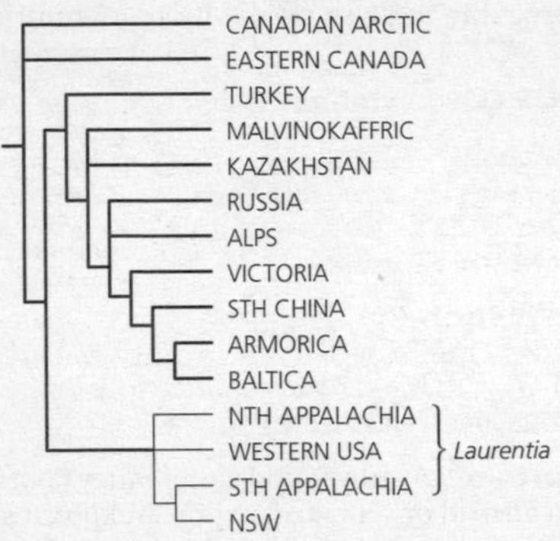

Areagram

and the way these have changed over time. An areagram shows the order in which geographic areas separated from one another.

area-restricted search A foraging pattern in which a consumer responds to an intake of food by slowing down its movement and remaining longer in the vicinity of the most recently located food item. This behaviour causes consumers to remain longer in areas where the density of food items is high than in areas where it is low.

arenaceous Sandy or *sand-like in appearance or texture. The term is applied to *clastic sedimentary rocks with a grain size 0.0625–2.00 mm. Three main groups of arenaceous rocks are recognized: quartz sandstones (quartzites), which contain 95 per cent quartz; arkoses, which have greater than 25 per cent feldspar; and greywackes, which essentially are poorly sorted sediments with rock (lithic) fragments in a mud matrix. *See also* ARKOSE.

arenite *See* SANDSTONE.

Arenosols A soil reference group in the FAO soil classification that comprises weakly developed, coarse-textured soils.

areolate Divided into small areas (areolae) by cracks or lines.

arête A knife-edged, steep-sided ridge that is found in upland areas that have been or are being glaciated. It is formed by the meeting of adjacent *cirque headwalls. It may be diversified by 'gendarmes' (abrupt rock pinnacles that have resisted frost shattering).

argic horizon **(argillic horizon)** A subsurface *soil horizon, at least one-tenth the thickness of the overlying horizon, that is identified by the illuvial (*see* ILLUVIATION) accumulation of silicate *clays. The amount of clay necessary is defined in comparison with the quantity in the overlying eluvial (*see* ELUVIATION) horizon, but it is at least 20 per cent more. *Cutans may be used to identify an argic horizon. The name is from the Latin *argilla*, meaning white clay.

argillaceous Applied to rocks which are *silt to *clay-sized sediments with grains less than 0.0625 mm in diameter. They account for more than 50 per cent of sedimentary rocks and most have a very high clay mineral content. Many contain a high percentage of organic material and can be regarded as potential source rocks for hydrocarbons.

argillans *See* CUTAN.

argillic horizon *See* ARGIC HORIZON.

argon-40 *See* POTASSIUM-ARGON DATING.

arid climate A climate with an *aridity index of 0.05–0.20.

aridic moisture regime The moisture balance of arid climates and soils, where the annual precipitation is less than the potential evaporation and transpiration, and where soil moisture status is normally less than *field capacity.

Aridisols An order of *mineral soils found in arid environments. These soils have very little organic matter in their surface *horizons, but may contain calcium carbonate or gypsum, and/or soluble-salt accumulations. The order includes infertile alkaline and saline soils of deserts. Except for a few tolerant species the vegetation is very sparse, making the soil prone to erosion.

aridity index An indication of moisture deficit. All climatic classifications include arid categories, defined either by

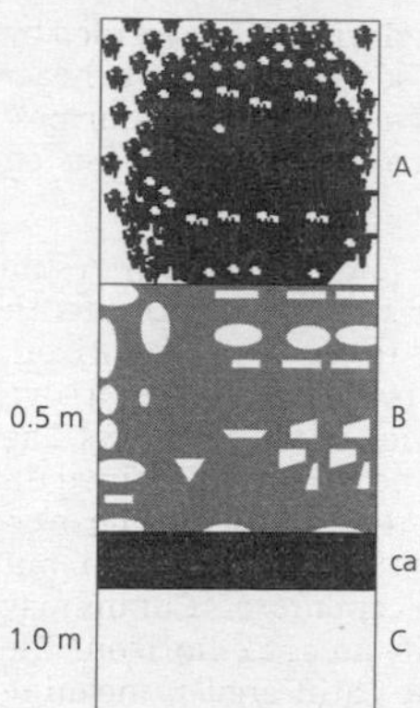

Aridisols

quantitative or, more usually, by mainly subjective criteria. C. W. *Thornthwaite first used the term 'aridity index' and calculated it as 100 × the water deficit/the potential evaporation. The most frequently used index is that devised by the United Nations Environment Programme to express the degree of climatic dryness. It is calculated by dividing the mean annual precipitation by the mean annual potential evaporation.

arillate Possessing an aril (i.e. a usually fleshy and often brightly coloured outgrowth from a seed), a diagnostic characteristic of the seeds of some flowering plants, especially tropical trees.

aristogenesis (directional evolution) An outmoded theory holding that evolution proceeds along a determined path. The modern view is that *natural selection does not direct evolution towards any particular kind of organism or physiological attribute, nor is there any mysterious inner guiding force. *See also* ENTELECHY; NOMOGENESIS; and ORTHOGENESIS.

arkose An *arenaceous rock that contains quartz and 25 per cent or more of feldspar. The feldspar is easily destroyed during transportation or chemical change, and the implication is that arkoses were deposited rapidly under fairly arid environmental conditions. Most were also deposited near to land and probably in close proximity to a granitic area.

Armillaria mellea *See* HONEY FUNGUS.

arousal **1.** The transition from the sleeping to the waking state. **2.** An increase in the responsiveness of an animal to sensory stimuli.

arrhenotoky Production of males from unfertilized eggs, a phenomenon that is characteristic of all Hymenoptera (ants, bees, wasps, etc.).

arroyo A gully found along valley floors in an arid or semi-arid region and possessing steep or vertical walls cut in fine-grained cohesive sediments. The floor is flat and usually sandy. Arroyos are found especially in the south-western United States, parts of India, South Africa, and around the Mediterranean.

artefact (artifact) **1.** A man-made object. **2.** Something observed that is not naturally present but that has arisen as a result of the process of observation or investigation.

artesian water *Groundwater that is confined in an *aquifer, but which may overflow on to the land surface via artificial boreholes or, sometimes, natural *springs, because of the high hydraulic head that may be developed in a confined aquifer. Artesian conditions are common when the aquifer has a synclinal form. The London Basin, England, provided artesian water during the nineteenth century from a chalk aquifer sealed by clays. The term is derived from the Artois region of north-western France.

artesian well (overflowing well) A well that flows at the surface without pumping, because it is sunk into a confined *aquifer with a hydraulic head (sometimes called the potentiometric or piezometric head) that lies above ground level. The aquifer is confined between two layers of impermeable rock or clay that are depressed, forming a basin. The water table lies below the ground surface at the top of the basin, but lower down the slope water draining downward by gravity is held under pressure. When a well is sunk into the aquifer water rises potentially to

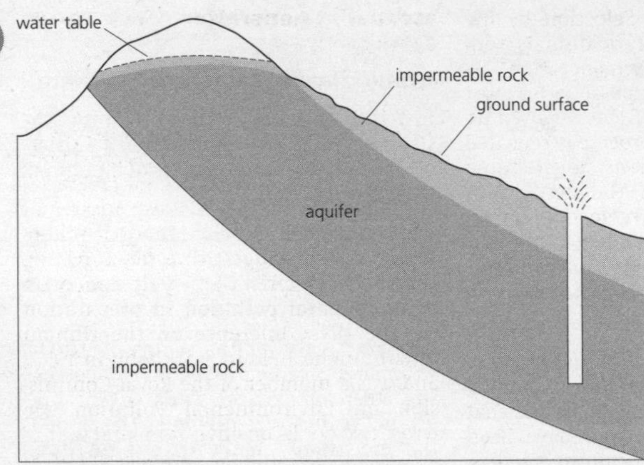

Artesian well

the level of the water table (see the diagram). *See* ARTESIAN WATER.

Arthropoda A highly diverse phylum of jointed-limbed animals, which includes the crustaceans, arachnids, and insects as the major components, as well as the classes Symphyla, Pauropoda, Chilopoda (centipedes), Diplopoda (millipedes), and the extinct trilobites and eurypterids. Arthropods comprise some 80 per cent of all animal species that have been described. They appeared first in the *Cambrian, already well diversified, implying an earlier, hidden history reaching back into the *Precambrian. Embryological evidence shows that they are a monophyletic group derived either from primitive polychaete worms or from ancestors common to both. Arthropods share several important features with annelid worms. The limbs of all arthropods are paired, jointed, and segmental, and the body has a chitinous exoskeleton. Primitively, the limbs and cuticular plates correspond to the segmentation of the body, but in many groups there is considerable loss and/or fusion of segments.

artifact *See* ARTEFACT.

artificial classification The ordering of organisms into groups on the basis of non-evolutionary features (e.g. the grouping together of plants according to the number and situation of their stamens, styles, and stigmas rather than their evolutionary relationships). As knowledge of shared characteristics and relationships has grown, artificial systems have been superseded by systems of *natural classification.

artificial freezing A method of controlling *groundwater and improving the strength of ground by pumping a refrigerant (e.g. calcium chloride or liquid nitrogen) through tubes in the ground; as the ground freezes around closely spaced tubes a continuous frozen zone may be formed.

artificial rain Rain, or increased rain, that is produced by seeding clouds artificially with 'dry ice' (frozen carbon dioxide), silver iodide, or other appropriate particles, which act as condensation nuclei. *See* CLOUD SEEDING.

artificial recharge A process whereby the amount of water in an *aquifer is supplemented by engineered as opposed to natural means. Artificial recharge may be through boreholes, purpose-built ponds, or simply by diverting more water on to the surface *catchment of the aquifer. Artificial recharge may be implemented as part of a conjunctive use scheme.

artificial selection Selection by humans of individual plants or animals from which to breed the next generation, because these individuals possess the most marked development of the required attributes. Typically the process is repeated in successive generations until those attributes are fixed in the descendent offspring. Such artificial selection can result in dramatic changes, like those that took place in the domestication of plants from their wild forebears. *See* GENETIC ENGINEERING.

asbestos Strictly, the fibrous variety of the mineral actinolite, but more generally any of a number of fibrous minerals that can be spun into yarn and have been mined commercially. Crocidolite (blue asbestos) is the fibrous variety of riebeckite, amosite of anthophyllite, and chrysotile of serpentine. Amosite, the most abundant form of asbestos, was widely used for insulation, construction, and in brake linings; crocidolite was used where resistance to acids was required. The inhalation of asbestos fibres over a prolonged period can cause the respiratory illness asbestosis in humans and it has also been reported in baboons and rodents living in the vicinity of asbestos mines. In the 1960s a correlation was found between the inhalation of crocidolite fibres and the incidence of mesothelioma (a form of lung cancer) and its use was restricted. Later, exposure to airborne fibres from other varieties of asbestos was also linked to lung cancer, and the use of all forms of asbestos is now forbidden or severely restricted in most countries.

asbestosis *See* ASBESTOS.

Ascencion and St Helena floral region Part of R. Good's (1974, *The Geography of the Flowering Plants*) *Palaeotropical floral kingdom, within the African subkingdom, comprising just two islands. This is probably the smallest floral region. There are five endemic (*see* ENDEMISM) genera, three of them being woody members of the Asteraceae. *See also* FLORAL PROVINCE and FLORISTIC REGION.

ascomycete A fungus of the subdivision Ascomycotina.

asexual generation *See* AGAMIC GENERATION.

Ashby, Eric (1904–92) A British botanist who held many public offices and professorships at the Universities of Sydney and Manchester. He was president, vice-chancellor, and chancellor (1970–83) of Queen's University, Belfast, master of Clare College, Cambridge, and vice-chancellor of Cambridge University (1967–9). Lord Ashby chaired the working party on environmental pollution in preparation for the UN Conference on the Human Environment, held in Stockholm in 1972, and was a member of the Royal Commission on Environmental Pollution. He wrote two books on environmental topics: *Reconciling Man with the Environment* (1978) and (with Mary Anderson) *The Politics of Clean Air* (1981).

aspergillosis A general term for any human or animal disease caused by a fungus of the genus *Aspergillus*. In humans the most common form of the disease is a lung infection which may lead to pneumonia; the disease is uncommon.

assart **1.** To grub up the trees and underwood of forest land in order to convert the area to arable or pasture use. **2.** Land cleared in this way. **3.** Private farmland formed by the clearance of part of a wood, common, or forest.

assemblage A collection of plants and/or animals characteristically associated with a particular environment that can be used as an indicator of that environment (e.g. in *geobotanical exploration). The term has a neutral connotation. Its use does not imply any specific relationship between the component organisms, whereas terms such as 'community' imply interactions.

assemblage zone (coenozone, faunizone) A stratigraphic unit or level of strata that is characterized by an assemblage of animals and/or plants. An assemblage zone is named after one or more of the distinguishing *fossils present, which are chosen without regard for their total time ranges, so that the assemblage is of purely environmental significance. *Compare* CONCURRENT RANGE ZONE.

a

assembly rules The principles that underlie the aggregation of species to make up a community.

assimilate **1.** The portion of the food energy consumed by an organism that is metabolized by that organism. Some food, or in the case of a plant some light energy, may pass through the organism without being used. **2.** To engage in *assimilation.

assimilation The incorporation of new materials, acquired by the digestion of food or *photosynthesis, into the internal structure of an organism.

assimilation efficiency The ratio of the amount of food absorbed (i.e. assimilated) to the total food ingested by an animal.

assisted migration **1.** The intentional establishment of populations or *metapopulations beyond the boundary of a species's historic *range for the purpose of tracking suitable *habitats through a period of changing climate. This might involve migration between islands, up mountain slopes, and between mountain tops. The term was coined in 2004 by Brian Keel. **2.** A technique used to establish a population of migratory animals in an area they did not formerly inhabit. For example, several groups of whooping cranes (*Grus americana*) raised in isolation from adult cranes have been taught to follow ultralight aircraft, finally travelling to the area in which they were to settle. Some have flown from the Necedah National Wildlife Refuge, Wisconsin, to the Chassahowitzka National Wildlife Refuge, Florida, a distance of 1 976 km.

association **1.** A learned connection between a type of event and a neutral stimulus with which it is paired (e.g. the sensation of hunger that may follow the chiming of a clock, denoting the time at which a meal is customarily presented). *See also* CONDITIONING. **2.** *See* PLANT ASSOCIATION.

association analysis A hierarchical method of classification (*see* HIERARCHICAL and NON-HIERARCHICAL CLASSIFICATION METHODS), which is divisive and *monothetic. The method uses χ^2 as a measure of association between pairs of species (attributes) found at a range of sample sites (individuals). The species or attribute with the highest overall sum of χ^2 values (i.e. the strongest links) with all other species is selected as the basis for subdivision into two groups of sites or individuals, either having or lacking that attribute. The process is then repeated for each new group until no further subdivision is required.

association measure Any measure of the link between two variables as shown by quantitative or qualitative data describing their characteristics. Links between continuous variables (e.g. mass and height or crop yield and rainfall) may be assessed statistically using *correlation and regression methods. Links between qualitative characteristics (e.g. aspect and slope form) may be assessed statistically using contingency tables and tests based on χ^2 or an appropriate non-parametric test. Sometimes continuous variables can be more conveniently tested in qualitative form, e.g. height as short, medium, or tall, especially when the link with some other qualitative variable is wanted (e.g. aspect and tree height).

associes A phytosociological term used in the Clementsian (American) and British traditions and generally implying a sub-climax community in a *successional sequence. *See* CLEMENTS, FREDERIC EDWARD.

assortative mating Sexual reproduction in which the pairing of male and female is not random, but involves a tendency for males of a particular kind to breed with females of a particular kind or the converse. If the two partners in each pair tend to be more alike than is expected by chance, then it is referred to as positive assortative mating. Assortative mating for some traits is common: examples include positive assortative mating for skin colour and height in humans, for height in wind-pollinated plants, and for flower colour and time of development to sexual maturity in plants and in many in-sects with one generation per year; and negative assortative mating for different plumage colour phases in Arctic skuas. *Compare* PANMIXIS.

asymmetric valley A valley that has one side steeper than the other, the opposing slopes having significantly contrasting angles. This contrast may be caused by geologic structure or variation in the nature and intensity of erosional (e.g. *periglacial) processes and there may be contrasts in the vegetation on the opposing slopes. Such valleys are common in past and present periglacial environments, where aspect has a significant effect on the nature of frost-based processes and on the depth of the *active layer.

astomatous Lacking a mouth or cytostome.

atavism The reappearance of a character after several generations, the character being the expression of a recessive gene or of complementary genes. The character, or the individual possessing this character, is sometimes referred to as a 'throw-back'.

Atlantic conveyor *See* GREAT CONVEYOR.

Atlantic North American floral region Part of R. Good's (1974, *The Geography of the Flowering Plants*) boreal kingdom. It is a large and extensive flora with 100–200 endemic (*see* ENDEMISM) genera. *Robinia* (false acacia) is a well-known example, and there is also *Franklinia*, a garden tree (not hardy in Britain) that was obtained from a single plant of a single species, and which is now extinct in the wild. As a whole the flora strongly resembles those of temperate Eurasia, China, and Japan. *See also* FLORAL PROVINCE and FLORISTIC REGION.

Atlantic Ocean One of the main oceanic areas of the world. It is relatively shallow, having an average depth of 3 310 m; and it is the warmest (average temperature 3.73°C) and most saline (average *salinity 34.9 parts per thousand) of the major oceans.

Atlantic Period A period in post-glacial times (i.e. post-*Devensian or *Flandrian) from about 7 500 years BP to 5 000 years BP which, according to pollen evidence (*see* POLLEN ZONE), was warmer than the present, and moist, with oceanic climatic conditions prevailing throughout north-western Europe. It corresponds to Pollen Zone VIIa, which throughout north-western Europe is characterized by the most *thermophilous species found in post-glacial pollen records. The *climatic optimum of the post-glacial, or current Flandrian *interglacial, is dated to the early Atlantic period. *Compare* BOREAL PERIOD.

atmometer An instrument that is used for measuring the rate of evaporation of water into air. It is normally in the form of an open-ended glass tube from which water can evaporate. By using an atmometer alongside a *potometer, it is possible to compare the rate of *transpiration from a plant with evaporation from a purely physical system.

atmophile Applied to the elements most typical of, and concentrated in, the Earth's atmosphere (e.g. H, C, N, O, I, and inert gases). They may occur in an uncombined state or combined (e.g. water (H_2O), carbon dioxide (CO_2) and methane (CH_4)).

atmosphere **1.** The air surrounding the Earth. The atmosphere has no precise upper limit, but for all practical purposes the absolute top can be regarded as being at about 200 km. The density of the atmosphere decreases rapidly with height, and about three-quarters of the mass of the atmosphere is contained within the lowest major layer, the *troposphere, whose depth varies between about 10 km and 17 km, being generally smaller further from the equator. **2.** A unit of pressure (abbreviation: atm.). Its value is approximately the average pressure of the atmosphere at sea level, the figure adopted being the pressure at sea level in the International Standard Atmosphere (760 mm of mercury, or 1013.25 mb). In SI units, 1 atm = 101 325 Pa. *See also* ATMOSPHERIC STRUCTURE.

atmospheric boil *See* SHIMMER.

atmospheric pressure The downward force exerted by the weight of the overlying *atmosphere, expressed per unit area in a given horizontal cross-section. Pressure varies throughout the atmosphere, owing to the distribution of mass; there are small diurnal variations of partly tidal and partly thermal origin, as well as bigger changes associated with the

a

passage of *depressions and *anticyclones. Atmospheric pressure is measured in millibars (mb), 1 mb being equal to 100 kilopascals (kPa). Measurements are usually made with a mercury barometer. The overall global average pressure at sea level is 1 013.25 mb, but as air is readily compressible, pressure decreases exponentially with altitude.

atmospheric structure The broadly horizontal layering of the *atmosphere, the layers being distinguished by differences in the rate of change of temperature with height, which either favour or discourage the development of vertical exchanges (convection). From the surface of the Earth upwards the layers are: (*a*) the *troposphere, in which convection is often prominent, especially over warm regions, extending to the *tropopause at a somewhat variable height, generally about 11 km over middle and higher latitudes and 17 km near the equator; (*b*) the *stratosphere, in which there is much less vertical motion, and which extends from the tropopause to about 50 km at the *stratopause; (*c*) the *mesosphere, in which there is once again more convection, extending from the stratopause to a height of about 80 km at the *mesopause; and (*d*) the *thermosphere, extending from the mesopause to the effective limit of the atmosphere, at about 200 km.

atmospheric 'window' The range of wavelengths (about 8.5–11 μm) at which radiation is only slightly absorbed by water vapour. Terrestrial radiation within this range may escape into space unless it is absorbed by cloud (water droplets can absorb in this range). *See also* 'GREENHOUSE EFFECT' and TERRESTRIAL RADIATION.

atoll A ring-shaped organic *reef that encloses or almost encloses a *lagoon and which is surrounded by the open sea. The reef may be built of coral and/or calcareous algae. An atoll is built on an existing structure such as an extinct, submerged volcano.

atrophy Of a structure, limb, organ, tissue, etc., to diminish in size.

Atterberg limits A series of thresholds which are observed when the water content of a soil is steadily changed. The 'contraction limit' occurs when sufficient water is added to a dry soil for contraction cracks to close. The addition of further water leads to plastic deformation at the 'plastic limit'. The 'liquid limit' occurs when just enough water is then added for the soil to behave like a liquid. Knowledge of these limits is important for understanding and predicting hill-slope failure. The limits were devised in 1911 by the Swedish soil physicist Albert Mauritz Atterberg (1846–1916).

atto- **(a)** Prefix used with *SI units to denote the unit $\times 10^{-18}$. The prefix is derived from the Danish or Norwegian *atten*, meaning eighteen.

aufwuchs *See* PERIPHYTON.

auger A tool used primarily for soil sampling, but also for sampling peats and other unconsolidated sediments. The simplest and most universal form has a screw head to bore the soil or sediment. Alternative auger heads are available for more specialized needs. Standard augers sample to one metre depth, but extension rods can be attached, enabling sampling at deeper levels.

Auger

austral An adjective from the Latin *australis*, meaning 'southern' or 'of the south'.

Australian faunal subregion A region that is distinguished by a unique marsupial (Marsupialia) fauna, including herbivores, carnivores, and insectivores. These evolved in isolation from the placental mammals (Eutheria) which now dominate the other continental faunas. In addition to marsupials there are also very primitive mammals (Monotremata), the spiny anteater and the platypus; and small rodents which are relatively recent (probably *Miocene) immigrants.

aut- A prefix derived from the Greek *auto-*, meaning 'self', that means 'self' or 'individual' (e.g. autotroph, 'self-feeder').

autapomorph An *apomorph character state that is unique to a particular species or lineage in the group under consideration.

autecology The ecology of individual organisms and populations, including physiological ecology, animal behaviour, and population dynamics. Usually only one or two species are studied. *See also* POPULATION ECOLOGY. *Compare* SYNECOLOGY.

authigenic Applied to mineral particles formed at or close to the surface, or to soils containing such particles.

auticidal control A type of *biological control in which the pest population contributes to an increase in its own mortality rate (e.g. the release of sterilized males leads to the laying of infertile eggs and, therefore, a reduction in the population).

autochory Dispersal of spores or seeds by the parent organism.

autochthonous Applied to material which originated in its present position (e.g. the plant material in a deposit, such as *peat, which actually grew where it is found, rather than being brought in by outside influences). *Compare* ALLOCHTHONOUS.

autoecious (monoxenous) Applied to a parasitic organism that can complete its life cycle in a single host species.

autogenic Applied to a *successional change owing to modification of the environment by vegetation (e.g. by producing *humus or providing shade). *See also* SUCCESSION. *Compare* ALLOGENIC.

autogrooming *Grooming of an animal by itself. *Compare* ALLOGROOMING.

autolysis The destruction of a cell or some of its components through the action of its own hydrolytic enzymes. It is a process that is particularly marked in organisms undergoing *metamorphosis.

automimicry The presence of a *polymorphism for palatability to predators. In the monarch butterfly, for example, unpalatability arises from the food plants which are chosen by the ovipositing female. After hatching, the feeding larvae take in substances that are toxic to birds but not to the insects themselves. This renders them unpalatable. The situation is used to advantage by palatable members of the species, which mimic the coloration of the unpalatable ones and so obtain protection. The polymorphism is frequency-dependent, and the unpalatable insects must be more abundant (and hence more frequently encountered by the bird predators) than the palatable insects.

autonomic Applied to behavioural responses (e.g. pallor and sweating in humans in response to fear) that are produced by the autonomic nervous system. Certain ritualized forms of behaviour used in communication may have originated as autonomic behaviour.

autophagy The digestion within a cell of material produced by the cell itself but which it no longer requires.

autopoiesis Compensation for environmental and genetic changes, as a result of which a taxon undergoes minimal morphological alteration (e.g. *Plethodon*, a salamander genus that has changed little morphologically over 60 Ma, despite frequent speciation and major changes to the environments its species inhabit, because of its behavioural flexibility and generalized tongue and teeth). *See also* CANALIZATION.

autopolyploid A *polyploid organism that originates by the multiplication of a single *genome (set of *chromosomes) such that all the chromosomes come from sets within one single *species. Autopolyploidy has been used commercially in breeding crop plants, including sugar beet

a

and tomatoes, to improve their vigour and growth, although the fertility of autopolyploids tends to be reduced.

autoradiograph A photographic print that is made by the action of a radioactive substance (e.g. carbon-14) upon a sensitive photographic plate.

autotetraploid An *autopolyploid organism with four similar *genomes. They occur either naturally by the spontaneous accidental doubling of a $2n$ genome to $4n$, or artificially, through the use of colchicine. They are present in many commercially important crop plants because, as with other *polyploids, they tend to be associated with increased size of the plants (through increased cell size, fruit size, stomata size, etc.).

autotomy The voluntary severance by an animal of a part of its body (commonly one of its own limbs), usually to escape capture by a predator that has seized that part. The part then regrows. Autotomy of the claws occurs in some Crustacea and of the tail in some Lacertilia (lizards). It is a defensive mechanism: in some lizards the detached tail continues to wriggle, distracting the predator while the lizard escapes.

autotroph **1.** An organism that is capable of synthesizing complex organic materials from simple inorganic substrates. The term includes photosynthetic autotrophs that use sunlight as a source of energy and chemosynthetic autotrophs that obtain energy from inorganic reactions (e.g. iron oxidation). **2.** An organism that uses carbon dioxide as its main or sole source of carbon. *Compare* HETEROTROPH.

autumnal *See* AESTIVAL; HIBERNAL; PREVERNAL; SEROTINAL; and VERNAL.

auxotroph An organism (usually a bacterium) that, as a result of a *mutation, has lost the ability to synthesize a particular substance essential for its growth (e.g. an amino acid or a sugar). The organism must be able to obtain this component from its environment if it is to survive and grow. *Compare* PROTOTROPH.

available nutrients Any elements or compounds in the soil solution that can be absorbed readily into plant roots, and that function as nutrients to growing plants. The available amount is usually much less than the total amount of that plant nutrient in the soil.

available relief The part of a landscape that is higher than the floors of the main valleys. It is therefore available for destruction by the agents of *erosion, controlled by the local base level. It is measured by the vertical distance between hilltops and valley floors.

available water In soil, the water that can be absorbed readily by plant roots. It is usually taken to be water held in the soil under a pressure of 0.3 to about 15 bars.

avalanche *See* MASS-WASTING.

avalanche wind A blast of often very destructive air ahead of a descending avalanche (*see* MASS-WASTING).

Aves *See* BIRDS.

avirulent Not virulent.

Avogadro constant **(Avogadro number)** The number of molecules, atoms, or ions in one mole of a substance: 6.02252×10^{23} per mol. It is derived from the number of atoms of the pure isotope ^{12}C in 12 grams of that substance and is the reciprocal of atomic mass in grams.

avoidance Behaviour that tends to protect an animal by reducing its exposure to hazard. Avoidance behaviour may be learned (e.g. when an animal does not enter an area in which a predator has been encountered, or when it does not eat an item that previously made it ill), or innate (e.g. when the young of some bird species utter distress calls and seek to hide when they see the shadow of a hawk-like object, although they have no experience of predatory birds).

avulsion The lateral displacement of a stream from its main channel into a new course across its *flood-plain. Normally it is a result of the instability caused by channel *aggradation. The avulsion of a stream into an adjacent valley may explain some cases of apparent *river capture.

axenic Applied to a culture of an organism that consists of one type of organism only (i.e. that is free from any contaminating organisms).

axial rift *See* MEDIAN VALLEY.

axial trough *See* MEDIAN VALLEY.

Azoic *See* ARCHAEAN and CRYPTOZOIC.

Azores high A semi-permanent anticyclonic region with subsiding air over the *Atlantic Ocean at around 30°N latitude. Movement of the system poleward in summer has a major impact upon the climate of Europe. The aridity of the Sahara Desert and the adjacent Mediterranean region is due to the subsidence of air in this high-pressure system. *See* AIR MASS and ANTICYCLONE.

B *See* BORON.

b *See* BAR (1).

backcross A cross of a *hybrid that is the progeny of two parental lines (i.e. the first filial generation) or *heterozygote with an individual whose *genotype is identical to that of one or other of the two parental individuals.

backing An anticlockwise shift of the direction of the wind. The reverse change is called *veering.

back mutation A reverse mutation (i.e. reversion) in which a mutant gene (called the non-wild-type form) reverts to the original standard form (the wild-type form).

backreef The area behind or to the landward of a *reef. This zone usually includes a *lagoon between the reef and the land.

backshore The part of a *beach that is above the level of normal high spring tides. This zone is usually dry; only when exceptionally high tides or storms occur does wave action influence this part of a beach. Characteristic plants of this area include *halophytes and plants with *xeromorphic adaptations.

backswamp An area of low, ill-drained ground on a *flood-plain away from the main channel. It stands slightly lower than adjacent *alluvial fans extending from the valley sides, and is below natural *levees that rise towards the main channel. It is a site of slow accumulation of silts and clays, usually inhabited by *marsh plants.

backwash The seaward return of water down a *beach. The process is affected by wave height and frequency, and by beach properties such as gradient and permeability. The general effect can be to steepen the beach profile.

Bacteria In *taxonomy, a *kingdom comprising 11 main groups of *prokaryotes. These are: purple (photosynthetic); *gram positive; *cyanobacteria; green non-sulphur; spirochaetes; flavobacteria; green sulphur; Planctomyces; Chalmydiales; Deinococci; and Thermatogales. Most bacteria are single-celled and most have a rigid *cell wall. Cell division usually occurs by *binary fission; *mitosis never occurs. Bacteria are almost universal in distribution and may live as *saprotrophs, *parasites, *symbionts, *pathogens, etc. They have many important roles in nature, e.g. as agents of decay and mineralization, and in the recycling of elements (such as nitrogen) in the *biosphere. Bacteria are also important to humans, e.g. as causal agents of certain diseases, as agents of spoilage of food and other commodities, and as useful agents in the industrial production of commodities such as vinegar, *antibiotics, and many types of dairy products. The oldest fossils known are of bacteria, from rocks in South Africa that are apparently 3 200 million years old. These must have been *heterotrophic bacteria, feeding off organic molecules dissolved in the oceans of that time. The first photosynthetic bacteria, of *anaerobic type, appeared a little later, about 3 000 Ma ago. In the widely used five-kingdom system of classification, Bacteria is one of the kingdoms and the only kingdom in the superkingdom Prokarya. The kingdom Bacteria comprises two subkingdoms: *Archaea and *Eubacteria. In the three-domain system of classification, the kingdom Bacteria is the only kingdom in the *domain Eubacteria.

bactericide Any agent which kills bacteria.

bacteriophage (phage) A type of *virus which infects bacteria. Infection with a bacteriophage may or may not lead to the death of the bacterium, depending on the phage and sometimes on conditions. A given bacteriophage usually infects only a single species or strain of bacterium. Phages can be found in most natural environments in which bacteria occur.

bacteriostatic Applied to an agent that prevents the growth of at least some types of bacteria without actually killing them.

bacteroid A modified bacterial cell, particularly of the type formed by species of *Rhizobium* within the root nodules of leguminous plants.

badlands The name originally applied to the intricately eroded plateau country of South Dakota, Nebraska, and North Dakota, but now widened to refer to any barren terrain that has been similarly intensively dissected. It is most common in areas of infrequent but intense rainfall and little vegetation cover.

Baermann funnel A device used to extract nematodes from a soil sample or plant material. A muslin bag containing the sample is submerged in water in a funnel sealed at the lower end by a rubber tube and clip. Being heavier than water, the nematodes pass through the muslin and sink to the bottom. After about 12 hours they can be collected by drawing off the bottom centimetre of water. The efficiency of the device is increased by gentle warming which immobilizes the nematodes.

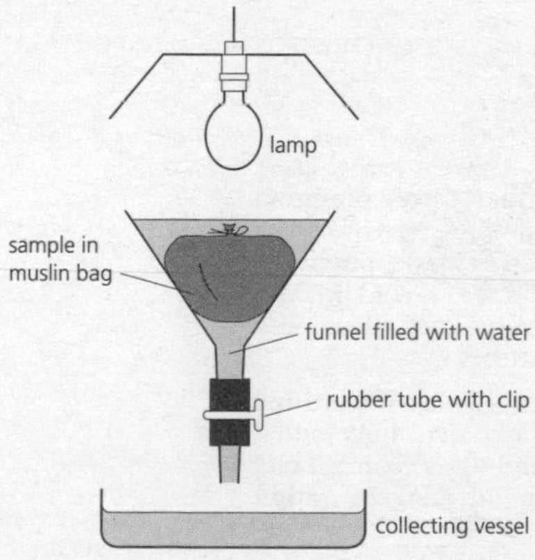

Baermann funnel

baguio The name given to a *tropical cyclone that forms in the vicinity of Indonesia and the Philippines (Baguio is the name of a town in Luzon, Philippines).

bahada *See* BAJADA.

Bailey's triple catch A method, devised in the 1950s by N. T. J. Bailey, for estimating the size of an animal population in which individuals are caught on two separate occasions, marked, released, and a third catch is made. Provided each sample contains more than 20 individuals, the population estimate (P) is given by $(a_2n_2r_4)/r_1r_3$, where a_2 is the number of marked individuals released from the second catch, n_2 is the number caught in the second sample, r_4 is the number of individuals in the third catch that were marked in the first catch, r_1 is the number of individuals marked in the first catch that were caught in the second, and r_3 is the number of individuals caught in the third sample that were marked in the second. If the samples number less than 20 individuals, an adjustment is made, giving: $P = (a_2(n_2 + 1)r_4)/((r_1 + 1)(r_3 + 1))$.

Bai-u season The principal rainy season of the south-east monsoon in southern and central Japan.

bajada (bahada) An extensive, gently sloping plain of unconsolidated rock debris resting against the foot of a mountain front in a semi-arid environment. Typically it is made of a number of coalescing *alluvial fans laid down by *ephemeral streams as their gradients lessen on leaving the mountain zone. Material is also supplied by the *weathering of the mountain front. Alternatively, it may comprise the alluvial accumulation on the lower part of a *pediment.

bakanae disease (foolish seedling disease) An important disease of rice caused by the fungus *Gibberella fujikuroi*. Early symptoms include excessive growth of the stem, apparently caused by gibberellins produced by the *pathogen.

balanced polymorphism A genetic *polymorphism that is stable and is maintained in a population by *natural selection, because the *heterozygotes for particular *alleles have a higher *adaptive value (i.e. fitness) than either *homozygote.

b

baleen In whales of the suborder Mysticeti, sheets of keratin which hang transversely from the roof of the mouth, their lower edges fringed with hairs which form a comb-like structure used to filter *plankton.

balloon sounding The use of balloons lighter than air to establish wind conditions in the upper air. Usually the balloons are tracked by radar, and instruments may be attached to the balloons to record temperature and humidity at given pressure levels. *See* RADIOSONDE, RAWINSONDE, and WIND SONDE.

Baltica *See* BALTIC SHIELD.

Baltic Shield (Baltica, Baltoscandia, Fennoscandia) A region underlain by *Precambrian rocks that covers the Kola Peninsula in western Russia, Finland, Sweden, and Norway. The region extends from the Arctic Ocean to the island of Bornholm, in the Baltic Sea off the southern tip of Sweden, and from the eastern coast of the Kola Peninsula to the westernmost tip of Norway. This continental mass once formed the south-eastern margin of the *Iapetus Ocean. The subduction of this ocean during the *Silurian and early *Devonian brought the Baltic Shield into contact with North America and Greenland. It separated again with the opening of the Atlantic Ocean.

Baltoscandia *See* BALTIC SHIELD.

bankfull flow The maximum amount of discharge (usually measured in m^3/s) that a stream channel can carry without overflowing. Its frequency of occurrence varies between streams, from a few times each year to once every few years. The water height at bankfull discharge is referred to as the 'bankfull stage'.

bankfull stage *See* BANKFULL FLOW.

banner cloud A motionless, flag-like cloud, commonly of lenticular (lens-like) shape, forming to the lee (eddy zone) side of a hill or mountain peak. The cloud extends downwind in a strong current of humid air. Many distinctive mountain peaks (e.g. the Matterhorn and Table Mountain) are associated with a characteristic banner cloud. *See also* LEE WAVE.

bar **1.** A unit of pressure approximately equal to one atmosphere (14 lb/in^2 in CGS units), and precisely equal to 10^5 Pa (105 N/m^2) in *SI units. The pressure of the atmosphere at sea level on average is very approximately one bar, or about 1 013 millibars (the bar commonly being divided into one thousand millibars, mb). **2.** (*a*) A low ridge of sand or shingle laid down by marine *aggradation in shallow water ad-

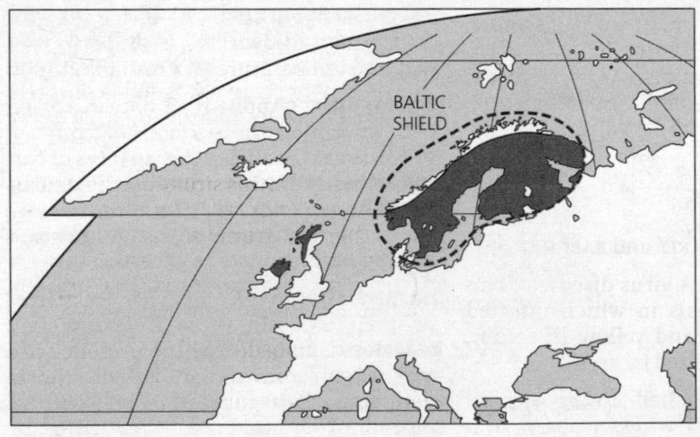

Baltic Shield

jacent to a coastline. There are several varieties: a bay bar joins the two flanks of a bay and may enclose a lagoon; an offshore or barrier bar runs parallel to a coastline and up to 40 km distant. (*b*) A rocky obstruction across a glaciated valley. *See* GLACIAL STAIRWAY. (*c*) A lobate river bed-form, typically constructed of gravel, often regularly spaced, and forming a riffle or shallow section. (*d*) Point bar: a low crescentic shoal on the convex side (inside) of a river bend, consisting of material that has been eroded from an outside bend, either opposite or upstream. Point-bar deposits consist of relatively coarse materials, often showing an upstream dip.

barachory (clitochory) Dispersal of spores or seeds by their own weight.

barat The local name for a fierce northwesterly wind common from December to February on the northern coast of Celebes.

barchan (adj. barchanoid) A crescent-shaped mobile *dune in a sand desert in which the wind blows predominantly from one direction. The dune moves by the erosion of sand from the windward slope and its accumulation on the steeper lee or slip slope, which stands at about 32°. Average velocities of dune movement are 10–20 m/yr.

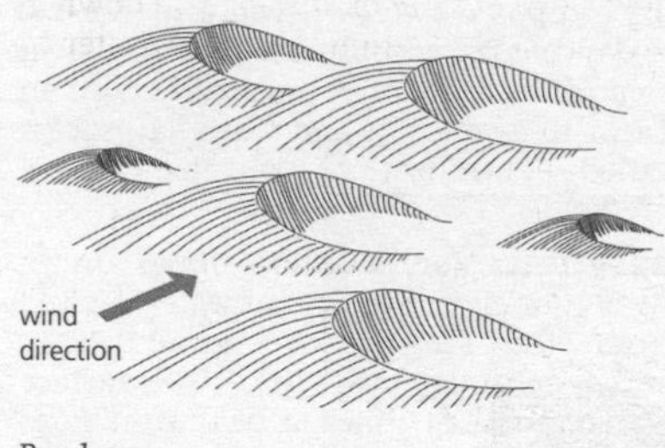

Barchan

barchanoid *See* AKLÉ and BARCHAN.

barley yellows A virus disease of barley and other cereals in which infected plants are stunted and yellow in colour. The virus is transmitted by aphids.

baroclinic 1. Applied to an atmospheric condition in which isobaric and constant-density surfaces are not parallel (e.g. in a frontal zone). **2.** Applied to a state in the ocean in which the surfaces of constant pressure intersect surfaces of constant density. In this situation, the water density gradient depends on water properties (temperature and *salinity) as well as pressure (depth). This can be contrasted with the *barotropic situation.

baroduric Capable of withstanding high pressures.

barograph A *barometer that gives a continuous recording of air pressure. It is based on an *aneroid instrument with levers attached to the vacuum chambers, and records a trace on a chart mounted around a clock drum.

barometer An instrument for the measurement of atmospheric pressure. The usual type is a mercury barometer, in which the atmosphere's pressure on a small reservoir of mercury supports a column of mercury in a vacuum tube, the open end of which is below the surface of the mercury in the reservoir. The column is on average about 76 cm (30 inches) high. Readings must be corrected to compensate for pressure variation due to gravitational anomalies and for thermal expansion or contraction of the mercury; therefore correction to a standard temperature is necessary. *See also* ANEROID BAROMETER; FORTIN BAROMETER; and KEW BAROMETER.

barophilic Living in or preferring environments subject to high pressures (e.g. deep-sea environments).

barothermograph A device for the continuous measurement of both pressure and temperature on a revolving chart.

barotropic Applied to a state in a water mass in which the surfaces of constant pressure are parallel to the surfaces of constant density. In this situation the density gradient depends on depth only, as in an isothermal freshwater lake. *Compare* BAROCLINIC.

barred basin A partially restricted sedimentary basin, where free movement of waters is impeded by the presence of a rock sill or sediment barrier. This restriction often results in anoxic or oxygenpoor waters, or, in arid areas, in evaporite deposition.

b

barren A *community of few and scattered plants that occupy less than half the available ground area. Barrens occur in the north-eastern United States and within the Arctic *tundra where they are typically dominated by a single plant species (e.g. *Dryas octopetala*, mountain avens); and often a single particle size predominates in the soil. *Compare* NORTH AMERICAN LOWLAND CONIFEROUS FOREST and PINE BARREN.

barren lands A term that was once used to describe the *tundra of northern Canada, a region characterized by sparse vegetation, a harsh climate, and *permafrost.

barrier A general term that describes a depositional feature standing on the seaward side of a coastline. *See* BARRIER BAR; BARRIER BEACH; BARRIER ISLAND; BARRIER REEF; BAYHEAD BARRIER; and BAYMOUTH BARRIER.

barrier bar A major *longshore bar of gravel or sand whose surface is below mean still-water level. Normally it is formed off a depositional coast of low gradient and with ample unconsolidated sediment. *See* BARRIER.

barrier beach A relatively small, shingle feature that protects a steep coast. *See* BARRIER.

barrier island An elongated ridge that may extend from a few hundred metres to 100 km along a coast, forming a segmented *barrier-bar complex and found between two tidal inlets. Barrier-island systems have a *lagoonal area on their landward side, and often have wind-blown *dunes and vegetation on the exposed (seaward) side of the *barrier. There are three main hypotheses to explain the origin of barrier islands: (*a*) the building up of submarine bars; (*b*) spit progradation parallel to the coast and segmentation by inlets; and (*c*) submergence of subaerial coastal beach ridges by a rise in sea level. Barrier islands are most common in areas of low tidal range.

barrier reef A *reef that trends parallel to a shore but is separated from it by a *lagoon. The reef-building organisms build up the structure to approximately the low-tide level. One of the finest examples is the Great Barrier Reef which lies off the north-western coast of Australia: it extends for about 1 900 km and is 30–160 km in width.

basal metabolic rate **(BMR)** The rate at which energy must be released metabolically in order to maintain an animal at rest. Since the BMR of different body tissues varies, the BMR for a particular animal is determined by the composition of its tissues; for a group of animals BMR is inversely proportional to body weight (i.e. small animals usually have a higher BMR than large ones).

basal sliding The process by which a temperate *glacier moves over its bed. It involves three mechanisms: relatively rapid creep in the basal layers; pressure melting, whereby ice under pressure melts on the up-glacier side of a small obstacle and the released water freezes on the down-glacier side; and slippage over a layer of water at the bed.

base According to the Brønsted–Lowry theory (proposed in 1923 by both J. N. Brønsted in Copenhagen and T. M. Lowry in Cambridge, who were working independently of one another), a substance that in solution can bind and remove hydrogen ions or protons. The Lewis theory (proposed by G. N. Lewis, also in 1923) states that it is a substance that acts as an electron-pair donor. A base reacts with an *acid to give a salt and water (a process called neutralization), and has a *pH greater than 7.0.

baseflow **(dry-weather flow)** In a stream or river, the flow of water derived from the seepage of *groundwater, and/or through-flow into the surface watercourse. At times of peak river flow, baseflow forms only a small proportion of the total flow, but in periods of *drought it may represent nearly 100 per cent, often allowing a stream or river to flow even when no rain has fallen for some time. *See also* INTERFLOW and SUBSURFACE FLOW.

base level A theoretical plane surface underlying a land mass, denoting the depth below which *erosion would be unable to occur. Sea level provides a base level on a regional scale. Local base levels may be provided by the base of a hill-slope, lakes, or by the junction between a tributary and the main river.

base pair **1.** Two *nucleotides on separate DNA strands that are connected through hydrogen bonds. **2.** A unit of measurement of a length of double-stranded DNA.

base saturation The extent to which the exchange sites of the soil's *adsorption complex are 'saturated' (or occupied) by exchangeable basic cations, or by cations other than hydrogen and aluminium, expressed as a percentage of the total *cation exchange capacity.

base surge A turbulent, dilute flow of ash and either water or steam, which expands radially as a collar-like cloud from the base of a vertically venting volcanic eruption column generated by the explosive interaction of magma and water. The surges are commonly cold and wet, consisting of ash mixed with water at temperatures below 100°C. With a high magma : water mass ratio the surges can become dry and hot, consisting of ash mixed with steam. Base surges can be very hazardous, as was the one that occurred during the eruption of Taal, Philippines, 1965. Radially expanding basal clouds observed in nuclear explosions are a type of base surge.

basic dye A dye that consists of an organic cation which combines with and stains negatively charged macromolecules (e.g. nucleic acids). It is used particularly for staining cell nuclei which contain nucleic acids.

basic grassland A vegetation that occurs on soils with an alkaline reaction, particularly those developed in chalk or limestone. In Britain, and north-western Europe generally, basic grasslands include some of the oldest areas of rough grazing, dating from Neolithic times. They are floristically rich and there are numerous characteristic herbs other than grasses.

basic rock Rock that has a relatively high concentration of iron, magnesium, and calcium, and with 45–53 per cent of silica by weight. Examples include gabbro, which is a coarse-grained basic intrusive rock, and basalt, which is a fine-grained basic volcanic (extrusive) rock. *See also* ALKALINE ROCK. *Compare* ACID ROCK and INTERMEDIATE ROCK.

basic soil A soil with a *pH greater than 7.0. *See* BASE. *Compare* ALKALINE SOIL.

basifugal movement *See* ACROPETAL MOVEMENT.

basin A depression, usually of considerable size, which may be erosional or structural in origin.

basin-and-swell sedimentation A form of sedimentation in a region of differential subsidence, where thin, condensed, sedimentary sequences are deposited on slowly subsiding highs or 'swells' and thicker, usually muddier, sediments accumulate in more rapidly subsiding basins between the swells.

basipetal Growing or developing from apex to base, so that the oldest parts are nearest the apex and the youngest are nearest the base. *Compare* ACROPETAL.

basipetal movement Movement of substances toward the basal region of the plant from the root and shoot apices. *Compare* ACROPETAL MOVEMENT.

basophilic Applied to a cell, its components, or products that can be stained by a *basic dye.

Bates, Henry Walter (1825–92) An English naturalist who, in 1848, accompanied A. R. *Wallace on an exploration of the Amazon, where he collected nearly 15 000 species of insects, 8 000 of which were new to science. His studies of them led him to propose the way in which the mimicry named after him can arise among unrelated species. He received enthusiastic support from *Darwin and Sir Joseph Hooker, and Darwin wrote an introduction to his only book, *The Naturalist on the Amazons*, published by John Murray in 1863.

Batesian mimicry Mimicking of brightly coloured, or distinctively patterned, unpalatable species by palatable species. This helps protect the mimics from predators. It is named after H. W. *Bates.

bathy- From the Greek *bathus*, meaning 'deep', used as a prefix applied to the oceans.

b

bathyal zone The oceanic zone at depths of 200–2 000 m, lying to the seaward of the shallower *neritic zone, and landward of the deeper *abyssal zone. The upper limit of the bathyal zone is marked by the edge of the *continental shelf. In marine ecology, it is the region of the *continental slope and rise. It may be geologically active, and include *trenches and *submarine canyons, with underwater erosion producing avalanches.

bathymetry The measurement of the depth of the ocean floor from the water surface; the oceanic equivalent of topography.

bathypelagic fish Deep-sea fish that live at depths of 1 000–3 000 m, where the environment is uniformly cold and dark. A swim-bladder is often absent, and many species (e.g. *Bathydraco scotiae*, dragonfish, and *Chiasmodon niger*, swallower) possess *photophores.

bathypelagic zone The part of the *pelagic zone of the ocean that extends from a depth of 700–1 000 m to 2 000–4 000 m. The temperature averages 10°C at the top of the zone and 4°C at the bottom. The pressure ranges from 7 to 40 MPa. *See* OCEAN DIVISIONS.

batrachotoxins A group comprising the most powerful poisons produced by any vertebrate animal; they are 250 times more toxic than strychnine and in tropical South America are used to poison the tips of blowdarts and arrowheads. The toxins are secreted by *Dendrobates* and *Phillobates* species of South American frogs and are also found in the skin and feathers of *Pitohui* and *Ifrita* species of birds found in New Guinea. The animals do not make the batrachotoxins themselves. Those in the New Guinea birds are acquired from *Choresine* species of beetles on which they feed. The beetles also occur in South America and may also be the source of the frog poison, although this has not been confirmed.

baumgrenze *See* TREE LINE.

bay bar *See* BAR.

bayhead barrier A *barrier beach protecting the head of a bay, but separated from it by a *lagoon.

bayhead beach A sand or shingle *beach in the low-energy environment at the head of a bay. It is typical of irregular coastlines in which bays and promontories alternate.

baymouth barrier A *barrier that partially encloses a bay at its entrance.

Bazin's average velocity equation An equation for calculating the average velocity of water flowing in an open channel, proposed in 1897 by H. E. Bazin (1829–1917). It relates the Chezy discharge coefficient (C) to the hydraulic radius (r) and a channel roughness coefficient ($k1$) by the formula: $C = 157.6/[1 + (k1/r1/2)]$. Other more complex formulae have also been proposed. *See* CHEZY'S FORMULA.

beach An accumulation of *sand and gravel found at the landward margin of the sea or a lake. The upper and lower limits approximate to the position of highest and lowest tidal water levels. The angle of slope and the sedimentary structures of a beach are related to the grain size of the beach materials, the nature of wave activity, and other sedimentary processes active in the area.

beach cusp One of a series of regularly spaced crescent-shaped structures forming local relief along a beach face. The horns or 'headlands' of the cusp are composed of coarse *sand or gravel and point seaward down the beach. The intervening troughs or 'bays' are made up of finer sand. The height of beach cusps is usually in the order of several centimetres, although larger examples have been described. The size and spacing of cusps appears to be related to the nature of waves breaking on the beach. The 'headlands' and 'bays' form distinct *microhabitats for benthic *microbiota.

beach drift The zigzag progression of sand and other debris along a *beach. Particles are driven obliquely up a beach by the *swash and are then returned down the steepest gradient of the beach by the *backwash. The combination of these two movements gives the zigzag progression. *See also* LONGSHORE DRIFT.

beach rock A cemented beach *sand deposit that develops within the intertidal zone by the precipitation of needle-like crystals of *aragonite in the *pore space between the grains. The cementation process is relatively rapid, taking as little as ten years for a lithified rock to develop. The precipitation of the cement is favoured by a warm climate, and may be aided by algal or bacterial action.

bean gall A red or pink, kidney-bean-shaped *gall, found on the leaves of *Salix* species (willow) and caused by the small sawfly *Pontania proxima* (family Tenthredinidae). A second brood laid in the early autumn by the adults that emerged in summer may overwinter in the gall. Sometimes the term 'bean gall' is used more generally to describe any red or brown gall caused by any *Pontania* species. Many galls may occur on a single leaf and, like other gall-causing species, the insects support a large community of *parasites and hyperparasites.

beat up A foresters' phrase meaning to replace dead trees with new ones, especially during the early years of the establishment or re-establishment of a plantation.

Beaufort scale A scale of values, from 0 to 12, for describing wind strength, as defined by Admiral Sir Francis Beaufort (1774–1857) in 1806 (the scale was accepted by the British Admiralty in 1838 and adopted by the International Meteorological Committee in 1874). Each wind force is recognized by its common effects on objects in the landscape (dust, flags, trees, etc.) and on people in the open, or on the state of the sea surface. The highest value on the Beaufort scale, force 12, describes all wind strengths exceeding 121 km/h as hurricanes, but this is inadequate for describing the winds generated by *tropical cyclones and *tornadoes, for which other scales have been devised. *See* FUJITA TORNADO INTENSITY SCALE and SAFFIR–SIMPSON HURRICANE SCALE.

becquerel (Bq) The *SI unit of radioactivity, being the activity of a nuclide that decays with an average of one spontaneous transition per second. The unit is named after the French physicist Antoine Henri Becquerel (1852–1908).

bed load (traction carpet, traction load) The coarser fraction of a river's total sediment load, which is carried along the bed by sliding, rolling, and *saltation. It constitutes 5–10 per cent of the total load.

bed roughness The surface relief developed at the base of a flowing fluid, comprising bedforms (form roughness) and particles projecting from the sediment carpet (grain roughness). Bed roughness is the quantifiable factor which expresses the frictional effect that the bed exerts on the flow. Surfaces whose bed-roughness elements do not project through the viscous sub-layer at the base of the flow are said to be smooth. Surfaces whose bed-roughness elements project through the sublayer are said to be rough.

bee bread A mixture of pollen, honey, and, in some species, plant oils and Dufour's gland exudates, that comprises the food of bee larvae.

beech bark disease A disease of beech trees (*Fagus*) caused by a variety of the fungus *Nectria coccinea*. Symptoms include yellowing of foliage and die-back; infected trees may be killed. The disease is spread from tree to tree by beech scale insects.

beeswax Wax secreted from glands beneath the abdominal terga or sterna of bees of the family Apidae, which is used in nest construction.

beetle analysis A technique for reconstructing past temperatures, using beetle remains that have been identified and *radiocarbon dated. Beetles respond more rapidly than plants to climatic change and many species occur only where the temperature lies within a well defined though rather broad range. Where a number of species occur together, the temperature tolerable for all of them is defined by the overlap of the tolerable ranges for each separate species, giving a much narrower range (the Mutual Climatic Range, MCR), which can be calculated for groups of coexisting fossil species from the known (or measurable) ranges for living beetles. The technique has been used (e.g. by T. C. Atkinson, K. R. Briffa, and

G. R. Coope, reported in 1987) to reconstruct temperature changes in Britain since 22 000 years BP.

beet yellows A virus disease of sugar beet, beetroot, etc. The leaves of infected plants become reddish-yellow, dry, and brittle. The virus is transmitted by aphids.

behavioural ecology **1.** The study of the behaviour of an organism in its natural habitat. **2.** The application of behavioural theories (e.g. *game theory) to particular activities (e.g. *foraging).

behavioural thermoregulation The maintenance of a constant body temperature by means of basking, sheltering, shivering, etc. *See* ECTOTHERM.

Beltian bodies Sausage-shaped organs at the tips of the leaves of certain *Acacia* species found in the African savannah that secrete oils and proteins as food for ants living in nests they have hollowed out in leaf bases; the ants also feed on nectar. In return for the food the ants defend the tree from attack by herbivores and cut away parts of adjacent plants that threaten to shade the acacia's leaves.

belt transect A strip, typically 1 m wide in herbaceous vegetation, that is marked out across a *habitat and within which species are recorded to determine their distribution in the habitat. *See* TRANSECT. *Compare* LINE TRANSECT.

Benedict's test A procedure devised by S. R. Benedict (1884–1936) that is used to detect the presence of a reducing sugar in solution. It depends upon the fact that an alkaline solution of copper (II) sulphate is reduced to insoluble copper (I) oxide by reducing sugars to give a red precipitate.

Benguela Current An oceanic water current that flows northward along the west coast of southern Africa between latitude 15°S and 35°S. It is distinguished by an area of cold upwelling water and is a relatively weak current, flowing at less than 0.25 m/s.

Bentham, George **(1800–84)** A British botanist, the nephew of the utilitarian philosopher Jeremy Bentham for whom he worked as secretary (1826–32), who prepared the *Flora Hongkongensis* (1861), seven volumes of the *Flora Australiensis* (1863–78), and what is considered his most important work, three volumes of *Genera Plantarum*, in collaboration with J. D. *Hooker (1862–83). His *Handbook of the British Flora* was first published in 1858, and its seventh edition in 1924 (the fifth and sixth editions were prepared by Hooker).

benthic fish Fish that live on or near the sea bottom, irrespective of the depth of the sea. Many benthic species have modified fins, enabling them to crawl over the bottom; others have flattened bodies and can lie on the sand; others live among weed beds, rocky outcrops, and coral reefs.

benthic zone The lowermost region of a freshwater or marine profile in which the *benthos resides. In bodies of deep water where little light penetrates to the bottom the zone is referred to as the benthic abyssal region and productivity is relatively low. In shallower (e.g. coastal) regions, where the benthic zone is well lit, the zone is referred to as the benthic *littoral region and it supports some of the world's most productive ecosystems.

benthos **(adj. benthic)** In freshwater and marine *ecosystems, the collection of organisms attached to or resting on the bottom sediments (i.e. *epifauna), and those which bore or burrow into the sediments (i.e. *infauna).

bentonite Montmorillonite-rich *clay that is formed by the breakdown and alteration of volcanic ash and volcanic *tuffs.

Bergen School The name used to distinguish the group of meteorologists (Vilhelm Bjerknes, his son Jacob Bjerknes, H. Solberg, and Tor Bergeron) who, working at the Bergen Geophysical Institute, in Norway, between 1917 and 1920, established the existence and role of *fronts and *air masses in the atmosphere.

Bergeron theory **(Bergeron–Findeisen theory)** A theory proposed around 1930 by T. Bergeron, and subsequently developed by W. Findeisen, that provides a mechanism for the growth of raindrops in ice/water cloud. It is based on the differential values for saturation vapour pressure over ice and supercooled water surfaces. At cloud temperatures of –12 to –30°C air can be saturated over ice but not over water particles, so evaporation can take place from the water droplets, and ice particles can grow by *deposition at the expense of water droplets. When they are large enough the ice particles can fall from the cloud, melting as they pass through lower, warmer air. The process depends on there being a mixture of ice and water, and so may operate in mid- and high-latitude cloud but not in all clouds (e.g. not in tropical clouds which are at temperatures above freezing). *See also* COLLISION THEORY and ICE NUCLEUS.

Bergmann, Karl Georg Lucas Christian (1814–65) An anatomist, physiologist, and biologist, Bergmann was born in Göttingen, Germany, and obtained his medical degree from Göttingen University in 1838. He became professor of anatomy and physiology at Rostock University. In 1847 he proposed the relationship between body size and climate known as *Bergmann's rule.

Bergmann's rule The idea that the size of *homoiothermic animals in a single, closely related, evolutionary line increases along a gradient from warm to cold temperatures (i.e. that races of species from cold climates tend to be composed of individuals physically larger than those of races from warm climates). This is because the surface area: body weight ratio decreases as body weight increases. Thus a large body loses proportionately less heat than a small one. This is advantageous in a cold climate but disadvantageous in a warm one. The rule was proposed in 1847 by the German biologist C. *Bergmann. Its universal validity has recently been called into question as studies of the musk ox in the Arctic have failed to demonstrate its operation. *See also* ALLEN'S RULE and GLOGER'S RULE.

bergschrund A wide and deep crevasse found between a *cirque glacier and its headwall. It forms when the glacier has developed to the stage at which it pulls away from the rock slope on its upper side. A series of small bergschrunds ('bergschrund crevasses') may form instead of the single feature.

berg wind A local wind, generically of the *föhn type, which blows offshore in southern Africa.

Beringia An area comprising the Bering Strait and adjacent Siberia and Alaska. At various times in the late *Mesozoic and *Cenozoic, the Strait was

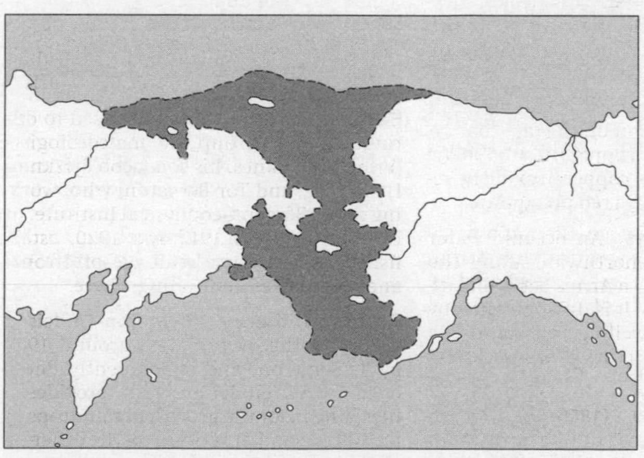

The shaded area indicates the probable extent of dry land

Beringia

dry land and so provided for plants and animals an important migration route between the Palaearctic and Nearctic biogeographical regions.

Bering land bridge A link (*see* LAND BRIDGE) between Siberia and Alaska that has existed intermittently during the *Cenozoic, permitting the migration of species between Eurasia and North America. It provided the means by which human invasion of the Americas from Asia took place at the close of the last glaciation. The Bering land bridge was the only route into North America once the direct link between Europe and North America had been broken by the opening of the Atlantic Ocean.

berm A ridge or nearly flat platform at the rear of a *beach and standing just above the mean high-water mark. Its distinguishing feature is a marked break of slope at the seaward edge.

Bermuda high An anticyclonic cell in the Bermuda region as a westward extension to, or displacement of, the *Azores high-pressure system.

Bernoulli equation An equation that describes the conservation of energy in the steady flow of an ideal, frictionless, incompressible fluid. It states that: $p_1/p_2 + gz + (v^2/2)$ is constant along any stream line, where p_1 is the fluid pressure, p_2 is the mass density of the fluid, v is the fluid velocity, g is the acceleration due to gravity, and z is the vertical height above a datum level. The equation was devised by the Swiss mathematician Daniel Bernoulli (1700–82).

beta decay The radioactive decay of an unstable atom by the emission of a negatively charged beta particle (negatron) from the nucleus, often accompanied by the emission of radiant energy (gamma rays). Beta decay may be regarded as the alteration of a neutron into a proton and an electron. As a result of beta decay the atomic number of the atom is increased by one, while the neutron number is decreased by one (e.g. $C_6^{14} \rightarrow N_7^{14}$).

beta-mesohaline *See* BRACKISH and HALINITY.

bet-hedging The behavioural response of a species population to a *K*-selecting environment in which occasional fluctuations in conditions affect the mortality rate of juveniles to a much greater extent than of adults. In such a situation adults release young into several different environments to maximize the chance that some will survive. *See also* *K*-SELECTION and R-SELECTION.

bevelled cliff A sea cliff whose upper slope has been trimmed to a relatively low angle by *Quaternary *periglacial processes, while the lower part is still steep as a result of recent marine activity. Such cliffs are common in south-west England.

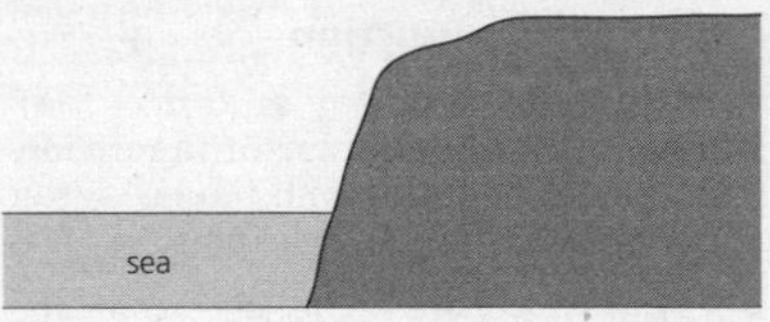

Bevelled cliff

biased gene conversion *Gene conversion where genetic information tends to flow more in one direction than in the other.

bicentric distribution The occurrence of a species or other *taxon in two widely separated places, but nowhere in between. For example, the tulip tree genus (*Liriodendron*) has species in eastern North America and in China, but none in between, and southern beeches (*Nothofagus* species) occur in New Guinea and Australasia, and in Chile, on opposite sides of the Pacific Ocean, but nowhere else. A number of plants endemic (*see* ENDEMISM) to Scandinavia occur only in both of two widely separated montane areas, possibly reflecting the former existence of *nunataks above the late *Pleistocene ice sheet. *Compare* UNICENTRIC DISTRIBUTION.

biennial Applied to a plant that lives for two years. During the first season food may be stored for use during flower and *seed production in the second year. If the plant is damaged, for example by insect attack, it may survive into a third year and flower again. This happens when the cinnabar moth (*Callimorpha jacobaeae*) attacks ragwort (*Senecio* species).

bifid Split in two.

bifurcation In a phylogenetic (*see* PHYLOGENY) tree, the dichotomous forking of an ancestral *branch which indicates a speciation event.

bifurcation ratio A dimensionless number denoting the ratio between the number of streams of one order (*see* STREAM ORDER) and those of the next higher order in a drainage network. It may be a useful measure of proneness to flooding: the higher the bifurcation ratio, the greater the probability of flooding.

big-bang reproduction *See* SEMELPARITY.

Biharian A faunal stage of the middle *Miocene that corresponds to the latter part of the *Calabrian marine stage. It is named from Bihari, Hungary.

bilateral symmetry The arrangement of the body components of an animal such that one plane divides the animal into two halves which are approximate mirror images of each other. Bilateral symmetry is associated with movement in which one end of the animal constantly leads. Sometimes bilateral symmetry is superimposed on another kind of symmetry (e.g. in some echinoids (Echinoidea) where basic pentameral (five-sided) symmetry has a bilateral symmetry superimposed upon it). *Compare* RADIAL SYMMETRY. Flowers can also possess a bilaterally symmetrical structure. For example, a pea flower is bilaterally symmetrical, being symmetrical about the vertical plane only.

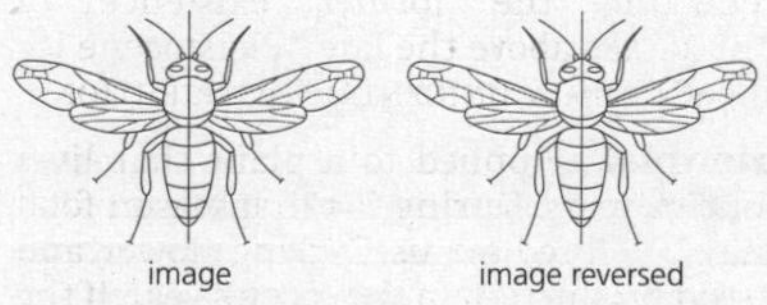

Bilateral symmetry

bilharzia *See* SCHISTOSOMIASIS.

billet A piece of wood that has been cut for fuel.

billow clouds Parallel rolls of cloud with distinct clear areas between the cloud bands, often associated with the cloud variety *undulatus.

bimodal distribution A distribution of data that is characterized by two distinct populations. For example, a bimodal grain size will be characterized by two particle size modes. A bimodal palaeocurrent distribution will exhibit two main current directions (not necessarily opposing directions, which would be termed a 'bipolar' distribution).

binary fission The division of one cell into two similar or identical cells; it is a common method of asexual reproduction in single-celled organisms.

binocular vision Vision that results from the ability of an animal to view an object using both eyes simultaneously, and which probably enables the animal to judge distances. In those mammals and birds which possess binocular vision, the orbits are directed forward. In some arboreal snakes, in which the orbits are directed laterally, binocular vision may result from the possession of eyes whose pupils are long, horizontal slits located eccentrically and close to the anterior margins of the orbits.

binomial classification *See* LINNAEUS.

bio- From the Greek *bios*, meaning 'human life', a prefix associating the word to which it is attached with living organisms or processes.

bioassay The use of living cells or organisms to make quantitative and/or qualitative measurements of the amounts or activity of substances.

biochemical oxygen demand *See* BIOLOGICAL OXYGEN DEMAND.

biochore A *biotic district (i.e. a distinctive *community and environment) and one of many precursors of the *ecosystem concept. The term is also used by *Raunkiaer to describe precise biological boundaries between major climatically determined plant regions and their main subdivisions, as identified by his life-form recording system (i.e. percentages of *phanerophytes, *chamaephytes, etc.).

biochronology The measurement of units of geological time by means of biological events. Biochronologists often derive their correlations from widespread and distinctive events in the biological history of the world, based on the first and last appearances of organisms.

biocide A natural or synthetic substance toxic to living organisms. Some ecologists advocate use of this term instead of *'pesticides', since most pesticides are also toxic to species other than the target pest species. Indirectly, pesticides may also affect non-target organisms detrimentally in many other ways (e.g. by loss of food species or loss of shelter) so that the effects of pesticides may be felt throughout a whole ecosystem. The term 'biocide' indicates this property more clearly than 'pesticide'.

bioclast A single, often broken, shell or fossil fragment.

bioclimatology The study of climate, with particular reference to the environments of living organisms, especially to those of agricultural plants and animals, and humans, together with the disease *vectors affecting humans and commercially important plants and animals.

biocoenosis **1.** The living part of a *biogeocoenosis, comprising the *phytocoenosis, the *zoocoenosis, and the microbiocoenosis. It is equivalent to the biotic component in an *ecosystem. **2.** *See* LIFE ASSEMBLAGE.

biodegradable Applied to substances that are easily broken down by living organisms.

biodeterioration Damage to materials that is caused by living organisms.

biodiversity A portmanteau term, which gained popularity in the late 1980s, used to describe all aspects of biological diversity, especially including *species richness, *ecosystem complexity, and genetic variation.

bio-energetics The study of energy transformations in living organisms, in particular the formation of ATP in photosynthesis and by other mechanisms, and its subsequent use in metabolic processes.

biofacies A rock unit, or an association of rock units, that is characterized by the presence of a fossil assemblage restricted to that particular facies and typical of a specific environment.

biogenesis The principle that a living organism can arise only from another living organism, a principle contrasting with concepts such as that of the spontaneous generation of living from non-living matter. The term is currently more often used to refer to the formation from or by living organisms of any substance (e.g. coal, chalk, or chemicals).

biogenetic law The early stages of development in animal species resemble one another, the species diverging more and more as development proceeds. The law was formulated by the Estonian embryologist of German origin K. E. von Baer (1792–1876).

biogenic Applied to the formation of rocks, traces (fossils), or structures as a result of the activities of living organisms (e.g. on the Isle of Wight, Britain, some of the calcium carbonate in rocks was formed from the remains of the fruits of the aquatic green alga (stonewort) *Chara* species).

biogeochemical cycle The movement of chemical elements from organism to physical environment to organism in more or less circular pathways. They are termed 'nutrient cycles' if the elements concerned are essential to life. The form and quantity of an element varies through the cycles, with amounts in the inorganic reservoir pools usually greater than those in the active pools. Exchange between the system components is achieved by physical processes (e.g. weathering) and/or biological processes (e.g. protein synthesis and decomposition). The latter form the vital negative-feedback mechanisms that regulate the cycles. Cycles may be described as varying from perfect to imperfect. A perfect cycle (e.g. the nitrogen cycle) has a readily accessible abiotic, usually gaseous, reservoir and many negative-feedback controls. In contrast the phosphorus cycle, which has a sedimentary reservoir accessed only by slow-moving physical processes, has few biological feedback mechanisms. Human activities can disrupt these cycles, leading to pollution. Theoretically, perfect cycles are more resilient than imperfect cycles.

biogeochemical exploration *See* GEOBOTANICAL EXPLORATION.

biogeochemistry The scientific study of the effects of living things on subsurface geology; or with the distribution and fixation of chemical elements in the *biosphere. Its principles are applied to the systematic collection and analysis of plants in the exploration for mineral deposits. It is also the study of the chemistry of organic sediments and of the chemical composition of fossils and fossil fuels. *See also* GEOBOTANICAL EXPLORATION.

biogeocoenosis A term equivalent to *ecosystem that is often used in Russian and Central European literature, and attributed to V. Sukachaev who is believed to have coined it in 1947. A biogeocoenosis comprises a *biocoenosis together with its habitat (the ecotope).

biogeographical barrier A barrier that prevents the migration of species. The various disjunctive geographical groupings of plants and animals are usually delimited by one or more such barriers which may be climatic, involving temperature and the availability of water, or physical, involving, for example, mountain ranges or expanses of sea water.

biogeographical province A biological subdivision of the Earth's surface, on the basis of taxonomic rather than ecological criteria, and embracing both faunal and floral characteristics. The hierarchical status of such a unit, and the total number of such units, varies from one authority to another. *See* FAUNAL REGION; FLORAL PROVINCE; and FLORISTIC REGION.

biogeographical region A biological subdivision of the Earth's surface that is delineated on the same general principles as a *biogeographical province but has superior taxonomic status. The provinces are grouped into regions, of which the following are generally recognized: Antarctic, Australasian, *Ethiopian, *Nearctic, *Neotropical, Oceanian, *Oriental, and *Palaearctic. A variant of this grouping has been proposed for mammals (*see* MAMMAL REGIONS). *See also* FAUNAL REGION; FLORAL PROVINCE; and FLORISTIC REGION.

biogeography The study of the geographical distribution of plants and animals at different taxonomic levels, past and present, the *habitats in which they occur, and the ecological relationships involved.

bioherm A build-up of largely *in situ* organisms that produces a *reef or mound of organic origin.

biohorizon **1.** An interface between strata at which some biostratigraphic change has occurred. **2.** A biostratigraphic marker bed.

biointermediate elements Elements that are partially depleted in surface waters as a result of biological activity. Four such elements are known: Ba, Ca, C, and Ra. *Compare* BIOLIMITING ELEMENTS and BIOUNLIMITING ELEMENTS.

biolimiting elements A few elements (N, P, and Si) which are almost totally depleted in surface waters, relative to deep water, by biological activity. As these elements are essential to living organisms, the depletion limits further biological production until the scarce elements are replaced, e.g. by *upwelling. *Compare* BIOINTERMEDIATE ELEMENTS and BIOUNLIMITING ELEMENTS.

biolithite A carbonate rock formed of organisms that grew and remained in place, comprising a rigid framework of organisms, together with associated debris. A *reef represents a typical biolithite.

biological amplification *See* BIOMAGNIFICATION.

biological clock An endogenous, physiological mechanism, whose exact nature has not been determined, that keeps time independently of external events, enabling organisms to determine and to respond to daily, lunar, seasonal, and other periodicities. Its existence has been inferred from the observation of organisms which retain rhythmic activity under constant conditions. *See also* CIRCADIAN RHYTHM.

b

biological conservation Active management to ensure the survival of the maximum *diversity of species, and the maintenance of genetic variety within species. The term also implies the maintenance of *biosphere functions, e.g. *biogeochemical cycling, without which the basic resources for life would be lost. Biological conservation embraces the concept of long-term sustained resource use or sustained yield from the biosphere, which may conflict with species conservation in some circumstances. Conservation of species and biological processes is unlikely to succeed without simultaneous conservation of *abiotic resources.

biological control In general, the control of the numbers of one organism as a result of natural predation by another or others. Specifically, the human use of natural predators for the control of pests, weeds, etc. (e.g. the control of rabbits in Australia by the introduction of the myxomatosis virus, and the control of citrus scale insects in California by the introduction of an Australian species of ladybird). The term is also applied to the introduction of large numbers of sterilized (usually irradiated) males of the pest species, whose matings result in the laying of infertile eggs (e.g. to control the screw-worm fly in the USA).

biological invasion *See* ECESIS.

biological magnification *See* BIOMAGNIFICATION.

biological oxygen demand (BOD, biochemical oxygen demand) An indicator of the polluting capacity of an effluent where pollution is caused by the take-up of dissolved oxygen by micro-organisms that decompose the organic material present in the effluent. It is measured as the weight (mg) of oxygen used by one litre of sample effluent stored in darkness at 20°C for five days.

biological productivity The productivity of organisms and ecosystems, as defined by primary, secondary, and community productivities. *See also* PRIMARY PRODUCTIVITY.

bioluminescence The production by living organisms of light without heat. Bioluminescence is a property of many types of organism (e.g. the fruit bodies of certain fungi, certain (mostly marine) bacteria, *dinoflagellates, and fireflies).

biomagnetism The influence of the magnetic field on living organisms, which is strong in some species. Certain *Spirillum* bacteria orient themselves along the geomagnetic lines of force. Higher organisms (e.g. snails, bees, birds, porpoises, and possibly humans) appear able to sense and utilize geomagnetic field directions for purposes of orientation. Strong magnetic fields may be detrimental to health in humans, but the effects of medium strength fields have not been studied. Magnetic fields are also used for scanning biological organisms.

biomagnification (biological amplification, biological magnification) An increase in the concentration of a chemically stable substance along a *food-chain, in extreme cases leading to physical ill effects.

biomass (standing crop) The total mass of all living organisms (*producers, *consumers, and *decomposers) or of a particular set (e.g. species), present in an *ecosystem or at a particular *trophic level in a *food-chain, and usually expressed as dry weight or, more accurately, as the carbon, nitrogen, or calorific content per unit area. Biomass is a quantity per unit area; productivity is a rate of biomass gain per unit area; *see* PRIMARY PRODUCTIVITY.

biome A biological subdivision that reflects the ecological and physiognomic character of the vegetation. Biomes are the largest geographical biotic communities that it is convenient to recognize. They broadly correspond with climatic regions, although other environmental controls are sometimes important. They are equivalent to the concept of major plant formations in plant ecology, but are defined in terms of all living organisms and of their interaction with the environment (and not only with the dominant vegetation type). Typically, distinctive biomes are recognized for all the major climatic regions of the world, emphasizing the adaptation of living organisms to their environment, e.g. *tropical rain-forest biome, *desert biome, *tundra biome.

biometrics *See* BIOMETRY.

biometry **(biometrics)** Quantitative biology, i.e. the application of mathematical and statistical concepts to the analysis of biological phenomena.

bionomic strategy The characteristic features of an organism or population (e.g. size, longevity, fecundity, range, and migratory habit) that give maximum fitness for the organism in its environment.

biophage *See* CONSUMER ORGANISM.

biophile Applied to those elements required by, or found in, living plants and animals, including C, H, O, N, P, S, Cl, I, Br, Ca, Mg, K, Na, V, Fe, Mn, and Cu.

biospecies A group of interbreeding individuals that is isolated reproductively from all other groups.

biosphere The part of the Earth's environment in which living organisms are found, and with which they interact to produce a steady-state system, effectively a whole-planet *ecosystem. Sometimes it is termed 'ecosphere' to emphasize the interconnection of the living and non-living components.

biosphere reserve One of a series of conservation sites designated by the United Nations Educational, Scientific, and Cultural Organization (UNESCO) in an attempt to establish an international network of protected areas encompassing examples of all the Earth's major vegetation and physiographic types. Biosphere reserves contain virgin vegetation, plus various kinds of cultural landscape, in the whole of which conservation is practised.

biostratigraphic unit A unit of strata that is characterized by a particular content of fossils which were deposited at the same time as the sediments and distinguish the unit from adjacent strata. A biostratigraphic unit may be of chronostratigraphic or environmental significance.

biostratigraphy The branch of stratigraphy that involves the use of fossil plants and animals in the dating and correlation of the stratigraphic sequences of rock in which they are discovered. A zone is the fundamental division recognized by biostratigraphers.

biostrome A layered, sheet-like accumulation of *in situ* organisms. It differs from a *bioherm in its geometry, lacking a mound-like or *reef-like form.

biosynthesis The formation of compounds by living organisms.

biota Plants and animals occupying a place together (e.g. marine biota, terrestrial biota).

biotelemetry A remote-sensing method for monitoring animal movements which uses small transmitters attached externally to the body of the animal or implanted within the body cavity. It enables the precise locations of animals to be followed, and may also monitor changes in heartbeat (e.g. in response to stress). It is especially useful for studying migratory animals in remote areas and is an important aid to the study of bird migration. *See also* RADAR TRACKING and RADIO TRACKING.

biotic Applied to the living components of the *biosphere or of an *ecosystem, as distinct from the *abiotic physical and chemical components.

biotic association A community of plants and animals. *See also* COMMUNITY and PLANT ASSOCIATION.

biotic climax *See* PLAGIOCLIMAX.

biotic factor The influence upon the environment of organisms owing to the presence and activities of other organisms (e.g. the casting of shade and competition), as distinct from a physical, *abiotic, environmental factor.

biotic indices *Indicator species, when used as a guide to the level of a particular *abiotic factor. For example, the presence of certain invertebrate groups in fresh water can be awarded a score that indicates the quality of the water. A scheme employing biotic indices has a scale ranging from 10 (clean water with diverse fauna) to 0 (grossly polluted water with no fauna or with only a few anaerobic organisms).

biotic potential **(intrinsic rate of natural increase)** The maximum reproductive potential of an organism, symbolized by the letter *r*. The difference between this and the rate of increase that actually occurs under field or laboratory conditions reflects the environmental resistance. *See also* LOGISTIC EQUATION.

biotope An environmental region characterized by certain conditions and populated by a characteristic *biota.

biotopographic unit **1.** A small *habitat unit with distinctive topography formed by the activities of an organism (e.g. an ant hill). **2.** A small topographic unit that, by its aspect, position, or other characteristics, generates a distinctive micro-environment for living organisms. Examples include solar or shade slopes, windward slopes, and similar units in sand-dunes.

biotroph A parasitic organism that obtains its nutrients from the living tissues of its host organism. *See* PARASITISM.

bioturbation The disruption of sediment by organisms, seen either as a complete churning of the sediment that has destroyed depositional sedimentary structures, or in the form of discrete and clearly recognizable burrows, trails, and traces (*see* TRACE FOSSIL).

biotype **1.** A naturally occurring group of individuals with identical *genomes. **2.** A physiological race (i.e. a group of individuals identical in structure but showing differences in physiological, biochemical, or pathogenic characters).

biounlimiting elements Elements (e.g. B, Mg, Sr, and S) that show no measurable depletion in surface waters compared to deep waters as a result of biological activity. *Compare* BIOLIMITING ELEMENTS and BIOINTERMEDIATE ELEMENTS.

biozone The total range of a given species defined within specific time limits or all the rocks laid down in the time interval during which that taxon existed.

bipectinate Resembling a comb in arrangement or shape, in which the 'teeth' occur on both sides of the main stem. The term is most commonly applied to insect antennae.

bipolar distribution **1.** The distribution of an organism that is found in the high latitudes of both hemispheres, e.g. the plant *Koenigia islandica* is found both in the Arctic and in Tierra del Fuego at the southern tip of South America. **2.** *See* BIMODAL DISTRIBUTION.

biradial symmetry The arrangement of the body components of an animal such that similar parts are located to either side of a central axis and each of the four sides of the body is identical to the opposite side but different from the adjacent side.

biramous Two-branched.

birds A class (Aves) of endothermic (*see* ENDOTHERM) vertebrates that are adapted for flight, bipedal walking or running, and, in some species, swimming on or below the surface of water; flightless species (ratites) are believed to have diverged from flying birds and subsequently to have lost their adaptations for flight. The bones are light and often tubular, sometimes strengthened by internal struts, the body is covered with feathers, the forelimbs are modified to form wings, and the jaws, which are usually long and slender, lack teeth and support a horny bill. Many of the bones contain extensions of the air sacs. Except in ratites, which lack it, a keel on the sternum provides attachment for powerful flight muscles. Birds are descended from archosaurian reptiles and retain a number of reptilian characteristics (e.g. the arrangement of the parts of the skull and the scaly covering of the legs and feet; feathers are also derived from scales). The skin is thin and lacks sweat glands. There are about 8 700 species, with a worldwide distribution.

bise The local name for a northerly winter wind affecting mountainous regions of southern France (in Languedoc the *bise noire* is associated with heavy cloud).

biserial Side by side.

bisexual Applied either to a species comprising individuals of both sexes, or to a hermaphrodite organism (in which an individual animal possesses both ovaries and testes and an individual plant possesses both stamens and pistils in the flower).

Biston betularia **(peppered moth)** A moth (order Lepidoptera) which has been the subject of classical studies of industrial melanism, where there is selective predation by birds upon light moths that are conspicuous against a dark, sooty background, and conversely upon dark moths (the *carbonaria* form) against a light, soot-free background.

bitter lake A saline lake which is rich in sulphates and carbonates, dominated by high concentrations of sodium sulphate.

bivalent A pair of *homologous chromosomes that are paired during the first stage of *meiosis.

black box system A system about whose structure nothing is known beyond what can be deduced from its behaviour. Statistical relationships between inputs to the system and outputs from it can be deduced by manipulating the inputs.

black earth *See* CHERNOZEM.

black ice A type of glazed frost (e.g. on roads or the superstructures of ships), caused when water falls on a surface that is below freezing temperature. The thin sheet of ice is rather dark in appearance and, unlike white hoar frost or rime, may be hard to see. *See also* CLEAR ICE.

blackleg The name given to a number of plant diseases in which symptoms include blackening of the base of the stem, often followed by the collapse of the stem. The diseases may be caused by any of several fungi or, less commonly, bacteria.

black mildew **(dark mildew)** A plant disease caused by a fungus belonging to the family Meliolaceae (order Dothideales). The name may also refer to the fungus itself. *Compare* SOOTY MOULD.

black smoker *See* HYDROTHERMAL VENT.

black-stem rust A disease which affects a wide range of cereals and other grasses. Symptoms include the appearance on stems and leaves of patches of reddish-brown spores. The causal agent is a fungus, *Puccinia graminis*; this organism requires two hosts to complete its life cycle, the other host being the barberry (*Berberis*).

blanket bog An *ombrogenous peat *bog community, typical of flat or moderately sloping areas in very wet, oceanic climates with high *humidity. In the British Isles, blanket bogs are widespread on the Pennine summits of England, in northwest Scotland, in Wales, and in parts of Ireland. In other parts of the world, this oceanic type of peatland is also found in Norway, Newfoundland, Tierra del Fuego, the Falkland Islands, and the Ruwenzori Mountains of East Africa.

blast disease A common disease of rice, caused by the fungus *Pyricularia grisea*. Lesions develop on the leaf-sheaths and on the stems, and the weakened stems are easily broken.

blastochory Dispersal of a plant by means of offshoots.

blastozooid A member of a colony of animals which are produced by asexual budding.

bleicherde Ash-grey soil that forms the leached layer in a *podzol soil.

blending inheritance Inheritance in which the characters of the parents appear to blend to form an intermediate state in the offspring, and in which there is no apparent segregation in later generations. The concept was proposed originally by biologists in the nineteenth century, including *Darwin, but later it was discredited as a model of inheritance after the results of *Mendel's experiments had been recognized.

blight A non-specific term applied to any of a wide range of unrelated plant diseases. The causal agent is usually fungus-like. *See* (for example) POTATO BLIGHT.

blind pores *See* EFFECTIVE POROSITY (2).

blizzard A storm of blowing snow with high winds and low temperatures.

Blizzards are a notable climatic feature of the northern and central parts of the USA in winter, and are related to depression tracks. In the USA, a blizzard is defined by the National Oceanic and Atmospheric Administration (NOAA) as a storm with winds of at least 56 km/h, temperatures below −6.7 °C, and enough falling or blowing snow to reduce visibility to less than 0.4 km. In a severe blizzard, wind speed is at least 72.5 km/h, temperatures below −12.2 °C, and visibility is close to zero.

blockfield **(felsenmeer)** A spread of coarse, angular, frost-shattered rock debris resting on a level or gently sloping upland surface, and found in present or former *periglacial environments. *See also* FROST WEDGING.

block glide The sliding movement of a large block of rock over a surface that has been lubricated. Alternatively, the block may be carried downslope by the plastic deformation of underlying material.

blocking In synoptic meteorology, the establishment in the mid-latitudes of a high-pressure system that interrupts or diverts for a considerable period the typically eastward movement of *depressions and other synoptic features in the zonal flow. Over western Europe, for example, blocking often forces depressions to move northward towards Scandinavia or southward over southern France and Spain.

blocking anticyclone *See* BLOCKING HIGH.

blocking high **(blocking anticyclone)** In *synoptic meteorology, an *anticyclone with deep circulation so placed as to interrupt or divert the eastward movement of the typical succession of low-pressure features (depressions) in the zonal flow of mid-latitudes. At such times the usual mid-latitude zonal flow is diverted into more meridional flow around the high-pressure area, where settled weather can then result. *See also* AIR MASS.

blood rain Reddish-coloured rainfall, caused by dust particles that have been lifted up from arid areas and carried long distances by the winds before they are washed out in precipitation. Saharan red dust sometimes occurs in rainfall over parts of Europe, even as far north as Finland.

bloom *See* ALGAL BLOOM.

blow-hole *See* COASTAL PROCESSES.

blow-out **1.** A wind-eroded section of a sand *dune that has been largely stabilized by vegetation. *Erosion results from a break in the vegetation cover, typically caused by overgrazing or recreational pressure. A parabolic dune may result. **2.** The sudden, often catastrophic release of oil and gas that results from the failure of equipment regulating the flow from a well.

blue Moon **(blue Sun)** The occasional appearance of the Moon or Sun when partly obscured by large particles in the atmosphere, as in dust storms, or following forest fires or great volcanic explosions. When the Moon or Sun is viewed through dust or smoke trails it usually appears to be very white, but when the suspended particles are predominantly of one size it sometimes appears blue, at other times green or orange. The phenomenon is attributed to diffraction, although no full explanation seems to be known. The smaller the particles, the more the colour of the Sun or Moon tends to the blue end of the spectrum. The phenomenon is believed to be more common in China than elsewhere. (In the neighbourhood of the 1883 Krakatoa volcanic explosion the Sun was seen as an azure blue sphere.)

blue-stain A common type of staining of freshly cut timber, caused by the growth of fungi (particularly species of *Ceratocystis*) in the sapwood.

blue Sun *See* BLUE MOON.

BMR *See* BASAL METABOLIC RATE.

BOD *See* BIOLOGICAL OXYGEN DEMAND.

bog A plant *community of acidic, wet areas. Decomposition rates in it are slow, favouring *peat development. In Britain and high northern latitudes typical plants include bog-mosses (*Sphagnum* species), sedges (e.g. *Eriophorum* (cottongrass) species), and heathers (e.g. *Calluna vulgaris* and *Erica tetralix*). Insectivorous plants (e.g. sundews, *Drosera* species) are especially characteristic; they compensate for low nutrient levels by trapping and digesting insects. Strictly the term 'bog' should be applied only to *ombrotrophic peatland, that is, those receiving water only from precipitation. These peatlands may be *raised bogs, *blanket bogs, or tropical bog forests, such as those found in the coastal regions of Sarawak. Sometimes the term bog is applied to other types of peatland, such as 'valley bog' and 'string bog', but these are *rheotrophic *mires and should be regarded as *fens rather than bogs.

bog forest A forest that develops as a *bog fills and dries. In boreal North America and Eurasia bog forests develop in post-glacial *kettle holes that filled with water and were later colonized by bog plants, especially *Sphagnum* mosses. As the sites dry out further, trees become established, eventually producing forests typically dominated by *Picea mariana* (black spruce), *Larix laricina* (tamarack), and *Pinus banksiana* (jack pine), with associated herbs and shrubs. In tropical regions bog forests develop in coastal regions; those in Asia are often dominated by *Shorea albida* (alan). *See also* OMBROGENOUS BOG.

bog peat *See* PEAT.

bole The trunk of a tree.

bolling The permanent trunk of a pollarded tree.

Bølling interstadial A relatively warm period that occurred towards the end of the last (*Devensian) glaciation in north-western Europe, and which is named after the type site in Denmark. The event took place about 13 000–12 000 *radiocarbon years BP. In Britain, the Bølling is not usually distinguished from the *Allerød. *See also* DRYAS.

bolochory Dispersal of spores or seeds by means of propulsive mechanisms.

bolson An *intermontane basin that extends from the divide of one block-faulted mountain to the divide of the adjacent mountain. The surface form is made up of mountain front, *bajada, *pediment, and *playa. It is classically described for the basin-and-range province of the western USA.

bomb calorimeter A device for measuring the energy content of material. The sample is burned in an oxygen-rich atmosphere inside a sealed chamber that is surrounded by a jacket containing a known volume of water. The rise in temperature of the water is recorded and used to calculate the amount of heat produced.

bonding Social behaviour that tends to keep individuals together as cohesive groups (e.g. as herds, flocks, or schools) or as pairs (as in parent-infant or mating pairs).

bone The skeletal tissue of vertebrates, which has a greater potential for preservation than cartilage, but is rarely found as intact skeletons. Bone consists of cells arranged regularly in a matrix mainly of collagen, heavily impregnated with calcium phosphate, which accounts for more than half the total weight. There are two main types: (*a*) endochondral bone, which forms the vertebrae and inner skull, develops from cartilaginous rudiments in the embryo; and (*b*) dermal bone, which develops directly in tissues beneath the skin without a cartilaginous precursor. It forms the scales in fish and the outer bones of the skull, the growth patterns of which are characteristic for each taxonomic group. Teeth are also derived from dermal bone, but have a denser structure. They are composed largely of dentine covered by hard enamel. Teeth are commonly preserved as *fossils and are of great diagnostic importance.

bootstrapping In *phylogenetics, a statistical procedure for testing the robustness of a particular tree *topology. Columns are randomly selected from an *alignment in order to construct a new alignment from which a new tree is calculated, and the process is repeated 100–1 000 times. A bootstrap score is then assigned to each *branch of the original tree. This score is equal to the number of randomly-generated trees containing that branch.

bor Open woodlands dominated by *Pinus sylvestris* (Scots pine) which occur on dry sand-plains within the closed-forest subzone of the *boreal conifer belt in

Russia. Similar vegetation types are known in Canada.

bora The local name for a cold and typically very dry wind from the north-east, blowing down from the mountains on the eastern side of the Adriatic Sea, which is most common in winter on northern Adriatic coasts. The wind is probably a consequence of continental high pressure in central Europe with low pressure to the south in the Mediterranean. It is often accompanied by much precipitation when associated with *depressions in the Adriatic. In other areas the name is used generically for cold *squalls moving downhill from uplands.

bore A very rapid rise of the tide, in which the advancing flood waters form a wave with an abrupt front. Bores occur in certain shallow *estuaries and river mouths where there is a large tidal range and suitably funnel-shaped regions (e.g. the Amazon, the Bay of Fundy, the Tsing Kiang River in China, and the Rivers Severn, Trent, and Ouse in England).

boreal Pertaining to the north (from Boreas, the Greek god of the north wind).

boreal climate The climate associated with the boreal (*taiga) forest zone of Eurasia, where it extends to 65–70°N in the west and 50°N in the east, and North America, where it extends from the fringe of the *tundra southwards to 55°N in the east. Winters are long and cold, with temperatures below 6°C for 6–9 months, and summers short, with temperatures averaging more than 10°C. Precipitation, as snow in winter, typically amounts to 380–635 mm per annum.

boreal forest The circumpolar, subarctic forest of high northern latitudes that is dominated by conifers. To the north it is bounded by *tundra and to the south by temperate, broad-leaved, deciduous forest, steppe, or semi-*desert. The American and Asian parts of the forest are floristically more diverse than the European parts.

Boreal Period A period in post-glacial (i.e. post-*Devensian or *Flandrian) times from about 8 800–7 500 years BP, which preceded the *climatic optimum of early Atlantic times (*see* ATLANTIC PERIOD). Pollen records (*pollen zone) typically show an increasing abundance of *thermophilous tree species and also indicate the drier, more continental conditions that characterized the ensuing Atlantic period. The early Boreal corresponds to late Pollen Zone V; otherwise the Boreal is linked with Pollen Zone VI, which is sometimes subdivided to give Zones VIa, VIb, and VIc, according to the most abundant tree *pollen represented. For Britain the Boreal Period is significant as the last period in post-glacial times in which Britain was joined to mainland Europe by a land bridge across the Dover Strait.

Boreal Realm An area in which the fauna or flora has northern affinities. At certain times in the *Cenozoic this implies a degree of coldness; in the *Mesozoic it is less specific, and is compared with the *Tethyan Realm.

boreal zone *See* CIRCUMBOREAL DISTRIBUTION.

bornhardt A rounded, often isolated hill developed in massive rock that is found in the humid tropics. Its shape is controlled by large-scale exfoliation or sheeting joints. Such hills are sometimes called 'sugar-loaf' hills, after the granitic dome of the Sugar Loaf, Rio de Janeiro, Brazil.

boron **(B)** An element that is commonly found as boric acid in the soil solution, and which is essential for healthy plant growth. It forms complexes with many organic molecules and is involved in phenolic metabolism and in membrane function, but its primary role is not yet understood.

boscus **(subboscus)** The wood or undergrowth produced by *coppice growth.

bottleneck A severe reduction in population size, often leading to a *founder effect. Bottleneck events are commonly followed by rapid population expansion in which the rate of loss of new lineages is greatly reduced, giving rise to *star phylogenies.

bottomset beds **1.** The part of a cross-bedded set of sediments that forms at the base of the down-current or lee-side of a dune- or ripple-form structure. **2.** Offshore clays formed at the base of a prograding deltaic sequence. *See also* DELTA and PROGRADATION.

bottom water The water mass that lies at the deepest part of the water column in the ocean. It is relatively dense and cold (e.g. the North Atlantic bottom water has a temperature of 1–2 °C). Bottom-water circulation is slow-moving, greatly influenced by sea-bed topography, and driven by differences in water density (*thermohaline circulation).

boulder clay A glacial deposit that consists of boulders of varying size in a clay-dominated matrix, laid down beneath a valley glacier or ice sheet. Typically it is unstratified and unsorted, and characterized by rock types derived from the country crossed by the depositing glacier. Boulder clay is an older term for what is now more usually known as boulder *till.

boundary current The northward- or southward-directed ocean-water current which flows parallel and close to a continental margin. Such currents are caused by the deflection of eastward- and westward-flowing currents by the continental land masses. Boundary currents on the western margins of ocean basins, such as the *Gulf Stream and the *Kuroshio Current, are deep, narrow, fast-moving currents; currents along the eastern boundaries, such as the *Canaries Current and the *California Current, tend to be relatively shallow, broad, diffuse and slow-moving.

boundary layer The layer of air that lies immediately adjacent to a surface and in which the atmospheric conditions are strongly conditioned by contact with the surface. The planetary boundary layer comprises the air between the surface and an average altitude of about 500 m, within which the air is strongly affected by surface conditions and wind speeds are reduced by friction with the surface. *See* FREE ATMOSPHERE.

boundary stratotype A specified rock section within which the time line ('golden spike', boundary zone) occurs that marks the standard demarcation between *chronostratigraphic units. In practice such time lines are usually based on either the appearance or the disappearance of a key species (*see* INDEX FOSSIL) or other taxon. Associated faunas and sediments may transgress zonal boundaries. The term 'boundary stratotype' has also been used in the sense of the time line itself.

boundary zone A time line that is based on either the appearance or the disappearance of a key species or fauna. Associated faunas and sediments may transgress a zonal boundary.

Bowen's ratio The ratio of sensible heat to latent heat transport from the ground to the atmosphere, which is generally calculated from the ratio of the vertical gradients of vapour pressure and temperature. It is used in the assessment of evaporation. It was described by I. S. Bowen in 1926.

box A short DNA sequence that performs a regulatory function with respect to a gene. Box sequences are usually under a high *functional constraint.

box model *See* COMPARTMENT MODEL.

BP Before the present (which is taken to be 1950). The initials should not be confused with BC. The year 1950 is taken as 'present' because the term BP is usually applied to radiocarbon dates (*see* RADIOCARBON DATING) and the atmospheric testing of atomic weapons has greatly modified the radiocarbon content of all organisms since about 1950, so this is taken as the datum horizon for this method.

Bq *See* BECQUEREL.

brachiation In some arboreal primates, a form of locomotion in which an animal swings hand over hand from branch to branch. In those species in which it is developed fully (species of gibbons (Hylobatidae) and apes (Pongidae)), apes are said to be able to move more quickly through the trees than a human can walk on the ground below.

brachypterous Applied to insects in which both pairs of wings are reduced.

bracket fungus Any one of a large group of fungi (order Aphyllophorales), many of which produce fruiting bodies that project from the trunks of trees. *Heterobasidion annosum* (*Fomes annosus*) is a parasite of coniferous trees that is the principal cause of decay in conifers.

brackish Applied to water that is *saline, but less so than sea water. According to the Venice system brackish waters are classified by the chlorine they contain and divided into zones. The zones, with their percentage chlorinity (mean values at limits), are: euhaline 1.65–2.2; polyhaline 1.0–1.65; mesohaline 0.3–1.0; alpha-mesohaline 0.3–0.55; beta-mesohaline 0.55–1.0; oligohaline 0.03–0.3; fresh water 0.03 or less.

bradycardia A condition in which the heart rate is reduced substantially.

bradytely An exceedingly slow rate of evolution, manifested by slowly evolving lineages which survive much longer than would normally be expected. Living fossils (e.g. the coelacanth) and groups that have remained relatively stable with time (e.g. opossums and crocodiles) represent the low end of the bradytelic range. *Compare* HOROTELY and TACHYTELY.

braided stream (braided channel, braided river) A stream whose plan form consists of a number of small channels separated by *bars. The bars may be vegetated and stable (e.g. the eyots (small islets) in the River Thames, England) or barren and unstable (as at glacial margins, where rapid changes occur).

brake An area of bracken (or other fern), scrub, or underwood.

branch The graphical representation of an evolutionary relationship in a phylogenetic (*see* PHYLOGENY) tree.

branchial Of the gills.

branching decay (dual decay) The decay of an isotope by different methods to two or more different end-members. For example, ^{40}K decays either to ^{40}Ar (12 per cent) by positron emission and electron capture, or to ^{40}Ca (88 per cent) by emission of a negative beta particle.

Brandenburg One of a series of *moraines which mark the southern limit of Weichselian ice (*see* DEVENSIAN), extending some 500 km across the north German Plain. A Russian equivalent extends a further 2 000 km into European Russia.

Brandon An episode in the *Upton Warren interstadial, which occurred within the Würm (*Devensian) Glacial.

brash A forester's term, meaning to cut off the lower branches of young trees, especially conifers, in plantations.

Braun-Blanquet, Josias (1884–1980) A Swiss ecologist who devised the most widely used European phytosociological method for the description of vegetation communities, and who was generally acknowledged as the foremost authority on such schemes. He developed the framework for his system in his doctoral study of the vegetation of the central Cévennes (1915), and the fully developed method was published in his authoritative textbook *Pflanzensoziologie* (Berlin, 1928) which he revised and updated most recently for a third, enlarged edition published in 1964. He worked at, and for many years was director of, the Station Internationale de Géobotanique Méditerranéenne et Alpine (SIGMA) at Montpellier, France. *See also* ZURICH–MONTPELLIER SCHOOL OF PHYTOSOCIOLOGY.

braunerde *See* BROWN EARTH.

Brazil Current Warm water that forms the oceanic *boundary current flowing southward along the Brazilian coastal margin. It is marked by its high *salinity (36.0–37.0 parts per thousand), its low velocity, and its shallowness (100–200 m) in comparison to its northern hemisphere equivalent, the *Gulf Stream.

breaker A wave that is collapsing or breaking as a result of the wave approaching the shore and reaching shallower water. The decreasing water depth causes the wave length and speed to decrease and the wave height to increase. Consequently wave steepness increases and the wave becomes unstable; it breaks when the wave height is about 0.8 times the water depth. Several types of breaker have been described, e.g. spilling breakers (in which the wave breaks forward, with broken water spilling down the front of the wave), and plunging breakers (in which the wave crest curls over a large air pocket and falls vertically into the trough).

breast height The standard height, 1.3 m above the ground, where foresters measure the girth of a tree in order to calculate its timber volume using *dimension analysis.

breckland A distinctive locality in south-western Norfolk and north-western Suffolk, England, with a mosaic of grass heath and *Pteridium aquilinum* (bracken) and *Calluna vulgaris* (heather or ling) communities. Forest once grew over the breckland. *Pollen analysis has demonstrated that the forest was cleared by Neolithic farmers.

breed An artificial mating group derived (by humans) from a common ancestor, usually for agriculture (e.g. domesticated animals and crop plants), or for genetic analysis, or for pleasure (e.g. cats and dogs).

breeding dispersal *See* DISPERSAL and PHILOPATRY.

breeding size The number of individuals in a population involved in reproduction during a particular generation. It does not include the non-breeding element of the population.

breeding true Producing offspring with *phenotypes for particular characters (i.e. organisms may breed true for some characters but not for others) that are identical to those of the parents. *Homozygous individuals necessarily breed true (unless *mutations arise), whereas *heterozygotes rarely do so.

breeze A relatively light wind, often of convective origin, or a particular local air movement (e.g. mountain, *land and sea breezes). On the *Beaufort scale breezes are classed as light (force 2, 6.4–11.2 km/h), gentle (force 3, 12.8–19.3 km/h), moderate (force 4, 20.9–28.9 km/h), fresh (force 5, 30.5–38.6 km/h), and strong (force 6, 40.2–49.8 km/h).

brezales *See* MEDITERRANEAN SCRUB.

brickearth A fine-grained, silty deposit found in south-eastern England. Of complex origin, it probably resulted from the reworking of *loess, either by hillslope washing or by redeposition in standing water.

brigalow scrub The semi-arid scrub vegetation that occurs in parts of Australia; *Acacia* species provide the main vegetation.

brine A concentrated solution of inorganic salts, formed by the partial evaporation of saline waters.

bristlecone pine Two pine species from California which are famous for their longevity and have been used to develop an exceptionally long, arid-site, tree-ring chronology. The oldest living specimens date back more than 4 600 years, but cross-dating these with remnants of dead bristlecone pines has extended the arid-site chronology to more than 8 200 BP. A 5 500-year chronology has been developed for bristlecone pines at the upper tree limit. These pines are also used to calibrate the *radiocarbon-dating method to allow for fluctuations in atmospheric $^{14}C:^{12}C$ ratios as revealed by measuring the $^{14}C:^{12}C$ ratios of individual *tree rings in the long, absolutely dated, tree-ring series. *Pinus longaeva* is the Great Basin bristlecone pine and *P. aristata* is the mountain bristlecone pine.

broad In East Anglia, England, the name given to a freshwater lake, usually fringed by reeds, which is connected to a slow-flowing river near to its estuary. It is derived from medieval peat diggings which subsequently flooded.

broad-leaved evergreen forest Forest dominated by broad-leaved, *evergreen tree species. North of the tropics, temperate, broad-leaved forests occur in the *formation types: (*a*) broad-leaved evergreen, and (*b*) broad-leaved *deciduous, the evergreen variants being restricted to the coastal plain of the Gulf of Mexico, southern Japan, and central China. South of the tropics, however, the broadleaved forests are mainly evergreen, the only exception occurring in Chilean Patagonia. This forest type requires plentiful, well-distributed rainfall.

bromatia The swellings that develop at the tips of hyphae of fungi cultivated by parasol ants; bromatia serve as food for the ants.

Brongniart, Adolphe Théodore (1801–76) A French botanist, who became a professor at the Muséum d'Histoire Naturelle, Paris, in 1831 and retained the position until his death. He was particularly noted for his work on the classification and distribution of fossil plants and their relationships with existing forms. In 1854 he founded the Société Botanique de France and was its first president.

Brongniart, Alexandre (1770–1847) Professor of mineralogy at the Muséum d'Histoire Naturelle, Paris, Brongniart collaborated with *Cuvier in his work on the mapping of the Paris basin and demonstrated that lower taxonomic orders of animals were found lower in the stratigraphic column, and hence gave evidence of progression in the succession of life.

brood **1. (noun)** All of the offspring that hatch from a single *clutch of eggs. **2. (verb)** To incubate eggs.

brood parasite (nest parasite) An animal which lays its eggs in the nest of a member of its own or another species. The recipient (host) raises the young which, in some species (e.g. *Cuculus canorus*, cuckoo), eject from the nest or kill the natural offspring of the host.

Brørup (Loopstedt) An *interstadial that occurred during the last (Weichselian) glacial (*see* DEVENSIAN). Named from a place in Jutland, it dates from about 60 000 years BP and is perhaps the equivalent of the *Chelford interstadial of the British Isles. Estimated mean July temperatures for this period are between 15 and 20 °C.

brotochory *See* ANDROCHORY.

Brown, Robert (1773–1858) A British botanist who contributed greatly to the adoption of a natural system of plant taxonomy, but who is best known for his discovery of Brownian motion (the continuous, random movement of very small (about 1 μm diameter) particles in a fluid that is caused by collisions with molecules of the fluid). He was the first person to distinguish between the *angiosperms and *gymnosperms, made a particular study of sexual processes in higher plants, originated the microscopical study of fossil plants, and strongly influenced the study of plant geography. In 1800–5 he took part in a botanical survey of the northern coast of Australia and returned to Britain with specimens of nearly 4 000 species, many of them unclassified. He was appointed librarian of the Linnean Society of London, where he remained until 1822, and from 1810 was librarian and botanist to Sir Joseph Banks, inheriting the use of his library and collections. These were transferred to the British Museum in 1827, with Brown as keeper, and when the botany department of the Museum was established in 1835, following the acquisition of the Sloane collection, Brown was appointed keeper, a position he held until his death.

brown clay *See* RED CLAY.

brown earth A freely draining soil-*profile type with only slight *horizons. It has a mull *humus in the surface horizon and very little differentiation of horizons below. Brown earths are well-weathered and slightly leached soils, with a *cambic horizon in the middle part of the profile (also known as braunerde and now included in the *Inceptisols of the *USDA Soil Taxonomy). Brown-earth types of soil are very productive and, although their natural *climax vegetation in humid, temperate latitudes is deciduous forest, they have been used extensively for agriculture. The soil type is common in southern England. *See also* ALFISOLS.

brown forest soil A little used soil-*profile term that has been applied to both acid *brown earths and *brown podzolics.

brown oak **1.** The wood from an oak tree (*Quercus*) that has been infected with *Fistulina hepatica* (beefsteak fungus). The wood has a rich, dark brown colour. Although mechanically weaker than wood from uninfected trees, it is nevertheless highly prized by cabinet-makers for decorative purposes. **2.** A disease of oak trees caused by *Fistulina hepatica*.

brown podzolic soil A freely draining, leached soil *profile that has developed acid surface *horizons, a *mor surface *humus, and a clearly visible enrichment of translocated iron oxide in the middle, or B, horizon. The profile has been leached to the early stage of *podzolization, identified by the movement down-profile of iron and aluminium compounds.

brown rot **1.** A very common disease of fruit caused by *Fungi of the genus *Sclerotinia* (usually *S. fructigena*). Fruits susceptible to it include apples, pears, cherries, plums, etc. Initially, soft brown patches appear on the fruit, often at a site of injury, and these gradually spread until the whole fruit is affected. Fungal conidia are produced from small, pale-coloured, cottony patches which typically appear on the rotting fruit. **2.** A type of timber decay in which the wood turns a reddish-brown colour and becomes cracked and eventually crumbly in texture. Fungi that cause brown rots are usually unable to break down the lignin component of wood.

browse wood Thin branches of trees that are fed to cattle and deer in winter.

Brückner cycle The tendency for the cyclical recurrence of runs of wet years at about 35-year intervals, with warmer and drier years in between, reported in 1890 by the German geographer and glaciologist Eduard Brückner (1862–1927). Much attention was paid to it for a time, but its operation is obscured by the greater magnitude of irregular year-to-year variations, and by cyclical recurrences of other periods.

Brunhes A palaeomagnetic *chron, characterized by normal magnetization (i.e. the geomagnetic North and South Poles are in the northern and southern hemispheres respectively), in which we are living at present. It began about 730 000 years ago. During the Brunhes there have been four reversed polarity excursions: Blake (110 000 BP); Lake Mungo (30 000 BP); Laschamp (20 000–18 730 BP); and Gothenburg (13 750–12 350 BP).

bryophyte A nonvascular plant belonging to the division Bryophyta. Some authors include mosses, liverworts, and hornworts in this division, but most now include only mosses, consigning the liverworts to the division Hepatophyta and hornworts to the division Anthocerotophyta.

Bryozoa **(Ectoprocta; moss animals)** A phylum of small, aquatic, colonial animals, related to the brachiopods, many of which possess a well-developed, *calcite skeleton which comprises microscopic, box-like divisions, each housing an individual animal possessing ciliated tentacles and a body cavity (coelom). Food is collected by the tentacles which surround the mouth. Reproduction takes place by asexual budding and by the release of larvae which give rise to new colonies. Bryozoans have occurred from the *Ordovician to the present day. Fossilized branched colonies are common in some rocks. They were important reef-builders in the *Phanerozoic, and underwent several great *adaptive radiations.

bubnoff unit A standard measure for describing the rates of geologic and geomorphologic erosional processes. One bubnoff unit (B) is equal to the removal of one micrometre per year, or one millimetre of surface material per thousand years. The unit is named after the Russian-born German geologist Serge von Bubnoff (1888–1957).

buccal Pertaining to the mouth (i.e. the buccal cavity); the adjective is derived from the Latin *bucca*, meaning 'cheek'.

buccal force pump A respiratory system, typical of amphibians, whereby air is forced into the lungs by raising the floor of the mouth while the valvular nostril is closed.

buccal incubation **(mouth brooding)** The incubation of eggs in the mouth of one of the parents. In a number of species of fish, the male or the female carries the fertilized eggs in the mouth until some time after hatching. During incubation the parent involved takes no food, and so may appear rather lean at the end of the incubation period.

buccopharynx The mouth and pharynx.

b

Buchan spells Several periods of the year, 'more or less well defined', when the normal seasonal rise or fall of temperature is halted or reversed for a time (e.g. May 9–14 is a cold period and December 3–14 a warm period). They are named after the Scottish meteorologist Alexander Buchan (1829–1907), whose analysis of temperature records for Scottish stations in the middle of the nineteenth century revealed them. Six cold and three warm periods of this type were identified by Buchan in 1869 from his examination of records covering several years. *See also* SINGULARITY.

buckminsterfullerene *See* CARBON.

'bucky balls' *See* CARBON.

buffer A solution of a weak acid and its conjugate weak base that resists changes in *pH which would otherwise result from the addition of an acid or base. The buffering properties of weak electrolytes in living organisms are of great importance, since most cells can survive only within narrow pH limits.

Buffon, Georges Louis Leclerc, Comte de (1707–88) A French naturalist and prolific author, whose 44-volume *Histoire Naturelle, générale et particulière* (first published between 1749 and 1804) was the first work to draw together disconnected items of information about natural history and present them in a coherent form.

Bufo marinus *See* CANE TOAD.

building phase A stage in the cyclical pattern of changes that is typical of many plant and *heathland communities, including forest grasslands and heaths. In grassland, the term refers to the accumulation of wind-borne particles to form a small hummock around any grass seedling that chances to invade a hollow, which in turn results from the erosion of a former hummock in the degenerate phase. In heathlands, the term describes the bushy phase of growth in individual *Calluna vulgaris* (heather) plants, lasting from about 7 to 13 years of age. At this stage net *primary production in the new shoots is at a maximum and the dense, bushy plant allows little light to reach the ground, hence it tends to suppress other species commonly associated with *C. vulgaris* in the pioneer, mature, and degenerate phases. *See also* MOSAIC EVOLUTION. *Compare* DEGENERATE PHASE; HOLLOW PHASE; MATURE PHASE; and PIONEER PHASE.

bujedales *See* MEDITERRANEAN SCRUB.

bulk density The mass per unit volume of soil (sampled as a clod or core), dried to constant weight at 105°C. *Compare* PARTICLE DENSITY.

bulking **1.** An increase in volume of material when it absorbs water; dry *sand or *clay may swell as much as 50 per cent. **2.** An increase in volume of solid rock or soil when broken.

bulk minerals *See* AGGREGATE.

bullate Blistered.

bunch grass prairie *See* PALOUSE PRAIRIE.

bundle-sheath cells *See* C_4 PATHWAY.

bunt **(stinking smut)** Common bunt is a seed-borne disease of wheat (and other grasses) caused by *Tilletia caries* or *T. foetida* (*T. laevis*). The grain in an infected ear of wheat is replaced by masses of fungal spores which have a characteristic fishy smell; the spores are released when the wheat crop is harvested. Dwarf bunt is a similar disease caused by *T. contraversa* (*T. brevifaciens*). Karnal bunt is caused by *Neovossia indica* (*Tilletia indica*); it occurs in northern India, Afghanistan, Iraq, and Pakistan.

bunya-bunya *See* DISJUNCT DISTRIBUTION.

buoyancy **1.** The upward force exerted on a body when it is immersed in a fluid. It occurs because an immersed body displaces its own volume of the fluid, reducing the weight of the body by the weight of the displaced fluid. If the body weighs more than the fluid it displaces it experiences negative buoyancy, and will sink. If the body weighs less than the fluid it displaces it experiences positive buoyancy and will rise. It is buoyancy due to the difference in density between hydrogen and helium and air that allows balloons and airships to remain airborne. A *parcel of air experiences buoyancy when it is surrounded by air of a different density. If the parcel of air is warmer than the surrounding air it will experience positive buoyancy, and negative buoyancy if it is colder than its surroundings. Buoyancy is calculated as $F/M = g[(\rho' - \rho)/\rho]$, where F is the buoyancy force, M is the mass of the parcel of air, ρ' is the density of the surrounding air, and ρ is the density of the parcel of air. *See* THERMAL. **2.** The greater powers of flotation provided by features such as air bladders possessed by aquatic plants or *algae. Both marine organisms such as *Fucus vesiculosus* (bladder wrack) and freshwater plants such as *Eichornia crassipes* (water hyacinth), which has inflated petioles, achieve buoyancy in this way.

buran The local name for a wind in Russia and central Asia which blows strongly from the north-east in both winter and summer. In winter it is very cold and associated with much snow, when it may strengthen to produce a blizzard called the 'purga'.

Burgess Shale A Middle Cambrian (540 Ma ago) rock horizon in British Columbia, Canada, that has yielded many exceptionally well-preserved fossils of metazoans, the remains of which were deposited in deep water or near a submarine fan. First discovered in 1909 by C. D. Walcott, the fossil fauna has been described many times. The fossils reveal a high level of *endemism, with many taxa that subsequently became extinct. The discovery of other well-preserved Cambrian faunas in China, Greenland, and Australia has helped with the interpretation of the Burgess fossils.

buried soil Soil covered by an *alluvial, colluvial (*see* COLLUVIUM), *aeolian, glacial, or organic deposit, being a product of a former period of *pedogenesis. In US usage a buried soil is defined as lying beneath 30–50 cm if the covering layer is more than half the thickness of the buried soil, otherwise beneath more than 50 cm.

'burst of monsoon' The onset of a marked change in weather conditions in the Indian subcontinent and southeast Asia, associated with the arrival of humid south-westerly winds which displace the hot, dry, pre-monsoon regime. The changed surface-level wind pattern is related to the establishment of a high-level easterly *jet stream.

bush **1.** Wilderness or uncleared land, contrasted with cultivated and settled land. **2.** An Australian term for forest. **3.** A *shrub.

bush veld *See* VELD.

butte A small, isolated, flat-topped hill that results from the erosion of near-horizontal strata. The diameter of the cap rock is less than the height of the landform above the surrounding country. Buttes are commonly found in semiarid regions dominated by *duricrust horizons. *See also* MESA.

buttress root A root that is similar to a *stilt root, but has a solid connection to the stem throughout its length. It is a means of providing stability to tall tropical trees in high winds. If the buttress root is detached from the trunk for part of its length it is known as a flying buttress root.

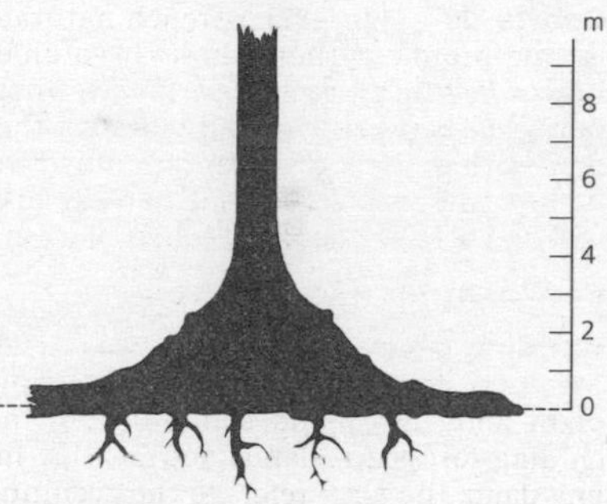

Buttress root

Buys Ballot's law A law enunciated in 1857 by the Dutch meteorologist Christoph Hendrik Buys Ballot (1817–90), professor of mathermatics at the University of Utrecht, that in the northern hemisphere the winds blow anticlockwise around centres of low pressure and clockwise around centres of high pressure. In the southern hemisphere both these tendencies are reversed.

buzz pollination The shedding of pollen from anthers that project from the flower, due to stimulation that occurs when the rapid vibration of a bee's wings resonates with the natural frequency of the anther structure. Some flowers, such as members of the Solanaceae (including nightshades and potato), have anthers projecting in this way.

byssoid Consisting of fine threads.

b-zone A biostratigraphic zone that is distinguished by benthonic (*see* BENTHOS) *fossils (e.g. brachiopods and trilobites). The term was proposed in 1965 by T. G. Miller. *Compare* P-ZONE.

^{14}C dating *See* RADIOCARBON DATING.

C_3 pathway The most common pathway of carbon fixation in plants. *See also* PHOTOSYNTHESIS.

C_4 pathway A pathway of carbon fixation, found most commonly in low-latitude plants, including many grasses, that are adapted to high temperatures and high light intensity. The first product formed as a result of the carboxylation by CO_2 of an acceptor molecule, phosphoenolpyruvate (PEP), is the four-carbon compound oxalo-acetate (OAA); hence the name C_4. OAA is converted to another four-carbon compound that leaves the mesophyll cells just below the leaf cells containing chlorophyll, passes through spaces between the cells (called plasmodesmata), and enters bundle-sheath cells that are tightly packed around the leaf veins. There the compound gives up its CO_2, which enters the light-independent stage of photosynthesis (*see* DARK REACTIONS). PEP carboxylase, the enzyme catalysing carboxylation, has no affinity for oxygen. Consequently C_4 plants suffer less than *C_3 plants from losses due to *photorespiration. Nevertheless, C_4 plants outcompete C_3 plants only under high light and high temperature conditions. Their efficiency is lower than C_3 plants in high latitudes. *Zea mays* (maize or corn) and *Saccharum officinarum* (sugar cane) are C_4 plants.

Ca *See* CALCIUM.

caatingas A thorn forest of semi-arid, tropical, north-eastern Brazil and similar or allied types of vegetation, including thorn woods and thorn bush. They are differentiated from *savannah woodlands by the paucity or absence of grass, which reflects a more prolonged dry period and smaller total amounts of rainfall. The name is also applied to physiognomically and structurally similar vegetation that occurs in patches through the Amazonian rain forest.

caducous Soon dropping off.

caecotrophy The passing of food through the alimentary canal twice. Rabbits and some rodents take soft faecal pellets directly from the anus at night and store them in the stomach to be mixed with food taken during the day, probably obtaining metabolites essential for digestion that are produced in the caecum. The animals will die if they are prevented from ingesting these pellets. *Compare* COPROPHAGY.

caespitose Growing in dense tufts.

Cainozoic *See* CENOZOIC.

Calabrian An early *Pleistocene marine stage, about 1.8–0.5 Ma ago, which is split between two land-mammal stages (Upper *Villafranchian and *Biharian).

calcareous Applied to substances containing calcium carbonate. *See also* CALCAREOUS OOZE and CALCAREOUS SOIL.

calcareous ooze A deep-sea, fine-grained, *pelagic deposit that contains more than 30 per cent calcium carbonate. The calcium carbonate is derived from the skeletal material of various *planktonic animals and plants (e.g. foraminiferan tests and coccoliths, which are calcitic, and pteropod tests, which are aragonitic). Calcareous ooze is the most extensive deposit on the ocean floor but is restricted to water depths less than about 3 500 m. *See also* CARBONATE-COMPENSATION DEPTH.

calcareous soil Soil that contains enough free calcium carbonate to effervesce visibly, releasing carbon dioxide gas, when treated with cold 0.1 *N* hydrochloric acid, and which could also be regarded as an *alkaline (basic) soil.

calcic horizon A *mineral *soil horizon with evidence of secondary calcium carbonate deposition which is more than 15 cm thick, with a calcium carbonate content of more than 15 per cent by weight, and with 5 per cent carbonate more than is in the parent material or horizons below it.

calcicole A plant species confined to, or most frequently found on, soils containing free calcium carbonate. *See also* ALKALINE SOIL. *Compare* CALCIFUGE.

C

calcicolous Applied to an organism that prefers to grow in, or can grow only in *habitats rich in calcium ('lime').

calcification The process of redeposition of secondary calcium carbonate from other parts of the soil *profile, which if sufficiently concentrated may develop into a *caliche, kunkar, or *calcic horizon (all similar and usually comprising more than 15 per cent weight of calcium carbonate in more than 15 cm soil thickness). Calcification involves limited upward and lateral movement of calcium salts in solution, and downward movement in less wet periods. When occasionally *leaching is deeper, some but not all of it may redissolve.

calcifuge A plant species not usually found on soils containing free calcium carbonate. *Compare* CALCICOLE.

Calcisols Soils that have a *calcic horizon within 125 cm of the surface. Calcisols are a reference soil group in the *FAO *soil classification.

calcispheres Small *calcite spheres, up to 50 μm in diameter, that are commonly found in *Palaeozoic limestones and believed to be of algal origin.

calcite A very common form of calcium carbonate ($CaCO_3$) that is the principal ingredient of many sedimentary rocks (e.g. limestones, marble, and chalk).

calcium **(Ca)** An element that is necessary for plant growth. In *eukaryotic cells it is found mainly in the apoplast, where it preserves membrane integrity and strengthens cell walls. It protects roots against the effects of low *pH, ion imbalance, and toxic ions, so that reduced root growth is a symptom of deficiency, as is 'die-back' of buds.

calcrete *See* CALICHE.

calcrete uranium A calcrete (*caliche) that is locally uraniferous and may constitute a workable ore (e.g. in western Australia and Namibia) because it has formed over a bedrock containing a high concentration of uranium.

caliche **(calcrete)** A carbonate *horizon (the K horizon) formed in a soil in a semi-arid region, under conditions of sparse rainfall (20–60 mm/yr) and a mean annual temperature of about 18°C, normally by the precipitation of calcium carbonate carried in solution. The soil profile develops over several thousand years, initially in the form of nodules (glaebules), more mature caliches taking a massive, laminar form. It may become cemented and indurated on exposure, when it gives rise to a tabular landscape. *See also* CALCIC HORIZON; DURIPAN; and PETROCALCIC HORIZON.

California Current A southward-flowing, eastern *boundary current that carries cool water down the western coast of North America from the North Pacific Current to join the North *Equatorial Current. It is a slow-moving and diffuse water mass.

calm A condition of general lack of wind, indicated by a wind speed of less than 1 knot (0.5 m/s). *See also* BEAUFORT SCALE.

calorie The energy required to raise one gram of water through one degree Celsius. One calorie is equivalent to 4.128 joules. The Calorie, used to measure the energy value of foods, is 1 000 calories.

calorific value The gross calorific value of a substance is the number of heat units that are liberated when a unit weight of that substance is burned in oxygen, and the residual materials are oxygen, carbon dioxide, sulphur dioxide, nitrogen, water, and ash. The energy content of biological materials has been expressed traditionally in calories (c) or kilocalories (C) per gram dry weight. Sometimes results are expressed more significantly in terms of ash-free dry weight (i.e. in terms of organic constituents only). Contemporary studies of ecological energetics express results in terms of the SI energy unit, the joule (4.182 J = 1 calorie).

calving The process whereby portions of a glacier's leading edge break off as icebergs into an adjacent body of water.

calvus The Latin word *calvus*, meaning 'bald' or 'stripped', used to describe a species of *cumulonimbus cloud in which upper protrusions form a more amorphous mass than appears from the cumuliform outlines. *See also* CLOUD CLASSIFICATION.

CAM *See* CRASSULACEAN ACID METABOLISM.

cambering The consistent *dip of strata towards valley centres, in conflict with the general regional dip. It is well displayed in the English Midlands, where ironstone beds above clays are cambered down as much as 30 m below their original level. It is probably due to large-scale structural disturbance when *permafrost thawed and the plastic clays allowed overlying massive beds to flow towards valleys.

cambic horizon A weakly developed *mineral *soil horizon of the middle part (B horizon) of soil *profiles, and one that has few distinguishing morphological characteristics except for evidence of weathering and sometimes of gleying. It is found in *brown earths and *gleys. The name is from the Latin *cambiare*, to change.

Cambisols Soils that have a *cambic or *mollic horizon above a B *soil horizon with a *base saturation lower than 50 per cent in the upper 100 cm, or an *andic, *vertic, or *vitric horizon with an upper boundary 25–100 cm below the surface. Cambisols are a reference soil group in the *FAO *soil classification.

Cambrian The first of six periods of the *Palaeozoic Era, from about 542–488.3 Ma ago, during which sediments deposited include the first organisms with mineralized skeletons. Common fossils include Brachiopoda, Trilobitomorpha, Ostracoda, and, late in the period, Graptolithina. The plant fossils are mainly of algae and fungi. Trilobites (arthropods) are important in the stratigraphic subdivision of the period. *See* BURGESS SHALE.

campanulate Shaped like a bell.

Campbell–Stokes sunshine recorder A device in which a glass sphere concentrates the Sun's rays on to a calibrated paper, where the resulting burns register the time and duration of sunshine. The instrument must be pre-set for a given latitude. This type of recorder has been in use since about 1880.

campo The general regional term applied to the tropical *savannah grasslands, with scattered, broad-leaved trees, of Brazil, developed on very poor soils, although, in the absence of fire, woodland or scrub would no doubt prevail. There are several variants of campo, including *campo cerrado and *campo sujo.

campo cerrado A tree-covered *savannah, which is one type of Brazilian *campo.

campo sujo A variety of *campo in which the aspect is fairly open and trees and shrubs are sparse.

Canada balsam resin A naturally occurring resin with optical qualities similar to those of glass, which is distilled from the bark of *Abies balsamea* (the balsam fir) and other North American *Abies* species. When heated to 160°C it becomes liquid and is used to cement specimens to glass microscope slides. Its use has been largely superseded by warm- or cold-setting epoxy resins which combine low viscosity (about 100 centipoise) with high shear strength (about 11.7×10^6 Pa), good adhesion, and a refractive index of 1.54.

canalization **1.** The holding of a developmental process within narrow bounds despite both genetic and environmental disturbing forces. Thus cells will progress along particular developmental pathways until they become differentiated into their final, adult forms. Development is such that all the different *genotypes have a standard *phenotype over the range of environments common to that species. *See also* AUTOPOIESIS. **2.** Regulating the flow of a river by such means as straightening its course, strengthening and/or raising its banks, constructing dams and locks, and building *levees, in order to prevent flooding or to make the river navigable. Effectively the river is converted to a canal, but one in which the water flows.

canalizing selection The elimination of *genotypes that render developing individuals sensitive to environmental fluctuations. Genetic differences may be revealed in organisms by placing them in a stressful environment or if a severe *mutation stresses the developmental system.

Canaries Current A cool, oceanic water current which flows south along the continental margin of Spain, Portugal, and West Africa. This slow-moving eastern *boundary current is the cause of frequent sea fogs off north-west Spain and Portugal. The water is further cooled by *upwelling of cold water off West Africa.

cancer cell A malignant cell within which the normal controls on growth, division, and cell-surface recognition have broken down. Such cells often give rise to cell masses (tumours) that can invade and destroy adjacent tissues.

candelabra tree *See* DISJUNCT DISTRIBUTION.

Candolle, Alphonse Louis Pierre Pyramus de (1806–93) The son of Augustin Pyramus de *Candolle, who, in 1842, succeeded his father as professor of natural history at the University of Geneva and completed his father's work by publishing 10 volumes of his *Prodromus* (one of them in collaboration with his own son). In his own *Géographie botanique raisonnée* (1855), *La Phytographie* (1880), and *Origine des plantes cultivées* (1833, published in English as *Origin of Cultivated Plants* in 1886), Candolle considered why particular species occur in some places but not others and contributed greatly to establishing the basis of modern biogeography.

Candolle, Augustin Pyramus de (1778–1841) A Swiss botanist who studied in Geneva and settled in Paris in 1796. At the request of the French Government he conducted a botanical and agricultural survey of the whole of France, the results of which were published in 1813. In 1816 he returned to the University of Geneva as professor of natural history and devoted the remainder of his life to developing his own system of botanical classification, which he tried to make more 'natural' than previous ones. He began his most famous work, *Prodromus systematis naturalis regni vegetabilis*, in 1824. Seven volumes were published before his death and a further 10 by his son, Alphonse de *Candolle.

cane blight A disease that affects the raspberry cane (*Rubus idaeus*). The leaves shrivel and die, dark patches appear on stem bases, and the bark splits open. The disease is caused by a fungus, *Leptosphaeria coniothyrium*, which enters the plant via wounds.

cane toad **(*Bufo marinus*)** A toad, native to North, Central, and South America, that was introduced to North Queensland, Australia, in June 1935 to control the grey-backed cane beetle and the frenchie beetle, two pests of sugar cane. The 100 imported individuals began unseasonal breeding almost immediately and within six months there were 60 000 toads. They spread through Queensland and by 2004 were advancing by 30–50 km/yr into the Northern Territory and 5 km/yr in northern New South Wales. The toads are large: a female may weigh 2.5 kg and be 26 cm long. Their skin secretions are highly toxic to vertebrates. They represent a hazard to domestic animals and wildlife and are classed as pests. Although cane toads eat grey-backed cane beetles and frenchie beetles when they encounter them, they were ineffective as control agents and the sugar-cane pests are now controlled by insecticides.

canine **1.** Pertaining to a dog. **2.** A conical, pointed tooth situated between the *incisors and *premolars, particularly well developed in carnivorous mammals which use their canine teeth to seize prey. In some mammals the canines are enlarged to form tusks, in others they are reduced or absent.

canker A general term for a localized disease of woody plants in which bark formation is prevented. Cankers may be caused by bacteria or fungi, or occasionally even by viruses.

cannibalism The eating by an animal of members of its own species. It is known to occur in nearly 140 species, most commonly among invertebrates, and often (but not always) in response to stress or a reduction in the availability of other food.

canopy The part of a woodland or *forest community that is formed by the trees. In complex forests, e.g. in *tropical rain

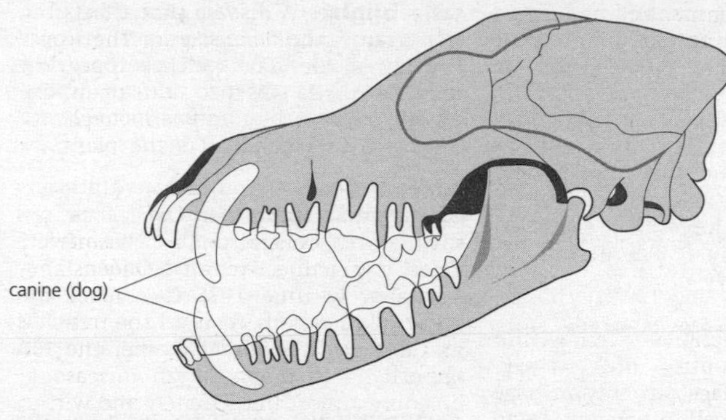

Canine

forest, the canopy is often arbitrarily subdivided into emergent, middle, and lower zones. The term may also be applied to the upper layer of shrub and scrub communities, or any terrestrial plant community in which a distinctive habitat is formed in the upper, denser regions of the taller plants. *Compare* GROUND LAYER.

cant The section of a *coppiced woodland that is cut in a particular year in the rotation. *See also* COUPE and HAG.

cap **1.** A thin, impermeable layer of soil particles covering the soil surface, caused by the impact of heavy rain on bare soil. **2.** *See* PILEUS.

capacity (of stream) The maximum load of solid particles that a stream is capable of carrying. It is largely a function of the *discharge, but influenced by particle size, in that a decrease in the size of particles involves an increase in the total load that can be transported. Ultimately a heavily loaded stream merges imperceptibly with a mud flow.

Cape floral region A region synonymous with R. Good's (1974, *The Geography of the Flowering Plants*) South African kingdom, and in relation to its size one that has perhaps the most remarkable and richest flora in the world. There are about 2 500 species, and an unusually high proportion of the genera and species (e.g. 520 species of *Erica*) are endemic (*see* ENDEMISM). Many well-known garden plants and greenhouse succulents originated in this floral region (e.g. many pelargoniums). *See also* FLORAL PROVINCE and FLORISTIC REGION.

capillarity *See* CAPILLARY ACTION.

capillary In the blood-circulation system of vertebrates, the narrowest blood vessel, with walls only one cell thick, through which molecules can pass to transfer oxygen and nutrients to cells and to remove waste products.

capillary action **(capillarity)** The process by which soil moisture may move in any direction through the fine (i.e. capillary) pores of the soil, under the influence of surface-tension forces between the water and individual soil particles. Soil moisture in this state is known as *capillary moisture. It exists as a film or skin of moisture on the soil particles, and may be drawn above the water-table by capillary action and into the plant roots by the process of *osmosis. 'Capillary conductivity' is now obsolete in US terminology.

capillary fringe *See* CAPILLARY ZONE.

capillary moisture **(capillary water)** Moisture that is left in the soil, along with hygroscopic moisture and water vapour, after the gravitational water has drained off. Capillary moisture is held by surface tension (known in the US as 'water potential') as a film of moisture on the surface of soil particles and *peds, and as minute bodies of water filling part of the pore space between particles. Curved water surfaces or menisci (singular: meniscus) form bridges across the pores at the boundaries between their water-filled and air-filled parts. Capillary moisture may move through the soil under the influence of surface-tension forces (*see* CAPILLARY ACTION) and some of it may be used by plants.

capillary water *See* CAPILLARY MOISTURE.

capillary wave A water wave whose wavelength is less than 1.7 cm and in which the primary restoring force is the surface tension of the water. The slightest of breezes may cause slight 'puckering' of the water surface, and the capillary waves so produced will be smoothed and flattened by the effects of surface tension.

capillary zone **(capillary fringe)** The zone immediately above the *water-table, into which water may be drawn upwards as a consequence of *capillary action. A typical height for the capillary fringe in clay with a pore radius of 0.0005 mm might be 3 m, compared with less than 10 cm in a fine sand with a pore radius of 0.02 mm.

capillatus The Latin *capillatus*, meaning 'with hair', used to describe a species of *cumulonimbus cloud with fibrous cirriform appearance in its upper parts. It is often associated with an *anvil or other protrusion in which wisps of cloud trail from the tip to give a hair-like appearance. Typically this cloud type brings *squalls with showers or thunderstorms. *See also* CLOUD CLASSIFICATION.

capsid The protein case of shell that encloses the *genome (RNA or DNA) in an individual *virus particle.

capsomere The protein 'building block' of a *virus *capsid.

capsule **1.** The gelatinous, outer surface layer of a prokaryotic cell (also known as the sheath in *cyanobacteria), which is composed primarily of polysaccharides. In *pathogenic bacteria, and possibly in others, it appears to serve a protective function against the defensive mechanism of the host, since such bacteria become non-infective in its absence. **2.** The *spore-bearing structure of a moss or liverwort. **3.** A dry fruit, normally *dehiscent.

carbamates A group of salts and esters of carbamic acid ($H_2N.COOH$) various members of which are used as insecticides, herbicides, and fungicides. They have a low toxicity to mammals, although some are harmful to non-target insects and fish, and they break down rapidly in the soil.

carbohydrate A generic term for molecules that are aldehyde or ketone derivatives of polyhydroxyl alcohols and that have a basic empirical formula of $C_x(H_2O)_y$, where x and y are variable numbers. They are normally classified as mono-, oligo-, or poly-saccharides, depending upon whether they are single sugars, short-chain molecules, or large polymers respectively.

carbon A non-metallic element, chemical symbol C, which is unique in the number of compounds it is able to form that contain chains or rings of carbon atoms. This ability to form large, complex molecules in which other elements are bonded to carbon atoms is exploited by all living organisms. The discipline of organic chemistry is essentially the study of cyclic carbon compounds. Carbon is extracted from gaseous carbon dioxide by plants during photosynthesis, is incorporated in living matter, and when organic matter decomposes its carbon is oxidized and so returned to the atmosphere as carbon dioxide. Pure carbon occurs naturally as diamond, graphite, and the amorphous carbon black, and recently an additional polymorph has been recognized called buckminsterfullerene ('bucky balls'), in which 60 or more atoms are arranged to form an approximately spherical shape, rather like a football. Charcoal produced by the destructive distillation of organic matter is also a relatively pure form of carbon.

carbon-14 *See* RADIOCARBON DATING.

carbonaria *See* *BISTON BETULARIA*.

carbonate compensation depth (CCD) The depth in the sea at which the rate of dissolution of solid calcium carbonate equals the rate of supply. Surface ocean waters are usually saturated with calcium carbonate, so calcareous materials are not dissolved. At mid-depths the lower temperature and higher CO_2 content of seawater cause slow dissolution of calcareous material. Below about 4 500 m waters are rich in dissolved CO_2 and able to dissolve calcium carbonate readily. Carbonate-rich sediments are common in waters less than 3 500 m depth, but are completely absent below about 6 000 m. *See also* CALCAREOUS OOZE.

carbonation A chemical *weathering process in which dilute carbonic acid, derived from the solution in water of free atmospheric and soil-air carbon dioxide, reacts with a mineral. The best-known example is the reaction between limestone (calcium carbonate) and carbonated water (carbonic acid) which yields calcium and bicarbonate ions in solution:

$$CaCO_3 + H^+ + HCO^-_3 \rightarrow Ca^+ + 2HCO^-_3$$

carbon cycle The movement of carbon through the surface, interior, and atmosphere of the Earth. Carbon exists in atmospheric gases, in dissolved ions in the *hydrosphere, and in solids as a major component of organic matter and sedimentary rocks, and is widely distributed. Inorganic exchange is mainly between the atmosphere and hydrosphere. The major movement of carbon results from *photosynthesis and respiration, with exchange between the *biosphere, atmosphere, and hydrosphere. Rates of exchange are very small, but over geologic time they have concentrated large amounts of carbon in the *lithosphere, mainly as limestones and *fossil fuels. This carbon was probably present as CO_2 in the primordial atmosphere. The burning of fossil fuels and the release of CO_2 from soil air through the clearance of tropical forests may eventually change the balance of the carbon cycle, although the climatic effects may be partly mitigated by the buffering action of the oceans; it is estimated that about 200 billion tonnes of CO_2 have been added to the atmosphere in this way since 1850. *See* 'GREENHOUSE EFFECT'.

carbon dating *See* RADIOCARBON DATING.

carbon dioxide **(CO_2)** The product of the complete oxidation of carbon and the compound most involved in the transport of carbon through the *carbon cycle. Carbon dioxide is utilized by *autotrophs in the process of *photosynthesis. When organic matter decomposes its carbon is oxidized to CO_2 and released into the atmosphere. Carbon that enters long-term 'storage', as carbonate rocks (e.g. limestone) or *fossil fuels, can be oxidized on exposure to oxygen (e.g. when fossil fuels

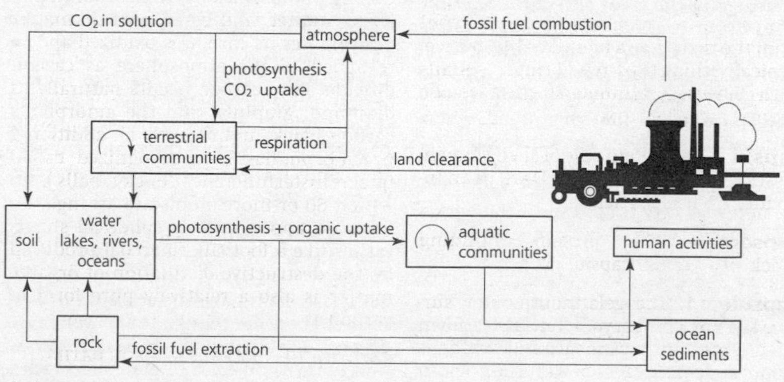

Carbon cycle

are burned). Apart from water vapour, carbon dioxide is the most important *greenhouse gas, absorbing long-wave radiation at wavelengths of about 5 μm and 18 μm.

carbon dioxide method Any non-destructive technique for measuring rates of *primary productivity which monitors changing carbon dioxide (CO_2) concentrations as a means of assessing the rate of carbon dioxide uptake in *photosynthesis and release in *respiration. A wide choice of carbon dioxide measuring techniques exists, some more applicable to laboratory than to field conditions. The most frequently used research methods include infrared gas analysis, conductivity measurements, and radiocarbon tracers. *See also* AERODYNAMIC METHOD.

Carboniferous The penultimate period of the *Palaeozoic Era, preceded by the *Devonian and followed by the *Permian. It began about 359.2 Ma ago and ended about 299 Ma ago. In Europe the lower part of the system is known as the Dinantian. It is divided into two stages and is characterized by marine limestones with a rich coral-brachiopod fauna. In contrast the upper part, the Silesian, which is subdivided into three stages, is noted for the deposition of terrestrial and freshwater sediments. North American geologists subdivide the Carboniferous System into two periods or subperiods. Of these the lower (359.2–318.1 Ma ago) is called the Mississippian and is the equivalent of the Dinantian stages (the Tournaisian and Visean) plus the lower part of the Silesian. The upper period, the Pennsylvanian (318.1–299 Ma ago), is the equivalent of most of the Silesian. During the Carboniferous very lush, swamp forests dominated the landscape in low-lying areas, where minor changes in sea level alternately exposed land supporting forest then inundated and buried the vegetation. The climate was very humid until the end of the period, when it became arid, conditions then selecting against seed-bearing plants. The forests were dominated by Lycopsida and Calamitaceae, some of which grew to the size of trees (e.g. *Lepidophloios* species grew up to 50 m tall) and the forest floor supported ferns and seed ferns. The first tetrapods (e.g. the amphibian *Ichthyostega*) appeared very early in the Carboniferous. The buried vegetation was compressed and changed through time to form the rich coal measures of southern Wales, England, Scotland, the USA, and many other areas worldwide, in which recognizable seed plants are common fossils.

carbon isotopes The naturally occurring isotopes of carbon, of which there are three: ^{12}C making up about 98.9 per cent; ^{13}C about 1.1 per cent; and ^{14}C whose amount is negligible, but which is detectable because it is radioactive. The relative abundance of these isotopes varies and the study of this variation is an important tool in geologic research, especially *radiometric dating. Carbon-isotope dating is a method of radiometric age-dating using the amount of the heavy, radioactive isotope carbon-14 remaining in organic matter. Carbon-14 has a half-life of 5 730 ± 40 years and the amount of the isotope present can be used to date materials up to about 50 000 years old (*see* RADIOCARBON DATING). Measurement of the ratio of carbon-13 to carbon-12 allows the recognition of carbonate precipitated from a variety of different sources. This ratio in plant-derived organic matter varies depending on which photosynthetic pathway was involved in its fixation (*C_3 or *C_4/CAM, *see* CRASSULACEAN ACID METABOLISM). The distinctive ratio persists in the tissues of *consumer organisms, allowing their dietary history to be reconstructed.

carbonization *See* FOSSILIZATION.

carcinogen A substance that can cause cancer.

Caribbean Current A warm, ocean-water current which flows westward through the Caribbean Sea, passes into the *Florida Current, and thus contributes to the *Gulf Stream. Its flow velocity averages 0.38–0.43 m/s.

Caribbean floral region Part of R. Good's (1974, *The Geography of the Flowering Plants*) *Neotropical floral kingdom. It comprises isthmian America and the West Indies. There are thought to be well in excess of 500 endemic genera (*see* ENDEMISM) and many thousands of endemic species. Some of the most diverse floras in the world are found here; for example, Cuba alone has roughly 8 000 species, many of which are endemic. A very large number of valuable economic plants and some garden plants are from this region. *See also* FLORAL PROVINCE and FLORISTIC REGION.

Carica *See* DISJUNCT DISTRIBUTION.

Caricaceae *See* DISJUNCT DISTRIBUTION.

carnassial In many *carnivores, a modification of *premolar or *molar teeth, commonly the lower first molar and the upper last premolar, giving them a scissor-like shearing action used for cutting flesh.

carnivore **1.** Any heterotrophic, flesh-eating animal. *See also* FOOD-CHAIN. **2.** A member of the mammalian order Carnivora.

carob *See* LOCUST (2).

carotene *See* CAROTENOID.

carotenoid A generic term for water-insoluble, polyisoprenoid pigments, which often function as accessory photosynthetic pigments in higher plants and photosynthetic bacteria, with absorption peaks between 450 and 480 nm. The group includes the carotenes, which are orange, and xanthophylls, which are yellow. During the *senescence of leaves, *chlorophyll breaks down faster than carotenoids and carotenoid colours are revealed. In the vertebrate liver carotene is changed into (i.e. is a precursor of) vitamin A.

carpal spur A sharp, horn-covered bone, situated on the carpus (wrist joint) of birds, and used in combat. It is found in some Anatidae (ducks, geese, etc.) and Charadriidae (plovers and lapwings), and in all Chionididae (sheathbills), Jacanidae (jacanas), and Anhimidae (screamers).

carr A locally variable term of Scandinavian origin, used to describe forested, wet, *rheotrophic habitats, usually with some *peat development but with neutral (not extremely acid) waters and good nutrient status. In East Anglia, England, they are typically alder woods (*Alnus glutinosa*), but where nutrient levels are less favourable willow (*Salix* species), especially *S. cinerea* (grey willow), may be the principal tree species. *See also* FEN. *Compare* BOG.

carrying capacity The maximum population of a given organism that a particular environment can sustain; the *K* (saturation) value for species populations showing *S-shaped population growth curves. It implies a continuing yield without environmental damage. It may be modified by human intervention to improve environmental potential (e.g. by applying fertilizers to range-land and reseeding it with nutritious grasses).

Cartesian projection A mapping technique in which every plane in the area being mapped is projected on to a plane in the map and every line on to a line. Each point in the area under study is identified by three values, representing its location

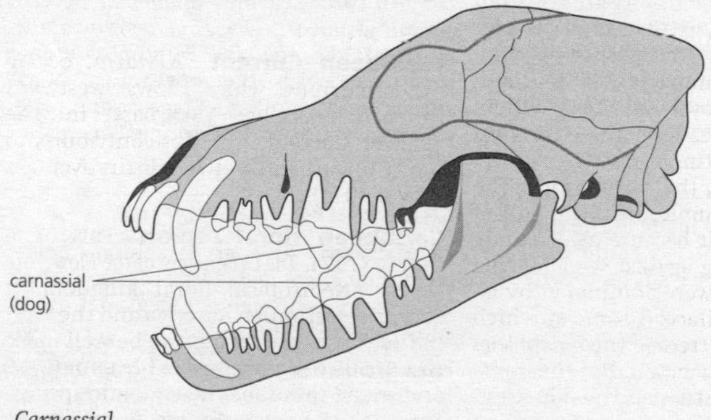

Carnassial

in relation to three mutually perpendicular axes (the Cartesian coordinates of the point). These coordinates are transformed mathematically into a homogeneous set of four coordinates which can then be plotted to produce a graphic representation (a map). The word 'Cartesian' is derived from the name of the French mathematician René Descartes (1596–1650).

cartilage In vertebrates, flexible skeletal tissue formed from groups of rounded cells lying in a matrix containing collagen fibres. It forms most of the skeleton of embryos and in adults is retained at the ends of bones, in intervertebral discs, and in the pinna of the ear; in Elasmobranchii (some sharks) calcified cartilage rather than true bone provides the entire skeleton.

caryopsis A dry, nut-like fruit typical of grasses, e.g. a cereal grain. It is an achene with the ovary wall united with the seed coat.

cascade effect A sequence of events in which each produces the circumstances necessary for the initiation of the next. **1.** In ecology, a *succession in which the organisms present at one stage provide resources that are exploited by those at the next. **2.** In geomorphology, the transfer of mass and energy through a chain of component subsystems, the output from one subsystem becoming the input for the next. An example is the *valley glacier, where the inputs of snowfall and rock debris from the slopes above and potential energy (derived from elevation) are cascaded through a sequence of climatic environments with a progressive reduction in mass and dissipation of energy, the output from the glacier being sediment and water which form the input to the *proglacial subsystem.

cast The preserved sediment infill of an impression or mould made in the top of a bed of soft sediment (e.g. flute cast (*see* FLUTE MARK), or casts of footprints or trails).

caste **1.** In social insects, the existence of more than one functionally different form (polymorphism) within the same sex in the same colony, characterized by morphological features, age, or both. In bees and wasps the female morphological castes are queens and workers. Within hive bees there are three castes, each performing a special function in the colony: non-reproductive workers (females which are usually sterile but help with the provision of food for the queen in the hive); drones (reproductive males); and queens (fertile females). In termites castes are not distinguished by sex: worker and soldier termites may be sterile males or females. There are several different kinds of workers among ants, all of which are sterile females, and ants also have a soldier caste. **2.** A system of social classification in humans, in which membership is determined culturally by birth and remains fixed; the group is ranked in a hierarchy of groups in the system.

caste determination In social insects, the process by which embryological development is influenced by physiological and environmental factors, giving rise to the various *castes.

castellanus From the Latin *castellum*, meaning 'castle', the name of a cloud species commonly associated with the upper parts of *altocumulus, *stratocumulus, *cirrus, and *cirrocumulus. Cumuliform, turreted protrusions extend in a linear fashion from the cloud top, producing a crenulate form. *See also* CLOUD CLASSIFICATION.

caste polyethism *See* POLYETHISM.

Castlecliffian A series in the *Quaternary of New Zealand, underlain by the Nukumaruan (the basal series), overlain by the *Holocene, and roughly contemporaneous with the uppermost *Calabrian and Emilian Stages, and the subsequent Upper *Pleistocene Series.

castle koppie *See* KOPPIE.

CAT *See* CLEAR-AIR TURBULENCE.

catabolism The part of cellular metabolism that encompasses the reactions that yield energy through the degradation of substrate molecules.

catadromous Applied to the migratory behaviour of fish that spend most of their lives in fresh water but travel to the sea in order to breed there (e.g. *Anguilla anguilla* (common eel) which breeds in the *Sargasso Sea). *Compare* ANADROMOUS; and POTAMODROMOUS. *See also* AMPHIDROMOUS; and DIADROMOUS.

catalysis The acceleration of a chemical or biochemical reaction that is brought about by the action of a catalyst. The catalyst, which is not affected by the overall reaction, acts by either lowering the activation energy of the reaction through a reorientation of molecules in collision, or by facilitating an alternative mechanism that has a different activation energy. Enzymes are naturally occurring catalysts which are universally present in all living cells. Enzymes are generally much more efficient and more specific than other catalysts.

catarrhine In primates, applied to nostrils that are close together and open downward. All Old World primates have catarrhine nostrils. *Compare* PLATYRRHINE.

catastrophic evolution (catastrophic speciation) A theory proposing that environmental stress might lead to the sudden rearrangement of chromosomes, which in self-fertilizing organisms may then give rise *sympatrically to a new species. Recent research suggests that at best this explanation applies only to some special cases.

catastrophism A theory that associates past geological change with sudden, catastrophic happenings. Early geologists, including William Buckland (1784–1856), *Cuvier, and Adam Sedgwick (1785–1873), claimed that catastrophism was a sound scientific theory. Although it met with considerable scorn in more recent times, many modern geologists acknowledge some degree of catastrophic change and so would describe themselves as 'neocatastrophists'. The collision of a bolide with the Earth at the Cretaceous/Tertiary boundary (65 Ma) is an accepted example of a singular catastrophic event that has had considerable influence on the Earth's history.

catchment The area from which a surface watercourse or a *groundwater system derives its water. Catchments are separated by *divides. A surface catchment area may overlie an *aquifer system, but may be unconnected with the aquifer rock itself if there are intervening impermeable aquicludes. In US usage a catchment is often termed a 'watershed'; in Britain the term 'drainage basin' is commonly used.

catena A topographic sequence of soils, of the same age and usually on the same parent material, which is repeated across larger landscape transects. Individual soil-profile types are related to site conditions and to position on a slope. The term was introduced in East Africa in the 1930s, and is mainly applicable in certain non-glaciated landscapes, particularly those with small, hilly relief (e.g. *loess areas).

cathemeral Applied to an organism that is equally active by day and by night.

cation An *ion that carries a positive electrical charge (e.g. the metallic element of salt compounds). A cation can combine with certain anions (which have negative charges).

cation exchange A process in which cations in solution are exchanged with cations held on the exchange sites of mineral and organic matter, particularly on the surfaces of colloids of clay and humus.

cation-exchange capacity (CEC) The total amount of exchangeable cations that a particular material or soil can adsorb at a given pH. Exchangeable cations are held mainly on the surface of colloids of clay and humus, and are measured in moles per gram of material or soil.

cation ordering The phenomenon, the extent of which is temperature-dependent, in which a *cation shows preference for the site it occupies because it provides greater chemical stability.

catotelm *See* ACROTELM.

caudal Pertaining to the tail. The word is derived from the Latin *cauda*, meaning 'tail'.

caudal fin The tail fin of a fish, used for steering, balancing, or locomotion.

caulescent In the process of becoming stalked.

cauliflorous Borne on the trunk; applied especially to tropical trees in which flower shoots grow from the main trunk. *Compare* RAMIFLOROUS.

cavernicolous Cave-dwelling.

cay A small, flat, marine island formed from coral-reef material or sand. The term is applied, for example, to the low-lying, sparsely vegetated islands off the coast of southern Florida.

CCD *See* CARBONATE COMPENSATION DEPTH.

CCL *See* CONVECTIVE CONDENSATION LEVEL.

CEC *See* CATION-EXCHANGE CAPACITY.

cecidium *See* GALL.

cecidization The formation of a plant *gall (especially by gall midges of the family Cecidomyidae).

cedar apple A *gall formed on juniper (*Juniperus*) trees infected with the *rust fungus *Gymnosporangium juniperivirginianae*.

celerity The velocity with which a wave advances. The celerity (c) of an ideal wave is related to its wavelength (λ) and frequency (f) by the wave equation $c = f\lambda$. The wave frequency (f) is the number of waves (n) passing a point in unit time (t), i.e. $f = n/t$. In deep water the wave celerity may be calculated by the equation: $c = (g\lambda/2\pi)^2 = 1.25\sqrt{\lambda}$, where λ is the wavelength in metres and g is the acceleration due to gravity (9.81 m/s). The speed of shallow-water waves may be calculated by the equation: $c = (gd)^{1/2} = 3.13\sqrt{d}$, where d is the depth of water in metres.

cell 1. The fundamental autonomous unit of plant and animal bodies, consisting of, at least, a cell membrane containing cytoplasm and nuclear material, but often having a more complex structure. Simple organisms are unicellular, but more complex organisms consist of many co-operating cells. Cells may be *eukaryote or *prokaryote. **2.** *See* MASS PROVISIONING.

cell culture (tissue culture) A mass of cells that have been derived either from a single cell or from a small group of cells from the same tissue or organ, and that is maintained *in vitro* using solid or liquid nutrient media.

cellulose A straight-chain, insoluble polysaccharide that is composed of glucose molecules linked by beta-1,4 glycosidic bonds. It is the principal structural material of plants, and as such is the most abundant organic compound in the world. It has also been found in certain sea squirts.

cellulytic Able to break open (lyse) cells.

cement A bone-like substance which coats the surface of that part of a tooth which is embedded in the jaw and which sometimes coats the enamel of the exposed part of the tooth.

cemented Applied to massive, infilled, and indurated *mineral soil: such soil has a hard and often brittle consistency because soil particles are joined together by cementing substances, e.g. calcium carbonate, silica, iron and aluminium oxide, or *humus. Cemented soil usually appears as a highly distinctive and resistant *horizon.

Cenozoic (Cainozoic, Coenozoic, Kainozoic) An era of geologic time that began about 65.5 Ma ago and continues to the present day. It includes the *Palaeogene, *Neogene, and *Pleistogene Periods; these are commonly grouped into the *Tertiary and *Quaternary Sub-Eras, although Tertiary and Quaternary are being abandoned as formal names. Molluscs and *microfossils are used in the stratigraphic subdivision of the era. The Alpine orogeny (episode of mountain-building) reached its climax during the Cenozoic Era.

centi- From the Latin *centum*, meaning 'hundred', a prefix (symbol c) used with *SI units to denote the unit $\times 10^{-2}$.

Central Australian floral region Part of R. Good's (1974, *The Geography of the Flowering Plants*) Australian kingdom, which accounts for most of the Australian continent, including all the central parts. The flora is imperfectly known, although the great majority of it is likely to be endemic (*see* ENDEMISM). Ecologically it coincides with extensive *thorn forest, with much *Acacia aneura* (mulga), *A. harpophylla* (brigalow), and *Eucalyptus hemiphloia* (mallee). The flora is poor, partly because of the extensive deserts and semi-deserts that account for most of the region, and partly because central Australia was submerged for much of the *Tertiary Period. *See also* FLORAL PROVINCE and FLORISTIC REGION.

central dogma The assertion that in DNA-based organisms information passes from DNA to RNA to protein.

Central European Sea *See* PARATETHYS.

central limit theorem The theorem stating that the arithmetic mean values for a series of similar-sized, fairly large samples ($n > 30$) taken from a large population will be approximately normally distributed about the true population mean (μ), irrespective of the actual distribution pattern of the individual counts.

centre of diversity **(gene centre)** The region where a particular *taxon exhibits greater genetic diversity than it does anywhere else. Many authorities believe that centres of diversity are also the *centres of origin of the taxa concerned.

centre of origin The region where a particular group of organisms is believed to have originated. Many authorities believe centres of origin are also *centres of diversity. For example, the tropical timber-tree family Dipterocarpaceae is strongly concentrated in the Malay Peninsula, Sumatra, and Borneo, although it ranges from Africa to New Guinea. It is difficult, however, to reconcile this explanation of centres of origin with the concept of *allopatric speciation.

centrifugal speciation The principle that new *species are likely to arise towards the centre of the range of the present species, rather than at the periphery. In practice, it is very often observed that primitive species are located on the edges of the distribution of a species-group or genus. *Compare* CENTRE OF DIVERSITY and CENTRE OF ORIGIN.

centripetal drainage pattern *See* DRAINAGE PATTERN.

cerebriform Resembling a brain; convoluted.

cerradão Areas of semi-evergreen woodland, in places with an almost closed *canopy, that occur in the *campo savannah grasslands of Brazil.

CFC *See* CHLOROFLUOROCARBON.

c.g.s. system A set of units of measurement derived from the metric system and based on the centimetre, gram, and second. It has now been largely replaced by the *SI (Système International d'Unités) system.

chaetotaxy **(setation, trichiation)** The arrangement of hairs on the body of an insect, which is often used as a taxonomic guide.

Chagas's disease A disease that can affect humans and other animals. It occurs chiefly in Central and South America. The causal agent is a protozoon, *Trypanosoma cruzi*. The pathogen is transmitted by blood-sucking bugs. Symptoms may include anaemia and various signs of heart, gland, and nervous-system involvement. The disease is named after the Brazilian physician Carlos Chagas (1879–1934).

chain An imperial measure, equal to 22 yards (20.11 m), that was a unit of length formerly used by foresters and land surveyors (10 sq. chains = 1 *acre).

chain response A sequence of behaviour in which each item produces a situation that evokes the next (e.g. in *courtship rituals the correct response of one partner is likely to evoke the next behaviour in the other).

Challenger expedition **(1872–5)** The first expedition to explore the deep oceans, led by John Murray, in the British naval ship HMS *Challenger*. With a staff of biologists, chemists, and geologists, the expedition surveyed the Atlantic, Indian, Antarctic, and Pacific Oceans, taking soundings and collecting specimens in dredges. Results of research into the material collected during the expedition were published between 1880 and 1895 as a long series of reports. The extent of the *Mid-Atlantic Ridge was first demonstrated by the crew of the *Challenger*. *See also* THOMSON, SIR CHARLES WYVILLE.

chalybeate Of natural waters, containing iron (chalybite is a synonym of siderite, an iron mineral, $FeCO_3$).

chamaephyte One of *Raunkiaer's lifeform categories, being a plant in which the perennating bud or shoot apices are borne very close to the ground. Four subcategories are recognized: (*a*) suffruticose chamaephyte, in which the aerial shoots die back partially at the onset of unfavourable conditions and buds arise on the lower, persistent stem portions; (*b*) passive chamaephyte, in which the aerial stems fall over as they die back, to give buds on horizontal axes near the ground; (*c*) active chamaephyte, which also produces buds on horizontal stems, but as the more normal growth form of the plant; and (*d*) cushion chamaephyte, which is essentially a compacted suffruticose chamaephyte. *Compare* CRYPTOPHYTE; HEMICRYPTOPHYTE; PHANEROPHYTE; and THEROPHYTE.

chanaral A type of thorny, scrub vegetation, in which chañar (*Gourliaea decorticans*) is common, that occurs in southern-central parts of South America.

channel **1.** The preferred linear route along which surface water and *groundwater flow is usually concentrated (although water can flow across wide, flat surfaces as sheet flow). It is commonly a linear, concave-based depression (e.g. river channel, submarine fan channel). The geometry may be sinuous, anastomosing, or straight, and with a widely variable width-to-depth ratio. *See* BRAIDED STREAM and MEANDER. **2.** A narrow seaway connecting two wider bodies of water (e.g. the English Channel).

channel fill The sediment infill of a *channel, produced either by the accretion of sediment transported by water flowing through the channel, or by the infilling of an abandoned channel. *See also* MULTISTOREY SANDBODY.

channelled wrack The common name for the brown seaweed *Pelvetia canaliculata*. The thallus is flattened and branched, the branches lacking a midrib, and having inrolled margins which form moisture-retaining channels. This plant can grow higher on the shore than can any other seaweed: at and above high-water mark. It is used sometimes as fodder for sheep and cattle or as manure.

chaos A theory derived from the observation that when the mathematical description of a system includes several nonlinear equations (i.e. equations that cannot be represented by straight lines on a graph), the future behaviour of that system may be unpredictable, because of wide variations that result from its sensitivity to very small differences in initial values supplied to any mathematical model. Chaos was first studied with reference to weather forecasting, but the theory has since been found to have many ecological implications (e.g. in studies of predator–prey relationships and population dynamics).

chaparral The *sclerophyllous vegetation of west California and adjacent regions. Like other sclerophyllous scrub in regions with Mediterranean-type climates, much of the chaparral has been derived by some disturbance, principally burning, of an earlier forest cover.

char **1.** A solid, carbonaceous residue, of high calorific value, derived from incomplete burning of organic material. It may be formed into briquettes and burned for fuel; if pure it can be used as a filter medium. Charcoal is made from wood or bone; coke, another char, is derived from *coal. *See* PYROLYSIS. **2.** The common name of four species of fish of the genus *Salvelinus* (family Salmonidae): arctic char (*S. alpinus*); brook trout (*S. fontinalis*); Dolly Varden trout (*S. malma*); and lake trout (*S. namaycush*).

character Any detectable attribute or property of the *phenotype of an organism. Defined heritable differences in the character may exist between individuals within a species.

character displacement The principle that two closely related species differ more in their morphological features where they occur together (i.e. are sympatric) than when they are separated geographically (i.e. are allopatric). *See* ALLOPATRY and SYMPATRY. *Compare* CHARACTER RELEASE.

characteristic species *See* KENNARTEN SPECIES.

character release The principle that a species will be able to exploit a greater range of habitats if it occurs in an environment from which a similar species with which it normally occurs is absent. *Compare* CHARACTER DISPLACEMENT.

character states The different versions of a *character that may exist in an organism. With DNA, at any particular base site there are four possible character states; A, T, G, or C (where A stands for adenine, T for thymine, G for guanine, and C for cytosine).

charade *See* DANCE LANGUAGE.

charcoal *See* CHAR.

Charpentier, Jean de **(1786–1855)** A Swiss superintendent of mines, Charpentier made extensive field studies in the Alps. Using evidence of erratic boulders and *moraines, he hypothesized that Swiss glaciers had once been much more extensive. His ideas were taken up and developed by *Agassiz.

chase **1.** In Britain, a royal forest that has passed into private ownership. **2.** A lane between two woods.

chattermark A small (less than 5 mm) crescentic scar typically found on the surface of rocks, of rock particles, and of rounded beach pebbles. It is a percussion fracture, produced when particles are thrown together in wind or water environments.

Chebotarev sequence An idealized sequence of chemical changes in *groundwater. As groundwater moves through rock its chemical composition normally changes. In general, the longer groundwater remains in contact with the *aquifer rocks the greater the amount of material it will take into solution. Changes in composition also occur with increasing depth of travel, as bicarbonate *anions, which dominate in many shallow groundwaters, give way to sulphate and then chloride anions, and calcium is exchanged for sodium. The sequence was first described by I. I. Chebotarev in 1955.

chela A prehensile claw or pincer (e.g. in Crustacea).

chelate **1. (adj.)** Pincer-like or claw-like. **2. (noun)** A ring structure formed as a result of the reaction of a metal ion with two or more groups on a ligand. Haemoglobin and chlorophyll are chelate compounds in which the metal ions are iron and magnesium respectively.

chelation An equilibrium reaction between a metallic ion and an organic molecule in which more than one bond links the two components. The metallic ion is termed the complexing agent, the chelating organic molecule the ligand. Chelation is a naturally occurring mechanism in soils, useful since it removes heavy-metal ions that are in solution in simple inorganic form where they may be directly toxic to plants or may interfere with the uptake of essential nutrients. Heavy-metal toxicity in wasteland will tend to be reduced by the application of organic material.

Chelford **1.** An *interstadial that occurred 65 000–60 000 years BP, during the *Devensian glaciation. **2.** Sections of *alluvial sands and organic muds containing tree remains from the Chelford interstadial, overlaid and underlaid by *till, that are exposed in sandpits between Chelford and Congleton, in Cheshire, England.

chemical evolution The process that is assumed to have led to the origin of life on Earth more than 3 300 Ma ago. The classical view is that lightning discharges in the Earth's early atmosphere produced the basic organic molecules from which, after they had dissolved in the oceans, life was assembled. How the assembly was achieved is still largely a matter of conjecture.

chemical oxygen demand **(COD)** An indicator of water or effluent quality, which measures oxygen demand by chemical (as distinct from biological) means, using potassium dichromate as the oxidizing agent. Oxidation takes 2 hours and the method is thus much quicker than a 5-day

BOD (*see* BIOLOGICAL OXYGEN DEMAND) assessment. Since the BOD : COD ratio is fairly constant for a given effluent, COD is used more frequently for routine monitoring of an effluent once this ratio has been determined.

chemical potential *See* WATER POTENTIAL.

chemical weathering The action of a set of chemical processes operating at the atomic and molecular levels to break down and re-form rocks and minerals. The results of chemical weathering are frequently new substances of reduced particle size, greater plasticity, lower density, and increased volume, compared with the original materials. Some of the important processes are solution, *hydration, *hydrolysis, *oxidation, *reduction, and *carbonation.

chemo-autotroph *See* CHEMOSYNTHETIC AUTOTROPH.

chemo-heterotroph A chemotrophic organism that obtains its carbon chiefly or solely from organic compounds.

chemo-lithotroph A chemotrophic organism that obtains its energy from the oxidation of inorganic compounds or elements.

chemo-organotroph A chemotrophic organism that obtains its energy from the metabolism of organic compounds.

chemoreceptor A sensory *receptor that responds to contact with molecules of chemical substances, producing the sensations of smell and taste.

chemosynthetic autotroph **(chemo-autotroph)** **1.** An *autotroph that is capable of synthesizing complex organic materials from inorganic reactions (e.g. iron oxidation). **2.** A chemotrophic organism that uses carbon dioxide as its main or sole source of carbon.

chemotaxis In a *motile organism or cell, a change in the direction of locomotion, made in response to a change in the concentrations of particular chemicals in its environment.

chemotroph An organism that obtains its energy from chemical reactions. *Compare* PHOTOTROPH.

chenier A beach ridge or sandy, linear mound that is built on a marsh area. It is at least 150 m broad, up to 3 m high, and up to 50 km long, and is typical of the Gulf Coast of America. A chenier is formed by the reworking of river-derived materials by waves. There are usually muddy, marshy zones to the front and rear of the chenier.

chenier plain An area of marine aggradation consisting of sandy ridges (*cheniers) separated by clay-rich depressions. Occasionally the ridges may be vegetated, as on the Gulf Coast of the USA. Chenier plains may be large: that on the north coast of South America is about 2 250 km long and up to 30 km wide.

chernozem **(black earth)** A freely draining soil *profile whose name is the Russian word for 'black earth'. Chernozems are associated with grassland vegetation in temperate climates, and identified by the deep and even distribution of *humus and of exchangeable *cations (calcium and magnesium) through the profile (included in *Mollisols of the *USDA Soil Taxonomy). Because of their richness in plant nutrients and their excellent *crumb structure, chernozems are among the most agriculturally productive soils in the world. Chernozems are a reference soil group in the *FAO *soil classification.

chert **1.** A variety of silica that lacks external evidence of crystal form. It is chalcedony (SiO_2) in a nodular or lens-like habit, formed in a sedimentary environment. Under magnification it is seen to be extremely fine-grained, and may be termed 'cryptocrystalline'. **2.** A fine-grained rock consisting of beds of cryptocrystalline silica, usually of *biogenic, volcanogenic, or diagenetic (*see* DIAGENESIS) origin.

chestnut blight **(chestnut canker)** A disease of the American chestnut (*Castanea dentata*) caused by the fungus *Endothia parasitica*. Invasion of the vascular cambium by the *pathogen results in wilting and the death of the tree. The introduction of the fungus to America from Asia around

1900 led to an epidemic, beginning on Long Island, New York, which by about 1935 had caused the wholesale destruction of the hitherto commercially important American chestnut.

chestnut canker *See* CHESTNUT BLIGHT.

chevron marks A linear pattern of small, V-shaped ridges formed by the dragging of an object over the surface of a viscous mud. The V-shapes close in the down-current direction, enabling the chevron marks to be used as a palaeocurrent indicator. *See* PALAEOCURRENT ANALYSIS.

Chezy's formula An empirical formula that relates river discharge (Q) to *channel dimensions and water surface slope. $Q = AC\sqrt{(rS)}$, where A is the cross-sectional area of the river, C is the Chezy discharge coefficient, r is the *hydraulic radius, and S is the slope of the water surface. This formula is useful for extending river-flow rating curves. Antoine Chezy (1718–98) was a French engineer, who devised the formula in 1798 while working on the design of a canal system to supply water to Paris.

Chile pine *See* DISJUNCT DISTRIBUTION.

chilling-sensitive plant A plant that is badly injured or killed by temperatures above freezing point, up to about 20°C.

chine A precipitous ravine that is found along eroding coastlines and is typically developed in the soft *Mesozoic and *Cenozoic sediments of southern England. It results from the rapid incision that occurs when a stream responds by downcutting to the rejuvenating effect of coastline retreat.

chinook A warm, dry, westerly wind of the *föhn type which blows on the eastern side of the Rocky Mountains. The quick onset of the wind and the sudden large rise in temperature is associated in spring with rapid melting of the snow.

chi-squared test **(χ^2 test)** A statistical test that is used to determine whether data obtained by sampling agree with those predicted hypothetically, and thus to test the validity of the hypothesis.

chitin A linear homopolysaccharide of N-acetyl-D-glucosamine that is found as a major constituent of fungal cell walls and as the major component of the cuticle of an insect, in which the molecules are linked to form chains that are arranged in layers. Depending on their orientation, these chains can be cross-linked to yield a very strong, lightweight material.

chlorine **(Cl)** An element which is necessary for normal plant growth. Its main role seems to be that of controlling *turgor, but it may also be involved in the light reaction of *photosynthesis. If plants are deficient in chloride ions wilting occurs and the young leaves become blue-green and shiny. Later they become bronze-coloured and chlorotic (*see* CHLOROSIS).

chlorinity A measure of the chloride content, by mass, of sea water. It is defined as the amount of chlorine, in grams, in 1 kg of sea water (bromine and iodine are assumed to have been replaced by chlorine). Chlorinity and *salinity are both measures of the saltiness of sea water. The relationship can be expressed mathematically, as salinity is equivalent to 1.80655 times the chlorinity.

chlorite *See* CLAY MINERAL.

chlorofluorocarbon **(CFC)** One of a range of chemically inert compounds in which a chlorinated hydrocarbon is treated with hydrogen fluoride (e.g. CFC-12 is CCl_2F_2, CFC-21 is $CHCl_2F$). CFCs have been used widely in fire extinguishers and as solvents, refrigerants, aerosol propellants, and in the manufacture of foam plastics. Their use is being phased out because of their capacity for absorbing long-wave electromagnetic radiation (i.e. they are 'greenhouse' gases) and because their chemical stability allows them to survive in the atmosphere long enough to enter the *stratosphere, where they are decomposed by ultraviolet radiation, liberating chlorine that is implicated in the thinning of the *ozone layer.

chlorophyll The green pigment in plants that functions in photosynthesis by absorbing radiant energy from the Sun, predominantly from blue (435–438 nm) and red (670–680 nm) regions of the spectrum. The light removes an electron from the chlorophyll molecule. This is used to produce either ATP or NADP (nicotinamide adenine dinucleotide phosphate) for carbon-dioxide fixation. Chlorophylls are magnesium-porphyrin derivatives, the principal variants in land plants being designated chlorophylls *a* and *b*, and in marine algae *c* and *d*.

chlorophyll method A method used for estimating the *primary productivity of *ecosystems. Experimental evidence shows that, with appropriate calibration, the *chlorophyll content of the community occupying a given area can form an index of the area's productivity. The method was pioneered for marine ecosystems, but has now been extended to other aquatic and to terrestrial communities.

chlorosis **(adj. chlorotic)** A symptom of disease or disorder in plants, which involves a reduction in or loss of the normal green coloration. Consequently, the plants are typically pale green or even yellow. Chlorosis is caused by conditions that prevent the formation of *chlorophyll (e.g. lack of light or a deficiency of iron or magnesium).

chlorotic *See* CHLOROSIS.

choanae Internal nostrils; the pair of openings that link the nasal cavity and throat in terrestrial vertebrates. They are believed to have evolved from the second, posterior nostrils in fish; in fish, water enters by the anterior nostrils and is expelled from the posterior nostrils. Posterior nostrils, or possibly choanae, have been discovered in the margin of the upper jaw of *Kenichthys*, a mid-Devonian fossil *tetrapodomorph fish from China that lived 395 Ma ago.

choice chamber A device that offers small invertebrates two or more contrasting environments (e.g. differing in humidity, temperature, illumination, or pH) and allows them to move freely into the one they prefer.

choke of grasses A disease which can affect many types of grass, particularly cocksfoot (*Dactylis glomerata*) and timothy (*Phleum pratense*). The causal agent is the fungus *Epichloe typhina*. A ring of white, felt-like mycelium develops around the flower-stalk and becomes yellow, then orange, as the fungal perithecia develop. Flowering is partially or completely suppressed.

-chory A suffix adopted in a little-used classification scheme of seed-dispersal mechanisms. *See* ANDROCHORY; ANEMOCHORY; AUTOCHORY; BARACHORY; BLASTOCHORY; BOLOCHORY; CRYSTALLOCHORY; ENDOZOOCHORY; ENTOMOCHORY; GYNOCHORY; HYDROCHORY; MYRMECOCHORY; PTEROCHORY; SAUROCHORY; and ZOOCHORY.

chroma One of the three variables (hue and value being the others) of colour; chroma measures the strength, wavelength purity, or saturation of colour. *See* MUNSELL COLOUR.

chromatography An analytical technique used for the separation of the components of complex mixtures, that is based on their repetitive distribution between a mobile phase (of gas or liquid) and a stationary phase (of solids or liquid-coated solids). The distribution of the different component molecules between the two phases is dependent on the method of chromatography used (e.g. gel-filtration, or ion-exchange), and on the movement of the mobile phase (which results in the differential migration and therefore separation of the components along the stationary phase).

chromatophore In many animals, a cell containing pigment granules; by dispersing or contracting such granules certain animals are able to change their colour.

chromosome A DNA-histone protein thread, usually associated with RNA, occurring in the nucleus of a cell. Although chromosomes are found in all animals and plants, bacteria and viruses contain structures that lack protein and contain only DNA or RNA: these are not chromosomes, though they serve a similar function. Chromosomes occur in pairs, which associate in a particular way during meiosis. Each species tends to have a characteristic number of chromosomes (e.g. 20 in maize, 23 in humans), found in most nucleated cells within most organisms. The presence of pairs of homologous chromosomes is referred to as the diploid state and is normal for the sexual phase of an organism. *Gametes (reproductive cells), and cells of the gametophyte (gamete-producing phase) of plants (*see* ALTERNATION OF GENERATIONS), however, are haploid with only one member of each pair in their nuclei. Usually chromosomes are visible only during mitosis or meiosis when they contract to form short thick rods coiled into a spiral. Each chromosome possesses chromomeres and a centromere, and some contain a nucleolar organizer. Chromosomes contain a line of different genes, a spindle attachment at some point along their length, and regions of heterochromatin, which stains strongly with basic dyes.

chromosome theory of heredity The unifying theory, advanced by W. S. Sutton in 1902, that *Mendel's laws of inheritance may be explained by assuming that *genes are located in specific sites on *chromosomes.

chron **1.** A small unit of geologic time, equivalent to the chronostratigraphic unit chronozone, that is usually based on fossil zonation. When used formally (e.g. Gilbert Chron) the initial letter is capitalized. **2.** (polarity chron) A single time interval of constant polarity of the geomagnetic field.

chronomere *See* CHRONOSTRATIGRAPHY.

chronosequence A sequence of related soils that differ in their degree of *profile development because of differences in their age. Chronosequences can be found in evolving landscapes such as those produced by deglaciation, volcanic activity, wind deposits, or sedimentation.

chronospecies **1.** (evolutionary species) According to one view of evolution (that of *phyletic gradualism), a new organism, or one of a series of new organisms, derived from an ancestor by a process of slow, steady, evolutionary change. **2.** A fossil species that occurs in strata formed at different times. **3.** *See* PALAEOSPECIES.

chronostratigraphy A branch of stratigraphy linked to the concept of time. In chronostratigraphy intervals of geological time are referred to as chronomeres. These may be of unequal duration. Intervals of geological time are given formal names and grouped within a Chronomeric Standard hierarchy. The formal terms are: eon, era, period, epoch, age, and chron. Of these, the last four are the equivalent of system, series, stage, and chronozone in the Stratomeric Standard hierarchy. The formal terms are written with initial capital letters when accompanied by the proper names of the intervals to which they refer. Some geologists hold that the term 'chronostratigraphy' is synonymous with *'biostratigraphy', but most agree that the two branches are separate.

chronozone The lowest-ranking chronostratigraphic unit. The duration of a chronozone is defined by a type section which may be based on a *biostratigraphic zone (in which case it would include all rocks laid down during that time interval regardless of *fossil content or rock type), or it may be defined on the basis of the time span of an existing lithostratigraphic unit. With increasing knowledge, chronozone boundaries may be found not to coincide with the boundaries of the stages to which they belong, and stages may more precisely be divided into substages. The geologic-time unit equivalent to achronozone is a *chron.

chrysotile *See* ASBESTOS.

ciguatera A type of poisoning caused by eating fish whose flesh contains ciguatoxin, a substance that is probably obtained initially from marine cyanobacteria eaten by small fish, in turn eaten by larger fish in which the toxin accumulates. The symptoms include nausea, vomiting, numbness of parts of the body, and even coma. Sometimes, therefore, it is probably wise not to eat certain species (e.g. mackerel, grouper, jack, and snapper).

cilia *See* CILIUM.

ciliates *See* PROTOZOA.

Ciliophora *See* PROTOZOA.

cilium **(pl. cilia)** A short, hair-like appendage, normally 2–10 μm long and about 0.5 μm in diameter, usually found in large numbers on those cells that have any at all. In certain protozoa, cilia function in locomotion and/or feeding. They generate currents in the fluid surrounding the cell by beating in a co-ordinated manner.

circadian rhythm The approximately 24-hourly pattern of various metabolic activities seen in most organisms. The rhythmic patterns may persist even when the organism is removed from exposure to 24-hour cycles of light and dark. In a natural habitat the rhythm is 24-hourly; in constant conditions it becomes slightly longer or shorter than 24 hours. The rhythm is thought to be controlled by an endogenous *biological clock, though little is known about the sites and modes of action of the mechanisms involved. The word is derived from the Latin *circa*, meaning 'about' and *dies*, meaning 'day'.

circalittoral zone The area of the *continental-shelf seabed that lies below the zone of periodic tidal exposure. It is approximately equivalent to the *sublittoral zone. *See also* LITTORAL ZONE.

circle of vegetation The highest classificatory unit used by the *Zurich–Montpellier school of phytosociology, this refers to a definite geographical region with a distinctive floristic element. This approximates to a major plant geographical region.

circularity ratio *See* DRAINAGE BASIN SHAPE INDEX.

circulation index *See* ZONAL FLOW.

circumaustral distribution The distribution pattern of organisms found around the high latitudes of the southern hemisphere in the *austral region, i.e. next to the Antarctic zone. An example of such a pattern is provided by the southern species of *Danthonia* (poverty grasses and wild-oat grasses).

circumboreal distribution The distribution pattern displayed by organisms around the high latitudes of the northern hemisphere in the boreal zone. The boreal zone lies next to the Arctic zone, and many circumboreal organisms are in fact also circumarctic. An example of a plant with both types of distribution is *Saxifraga oppositifolia* (purple saxifrage).

circumpolar distribution Denotes the range of organisms distributed around the North or South Poles. Strictly the term should be reserved for organisms that extend into the polar regions. *Eutrema edwardsii* is a good example of a flowering plant with such a distribution around the North Pole; an animal example is the reindeer or caribou (*Rangifer tarandus*).

cirque **(corrie, cwm)** A half-open, steep-sided hollow in a mountain region that has been, or is being, glaciated. Its form is due to a combination of glacial scouring (which deepens the floor and often produces a reversed long-profile gradient), and glacial erosion by basal sapping and *frost wedging (which acts on the cirque's head and side walls).

cirque glacier A relatively small body of ice, *firn, and snow, occupying an arm-chair-shaped hollow in bedrock. It is generally wide in relation to its length. It is actively supplied by drifting snow and therefore shows vigorous glacial erosion, involving rotational sliding.

cirri *See* CIRRUS.

cirrocumulus A cloud type comprising shallow, high-level cloud in stable air. The form is sheeted or layered with small-scale billows or ripples. As freezing of the super-cooled droplets of young cloud occurs, so the fine-patterned features of the cloud become more diffuse. *See also* CLOUD CLASSIFICATION.

cirrostratus A cloud type that comprises a semi-transparent veil of fibrous or smooth appearance, often covering the sky, and identified by halo phenomena. When associated with a cold front, the cloud sheets have distinct edges. *See also* CLOUD CLASSIFICATION.

cirrus **(plural cirri)** A Latin word meaning 'a tuft or lock of hair'. **1.** A cloud type that comprises high-level, banded clouds in fibrous filaments aligned approximately along their line of movement. *See also* CLOUD CLASSIFICATION. **2.** In certain ciliate protozoa, an organelle formed by the fusion of a group of cilia (*see* CILIUM), which usually functions in locomotion or feeding. **3.** In many invertebrates, a slender bodily appendage, often resembling a tentacle. In polychaete worms (Polychaeta) it bears cilia (*see* CILIUM). **4.** In some flatworms and trematodes, an *eversible copulatory organ. **5.** In some fishes (e.g. Creediidae) a tuft of skin on the jaw.

Cl *See* CHLORINE.

clade In *cladistics, a lineage branch that results from a dichotomous splitting in an earlier lineage. A split produces two distinct new taxa, each of which is represented as a branch in a phylogenetic diagram. The term is derived from the Greek *klados*, meaning 'twig' or 'branch'.

cladism *See* CLADISTICS.

cladistics (cladism, phylogenetic systematics) A special taxonomic system, founded in 1966 by W. Hennig, that is applied to the study of evolutionary relationships. It proposes that common origin can be demonstrated by the shared possession of derived characters, characters in any group being either *primitive or derived. In the *cladograms used to portray these relationships, *cladogenesis always creates two equal sister groups: the branching is dichotomous. Thus each pair of sister groups constitutes a *monophyletic group with a common stem *taxon, unique to the group, and a parent taxon always gives rise to two daughter taxa which must be given different names from each other and from the parent, so the parent species ceases to exist. Monophyletic groups are deduced by identifying *synapomorphic character states. *See also* AREA CLADISTICS.

cladogenesis In *cladistics, the development of a new *taxon as a result of the splitting of an ancestral lineage to create two equal sister groups, taxonomically separated from the ancestral taxon. *See also* ADAPTIVE RADIATION.

cladogram A diagram that delineates the branching sequences in an evolutionary tree. *See also* DENDROGRAM.

claspers The pelvic fins of male sharks and rays, modified to serve as copulatory organs.

classical conditioning *See* CONDITIONING.

classification Any scheme for structuring data that is used to group individuals or sometimes attributes. In ecological and taxonomic studies especially, quite sophisticated numerical classification schemes have been devised, and the methods developed in these disciplines are being applied increasingly in other fields, notably pedology and palaeontology. Various classificatory strategies may be used (e.g. *hierarchical or non-hierarchical; and where hierarchical schemes are used, these may be divisive or *agglomerative, and *monothetic or *polythetic, with the divisive polythetic approach being considered the optimum strategy). *Compare* ORDINATION METHODS.

clast A fragment of rock that has broken away from the main rock mass due to erosion and been deposited in a new location. Clasts may be of any size from small grains to boulders.

clastic Applied to the texture of sedimentary rocks that have broken into fragments (clasts).

clavate Club-shaped; thicker towards one end.

clay **1.** A soil separate comprising mineral particles less than 2 μm in diameter. **2.** In the Udden–Wentworth scale, particles less than 4 μm in diameter. **3.** A class of soil texture, usually containing at least 20 per cent by weight of clay particles. *Compare* CLAY MINERAL.

clay films *See* CUTAN.

clay mineral A member of a group of chemically related hydrous aluminium silicates, which generally occur either as very small platy or fibrous crystals. They have a layered structure and the ability to take up and lose water readily. It is difficult to distinguish various clay minerals, and therefore geologists employ sophisticated techniques (e.g. X-ray diffraction analysis and electron microscopy). The most important clay minerals belong to the kaolinite, hydrous mica, smectite, chlorite, vermiculite, and talc groups.

clay pan *See* PAN.

clayskins *See* CUTAN.

cleaner fish A fish (e.g. the cleaner wrasse *Labroides dimidiatus*) that removes ectoparasites (*see* PARASITISM) from other fish (which are usually much larger in size). Cleaner fish have conspicuous colour markings which make them easily recognizable to their temporary 'host', allowing them to come close with little risk of being eaten. About 45 species of cleaner fish are known.

cleaning station A *territory that is defended by a member of certain species of *cleaner fish and to which their customers go to have their parasites and dead surface tissue removed. At the cleaning station, cleaner fish and their customers recognize one another by performing *displays.

clear-air turbulence (CAT) The variable pattern of up- and down-draughts, or turbulence, sometimes occurring in the absence of any cloud. It is caused by strong *wind shear, especially associated with *jet streams in the upper *troposphere and lower *stratosphere. The phenomenon is significant for aircraft.

clear ice (glaze) A layer of transparent ice formed on objects near the ground, or on aircraft in flight, by freezing rain.

cleidoic egg An *egg that is enclosed by a shell which effectively isolates it from the outside environment and prevents the loss of moisture (i.e. the egg of a land-dwelling animal).

Clements, Frederic Edward (1874–1945) An American botanist and ecologist whose *Research Methods in Ecology* (1905) was the first work to deal fully with field experiments. Clements maintained that a plant community develops by constant adjustments of the relationships among species (i.e. a *succession) to an optimal state, the *climax community, and that a plant community may in some sense be regarded as a superorganism. *See also* TANSLEY, SIR ARTHUR GEORGE.

climacteric In plants, the phase of increased respiration found at the ripening of *fruit and at *senescence.

CLIMAPP *See* CLIMATE/LONG-RANGE INVESTIGATION MAPPING AND PREDICTIONS PROJECT.

climate The average weather conditions experienced at a particular place over a long period (usually more than 70 years).

climate classification The grouping of climates into broad types according to their shared characteristics. There are three principal approaches to the task. (*a*) Generic classification is based on levels of temperature and aridity as these relate to vegetation boundaries. Aridity is usually expressed as 'effective precipitation', which is calculated as the ratio of rainfall to temperature. Climatic types are defined by the response of flora to them. The *Köppen system, with its modifications, uses this approach. (*b*) Classifications based on the moisture budget and 'potential evapotranspiration' (i.e. the maximum moisture that will be transferred from the ground to the atmosphere, provided that sufficient moisture is available) which do not rely on vegetation boundaries. The *Thornthwaite system uses this approach. (*c*) Genetic (i.e. pertaining to its origin) classification, based on factors related to the atmospheric circulation of major winds and *air masses, and on other factors that cause climate, is used in the systems of H. Flohn (1950) and A. N. *Strahler.

Climate/Long-range Investigation Mapping and Predictions Project (CLIMAPP) An integrated project to study the climatic history of the *Quaternary, conducted by a team of scientists engaged in Earth and ocean research. Since 1971 the administrative base for the project has been at Columbia University, New York.

climate modelling The construction of mathematical descriptions of the *general circulation of the global atmosphere for the purpose of predicting future climates. Because of their complexity, such models can be made and manipulated only with the largest, fastest, and most powerful computers and, even then, certain phenomena (e.g. cloud formation and precipitation, which occur on scales smaller than the models can accommodate, and the relationship between oceanic heat transport and the atmosphere, which is not well understood) must be greatly simplified or described by simple assumptions. *See* GLOBAL WARMING and 'GREENHOUSE EFFECT'.

climatic climax A plant community that is in equilibrium with a zonal climate. *See also* CLIMAX THEORY.

climatic geomorphology That branch of *geomorphology which deals with the effects of climate on geomorphological processes and consequently on the character of land-forms. It has included the identification of climatically controlled zones and has attempted to define provinces with distinctive denudational processes.

climatic optimum The period of highest prevailing temperatures since the last ice age. In most parts of the world it occurred about 4 000–8 000 years ago.

climatic zone A region or zone that is characterized by a generally consistent climate. Climatic zones approximate to distinct latitude belts around the Earth. The principal ones are the humid tropical; subtropical arid and semi-arid; humid temperate; boreal (northern hemisphere) or sub-Arctic/sub-Antarctic; and polar zones.

climatostratigraphy The study of geologic-climatic units, which are climatic episodes defined in *Quaternary rocks. These record the effects of climate (i.e. the type of biota, soils, etc.), but not the climate itself.

climax The final stage in a plant *succession in which the vegetation attains equilibrium with the environment and, provided the environment is not perturbed, the plant community becomes more or less self-perpetuating. Subsequent changes occur much more slowly than those during earlier successional stages. The climax can be defined in productivity terms as the point in time when the gross *primary productivity of the *ecosystem is matched by the total ecosystem *respiration.

climax adaptation number A value, in the range 1–10, assigned to species in the *Wisconsin ordination scheme and based on the species' *importance values. Stands with the same leading dominants are grouped together, and importance values for all species in the group are recalculated. By comparing the changing importance of species in groups with different leading dominants, the leading dominants and then other species can be placed in *phytosociological order. *See also* DENSITY–FREQUENCY–DOMINANCE.

climax community (climax vegetation) The plants that inhabit an area within which the final stage of a *succession has been reached (*see* CLIMAX). The community is self-perpetuating, except that changes may occur very slowly and over a time-scale that is extensive compared with the rapid and dramatic changes during the early stages of succession. *See also* CLIMAX THEORY.

climax theory A theory that embraces the idea of successional development to an optimum sustainable vegetation community that is in equilibrium with its environment. Two conceptual models of climax vegetation exist: monoclimax (F. E. *Clements, in 1904 and 1916) and polyclimax (A. G. *Tansley, in 1916 and 1920). The essential difference between them lies in the time-scale envisaged for the development of the climax. Monoclimax envisages the development of vegetation and soil toward a definite end-point controlled by climate. Many ecologists consider this approach unworkable in practice, since equilibrium can never be reached owing to variations in climate over long periods of time (e.g. the known climatic changes in historical and post-glacial times in mid-latitudes). Evidence from climatically relatively uniform and stable tropical areas shows different, equally persistent, woodland communities *edaphically controlled. Insistence on a single uniform climax community for a major climatic region necessitates the introduction of an extensive terminology for subclimax communities, so tending to obscure an otherwise useful and simple concept. The alternative (polyclimax) approach proposes successional development to equilibrium with the environment, which may be controlled by climate or by some other factor (e.g. soil or fire) as often occurs in practice. *See also* CLIMAX COMMUNITY and SUCCESSION.

climax vegetation If a bare surface becomes available for colonization by plants, the vegetation will pass through a *succession of so-called *seral stages as more complex communities replace the earlier, simpler ones. Ultimately a state of equilibrium or climax is reached, which may reflect predominantly the control of climate, soil, or humans. *See also* CLIMAX COMMUNITY and CLIMAX THEORY.

climograph A two-dimensional, graphic representation of one major climatic element against another (e.g. rainfall against temperature). Typically the average values for each month are plotted, and the points joined together sequentially to give a dodecagon of a shape that is characteristic for a particular area. This forms an easily prepared basis for quick visual comparison of the climates of different areas. Uses include the assessment of the suitability of a new environment for a proposed species introduction. *See also* WALTER CLIMATE DIAGRAM.

climosequence A sequence of soil *profiles, usually on the same parent material, which differ from each other in their profile development because of local or site differences in climatic conditions. Climosequences can be found along mountain slopes in certain highland areas.

climotope The climatic component of the *habitat or *ecotope of a *biogeocoenosis.

clinal speciation A type of *allopatric speciation in which a geographic barrier falls across a *cline, cutting a *species (which already shows some variation) into two segments that continue to diverge in their new isolation.

cline A gradual change in gene frequencies or *character states within a species across its geographic distribution. *Compare* AREA-EFFECT SPECIATION.

clino- A prefix derived from the Greek *klino*, meaning 'sloping' or 'inclined'.

clinosequence A sequence of soils in which soil- *profile development is related to the angle of slope of the soil surface. Clinosequences can be found on landforms such as escarpments and *drumlins with varying surface angles of slope.

clint *See* LIMESTONE PAVEMENT.

clitochory *See* BARACHORY.

clitter A local name for the gently sloping spreads of coarse, often angular rock debris on the slopes of Dartmoor, south-west England. It was produced when blockfields developed on the Dartmoor granite during the *periglacial phases of the *Pleistocene and the exposure of *tors, but the rock debris is now stable and largely vegetated.

clod A compact, coherent block of soil, found *in situ* when soil is broken up by digging or ploughing. Clods are of varied sizes.

clonal dispersal A method of plant *dispersal in which the plant produces *stolons or *rhizomes from which new plants develop, the new plants being genetically identical to one another and to the parent (i.e. they comprise a *clone). Bracken (*Pteridium aquilinum*) and American quaking (or trembling) aspen (*Populus tremuloides*) are among the plants that disperse in this way. Such clones often cover a large area (nearly 14 ha in the case of one clone of *P. tremuloides*) and sometimes achieve great longevity (1 400 years in the case of a *P. aquilinum* clone). *See also* GUERILLA GROWTH FORM and PHALANX GROWTH FORM.

clone A group of genetically identical cells or individuals, derived from a common ancestor by asexual mitotic division. If a section of DNA is engulfed into the *chromosome of a bacterium, phage, or plasmid vector and is replicated to form many copies, each copy is referred to as a DNA clone.

closed canopy The condition in which the crowns or canopies of individual trees overlap to form a virtually continuous layer. *See also* CANOPY.

closed population A population in which there is a barrier to *gene flow. *Compare* OPEN POPULATION.

cloud amount The proportion of the sky that is seen to be covered by cloud. Nowadays it is commonly measured in 'oktas' or eighths of the sky covered, but sometimes it is quoted as a percentage or as tenths.

cloud base The undersurface of cloud, representing the *condensation level of water droplets or ice.

cloud classification The system by which clouds are accorded formal names based on their form, altitude, and the physical processes generating them. The World Meteorological Organization (*International Cloud Atlas*, 1956) classifies ten genera in three major groups (cumulus or heap clouds, stratus or sheet clouds, and cirrus or fibrous clouds) by criteria essentially based on cloud form. Some of the genera are subdivided according to variations in internal shape and structure, to give 14 species. Additional or supplementary features (e.g. transparency, arrangement, and characteristics of growth) are defined by Latin names as variants and accessory types of cloud. The cloud genera, with their abbreviations, are: cirrus (Ci), cirrocumulus (Cc), cirrostratus (Cs), altocumulus (Ac), altostratus (As), nimbostratus (Ns), stratocumulus (Sc), stratus (St), cumulus (Cu), and cumulonimbus (Cb). Clouds may also be referred to, according to their composition, as water or ice clouds; combinations of these are called mixed cloud.

cloud droplet The liquid component of clouds, occurring as water droplets with an average size of 10 μm. In a non-rain cloud the droplets are suspended at a near-constant level, because air friction approximately balances the gravitational force.

cloud forest Those tropical *montane forests that are shrouded in cloud for much of the day; usually they are located more than 1 000 m above sea level. The moisture encourages profuse *epiphytic growth, especially of ferns, filmy ferns, lycopods, and bryophytes.

cloud seeding The introduction of nuclei, e.g. silver-iodide crystals or solid carbon dioxide (dry ice), into clouds composed of supercooled water droplets, in an attempt to induce precipitation. Dry ice introduced (at −80 °C) from the air into cloud lowers the air temperature so that (particularly at temperatures below −40 °C) some of the supercooled water droplets are converted into ice crystals, which then grow by collisions with further droplets. Silver iodide (which has a crystal structure similar to that of ice), introduced from the air or ground, is the substance most commonly used in seeding: its crystals act as ice nuclei. Other substances (e.g. common salt or fine water droplets) may also be used to encourage coalescence. Natural seeding may be significant in cases where ice crystals from a high 'releaser' cloud (e.g. *altostratus or *cirrostratus) fall into a supercooled water 'spender' cloud (e.g. *nimbostratus) and encourage ice-crystal growth.

cloud street A band (or bands) of (usually) *cumulus cloud parallel to the wind direction in a sky that is otherwise more or less clear of cloud. Streets may form in an *air mass at a sharply demarcated convection layer, with separate *thermals producing parallel streets.

clubroot **(finger and toe)** A serious disease of cruciferous (Brassicaceae) plants, e.g. cabbages and broccoli. Symptoms include *wilting of plants in bright sunshine, and the leaves may take on a bluish or reddish coloration. The roots show characteristic *gall-like swellings. The disease is caused by the fungus *Plasmodiophora brassicae*. This organism can survive in the soil for many years, even in the absence of host plants.

clutch All the eggs deposited during one episode of laying.

co-adaptation The development and maintenance of advantageous genetic traits, so that mutual relationships can persist (i.e. both parties evolve adaptations that increase the effectiveness of the relationship). Predator-prey and flower-pollinator relationships often exhibit examples of co-adaptation, which is an aspect of *co-evolution. For example, the relationship between the ant *Pseudomyrex ferruginea* and the plant *Acacia hindsii* is obligatory and dependent on co-adaptations. The ant is active 24 hours a day (which is unusual for ants) and thereby provides continuous protection for the acacia. In a similar evolutionary gesture, the acacia bears leaves throughout the year (most related species lose their leaves), providing a continuous source of food for the ants. *See also* BELTIAN BODIES.

coagulation The clumping together of colloidal particles to form a large mass; it may be caused by heating (e.g. the cooking of an egg causes the albumen (the 'white') to solidify) or by the addition of ions that neutralize the electrical charge which stabilizes the colloid.

coal A carbon-rich mineral deposit formed from the remains of fossil plants. These are deposited initially as *peat, but burial and increase in temperatures at depth bring about physical and chemical changes. The process of 'coalification' results in the production of coals of different ranks ('coal series'), from peat, through the bituminous coals and lignite to anthracite. Each rank marks a reduction in the percentage of *volatiles and moisture, and an increase in the percentage of carbon. They are termed 'woody' or 'humic' coals if formed from fragments of trees or bushes. If the major constituents of coal are pollen grains and/or finely divided plant debris, the term 'sapropelic coal' is used.

coalescence The process by which the size of cloud water droplets increases as larger drops (more than 19 μm) fall and collide with smaller drops. Ice crystals in frozen cloud tops may also fall and grow by coalescence with drops below, ultimately melting into raindrops. In cumuliform cloud, turbulence may encourage this process of drop growth by the upthrust of small droplets to overtake and coalesce with other droplets. Repetition of the process can eventually produce rain-sized droplets. Factors favouring coalescent growth are high moisture content, water cloud of large vertical extent, and strong upward turbulence. *See also* BERGERON THEORY and COLLISION THEORY.

coalescence theory A theory stating that all *alleles of a gene must have descended from a single allele, the coalescent. The theory applies to neutral or nearly neutral genes, usually in a population of constant size where random mating occurs. The coalesence time (i.e. the time to the most recent common ancestor) depends on the population size and generation time.

coalescence time *See* COALESCENCE THEORY.

coalescent *See* COALESCENCE THEORY.

coalification The process by which *coal is formed.

coal maceral The elementary and microscopic constituent of *coal. There are a number of different types. Alginite is formed from algal remains; collinite from cell infillings; cutinite from plant *cuticles; and fusinite from woody material. Micrinite is opaque, granular, and has no cell structure. Resinite consists of small ellipsoids or spindles of resin, a cell-infilling material; sclerotinite of variously sized round or oval bodies of irregular structure, which may have been fungi or spores; sporinite of spore exines, usually flattened parallel to stratification; and telinite from cell-wall material.

coal measures A series of sedimentary rocks in which seams of *coal lie between strata of other material (e.g. clay or sandstone).

coal series *See* COAL.

coastal onlap The deposition and advance of coastal non-marine and *littoral deposits further and further inland, owing to a relative rise in sea level. If, subsequently, relative sea level falls, the base level is lowered and *erosion probably occurs at the top of the sequence. At the next relative sea-level rise, coastal onlap will recommence but from a lower level. This downward shift in coastal onlap is thus an indication that there has been a relative fall in sea level.

coastal processes The set of mechanisms that operate along a coastline, bringing about various combinations of *erosion and deposition. A cliffed coastline is affected by slope processes and by wave activity. Both agents give rise to distinctive land-forms, including the *geo, the *bevelled cliff, and the 'blow-hole' (a chamber with a relatively narrow exit at the top of the cliff, from which water and spray are forced when waves are driven against the coast). A low coastline is largely affected by processes in the *surf zone, where most work is done by shoaling and breaking waves, and in the offshore zone, where tidal currents are the chief agents of sediment transfer.

coccidiosis Any disease caused by a protozoon of the suborder Eimeriina. These organisms seldom cause disease in humans, but they may cause important economic losses among domestic animals. The disease usually affects the intestine, and symptoms often include diarrhoea with bloody stools.

coccolithophorids A family (phylum Prymnesiophyta) of unicellular, marine, planktonic protists which are, at least at some stage in their life cycle, covered in calcareous plates (*coccoliths) embedded in a gelatinous sheath. They are spherical or oval, and less than 20 μm in diameter. They range in age from the Upper *Triassic to *Holocene, although dubious examples have been described from the Upper *Precambrian and *Palaeozoic.

coccoliths The microscopic calcareous plates or discs, often oval and commonly intricately patterned and ornamented, that occur as part of the protective covering of a group of the unicellular algae called *coccolithophorids. Coccoliths are a major component of the modern deep-sea *calcareous oozes, and were especially abundant in the *Mesozoic, particularly the *Cretaceous Period, in which they became a major component of the Chalk lithology.

COD *See* CHEMICAL OXYGEN DEMAND.

coding codon A *nucleotide triplet (i.e. a codon) in *messenger-RNA which codes for an *amino acid to be included at a particular point in a forming *polypeptide.

codon *See* CODING CODON.

codon bias The disproportional use of one or more members of a *codon family.

codon family All the codons (*see* CODING CODON) that code for the same *amino acid.

coefficient of interference *See* MUTUAL INTERFERENCE.

coenobium An algal colony consisting of a definite number of cells in a specific arrangement. The colony as a whole behaves as an individual organism.

coenocline A gradient of communities (e.g. in a transect from the summit to the base of a hill), reflecting the changing importance, frequency, or other appropriate measure of different species populations. The term is applied most widely to vegetation records. *See also* ECOCLINE. *Compare* COMPLEX GRADIENT.

Coenozoic *See* CENOZOIC.

coenozone *See* ASSEMBLAGE ZONE.

co-evolution A complementary *evolution of closely associated species. The interlocking *adaptations of many flowering plants and their pollinating insects provide some striking examples of co-evolution. In a broader sense, predator-prey relationships also involve co-evolution, with an evolutionary advance in the predator, for instance, triggering an evolutionary response in the prey. *See also* CO-ADAPTATION and GENE-FOR-GENE CO-EVOLUTION.

cognition The mental processes that are presumed to be occurring within an animal but which cannot be observed directly. Cognition may be important in *insight learning.

cognitive map A mental model (or map) of the external environment which may be constructed following *exploratory behaviour.

cohesion The ability of particles to stick together without dependence on interparticle friction. In soils, cohesion is due to the shearing strength of the cement or film of water that separates individual grains.

cohort **1.** A group of individuals of the same age. **2.** In plant taxonomy, a little-used term meaning a group of related families. **3.** In animal taxonomy, a group of orders.

C

col **1.** A high pass or saddle in a ridge. It may mark the line of a former stream valley or of a former glacier, and so provides evidence of an early stage in the development of the landscape. **2.** The saddle region of the atmospheric-pressure field between two high- and two low-pressure centres.

cold front A boundary between dense, cold air and the warmer air ahead of it, which the cold air tends to undercut as it advances. The gradient of the upper surface of the cold air may be steep (e.g. 1 : 50). Along this steep front violent upwelling and *instability result in high *cumulonimbus cloud with rain and thunder. Weather changes occur with the passage of a cold front, sometimes including a pronounced temperature fall, a rise in pressure, and wind *veering (often with *squalls) to northerly or north-westerly (in the northern hemisphere). The passage of the front commonly brings clearer, brighter weather, but the unstable air may produce showers.

cold glacier *See* GLACIER.

cold low Basically, a *non-frontal depression that is typically associated with circulation in cold *air masses in the mid-*troposphere over the north-eastern USA and north-eastern Siberia, though it sometimes also occurs over the oceans in air that has emerged from the Arctic. Cold lows, marked by more or less concentric *isotherms around the core of the low, may originate along the Arctic coast as a result of strong vertical uplift and *adiabatic cooling in *occlusions. They do not necessarily influence surface weather conditions, but when they occur over warmer surfaces in middle latitudes, convection develops strongly. *See also* CUT-OFF LOW.

cold pole The place in each hemisphere that records the lowest mean temperature. The southern hemisphere cold pole is at Vostok Station, Antarctica, where the mean temperature in the coldest month (August) is −67.6°C and the annual mean temperature is −55.1°C, but where a temperature of −89.2°C was recorded on 21 July 1983. This is the lowest temperature ever recorded on Earth. There are two cold poles in the northern hemisphere. One is at Verkhoyansk, Siberia, where the mean temperature in the coldest month (January) is −50.3°C, the mean annual temperature is −17.2°C, and the lowest temperature ever recorded was −67°C. The other cold pole is at Snag, Yukon, where the mean temperature in the coldest month (January) is −28.1°C, the mean annual temperature is −5.8°C, and the lowest recorded temperature is −63°C.

cold sector The zone of colder air that surrounds the narrowing wedge of warm air, the warm sector, in a developing depression. At the *occlusion stage the whole of the surface layer forms a cold sector, with the warm air lifted off the surface.

cold wave The conditions associated with air of continental polar origin, often dominated by an *anticyclone behind a *cold front, that moves south into central and eastern parts of the USA. Cold waves are defined as a fall of 11°C or more to a minimum base (−18°C in northern, central, and north-eastern regions) within a 24-hour period. In southern states (Florida, California, and the Gulf Coast) the minimum fall is 9°C and the base minimum 0°C.

coliform bacteria Bacteria found in the intestinal tracts of mammals. They are rod-shaped, *facultative anaerobes that obtain energy through the fermentation of lactose sugar, a process that produces acid and gas. They do not produce spores. The group includes *Escherichia coli*, *Enterobacter aerogenes*, and *Klebsiella pneumoniae*.

coliform count A count made of the numbers of *coliform bacteria present as part of most standard analyses of water intended for potable use. The number of organisms present is normally expressed per 100 ml of water.

colinearity The correspondence between the DNA sequence of intronless genes (*see* INTRON) and the *amino acid sequence of the encoded proteins.

collector In a freshwater ecosystem, an organism which feeds on small particles (less than 2 mm). Collector–gatherers feed on particles of dead organic matter found on debris and sediments; collector–filterers abstract particles from the flowing water.

collinite *See* COAL MACERAL.

collision theory A theory that accounts for the growth of water droplets in cloud to produce raindrops, based on the mechanisms of collision, coalescence, and 'sweeping'. It holds that larger drops, with terminal velocities increasing in proportion to their diameter, fall faster than smaller drops, and collide with them. The probability of collision depends on the spacing of the drops in the cloud (i.e. on the mean free path) and on the relative sizes of droplets. For example, if some drops are up to 50 μm diameter in cloud consisting mainly of droplets smaller than this, collisions can be frequent. Such collisions can lead to coalescence, and an overall increase in size to produce particles of raindrop size. 'Sweeping' is an ancillary process whereby small drops that are swept into the rear of larger drops may be absorbed. These mechanisms are believed to be entirely responsible for rainfall from tropical convection cloud, as well as playing a part in other clouds, including those of mid-latitudes. *See also* BERGERON THEORY.

colloid **1.** A substance that is composed of two homogenous phases, one of which is dispersed in the other. **2. (pedol.)** Soil colloids are substances of very small particle size, either mineral (e.g. clay) or organic (e.g. humus), which therefore have a large surface area per unit volume. Colloids usually provide surfaces with high *cation exchange capacity, and also exhibit an instability controlled by soil chemistry.

colluvium Weathered rock debris that has moved down a hill slope either by creep or by surface wash.

colonization The successful establishment of an invading species in a *habitat.

colostrum Milk that is secreted in the first few days after parturition. It differs in its physical and biological properties from that secreted later, containing a large number of globulins that represent all the *antibodies in the maternal blood, and therefore conferring temporary passive immunity upon the new-born mammal.

colpus **(pl. colpi)** A germinal groove or aperture on the surface of a pollen grain, which is elliptical or approximately rectangular in shape and at least twice as long as it is wide. The shape and arrangement of colpi are diagnostic in the identification of *pollen.

comber A deep-water wave that has a breaking crest blown forward by a strong wind. The term is also applied to a long-period *spilling breaker.

comfort behaviour Behaviour (e.g. yawning, scratching, and *grooming) that increases the physical comfort of an animal.

comfort zone The range of atmospheric temperature and humidity within which the human body feels and works comfortably and efficiently, typically 19–24°C. Beyond these limits corrective adaptations are necessary to produce bodily comfort.

commensal *See* COMMENSALISM.

commensalism The interaction between species populations in which one species, the commensal, benefits from another, sometimes called the host, but this other is not affected. For example, a hydroid (*Hydractinia echinata*) living on a whelk shell occupied by a hermit crab is carried by the crab to sites where it can feed, but it does not deprive the crab because the two species have different food requirements. *Compare* MUTUALISM and PARASITISM.

Commons, the The concept that the major resources of the planet (land, air, and water) are commodities to which all people have equal right of access and use, and which no one has a right to spoil. The concept was popularized following the publication of an article, 'The Tragedy of the Commons', by Garrett Hardin, in *Science* in 1968, Hardin having derived it from a pamphlet by William Forster Lloyd (1794–1852) published in 1833. *See also* MALTHUSIANISM.

communication The transfer of information from one animal to another through the sense organs, resulting in behavioural changes that have survival value to one or both of the animals.

community A general term applied to any grouping of populations of different organisms found living together in a particular environment; essentially, the *biotic component of an ecosystem. The organisms interact (by *competition, *predation, *mutualism, etc.) and give the community a structure. Globally, the *climax communities characteristic of particular regional climates are called *biomes. Plant ecologists often use the term to cover merely the botanical components of a total biotic community. *See also* INDIVIDUALISTIC HYPOTHESIS.

community ecology An approach to ecological study which emphasizes the living components of an ecosystem (the *community). Typically it involves description and analysis of patterns within the community, employing methods of classification and ordination, and examines the interactions of community members, e.g. in the partitioning of resources and in succession. *See also* SYNECOLOGY.

companion species A term formerly used in the *Braun-Blanquet *phytosociological scheme to mean 'indifferent species'. Companion species at the association level sometimes emerge as characteristic (*kennarten or *faithful) species when communities are classified at higher levels, such as alliances or orders.

compartment model (box model) A modelling approach which emphasizes the quantities and materials in different compartments of a system, and which may also express connections between compartments by some form of transfer coefficient. The approach is frequently used for studies of whole *ecosystems.

compass orientation The ability to head in a particular compass direction without reference to landmarks possessed by many migratory birds (e.g. starling, *Sturnus vulgaris*), other vertebrates (e.g. some fishes, leatherhead and loggerhead turtles, *Caretta caretta* and *Derrunchelys coriacea* respectively), and some insects (e.g. bees, *see* DANCE LANGUAGE).

compensation level The depth at which light penetration in aquatic ecosystems is so reduced that oxygen production by photosynthesis just balances oxygen consumption by respiration. Generally this implies a light intensity of about 1 per cent of full daylight.

compensation light intensity The light intensity at which the amount of oxygen released by plants through *photosynthesis is equal to the amount absorbed through *respiration.

compensation point **1.** The light intensity at which plants do not accumulate carbon through *photosynthesis, because the rate at which they fix carbon is equal to the rate at which they release carbon by *respiration. **2.** At a given light intensity, the atmospheric concentration of carbon dioxide at which the rate of carbon fixation by photosynthesis is equal to the rate at which carbon is released by respiration.

competition The interaction between individuals of the same species (intraspecific competition), or between different species (interspecific competition) at the same *trophic level, in which the growth and survival of one or all species or individuals is affected adversely. The competitive mechanism may be direct (active), as in *allelopathy and *mutual inhibition, or indirect, when a common resource is scarce. Competition leads either to the replacement of one species by another that has a competitive advantage, or to the modification of the interacting species by selective adaptation (whereby competition is minimized by small behavioural differences, e.g. in feeding patterns). Competition thus favours the separation of closely related or otherwise similar species. Separation may be achieved spatially, temporally, or ecologically (i.e. by adaptations in behaviour, morphology, etc.). The tendency of species to separate in this way is known as the *competitive exclusion or Gause principle. Some ecologists differentiate between *interference competition (where space is substituted for a resource) and *exploitation competition (where organisms compete for a resource by enhancing their efficiency in gaining access to it).

competitive exclusion principle (exclusion principle, Gause principle) The principle that two or more resource-limited species, having identical patterns of resource use, cannot coexist in a stable environment: one species will be better adapted and will out-compete or otherwise eliminate the others. The concept was derived mathematically from the *logistic equation by Lotka and Volterra (*see* LOTKA–VOLTERRA EQUATIONS), working independently, and was first demonstrated experimentally by G. F. *Gause in 1934, using two closely related species of *Paramecium*. When grown separately, both species populations showed normal *S-shaped growth curves; when grown together, one species was eliminated.

competitive release The expansion of the *range of a species when a competitor for its *niche is removed.

competitive strategy *See* GRIME'S HABITAT CLASSIFICATION.

competitor In the classification of *plant strategies proposed by J. P. Grime, a plant species that exploits conditions of low stress and low disturbance. *Compare* RUDERAL and STRESS-TOLERATOR.

complementary resources Two or more resources that can substitute for one another and, when taken together, augment one another, so that the consumer requires less of them when taken together than when taken separately. *Compare* ANTAGONISTIC RESOURCES.

complex gradient A gradient of environmental factors linked in a complex fashion (e.g. the interrelated changes in rainfall, wind speed, and temperature, found along a transect from high to low elevation). *See also* ECOCLINE. *Compare* COENOCLINE.

concerted evolution The maintenance of homogeneity between two or more genes of a multigene family within a population, such that there is little variation within a population, but variation accumulates between reproductively isolated populations.

concrescent Applied to unified, or coalesced organs that are ordinarily separate, e.g. petals may be concrescent with each other or with stamens (which are then termed epipetalous).

concretion A localized concentration of material (e.g. calcium carbonate or iron oxide) in the form of a nodule of varying size, shape, or colour.

concurrent range zone (overlap zone) A body of strata which is characterized by the overlapping stratigraphic range of two or more taxa, selected as diagnostic, and after which the zone is named. Concurrent range zones are very widely used in the time correlation of strata. *See also* OPPEL ZONE. *Compare* ASSEMBLAGE ZONE.

condensation level The atmospheric level at which condensation occurs as a result of convection, the lifting of air (e.g. *orographic lifting), or vertical mixing. *See also* LIFTING CONDENSATION LEVEL and CONVECTIVE CONDENSATION LEVEL.

condensation nucleus A small particle of an atmospheric impurity (e.g. salt, dust, or smoke), which provides surfaces for the condensation of water. Condensation nuclei vary in size from about 0.1 μm to more than 3 μm. Some nuclei (e.g. salt and acid particles) can encourage condensation at a *relative humidity well below 100 per cent. *See also* AITKEN NUCLEUS; CONDENSATION LEVEL; and HYGROSCOPIC NUCLEUS.

conditional instability An atmospheric condition in which otherwise stable air, on being forced to rise (e.g. over a mountain barrier), cools at a rate less than that at which the temperature drops with height in the surrounding air. The rising air therefore becomes warmer than the surrounding air, and so continues to rise. This lesser rate of fall of temperature in the rising air is owing to the condensation that occurs, as this is accompanied by a release of the latent heat of condensation. In such cases the instability is thus conditional upon the *relative humidity of the rising air. *See also* INSTABILITY.

conditional stimulus *See* CONDITIONING.

conditioning (classical conditioning; Pavlovian conditioning) A form of learning in which an animal comes to associate an unconditional (significant) stimulus (e.g. the smell of food) with a conditional (neutral) stimulus (e.g. a sound), so that the previously conditional stimulus evokes a response that is rarely identical to the unconditional response but is nevertheless appropriate to the unconditional stimulus. The method for studying this form of conditioning is derived from the work of I. P. *Pavlov. *Compare* OPERANT CONDITIONING.

cone of depression The region shaped like an inverted cone, in which the *water-table is drawn down or depressed in the vicinity of a borehole from which *groundwater is being abstracted by pumping.

confidence limits In statistics, the probability that any value in a set of data will fall within a given range of the mean. Confidence limits of 95 per cent are given approximately by the mean plus or minus twice the *standard error of the mean.

confined aquifer *See* AQUIFER.

conflict The condition in which *motivations urge an animal to perform more than one activity at a time. Conflict may lead to one motivation becoming dominant or to unresolved, *ambivalent behaviour, *displacement activity, *intention movements, or *redirected behaviour.

congelifraction *See* FROST WEDGING.

congeliturbation *See* GELITURBATION.

congestus The Latin *congestus*, meaning 'piled up', used to describe a species of deep, bulging, *cumulus cloud, the upper part of which resembles a cauliflower. *See also* CLOUD CLASSIFICATION.

connate From the Latin *connatus*, meaning 'born together'. **1.** Applied to similar organs (e.g. leaves or petals) that are joined together. *Compare* ADNATE. **2.** Applied to water that has remained trapped in a sedimentary rock since the original sediments were laid down in that water, prior to lithification. Connate water may be very old and saline.

connectivity A measure of the degree to which landscape units are linked to one another. For example, hedges that have intact and frequent lateral branches have a high degree of connectivity.

consanguinity **(adj. consanguineous)** A genetic relationship in which individuals share at least one ancestor in the preceding few generations. Matings between related individuals may reveal deleterious recessive alleles. For example, first cousin marriages among humans account for about 18–24 per cent of albino children and 27–53 per cent of children with Tay–Sachs disease, both of which are rare recessive conditions.

consensus sequence In an *alignment of homologous (*see* HOMOLOGY) sequences of DNA or *amino acids, that sequence which represents the most common *character state at each site.

consequential dormancy *See* DORMANCY.

consequent stream A stream whose course is consequent upon the shape of a newly emerged land surface. Its course shows no necessary relationship with the underlying geologic structure, although older usage tended to restrict the term to a stream flowing in a downdip direction across gently inclined strata.

conservation The maintenance of environmental quality and resources or a particular balance among the species present in a given area. The resources may be physical (e.g. fossil fuels), biological (e.g. tropical forest), or cultural (e.g. ancient monuments). In modern scientific usage conservation implies sound *biosphere management within given social and economic constraints, producing goods and services for humans without depleting natural ecosystem diversity, and acknowledging the naturally dynamic character of biological systems. This contrasts with the preservationist approach which, it is argued, protects species or landscapes without reference to natural change in living systems or to human requirements. *See also* BIOLOGICAL CONSERVATION.

conservative substitution The substitution of one *amino acid for another with similar chemical properties. *Compare* RADICAL SUBSTITUTION.

consistence **(consistency)** The resistance of soil to physical impact such as ploughing, digging, or handling. It is controlled by the degree of adhesion between soil particles. It is described when dry as loose, soft, or hard, when moist as loose, friable, or firm, and when wet as sticky or plastic.

consistency *See* CONSISTENCE.

consociation A phytosociological term of the British and American traditions, meaning a *community with a single dominant species (e.g. an oak or beech wood). The term is also used by the *Uppsala school. (A community with several dominants is a *plant association.)

consocies A little-used term, derived from the Clementsian scheme for vegetation classification, which refers to a seral community in the succession to a *climax community with *consociation status. *See also* PHYTOSOCIOLOGY.

consolidated species list A list of all the organisms recorded in a sampling area, usually arranged in groups (e.g. taxonomically or by habitat type).

conspecific Applied to individuals that belong to the same species.

constancy *See* CONSTANT SPECIES.

constant head permeameter *See* PERMEAMETER.

constant site A site within a DNA sequence that is occupied by the same nucleic acid in all the sequences being compared.

constant species **(constancy)** In *phytosociology, a species common to a particular association or *community, but not necessarily confined to that community; a species of wide ecological amplitude compared with a *faithful species. A species of high constancy would be present in all, or almost all, of a series of *relevés or field samples that describe an association or community.

constellation diagram A representation of species affinities based on *χ^2 as a measure of the association between species. The reciprocal of the χ^2 value for each species pair is used to plot the diagram, so that highly positively associated species with large χ^2 values are positioned closely together. Thus clusters of similarly distributed species may emerge, while the transitional affinities of a species with those of another main focal area or cluster will be evident. The constellation diagram exemplifies a simply calculated ordination method.

C

constructive wave A wave that leads to the build-up of a beach, owing to the *swash of the wave being more effective in moving material than the *backwash. Usually, constructive waves are associated with low-energy conditions and a gentle offshore gradient.

consumer In the widest sense, a *heterotrophic organism that feeds on living or dead organic material. Two main categories are recognized: (*a*) macroconsumers, mainly animals (*herbivores, *carnivores, and *detritivores), which wholly or partly ingest other living organisms or organic particulate matter; and (*b*) microconsumers, mainly bacteria and fungi, which feed by breaking down complex organic compounds in dead protoplasm, absorbing some of the decomposition products, and at the same time releasing inorganic and relatively simple organic substances to the environment. Sometimes the term 'consumer' is confined to macroconsumers, microconsumers being known as *'decomposers'. Consumers may then be termed 'primary' (herbivores), 'secondary' (herbivore-eating carnivores), and so on, according to their position in the food-chain. Macroconsumers are also sometimes termed phagotrophs or biophages, while microconsumers correspondingly are termed *saprotrophs or *saprophages. *Compare* PRODUCER.

consummatory In animals, applied to behaviour associated with the achievement of a goal (e.g. the eating of food) as distinct from *appetitive behaviour (e.g. searching for food). *See also* QUIESCENCE.

contagious distribution *See* OVERDISPERSION.

contessa del vento A type of cloud that typically has a rounded base and a bulging upper surface. Sometimes a number of separate discs of this cloud-form extend one above another. Such cloud occurs on the lee side of distinct mountains within an eddy zone. In the case of Mt Etna the cloud is related to a westerly air stream.

C

contest competition *Competition for a resource that is partitioned unequally, so that some competitors obtain all they need and others less than they need (i.e. there are winners and losers). *Compare* SCRAMBLE COMPETITION.

contiguous grids A system of adjacent *quadrats. *See also* GRID ANALYSIS OF PATTERN.

continental drift The hypothesis proposed around 1910 to describe the relative movements of continental masses over the surface of the Earth. A major theorist of continental drift, and certainly the one who gave the hypothesis scientific plausibility, was Alfred Wegener (1880–1930). His work was based on qualitative data, but has been vindicated in recent years by the development of the plate tectonics theory, which has provided geologists with a viable mechanism to account for continental movements.

continental freeboard The average level of the sea surface relative to the continents.

continentality A measure of how the climate of a place is affected by its remoteness from the oceans and oceanic air. The difference between the average temperatures prevailing in January and July is most often quoted as an indicator of this.

continental rise A smooth-surfaced accumulation of sediment which forms at the base of the *continental slope. The surface of the rise is gently sloping with gradients between 1:100 and 1:700. The width of the rise varies but is often several hundred kilometres. Two types of deposit lead to the formation of rises: *turbidites laid down by *turbidity currents flowing down the continental slope; and *contourites laid down by *contour currents flowing along the rise at the base of the continental margin.

continental shelf The gently seaward-sloping surface that extends between the shoreline and the top of the *continental slope at about 150 m depth. The average gradient of the shelf is between 1:500 and 1:1 000 and, although it varies greatly, the average width is approximately 70 km. Five major types of shelves may be recognized: (*a*) those dominated by tidal action; (*b*) those dominated by wave and storm action; (*c*) those dominated by carbonate deposition; (*d*) those subject to modern glaciation in Arctic areas; and (*e*) those floored by relict sediments which constitute up to 50 per cent of the total shelf area.

continental-shelf waves Vorticity waves that are produced in a *continental-shelf area where there is a sea-bed slope. In the northern hemisphere, if a water column is displaced into shallower water it develops negative relative vorticity or anticyclonic motion; if displaced into deeper water it will develop positive relative vorticity or cyclonic motion. The net result is for shelf waves to progress in a poleward direction along the west coasts of continents, or towards the equator along east coasts.

continental slope The relatively steeply sloping surface that extends from the outer edge of a *continental shelf down to the *continental rise. The total relief is substantial, ranging from 1 km to 10 km, but the slope is not precipitous and ranges from 1° to 15° of slope (average 4°). Along many coasts of the world the slope is furrowed by deep submarine canyons, terminating as fan-shaped deposits at the base.

continental South-east Asia floral region Part of R. Good's (1974, *The Geography of the Flowering Plants*) *Palaeotropical floral kingdom, which lies within the Indo-Malaysian subkingdom. Floristically it is a poorly documented region, transitional between the rich floras of China to the north and Malaysia to the south. There are relatively few endemic genera (*see* ENDEMISM), probably around 250, though the proportion of endemic species is thought to be high. This region is very likely the source of such important crop plants as rice, tea, and citrus fruit. *See also* FLORAL PROVINCE and FLORISTIC REGION.

contingency table (two-way table) A table of data for two methods of classification of the same individuals (e.g. leaf shape and flower structure or hair colour and eye colour). This type of data can then be analysed statistically for association between these properties using a *χ^2 test.

continuous distribution Data that yield a continuous spectrum of values (e.g. the height of a plant, the wing length of a bird, or the weight of a mammal).

continuous plankton recorder A device for sampling marine *plankton, comprising a screen of silk gauze that is wound from a storage spool, across a second spool, and on to a take-up spool contained in a tank of formalin. The screen mechanism is enclosed in a torpedo-shaped cylinder with stabilizing fins, which is towed behind a vessel at a known speed and depth and powered by a turbine turned by the flow of water. Water enters through an aperture at the forward end of the casing, flows through the moving screen, which collects and stores the plankton, and leaves through an exit aperture at the rear. The device was invented by Sir Alister *Hardy.

continuous variation An assemblage of measurements of a phenotypic character which form a continuous spectrum of values. The continuity of a *phenotype is a result of two phenomena: (*a*) each phenotype does not have a single phenotypic expression but a norm of reaction that covers a wide phenotypic range; (*b*) there may be many segregating loci whose alleles make a difference to the phenotype being observed.

continuum The idea that vegetation is continuously variable and cannot be classified into discrete entities, since it shows gradual change in response to environmental change. Such change may be analysed using *ordination methods (e.g. the 'continuum approach' of the *Wisconsin ordination scheme).

continuum index A measure of the total environment of a stand of trees, expressed in terms of species composition and their relative abundance. In the *Wisconsin ordination scheme the index is calculated by multiplying the importance value of each species in each sample by the adaptation number, and summing these results for each stand. *See also* CLIMAX ADAPTATION NUMBER.

contour current An undercurrent, typical of the *continental rise, which flows along the western boundaries of ocean basins. Such currents occur particularly in regions in which density stratification is strong because of the supply of cold waters originating near the poles. A well-known example is the Western Boundary Undercurrent, which hugs the continental rise of eastern North America. Contour currents are persistent, slow-moving (velocity 5–30 cm/s) flows capable of transporting mud, silt, and sand.

contour diagram A stereographic *equal-area net on which orientation data (i.e. the azimuths of structural features) are plotted as lines or points and then joined to form contours linking areas of equal density of data, thus providing a visual description of the range and concentration of the data. There are several ways to construct such diagrams manually or by computer, and there are statistical tests for evaluating the significance of the densities revealed.

contourites Sediments that have been deposited by *contour currents on the *continental rise. The sediments are thin-bedded silts, sands, and muds. The sands are well sorted, laminated, or cross-laminated, with many internal erosion surfaces, and there are concentrations of heavy minerals. Both bottom and top contacts of the beds are sharp, and the beds lack great lateral continuity.

contraction limit *See* ATTERBERG LIMITS.

controlled pollination A practice used in plant hybridization, in which pistillate flowers are enclosed in bags (usually of muslin) to protect them from unwanted pollen and, when they are in a receptive condition, the flowers are dusted with pollen of the required type. This is commonly performed in horticulture and in genetic experiments.

convection **1.** Vertical circulation within a fluid that results from density differences caused by temperature

C

variations. Convection currents occur in the oceans when a water mass that is denser than the water below it sinks and is replaced by lighter, warmer water. **2.** In meteorology, the process in which air, having been warmed close to the ground, rises. The convective uplift of air parcels is one of the main processes leading to condensation and cloud formation. *See also* DISHPAN EXPERIMENT; FORCED CONVECTION; HADLEY CELL; INSTABILITY; and STABILITY. **3.** Within the Earth, the release of radiogenic heat results in convective motions causing tectonic plate movements. The location and configuration of the *convective cells is uncertain, but they appear to be mantle-wide and marked by most heat loss along the mid-ocean ridges. The difference in temperature between upgoing and downgoing convective limbs within the mantle may be only 1–2°C. In the upper oceanic crust, heat loss is mainly by convective circulation systems combined with thermal conduction.

convective cell The pattern formed in a fluid when local warming causes part of the fluid to rise, and local cooling causes it to sink again elsewhere. The atmosphere in low latitudes forms such cells, known as *Hadley cells, as warm equatorial air rises, moves away from the equator and cools, and then descends over the subtropics.

convective condensation level The level at which surface air will become saturated when rising by *convection. *See also* LIFTING CONDENSATION LEVEL.

convective instability *See* POTENTIAL INSTABILITY.

convergence 1. The situation in which, over a given lapse of time, more air flows into a given region than flows out of it. It is commonly accompanied by confluence of the streamlines, but may be caused by differences of velocity (e.g. where the wind comes against a coast or a mountain wall). Surface friction can produce convergence. *Compare* DIVERGENCE. **2.** The point, line, or region where two oceanic water masses or surface currents meet. This leads to the denser water from one side sinking beneath the lighter water of the other side.

convergent evolution The development of similar characteristics in organisms which are unrelated (except through distant ancestors) as each adapts to a similar way of life. Sharks (fish), dolphins (mammals), and ichthyosaurs (extinct reptiles) provide good examples of convergence in the aquatic habitat. Similarities in appearance and behaviour between many marsupial and placental mammals (e.g. Tasmanian 'wolf', native 'cat', and marsupial 'mouse') have arisen by convergence. The wings of birds and bats, which are modified forelimbs, have arisen by convergent evolution in unrelated organisms that have adapted to flight. *Compare* DIVERGENT EVOLUTION and PARALLEL EVOLUTION.

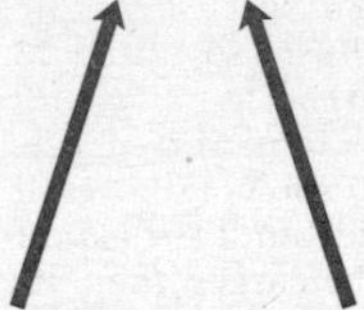

Convergent evolution

convergent substitution The substitution by the same *nucleotide of two different nucleotides in homologous sequences at the same site.

convex slope *See* SLOPE PROFILE.

Cooksonia One of the earliest of vascular land plants, known from the late *Silurian and early *Devonian (400 Ma ago), which is believed to be ancestral to all vascular plants. A few centimetres tall, it was upright, dichotomously branching, produced thick-walled *spores, possessed a *cuticle and stomata to control the passage of gases, and an underground rooting portion, the rhizome. Long suspected of being a vascular plant, the presence of a variety of conducting elements was confirmed in the Lower Devonian *C. pertoni* in 1992 (by D. Edwards, K. L. Davies, and L. Axe). Two species are known: *C. pertoni* and *C. hemispherica* (Silurian), the latter being the first representative of the order Rhyniopsida. *Cooksonia* may share a common ancestor with the club mosses or be ancestral to them.

co-operation In animals, mutually beneficial behaviour that involves several individuals (e.g. collaborative hunting and care of the young). Co-operation may involve *altruism. Co-operation among members of different species is usually called '*symbiosis'.

Cope's rule In 1871, the American palaeontologist Edward Drinker Cope (1840–97) noted a phylogenetic trend towards increased body size in many animal groups, including mammals, reptiles, arthropods, and molluscs. This came to be known as Cope's rule. It remained unchallenged until a study of more than 1 000 insect species in 1996 and was finally disproved in 1997, by a study in which David Jablonski made more than 6 000 measurements on 1086 species of Late *Cretaceous fossil molluscs spanning 16 million years and found that as many lines led to decreased size as to increased size. Evolutionary lineages show no overall tendency to greater size, but if the extant survivor happens to be larger than its immediate ancestor (e.g. the horse) this coincidence appears to validate Cope's rule. Cope also formulated a second rule, that species characterized by larger body size are more likely to go extinct.

copper (Cu) An element that is required in small amounts by plants, although high concentrations of it can be toxic. It is found bound to proteins and is involved in oxidation-reduction reactions, especially those involving molecular oxygen. The signs of copper deficiency can be very varied: leaves may become chlorotic (*see* CHLOROSIS) or dark green; the bark of trees may blister; shrubs may become very bushy.

copper moss A moss that grows on copper-rich rocks. Such mosses, including species of *Mielichhoferia*, *Dryptodon*, and *Merceya*, can survive levels of copper far in excess of those lethal to other mosses.

coppice **1.** A traditional European method of woodland management and wood production, in which shoots are allowed to grow up from the base of a felled tree. Trees are felled in a rotation, commonly of 12–15 years. A coppice may be large, in which case trees, usually ash (*Fraxinus*) or maple (*Acer*), are cut, leaving a massive stool from which up to 10 trunks arise; or small, in which case trees, usually hazel (*Corylus*), hawthorn (*Crataegus*), or willow (*Salix*), are cut to leave small, underground stools producing many short stems. The system provides a continuous supply of timber for fuel, fencing, etc., but not structural timber. In Britain, coppicing is largely abandoned now, except for conservation purposes, since high labour costs and alternative fuels and materials render the practice unprofitable. One consequence of coppicing is that the stool enlarges because each subsequent growth of shoots occurs on its outside. The diameter of a stool is thus directly related to its age. **2.** The smaller trees and bushes that regenerate from cut stumps and occasionally (e.g. in *Ulmus* species) from root suckering. **3.** An area of land in which underwood and timber is or was grown. **4. (copse)** Any type of wood in which the shrub layer predominates and is periodically coppiced. **5.** The action of cutting coppice.

coppiced scrub The regrowth of trees resulting from irregular *coppicing.

coppice shoot A shoot that arises from a bud at the base of the *coppice stump of a tree that has been cut near the ground.

coppice stump (moot, stool) The remnant of a tree that has been cut, usually to or near to ground level, and from which coppice growth develops.

coppice-with-standards A *coppice system in which scattered trees, typically oak (*Quercus robur*) are allowed to grow to their full height (standards) for use as structural timber, while the understorey, commonly of hazel (*Corylus avellana*), is coppiced.

coprolite Fossilized droppings or excreta (i.e. a fossilized faecal pellet). Coprolites may have distinctive shapes or markings which can provide information regarding the structure of the animal's alimentary canal, and analysis of the contents may also reveal its diet.

coprophagy Feeding by the ingestion of faecal pellets that have been enriched by microbial activity during exposure to the external environment. *Compare* CAECOTROPHY.

coprophilous Growing on or in dung.

copse **1.** *See* COPPICE (4). **2.** In modern usage, any small wood, particularly a detached wood that is isolated from other woodland.

copse-bank A boundary between areas of land made from an earth bank, especially where the boundary separates an intermittently enclosed *coppice within a *wood-pasture system.

copulation Sexual behaviour that is most closely associated with the fertilization of the egg by sperm (hence *in copula*, sometimes written incorrectly as *in copulo*).

copy number **1.** The number of times a segment of DNA is repeated throughout a *genome. **2.** The number of copies of a *plasmid that are present in a cell.

coral growth lines Minute growth lines on the outer surfaces of all corals that have a calcified outer wall. The carbonate is secreted by symbiotic algae (*zooxanthellae) which, responding to day and night, create a series of diurnal growth increments. Studies on *Devonian isolate corals indicate a 400-day year and, therefore, a 22-hour day. Post-Devonian data confirm a near-linear deceleration of the Earth's rotational velocity towards the present 24-hour day. The number of daily growth lines can therefore be used in a crude way to calibrate the geologic record.

corallite The skeleton formed by an individual coral polyp, which may be either solitary or part of a colony.

corallum The skeleton of a colonial coral, made up from individual *corallites.

coral reef A massive, wave-resistant structure, built largely by coral, and consisting of skeletal and chemically precipitated material. Coral reefs extend over an area of more than 175 × 10^6 km^2 in tropical and subtropical seas, being best developed where the mean annual temperature is 23–25°C; they do not develop significantly at less than 18°C. Surface illumination is important and reefs do not grow in regions of high sedimentation, their skeletal formation depending on the activity of symbiotic algae and *zooxanthellae.

coral spot The common name for the coral-pink conidial pustules formed by the fungus *Nectria cinnabarina* on twigs, branches, etc.

cord A stack of wood, usually measuring 2.54 m long by 1.27 m high, but with local variations. It is usually composed of small-diameter material. A cord of this size measures about 3.6 m^3 when stacked, and contains 2–2.8 m^3 of wood. A 'short' cord is equal to half a cord.

cordillera A Spanish word meaning 'little cord'. **1.** A broad assemblage of mountain ranges belonging to orogenic belts (*see* OROGENY) of different ages, which formed originally at destructive plate margins. **2.** A system of mountain ranges, together with their related plateaux and intermontane basins. For example, the Cordillera of North America includes all the mountain ranges and plateaux west of the Great Plains and the Mexican lowlands. **3.** A subsidiary complex of ranges within a mountain system, e.g. the eastern and western cordillera of the Andes. **4.** An individual range (e.g. the Cordillera Patagonica of the southern Andes).

cordwood A small-diameter wood that has been cut into lengths suitable for use as domestic fuel or for making charcoal.

core area Part of a *range in which an animal or group of animals may rest securely, in which young may be raised, and to which in some species food may be taken to be eaten. It is likely to be defended against members of the same species which do not share the range.

Cor F *See* CORIOLIS EFFECT.

coriaceous Having a leathery texture.

Coriolis effect **(CorF)** A deflection of the path of bodies moving across the Earth's surface that is due to the rotation of the Earth beneath them. It gives the appearance of a force, hence its abbreviation, but a body affected by it is not 'pushed';

the deflection results from the difference between the velocity of the body and that of the Earth. The Coriolis effect causes bodies, including moving air and ocean currents, to be deflected to the right in the northern hemisphere and to the left in the southern hemisphere. The magnitude of CorF is related to the speed of the moving body and its latitude. The relationship of CorF to latitude is given by the Coriolis parameter: $2\Omega \sin \Phi$, where Ω is the angular velocity of the Earth (7.29×10^{-5} rad s^{-1}) and Φ is the latitude. The magnitude of CorF is given by: CorF $= 2\Omega \sin \Phi v$, where v is the speed. At the equator $\Phi = 0$, therefore $\sin \Phi = 0$. At the North and South Pole $\Phi = 90$, therefore $\sin \Phi = 1$. Consequently the magnitude of the Coriolis effect is 0 at the equator and reaches its maximum over the poles. The effect was first described in 1835 by the French physicist, mathematician, and engineer Gaspard Gustave de Coriolis (1792–1843).

cork In woody plants, a layer of protective tissue that forms below the epidermis. It comprises dead cells, derived from the cork cambium (phellogen), and coated with a waxy substance (suberin) that renders them waterproof. Cork develops abundantly in the bark layer of certain plants (e.g. *Quercus suber*, cork oak), and is removed for commercial use.

corm In plants, an underground storage organ formed from a swollen stem base, bearing adventitious roots and scale leaves. Often it is renewed annually, each new corm forming on top of the preceding one. It may function as an organ of vegetative reproduction or in perennation.

corona Coloured rings of lights, typically from blue inside to red outside, which sometimes appear to surround the Sun or Moon. The effect is created by the diffraction of light by spherical water drops in such clouds as *altocumulus. *Compare* HALO.

corrasion *See* ABRASION.

correlated response A change in a character which occurs as an incidental consequence of selection for an apparently independent character (e.g. selection for increased bristle number in *Drosophilia* may also result in reduced fertility).

correlation A statistical association between variables, such that changes in one variable are associated with changes in others.

correlation coefficient A statistic that is used to measure the degree of relationship between two variables.

correlogram A graph showing the strength of correlations in data at different time intervals and thereby exposing the existence and phases of cycles.

corridor dispersal route As originally defined by the American palaeontologist G. G. *Simpson in 1940, a corridor is a *migration route that allows more or less uninhibited faunal interchange. Thus many or most of the animals of one faunal region can migrate to another one. A dispersal corridor has long existed between Western Europe and China via central Asia.

corridor farming *See* AGROFORESTRY.

corrie *See* CIRQUE.

corticolous Growing on or in tree bark.

cosmic radiation Ionizing radiation from space, consisting principally of protons, alpha particles, and 1–2 per cent heavier atomic nuclei, as well as some high-energy photons and electrons. On encountering the Earth's atmosphere, secondary radiation is produced, mainly gamma rays, electrons, pions, and muons. Three sources are identified: (*a*) galactic cosmic rays, from outside the solar system, with energies in the range 1–10 GeV per nucleon; (*b*) solar cosmic rays, mainly associated with solar flares, with energies in the range 1–100 MeV per nucleon; and (*c*) solar wind, with energies of about 1 000 eV per nucleon.

cosmopolitan distribution A distribution of an organism that is worldwide. Apart from *weeds, commensal animals (*see* COMMENSALISM), and some of the lower groups of *cryptogams, there are relatively few organisms that occur on all six of the widely inhabited continents. *See* BICENTRIC DISTRIBUTION; DISJUNCT DISTRIBUTION; and UNICENTRIC DISTRIBUTION.

cotidal line A line joining points at which given tidal levels (such as mean high water or mean low water) occur simultaneously. The lines are shown on certain hydrographic charts. The same information is contained in tide tables, where the data are given as differences from the times of high or low water at a 'standard port'.

coupe **1.** An area that is to be cut in a particular year of a *coppice rotation. **2.** An area of woodland that is to be or has been clear-felled.

courtship The behaviour that precedes the sexual act and involves *displays and posturings, usually by the male partner.

courtship feeding The presentation of food, or food-like objects, usually by a male to a female during *courtship. The gift is sometimes of nutritional significance but more usually the presentation is an act of appeasement, and it is often highly ritualized (*see* RITUALIZATION).

Couvinian *See* DEVONIAN.

covalent bond A bond in which a pair (or pairs) of electrons is shared between two atoms. The bond is often represented by drawing a single line between the symbols of the two atoms that have bonded together. Sometimes the bonding is between atoms of different elements (e.g. hydrogen chloride, H–Cl), and sometimes between atoms of the same element (e.g. fluorine, F–F). The name 'molecule' is used to describe any uncharged particle containing covalently bonded atoms. *See also* HYDROGEN BOND, IONIC BOND, and METALLIC BOND.

cover In descriptions of plant communities, the proportion of ground, usually expressed as a percentage, that is occupied by the perpendicular projection down on to it of the aerial parts of individuals of the species under consideration. The most widely used visual scales are the *Domin scale and the *Braun-Blanquet five-point scale. A more objective estimate may be obtained using a *pin-frame or *point-quadrat.

cover-abundance measure A linked scheme for estimating cover visually. It is based on percentages at the top end, but uses abundance estimates for species with low-cover values. The most widely known is the *Domin scale.

covered smut A disease of cereals caused by *smut fungi (order Ustilaginales) in which the fungal spore masses are enclosed within the glumes of the cereal and are not released until threshing. *Compare* LOOSE SMUT.

cover-sociability scale A method for recording vegetation, devised by *Braun-Blanquet, in which two scales are combined. The first describes the number and cover of a species, ranging from presence through five scales of cover; the second indicates the spatial arrangement of the individuals concerned (e.g. as 'isolated' or 'clumped').

crachin Condensation as low cloud or fog, often with drizzle, which is frequent in spring in coastal areas of the Gulf of Tonkin and southern China. The weather results from the *advection of warm air over a cold surface, or the mixing of *air masses at the surface.

crag Shelly sand.

crag and tail A land-form consisting of a small rocky hill (crag) from which there extends a tapering ridge of unconsolidated debris (tail). The crag is a residual feature left by selective glacial erosion, while the tail is drift-deposited by ice on the lee side of the obstacle.

crassulacean acid metabolism (CAM) A method of carbon dioxide fixation that conserves water in certain succulent, drought-resistant plants. At night, when the external temperature and therefore the evaporation rate are low, the leaf stomata open allowing carbon dioxide to enter leaf cells where it is incorporated in an organic acid. During the day the stomata remain closed, conserving moisture, while the acids are decarboxylated and the carbon dioxide is used for the *dark reactions of *photosynthesis. The initial fixation of carbon dioxide results in the formation of compounds with four carbon atoms; this process uses the enzyme phosphoenolpyruvate carboxylase, as in C_4 photosynthesis (*see* C_4 PATHWAY). In CAM, however, there is no spatial movement of this product prior to processing by the Calvin cycle. This pathway of carbon fixation was first observed in members of the Crassulaceae (stonecrops and houseleeks), hence the name.

creationism A modern variant of *special creation, in which it is maintained that all 'kinds' of organisms were created during one week, 6 000–10 000 years ago, exactly as is stated in the biblical Book of Genesis. Creationism involves a rejection not only of the concept of evolution but also of the whole of geology and radiometric dating. *See also* CREATION 'SCIENCE'.

creation 'science' The extreme form of *creationism, in which it is maintained that science does not refute the Genesis stories of the creation and the flood, but confirms them. Many analyses of creation 'science' in recent years have shown that it engages in twisting or ignoring scientific evidence, as well as special pleading, and that it is scientifically dishonest.

creep **1.** The slow, downslope movement of the *regolith over hill-slopes, caused by gravity. The necessary disturbance of the regolith may be due to freezing and thawing, to expansion and contraction (resulting from temperature change or from wetting and drying), to additional weight and lubrication by water, or to the activities of burrowing animals. **2.** *See* CREEP MECHANISMS.

creep mechanisms The mechanisms by which materials deform at the Earth's surface or, more commonly, at depth. These can be: (*a*) cataclastic, in which individual grains or fragments physically rotate or glide past one another; (*b*) dislocational, by a gliding motion along crystalline dislocations; (*c*) grain-boundary sliding; (*d*) recrystallization; and (*e*) the diffusion of individual atoms. Each process is dependent on the degree, temperature, and duration of the stress.

creep strength The strength of a rock which is undergoing long-term *creep processes. It is the threshold value, beyond which creep gives way to permanent rupture.

Crenarchaeota The less derived (*see* APOMORPH) of the *Archaea, consisting principally of extreme *thermophiles and *psychrophiles. Members of the Crenarchaeota show a greater genetic similarity to *eukaryotes than to members of the *Eubacteria. In the widely used five-kingdom system of classification, the Crenarchaeota form a group within the Archaea subkingdom of the *Bacteria. In the three-domain system they comprise a kingdom within the *domain Archaea.

crenate With a round-toothed or scalloped edge or margin.

crenulate Finely notched.

crepuscular Of the twilight; applied to animals that are active at dusk.

crepuscular rays Beams of sunlight that are made visible by haze in the atmosphere and are seen where rays penetrate gaps in clouds such as *stratocumulus (this effect is called 'Jacob's ladder'). In other cases rays from a low Sun diverge upwards above cumuliform cloud.

Cretaceous The third of the three periods that comprise the *Mesozoic Era. The Cretaceous lasted from 145.5 to 65.5 Ma ago; its end is defined by the mass extinction of many invertebrate and vertebrate stocks associated with a bolide impact. The period is noted for the deposition of the chalk of the White Cliffs of Dover, England, much of the chalk being derived from the calcareous plates (*coccoliths) of marine *algae. Angiosperms, which arose during the *Jurassic, came to dominance during the Cretaceous at the expense of such groups as cycads and pteridosperms. Woody species evolved during the Cretaceous and around 65 Ma stratified forests appeared and there was an increase in the variety of fruits.

crevasse **1.** A deep fissure in the surface of a glacier, caused when tensile stress overcomes the shear strength of ice in the brittle upper few metres. Appropriate tensile stresses are typically developed when a glacier moves over a convex slope. **2.** A breach in a *levee along the bank of a river, through which flood water may flow and produce crevasse splays. (*See* CREVASSE DEPOSIT (2)).

crevasse deposit **1.** The gravelly or sandy sediment infill of a *crevasse in glacial ice. **2. (crevasse splay)** The deposit generated by a river crevasse event. It is sheet-like in geometry, thinning away from the side of the breach in the river bank, and characterized by rapidly deposited sands, fining upwards to a muddy top, produced by the waning flow of a flood event.

crevasse splay *See* CREVASSE DEPOSIT.

critical erosion velocity *See* CRITICAL VELOCITY.

critical flow The flow that occurs when the flow velocity in a river channel equals the wave velocity generated by a disturbance or obstruction. In this condition the *Froude number (Fr) = 1. When the wave velocity exceeds the flow velocity (i.e. Fr is less than 1) waves can flow upstream, water can pond behind an obstruction, and the flow is said to be subcritical or tranquil. When Fr is greater than 1 waves cannot be generated upstream and the flow is said to be supercritical, rapid, or shooting. In this condition a standing wave is formed over obstructions in the river bed. In nature, supercritical flow is found only in rapids and waterfalls, but it is often created artificially by weirs and *flumes, with the aim of measuring discharge.

critical species *See* RARITY.

critical velocity (critical erosion velocity) The minimum velocity of a flowing fluid that is required in order to entrain a particle.

crocidolite *See* ASBESTOS.

Cromerian 1. A northern European *interglacial stage dating from about 0.8 Ma to 0.5 Ma ago. It coincides approximately with the *Günz/Mindel Interglacial of the Alps. **2.** Temperate deposits of Middle *Pleistocene age found at West Runton, Norfolk, England. Estuarine sands and silts and freshwater peat (Upper Freshwater Bed), with a temperate-forest flora, are succeeded by glacial deposits. It is not possible to correlate them with continental European deposits.

Cromwell Current *See* EQUATORIAL UNDERCURRENT.

cross-breeding Usually, *outbreeding or the breeding of genetically unrelated individuals. In plants this may entail the transfer of pollen from one individual to the stigma of another of a different *genotype.

cross-dating The matching of *tree-ring width patterns and other properties among the trees and fragments of wood from a particular area. This enables the year in which each ring was formed in living trees and recent stumps to be determined accurately, the presence of false rings or the absence of rings in individual specimens being made apparent. By matching ring series from living specimens with those from older (e.g. constructional) timbers, the chronology may be extended backwards (a dating procedure known as *dendrochronology).

crossing over The exchange of genetic information between two *homologous chromosomes. *Compare* UNEQUAL CROSSING OVER.

Crotonian *See* QUATERNARY.

crotovina *See* KROTOVINA.

crown gall A common and widespread plant disease which can affect a very wide range of woody and herbaceous plants. The disease is caused by the bacterium *Agrobacterium tumefaciens*. *Galls are formed at the crown (the junction of the stem and the root) or, less commonly, on roots, stems, or branches of infected plants. On the herbaceous plants the galls are usually soft, while on woody plants they tend to be hard. On trees, crown galls may reach sizes of 1 metre or more across.

crumb structure A type of soil structure in which the structural units or peds have a spheroidal or crumb shape. Crumb structure is often found in more porous than granular organo-mineral surface *soil horizons, and provides optimal *pore space for soil fertility.

Crustacea (crabs, lobsters, shrimps, slaters, woodlice, barnacles) A diverse subphylum of *Arthropoda, comprising animals which have two pairs of antennae, one pair of mandibles, and two pairs of maxillae. The limbs are *biramous, and adapted for a wide range of functions. Most of the 31 400 species are marine, but there are many freshwater species, and a relatively small number have invaded the land. A few marine species are parasites of other Crustacea and one group (Cyamidae) are ectoparasites of whales (whale lice). The first crustaceans are known from *Cambrian rocks.

cryergic Applied to the work of ground ice and, therefore, to *frost heaving, *frost wedging, and thaw processes. The term has been used as a synonym for *periglacial.

cryic *See* PERGELIC.

cryonival Applied to the set of geomorphological processes that comprise *cryergic and *nival mechanisms.

cryopediment A bench-like landform, cut indiscriminately across bedrock, and confined to past or present *periglacial environments. Its position on the lower part of a hill-slope is the main criterion for distinguishing it from a *cryoplanation terrace. It is the periglacial analogue of the warm desert *pediment.

cryophilic Applied to organisms preferring to grow at low temperatures.

cryoplanation The reduction of relief to a gently undulating land surface under *periglacial conditions. It is equivalent to extensive *altiplanation.

Cryosols Soils that have a *permafrost layer within 100 cm of the surface. Cryosols are a reference soil group in the *FAO *soil classification.

cryoturbation *See* GELITURBATION.

crypsis *See* CRYPTIC COLORATION.

crypt- A prefix derived from the Greek *kruptos*, meaning 'hidden'.

cryptic coloration (crypsis) Coloration that makes animals difficult to distinguish against their background, so tending to reduce predation. The effect of cryptic coloration may be to cause the appearance of the animal to merge into its background (e.g. the absence of all colour in some *pelagic fish larvae) or to break up the body outline (e.g. the spotted patterns of many bottom-dwelling flat fish). Both effects often occur in the same animal.

cryptic species Species which are apparently identical phenotypically (*see* PHENOTYPE), often to the point where individuals of such species are themselves unable to make the distinction, but that are incapable of producing hybrid offspring.

cryptobiosis Dormancy, used, for example, in relation to microbial spores which may show no signs of life for extended periods of time.

cryptogam A plant that reproduces by spores or *gametes rather than seeds (i.e. an alga, bryophyte, or pteridophyte).

cryptophyte One of *Raunkiaer's life-form categories, being a plant in which the perennating bud lies below the ground or water surface. The groups geophytes, helophytes, and hydrophytes are distinguished by the environment in which the perennating bud is found (i.e. land, marsh, and water respectively). *Compare* CHAMAEPHYTE; HEMICRYPTOPHYTE; PHANEROPHYTE; and THEROPHYTE.

cryptozoa Invertebrate animals, large enough to be visible to the naked eye, which live between litter and the soil.

Cryptozoic (Archaeozoic, Azoic) The name formerly given to the time from the origin of the Earth, about 4567 Ma to the evolution of the first shelled animals at the beginning of the *Cambrian, at about 542 Ma, and consisting of three eras, *Hadean, *Archaean, and *Proterozoic. These are now ranked as eons, although the name Hadean is used only informally, and the name Cryptozoic has been abandoned.

crystallochory Dispersal of spores or seeds by glaciers.

CTD An instrument for measuring sea-water conductivity (from which *salinity can be calculated), temperature, and depth (actually, pressure). A sensor unit is lowered through the water on the end of an electrical conductor cable which transmits the information to indicating and recording units on board ship.

Cu *See* COPPER.

cuesta An asymmetric land-form that consists of a steep *scarp slope and a more gentle dip (or back) slope. It is typical of areas underlain by strata of varying resistance that are dipping gently in one direction, and is intermediate between the flat-topped *mesa and *butte and the more symmetric ridge form of the hog's back.

cultivar Any variety or strain of plant which has been produced by horticultural techniques and is not normally found in wild populations (e.g. bread wheat, *Triticum aestivum*).

cultivation Tilling the soil by ploughing, digging, draining and/or smoothing. It is done in the course of seeding, transplanting, loosening soil, controlling weeds, or incorporating residues.

cultural Applied to the transmission of information from one generation of animals to another by non-genetic means (e.g. by *imprinting or *imitation). Behaviour may be modified culturally, so that while individuals usually behave in a traditional way, traditional behaviour in the same species may vary widely from one population to another.

cultural landscape A landscape that is essentially dominated by the effects of human activity, such as arable or pastoral farming, or forestry.

cultural eutrophication *See* EUTROPHICATION.

cultural evolution Evolution that is based on the transmission of information other than genetically (i.e. culturally). In humans, such transmission embraces customs, beliefs, and the acquisition and communication of knowledge. Adaptation by cultural change can be far more rapid than by genetic alteration.

culture **1.** A population of microorganisms or of the dissociated cells of a tissue grown, for experiment, in a nutrient medium: they multiply by asexual division. **2.** The transfer of behavioural traits between individuals in a non-genetic manner (i.e. the traits are not inherited genetically although they may be passed from parent to offspring by verbal or visual *communication). Culture is most developed in primates, particularly humans, but may also occur in other organisms such as social insects.

cumec *See* DISCHARGE.

cumulative percentage curve A graphical plot in which size classes are plotted against the percentage frequency of the class plus the sum of the percentages in preceding size classes. When plotted on normal graph paper, the cumulative frequency curve resembles an S-shape. When plotted on a normal-probability scale, the cumulative percentage data appear in a series of straightline segments, each with a different gradient.

cumuliform Heaped; the term is most commonly applied to clouds. *See* ALTOCUMULUS; CIRROCUMULUS; CUMULONIMBUS; CUMULUS; and STRATOCUMULUS.

cumulonimbus From the Latin *cumulus*, meaning 'heap' and *nimbus* meaning 'cloud', the name of a cloud of bulging, dense form, often towering to great height in unstable air. Young clouds have distinctive fibrous or lined features; older, glaciated types, with abundant ice crystals, are lustrous. Typically, the upper parts are spread into anvil or plume features. The cloud base is dark and usually gives rise to precipitation, often with *virga. *See also* CLOUD CLASSIFICATION.

cumulus The Latin *cumulus*, meaning 'heap', used as a name for a dense, isolated, and clearly defined cloud with vertical growth in bulges or domes, and a flattened, darker base. Sharply outlined, bulging cloud tops indicate vigorous growth. Occasionally a more ragged form occurs. *See also* CLOUD CLASSIFICATION.

cupola *See* RAISED BOG.

cupulate Cup-shaped.

curare A plant extract containing *alkaloids that block the passage of nerve impulses at synaptic junctions by competing with acetyl-choline for receptor sites on the post-synaptic membrane. It is used by some South American Indians as an arrow poison. Curare is obtained from the bark of *Strychnos toxifera* and the root of *Chondrodendron tomentosum*. Curare is also used as a muscle-relaxant in surgery.

current competition Competition in which species restrict one another to niches that are smaller than they would occupy were competitors absent. *See* COMPETITIVE RELEASE.

current meter An instrument for measuring the speed of flow in a watercourse. The most common type of current meters relate current speed to the rate at which an impeller is rotated by the flowing water.

cushion chamaephyte *See* CHAMAEPHYTE.

cushion plant A plant that has small, hairy, or thick leaves borne on short stems and forming a tight hummock. The habit is an adaptation to cold, dry, or windy conditions.

cuspate foreland A large, triangular area of coastal deposition, which is dominated by many shingle ridges and is often terminated on the landward side by poorly drained terrain. It is the result of a long episode of local marine aggradation under wave advance from two dominant directions. Dungeness, on the south coast of England, is a typical example.

cuspidate Having a sharp tip or point.

customer A fish that allows *cleaner fish to remove parasites, dead tissue, and other unwanted matter from its body surface and mouth.

cutan (clay films, clayskins, argillans, tonhäutchens) A deposited skin or coating of material on the surfaces of peds and stones, which is usually composed of fine, *clay-like soil particles which have been moved down through the soil.

cuticle In plants, a thin, waxy, protective layer covering the surface of the leaves and stems. In animals, a layer covering, and secreted by, the epidermis. In invertebrates, it is mainly protective against mechanical or (in endoparasites, *see* PARASITISM) chemical damage.

cutinite *See* COAL MACERAL.

cut-off (ox bow) A section of a river channel that no longer carries the main discharge. Its abandonment results from meander development associated with lateral channel migration across a *floodplain. Channel length is shortened by contact at the neck of a loop, the remainder of which becomes separated.

cut-off high An *anticyclone that is isolated from the main subtropical belt of high pressure and around which the main flow of the upper westerlies is diverted, causing a *blocking situation in middle latitudes.

cut-off low A *cold low in midlatitudes (occasionally almost in subtropical latitudes) where it is cut off from the main subpolar belt of low pressure. Sometimes a cut-off low occurs with a *cut-off high over the higher latitudes, typically in *blocking situations. In summer, weather associated with such slow-moving lows is unsettled and thundery.

Cuvier, Georges Léopold Chrétien Fréderic Dagobert, Baron (1769–1832) A French naturalist who was one of the founders of the disciplines of comparative anatomy and palaeontology. His studies of marine fauna led to detailed work on the structure and classification of Mollusca and the anatomy of fishes. Later he turned his attention to fossil vertebrates and then to living members of those groups, eventually describing the structure of all known animal groups. He believed each species and each organ was created for a particular purpose and extinctions were caused by catastrophes, the affected areas being recolonized by immigration.

cwm *See* CIRQUE.

cyanobacteria A large and varied group of bacteria which possess *chlorophyll *a* and which carry out *photosynthesis in the presence of light and air, with concomitant production of oxygen. They were formerly regarded as algae (division Cyanophyta) and were called 'blue-green algae'. Fossil cyanobacteria have been found in rocks almost 3 000 Ma old and they are common as *stromatolite colonies in rocks 2 300 Ma old. They are believed to have been the first oxygen-producing organisms and to have been responsible for generating the oxygen in the atmosphere, thus profoundly influencing the subsequent course of evolution. The organisms may be single-celled or filamentous, and may or may not be colonial. Some are capable of a gliding motility when in contact with a solid surface. Many species can carry out the fixation of atmospheric nitrogen. Cyanobacteria are widely distributed and are found in freshwater and marine environments, on soil, on rocks, and on plants as *epiphytes or as *symbionts. Their taxonomy is confused, because most of the phycological genera and species (defined when they were regarded as algae) are now known to be based on unreliable characteristics (e.g. characteristics that may depend on conditions of growth) and they have been redefined or abandoned according to bacteriological criteria.

C

cyanogenic Applied to plants that emit hydrogen cyanide when cut or bruised. Some *genotypes of bird's foot trefoil (*Lotus corniculatus*) are cyanogenic but not all. Bracken (*Pteridium aquilinum*) is cyanogenic in all except its very young stages.

Cyanophyta *See* CYANOBACTERIA.

cyanophyte A cyanobacterium (blue-green alga). *See* CYANOBACTERIA.

cybernetics The study of communications systems and of system control in animals and machines. In the life sciences, it also includes the study of feedback controls in *homoeostasis.

cycling pool *See* ACTIVE POOL.

cyclogenesis The formation and strengthening of cyclonic air circulation, which tends to form or deepen depressions. The process is associated with upper-air divergence over or near the frontal zone.

cyclolysis The processes of dissipation of cyclonic air circulation around a depression, or the weakening of lesser cyclonic features.

cyclomorphosis Seasonal changes in body shape found in rotifers (phylum Rotifera), and in cladoceran Crustacea. In cladocerans (e.g. *Daphnia* species) the changes in shape involve the head, which is rounded from midsummer to spring and then progressively becomes helmet-shaped from spring to summer, reverting to the rounded shape by midsummer. The process is poorly understood and may be the result of genetic factors interacting with external conditions, e.g. temperature or day length, or, as in rotifers, the result of internal factors alone.

cyclone **1.** The name given to a *tropical cyclone that develops in the Indian Ocean and Bay of Bengal. Cyclones usually travel north, on tracks that carry them over Bangladesh. **2.** *See* DEPRESSION.

Cylicomorpha *See* DISJUNCT DISTRIBUTION.

cytoplasmic inheritance A non-Mendelian (extra-chromosomal) inheritance via genes in cytoplasmic organelles. Examples of such organelles are viruses, mitochondria, and plastids.

cytosine A *pyrimidine base found in nucleic acids. *See also* ADENINE, GUANINE, and THYMINE.

dactylozooid In some colonial coelenterates, a defensive or protective polyp. The mouth, tentacles, and enteron are reduced or lost, so that the more extremely specialized dactylozooids have the form of a club bearing many *nematocysts.

dagalas *See* KIPUKA.

damping-off A disease of young seedlings, in which the stems decay at ground level and the seedlings collapse. It may be caused by any of a number of fungi (including, e.g., species of *Pythium* and *Rhizoctonia*).

dance language **(bee dance)** The ritualized behaviour (first recorded by Aristotle and described in detail by Karl von *Frisch) by which a returned worker honey-bee (*Apis* species) communicates the quality and source of food to other workers. There are two dances, both performed in the darkness of the nest or hive, on the vertical wax comb. The 'round' dance conveys the presence and quality of a food source close to the nest; no directional information is given. It comprises runs in a small circle, with regular changes of direction. The more frequent the change of direction, the greater the profitability of the food source (i.e. the greater its calorific value). Workers of foraging age pay particular attention to the pollen carried by the dancer and constantly *antennate her. Presumably they identify the source of the pollen by its characteristic scent and are then recruited to exploit the nearby source of food. The 'waggle' dance conveys a wider range of information, relating to the distance and direction of the food source, as well as its quality. It refers to more distant sources of food than does the round dance, and often recruits workers to exploit flowers several kilometres from the nest. The waggle dance comprises runs in a flattened figure-of-eight pattern, with the waggle action performed in the straight run between the two rounds of the figure. The duration and vigour of the waggle are scaled to indicate distance, as is the loudness and duration of the buzzing associated with the waggle. Direction is indicated by the angle by which the straight run deviates from the vertical, the angle being equal to the angle between the direction of the food source and the sun as seen at that time from the nest entrance. Workers recruited by this dance will fly on a course that maintains this angle until they reach the food source, which they recognized by olfactory cues picked up earlier by antennation of the dancing worker in the hive. The dance language can be regarded as a ritualized enactment or 'charade' of the foraging flight made by the dancing worker. The language varies slightly between one geographic race of honey-bees and another, and thus resembles birdsong and human languages in having local dialects.

darcy The unit of intrinsic *permeability, used particularly in the oil industry. One darcy is equal to $0.987 \times 10^{-12}\,m^2$. The unit is named after the French engineer Henri Philibert Gaspard Darcy (1803–58).

Darcy's law A description of the relationships among the factors that determine *groundwater flow, expressed as an equation. At its simplest, Darcy's law states that $Q = kIA$, where Q is the groundwater flow, k the *hydraulic conductivity of the rock, I the hydraulic gradient (i.e. gradient of the hydraulic head), and A the cross-sectional area through which flow occurs.

dark bottle A bottle covered with tape, or similarly adapted to exclude light: it is used to monitor respiration rates in aquatic productivity experiments. *See also* OXYGEN METHOD.

dark mildew *See* BLACK MILDEW.

dark reactions Photosynthetic reactions (*see* PHOTOSYNTHESIS) that involve the reduction of carbon dioxide in the Calvin cycle and which can take place in darkness, provided there is sufficient ATP and NADPH. See illustration overleaf.

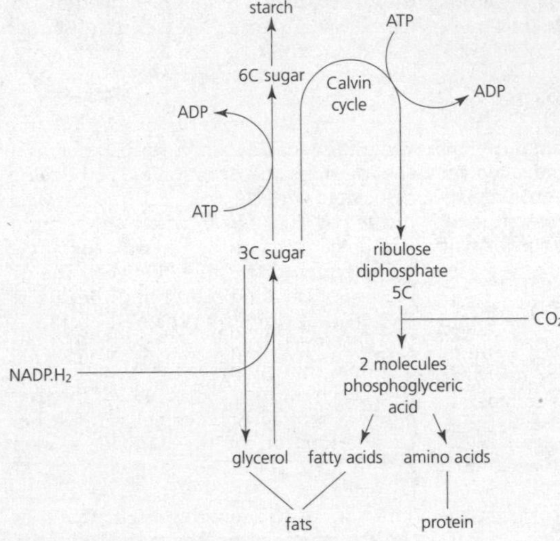

Dark reactions

Darling, Sir Frank Fraser **(1903–79)** A British ecologist and animal geneticist whose 1969 Reith Lectures for the BBC, 'Wilderness and Plenty' (published in 1970 with the same title) was influential in drawing attention to the relationship between humans and the natural environment. His ecological and social study in Scotland was published in 1955 as *West Highland Survey* and became a classic, as did his natural history book, *The Highlands and Islands*, written in collaboration with J. Morton Boyd and published in 1964. Darling was knighted in 1970, served on the Royal Commission on Environmental Pollution from 1970 to 1973, and from 1959 to 1972 was vice-president of the Conservation Foundation, based in Washington, USA.

darwin A measure of evolutionary rate (introduced by J. B. S. Haldane in 1948), given in exponential units of change over time, such that 1 darwin = 1/1 000 of the genome changed per 1 000 years.

Darwin, Charles Robert **(1809–82)** The English naturalist who is remembered mainly for his theory of evolution, which he based largely on observations made in 1832–6 during a voyage around the world on HMS *Beagle*, which was engaged on a mapping survey. In 1858, prompted by and published together with a paper by Alfred Russel *Wallace (who had reached similar conclusions independently), he published in the third volume of the *Journal of the Linnean Society* a short paper, 'On the tendency of species to form varieties: and on the perpetuation of varieties and species by natural means of selection', and in 1859 he published a longer account in his book, *On the Origin of Species by Means of Natural Selection*. In this he presented powerful evidence suggesting that change (evolution) has occurred among species, and proposed natural selection as the mechanism by which it occurs. The theory may be summarized as follows: (*a*) The individuals of a species show variation. (*b*) On average, more offspring are produced than are needed to replace their parents. (*c*) Populations cannot expand indefinitely and, on average, population sizes remain stable. (*d*) Therefore there must be competition for survival. (*e*) Therefore the best-adapted variants (the fittest) survive. Since environmental conditions change over long periods of time, a process of natural selection occurs which favours the emergence of different variants and ultimately of new species (the 'origin of species'). This theory is known as Darwinism. The subsequent discovery of chromosomes and genes, and the development of the science of genetics, have led to a better understanding of the ways in which variation may be caused. Modified by this modern knowledge, Darwin's theory is sometimes called 'neo-Darwinism'.

Darwinian fitness *See* ADAPTIVE VALUE.

Darwinism The theory of *evolution by *natural selection, often used incorrectly as a synonym for the theory of evolution itself (a concept that was described by Aristotle and debated in classical literature). The term 'neo-Darwinism' is often used to denote the 'new synthesis' (i.e. *synthetic theory).

Darwin's finches Fourteen species of Geospizinae that are endemic (*see* ENDEMISM) to the Galápagos Islands. There are only six species of all other passerine birds and one species of cuckoo on the islands. Thus it is inferred that an ancestor of the finches arrived on the islands before other birds and then underwent adaptive radiation. Each species has evolved a distinctive beak type, and feeds on food not eaten by the other species.

dasmatrophy A mode of feeding in which an organism releases a substance that punctures the tissues of its prey, allowing it to extract the contents. Dasmatrophy has been observed in a toxic, photosynthetic member of the *phytoplankton.

dating methods The methods used to determine the relative or *absolute age of rocks, *fossils, or remains of archaeological interest. A relative time scale, constructed in the last century, is based on correlations between palaeontological and stratigraphic data. The rate at which sediments accumulate can also be used for dating (*see* VARVE). Absolute dating relies on the decay of radioactive isotopes of elements present in the material to be dated (*see* DECAY CONSTANT; DECAY CURVE; DECAY SERIES; ISOTOPIC DATING; RADIOCARBON DATING; and RADIOMETRIC DATING).

datum level A surface or level which is regarded as a base from which other levels can be counted (i.e. a datum). For example, sea level is often used as a datum level against which the height of land and the depth of the sea bed are measured.

Davis, William Morris (1850–1934) An American geologist and geographer, who taught at Harvard University from 1876 to 1912, and who is recognized as the

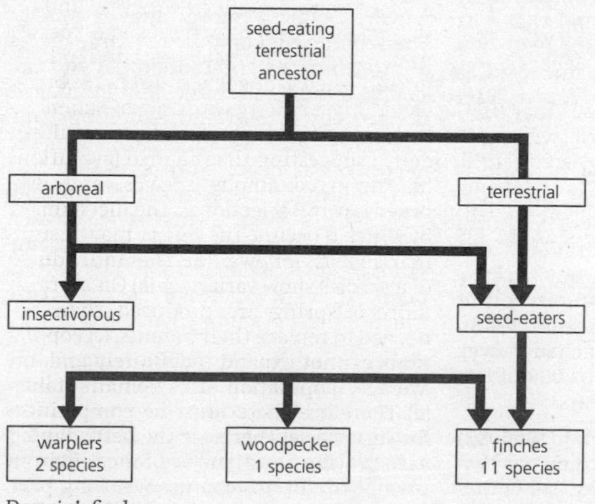

Darwin's finches

founder of modern *geomorphology. He described the action of rivers in the shaping of landscapes and developed the concept of the *Davisian cycle of erosion.

Davisian cycle The orderly series of stages through which it is suggested that land-forms pass from their initial uplift as hills or mountains until erosion finally reduces them to plains. The main stages are youth, when hill gradients are steep and river profiles irregular, maturity, when rivers are incising into the surface more slowly and their profiles are smoother, and old age, when the landscape has a gently rolling surface or is a *peneplain. This framework for the study of land-forms is now little used.

day degrees The departure of the average daily temperature from a defined base (e.g. the minimum recognized temperature for the growth of a plant species). The number of day degrees may be totalled to assess the accumulated warmth of a particular year's growing season for crops. *See also* ACCUMULATED TEMPERATURE; AGROMETEOROLOGY; and MONTH DEGREES.

day length *See* PHOTOPERIOD.

day-neutral plant A plant in which flowering is independent of the *photoperiod (e.g. dandelion). *Compare* LONG-DAY PLANT and SHORT-DAY PLANT.

DDT (dichlorodiphenyltrichloroethane) An *organochlorine compound that was first synthesized in 1874. In 1939 the Swiss chemist Paul Hermann Müller (1899–1965) discovered its insecticidal properties, for which he was awarded the 1948 Nobel Prize in Physiology or Medicine. DDT was used widely to control arthropod parasites of humans and then as an agricultural and horticultural insecticide. Its use was curtailed, and abandoned in many countries, because the chemical stability that allowed it to remain effective for long periods also caused its accumulation along *food-chains and because target insects were acquiring resistance to it.

death assemblage (thanatocoenosis) An assemblage of fossils of organisms that were not associated with one another during their lives. The remains were brought together after death, often by the action of currents. *See also* TAPHONOMY.

death feigning *See* THANATOSIS.

débâcle The break-up of river ice in spring over northern Eurasia and North America. The onset of thaw begins in March in low latitudes, and later further north.

Debenham level A simple device for measuring the vertical distance between two points; it consists of a clear container (e.g. a plastic can), a piece of clear plastic tubing about 10 m long, wire to attach the tubing to the container, and a clip. The container is filled with water and one end of the tubing inserted almost to the bottom and fixed to the mouth. Water is siphoned into the tubing to within about 50 cm of its further end and that end is sealed with the clip. The tubing is extended downslope over the surface, its end raised until the water level in the tubing is higher than that in the container, and the clip is removed, allowing the water level in the tubing to fall to that in the container. The difference in the height above the surface of the water levels in the tubing and the container represents the vertical distance between the two sites. The method was devised by the biologist C. Wood-Robinson and first described in 1981.

debris flow A slow-moving flow in which sediment moves gravitationally. The sediment is composed of large pieces of broken rock (clasts) supported and carried by a mud and water mixture. Debris flows occur as overland and submarine mudflows, and as submarine deep-sea deposits. The deposits are poorly sorted and internally structureless; typically they have a pebbly, mudstone texture. Debris-flow deposits cover many thousands of square kilometres of the *abyssal plain after originating on the *continental slope as slumped material.

debris slide A shallow landslide within rock debris, characterized by a displacement along one or several surfaces within a relatively narrow zone. It may take place as a largely unbroken mass or may be disrupted into several units, each consisting of rock debris.

deca- The Greek *deca*, meaning 'ten', used as a prefix (symbol da) with *SI units to denote the unit × 10.

decay constant The probability that an atom of a radioactive isotope will decay within a stated period of time. Radioactive decay involves only the nucleus of the parent atom, and thus the rate of its decay is independent of all physical and chemical conditions (e.g. pressure, temperature, etc.) and follows an exponential law. The fundamental equation describing the rate of disintegration may be written as: $-(dN/dt) = \lambda N$, where λ is the decay constant, representing the probability that an atom will decay in unit time t, and N is the number of radioactive atoms present. It is a fundamental assumption in geochronology that λ is a constant and that the only alteration in the amount of daughter or parent in the system is due to radioactive decay. The constant λ is usually expressed in units of 10^{-10} per year (e.g. ^{235}U is 9.72, ^{40}K is 5.31, ^{87}Rb is 0.139, and ^{238}U is 1.54). The total lifetime of a radioactive parent in a given system cannot be specified; in theory it is infinite. It is a simple matter, however, to specify the time for half of the radioactive parent atoms in a system to decay. This is called the 'half-life' (T), which is related to the decay constant by the expression $T = 0.693/\lambda$. *See also* DECAY CURVE.

decay curve A graphic representation of the rate of decay of a radioactive isotope of an element. Radioactive disintegration is exponential. If half the parent nuclide remains after one time increment, one-quarter will remain after the next (identical) time increment, and so on. A plot of the surviving parent atoms against time in half-lives (*see* DECAY CONSTANT) gives a decay curve that approaches the zero line asymptotically. In theory it should never attain zero.

decay series The sequence of radioactive daughter nuclides that are formed by the radioactive decay of a parent nuclide to a final, stable daughter nuclide. For example, uranium has three naturally occurring isotopes, ^{238}U, ^{235}U, and ^{234}U, all of which are radioactive. Thorium exists mainly as one isotope (^{232}Th) which is also radioactive. In addition, five radioactive isotopes of thorium occur in nature as short-lived, intermediate daughters of ^{238}U, ^{235}U, and ^{232}Th. Each of these isotopes is the parent of a chain (decay series) of radioactive daughters, each ending with stable isotopes of lead. Decay of ^{238}U gives rise to the 'uranium series' which includes ^{234}U as an intermediate daughter and ends in stable ^{206}Pb. The decay of ^{235}U gives rise to the 'actinium series' which ends in stable ^{207}Pb. The decay of ^{232}Th results in the emission of six alpha and four beta particles (*see* ALPHA DECAY and BETA DECAY) leading to the formation of stable ^{208}Pb.

deci- From the Latin *decimus*, meaning 'tenth', a prefix (symbol d) used with *SI units to denote the unit × 10^{-1}.

deciduous Applied to parts of a plant or animal that are shed seasonally (e.g. deer antlers, leaves of certain plants), to trees that shed their leaves seasonally, and to the perianth of a flower if this is shed after fertilization. In trees, this is not an indicator of taxonomic status; although deciduous trees are generally angiosperms (flowering plants), some (e.g. larch) are gymnosperms.

deciduous summer forest One of the two kinds of broad-leaved forest in the middle latitudes, the other being broad-leaved and evergreen (*see* EVERGREEN FOREST). The deciduous summer forest is by far the more important of the two in the northern hemisphere, and is absent from the southern hemisphere. The deciduous nature of the forest is believed to be principally an adaptation to drought in winter, when the soil is frozen.

declination **1.** The angle between magnetic North and true (geographic) North. If the compass needle points to the east of true North the declination is said to be

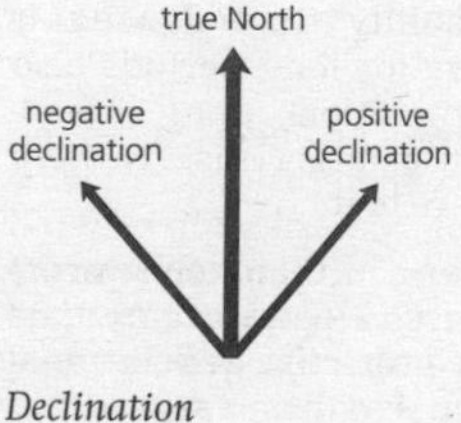

Declination

positive; if the needle points to the west of true North the declination is said to be negative. **2.** The angle between a celestial object and the celestial equator.

decomposer A term that is generally synonymous with 'microconsumer'. In an ecosystem, decomposer organisms (mainly bacteria and fungi) enable nutrient recycling by breaking down the complex organic molecules of dead protoplasm and cell walls into simpler organic and (more importantly) inorganic molecules which may be used again by *primary producers. Recent work suggests that some macroconsumers may also play a role in decomposition (for example, detritivores, in breaking down litter, speed its bacterial breakdown). In this sense 'decomposer' has a wider meaning than that traditionally implied. *See also* CONSUMER ORGANISM.

decorticate To remove the bark from a woody stem or branch; a stem or branch from which the bark has been stripped.

decurved Curved or bent downwards (e.g. as in the bill of a bird).

deepening In meteorology, a lowering of the central pressure in a depression. *Compare* FILLING.

deep scattering layer **(DSL)** A sound-reflecting layer in ocean waters, consisting of a stratified, dense concentration of zooplankton and fish. Such organism-rich layers, which cause scattering of sound as recorded on an echo sounder, may be 50–200 m thick.

deflation The removal of material from a land surface by *aeolian processes. It is most effective where extensive unconsolidated materials are exposed (e.g. on beaches, and on dry lake and river beds). The very large, enclosed hollows of many *deserts (e.g. the Qattara Depression of the Egyptian Sahara) may be owing to deflation.

deflation hollow An enclosed depression produced by wind erosion. It may be found both in hot *deserts, where wind may scour a hollow in relatively unconsolidated material, and in more temperate regions, where a protective vegetational cover has been removed from a sand dune.

deflected climax A vegetation *climax that is maintained by the activities of other organisms (e.g. by browsing, grazing, mowing, etc.). *See* PLAGIOCLIMAX.

defoliation The process of leaves being removed from a plant.

deforestation **(disafforestation)** **1.** The permanent clear-felling of an area of forest or woodland. On steep slopes this can lead to severe soil erosion, especially where heavy seasonal rains or the melting of snow at higher levels cause sudden heavy flows of water. In the humid tropics it may also lead to a release of carbon dioxide from the soil (due partly to the loss of gases as soil structure deteriorates and partly to the decomposition of organic material, including tree roots). **2.** A legal process whereby an area of forest land ceases to be regarded as forest under the terms of *forest law.

de Geer moraine A type of *moraine landscape that consists of a series of separate, narrow ridges trending parallel to a former ice front, and which can form annually. The ridges may be up to 300 m apart and up to 15 m high. Each ridge consists typically of a *till core, capped by a layer of partly rounded boulders. This landscape may have formed beneath the grounded part of an *ice sheet that extended into a lake or sea. It was first described by the Swedish geologist Gerard Jacob de Geer (1858–1943).

degenerate Applied to parts of the body, or stages in the life cycle of an organism, which have become greatly reduced in size, or have disappeared entirely, in the course of evolution.

degenerate code A genetic code in which the number of *sense codons exceeds the number of *amino acids, with the consequence that some amino acids are coded by more than one codon. This results in two-fold degenerate sites in DNA coding regions where one possible *nucleotide change is *synonymous and the other two are nonsysonymous, and four-fold degenerate sites where all nucleotide changes are synonymous.

degenerate phase The stage in the cyclical pattern of changes typical of grassland and *heathland *communities. In grassland, the term refers to the extended colonization by lichens of the small, grassy hummocks of the mature phase. As the frequency of soil-binding grass plants declines, erosion of the hummocks starts and a new cycle begins. In heathland, the term refers to *Calluna vulgaris* plants that are 20, 30, or more years old. A gap appears and widens gradually in the centre of the *C. vulgaris* bush, which becomes very straggly, with new shoots confined to the branch tips. In the gap and on the bare central stems lichens, especially *Cladonia* species, and mosses colonize. Eventually new *C. vulgaris* seedlings invade the central gap and a new cycle begins. *Compare* BUILDING PHASE; HOLLOW PHASE; MATURE PHASE; and PIONEER PHASE.

degradative succession (heterotrophic succession) A *succession that occurs on dead organic matter over a relatively short time-scale (months to years). *Detritivores feed in sequence, each group releasing nutrients that are utilized by the next group in the sequence until the resources are exhausted.

degree-day *See* ACCUMULATED TEMPERATURE.

dehiscent Bursting or splitting open at maturity. Usually used of a fruit that bursts open to release its seeds (e.g. a pea pod).

dehydration The removal of water, especially from a chemical compound by heat, sometimes in the presence of a catalyst or dehydrating agent (e.g. sulphuric acid), or the removal of water from crystals, oil, etc., commercially by distillation, chemicals, or heat.

deimatic behaviour Intimidating behaviour in animals which serves to warn off potential predators. It may be bluff (e.g. the inflation of the lungs in some toads, which increases their size) or may precede an attack (e.g. when skunks rise on their forelegs prior to spraying an evil-smelling liquid).

delayed flow Water that reaches river channels after flowing through subsurface routes and also from *groundwater. It contrasts with *surface run-off; together, surface run-off (quick flow) and delayed flow make up the total of river flow.

deleterious mutation A *mutation that lowers the *fitness of its carriers.

deletion The removal of one or more *nucleotides from a DNA sequence.

delta A discrete protuberance of sediment formed where a sediment-laden river enters an open body of water, such as the ocean, a partially enclosed sea, a *lagoon, or a lake, delivering sediment faster than *coastal processes can remove it. There is a reduction in the rate of flow in the river, resulting in rapid deposition of sediment and a characteristic coarsening upwards of sediments. Although the river supplies the material for building the delta its shape is controlled by a variety of factors, including climate, water discharge, sediment load, rate of subsidence of the sea or lake floor, and the nature of the river-mouth processes (particularly tidal and wave energy). Various attempts have been made to classify delta types. One system divides deltas into three classes: river-dominated (e.g. the Mississippi and Po); tide-dominated (e.g. the Ganges and Mekong); and wave-dominated (e.g. the Rhône and Nile). Each of these broad divisions includes many variants. For example, the Mississippi Delta has a highly indented coastline and the river ends with several distributary channels, forming a 'bird foot' shape, whereas the Ebro coastline, also river-dominated, is smooth and has a single river channel that protrudes into the sea. The Nile has a smooth shoreline and two distributary mouths

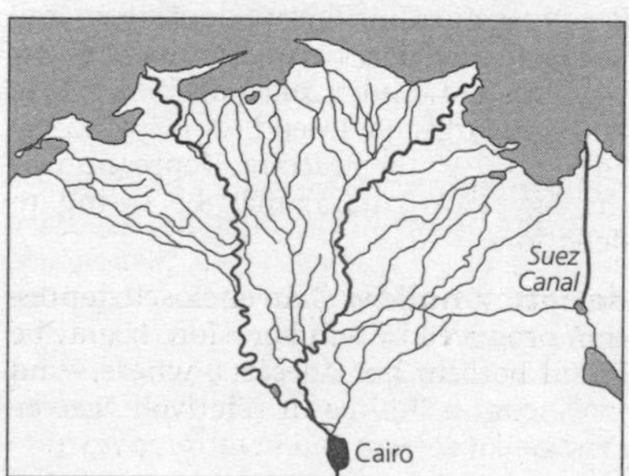

Delta

that protrude slightly into the sea, whereas the Niger, which is also wave-dominated, also has a smooth coastline but with many river mouths protruding slightly into the sea. *See also* GILBERT-TYPE DELTA.

d

delta front The sloping portion of a *delta, developed offshore from the bar at the mouth and passing at its toe into the *pro-delta. Delta fronts are the site of active, and often rapid, sedimentation.

delta plain The lowland area that lies to the landward side of a *delta front. It is covered by lakes, *salt marshes, freshwater *marshes, and *tidal flats crossed by one or more distributary channels.

deme A spatially discrete interbreeding group of organisms with definable genetic or cytological characters (i.e. a subpopulation of a species). There is very restricted genetic exchange, if any, with other demes, although demes are usually contiguous with one another, unlike subspecies or races which are generally isolated by some geographical or habitat barrier. All possible male and female pairings within a deme have an equal chance of forming, for one breeding season at least. Populations which fulfil only one of the two key criteria (i.e. very occasional or no cross-breeding and free pairing) are also referred to as demes by some authors.

demersal Applied to fish that live close to the sea floor, e.g. the cod (*Gadus morrhua*), hake (*Merluccius merluccius*), and saithe (*Pollachius virens*).

demography The statistical study of the size and structure (e.g. with regard to age or sex distribution) of populations and of changes within them. The word is derived from the Greek *demos*, meaning 'people' and *graphe*, 'writing'.

denaturation Reversible or irreversible alterations in the biological activity of proteins or nucleic acids, which are brought about by changes in structure other than the breaking of the primary bonds between amino acids or nucleotides in the chain. This may be accomplished by changes of solvent, *pH, or temperature, or through the physical abuse of the molecules.

dendritic Branching many times, like the branches of a tree.

dendritic drainage A *drainage pattern that may develop on homogeneous rock, which has a shape resembling the pattern made by the branches of a tree or the veins of a leaf.

dendrochronology (tree-ring analysis) 1. The science of dating by means of tree rings. **2.** All aspects of the study of annual growth layers in wood. **Pinus longaeva* and *P. aristata* are widely used in North America; in Europe, much information is derived from oak (*Quercus* species).

dendroclimatology The branch of *dendrochronology that deals with the relationships between the annual growth increment in tree trunks and climate, and especially with the reconstruction of past climates from dated tree-ring series. It is assumed that by studying present tree-ring patterns in relation to climate, older tree-ring chronologies may be used to indicate the climatic conditions experienced before detailed climate and weather records were kept. The value of such studies may be limited because of environmental influences (*see* DENDROECOLOGY).

dendroecology A branch of *dendrochronology that deals with the relationships between patterns in dated *tree-ring series and all the ecological factors that may influence those patterns. As well as climate, it considers *competition, *predation, etc.

dendrogeomorphology The use of dated tree-ring series to study land-forms and geomorphological processes.

dendrogram A diagram that represents relationships among groups of taxa, with the highest taxon at the base of a vertical line from which lower taxa branch at appropriate levels. There are two principal types: (*a*) the *phenogram, which is based solely on similarities in *phenotypes; and (*b*) the *cladogram.

dendrograph An instrument that provides a continuous record of the circumference of a tree stem (often called a 'girthing'). It is especially useful for recording *diurnal changes in stem size related to differences in hydration. There are several designs. *See also* DENDROMETER.

dendrohydrology The use of dated tree-ring series to study hydrological questions, especially relating to the periodicity of river flow and flooding.

dendroid **1.** In corals, applied to a colony formed by the irregular branching of *corallites. The individual corallites are separated from one another but may be connected by tubules. **2.** In graptolites (Graptolithina), applied to a bushy colony formed by irregular branching of the stipes.

dendrometer An instrument for measuring the size of tree stems. Most commonly it consists of a thin metal band with a vernier scale: it allows precise measurement of the stem circumference.

Denekamp **(Zelzate)** A Middle *Devensian *interstadial from the Netherlands, dating from about 30 000 years BP, when the estimated mean July temperature for the Netherlands was 10 °C. The Denekamp interstadial, together with the *Moershoofd and *Hengelo interstadials, is perhaps the equivalent of the *Upton Warren interstadial of the British Isles.

denitrification The conversion of nitrate or nitrite to gaseous products, chiefly nitrogen (N_2) and/or nitrous oxide (N_2O), by certain types of bacteria (*denitrifying bacteria). Denitrification occurs mainly under *anaerobic or *micro-aerobic conditions. It is also usually associated with higher pH conditions.

denitrifying bacteria Bacteria which can carry out *denitrification; they occur, for example, in soil and in freshwater and marine environments, and include, for example, certain species of *Bacillus*, *Hyphomicrobium*, *Paracoccus*, *Pseudomonas*, and *Thiobacillus*.

density Mass per unit volume, expressed in *SI units as kilograms per cubic metre (kg/m^3).

density current A current that is produced by differences in density. Where a flow of sea water has a greater density than that surrounding it, the denser water will dive beneath the less dense water. The density of sea water is affected by temperature, *salinity, and the content of suspended sediment. Most deep and bottom ocean currents are density currents. *See also* TURBIDITY CURRENT.

density dependence The regulation of the size of a population by mechanisms themselves controlled by the size of that population (i.e. *environmental resistance factors), whose effectiveness increases as population size increases. Density dependence may act through increasing mortality within a population or by decreasing fecundity. *See also* S-SHAPED GROWTH CURVE.

density determination **(specific gravity determination)** The determination of the density of a substance, using one of several techniques. (*a*) With a Jolly's spring balance, the sample is first weighed in air, then immersed in water and reweighed, and the density is calculated by comparing the two weights. (*b*) The Berman balance is a very sensitive torsion balance, used for determining the density of small samples, and is based on the same principle as the Jolly balance. (*c*) A pycnometer is used to determine the density of soils or powders. It consists of a small bottle fitted with a ground-glass stopper with a capillary opening. The specimen is placed in the pycnometer, weighed, and then the bottle is filled with water and reweighed. (*d*) Heavy liquids can also be used. Liquids which are relatively dense, such as bromoform (sp. gr. 2.89) and methylene iodide (sp. gr. 3.33), may be mixed with acetone (sp. gr. 0.79) to produce a series of liquids of known density. If a sample is introduced into one of these liquids and it neither rises nor sinks, its density is the same as that of the liquid. These heavy liquids are often toxic, and great care must be taken when using them. Gloves and face masks should always be worn.

density–frequency–dominance **(DFD measure)** A combined abundance estimate used in early North American ordination schemes but now rarely used. Usually expressed as relative values, relative density is the number of a given species expressed as a percentage of all species present, relative frequency is the frequency of a given species expressed as a percentage of the sum of frequency values for all species present, and relative dominance is the basal area of a given species expressed as a percentage of the total basal area of all species present. These three measures are summed to give the importance value, which may lie between 0 and 300.

d

density independence *See* J-SHAPED GROWTH CURVE.

density measure An estimate of the abundance of a particular plant species, as the number of individuals per unit area. In practice, problems arise with those plants, including many grasses, for which identification of single individuals is difficult.

dentate Toothed or serrated.

denticulate Having very small teeth or serrations.

dentine A bone-like substance, lacking cell bodies, which consists mainly of calcium phosphate in a fibrous matrix.

denudation From the Latin *denudare*, meaning 'to strip bare', a general name for the processes of *weathering, transport, and *erosion.

denudation chronology A branch of landform studies that deals with the historical development of landscapes by *denudation, especially during pre-*Quaternary time. Evidence for developmental stages is provided by studies of erosion surfaces and their mantling deposits, drainage patterns, stream long-profiles, and geologic structures.

deoxyribonucleic acid *See* DNA.

depensation In some animal species (e.g. whales) an increase in *parasitism and *predation that occurs when the size of a population falls below a threshold. This further depletes the population, reducing the likelihood that it will recover.

depocentre The site of maximum deposition within a sedimentary basin, where the thickest development of the sedimentary sequence is found.

deposit feeder An animal that lives on or in the sediment of the sea floor, and ingests mud rich in organic material in order to obtain nutriment.

deposit gauge A device for collecting solid and liquid atmospheric pollutants.

depositional remanent magnetization (DRM) The magnetization acquired by a sediment during deposition. This is predominantly *detrital remanent magnetization, but, used more widely, the term covers all magnetization acquired at such a time.

depression An enclosed area of low pressure revealed by the pattern of pressure distribution. Depressions are also described as cyclones or cyclonic systems, and have a characteristic pattern of wind circulation (anticlockwise around low-pressure areas in the northern hemisphere). Mid-latitude depressions are associated with the convergence of polar and tropical *air masses along a frontal zone: this commonly becomes deformed, and each air mass in turn advances over parts near (especially south of) the depression path, bringing first a warm and then a cold front.

deranged drainage An original drainage pattern that has been deranged or disturbed by the intervention of external factors (e.g. tectonic activity or glaciation).

derived *See* APOMORPHIC.

dermatophyte A fungus that lives *parasitically on skin.

desalination A series of processes whereby salt water is rendered potable.

desert An area within which the rate of evaporation exceeds the rate of precipitation for most of the time. The rate of evaporation depends on temperature but a desert is likely to form within any temperature range if the average precipitation is less than 250 mm/yr, and is typically very erratic. Plants and animals are either absent or sparsely distributed, and they are adapted to long droughts or to a lack of access to free water.

desert biome The characteristic *biotic *community of arid regions, generally defined as areas with rainfall of less than 250 mm/yr. Typically, such areas have high evaporation rates and a large *diurnal temperature range. Organisms commonly show adaptations to drought and heat (e.g. water storage in succulents such as cacti, and the frequent use among desert mammals of the burrowing habit). A high percentage of *ephemeral species is characteristic of desert floras.

desertification The process of desert expansion or formation, which may occur as a direct consequence of climatic change (e.g. shifts in the location of the major planetary pressure and wind systems), as the result of poor land-use policy (e.g. overgrazing), or owing to some complex interaction of these factors (e.g. overgrazing leading to *albedo change, favouring climatic change in the form of increased dryness).

desert pavement A thin surface covering of gravel and stones found in many desert areas, left after *erosion by wind and water has removed the finer soil materials.

desiccation **1.** A long-term loss of water that is associated with regional climatic change. **2.** The drying-out of an organism that is exposed to air.

desiccation cracks **(mud-cracks, shrinkage cracks, sun-cracks)** The polygonal-shaped cracks developed in mud which has dried out in a terrestrial environment. They are most often preserved when loose sand infills the cracks and then buries the desiccated mud surface.

desquamation *See* WEATHERING.

destructive wave A relatively high-energy, shallow-water wave that causes degradation of a beach by moving more material seawards than landwards, thus having a net erosional effect on the adjacent beach. It is characterized by high frequency, which implies that the *swash is impeded by the *backwash from the preceding wave. The backwash is more effective than the swash in moving material, and waves having this character are usually steep and associated with onshore winds. The erosional effect is also enhanced by the near-vertical plunge on breaking. It is favoured by a relatively steeply sloping offshore zone.

deterministic model A mathematical representation of a system in which relationships are fixed (i.e. taking no account of probability), so that any given input invariably yields the same result. *Compare* STOCHASTIC MODEL.

detrital Applied to material derived from the mechanical breakdown of rock by the processes of *weathering and *erosion, or to debris of biological origin.

detrital pathway **(detritus food-chain)** Most simply, a *food-chain in which the living primary producers (green plants) are not consumed by grazing *herbivores, but eventually form litter (detritus) on which decomposers (microorganisms) and detritivores feed, with subsequent energy transfer to various levels of carnivore (e.g. the pathway: leaf litter → earthworm → blackbird → sparrowhawk). Detritus from organisms at higher trophic levels than green plants may also form the basis for a detrital pathway, but the key distinction between this and a grazing pathway lies in the fate of the primary producers.

detrital remanent magnetization (DRM) The magnetization of a sediment acquired as ferromagnetic detrital grains become aligned by the geomagnetic field as they fall through the water and become part of the sediment on the bottom of a lake, pond, river system, or the sea.

detritivore **(detritus feeder)** A *heterotrophic animal that feeds on dead material (*detritus). The dead material is most typically of plant origin, but it may include the dead remains of small animals. Since this material may also be digested by decomposer organisms (fungi and bacteria) and forms the *habitat for other organisms (e.g. nematode worms and small insects), these too will form part of the typical detritivore diet. Animals (e.g. the hyena) that feed mainly on other dead animals, or that feed mainly on the products (exuviae, e.g. dung), of larger animals, are termed scavengers. *See also* FOOD-CHAIN; *compare* CARNIVORE and HERBIVORE.

detritus Litter formed from fragments of dead material (e.g. leaf litter, dung, moulted feathers, and corpses). In aquatic habitats, detritus provides habitats equivalent to those which occur in soil *humus.

detritus agriculture The planned production of detritus as a source of food (e.g. silage production). Detritus agriculture is thought by some workers to have great potential as an alternative to harvesting the products of the *grazing pathway (i.e. animal products). They argue that the economic and environmental costs of microbial conversion of the detritus to palatable human food would be less than the costs of pest and disease control, etc.

detritus feeder *See* DETRITIVORE.

detritus food-chain *See* DETRITAL PATHWAY.

Devensian The last glacial stage in Britain (named after a Celtic tribe, the Devense), which lasted from around 70 000 years BP (possibly earlier) to about 10 000 years BP. It is approximately synchronous with the *Wisconsinian Stage in North America, the Weichselian Glaciation in northern Europe, and the Würm Glaciation in the Alps. The Devensian was preceded by the *Ipswichian interglacial, and followed by the *Flandrian (i.e. present) interglacial. The Devensian was unusual because of the high rate of megafaunal extinction with which it ended; these extinctions are sometimes attributed to human activities, but the matter is controversial. *See also* ALLERØD; BØLLING; CHELFORD INTERSTADIAL; DRYAS; LATE GLACIAL; LOCH LOMOND STADIAL; UPTON WARREN INTERSTADIAL; and WINDERMERE INTERSTADIAL.

Devonian The fourth of the six periods of the *Palaeozoic Era, preceded by the *Silurian and followed by the *Carboniferous. It began about 416 Ma ago and ended about 359.2 Ma ago. In Europe there are both marine and continental facies present, the latter being commonly known as the Old Red Sandstone. Although originally described from the type area in Devon, England, the marine Devonian is subdivided stratigraphically into stages established in other parts of the world. These stages are: the Lochkovian (416–411.2 Ma, Czech Republic); Praghian (411.2–407 Ma, Czech Republic); and Emsian (407–397.5 Ma, Uzbekistan) of the Lower Devonian; the Eifelian (397.5–391.8 Ma, Germany) and Givetian (391.8–385.3 Ma, Morocco) of the Middle Devonian, and the Frasnian (385.3–374.5 Ma, France) and Famennian (374.5–359.2 Ma, France) of the Upper Devonian. The subdivision of the marine deposits is based on lithologies and the presence of an abundant invertebrate fauna including goniatites (Ammonoidea) and spiriferid brachiopods (Spiriferida). The continental Old Red Sandstone deposits contain a fauna of jawless fish and plants belonging to the primitive psilophyte group. As a result of the Caledonian orogeny of late Silurian times, much of the British Isles was covered with continental red-bed facies. Fossils of vascular plants are abundant in Devonian beds; the Rhynie chert flora (Middle Devonian) consists of well-preserved psilophytes. Insects are also present within the Devonian and may have their origin in the preceding Silurian.

dewatering The removal of *groundwater to reduce flow-rate or diminish pressure. Methods used depend on the *permeability of the ground, the proximity of hydrogeologic boundaries (*see* HYDRAULIC BOUNDARY), the *storage coefficient of the soil, pressure, and *hydraulic gradient. Methods used include *abstraction by wells, *electro-osmosis, sumps and drains, vertical drains, or exclusion by *grouting, compressed air, or freezing techniques. Dewatering is usually undertaken to improve conditions in surface excavations and to help construction work at or near the surface.

dew-point The temperature, with constant pressure and water-vapour content, to which air must be cooled for saturation to occur. Dew-point may be determined from the dew-point *hygrometer which indicates initial condensation on a cooled surface.

dextrorse Growing or arising in a righthanded or clockwise spiral from the point of view of an observer. The term is applied to a climbing-plant stem, a chain of spores, etc.

DFD measure *See* DENSITY–FREQUENCY–DOMINANCE.

diachronous Differing in age from place to place.

diadromous Applied to fish that regularly migrate between the sea and freshwater systems. *Compare* AMPHIDROMOUS. *See also* ANADROMOUS; CATADROMOUS; and POTAMODROMOUS.

diagenesis All the changes that take place in a sediment at low temperature and pressure following deposition. With increasing temperature and pressure, diagenesis grades into metamorphism. Diagenetic processes such as compaction, dissolution, *cementation, replacement, and recrystallization are the means by which an unconsolidated, loose sediment is turned into a sedimentary rock (e.g. sand into sandstone, or peat into coal).

diageotropism A tropic response (*see* TROPISM) of a plant organ in which it takes up a position at right angles to the direction of the force of gravity. An example is the ivy-leaved toadflax (*Cymbalaria muralis*), which climbs vertical stone walls and has diageotropic fruits. The seeds are thus pushed laterally into crevices in the wall.

diagnostic horizon A soil layer that contains a combination of characteristics typical of that kind of soil.

dialect *Vocalizations among a *population of animals that differ from those of another population of the same species. There are many dialects in bird-song.

dialysis The separation of dissolved crystalloids from colloidal macromolecules by means of a partially permeable membrane that allows the passage of the former but not of the latter.

diamond dust **(ice prisms)** Minute ice crystals that form in extremely cold air. They are so small as to be barely visible and seem to hang suspended, twinkling as they reflect sunlight.

diapause A temporary cessation that occurs in the growth and development of an insect. Insects can enter the diapause state as eggs, larvae, pupae, or as adults. Diapause is frequently associated with seasonal environments, the insect entering it during the adverse period, and breaking from it when more favourable conditions return. *See also* DORMANCY.

diaphototropism A tropic response (*see* TROPISM) of a plant organ in which it grows at right angles to the stimulus of light.

diaspore A *spore, *seed, or other structure that functions in dispersal; a *propagule.

diastem A very small break in a conformable succession of strata, indicated only by a bedding plane, and representing a brief interruption in the deposition of sediments, with little or no *erosion having occurred. Diastems may be very localized, with deposition having occurred elsewhere.

diastema A naturally occurring gap (i.e. not one resulting from the removal of a tooth) in the tooth row of a mammal, most commonly between the incisors and the first premolar.

diatom A *protist, belonging to the division Bacillariophyta, in which the cell wall (frustule) is composed of silica and consists of two halves, one of which overlaps the other like the lid of a box. Frustules are often delicately ornamented. Most diatoms are unicellular, but some are colonial or filamentous. Most are photosynthetic, but some species lack *chlorophyll and live *heterotrophically among decaying marine algae. Pennate (i.e. bilaterally symmetrical) diatoms occur in both freshwater and marine habitats; centric diatoms (i.e. radially symmetrical) occur predominantly as part of the marine *plankton. There are more than 10 000 species. *See* DIATOMACEOUS EARTH; DIATOMITE; and DIATOM OOZE.

diatomaceous earth **(kieselguhr)** A deposit composed of fossil diatoms (Bacillariophyceae), which is mined for many industrial uses: as a mild abrasive in metal polishes, as a filtering medium (e.g. in sugar refineries), for insulation of boilers and blast furnaces, etc., and under the name 'kieselguhr' as a vehicle for explosives. Vast deposits of diatomaceous earth (called '*diatomite') are mined at Lompoc, California.

diatomite A diatom-rich *sediment, which has been laid down in a lacustrine or deep-sea environment. The diatom cell wall is made from silica, therefore the sediment is siliceous.

diatom ooze A soft, siliceous, deep-sea deposit composed of more than 30 per cent (by volume) diatom cell walls. Diatomaceous sediments predominate in the high latitudes both around the coast of Antarctica and in the North Pacific, but in the North Atlantic the diatom content is overwhelmed by the terrigenous sediment derived from the adjacent continents. *See also* RADIOLARIAN OOZE.

diatropism A tropic response (*see* TROPISM) of a plant organ in which it grows at right angles to a stimulus.

diazotroph An organism capable of utilizing ('fixing') atmospheric nitrogen. *See also* NITROGEN FIXATION.

dichotomize To split into two equal parts.

dieldrin An *organochlorine insecticide that was formerly used as a seed dressing. In the 1950s seed-eating birds were exposed to lethal doses and predatory birds accumulated doses ingested from their prey, which caused them to lay eggs with shells too thin to protect the embryo adequately. Restrictions on the use of dieldrin, introduced in the 1960s, were followed by a recovery in populations of birds of prey.

differential resource utilization (resource partitioning) The situation in which ecologically similar species sharing the same habitat exploit different resources, or the same resources but in different ways, thereby avoiding competition.

differential species In *phytosociology, species which seem to be mutually *exclusive in a comparison of two *community types, but which examination of other communities reveals are not necessarily characteristic for either community. *See also* KENNARTEN SPECIES.

differentially permeable membrane A membrane that allows the passage of small molecules but not of large molecules. *See* SEMI-PERMEABLE.

differentiation, theory of On the basis of the different distribution patterns of the families, genera, and *species of flowering plants, the suggestion by some early plant geographers that *evolution in this group proceeded from the family level downward into genera and species, rather than the other way around.

diffusion The movement of molecules or ions from a region of higher to one of lower solute concentration as a result of their random thermal movement. It is an important means of transport within cells.

digitigrade Applied to a *gait in which only the digits make contact with the ground, the hind part of the foot remaining raised (e.g. in cats and dogs).

dimension analysis The detailed measurement of plant dimensions in productivity studies. Control studies enable estimation of allometric (*see* ALLOMETRY) relationships between external dimensions and dry-matter production, so that in future studies destructive sampling is unnecessary. Dimension analysis has particular relevance to forest productivity studies and has been much used.

dimictic Applied to a lake in which two seasonal periods of free circulation occur, as is typical of lakes in mid-latitude climates. In summer, thermal stratification occurs as surface waters are warmed and cease to mix with the denser, colder, deep waters. In winter, when they cool to below 4°C, surface waters expand, so becoming less dense than warmer waters beneath them, giving a reverse stratification. Free circulation through the depth of the lake is possible only in spring, when the surface temperature rises to above 4°C, and the water becomes heavier than that beneath and so sinks and mixes, and in autumn, when the surface (*epilimnion) waters cool to the temperature of the deep (*hypolimnion) waters.

dimidiate **1.** Divided into two. **2. (Of a fungal fruit body)** Semicircular in outline.

dimorphism **1.** The presence of one or more morphological differences that divide a species into two groups. Many examples come from sexual differences of particular traits, such as body size (males are often larger than females), plumage (male birds are usually more colourful than females), and types of flowers in *dioecious plants. These result from sex-linkage of the genes coding for the particular trait. However, some dimorphism, such as the aerial or submerged leaves of some aquatic plants (e.g. water crowfoot), may not be sex-linked. *See also* SEXUAL DIMORPHISM. **2.** *See* POLYMORPHISM.

Dinantian *See* CARBONIFEROUS.

dinoflagellates A division (Dinomastigota) of *protists that are *heterotrophs but closely allied to brown *algae and diatoms (they are were formerly classified as algae). Many have brown or yellow chromoplasts containing xanthophyll and chlorophylls *a* and *c*; others are colourless. Typically, dinoflagellates have two flagella, one propelling water to the rear and providing forward motion, attached just behind the centre of the body and directed posteriorly, the other causing the body to rotate and move forwards, forming a transverse ring or spiral of several turns around the centre of the body. Some dinoflagellates are naked, others are covered with a membrane or plates of cellulose. Many species are capable of emitting light, and these are the main contributors to bioluminescence in the sea. Most are planktonic, some in fresh water but most in marine environments, and some live in *symbiosis with animals (e.g. the flatworm *Amphiscolops*, sea anemones, and corals) with which they exchange nutrients. Some are colonial. There are many species.

dioecious Possessing male and female flowers or other reproductive organs on separate, unisexual, individual plants. *Compare* MONOECIOUS.

dioestrus The period between two *oestrus cycles in a female mammal.

dioxin A member of a range of about 300 compounds produced as by-products of certain industrial chemical processes and also by the incomplete incineration of chlorinated hydrocarbon compounds, especially *polychlorinated biphenyls, when the incineration temperature is lower than about 1200°C. The name 'dioxin' usually refers to one member of the group, 2,3,7,8-tetrachlorodibenzo-*p*-dioxin ($C_{12}H_4Cl_4O_2$), usually known as TCDD, severe exposure to which can cause chloracne in humans.

dip The angle by which a plane feature inclines from the horizontal.

diphyodont Applied to vertebrates in which a set of deciduous teeth (milk teeth) is shed and replaced by a second set of (permanent) teeth.

diploid Applied to a cell with two *chromosome sets, or an individual with two chromosome sets in each cell (excluding the *sex chromosomes, which may or may not be represented twice, according to the sex of the individual). A diploid state is written as $2n$ to distinguish it from the *haploid state of n. Almost all animal cells (except *gametes) are diploid.

diploidy The *diploid condition.

dipole Having a different electromagnetic charge (i.e. pole) at each end (e.g. a molecule with an uneven charge distribution, one pole having a net negative charge, the other a net positive charge).

dipole field The field that results from the presence of two oppositely polarized magnetic poles. The term is usually applied to the geomagnetic dipole field or that associated with a magnetic anomaly.

dipole moment *See* POLAR MOLECULE.

dip pole A point, usually on the Earth's surface, where the magnetic vector is vertical (i.e. the *inclination is 90°). A positive (downward) inclination is a positive (north) dip pole, and a negative (upward) inclination is a negative (south) dip pole. Many dip poles exist and these are not the same as the two geomagnetic poles.

direct circulation A circulation system in which lighter air rises while denser air descends, leading to the conversion of potential energy to kinetic energy. Land and sea breezes are examples of such circulation.

directed speciation A speciational trend recognized in plants: the *species do not conform to a continuum of *adaptive types, but rather to a stepwise succession of distinct species. The evolutionary significance of such a trend is unclear. It may relate to differential survival as opposed to differential speciation along an environmental gradient, i.e. to directional species selection rather than directed speciation.

d

directional evolution *See* ARISTOGENESIS.

directional selection A selection that operates on the range of *phenotypes for a particular characteristic existing in a population, by moving the mean phenotype towards one phenotypic extreme. Directional selection usually occurs in response to a steady change in environmental conditions, with a consequent shift in selection pressure such that the frequency of particular *alleles will change in a constant direction. It is often used in agriculture and horticulture to produce a shift in the population mean of a trait derived by humans. For example, the breeder might select for cows that yield more milk or plants that fruit only in a particular season. *Compare* DISRUPTIVE SELECTION and STABILIZING SELECTION.

disafforestation *See* DEFORESTATION.

disassortative mating Mating between individuals of unlike phenotypes. *Compare* ASSORTATIVE MATING.

discharge A measure of the water flow, expressed as volume per unit time, at a particular point (e.g. a river *gauging station, sewage works, or *groundwater abstraction well. Various units of measurement are in common use, depending on the nature of the discharge being measured. River flow may be expressed in cubic metres per second (misleadingly called 'cumecs' in some literature), while borehole flows may be more conveniently expressed as litres per second.

discharge hydrograph *See* HYDROGRAPH.

disclimax In the monoclimax model of vegetation development (*see* CLIMAX THEORY), a plant community replacing the climax community following an environmental disturbance, e.g. the introduction and maintenance of grazing pressure. 'Disclimax' is analogous to the more widely used terms *plagioclimax and/or biotic climax.

discontinuous distribution *See* DISJUNCT DISTRIBUTION.

discordant drainage A drainage pattern that runs across (i.e. is discordant to) the geologic structure.

discrimination Differential responsiveness to stimuli when several stimuli are presented simultaneously.

dish-pan experiment A method for simulating the convective cell and other global atmospheric motions. A shallow layer of water in a round vessel is heated at the edge, to represent the equator, and cooled at the centre, to represent the polar region. As the dish-pan is rotated, various motions are produced in response to the radial temperature gradient and to the speed of rotation. At lower speeds the flow pattern is zonal, but with faster rotation large waves develop and incorporate closed circulations. *See also* HADLEY CELL and ROSSBY WAVES.

disjunct distribution (discontinuous distribution) The occurrence of closely related species in a limited number of locations separated by oceans. For example, the family Caricaceae comprises four genera and about 30 species of trees, one of which is *Carica papaya* (papaya or pawpaw). Members of the family are most abundant in South America, with some in Central America, but one genus, *Cylicomorpha*, grows in tropical Africa. *Araucaria* species (*A. araucana* is the monkey puzzle tree or Chile pine and *A. angustifolia* is paraná pine or candelabra tree) are native to South America, where they form forests. *A. bidwilli* (bunya-bunya) occurs naturally in New Guinea, north-eastern Australia, and on islands in the South Pacific Ocean. Disjunct distribution is evidence supporting *continental drift.

dispersal The tendency of an organism to move away, either from its birth site (natal dispersal) or breeding site (breeding

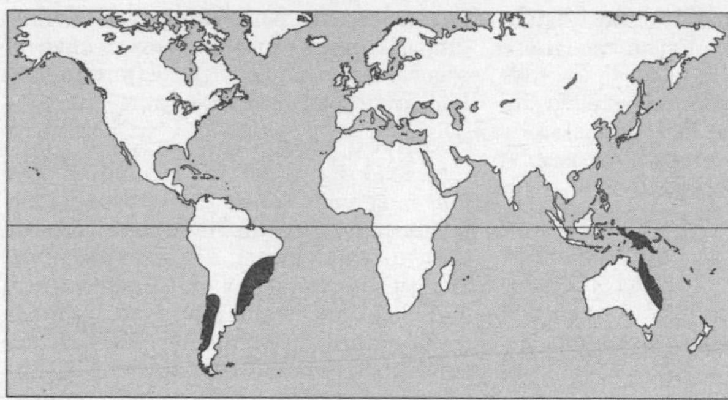

Disjunct distribution

dispersal): the opposite of *philopatry. Rates of regional dispersal depend on the interaction of several factors, notably the size and shape of the source area, the dispersal ability of the organisms, and the influence of such other environmental factors as winds or ocean currents. Dispersal may be passive (e.g. of winged seeds or ballooning spiderlings), active (e.g. of many mammals), passive but involving an active agent (e.g. seeds carried on the coats of mammals), or *clonal; in practice these categories are difficult to define precisely. Mathematical modelling using these factors has practical applications in the design of nature reserves, and provides an insight into the present distribution of organisms.

dispersal barrier **(ecological barrier)** An area of unfavourable *habitat separating two areas of favourable habitat, e.g. oceans in the case of terrestrial organisms, or a cereal *monoculture in the case of woodland organisms.

dispersal biogeography The term now applied to the traditional school of *biogeography, regarding organisms as arising in a *centre of origin and spreading out, utilizing *corridor dispersal *sweepstake routes, etc.

dispersal mechanism The characteristic adaptation for dispersal which forms part of the reproductive strategy of many slow-moving or *sessile organisms. It is most characteristic of the dispersal of spores, seeds, and fruit from plants, but is also found in other organisms, especially for the dispersal of larvae. Typical examples are the hooked seeds and fruits that attach themselves to the coats of animals.

dispersion **1.** In statistics, the internal pattern of a population (i.e. its distribution about the mean value). In spatial statistics, the pattern relative to some specific location, or of individuals relative to one another (e.g. clumped or random). **2.** In *pedology, the process of separating soil particles (as in *aggregates) from each other so that they may react as individual particles. Aggregates or *peds of soil particles are destroyed by dispersion (and formation is initiated by *flocculation). **3.** The spreading of a body of water as it flows. Lateral dispersion is the widening of the path taken by *groundwater as it flows from a known point of origin through a rock matrix, owing to its movement around individual mineral grains within the main rock body. Unless it flows in well-defined fissures or fractures, water does not travel through a rock in a straight line but is forced to flow across a widening front because of the granular nature of the rock matrix. Longitudinal dispersion is the spreading out of a body of water along its own flow path, due to the differences in water velocities in larger and smaller pores of the rock. Both modes of dispersion are normally observed by means of tracers. Lateral and longitudinal dispersion also occur in river *channels, where they are caused by differences in flow velocity across the channel and between the water surface and the bed, and also by random fluctuations in velocity caused by turbulent eddies. **4.** *See* SWELL.

d

dispersion coefficient The measure of the spread of data about the mean value, or with reference to some other theoretically important threshold or spatial location, e.g. the standard deviation. *See also* DISPERSION; OVERDISPERSION; and UNDERDISPERSION.

disphotic zone *See* APHOTIC ZONE.

displacement activity The behaviour of an animal which is or appears to be irrelevant to the situation in which it occurs and which may interrupt other activity. It may result from conflicting motivation, in which two motivations cancel one another, allowing a less urgent motivation (e.g. feeding) to dominate behaviour; or it may arise because the animal is prevented from attaining a goal, and the consequent *frustration causes the attention to be switched to another stimulus to which it then responds.

display Stereotyped behaviour, involved in communication, which is largely acquired genetically. It may be associated with *courtship, in which physical characteristics (plumage, antlers, etc.) are exhibited by an animal as a means of attracting and securing the co-operation of a sexual partner, *deimatic, a threat (e.g. in a bird establishing a *territory), or cryptic (i.e. making the animal more difficult to see).

disruptive coloration In an animal, a colour pattern that is thought to disrupt the perceived contour of the body or parts of the body, thereby making the animal more difficult to see.

disruptive selection A selection that changes the frequency of alleles in a divergent manner, leading to the fixation of alternative alleles in members of the population. The result after several generations of selection should be two divergent *phenotypic extremes within the population. This may be achieved, for example, by selecting seeds from the longest and shortest ears of corn in a population over a number of generations. *Compare* DIRECTIONAL SELECTION and STABILIZING SELECTION.

dissolved load The part of a river's total load that is carried in solution. Five ions normally constitute approximately 90 per cent of the dissolved load: chloride (Cl^-), sulphate (SO_4^{2-}), dissolved bicarbonate (HCO_3^-), sodium (Na^+), and calcium (Ca^{2+}). Generally the load is at its maximum concentration during low-discharge conditions when *groundwater is the main source of flow.

dissolved oxygen level The concentration of oxygen held in solution in water. Usually it is measured in mg/l (sometimes in $\mu g/m^3$) or expressed as a percentage of the saturation value for a given water temperature. The solubility of oxygen varies inversely with temperature; this is important, because the warmer the water the larger the proportion of dissolved oxygen that is used by *poikilotherms. The dissolved oxygen level is an important first indicator of water quality. In general, oxygen levels decline as pollution increases.

distal **1.** Applied to the region of an organ that is furthest from the point by which it is attached to an organism. **2.** Applied to a depositional environment sited at the furthest position from the source area, and generally characterized by fine-grained sediments.

distance matrix A matrix of genetic distances between the *homologous sequences of DNA of *amino acids being compared. Such matrices can be used to construct *distance-matrix *phylogenetic trees.

distance-matrix tree A *phylogenetic tree constructed solely on the basis of genetic distances calculated in a *distance matrix, as opposed to being constructed directly from the *character states present in an *alignment, as in a *maximum-parsimony tree or *maximum-likelihood tree.

distichous In two ranks.

distributary channel A natural stream channel that branches from a trunk stream which it may or may not rejoin. It occurs typically on the surface of an *alluvial fan or delta, where it may be part of a complex, fan-shaped network that distributes the discharge and sediment load of the main channel among many small distributory channels, between which a variable assemblage of bays, lakes, *tidal flats, or *marshes may exist. The larger distributory channels show *crevasse splays and *levees. *See* ANABRANCHING CHANNEL.

distribution **1.** The geographical area (i.e. *range) within which a taxon or other group of organisms occurs. **2.** The arrangement of organisms within an area. **3.** In statistics, a particular arrangement of data, as in normal, log-normal, and Poisson distribution.

disturbance A general term for low-pressure features (e.g. a *depression or *trough). Disturbances commonly appear as waves in the major air flows in the mid-*troposphere (e.g. the equatorial easterlies, the prevailing westerlies over middle latitudes, and the *trade winds).

diurnal **1.** During daytime (as opposed to nocturnal), as applied to events that occur only during daylight hours or to species that are active only in daylight. **2.** At daily intervals, as applied to such daily rhythms as the normal pattern of waking and sleeping, leaf or flower opening and closing, or the characteristic rise and fall of temperature associated with the hours of light and darkness. *See also* CIRCADIAN RHYTHM.

diurnal curve method A technique for measuring oxygen production in aquatic *ecosystems as a means of assessing gross *primary or community productivity. Dissolved oxygen measurements are taken throughout a 24-hour period so that oxygen production by day and use at night by the aquatic community can be assessed.

diurnal temperature variation Daily variations in temperature at a particular place, related to the local radiation budget. In mid-latitudes, for example, maximum temperatures usually occur after noon and minimum temperatures in the early morning. The range varies according to location, with high variation in continental areas and low variation in maritime areas. The diurnal range in equatorial areas exceeds the annual variation in average temperature.

divergence **1.** In meteorology, the situation in which air moves outward from a region of high pressure. The effect over time is to reduce pressure at the centre. Divergence near the surface is associated with the subsidence of air; consequently the high pressure that causes low-level divergence draws air downward, producing high-level *convergence. Diverging air turns in a clockwise direction in the northern hemisphere and anticlockwise in the southern hemisphere. **2.** A horizontal flow of water in different directions away from a common centre or line. A particular example of divergence in the oceans is seen in areas of *upwelling.

divergent evolution The situation in which descendants of an ancestral group of organisms split into two or more groups that become increasingly different as time passes. Genetic separation and differentiation occurs to such an extent that distinct derivative taxa may result. Divergence may be at the *species, genus, family, order, or higher level. For example, gymnosperms and angiosperms arose from a stem group and subsequently diverged.

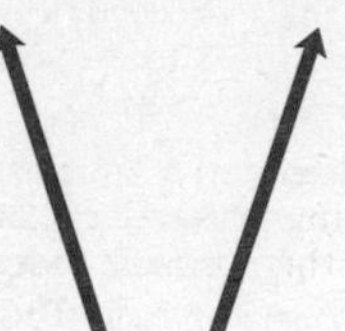

Divergent evolution

diversification An increase in the diversity of distinct types in one or more taxonomic categories (i.e. species, genus, etc.). *Phanerozoic, well-skeletonized marine invertebrates provide an illustration: their diversity at phylum level remains much the same throughout, whereas at family level there is a peak at the mid-*Palaeozoic and a trough at the Permo-*Triassic boundary; after this there is a steady increase to a second, higher peak in the *Cenozoic.

diversity Most simply, the species richness of a *community or area, though it provides a more useful measure of community characteristics when it is combined with an assessment of the relative abundance of species present. Diversity in ecosystems has been equated classically with stability and *climax communities. Such generalizations may be criticized on many counts, however.

diversity index The mathematical expression of the species *diversity of a given *community or area, which includes due allowance for the relative abundance of different species present. Such indices (e.g. the *Shannon–Wiener index) are generally considered an important means for comparisons of community structure and stability. A different and specialized case of a diversity index is the *'biotic index' used in water-pollution studies.

diversivore *See* OMNIVORE.

divide The boundary between separate *catchment areas or drainage basins. It is normally marked topographically by high ground. In British usage, a divide is sometimes called a watershed, but watershed has a different sense in US usage.

divisive method A system of hierarchical classification that proceeds by subdividing the whole into successively smaller and more homogeneous units.

DNA (deoxyribonucleic acid) A *nucleic acid, characterized by the presence of the sugar deoxyribose, the *pyrimidine bases cytosine and thymine, and the *purine bases adenine and guanine. It is the genetic material of organisms, its sequence of paired bases constituting the genetic code.

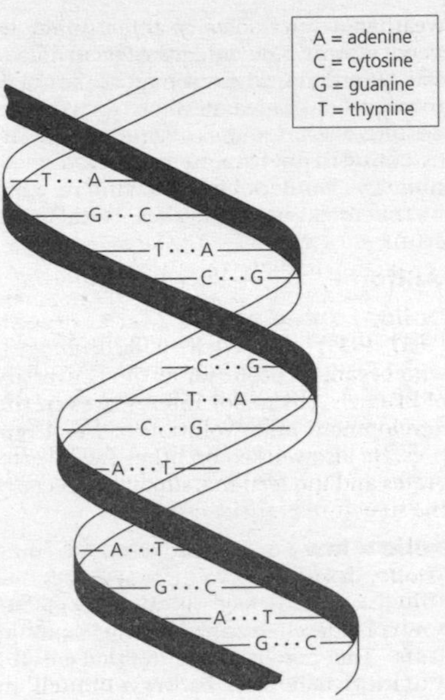

DNA

DNA fingerprint A set of DNA-based *characters which are in a combination unique, or nearly so, to an individual organism. *See also* SATELLITE DNA.

Dobson unit (DU) The unit of measurement used to report the concentration of a gas that is present in the atmosphere or in some part of the atmosphere. It is most often used to measure ozone concentration in the stratosphere (*see* OZONE LAYER). The unit is the thickness in millimetres of the layer the gas would occupy if all other gases were removed and the gas in question were subjected to sea-level atmospheric pressure. For example, 1 DU of ozone is equal to a layer 0.01 mm thick. The unit is named after Gordon Miller Bourne Dobson (1889–1976), the British physicist who spent many years studying the ozone layer.

doctor, the A local name for the West African *harmattan wind; it refers to the health-giving properties of this wind.

Dokuchaev, Vasily Vasilievich (1840–1903) A Russian soil scientist, who became director of the Kharkov Institute of Agriculture and Forestry. He studied the formation of soils, especially the *chernozem, developing a *soil classification and making soil maps.

doldrums Sea areas in which the winds are light and variable and sailing ships are likely to be becalmed, although the calm weather is periodically interrupted by fierce storms. The doldrums occur inside the *Intertropical Convergence Zone, in three principal areas of the eastern Pacific, western Pacific, and eastern Atlantic. The extent of these areas varies with the seasons. The word 'doldrums' came into use in this sense in the middle of the 19th century.

doliform Barrel-shaped or jar-shaped.

Dollo, Louis Antoine Marie Joseph (1857–1931) A French palaeontologist, who became a professor at the University of Brussels. His major interest was in the development and evolution of fossil reptiles. He also worked on other fossil vertebrates and modern marsupials, especially the structure of their limbs.

Dollo's law A law proposed by Louis *Dollo, describing evolutionary irreversibility: once regarded as inevitable, but now considered to apply mainly in special cases. The potential for further useful mutation may well be very limited in highly specialized organisms, since only those mutations that will allow the organism to continue in its narrow *niche will normally be functionally possible. In such cases there is therefore a self-perpetuating, almost irreversible, evolutionary trend, so much so that it is regarded virtually as a law, Dollo's law. The trend results from steady directional selective pressure, or *orthoselection reinforced by specialization.

dolomite **(dolostone)** A limestone rock that contains at least 90 per cent of the mineral dolomite, which is a mixture of calcium and magnesium carbonates ($CaMg(CO_3)_2$). It is used as a building stone and to make bricks for furnaces.

dolostone *See* DOLOMITE.

domain The highest taxonomic category in a classification system based on comparisons of ribosomal RNA. There are three domains: *Archaea; *Eubacteria; and *Eukarya. Each domain comprises organisms that are not closely related genetically to members of either of the other domains. *See also* KINGDOM.

domestication The selective breeding by humans of plant and animal *species in order to accommodate human needs. Domestication also requires considerable modification of natural *ecosystems to ensure the survival of, and optimum production from, the domesticated species (e.g. the removal of competing weed species when growing cereal crops).

dominance The possession of high social status within an animal group that exhibits social organization; it is often achieved and sustained by *aggression toward inferior individuals. *See* PECKING ORDER.

dominant In ecology, the species having the most influence on *community composition and form. Sometimes the term is also used to refer to the largest and/or most abundant species in the community.

Domin scale A system devised by K. Domin for describing the cover of a species in a vegetation *community. The scale ranges from simple presence through 10 grades of linked *cover-abundance measures. The scheme is based on the original (1927) five-point cover scale of *Braun-Blanquet, but the finer subdivisions allow more detailed interpretation. *See also* COVER-SOCIABILITY SCALE.

Donau A period of glaciation which occurred at about the beginning of the *Pleistocene. It may correspond to the Nebraskan Stage in North America. Evidence of it has largely been removed by later glacial episodes.

Donau/Günz interglacial A *Pleistocene *interglacial stage of the Alpine areas that may be equivalent to the *Waalian of northern Europe and the *Aftonian of North America.

dormancy **(hypobiosis)** A resting condition with reduced metabolic rate. This is found in non-germinating seeds and non-growing buds. Dormancy is predictive if it protects the organism against adverse conditions and occurs before their onset. Predictive dormancy most commonly occurs in environments that undergo regular seasonal change; in animals it is often called *diapause, in plants innate dormancy. Consequential (secondary) dormancy commences after the onset of adverse conditions.

dorsal Towards the upper surface of an organism or nearest to the back; in vertebrates nearest the spinal column. The opposite of ventral. In a plant, abaxial (i.e. facing away from the stem).

dorsiventral With upper and lower sides differing in structure.

double fertilization In flowering plants, the union of one sperm nucleus with the egg nucleus and of a second sperm nucleus with the two polar nuclei to form a triploid endosperm nucleus. The male gametophyte (the pollen grain and pollen tube) contains three sperm nuclei, but one (the vegetative nucleus) degenerates once double fertilization has been accomplished.

double planation The development of two erosion surfaces, one below the surface produced by *weathering, the other at the surface produced by wind and water *erosion. The phenomenon occurs in tropical areas and its rate of development is determined by climate and local conditions. It can contribute to the production of an *etchplain.

down The name applied to *grassland in the lowland zone of Britain, which has been created and maintained by grazing. Typically such grassland occurs on chalk and limestone hills, but occasionally it is found on acidic rocks, such as the Old Red Sandstone of the Gower Peninsula in southern Wales.

downstream In the direction of the 3′ end of a DNA sequence as measured from a reference site.

downwelling **(sinking)** The downward movement of surface waters caused by the convergence of different water masses in the open ocean, or where surface waters flow towards the coast. An example of the latter is found along the Washington–Oregon coast in winter. The Antarctic convergence zone is a major downwelling region in the Southern Ocean and is the source of *Antarctic intermediate waters.

downy mildew Either a fungus of the order Peronosporales or a plant disease caused by such a fungus. The leaves of infected plants typically show yellowish spots or patches, with whitish or purplish mould on the underside. A wide range of plants may be affected.

draa A high (more than 300 m) sand ridge or chain of *dunes in the Sahara, lying about 0.5–5 km from its nearest neighbour and moving at 2–5 cm per year. It is the largest land-form of the *erg or sand desert. A star-shaped dune, or *rhourd, is developed at the site where two draa chains cross.

drainage **1.** The passage of water over and through the land surface, ultimately towards the sea. *See* DENDRITIC DRAINAGE; DERANGED DRAINAGE; DISCORDANT DRAINAGE; DRAINAGE DENSITY; DRAINAGE PATTERN; INCONSEQUENT DRAINAGE; and SUPERIMPOSED DRAINAGE. **2.** The process of removing the gravitational water from soil, using artificial or natural conditions, such that freely moving water can drain, under gravity, through or off soil. *See* MOLE DRAIN and TILE DRAIN.

drainage basin *See* CATCHMENT.

drainage basin morphometry The measurement of the characteristics of the surface form of a drainage basin (*catchment), and of the arrangement and organization of the associated river network. Properties such as area, shape, gradient, and relief are important elements of form (*see* DRAINAGE BASIN SHAPE INDEX and DRAINAGE BASIN RELIEF RATIO), while the stream network is investigated through a study of its components and of the ways in which they are related. *See* DRAINAGE NETWORK ANALYSIS.

drainage basin relief ratio An index (Rh) of the relief characteristics of a drainage basin. It is expressed as $Rh = H/L$, where H is the difference in height between the highest and lowest points in the basin and L is the horizontal distance along the longest dimension of the basin parallel to the main stream line. The ratio can be positively correlated with the rate of sediment loss from a basin.

drainage basin shape index A measure of the shape of a drainage basin (*catchment), normally expressed as the ratio between two dimensions of the basin being considered. One such measure is the circularity index (or ratio), C, expressed as $C = A_b/A_c$, where A_b is the area of the basin and A_c is the area of a circle with the same length of perimeter as the basin. Another index is the form factor, F, expressed as $F = A/L$, where A is the area of the basin and L is its length. Such indices may help in forecasting the flood potential of a basin.

drainage density A measure of the average spacing between the streams draining an area. It is obtained by dividing the total length of streams by drainage area. Its magnitude is affected by factors such as the amount of rainfall, permeability of the ground surface, and age. *See* DRAINAGE.

drainage morphometry (morphometric analysis, Horton analysis) The calculation of a range of dimensionless drainage network relationships (*see* BIFURCATION RATIO, FOR EXAMPLE), based on a system of stream ordering (i.e. the numerical ranking of channel segments within a channel network). It was proposed by R. E. Horton in 1945.

drainage network *See* DRAINAGE PATTERN.

drainage network analysis The study of the way in which the pattern of streams in a drainage basin is organized. Classical work focused on the relations between the components of a network (e.g. on the link between the importance (or 'order') of a stream segment and its frequency). Certain 'laws' of drainage network composition were derived. The modern approach emphasizes the importance of random processes in an explanation of these 'laws' and is more concerned with the density of the drainage network (drainage density = total length of channels/drainage area).

drainage pattern The spatial relationship between individual stream courses in an area. The resulting pattern often reflects the underlying rock type and structure, and several varieties are recognized. A dendritic pattern is the most common, characterized by a randomly branched arrangement. It is not structurally controlled and is developed on a homogeneous rock (e.g. clay). A trellis pattern consists of subparallel streams, usually aligned along the geologic strike, joined at right angles by tributaries. A rectangular pattern is dominated by right-angled bends, and reflects control by joints or faults. A centripetal pattern consists of stream courses converging into a central depression. The drainage network is the drainage pattern viewed geometrically.

drainage wind *See* KATABATIC WIND.

drawdown The lowering of the *water-table or *potentiometric surface, normally as a result of the deliberate extraction of *groundwater.

drift The British Geological Survey has used this term to refer to all superficial (i.e. drift) deposits (*see* DRIFT MAP). Sometimes the term has been used to describe any sediment laid down by, or in association with,

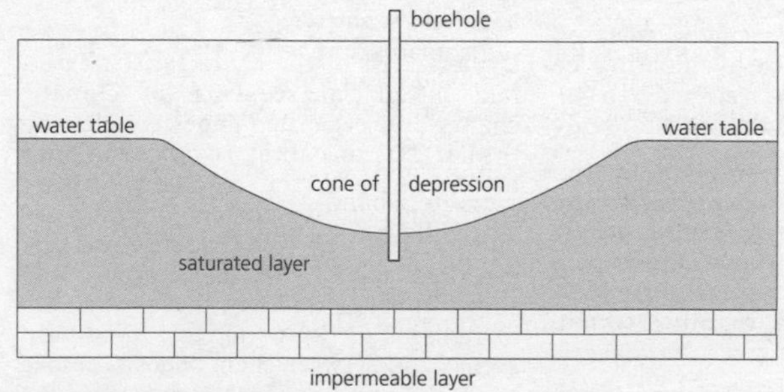

Drawdown

the activity of glacial ice and it is often widened to include related submarine and lacustrine deposits. The word was introduced by C. Lyell (1797–1875), who suggested that glacial deposits were laid down by melting icebergs which drifted across an ice-age sea covering Britain. This old term is now largely superseded by more recent classifications.

d

drift map A geologic or geomorphological map which shows the distribution of more recent glacial, *fluvial, fluvioglacial, *alluvial, and marine sediments (i.e. all superficial deposits). Depending on the distribution and extent of *drift, the map may show a combination of solid and drift exposures.

drive An outdated term used formerly to describe a type of motivation in animals, the psychological 'force' that leads to physical action. The term was abandoned because 'force' in the sense of 'physical energy' plays no direct part in psychological processes, and because the attempt to attribute a drive to each aspect of behaviour led to an uncontrollable proliferation of drives.

drizzle Precipitation of very small (200–500 μm) water droplets generated by coalescence at the base of cloud such as *stratus. *See also* CRACHIN.

DRM **1.** *See* DEPOSITIONAL REMANENT MAGNETIZATION. **2.** *See* DETRITAL REMANENT MAGNETIZATION.

drone The male of ants, bees, and wasps, whose only function is to mate with fertile females: the drone contributes nothing to the maintenance of the colony.

dropstone A rock fragment, released by melting from the base of a floating ice sheet or glacier, or a bomb ejected by a volcano, which falls through the water body to settle in muddy sediment.

drought A period during which rainfall is either totally absent or substantially lower than usual for the area in question, so that there is a resulting shortage of water for human use, agriculture, or natural vegetation and fauna.

drought cycle A temporary and repetitive phase of drier conditions in an otherwise favourable environment (e.g. the 22-year drought cycles of North American grasslands).

Drude, Carl Georg Oscar (1852–1933) A German botanist and biogeographer, who described vegetation types in terms of *formations. He published many works, including *Handbuch der Pflanzengeographie* (1890) and *Oekologie der Pflanzen* (1913), and collaborated with H. E. *Engler to produce *Die Vegetation der Erde*, which began publication in 1896. Drude was professor of botany and director of the botanic garden at the University of Dresden.

drumlin A smooth, streamlined, oval-shaped land-form, one end of which is blunt and the other tapered. Drumlins may occur singly, but more commonly they are found within a large group, called a 'drumlin field' or 'drumlin swarm'. Usually they are composed of till or boulder clay, but occasionally they are composed largely of solid rock (hence 'rock drumlin'). They are believed to be formed beneath the outer zone of an expanding ice sheet, during a major advance: they result from the selective deposition of material that is then streamlined by the moving ice. The long axis of a drumlin lies parallel to the direction of the advance.

drumlin field *See* DRUMLIN.

drumlin swarm *See* DRUMLIN.

dry adiabatic lapse rate The rate at which dry (i.e. unsaturated) air cools when rising adiabatically through the atmosphere as a result of the utilization of energy in expansion. It is 9.8°C/km. *See also* INSTABILITY; SATURATED ADIABATIC LAPSE RATE; and STABILITY.

Dryas Part of the characteristic threefold late-glacial sequence of climatic change and associated deposits following the last (*Devensian) ice advance and prior to the onset of the markedly warmer conditions of the current (*Flandrian) *interglacial. The type sequence was first described for Allerød in Denmark, and shows upper- and lower-clay deposits rich in remains of *Dryas octopetala* (mountain avens), and between them deposits of lake mud with remains of cool temperate flora, e.g. tree birches. The colder Dryas phases mark times of cold, *tundra-like conditions throughout what is now temperate Europe. The threefold Dryas–Allerød–Dryas sequence forms Pollen Zones I, II, and III of the widely accepted late and postglacial chronology of Europe. The basal, older Dryas deposit forms Zone I; the Allerød Zone II; and the younger Dryas Zone III. In north-western Europe, Pollen Zone I is subdivided into a, b, and c. Zone 1b represents a proposed *Bølling interstadial, with Zones 1a and 1c referred to as Oldest and Older Dryas respectively.

dry-bulb thermometer A thermometer that registers normal air temperature. It may be used in conjunction with a *wet-bulb thermometer: the *relative humidity can be found by measuring the depression of temperature registered by the wet bulb. *See also* HYGROMETER and PSYCHROMETER.

dryfall The delivery of nutrients to surface communities as molecules or particles deposited on to surfaces from dry air. *Compare* WETFALL.

dry ice Solid carbon dioxide. It is used in *cloud seeding to cool air in supercooled clouds by *sublimation at low temperatures. This can generate many ice crystals for further ice nucleation. Solid carbon dioxide sublimes at 195 K (−78°C).

dry-matter production The expression of plant or animal productivity in terms of the dry weight of material produced per unit area during a specified time period. It is a more easily achieved, though technically less accurate, measure of organic *production than are *calorific values; in the latter, the inorganic (ash) component can be separated.

dry rot **1.** Any of several plant diseases which are characterized by the formation of dry, shrivelled lesions; they are usually due to fungal infection. **2.** A serious type of timber decay in buildings, caused by the fungus *Serpula lacrymans*. Typically, infected timber develops longitudinal and cross-grain cracking and bears a surface growth of whitish mycelium; leathery fruit bodies bearing rust-coloured spores may appear.

dry season A period each year during which there is little precipitation. In tropical climates (e.g. over much of India) the dry period is often in the winter season. In places in very low latitudes two dry seasons may occur each year, between the northward and southward passage of the equatorial rains. In subtropical, Mediterranean, and west-coast climates, the dry season is in the summer.

dry-subhumid climate A climate with an *aridity index of 0.50–0.65.

dry valley A linear depression that lacks a permanent stream but shows signs of past water *erosion. It is a common land-form in areas underlain by permeable rock (e.g. the chalk of southern England). The dry valley was eroded during an episode of surface drainage, perhaps due to *permafrost conditions, to greater precipitation, or to a higher *water-table.

dry-weather flow *See* BASEFLOW.

DU *See* DOBSON UNIT.

dulosis **(adj. dulotic)** The slave-making behaviour of certain parasitic ant species, in which the workers raid nests of other species and remove pupae. These are then reared as slaves in the nest of the dulotic ant species, and pass on foraged food to their captors.

dune A land-form produced by the action of wind on unconsolidated sediment, normally sand. Aeolian dune forms range from small ripples less than 1 cm in height to the *draa forms of the Sahara, which rise to more than 300 m. Such dunes may be divided into three basic categories: *barchans; longitudinal or 'seif' dunes, which parallel the wind direction; and transverse dunes, which are aligned normally to the dominant wind. Transverse dunes are initial forms on sandy coastlines in temperate regions. They migrate inland and may be eroded locally by the wind to form a damp hollow or *'dune slack'. The enclosing crescentic dune is a 'parabolic' dune whose form reverses that of the barchan. *See also* AKLÉ DUNE; DUNE BEDFORM; and STAR DUNE.

dune bedform **('megaripples')** Mounds or ridges of sand which are asymmetrical and are produced subaqueously by flowing water. The external morphology is similar to the smaller 'ripple' and larger *'sand wave', with a gently sloping, upstream side (stoss), and a steeper downstream side (lee). The crestline elongation extends transverse to the flow direction and is sinuous or lunate in plan. The height varies between 0.1 m and 2 m, while the wavelength (spacing) between dunes is 1–10 m. Size and growth are limited by water depth and, in general, dune height is less than one-sixth of the flow depth. The down-current migration of dune bedforms leads to the formation of cross-bedding in sediments.

dune slack A flat-bottomed, hollow zone within a sand-dune system that has developed over impervious strata. The slack may result from erosion or *blow-out of the dune system, and the flat base level is therefore close to or at the permanent *water-table level. Characteristically, dune slacks have rich, marshy flora, with *Salix* species (willows) as typical woody colonizers. Water draining into the slack is often rich in calcium, being derived from the calcium carbonate of mollusc shells mixed in the dune sand.

duplication The presence or creation of two copies of a DNA segment; usually the term is preceded by the unit of duplication (e.g. *exon duplication, gene duplication, partial gene duplication).

duplicatus The Latin *duplicatus* meaning 'doubled', used to describe a type of cloud with overlapping layers, typified by such genera as *stratocumulus, *altocumulus, *altostratus, *cirrostratus, and *cirrus. *See also* CLOUD CLASSIFICATION.

duric horizon A *soil horizon containing cemented silica. The name is from the Latin *durum*, meaning hard.

duricrust A weathered soil deposit, found especially in subtropical environments, which may ultimately develop into a hardened mass. A range of types occurs, each distinguished by a dominant mineral. Ferricrete and alcrete are dominated by sesquioxides of iron and aluminium respectively, silcrete by silica, and *caliche (calcrete) by calcium carbonate. *See also* DURIPAN.

du Rietz, Gustaf Einar (1895–1967) A Swedish ecologist, who, in 1934, became professor of plant ecology at the University of Uppsala. He contributed to several branches of biogeography and was a leading figure in the *Uppsala school of phytosociology.

duripan A mineral *diagnostic *soil horizon which is cemented by silica and so will not slake or fall apart in water or hydrochloric acid. It may contain secondary cement (e.g. carbonates and iron oxide). Where duripans are exposed on the soil surface, they are called *duricrust. *Compare* CALICHE.

Durisols Soils that have a *duric horizon within 100 cm of the surface. Durisols are a reference soil group in the *FAO *soil classification.

dust-bowl An area of the Great Plains region, USA, where a combination of drought and inappropriate farming practices, especially an expansion of wheat production, led to severe *deflation and soil erosion during the middle 1930s. More generally, any region where deflation of cultivable land occurs.

dust devil (dust whirl, sand pillar) A very localized *whirlwind in a desert area, where strong convection uplifts dust and sand often to a height of a few tens of metres.

dust storm A storm in which dust is blown up from the ground. It occurs when the wind speed exceeds a critical value (commonly 24–48 km/h) that depends on quantity of dust present and the size, shape, dampness, and specific gravity of the dust particles. Dust may be carried to a height of 1 500–1 800 m or more. The *haboob and *khamsin are types of dust storm.

dust whirl *See* DUST DEVIL.

Dutch elm disease A devastating disease which can affect all species of elm (*Ulmus*). The causal agent is *Ceratocystis ulmi*, a fungus originating in Asia but first identified in Holland. The fungus develops and spreads in the xylem vessels; tyloses are formed. Symptoms include wilting, with curling and yellowing of foliage, followed by the rapid death of branches or the whole tree. The fungus is spread from tree to tree by elm-bark beetles (commonly *Scolytus* species). In the 1960s a new and more virulent strain of the *pathogen was introduced into Britain on logs imported from Canada; since then many millions of elms have been killed, dramatically changing the landscape in many regions of Britain. The disease has also been proposed as the cause of the *elm decline.

Du Toit, James Alexander Logie (1878–1948) A South African geologist who made field studies of the provinces of *Gondwana and found extensive evidence for *continental drift. He published his ideas in *Our Wandering Continents: An Hypothesis of Continental Drift* (1937).

dysgenic Genetically deleterious.

dysphotic zone The region of the *photic zone that lies below the *compensation level, and within which light penetration is such that oxygen production by photosynthesis is exceeded by oxygen consumption by respiration. *Compare* EUPHOTIC ZONE.

dystrophic Applied to a lake that is usually shallow, rich in humus giving its water a brown colour, with variable amounts of nutrients, and with the deeper water often depleted of oxygen. A dystrophic lake was proposed (by A. Thienemann in 1925) as one of three categories of standing water, the others being described as *oligotrophic and *eutrophic, with *mesotrophic water comprising an intermediate category.

E *See* EXA-.

-eae In plant *taxonomy, the suffix used to indicate a *tribe.

earthflow A flow of unconsolidated material down a hill-slope, normally resulting from an increase in pore-water pressure, which reduces the friction between particles. Flow velocities vary from slow, when behaviour is plastic, to rapid, when behaviour is more liquid, and reflect variations in water content. Dry flows may occur when an earthquake shock breaks inter-granular bonds.

earthquake Motion of the Earth, which usually results from the sudden release of stress. Earthquakes are usually classified in terms of their depth: shallow ones are less than 70 km depth; intermediate 70–300 km; and deep more than 300 km. No earthquakes are known below 720 km depth. *See* EARTHQUAKE PREDICTION; FORESHOCK; MERCALLI SCALE; and RICHTER SCALE.

earthquake prediction The provision of advance warning of an *earthquake. Most predictions are based on attempts to determine a stress increase prior to rock rupture. This may involve geodetic measurements to monitor relative motions, changing elevation, etc., or phenomena resulting from stress accumulation (changes in the magnetization, temperature, gas release, etc.), some of which may affect animals. So far, most methods indicate only an increasing probability of seismic activity and cannot be used to predict an actual occurrence, other than the use of *foreshocks, often only minutes prior to a major main earthquake, but such small earthquakes do not necessarily lead to major activity. Quiescence within an active seismic area can indicate either a gradual increase in stress or that stress release is taking place gradually.

East African steppe floral region Part of R. Good's (1974, *The Geography of the Flowering Plants*) African subkingdom, which lies within his *Palaeotropical floral kingdom. The flora includes about 150 endemic (*see* ENDEMISM) genera, nearly all of which are small. They include *Saintpaulia*, one species of which, *S. ionantha* (African violet), is now a popular house plant. *See also* FLORAL PROVINCE and FLORISTIC REGION.

East Australian Current An oceanic water current that flows along the east coast of Australia. This narrow (100–200 km wide) current forms the westerly part of the anticyclonic circulation in the South Pacific. The flow velocity varies in the range 0.3–0.5 m/s. It is an example of a western *boundary current.

easterly wave A type of weak *trough in a tropical easterly airflow, which has a wavelength generally of 2 000–4 000 km. Such waves occur, for example, over West Africa and in the Caribbean area, where they develop in summer and autumn with a weak or absent *trade-wind inversion, and in the central Pacific, when the equatorial trough is displaced northwards. Disturbances in such waves often lead to tropical storms, which vary in intensity and sometimes develop into *hurricanes. Most hurricanes begin as easterly waves.

ebb tide Falling *tide: the phase of the tide between high water and the succeeding low water. *Compare* FLOOD TIDE.

Eburonian A northern European stage, dating from about 1.6 to 1.4 Ma ago, that is associated with a period of glaciation at about the beginning of the *Pleistocene. It may correspond to the *Donau stage in the Alpine area and the Nebraskan stage in North America.

eccritic temperature The body temperature an *ectotherm prefers, often maintaining it by alternately basking and seeking shade.

ecdysis The periodic shedding of the exoskeleton by some invertebrates, or of the outer skin by some amphibians (*see* AMPHIBIA) and *reptiles.

ecesis (biological invasion) The ability of some migrating plant species, having arrived at a new site, to germinate, grow, and reproduce successfully, while others fail to become established in the new environment. It represents the third in a series of six phases in plant *succession. The associated verb is to ecize.

echinulate Covered with small points or spines.

echo-location The detection of an object by means of reflected sound. Bats, some cetaceans (whales and dolphins), and other animals use echo-location for purposes of orientation and the pursuit of prey. Some nocturnal birds, such as the oilbird (*Steatornis caripensis*) of South America, use echo-location for navigating in the dark. In addition to its vocalizations of screams, squawks, and clucking sounds, the oilbird navigates at night by means of short clicking sounds.

ecize *See* ECESIS.

ecliptic The plane of the orbit of the Earth around the Sun. It forms an angle of 23°27′ with the Earth's equator. The orbits of the planets all lie within 3.4° of this plane, except for those of Pluto (17.2°) and Mercury (7°).

eco- A prefix derived from the Greek *oikos*, meaning 'house' or 'dwelling place'.

ecocline (ecological gradient) A gradation from one *ecosystem to another when there is no sharp boundary between the two. It is the joint expression of associated *community (*coenocline) and complex environmental gradients.

ecological amplitude The range of *tolerance of a species, diagrammatically forming a bell-shaped curve. Species with a narrow ecological amplitude often form good *indicator species.

ecological and phytosociological distance *See* AFFINITY INDEX.

ecological backlash The unexpected and detrimental consequences of an environmental modification (e.g. dam construction) which may outweigh the gains anticipated from the modification scheme.

ecological barrier *See* DISPERSAL BARRIER.

ecological efficiency The ratio between energy flows measured at different points in a food-chain, usually expressed as a percentage. Many approaches have been devised to relate different aspects (e.g. intake, *assimilation, and *production). Two main categories of efficiencies are studied: (*a*) those of energy transfer between different *trophic levels; and (*b*) those of energy transfer within a single trophic level.

ecological energetics The study of energy transformations within *ecosystems.

ecological factor *See* LIMITING FACTOR.

ecological genetics The study of genetics with particular reference to variation on a global and local geographic scale.

ecological gradient *See* ECOCLINE.

ecological indicator Any organism or group of organisms indicative of a particular environment or set of environmental conditions. For example, *lichens may be used as indicators of air *pollution and fossil assemblages as indicators of past environments. *See also* BIOTIC INDICES and INDICATOR SPECIES.

ecological isolation The separation of groups of organisms as a result of changes in their *ecology or in the environment in which they live. This is one of the processes leading to *speciation, since there will be a restriction in the movement of genes between groups thus separated, and changes in gene frequencies may occur, owing to local selection or drift, until eventually the groups may be so divergent that reproductive barriers exist.

ecological niche *See* NICHE.

ecological pyramid (Eltonian pyramid) A graphical representation of the *trophic structure and function of an *ecosystem. The first *trophic level, comprising *producer organisms (usually green plants), forms the base of the pyramid, with succeeding levels of *consumers arranged above it to the apex. Each level is represented by a horizontal bar; the bars are of equal thickness but of widths that vary to indicate the magnitude at that level. There are three types of pyramids: the *pyramid of numbers, the *pyramid of biomass, and the *pyramid of energy. Usually the bar at the base is the widest, with progressively narrower bars above it, creating the pyramid shape, but the concept is sometimes reversed in aquatic ecosystems. For example, the pyramid of biomass in an aquatic system stands on its head, having a higher biomass of predatory fish than of either primary producers or grazers. The alternative name for the concept is taken from the name of Sir Charles *Elton, FRS, the British ecologist who devised it.

ecological system *See* ECOSYSTEM.

ecologism **1.** The use of ecological terminology or simplistic interpretations of ecological concepts in support of political or moral arguments. **2.** Any supposedly ecological expression that is so used.

ecology The scientific study of the interrelationships among organisms and between organisms, and between them and all aspects, living and non-living, of their environment. Ernst Heinrich *Haeckel is usually credited with having coined the word 'ecology' in 1866.

ecomorph A small difference in form or colour that distinguishes populations of a species that have become reproductively isolated fairly recently. Ecomorphs represent adaptations to very local environmental variations and may form a series (e.g. among anolis lizards, all of the species *Anolis sagrei*, descended from populations that were introduced experimentally to 14 very small islands in the Bahamas in the 1970s, which differ in the length of their hind legs according to the structures on which they rest).

economic injury level **(EIL)** The level of pest infestation below which the cost of further reducing the pest population exceeds the additional revenue or value of other benefits such reduction would achieve. In the case of *vectors of life-threatening diseases, the value of saving one human life may be thought to exceed the (probably very high) cost of eliminating the vector totally. This is rarely the case where the damage is purely economic (e.g. to a farm crop or stored goods). If the benefits under consideration are social (e.g. relative freedom from clouds of insects which are a nuisance but otherwise harmless) EIL is known as the aesthetic injury level.

eco-organ A term used in some modern attempts to devise a system for describing vegetation types on the basis of life-form rather than species composition. An eco-organ is a characteristic morphological feature which reflects adaptation to external environmental conditions. Sunken stomata are one example; narrow, needle, or divided leaves are another. *Compare* LIFE-FORM and RAUNKIAER.

ecophysiology The scientific study of the physiological adaptation of species to their environments.

ecosphere *See* BIOSPHERE.

ecostratigraphy The study of the occurrence and development of fossil communities throughout geologic time, as evidenced by *biofacies, with particular reference to its relevance in stratigraphic correlation and other fields, such as *biogeography and basin analysis.

ecosystem **(ecological system)** A term first used by A. G. *Tansley (in 1935) to describe a discrete unit that consists of living and non-living parts, interacting to form a stable system. Fundamental concepts include the flow of energy via *food-chains and food-webs, and the cycling of nutrients biogeochemically (*see* BIOGEOCHEMICAL CYCLE). Ecosystem principles can be applied at all scales. Principles that apply to an ephemeral pond, for example, apply equally to a lake, an ocean, or the whole planet. In Russian and central European literature 'biogeocoenosis' describes the same concept.

ecosystem management The active and purposeful manipulation of an *ecosystem in order to exploit its productivity (*see* PRIMARY PRODUCTIVITY) or to enhance its *biodiversity and conservation value.

ecosystem respiration The total *respiration rate per unit area of an *ecosystem, including the respiration of *producers, *consumers, and *decomposers.

ecotone **1.** A narrow and fairly sharply defined transition zone between two or more different *communities. Such edge communities are typically species-rich. Ecotones arise naturally, e.g. at land–water interfaces, but elsewhere may often reflect human intervention (e.g. the agricultural clearance of formerly forested areas). *See* EDGE EFFECT. **2.** A physical gradient that causes a gradual change in biological composition in response to physical factors.

ecotope The habitat component of a *biogeocoenosis.

ecotourism Travel to an area of ecological, geographical, or natural history interest, with the intention of study and education rather than pure recreation. Ecotourists seek to avoid bringing additional pressures upon the region that they visit, and are concerned to ensure that both local human cultures and their environment are enhanced rather than damaged by their activity.

ecotron *See* MICROCOSM.

ecotype A locally adapted *population of a widespread species. Such populations show minor changes of morphology and/or physiology, which are related to *habitat and are genetically induced. Nevertheless they can still reproduce with other ecotypes of the same species. *Heavy-metal-tolerant ecotypes of common grasses (e.g. *Agrostis tenuis*) are an example.

ecto- The Greek *ekto*, meaning 'outside', used as a prefix in the same sense.

ectomycorrhiza A type of *mycorrhiza in which the fungal hyphae do not penetrate the cells of the root, but cover the root and grow between the root cells. This type of mycorrhiza is common in forest trees.

ectoparasite *See* PARASITISM.

Ectoprocta *See* BRYOZOA.

ectotherm An animal that maintains its body temperature within fairly narrow limits by behavioural means (e.g. basking or seeking shade). Terrestrial reptiles are ectotherms. *Compare* POIKILOTHERM.

edaphic Of the soil, or influenced by the soil. Edaphic factors that influence soil organisms are derived from the development of soils and are both physical and biological (e.g. mineral and humus content, and *pH).

edaphotope The soil component of the ecotope (*habitat) of a *biogeocoenosis.

eddy The motion of a fluid in directions differing from, and at some points contrary to, the direction of the larger-scale current. In air, eddies vary in size from small-scale turbulence (which can transport dust and diffuse pollutants) to large-scale movements (e.g. cyclone and *anticyclone cells) within the general global circulation of the atmosphere.

eddy viscosity A coefficient that relates the average shear stress within a turbulent flow of water or air to the vertical gradient of velocity. The eddy viscosity depends on the fluid density and distance from the river bed or ground surface. The concept of eddy viscosity is fundamental to the *von Karman–Prandtl equation describing the velocity profile in turbulent flow, and is important in determining rates of evaporation or cooling by wind, and the shear stress exerted by rivers on moving particles on their beds.

edge effect The change in the number of species occurring in the zone where two *habitats are in contact. Since this zone may contain biotic elements from both habitats and some unique to itself it may be rich in species, but because those species are ill-adapted to the immediately adjacent habitat, the rate of local extinction is usually high at edges. Predation, in particular, is greatest at a habitat edge. The effect occurs because the overlap region supports some species from both adjacent ecosystems (indicated in the diagram by differently marked circles and squares) and some peculiar to itself

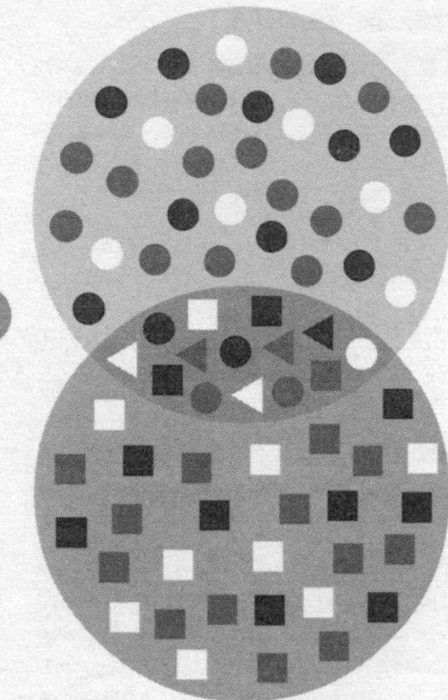

Edge effect

(triangles in the diagram). Ecologists now regard the edge effect as a sign of ecological deterioration. The fragmentation of habitats causes an increase in edge areas, but a decrease in the internal areas of *ecosystems, leading eventually to a loss of species from all affected ecosystems and an increase in edge species, which are usually commonplace.

Ediacaran The last (i.e. youngest) faunal stage of the *Precambrian; it is marked by the presence, at localities all over the world, of the distinctive Ediacaran fauna, named after a site at Ediacara in South Australia. Ediacaran fossils come from a shallow, littoral, marine environment and the animals appear to have been stranded on mud-flats or in tidal pools. About 30 genera are known and include: medusoids (jellyfish, e.g. *Medusina mawsoni* and *Medusinites*); pennatulaceans (soft corals, e.g. *Charniodiscus*); and annelid worms (e.g. *Dickinsonia* and *Spriggina*). There are also ovoid or discoid forms of unknown affinity. Since the fossils were first described by Martin Glaessner in 1961, similar faunal types of equivalent age have been found elsewhere in the world.

Eemian An *interglacial stage in northern Europe, dating from about 100 000 years BP to about 70 000 years BP, which may be the equivalent of the *Riss/Würm interglacial of the Alpine area and the *Ipswichian of the East Anglian succession.

effect hypothesis A model proposed in 1980 by the palaeontologist Elisabeth Vrba to account for *evolutionary trends. She proposed that a species, occupying a restricted ecological *niche, would continually give rise to daughter species by *punctuated equilibrium. These new species would have a variety of characteristics, but because of the features of the particular ecological niche, only species that possessed a particular suite of characters would survive. The surviving species would speciate in their turn, with the same result, and at each level the lineage appears to be 'pushed' further and further in a given direction. *Compare* DIRECTED SPECIATION.

effective population size The average number of individuals in a population that actually contribute genes to succeeding generations by breeding. This number is generally rather lower than the observed, censused, population size, being reduced by the following factors: (*a*) a higher proportion of one sex may mate; (*b*) some individuals will pass on more genes by having more offspring in a lifetime than others; (*c*) any severe past reduction in population size may result in the random loss of particular *genotypes.

effective porosity **1.** The proportion of the total *pore space in a rock which is capable of releasing its contained water. Clay, for example, may have a total porosity of 50 per cent or more, but little if any of the water contained in these pores may be released, because of the retentive forces (e.g. surface tension) that hold it within the rock. **2.** The proportion of the pore space through which *groundwater flow occurs. For example, in fractured rocks the majority of flow occurs in the fractures, and intergranular pore water may be almost static. In porous rocks some pores may have only one connection with the general pore space ('blind' pores) and so contain only static water.

effective precipitation Net precipitation after losses by evaporation. As higher temperatures increase evaporation, an index of effective precipitation derived from a temperature : precipitation ratio has been used as a criterion for some systems of climate classification (e.g. those of *Köppen in 1936 and *Thornthwaite in 1948).

effective stress In soil, the pressure between grains at their points of contact; it is at equilibrium in saturated soil. Effective stress equals total pressure minus the neutral pressure of water in pores. During consolidation, effective stress increases and reaches a maximum at complete consolidation before shear failure occurs.

effective temperature The temperature of a planetary surface in the absence of an atmosphere. The effective temperature of Earth is some 35–40°C lower than the actual Earth surface temperature (approximately 15°C) owing to the *'greenhouse effect' of the Earth's atmosphere.

effused Loosely or irregularly spreading.

egestion The expulsion by an organism of waste products that have never been part of the cell constituents.

egg In egg-laying animals (*see* OVIPARY and OVOVIVIPARY; *compare* VIVIPARY), the fertilized ovum and the embryo into which it develops, enclosed within a protective egg membrane. The eggs of animals that are laid out of water have an impermeable outer covering (e.g. the shells of eggs laid by birds and *reptiles) that protects them against drying.

egg membrane *See* EGG.

E_H *See* REDOX POTENTIAL.

EIA *See* ENVIRONMENTAL IMPACT ASSESSMENT.

Eichhornia A genus of herbs (family Pontederiaceae) that includes *E. crassipes*, water hyacinth, a floating plant with bladder-like swollen petioles, which is a noxious weed in many parts of the tropics. There are seven species occurring in warm and tropical America.

Eifelian *See* DEVONIAN.

eigen value **(latent root, λ)** The components (latent roots) derived from the data which represent that variation in the original data accounted for by each new component or axis.

eigen vector **(latent vector)** The loading of an attribute or variable on a component, as measured by the correlation between the original variable and the new component.

EIL *See* ECONOMIC INJURY LEVEL.

Ekman depth *See* EKMAN SPIRAL.

Ekman spiral A theoretical model which explains the currents that would result from a steady wind blowing over an ocean of unlimited depth and extent, which was proposed by the Swedish oceanographer Vagn Walfrid Ekman (1874–1954). In the northern hemisphere, the surface layer of the water would flow at an angle of 45° to the right of the wind direction. Water at increasing depths would flow in directions more to the right, until, at a depth known as the Ekman depth, the water would move in a direction opposite to that of the wind. The Ekman depth varies with latitude but is of the order of 100 m in mid-latitudes. The velocity of the water flow decreases with depth throughout the spiral. In the northern hemisphere, the net water transport is at 90° to the right of the wind direction, and is known as the Ekman transport. A similar spiral effect occurs in air, between the surface and the top of the *boundary layer, when the changing direction of the wind with height is plotted on a two-dimensional surface. Air moving under the influence of the *pressure-gradient force (PGF) is deflected by the *Coriolis effect (CorF). The magnitude of the CorF is proportional to the wind speed. At the surface, friction reduces the wind speed and therefore the CorF, thereby increasing the influence of the PGF and causing the wind to blow across the *isobars at an angle of 10–20° over the sea and 25–35° over land. Friction decreases with increasing height above the surface, increasing wind speed and CorF, and reducing the angle at which the wind crosses the isobars. Above the boundary level the wind is not affected by surface friction and the wind is *geostrophic, blowing parallel to the isobars.

Ekman transport *See* EKMAN SPIRAL.

elbow of capture *See* RIVER CAPTURE.

electrode potential *See* REDUCTION POTENTIAL.

electrolocation The detection of an object by its distortion of a weak electrical field generated and sensed by the animal (e.g. by electric fishes).

electromagnetic location The detection of an object by its distortion of the Earth's electromagnetic field. *See* ELECTROMAGNETIC SENSE.

electromagnetic sense The mechanism by which some fish (e.g. *Scyliorhinus*, dogfish) and many invertebrates are able to orientate themselves within an electric field that they themselves generate, or by disturbances that they detect in the Earth's electrical or magnetic fields. In some fish the electric pulses are altered by individuals to prevent interference with neighbours, and are used as a means of communication.

electromagnetic wave A wave comprising an electrical and a magnetic component which are at right angles to one another but are in phase and have the same frequency. The electrical component represents the electrical field strength and the magnetic component the magnetic flux density. Electromagnetic waves travel at the speed of light (about 2.998×10^5 km/s in free space).

electrometer *See* MASS SPECTROMETRY.

electronegativity **1.** The tendency to form negative *ions, measured by combining ionization-potential and electron-affinity values for an element to find the degree to which its atoms attract electrons. **2.** The ability of an atom to attract electrons, usually in non-metallic, acid-forming elements. Elements with sharply constrasting electronegativities tend to form ionic compounds (i.e. with ionic bonds), e.g. NaCl, where Na and Cl have electronegativities of 0.9 and 3.0 respectively. Elements with similar electronegativities are likely to form covalent bonds (e.g. CH_4 (methane), where C and H have electronegativities of 2.5 and 2.1 respectively).

electro-osmosis The phenomenon whereby some fine-grained sediments with low *permeability expel pore water when an electric current is passed through them. This is sometimes exploited to reduce *groundwater by passing currents between anodes and cathodes.

electrophoresis The migration, under the influence of an electric field, of charged particles within a stationary liquid. The liquid may be a normal solution or held upon a porous medium (e.g. starch, acryl-amide gel, or cellulose acetate). The rate at which migration occurs varies according to the charge on the particle and also its size and shape. The phenomenon is exploited in a variety of analytical and preparative techniques employed in studies of macromolecules.

electropositive element An element whose electrode potential is more positive than that of the standard hydrogen electrode which is assigned an arbitrary value of zero. Electropositive elements tend to lose electrons and form positive *ions (e.g. the univalent alkali metals Li^+, Na^+, K^+, etc., and the divalent alkaline-earth metals Be^{2+}, Mg^{2+}, and Ca^{2+}). *Compare* ELECTRONEGATIVITY.

electrovalent bond *See* IONIC BOND.

elevation head *See* ELEVATION POTENTIAL ENERGY.

elevation potential energy The energy possessed by a mass (e.g. a body of water) by virtue of its being raised above a particular datum point, usually taken as either sea level or local ground level. The energy may be released when the mass is allowed to fall to a lower level, and may be harnessed (e.g. in the case of water by powering a turbine in a hydroelectric scheme). The elevation head is the energy possessed by a unit weight of water at a point, owing to this cause. *See also* BERNOULLI EQUATION; DARCY'S LAW; and PRESSURE HEAD.

elfin woodland A *facies of tropical upper *montane or subalpine forest, with dwarfed and gnarled trees, which grows on extreme or exposed sites. The term is sometimes applied outside the tropics, although in this context it is more usual to apply the name *kampfzone or *krummholz.

elm decline The sharp decrease in the amount of elm (*Ulmus*) pollen that occurs throughout north-western Europe in soil horizons radiocarbon-dated within 100–200 years of 5 000 BP. Associated changes in vegetation vary from one site to another, as do the elm species involved (e.g. *U. glabra* in northern Britain and possibly *U. minor* in Denmark). It has been suggested that the decline is linked to the activities of early farmers, but it may have been caused by *Dutch elm disease.

El Niño A weakening of the Equatorial Current, allowing warm water to accumulate off the S. American Pacific coast; it is associated with the *southern oscillation (these two effects are known collectively as an El Niño–Southern Oscillation or ENSO event) and with climatic effects throughout the Pacific region. A similar phenomenon may also occur in the Atlantic. Approximately once every seven years, during the Christmas season (the name refers to the Christ child), prevailing trade winds weaken and the *Equatorial countercurrent strengthens. Warm surface waters, normally driven westward by the wind to form a deep layer off Indonesia, flow eastwards to overlie the cold waters of the *Peru Current. In exceptional years (e.g. 1953, 1972–3, 1982–3, and 1997–8) the extent to which the upwelling of the nutrient-rich cold waters is inhibited causes the death of a large proportion of the plankton population and a consequent decline in the numbers of surface fish.

ELR *See* ENVIRONMENTAL LAPSE RATE.

Elsterian A glacial period in northern Europe, dating from about 0.5 Ma to 0.3 Ma ago, which is probably equivalent to the *Mindel glaciation of the Alpine area.

Elton, Charles Sutherland (1900–91) A British zoologist who studied animal communities and made major contributions to the development of ecological studies in general and of animal ecology in particular. One of his early books, *Animal Ecology* (first published in 1927), was very influential. He emphasized the importance of conserving species and habitats. He proposed the existence of *niches, occupied by species at particular points in a *food-chain. From 1932 until 1967 he was director of the Bureau of Animal Population at the University of Oxford. Elton was elected a Fellow of the Royal Society in 1953 and was awarded many scientific prizes.

Eltonian pyramid *See* ECOLOGICAL PYRAMID.

elutriation A method of analysing the size of grains, in which the finer particles are separated from coarser and heavier ones by the use of a rising current of air or water. This carries the light particles upwards and allows the heavier grains to sink. By controlling the velocity of the flow, grains of different sizes can be separated.

eluviation The removal of soil materials from surface *soil horizons, in suspension or in solution, and their partial deposition in lower horizons. Removal in solution is called *leaching and hence the term 'eluviation' is often limited in use to removal in suspension.

emarginate Having a notch or notches at the edge.

emergence marsh The upper zone of a *salt-marsh, from the general mean high-water level to the mean high-water level of spring tides. Typically it has fewer than 360 submergences per annum, often with less than an hour of submergence daily during sunlight. The minimum period of continuous exposure may exceed 10 days. *Compare* SUBMERGENCE MARSH.

emergent aquatic An aquatic herb that is rooted below the water surface, but with its shoot and/or leaves above the water. Examples include bulrushes and reeds.

emergents The individual trees, or clumps of trees, which stand prominently higher than the top of the continuous *canopy of many lowland tropical rain forests.

Emilian *See* CASTLECLIFFIAN and QUATERNARY.

Emsian *See* DEVONIAN.

emu Electromagnetic units in the *c.g.s. units (gauss, oersteds, etc.). These have now been replaced by *SI units of ampere per metre, weber per metre, and tesla.

e

en- The French *en*, meaning 'in', used as a prefix meaning 'in', 'into', or 'inside'.

enation The outgrowth from a leaf in a plant infected with a certain type of virus.

enation theory The theory that accounts for the origin of the fern leaf by suggesting that it arose from the development of simple outgrowths (enations). Any such theory has to account for the large, branching fronds of a fern with branching veins (a 'megaphyll') and also for small leaves ('microphylls') with a single median vein.

endangered species *See* RARITY.

endemic *See* ENDEMISM.

endemism The situation in which a species or other taxonomic group is restricted to a particular geographic region, owing to factors such as isolation or response to soil or climatic conditions. Such a taxon is said to be endemic to that region. The size of the region in this context will usually depend on the status of the taxon: thus a family will be endemic to a much larger area than a species, all other things being equal. Reference is frequently made to 'narrow endemics' (i.e. taxa with markedly restricted ranges). Some of these are evolutionary relics, such as the maidenhair tree (*Ginkgo biloba*), a single species within a single genus which is confined to Chekiang Province, China, where it was discovered in 1758. Endemics are always native to the region in which they are found and to which they are confined. They may be neoendemics, in which case they have evolved recently and may be restricted in their distribution simply because they have not yet had time to spread further, or palaeoendemics, in which case they have a long evolutionary history and their confinement is caused by barriers to dispersal.

end moraine *See* MORAINE.

endo- A prefix meaning 'internal', derived from the Greek *endon*, meaning 'within'.

endobiotic Growing within a living organism.

endobyssate Applied to the habit of specific bivalves (Bivalvia) that live in sediment. In contrast to *epibyssate forms, the byssus (strands, usually of tough, horny thread) of these animals is used to anchor the animal within a burrow or boring.

endocytosis The mechanism by which a cell engulfs large material by the invagination of the cell membrane to form a vesicle or vacuole.

endogenetic *See* EXOGENETIC.

endogenous infection Infection with a micro-organism that is normally resident in the body so infected.

endolithic Growing within or between stones (e.g. the thalli of certain *lichens). Some desert lichens and *protists avoid desiccation by living beneath the surface of a rock.

endomycorrhiza A type of *mycorrhiza in which the hyphae of the fungus actually penetrate the cells of the root. Endomycorrhizas are found in a wide range of plant types.

endonuclease An *enzyme that will break a *nucleotide sequence.

endoparasite *See* PARASITISM.

endophloeodal Growing within bark.

endorheic lake A lake that loses water only by evaporation (i.e. no stream flows from it). *Compare* EXORHEIC LAKE.

endospore A type of bacterial resting cell that develops within a vegetative cell under certain conditions. Endospores are extremely resistant to adverse environmental conditions.

endosymbiont A *symbiont that lives within the body of its host. *Compare* EXOSYMBIONT.

endotherm An animal that is able to maintain a body temperature that varies only within narrow limits by means of internal mechanisms (e.g. dilation or contraction of blood vessels, sweating, panting, and shivering). Birds and mammals are endotherms. *Compare* HOMOIOTHERM.

endotoxin A component (lipopolysaccharide) of the walls of certain (Gramnegative) bacteria: it is toxic to animals, including humans. *See* GRAM STAINING.

endozoochory Dispersal of spores or seeds by animals after passage through the gut.

energy budget **1.** A comparison between the amount of energy that enters the body of an animal, or a particular *trophic level, and the amount of energy that leaves that animal or level. **2.** *See* RADIATION BUDGET.

energy flow The exchange and dissipation of energy along the *food-chains and food-webs of an *ecosystem.

energy of activation *See* ACTIVATION ENERGY.

englacial Contained within the interior of a *glacier, as opposed to being at its base (subglacial) or on its surface (supraglacial). Normally the term is applied to meltwater or *drift.

Engler, Heinrich Gustav Adolf (1844–1930) A German taxonomist and biogeographer who helped to develop a system for classifying plant families and genera. He collaborated with C. G. O. *Drude to produce *Die Vegetation der Erde*, which began publication in 1896. Engler was professor at the museum and botanical gardens in Berlin-Dahlem.

engram A memory trace stored in the brain.

enriched uranium Uranium that contains up to 3 per cent ^{235}U. It is produced industrially for use in some nuclear reactors. Natural uranium contains as little as 0.7 per cent ^{235}U.

enrichment **1.** In forestry, the planting of young trees within a forest, commonly after depletion by timber extraction. **2.** In nuclear engineering, the processing of natural uranium to increase, for power generation commonly to about 3 per cent, the proportion of ^{235}U to ^{238}U.

ensiform Sword-shaped.

ensilage *See* SILAGE.

ENSO *See* EL NIÑO.

entelechy An outmoded theory holding that evolution proceeds by the realization of that which was always potential. The word is from the Greek *entelekheia*, meaning 'become perfect'. *See also* ARISTOGENESIS; NOMOGENESIS; and ORTHOGENESIS.

enterotoxin A type of *toxin, produced by certain bacteria, which affects the function of the intestinal mucosa, causing diarrhoea, gastroenteritis, etc. in animals.

enthalphy **(*H*)** The heat content per unit mass of a substance, measured as the internal energy plus the product of its volume and pressure.

Entisols An order of embryonic *mineral soils, including those that have no distinct pedogenic *horizons. Representing only the initiation of soil-*profile development, Entisols are common on recent flood-plains, steep eroding slopes, stabilized sand-dunes, and recent deep ash or wind deposits.

entognathous Eversible and contained within a small pocket.

entomochory Dispersal of spores or seeds by insects.

entomopathogenic Capable of causing disease in insects.

entomophilous Applied to flowering plants with floral parts that are adapted to pollination by insects.

entrainment The process by which air from the environment outside a growing cloud is caught into the rising convective current within the cloud and mixed with the cloudy air. This is significant in that it reduces the *buoyancy of the rising current and causes cloud growth by reason of the cooler, drier air which is introduced. This also causes some evaporation of cloud droplets. When very dry air is introduced, entrainment can produce rapid dissipation of the cloud.

entrenched meander *See* MEANDER.

entropy **1.** A measure of disorder or unavailable energy in a thermodynamic system; the measure of increasing disorganization of the universe. **2.** *See* LEAST-WORK PRINCIPLE and LEAST-WORK PROFILE.

environment The complete range of external conditions, physical and biological, in which an organism lives. Environment includes social, cultural, and (for humans) economic and political considerations, as well as the more usually understood features such as soil, climate, and food supply.

environmental geology The study of the problems that result from natural hazards and human exploitation of the natural environment. The geologic techniques used include those of engineering geology, economic geology, hydrogeology, etc., as applied to waste disposal, water resources, transport, building, mining, and general land use.

environmental impact assessment **(EIA, environmental impact statement)** An attempt to identify and predict the impact on the biogeophysical environment and on human health and well-being of proposed industrial developments, projects, or legislation. EIA also aims to devise easily comprehended, universally applicable schemes for communicating the results of the assessment.

environmental impact statement *See* ENVIRONMENTAL IMPACT ASSESSMENT.

environmentalist **1.** One who holds that damage to the natural environment resulting directly from human activity is so severe as to present a challenge to the survival of many *habitats and ultimately perhaps to the continuance of life on Earth, and can be redressed only by major reforms of the way people live and industries function. **2.** One who holds that the environment exerts a more important role than heredity in determining the characteristics of individuals (i.e. favouring nurture in the debate over the relative contributions of *nature and nurture).

environmental lapse rate **(ELR)** The rate at which air temperature decreases (lapses) with height at a particular place and time in air that is unaffected by *adiabatic warming or cooling. ELR varies greatly in different regions of the world, in different air streams, and at different times of day and year. Where the ELR is negative (i.e. air temperature increases with height) a *temperature inversion is said to exist. ELR is calculated from observation, by measuring the difference between the sea-level air temperature and the temperature at the *tropopause, and dividing that difference by the height of the tropopause to give a value in degrees C per kilometre or degrees F per 1 000 feet. The average ELR is 6.5°C/km, based on a sea-level temperature of 15°C, a tropopause temperature of −56.5°C, and a tropopause height of 11 km.

environmental resistance The sum total of the environmental *limiting factors, both *biotic and *abiotic, which together act to prevent the *biotic potential of an organism from being realized. Such factors include the availability of essential resources (e.g. food and water), predation, disease, the accumulation of toxic metabolic wastes, and, in some species, behavioural changes due to stress caused by overcrowding. *See also* LOGISTIC EQUATION.

environmental science The study of environments. This may be interpreted fairly strictly as the physical environment, or may include the biological environment of an organism; or, in its widest sense, it may also consider social, cultural, and other aspects of the environment.

environmental variance The portion of *phenotypic variance that is due to differences in the environments to which the individuals in a population have been exposed. The total amount of variance observed among individuals in a population will be made up by an environmental component, determined by environmental variation, and a genetic component (*see* GENETIC VARIANCE), determined by the variation that is inherited.

enzootic Applied to a disease of animals that is restricted to a given geographical locality.

enzyme A molecule, wholly or largely protein, produced by a living cell, which acts as a biological catalyst. Enzymes are present in all living organisms, and through their high degree of specificity exert close control over cellular metabolism.

Eocambrian A little-used term employed in the description of sequences of unfossiliferous rocks that were deposited at the end of the *Precambrian.

Eocene A *Tertiary epoch, from about 55.8–33.9 Ma ago, which began at the end of the *Palaeocene and ended at the beginning of the *Oligocene. It is noted for the expansion of mammalian stocks (horses, bats, and whales appeared during this epoch), and the local abundance of nummulites (marine protozoans of the Foraminiferida). In southern Britain a humid, subtropical climate allowed rain forests to flourish. The name means 'dawn of the new' and is derived from the Greek *eos*, meaning 'dawn', and *kainos*, meaning 'new'.

eolian *See* AEOLIAN.

eon **1.** The largest unit of geologic time, incorporating a number of *eras. The equivalent chronostratigraphic unit is the (little used) eonothem. Originally, in 1930 G. H. Chadwick proposed two eons. The younger was the *Phanerozoic Eon (the time of evident life), comprising the *Cenozoic, *Mesozoic, and *Palaeozoic Eras, and this name is still used (as are the names of the eras). The name suggested for the preceding eon was the *Cryptozoic (the time of hidden life). This eon was also known as the Archaeozoic (the time of most ancient life), but it was most commonly known as the *Precambrian. The names Cryptozoic, Archaeozoic, and Precambrian have been abandoned and are no longer used formally, although the name Precambrian is often used informally. The time previously known as the Cryptozoic or Precambrian is now divided into three eons: the *Proterozoic (2 500–542 Ma); the *Archaean (3 800–2 500 Ma); and the *Hadean for the time from the formation of the Earth until 3 800 Ma. Hadean is not a formal term, however, because its base cannot be dated reliably. **2.** A time unit equal to one billion (10^9) present Earth years.

eonothem *See* EON.

epeiric sea **(epicontinental sea)** A shallow sea which extends far into the interior of a continent (e.g. Hudson Bay and the Baltic Sea). The term also denotes shallow sea areas that cover the *continental shelf and are partially enclosed (e.g. the North Sea).

epeirogenesis The large-scale upward or downward movements of continental or oceanic areas. Epeirogenic movements should not be confused with the more dynamic mountain-building episodes of an *orogeny.

ephemeral **1.** Short-lived, or of brief duration (e.g. the life of a mayfly, Ephemeroptera). **2. (ephemerophyte)** A plant that completes its life cycle very rapidly. In favourable environments ephemerals may germinate, bloom, and set seed several times during a single year. Many weed species are ephemerals. Ephemerals are also particularly characteristic of *desert environments. The word is derived from the Greek *ephemeros*, meaning 'lasting for only a day'. One type of ephemeral is the winter annual, often found in summer-dry locations such as sand dunes. These germinate in the moist conditions of autumn, grow as a small rosette through the winter, flower and fruit in the early spring, then enter a dormant state as a seed through the hot drought of summer.

ephemeral stream A stream which flows only after rain or snow-melt and has no *baseflow component. A desert *wadi may form an ephemeral stream.

ephemerophyte *See* EPHEMERAL.

epi- The Greek preposition *epi*, used as a prefix meaning 'upon', 'above', or 'in addition to'.

epibiontic **(noun epibionty)** Applied to old endemic (*see* ENDEMISM) taxa which now have a restricted distribution although formerly they were more widespread.

epibionty *See* EPIBIONTIC.

epibiotic Growing on the surface of a living organism.

epibole *See* ACME ZONE.

epibyssate Applied to animals that use the byssus (strands, usually of tough, horny threads) to anchor themselves to rock or seaweed. *Compare* ENDOBYSSATE.

epicontinental sea *See* EPEIRIC SEA.

epifauna Benthic organisms (*see* BENTHOS) that live on the surface of the seabed, either attached to objects on the bottom or free-moving. They are characteristic of the intertidal zone. *Compare* INFAUNA.

epigamic Applied to a character of an animal that serves to attract or stimulate members of the opposite sex during courtship. Examples are the distinctive coloration of some male birds and fish, the colourful crest of the male crested newt, and the song of birds. Such characters may also have other functions such as discouraging potential competitors for mates or food resources.

epigeal (epigean) Growing or occurring above ground, commonly with reference to a mode of seed *germination in which the cotyledons are carried above the soil on an axis (the hypocotyl). *Compare* HYPOGEAL.

epigean *See* EPIGEAL.

epigene Produced or occurring at the Earth's surface. The term is used especially in relation to the processes of *weathering, *erosion, and deposition. 'Epigenetic drainage' is sometimes used as a synonym for 'superimposed drainage'.

epigenesis The hypothesis that an organism develops by the new appearance of structures and functions. An alternative hypothesis (termed 'preformation') is that development of an organism occurs by the unfolding and growth of characters already present in the egg at the beginning of development.

epigenetic drainage Superimposed drainage. *See also* DRAINAGE and EPIGENE.

epigenetics The study of the mechanisms by which genes bring about their *phenotypic effects.

epilimnion The upper, warm, circulating water in a thermally stratified lake in summer. Usually it forms a layer that is thin compared to the *hypolimnion.

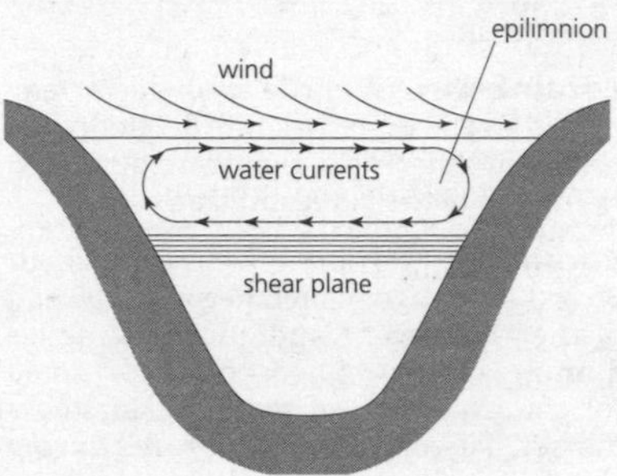

Epilimnion

epilithic Growing on or attached to the surfaces of rocks or stones.

epinastic growth (epinasty) Differential growth of the upper or *adaxial part of a plant organ. A well-known example is the growth of the upper side of the petiole (i.e. the stalk by which the leaf is attached to the stem) causing the leaf to be bent downwards. Auxin and ethylene are involved in this response, which may also be induced by application of *herbicides. It may be a symptom of disease, in which there is a pronounced and atypical bending downwards of the leaves.

epinasty *See* EPINASTIC GROWTH.

epineustic *See* ZOONEUSTON.

epineuston The organisms (*zooneuston) living in the upper part of the surface film of water. *Compare* HYPONEUSTON and NEUSTON.

epipedon A subsurface *soil horizon.

epipelagic zone The part of the *pelagic zone of the ocean that extends from the surface to a depth of about 200 m, coinciding with the *photic zone. *See* OCEAN DIVISIONS.

epipelic Growing on mud.

epipetalous Borne on the petals, by concrescence. *See also* CONCRESCENT.

epiphloeodal Growing on or attached to the surface of tree bark.

epiphyllous (folicolous) Growing on leaves.

epiphyte A plant that uses another plant, typically a tree, for its physical support, but which does not draw nourishment from it. Well-known examples include *Platycerium* (staghorn fern) and many members of the Bromeliaceae and Orchidaceae. Epiphytes are a conspicuous feature of many kinds of tropical rain forest. Even the epiphytes of tropical rain forests need to be adapted to drought because of their total dependence on rainfall for a water supply, hence the water-collecting systems of bromeliads and the occurrence of some cacti as epiphytes, e.g. *Zygocactus* (Christmas cactus).

epiphytotic An outbreak of disease (an epidemic) in a population of plants.

episepalous Borne on the sepals, by concrescence (*see* CONCRESCENT).

episodic evolution An overall picture of evolution, reflected in the fossil record, that is characterized by extinction events and succeeding phases of rapid evolutionary innovation (i.e. evolution is episodic). However, the term has recently acquired other connotations, and tends to be linked with *punctuated equilibrium.

epistasis The situation in which an *allele of one gene (the epistatic gene) prevents the expression of all allelic alternatives of another gene.

epixylous Growing on wood.

epizoic Applied to non-parasitic animals that attach themselves to the outer surface of other animals.

epizootic An outbreak of disease (an epidemic) in a population of (non-human) animals.

epoch One of the intervals of geological time recommended by the International Subcommission on Stratigraphic Terminology. An epoch is ranked as a third-order time unit, and is the equivalent of the chronostratigraphic unit 'series'. Several epochs form a period, several periods an era. When used formally, the initial letter is capitalized (e.g. Early Devonian Epoch).

equal-area net (Schmidt–Lambert net) A system of coordinates providing a two-dimensional representation of a three-dimensional object, which is used to display structural data graphically.

Equatorial Countercurrent *See* EQUATORIAL CURRENT.

Equatorial Current An oceanic current which flows in an east-west direction in the equatorial regions of all the oceans. The broad (1 000–5 000 km) westward-flowing currents (North and South Equatorial Currents) are separated by a comparatively narrow eastward-flowing countercurrent (the Equatorial Countercurrent). The flow tends to be limited to the upper 500 m of the water column and the velocity is 0.25–1.0 m/s. The strength and position of the equatorial currents are controlled by the overlying wind system.

equatorial trough A shallow low-pressure zone around the equator, where the *trade-wind systems tend to converge. *See also* DOLDRUMS and INTERTROPICAL CONVERGENCE ZONE.

Equatorial Undercurrent (Cromwell Current) A shallow, subsurface, eastward-flowing current in the central Pacific Ocean between 1.5°S and 1.5°N. The current is 300 km wide and flows at up to 1.5 m/s at depths of 50–300 m. The surface waters in this area flow in the opposite direction.

equifinality The theory that the members of morphologically similar land-forms may each have been produced by a different process or sequence of processes. For example, an armchair-shaped hollow may have been produced by a *cirque glacier, by a rotational landslide, by *nivation, or by spring-sapping. Their present-day similarity may make the origin of such land-forms difficult to determine.

equilibrium line (firn line) The line on a *glacier that divides the accumulation zone, where the glacier is acquiring more ice through precipitation than it loses through *ablation, from the ablation zone. It marks the level at which the glacial mass remains constant.

equilibrium species A species in which competitive ability (*see* COMPETITION), rather than *dispersal ability or reproductive rate, is the chief survival strategy: competition is the typical response to stable environmental resources. In unstable or extreme environments (e.g. deserts) equilibrium species survive unfavourable periods by living on stored food resources and reducing life processes to a minimum. *Compare* FUGITIVE SPECIES.

equinoctial gale A term derived from the popular misconception that gales are more frequent at periods close to the equinoxes.

equivalence point *See* MINIMAL AREA.

era A first-order geological time unit composed of several periods. The *Mesozoic Era, for example, is composed of the *Triassic, *Jurassic, and *Cretaceous Periods, dating from approximately 251 Ma to 65.5 Ma. When used formally the initial letter is capitalized. The equivalent chronostratigraphic unit is the erathem.

erathem *See* ERA.

erg **1.** A 'sand sea' in a hot desert. It is a feature of the Sahara, where sand is typically accumulated in wide shallow basins as *alluvial and *lacustrine deposits derived from adjacent rocky desert during the *Cenozoic. It is often very large: the Grand Erg Oriental covers 196 000 km^2 in Algeria and Tunisia. **2.** The unit of energy or work in the *c.g.s. system; 1 erg = 10^{-7} J.

ergatogyne A form intermediate between a worker and a queen which occurs in some ant species. The gaster (abdomen) is larger than that of a worker, but the head and thorax are worker-like. It can function as an additional reproductive or replace the queen.

ergot **(St Anthony's fire)** A disease which affects many grasses, including cereals. It is caused by the fungus *Claviceps purpurea*; the conspicuous, hard, black sclerotia of the fungus replace the grain in the spikelets of an infected plant. These sclerotia contain *alkaloids which can cause severe poisoning or even death if ingested by humans or other animals.

ericeto *See* MEDITERRANEAN SCRUB.

erosion **1.** The part of the overall process of *denudation which includes the physical breaking down, chemical solution, and transportation of material. **2.** The movement of soil and rock material by agents such as running water, wind, moving ice, and gravitational creep (or mass movement). *See* BUBNOFF UNIT.

erosion rate The rate at which geomorphological processes wear away land surfaces. Rates vary widely, depending on both processes and environments, as shown by the following figures: glacial *abrasion, 1 000 B (*bubnoff units); soil creep (under a temperate, maritime climate), 1–5 B; *solifluction (cold climate), 25–250 B; slope wash, 2–200 B; solutional loss, 2–100 B; sea cliff retreat (rocks of medium hardness), 4 000 B; soil erosion resulting from human activity, 2 000–8 000 B.

erosion surface **1. (planation surface)** A gently undulating land surface which cuts indiscriminately across underlying geologic structures and is the end-product of a long period of *erosion. Different types are produced in different environments. Among the more important are the *peneplain (humid temperate), *etchplain (humid tropical) and *pediplain (semi-arid). Marine processes may also cut a marine erosion surface. **2.** An irregular surface cut into rock or sediment by the eroding action of flowing water, ice, or the wind.

erratic A glacially transported rock whose lithology shows that it could not have been derived from the local country rock. An example is the occurrence near Snowdonia, Wales, of granites derived from Scotland. An 'indicator' is an erratic whose origin can be located precisely.

errors Generally, the deviation of measured values from their true values. Such errors may be random or systematic. Random errors should have a *Gaussian (normal) distribution about the arithmetic mean of the measurements and this should approach the true value as the number of measurements increases. Systematic errors are consistent differences between the true value and a set of measurements, such that their arithmetic mean is displaced from the true value. Thus random errors determine the precision of a set of measurements, while systematic errors limit their accuracy.

eruciform Resembling a caterpillar (i.e. having an approximately cylindrical body with prolegs in the hind region and true thoracic legs). The word comes from the Latin *eruca*, meaning 'caterpillar'.

erumpent Bursting through.

erythrism In mammals, the possession of red hair, caused most commonly by a lack of black pigment (eumelanin) which allows the red pigment (phaeomelanin) to dominate.

escape reaction A form of defensive behaviour that occurs when an animal detects a predator or the predator attempts to capture it. Many animals have specialized forms of escape (e.g. flying fish disappear from the view of the predator by leaping from the water).

esker A long, sinuous, steep-sided, narrow-crested ridge which consists of cross-bedded sands and gravels. It is laid down by glacial meltwater either at the retreating edge of an ice sheet, or in an ice-walled tunnel. Subglacially engorged eskers are deposited in *subglacial or *englacial ice tunnels; to be preserved these tunnels must have been in *stagnant ice.

ESS *See* EVOLUTIONARY STABLE STRATEGY.

essential element A chemical nutrient that is vital for the successful growth and development of an organism. Elements needed in relatively large amounts are termed macronutrients; those needed only in small or minute quantities are called micronutrients or trace elements. Macronutrient elements include carbon, hydrogen, oxygen, nitrogen, sulphur, phosphorus, potassium, magnesium, and calcium. Important micronutrient elements include iron, manganese, zinc, copper, boron, molybdenum, and cobalt.

established cell line Cells derived from a primary culture, which may be subcultured indefinitely *in vitro*.

ester A compound that is formed as the condensation product (i.e. water is removed) of an acid and an alcohol; water is formed from the OH of the acid and the H of the alcohol.

estivation *See* AESTIVATION.

estovers (common of) 1. The right to gather lying and dead standing wood, usually up to a stated size, for firewood on common land. Dead standing wood and branches can be obtained 'by hook or by crook', i.e. by the use of a blunt tool. **2.** Firewood, timber, and other materials taken from common land as a privilege.

estrus cycle *See* OESTRUS CYCLE.

estuary A coastal body of water which has a free connection with the open sea and where fresh water, derived from land drainage, is mixed with sea water. Estuaries are often subject to tidal action and, where tidal activity is large, ebb and flood tidal currents tend to avoid each other, forming separate channels. In estuaries where tidal activity is small, the invading dense sea water may flow under the lighter fresh water, forming a *salt wedge. A positive estuary is one in which surface salinities are lower within the estuary than in the open sea, owing to freshwater inflow exceeding outflow caused by evaporation. A negative estuary is one in which evaporation exceeds freshwater inflow and therefore hypersaline conditions exist in the estuary. Normally an estuary is the result of valley drowning by the post-glacial rise in sea level. The action of tidal currents on the large amount of available sediment may give rise to a range of mobile bottom forms, including ebb and flood channels, sandbanks, and *sand waves.

-etalia *See* ZURICH–MONTPELLIER SCHOOL OF PHYTOSOCIOLOGY.

etchplain A plain produced in a tropical or subtropical environment as a result of a phase of deep chemical *weathering during tectonic stability, followed by one of *erosion in which the weathered debris is stripped away. Etchplain relief reflects differences in the resistance of the bedrock to weathering and involves *double planation.

-etea *See* ZURICH–MONTPELLIER SCHOOL OF PHYTOSOCIOLOGY.

etesian winds The Greek name for north-easterly, easterly, northerly, or north-westerly winds which blow between May and September in the Aegean Sea. The equivalent Turkish term is meltemi.

e

ethanol (ethyl alcohol) A colourless liquid (CH_3CH_2OH) that is produced naturally as a by-product of the fermentation of sugary liquids during *anaerobic respiration. Ethanol mixes completely with water and pure ethanol absorbs water vapour. Many organic compounds are soluble in ethanol and some gases are more soluble in it than they are in water.

ethene (ethylene) A gas (C_2H_4) that is produced naturally by plants, and which functions as a hormone in the control of such processes as germination, cell growth, fruit ripening, *senescence, and *abscission. It is also involved in the response of a plant to gravity and to stress.

Ethiopian faunal region An area that corresponds with sub-Saharan Africa, although it is not completely separated from the neighbouring *faunal regions; generally it is taken to include the Southwest corner of the Arabian peninsula. Varied game animals are found there, and there are a number of endemic (*see* ENDEMISM) families, including various rodents and shrews, giraffe, hippopotamus, aardvark, tenrec, and lemurs.

ethocline A progressive change in the pattern of behaviour of a group of closely related organisms, the more complicated patterns often being associated with more recently evolved members of the group, which evolved from members exhibiting a more primitive pattern. For example, variations in *courtship behaviour among species of *Hilara* flies (Empididae) suggest an evolutionary lineage: (*a*) a male searches for a female and courts her in isolation; (*b*) a male captures a fly, presents it to a female, and she consumes it before mating; (*c*) a male captures a fly, joins a swarm of males, presents the fly to a female, and she consumes it before mating; (*d*) as (*c*), but the male adds a few strands of silk to the fly before joining the swarm; (*e*) as (*d*), but the male covers the fly in silk; (*f*) as (*e*), but the male eats the edible portion of the fly, leaving the husk wrapped in silk; (*g*) males of non-carnivorous species use dried insect fragments or petals, around which they weave a silken balloon before joining the swarm; (*h*) the male makes only a balloon (i.e. containing nothing) before joining the swarm.

ethogram A detailed record of an animal's behaviour.

ethological isolation The failure of individuals of species with recent common ancestry (i.e. related species or semi-species) to produce hybrid offspring because differences in their behaviour prevent successful mating taking place.

ethology (animal behaviour) The scientific study of the behaviour of animals in their normal *environment, including all the processes, both internal and external, by which they respond to changes in their environment.

ethyl alcohol *See* ETHANOL.

ethylene *See* ETHENE.

etiolation The state of plants that have been grown in the dark: they are not green, having little or no *chlorophyll, and they have very long internodes and rudimentary leaf growth. These features associated with etiolation ensure, under natural conditions, that the shoot is carried towards the light as rapidly as possible.

-etosum *See* ZURICH–MONTPELLIER SCHOOL OF PHYTOSOCIOLOGY.

-etum A suffix used in the *Zurich–Montpellier phytosociological scheme to denote a *community of association rank. It is applied to the generic name of the most characteristic species, i.e. that which differentiates the association from others in the same alliance. A *heathland community typified by *Calluna vulgaris* thus becomes a Callunetum. *See also* DIFFERENTIAL SPECIES.

eu- The Greek *eu*, used as a prefix meaning 'well', 'good', etc. It is used in ecology to denote, in particular, enrichment or abundance, e.g. *eutrophic, nutrient-rich; *euphotic, light-rich.

Eubacteria In the widely used five-kingdom system of *taxonomy, one of the subkingdoms in the *kingdom *Bacteria, containing organisms that need to be distinguished from members of the other subkingdom, the *Archaea. The Eubacteria comprises all of the 'true' bacteria, ranked as three divisions: Gracilicutes; Tenericutes; and Firmicutes. Division Gracilicutes comprises the proteobacteria (the largest group, including such organisms as *Escherichia coli*), spirochaetes, *cyanobacteria, and other groups. In the three-domain classification system, one of the three *domains, containing the single kingdom Bacteria.

Eucaryota An alternative name, now seldom used, for the *Eukarya.

eugenics A postulated method of improving humanity by altering its genetic composition. It would involve encouraging the breeding of those presumed to have desirable genes (positive eugenics), and discouraging the breeding of those presumed to have undesirable genes (negative eugenics). It relies upon the notion of 'desirable' and 'undesirable' *genotypes and assumes that selection for these can be particularly applied. Neither concept is generally accepted as valid, the former because it is often a value judgement that will vary between individuals, societies, and circumstances; the latter because genotypes are often present in the heterozygous state and so are not readily detected. Every human is heterozygous for several different deleterious recessive genes, so that, if these were detectable, *no one* would be allowed to breed under such a programme.

euhaline *See* BRACKISH and HALINITY.

Eukarya (Eukaryota) In *taxonomy, the group that includes all *eukaryotes. In the widely used five-kingdom classification system, the Eukarya is ranked as a superkingdom containing the kingdoms *Protoctista, *Animalia, *Fungi, and *Plantae. In the three-domain classification system, Eukarya is the *domain containing the kingdoms Protoctista, Animalia, Fungi, and Plantae.

Eukaryota *See* EUKARYA.

eukaryote An organism whose cells have a distinct nucleus enveloped by a double membrane, and other features including double-membraned mitochondria and 80S ribosomes in the fluid of the cytoplasm (i.e. all protists, fungi, plants, and animals). The first eukaryotes were almost certainly green algae (Chlorophyta) and what appear to be their microscopic remains appear in *Proterozoic sediments dating from a little less than 1 500 Ma ago.

Eulerian current measurement A technique for measuring the direction and speed of water movement at a series of fixed points. A current-measuring device is held at a fixed point and as the water flows past its speed is measured. A series of measurements taken at different times or places may be plotted on a map as individual current vectors or streamlines. A number of devices exist for Eulerian current measurement, the most common being the propeller-type current meter. It is named after the Swiss mathematician Leonhard Euler (1707–83).

eulittoral zone The *habitat formed on the lower shore of an aquatic *ecosystem, below the *littoral zone. In marine ecosystems, it is the main area of the littoral zone lying below the *littoral fringe. Barnacles (*Balanus* and *Chthamalus* species) are the most characteristic animals on rocky shores, with mussels and oysters also typical; the green algae *Enteromorpha* and *Ulva* species are also common. On sandy shores burrowing animals are common, e.g. shrimps, crabs, and polychaete worms (such as the lug worms, *Arenicola species*).

eumelanic Black or brown, as applied to the colour of mammalian hair. *Compare* PHAEOMELANIC.

euphotic depth In a lake, the depth at which net *photosynthesis (i.e. carbon dioxide uptake by photosynthesis minus carbon dioxide release by respiration) occurs in a light intensity about 1 per cent of that at the surface.

euphotic zone The upper, illuminated zone of aquatic *ecosystems: it is above the *compensation level and therefore the zone of effective *photosynthesis. In marine ecosystems it is much thinner than the deeper *aphotic zone, typically reaching 30 m in coastal waters but extending to 100–200 m in open ocean waters. In freshwater ecosystems it is subdivided into *littoral (shallow edge) and *limnetic zones. The term is occasionally used of the upper, well-lit strata of architecturally complex terrestrial ecosystems, such as rain forests. *Compare* APHOTIC ZONE; DYSPHOTIC ZONE; PHOTIC ZONE; and PROFUNDAL ZONE.

Euramerica The continental mass which resulted from the fusion of north-western Europe and North America during the Caledonian orogeny. This cratonic area subsequently fused with *Angara and *Gondwana during the Variscan orogenic event to form *Pangaea.

Eurosiberian floral region Part of R. Good's (1974, *The Geography of the Flowering Plants*) *Boreal Realm: an extensive region, in which the flora of the western part is richer than that of the eastern part. Whereas Europe has roughly 150 endemic genera (*see* ENDEMISM), Siberia apparently has only about 12, almost all of which are *monotypic. The contrast between the two subregions reflects primarily the different climatic regimes. *See also* FLORAL PROVINCE and FLORISTIC REGION.

eury- A prefix derived from the Greek *eurus*, meaning 'wide', which is used in ecology to describe species with a wide range of tolerance for a given environmental factor (e.g. 'euryecious', having a wide range of habitats). *Compare* STENO-.

Euryarchaeota **1.** In the widely used five-kingdom system for classifying organisms, a phylum within the subkingdom *Archaea in the *kingdom *Bacteria. **2.** In the three-domain classification, the more derived (*see* APOMORPH) of the two kingdoms (sometimes called subdomains) within the *domain Archaea. The Euryarchaeota contains a broad range of *phenotypes, including *methanogens, *halophiles, *hyperthermophiles, *acidophiles, and *sulphur-reducing organisms. Genetically, members of the Euryarchaeota are more different from members of the domains *Eukarya and *Eubacteria than are members of the *Crenarchaeota.

euryecious Having a wide range of habitats.

euryhaline Able to tolerate a wide range of degrees of salinity.

euryoxic Able to tolerate a wide range of concentrations of oxygen.

euryphagic Using a wide range of types of food.

eurythermal Able to tolerate a wide temperature range.

eurytopic Able to tolerate a wide range of a number of factors.

eusocial Highly social (e.g. termites, ants, honey bees, and, the only known example among mammals, naked mole rats (*Heterocephalus glaber*)).

eustatic Applied to the worldwide changes in sea level caused either by tectonic movements or by the growth or decay of glaciers (*glacioeustasy or glacioeustastism).

eutrophic Originally applied to nutrient rich waters with high *primary productivity but now also applied to soils. Typically, eutrophic lakes are shallow, with a dense plankton population and well-developed *littoral vegetation. The high organic content may mean that in summer, when there is stagnation caused by thermal stratification, oxygen supplies in the *hypolimnion become limiting for some fish species (e.g. trout). *Compare* OLIGOTROPHIC.

eutrophication The process of nutrient enrichment (usually by nitrates and phosphates) in aquatic ecosystems, such that the productivity of the system ceases to be limited by the availability of nutrients. It occurs naturally over geological time, but may be accelerated by human activities (e.g. sewage disposal or land drainage): such activities are sometimes termed 'cultural eutrophication'. The rapid increase in nutrient levels stimulates *algal blooms. On death, bacterial decomposition of the excess algae may deplete oxygen levels seriously. This is especially critical in thermally stratified lakes, since the decaying algal material typically sinks to the *hypolimnion where, in the short term, oxygen replenishment is impossible. The extremely low oxygen concentrations that result may lead to the death of fish, creating a further *oxygen demand, and so leading to further deaths.

euxinic Applied to an environment in which the circulation of water is restricted, leading to reduced oxygen levels or anaerobic conditions in the water. Such conditions may develop in swamps, barred basins, stratified lakes, and *fiords. Euxinic sediments are those deposited in such conditions, and are usually black and organic-rich.

evagination **1.** Turning inside out. **2.** The release of the contents of membranaceous vesicles to the exterior.

evanescent Temporary; soon disappearing.

evaporation *See* EVAPOTRANSPIRATION.

evaporation pan A broad, shallow, water-filled pan of standard size that is used to obtain an estimate of evaporation losses by monitoring the amount of water within it. The class A pan used by the US Weather Bureau is 122 cm in diameter and 25 cm deep.

evaporimeter *See* LYSIMETER.

evapotranspiration A combined term for water lost as vapour from a soil or open water surface (evaporation) and water lost from the surface of a plant, mainly via the stomata (*transpiration). The combined term is used since in practice it is very difficult to distinguish water vapour from these two sources in water-balance and atmospheric studies.

evapotron An instrument developed in Australia to measure the extent and direction of vertical air eddies which are involved in the vertical transfer of water vapour, and thus to provide a direct measurement of evaporation rates over short periods of time.

event deposit *See* STORM BED.

event recorder An instrument that is used to record animal behaviour. The operator has a keyboard, similar to that of a small electronic organ, which permits a very large number of signals to be produced by pressing keys or combinations of keys. The signals are transmitted to a recording tape from which they can be read by a suitably programmed computer.

event stratigraphy A term first proposed by D. V. Ager in 1973 for the recognition, study, and correlation of the effects of significant physical events (e.g. marine transgressions, volcanic eruptions, geomagnetic polarity reversals, climatic changes) or biological events (e.g. extinctions) on the stratigraphic record of whole continents or even of the entire globe. It is argued that by correlating these effects, as they are evidenced in the sedimentary record, it will be possible to define truly synchronous horizons, thus leading to greater resolution and a more accurate chronostratigraphic scale. In 1984 A. Seilacher suggested the term 'event stratinomy' for the study of events at the level of individual beds.

event stratinomy *See* EVENT STRATIGRAPHY.

Everglades The low, flat plains area of southern Florida, USA, which is subject to periodic freshwater flooding. In summer the area becomes swampy, but in winter it is extremely dry. *Cladium effusum* is widespread. Scattered trees include palms, and pines occur on higher ground. Within the same area small, isolated, subtropical rain-forest communities, locally termed hammocks, are found. It is the habitat of the swamp cypress (*Taxodium distichum*), which has conspicuous knee roots. The diverse animal population includes the manatee (*Trichecus manatus*), Florida panther (*Felis concolor coryii*), alligator (*Alligator mississipiensis*), red-shouldered hawk (*Buteo lineatus*), and water moccasin or cottonmouth snake (*Agkistrodon piscivorus*). The coastal areas are renowned for *mangrove forest.

evergreen Applied to a tree or shrub that has persistent leaves, and whose crown is never wholly bare. Although the entire plant remains green throughout the year, each leaf has a limited life span, but is physically tougher and usually longer-lived than a deciduous leaf. Evergreen leaves have the advantage that where nutrients such as nitrogen are in short supply their longer life span allows a more efficient use of the limited resource.

evergreen forest A forest in which there is no complete, seasonal loss of leaves (i.e. trees shed old leaves and produce new ones partially, and sometimes throughout the year, rather than during particular periods). The trees may be softwood conifers or broad-leaved *hardwoods. The distribution of these forests ranges through *boreal, middle, and tropical latitudes. The northern coniferous forests and the equatorial rain forests (*see* TROPICAL RAIN FORESTS) are the most extensive evergreen forests. Evergreen coniferous forests predominate where the growing season is less than half the year; *broad-leaved evergreen forest is found in regions that lack a prolonged dry season.

evergreen mixed forest A forest in which the dominant trees are both *evergreen broad-leaved hardwoods and conifers. Such forests are particularly well developed in the southern hemisphere, in South Africa, Tasmania, New Zealand, and Chile. In the northern hemisphere, eastern Asia and the Mediterranean basin once contained large tracts of such forest, but most of it has long been cleared.

eversible Capable of being everted (turned inside out).

evolution Change with continuity in successive generations of organisms (i.e. 'descent with modification', as *Darwin called it). The phenomenon is amply demonstrated by the fossil record, for the changes over geological time are sufficient to recognize distinct eras, for the most part with very different plants and animals. *See also* MACRO-EVOLUTION; MICROEVOLUTION; NATURAL SELECTION; PHYLETIC EVOLUTION; PHYLETIC GRADUALISM; PHYLOGENY; and PUNCTUATED EQUILIBRIUM.

evolutionary allometry *See* ALLOMETRY.

evolutionary determinism The change in gene frequencies by directed or deterministic processes, in contrast with change due to random or stochastic processes. The relative importance of the two kinds of change in evolutionary development is uncertain.

evolutionary lineage A line of descent of a taxon from its ancestral taxon. A lineage ultimately extends back through the various taxonomic levels, from the species to the genus, from the genus to the family, from the family to the order, etc.

evolutionary rate The amount of evolutionary change that occurs in a given unit of time. This is often difficult to determine, for several reasons: for example, should the unit of time be geological or biological (the number of generations)? How should morphological change in unrelated groups be compared? In practice it is necessary to adopt a pragmatic approach, such as the number of new genera per million years. *See also* DARWIN.

evolutionary species *See* CHRONOSPECIES.

evolutionary stable strategy (ESS) In the application of *game theory to evolutionary studies, moves in the game that cannot be beaten; i.e. traits or combinations of traits cannot be replaced by any invading mutant. Evolutionary stable strategy theory has proved very useful in the analysis of certain types of animal behaviour.

evolutionary trend A steady change in a given adaptive direction, either in an *evolutionary lineage or in a particular attribute (e.g. height of shoot). Such trends are often apparent in unrelated taxa. Formerly they were attributed to *orthogenesis; now *orthoselection or the contending theory of species selection are invoked. *See also* EFFECT HYPOTHESIS.

evolutionary zone *See* LINEAGE ZONE.

ex- The Latin *ex*, meaning 'out of', used as a prefix meaning 'out' or 'not having'.

exa- **(E)** Prefix used with *SI units to denote the unit $\times 10^{18}$.

exafference The stimulation of an animal as a result of factors external to the animal itself. *Compare* REAFFERENCE.

exaptation A characteristic that opens up a previously unavailable *niche to its possessor. The characteristic may have originated as an *adaptation to some other niche (e.g. it is proposed that feathers were an adaptation to thermoregulation, but opened up the possibility of flight to their possessors), or as a *neutral mutation.

excess baggage hypothesis The proposition that organs for which an organism no longer has any use (e.g. functional eyes in some cave-dwelling fish and excess brain tissue in some vertebrates) will be selected against because the energy cost of maintaining them reduces fitness (i.e. they are 'excess baggage'). The hypothesis also applies to genetically engineered organisms. The cost they incur by carrying inserted genes may reduce their fitness if they are released into the environment, so reducing the risk that release will prove environmentally harmful.

exchangeable ions Charged ions that are adsorbed on to sites, oppositely charged, on the surface of the *adsorption complex of the soil (mainly clay and humus colloids). Exchangeable ions can replace each other on this surface, and are also available to plants as nutrients. Although *cations (e.g. calcium and magnesium) are the most common, exchanging at negatively charged sites, some *anions (e.g. sulphate and phosphate) do exchange at positively charged sites. *See* ANION-EXCHANGE CAPACITY and CATION-EXCHANGE CAPACITY.

exchange capacity The total ionic charge of the *adsorption complex in the soil which is capable of adsorbing *cations or *anions.

exchange pool *See* ACTIVE POOL.

exclusion principle A principle introduced in 1934 by the Russian ecologist G. F.*Gause (1910–86), that two species cannot coexist in the same locality if they have identical ecological requirements.

exclusive species In *phytosociology, the optimum fidelity class 5 (*see* FAITHFUL SPECIES), i.e. a species that is confined completely or almost completely to a particular *community. *Compare* ACCIDENTAL SPECIES; INDIFFERENT SPECIES; PREFERENTIAL SPECIES; and SELECTIVE SPECIES.

excurrent Of, for example, a leaf: having a midrib that projects beyond the tip.

exhumed topography An ancient land-form or landscape that had been buried beneath younger rocks or sediment and that is exposed by their subsequent *erosion.

exine The outer, decay-resistant coat of a *pollen grain, composed of sporopollenin, an inert polymer (possibly the toughest of all polymers of biological origin). The exine is characteristic for different plant families and genera, and sometimes even for different species. Hence it forms the basis for the identification and quantitative analysis of the vegetation composition of *peats and other suitable sedimentary deposits dating back many thousands of years. *See also* PALYNOLOGY.

exogenetic Applied to the various processes of erosion, transport, and deposition that take place at the Earth's surface. Usually the term is used in contrast with the 'endogenetic' or internal mechanisms.

exon That part of a gene which appears in the mature RNA transcript.

exon insertion The incorporation in one gene of an *exon from another.

exon shuffling Exon *duplication and *exon insertion.

exopterygote Applied to an insect in which the wings develop externally and gradually. The insect undergoes no pupal stage or rapid metamorphosis, and its young are called 'nymphs'.

exorheic lake A lake that has one or more outflow streams. *Compare* ENDORHEIC LAKE.

exosphere The outer region of the upper atmosphere extending from a base of 500–750 km altitude. The zone has a very low concentration of gases (mostly molecules of oxygen, hydrogen, and helium, although some are ionized). Gases can escape from it into space, as molecular collisions are much reduced because of the low gas density. The exosphere and much of the underlying *ionosphere form part of the *magnetosphere.

exosymbiont A *symbiont that lives on the exterior of the body of its host. *Compare* ENDOSYMBIONT.

exotherm *See* POIKILOTHERM.

exotic species An introduced, non-native species.

exotoxin A toxin that is secreted by a living organism.

experimental error An error that arises because of variation between experimental samples. It may be attributed to differences in materials and/or techniques, rather than to real differences (e.g. in growth). Experimental error must be monitored in any statistical comparison of experimental data.

exploitation competition (exploitative competition) *Competition occurs where two species require the same resource and that resource is in short supply. Whichever of the two species is more efficient in accessing the resource is more likely to succeed. This is known as exploitation competition, the alternative being *interference competition.

exploitative competition *See* EXPLOITATION COMPETITION.

exploratory behaviour A form of *appetitive behaviour that may be goal-orientated (e.g. the search for food or nesting material) or concerned with the examination of areas or articles with which an animal is unfamiliar, in which case the behaviour often exhibits signs of *conflict.

exploratory learning A form of *latent learning in which an animal learns by exploring new surroundings, without apparent reward or punishment.

explosive evolution A sudden diversification or *adaptive radiation of a group. The term is an old one, rarely used nowadays. Phases of explosive evolution have occurred in all the higher taxonomic groups, i.e. in genera, families, orders, and classes. An example at the ordinal level is the diversification of the mammals at the dawn of the *Cenozoic Era.

exponential A mathematical function that varies as the power of a particular quantity (i.e. a rate of change (increase or decrease) that is calculated as a fixed percentage of the starting value and, therefore, in which the logarithms of a value change linearly in any given period of time). The amount of change in a particular period is calculated as the starting value plus the interest accrued in preceding periods, multiplied by the rate of change (e.g. for a quantity, starting at 100, that grows exponentially by 10 per cent in each unit of time: 100, 110 (100 + 10 per cent of 100), 121 (110 + 10 per cent of 110), 132.1 (121 + 10 per cent of 121), . . .). Fears have been expressed (e.g. by some environmentalists) that the size of the human population may increase exponentially and the demands such growth imposes on essential resources may cause stocks of these to be depleted exponentially. In biological (and economic) systems, however, exponential rates of change are never sustained (*see* J-SHAPED GROWTH CURVE and S-SHAPED GROWTH CURVE).

expressivity The degree to which a particular *genotype is expressed in the *phenotype.

exserted Protruding (e.g. of stamens projecting beyond a corolla, or of a moss capsule projecting beyond the leaves of the gametophyte).

exsiccation Dehydration of an area by a process (e.g. drainage) in the absence of changes in precipitation levels. Draining of marshlands and deforestation are examples of processes that can lead to exsiccation. *See also* DESICCATION.

extant Applied to a *taxon some of whose members are living at the present time. *Compare* EXTINCT.

extinct Applied to a *taxon no member of which is living at the present time. *Compare* EXTANT.

extinction **1.** The elimination of a *taxon. The term can be used of the local loss of a species or a population. This may take place in several ways. In the simplest case the taxon disappears from the record and is not replaced. Alternatively, one taxon may replace another, the earlier group consequently disappearing. Thus there is a process of either subtraction or substitution. Extinction generally takes place at particular times and places, but there are recurring periods when episodes of mass extinction have taken place. Environmental catastrophe, occurring for whatever reason, removes many groups from the environment and ecosystems collapse. Eventually new forms appear and evolution resumes. It would appear that periods of mass extinction control the pattern of evolution. **2.** The loss of learned patterns of behaviour from the repertoire of an animal to which they have become irrelevant.

extinction point The lowest percentage of the intensity of full daylight in which a given species occurs naturally.

extirpation The bringing of a species to *extinction within a part of its range.

extraction *See* ABSTRACTION.

extremophile A micro-organism (domain *Archaea) that thrives under extreme environmental conditions of temperature, pH, or salinity. *See also* ACIDOPHILE, ALKALIPHILE, HALOPHILE, HYPERTHERMOPHILE, PSYCHROPHILE, and THERMOPHILE.

extremozyme One of a range of enzymes, present in *extremophiles, that continue to function at temperatures, salinities, acidities, or alkalinities at which other enzymes would fail.

'eye' of storm The central part of a tropical cyclone, where winds are light, skies are generally clear, and there is a slight, horizontal pressure gradient. The diameter of the 'eye' averages 20 km but in a large cyclone can be 40 km or more. The 'eye' is an area of some air subsidence which produces *adiabatic warming.

eyot *See* BRAIDED STREAM.

facies **1.** Aspect or appearance. **2.** The sum total of the features that reflect the specific environmental conditions under which a given rock was formed or deposited.

facies association A group of sedimentary *facies that are used to define a particular sedimentary environment. For example, all the facies found in a *fluviatile environment may be grouped together to define a fluvial facies association.

facies fossil A *fossil organism that is restricted to a particular rock, reflecting the original environment of deposition.

facilitation **1.** The intensification of a behaviour that is caused by the presence of another animal of the same species. **2.** The process whereby a plant so modifies a *habitat as to allow other species to invade (as in a *succession). For example, the grass *Elytrigia juncea.* facilitates the invasion of *Ammophila arenaria* (marram grass) in the early stages of sand-dune formation. If the invader is more efficient in the new circumstances, this may lead to the decline and eventual exclusion of the facilitating species. *Compare* INHIBITION.

factor **1.** In statistics, one of a pair or series of numbers which yield a given product when multiplied together. **2. (ecological factor)** *See* LIMITING FACTOR.

facultative Applied to organisms that are able to adopt an alternative mode of living. For example, a facultative anaerobe is an aerobic organism that can also grow under anaerobic conditions.

facultative mutualism *See* PROTOCOOPERATION.

facultative parasite *See* PARASITISM.

Fahrenholz's rule The principle that the phylogenies of parasites and their hosts generally evolve in parallel.

fairy ring A circle of dark-green grass (in a lawn or field) in which toadstools may be found. The circle is formed as a result of the radial growth of a fungus through the soil, away from the centre of the ring; as the fungal mycelium grows it deprives grass roots of nutrients, but as it dies and decomposes the release of nutrients stimulates the growth of the grass, producing the dark coloration. Fairy rings are often formed by *Marasmius oreades*.

faithful species **(fidelity)** In *phytosociology, a species confined, or nearly so, to a particular association. The species may or may not be constant to the association fidelity, and therefore faithful or characteristic species can be determined satisfactorily only if a full knowledge of other *community types is available. Five fidelity classes are commonly distinguished: (*a*) *accidental; (*b*) *indifferent; (*c*) *preferential; (*d*) *selective; (*e*) *exclusive. *See also* COMPANION SPECIES; KENNARTEN SPECIES; and PLANT ASSOCIATION.

falcate Curved, as a sickle.

falcato-secund Curved to one side.

falling head permeameter *See* PERMEAMETER.

fall-stripes *See* VIRGA.

false cirrus *See* SPISSATUS

false colour In remote sensing techniques, the display of data collected in a number of different wavelengths, usually longer or shorter than those perceptible to the naked eye. Typical false-colour images include infrared data, which are often displayed as visible red. Thus green vegetation, which is highly reflective in the infrared, typically appears red on a false-colour image.

false rings *See* TREE RING.

Famennian *See* DEVONIAN.

FAO The Food and Agriculture Organization of the United Nations.

farinose Floury or powdery in appearance or texture.

fasciation A malformation in plants in which shoots tend to be thick and flattened and may occur in masses. Fasciation is sometimes caused by infection with the bacterium *Rhodococcus fascians*.

fast breeder reactor A nuclear reactor that uses fast neutrons to convert uranium to plutonium and which creates more fuel than it uses. A chain reaction is set up in which ^{238}U, with an initial charge of ^{239}Pu to begin the fission process, discards fast neutrons, each of which induces fission in another nucleus. Excess ^{239}Pu is produced where plutonium production exceeds rate of fission. In fast breeder reactors, about 60 per cent of the fissionable material in the fuel elements is converted to useful energy compared to 0.5–1 per cent in burner reactors. Plutonium can be recycled, but other radioactive products are waste. *See* RADIOACTIVE WASTE.

fathom A unit of water-depth measurement, originally 6 feet, equal to 1.83 m.

fatigue *See* ACCOMMODATION.

fauna **(adj. faunal, faunistic)** The animal life of a region or geological period.

faunal *See* FAUNA.

faunal province *See* FAUNAL REGION.

faunal region **(faunal province, faunal realm, faunal zoogeographic kingdom)** A biological division of the Earth's surface (i.e. a large geographical area) that contains a fauna more or less peculiar to it. The degree of distinctiveness varies with the region concerned and reflects partly climate and partly the existence of barriers to migration. The number of regions recognized varies from one authority to another, but a minimum of six are recognized: *Australian, *Ethiopian, *Nearctic, *Neotropical, *Oriental, and *Palaearctic. In defining a region, greatest emphasis is given to mammals.

faunal succession The principle, first recognized at the beginning of the nineteenth century by William Smith (1769–1839), that different strata each contain particular assemblages of *fossils by which the rocks may be identified and correlated over long distances; and that these fossil forms succeed one another in a definite and habitual order. This law, together with the law of superposition of strata (i.e. that sedimentary strata are deposited sequentially, so that in an undisturbed sequence each stratum is younger than the one beneath it), enables the relative age of a rock to be deduced from its content of fossil faunas and floras.

faunal zoogeographic kingdom *See* FAUNAL REGION.

faunistic *See* FAUNA.

faunizone *See* ASSEMBLAGE ZONE.

Fe *See* IRON.

fecundity *See* FERTILITY.

feedback loop **(feedback mechanism)** A control device in a system. Homoeostatic systems have numerous negative-feedback mechanisms which tend to counterbalance positive changes and so maintain stability. For example, denitrifying bacteria counteract the effects of nitrogen-fixing bacteria. Positive feedback reinforces change and in natural systems may result in radical environmental alteration. For example, an exceptionally cool summer in high mid-latitudes of the northern hemisphere may impede the melting of snow, leading to unusually high albedo, which reduces absorption of solar energy, leading to further cooling, etc.

feedback mechanism *See* FEEDBACK LOOP.

feedback regulation The process by which the product of a *metabolic pathway influences its own production by controlling the amount and/or activity of one or more enzymes involved in the pathway. Normally this influence is inhibitory.

feeding All behaviour that involves the obtaining, manipulation, and ingestion of food. *Compare* FORAGING.

fell An area of open mountainside with low-lying vegetation. The word is derived from the Old Norse *fiall*, meaning 'hill', and survives in a number of place names in northern England, probably as a result of Viking settlement.

fell field An area within the *tundra belt of frost-shattered stony debris with interstitial fine particles, which supports various plant species in a mixed *community. The vegetation is sparse, however, and typically occupies less than half the ground. Frequently fell fields display patterned-ground phenomena, due to freeze-thaw activity in the soil.

femto- From the Danish and Norwegian *femten*, meaning 'fifteen', a prefix (symbol f) used with *SI units to denote the unit $\times 10^{-15}$.

fen An area of wetland vegetation that receives its water by both rainfall and groundwater flow (rheotrophic), and in which the summer *water-table is at or below the surface of the sediment. Fens are peat-forming *ecosystems (mires) and can be divided into rich fens, which are usually neutral to acid in their reaction and are often fed by streams draining *limestone rocks, or poor fens, in which case their water is slightly acid and the concentration of dissolved nutrients is low. Some fens are erroneously given the label *bog (e.g. string bogs and valley bogs), a term that should be reserved for *ombrotrophic mires.

fenestrae **1.** Pits found on the surfaces of certain pollen grains. **2.** Irregular cavities found in muddy *intertidal to *supratidal carbonate sediments. They take a number of forms: birdseye fenestrae (irregular, 'birdseye'-shaped cavities, usually 1–5 mm across, formed by gas entrapment in the sediment); laminoid fenestrae (long, thin cavities, parallel to the sediment laminae, formed particularly in algal, laminated muds, and produced by the decay of organic material); and tubular fenestrae (cylindrical, near vertical tubes, formed by burrowing organisms or plant rootlets). Fenestral cavities may become filled with sparry calcite (sparite). If they remain unfilled, the fenestrae are responsible for the development of fenestral porosity in the sediment. The term comes from *fenestra* (pl. *fenestrae*), the Latin for an opening or window.

fenestrate A term describing a form of sculpturing on certain pollen grains (e.g. the Lactucae tribe of the Asteraceae) which have *fenestrae.

fenestrated Perforated with small openings or transparent areas.

fen peat *See* PEAT.

Fennoscandia *See* BALTIC SHIELD.

feral From the Latin *ferus*, meaning 'wild', an adjective applied to a wild or undomesticated organism. In particular, the term is applied to wild strains of an otherwise domesticated species or to an organism that has reverted to a wild condition following escape from captivity, etc. Some authors make these distinctions: wild species, subject to natural selection only; domestic species, subject to selection by humans; and feral species, formerly domestic species which are now, as escapees, subject once again to natural selection.

ferrallization Part of the *leaching process found in tropical soils, by which large amounts of iron and aluminium oxides accumulate in the B *horizon of such soils as *red podzolic soils.

Ferralsols Soils that have an iron-rich B *soil horizon with distinctive red mottling that is more than 15 cm deep. This horizon is highly weathered (*see* WEATHERING) and contains high concentrations of iron and aluminium. Ferralsols are a reference soil group in the *FAO *soil classification.

ferricrete *See* DURICRUST.

ferruginous Resembling rust, containing rust, or rust-coloured.

fertility **1. (fecundity, fruitfulness)** The reproductive capacity of an organism, i.e. the number of eggs that develop in a mated female over a specified period. It is usually calculated at the stage when this number is readily observable (i.e. in *oviparous animals when eggs are laid and in *viviparous animals when young are born), although strictly speaking it applies from the time that fertilization occurs. Sometimes the term 'fertility' is applied only to the production of fertilized eggs (ova), while 'fecundity' is used for the production of offspring, so excluding those embryos which fail to develop. **2.** The condition of a soil relative to the amount and availability to plants of elements necessary for plant growth. Soil fertility is affected by physical elements (e.g. supply of moisture and oxygen) as well as by the supply of chemical plant nutrients.

fertilization **1.** The union of two *gametes to produce a zygote, which occurs during sexual reproduction. Fertilization involves the fusion of two haploid nuclei containing genetic material from two distinct individuals (cross-fertilization) or from one individual (self-fertilization). The resulting zygote then develops into a new individual. Most aquatic animals, e.g. echinoderms, achieve fertilization externally, gametes uniting outside the body of the parents. Some other animals, particularly terrestrial species, have internal fertilization, with the union of gametes inside the female. Some lower plants (e.g. mosses) release their male gametes externally, which then swim like spermatozoa to the female gamete. Most higher plants have the male gamete released internally from the *pollen grain directly to the female gamete and *double fertilization occurs. **2.** The application of plant nutrients (i.e. fertilizers) to land in order to promote the growth of desired plants.

fetch **1.** The length of water surface over which the wind blows in generating waves. Together with wind velocity and duration, this determines wave height. Many features of coastal deposition tend to become orientated normally to the direction of maximum fetch. **2.** The distance over which an airstream has travelled across sea or ocean.

Fibonacci series A mathematical series that is obtained by adding the two previous elements of the series, as in 1, 2, 3, 5, 8, 13, 21, etc. The interlocking spirals found in many plant structures, such as the fruit arrangement in a sunflower head, or the sections on a pineapple, are described by this series. The series was discovered by the Italian mathematician Leonardo Fibonacci (*c.*1180–*c.*1250), who was also influential in introducing Indo-Arabic numerals to Europe. Fibonacci lived in Pisa.

fibratus The Latin *fibratus*, meaning 'fibrous', used to describe a species of separate cloud or cloud veil which has rather curved elements, but without hooks. *See also* CLOUD CLASSIFICATION.

fibril A cloud trail observed in *cumulonimbus, where *drizzle-sized droplets are large enough for their terminal velocities to allow them to depart from the main cloud body.

fidelity *See* FAITHFUL SPECIES.

fiducial point **1.** The temperature at which the atmospheric-pressure scale of a particular barometer reads correctly. The temperature at which this is so in latitude 45° is called the standard temperature of that barometer. At other temperatures and other latitudes, corrections must be applied. **2.** The fixed point (indicated by a pointer) that is the zero of the scale of a *Fortin barometer.

field capacity Water that remains in soil after excess moisture has drained freely from that soil. Usually it is measured as a percentage of the weight or volume of oven-dry soil.

field layer The herb and small shrub layer of a plant *community. *Compare* CANOPY; GROUND LAYER; and SHRUB LAYER.

filiform Thread-like; long and slender.

filling In *synoptic meteorology, an increase in pressure at the centre of a *depression.

film water *See* PELLICULAR WATER.

filter route A term introduced by the American palaeontologist G. G. *Simpson in 1940 to describe a faunal migration route across which the spread of some animals is very likely, but the spread of others is correspondingly improbable. The route thus filters out part of the fauna, but permits the rest to pass. *Deserts and mountain ranges provide examples of filter routes.

finger and toe *See* CLUBROOT.

finite resource (non-renewable resource) A resource that is concentrated or formed at a rate very much slower than its rate of consumption and so, for all practical purposes, is non-renewable. *Compare* RENEWABLE RESOURCE.

fiord (fjord) A long, narrow, deep, U-shaped coastal inlet which usually represents the seaward end of a glaciated valley that has been partially submerged. The water depths often exceed 1 000 m except near the mouth, where a bar or sill may be present.

fire-blight An important disease which affects many trees of the Rosaceae (e.g. apple, pear, and hawthorn). Infected trees have the appearance of having been scorched by fire; the tree may be killed within months. The causal agent is the bacterium *Erwinia amylovora*. The disease occurs in North America and in many parts of Europe; it is notifiable in Britain.

fire climax (pyroclimax) A *climax community for which fire is the dominating control factor, as in the long-leaf pine forests of the USA. Fire is also thought to be an important determining factor, interacting in complex ways with grazing and trampling pressure, in the formation of the major *grassland areas, e.g. *prairies and *savannah. Fire is also a major influence, with grazing, on the development and maintenance of *heathland communities.

fire scar A scar that is often found in the annual rings of a tree that has been subjected to fire but has survived. The scar results from fire damage and is visible in a cross section through the trunk. Fire scars make it possible to assign dates to past fires and to calculate fire frequencies.

firn (névé) Snow that has survived a summer melting season. It is an intermediate material in the conversion of snow to glacial ice. Normally it is granular, owing to the partial melt. The obsolete German word *firn* means 'of last year'; *Firner* means 'everlasting snow' or 'snow that remains from last year'.

firn limit *See* FIRN LINE.

firn line (annual snow line, firn limit) A line on a *glacier that marks the upper limit to which winter snowfall melts during the summer *ablation season. It is often clearly marked, and on many glaciers separates hard, blue ice below from snow above. *See* EQUILIBRIUM LINE.

firn wind (glacier wind) A downhill airflow which develops over a *glacier during the day, usually in summer. The greater air density over the glacier than over the surrounding surfaces causes this air to sink.

fish Broadly speaking, any *poikilothermic, legless, aquatic vertebrate that possesses a series of gills on each side of the pharynx, a two-chambered heart, no internal nostrils, and at least a median fin as well as a tail fin. If the lampreys and hagfish (Agnatha) are excluded, this definition includes the sharks and rays (Chrondrichthyes), in which the skeleton is cartilaginous, as well as the bony fish (Osteichthyes). Some consider, however, that only the bony fish should be classed as real fish.

Fisher, Sir Ronald Aylmer (1890–1962) The British statistician and geneticist who demonstrated mathematically that *Mendel's laws must lead to the results observed and that these laws accorded well with *Darwins's theory of evolution by natural selection. He pioneered the application of statistical methods to data derived from genetic and ecological studies. He worked at Rothamsted Experimental Station from 1919 until his appointment as professor of genetics at the University of Cambridge in 1943. In 1959 he joined the Commonwealth Scientific and Industrial Research Organization (CSIRO) in Australia.

Fisher's fundamental theorem The principle that the increase in fitness over one generation is equal to the additive genetic variance in fitness during the same time. Selection maximizes fitness, thus reducing differences in fitness and tending to reduce the heritability of traits connected to fitness (although there are exceptions).

fish lice Members of two orders of Copepoda, the Caligoida and Lernaeopodida, which are highly modified ectoparasites of marine and freshwater fish and whales.

fission **1.** The asexual reproduction of a cell by division. **2.** The splitting of a heavy atomic nucleus by collision, with the ejection of two or more neutrons, and the release of energy.

fitness **1.** In ecology, the extent to which an organism is well adapted to its environment. The fitness of an individual animal is a measure of its ability, relative to others, to leave viable offspring. **2.** **(Darwinian fitness)** *See* ADAPTIVE VALUE.

fixation **1.** A soil process by which certain nutrient chemicals required by plants are changed from a soluble and available form into a much less soluble and almost unavailable form. **2.** The first step in making permanent preparations of tissues for microscopic study, by killing cells and preventing subsequent decay with as little distortion of structure as possible. Examples of fixatives are formaldehyde and osmium tetroxide, often used as mixtures. **3.** A term applied to 100 per cent gene frequencies when all members of a population are homozygous for a particular *allele at a given *locus (i.e. there is no *polymorphism). **4.** A biological process by which inorganic molecules can be incorporated into organic molecules within living organisms. In *photosynthesis the green plant fixes carbon dioxide from the atmosphere, and several microbes are capable of fixing atmospheric nitrogen into amino acids and hence proteins (*see* NITROGEN FIXATION).

fixation probability The probability that a particular *allele will become fixed within a population.

fixation time The time taken for a new *allele to become fixed within a population.

fixed-action pattern Apparently stereotyped behaviour, exhibited or capable of being exhibited by all members of a species or higher taxonomic group, which may be used to achieve more than one objective, which may be innate or learned, and whose acquisition may be affected by environmental factors. Examples include the calls of certain birds: these are influenced by sounds heard by the birds early in their lives, but once acquired the songs are performed in a stereotyped way.

fjord *See* FIORD.

flabelliform Shaped like a fan.

flaccid Wilted (*see* WILTING). The state of cells when they are short of water.

flacherie A disease of the silkworm (*Bombyx mori*), which is apparently caused by a virus.

Fladbury A site in Hereford and Worcester, England, where palaeontological and palaeozoological evidence suggests there was a barren, tundra-like landscape during a cold period following the *Upton Warren interstadial.

flagellates *See* PROTOZOA.

Flandrian A local geological term used in the British Isles to describe the current interglacial, starting about 10 000 years ago. Most geologists regard the Flandrian as a term equivalent to the *Holocene Epoch, representing the second epoch of the *Quaternary Period, following the *Pleistocene Epoch. It can reasonably be argued, however, that the current interglacial is not sufficiently distinct from previous interglacials within the Pleistocene to merit an elevation to the level of epoch. On the other hand the dominance and impact of the human species in this interglacial could be regarded as a reason for such a distinction. In Europe the warmest *stage occurred during Atlantic times, about 6 000 BP, when warmth-loving trees dominated the landscape (the Hypsithermal is the equivalent North American *climatic optimum). No consensus view exists as to when the ice will advance again, bringing extremely cold conditions to high mid-latitudes, nor as to how quickly these conditions may arise.

flanking sequence The untranscribed regions at the 5′ or 3′ ends of a transcribed gene.

flaser bedding A form of bedding, characterized by cross-laminations draped with silt or clay, which forms in environments where flow strengths fluctuate considerably, thus permitting the transport of sand in ripples, followed by low-energy periods when mud can drape the ripples.

flash flood A brief but powerful surge of water that flows either over a surface ('sheet flood') or down a normally dry stream channel ('stream flood'). Usually it is caused by heavy convectional rainfall of short duration, and is typical of semi-arid and *desert environments.

flehmen Behaviour, probably with a sexual significance, exhibited by felids, goats, and some other mammals when examining a scent mark (*see* SCENT MARKING). Having sniffed the mark intensively, the animal raises its head with its mouth partly open and upper lip slightly drawn back, stares fixedly to its front, and breathes slowly. Sniffing followed by flehmen may occur more than once at a single encounter with a scent mark.

flexuous Wavy or bending in a zigzag manner; the term is usually applied to a stem or other axis.

floating chronology A tree-ring chronology for a particular area that does not overlap with chronologies from living trees in that area, and therefore cannot be dated absolutely.

floccose Having a loose, woolly, or cottony surface.

flocculation A process in which *clay and other soil particles adhere to form larger groupings or *aggregates, thereby coarsening the soil texture and making heavier soils easier to cultivate. The reverse of this process is known as *dispersion.

floccus The Latin *floccus*, meaning 'tuft', used to describe filaments or hairs borne in tufts and having a woolly appearance. In meteorology, a species of cloud with a tufted appearance, the lower parts of which are rather ragged, often with *virga. The species is most associated with *cirrus, *cirrocumulus, and *altocumulus. *See also* CLOUD CLASSIFICATION.

flocking Among birds, the formation of a group with a social organization. In some species flocking is accompanied by communal nesting and roosting, in others it occurs only outside the breeding season. About half of all known species of birds exhibit flocking behaviour at some stage in their life cycles.

flood forecasting A technique which uses the known characteristics of a river basin to predict the timing, discharge, and height of flood peaks resulting from a measured rainfall, usually with the objective of warning populations who may be endangered by the flood. *Compare* FLOOD PREDICTION.

flood-plain The part of a river valley that is made of unconsolidated, river-borne sediment and is periodically flooded. It is built up of relatively coarse debris left behind as a stream channel migrates laterally, and of relatively fine sediment deposited when *bankfull flow discharge is exceeded.

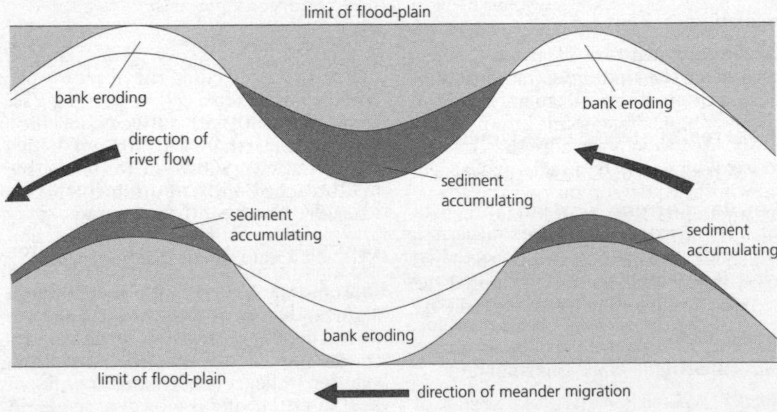

Flood-plain

flood prediction The study of rainfall patterns, catchment characteristics, and river hydrographs to predict the future average frequency of occurrence of flood events. Flood predictions seek to estimate the probable discharge which, on average, will be exceeded only once in any particular period, hence the use of such terms as '50-year flood' and '100-year flood'. *Compare* FLOOD FORECASTING.

flood tide The rising tide: the phase of the tide between low water and the next high tide. *Compare* EBB TIDE.

flood zone *See* ACME ZONE.

flora **(adj. floral, floristic)** All the plant species that make up the vegetation of a given area. The term is also applied to assemblages of fossil plants from a particular geological time, or from a geographical region in a former geological time. Examples of all three types of usage, respectively, are: British flora, *Carboniferous flora, and *Gondwana flora.

floral *See* FLORA.

floral kingdom *See* FLORAL PROVINCE.

floral province **(floral kingdom, floral realm, floral region)** A major geographical grouping of plants, especially flowering plants, identified on the basis of distinctiveness, particularly with regard to the degree of *endemism at family and generic level. Some authorities distinguish between kingdoms or realms, which are accorded the highest status, and provinces, which have lesser status and go to make up the kingdoms. *Compare* FLORISTIC REGION.

floral realm *See* FLORAL PROVINCE.

floral region *See* FLORAL PROVINCE.

Florida Current Part of the *Gulf Stream that extends from the southern tip of Florida to Cape Hatteras, North Carolina. It is a fast-flowing (1–3 m/s), narrow (50–75 km wide), and deep current, still evident at depths of 2 000 m, where velocities of up to 10 cm/s have been measured. It is an example of a western *boundary current.

floristic *See* FLORA.

floristic region The spatial categorization of plants recognized in terms of floristic composition. The term is often used synonymously with *floral province, kingdom, or realm, depending on the authority.

floristics *See* PHYTOGEOGRAPHY.

flow till Sediments which flow after they have been deposited by *ablation. *See* TILL.

flume **1.** A short section of artificial channel constructed in a river in order to create a constriction in which *critical flow will be established, allowing the discharge to be calculated from the water depth. **2.** An experimental channel used for studying relationships between sediment movement and flow conditions. There are a number of different flume designs, but most flumes are capable of carrying water at variable depths and velocities, either in a unidirectional flow or generating waves. Flume studies have been responsible for establishing important relationships between the grain size and *erosion velocities and stability fields for the various sediment bed-forms.

fluorescence A kind of luminescence, in which an atom or molecule emits radiation when electrons within it pass back from a higher to their former, lower energy state. The term is restricted to the phenomenon in cases where the interval between absorption and emission is very short (less than 10^{-3} s).

fluorescent-antibody technique A method for detecting the location of a specific antigen in a cell by staining a section of the tissue with antibody (specific to that antigen) that is combined with a fluorochrome (a substance that fluoresces in ultraviolet light). The antigen is located wherever fluorescence is observed.

fluorosis A disease that results from the ingestion of fluorine in amounts that substantially exceed bodily requirements. Fluorine is involved in the formation of teeth and bones; in excess, it leads to the thickening of bones, sometimes to such an extent that joints stiffen and the increased weight of the skeleton makes it difficult for the head to be raised, and the softening and staining of teeth. The complaint has been recorded among ruminants grazing near factories (e.g. brick works and aluminium smelters) emitting fluorine salts (fluorides), which are deposited on pasture, where they may stimulate luxuriant growth. Fluorides are highly soluble in water and the pollution of pasture occurs only immediately downwind of the source.

f

flute cast *See* FLUTE MARK.

fluted moraine A ground *moraine surface which shows streamlined ridges and grooves trending at right angles to the ice front. Individual ridges are generally less than 1 km long and less than 10 m high. They may form as a consequence of high vertical ice pressure forcing a *subglacial, plastic *till up into the low-pressure zone down-glacier of a large boulder.

flute mark A tongue-shaped scour cut into mud by a turbulent flow of water. The tongue is deepest at the up-current end and the flute can thus be used in a *palaeocurrent analysis. If the flute is infilled by sediment a flute cast will be preserved in the base of the overlying bed. Although once believed to be diagnostic of *turbidite deposition, flutes can form in any setting where water flows strongly over soft mud.

fluvial Pertaining to a river.

fluvial processes The set of mechanisms that operate as a result of water flow within (and at times beyond) a stream channel, bringing about the *erosion, transfer, and deposition of sediment. The erosional processes include: the displacement of bed particles through drag and lift forces; abrasion, causing the wearing away of bed and banks as mobile sediment is dragged against them; and bank collapse, a consequence of hydraulic activity. Transport processes include the transfer of material in solution and suspension, and by *saltation. Depositional processes act when the immersed weight of a particle is greater than the force driving it down-channel.

fluviatile Applied to sediments of fluvial (river) origin.

fluvic horizon A *soil horizon that is dark in colour and usually rich in *pyroclastic material. The name is from the Latin *fluvius*, meaning river.

Fluvisols Soils that have formed on recent *alluvial deposits and have a *fluvic horizon extending from 25 cm below the surface to a depth exceeding 50 cm. Fluvisols are a reference soil group in the *FAO *soil classification.

flux study An examination of the productivity and *respiration of an entire *ecosystem by tracing changes in atmospheric gas (usually carbon dioxide) concentrations. Such studies are complicated by turbulence in the atmosphere and require the construction of detailed computer models of gas movements in order to calculate changes in gas levels. *See* AERODYNAMIC METHOD.

foehn wall *See* FÖHN WALL.

foehn wind *See* FÖHN WIND.

fog A condition of atmospheric obscurity near the ground surface, caused by the suspension of minute water droplets in conditions of near saturation of the air. The formation of fog is aided by any concentration of smoke particles, as these act as condensation nuclei and may cause fog at levels of humidity below saturation point. Visibility is below 1 km. The cause of condensation of the water droplets may be radiation cooling of the ground, advection of warm air over a cold ocean or cold ground, or conditions at a *front. *See also* SMOG.

föhn wall **(foehn wall)** A mass of cap cloud and associated precipitation over windward slopes and parts of leeward slopes on mountain barriers, resulting from the föhn effect (*see* FÖHN WIND).

föhn wind **(foehn wind)** A generic term for warm, dry winds in the lee of a mountain range. It was originally used in the European Alps. After cooling on the windward ascent at the *saturated adiabatic lapse rate of 0.5 °C/100 m, with the resulting condensation and precipitation, the air descending on the leeward side of the mountain range is warmed through compression at the *dry adiabatic lapse rate of 1 °C/100 m. This produces a warming wind on the lee side, with higher temperatures than occurred in the same air on the upslope side of the mountains.

folicolous *See* EPIPHYLLOUS.

foliose Leaf-like; made up of thin flat lobes (e.g. the thalli of certain types of lichen).

Fomes annosus **(*Heterobasidion annosum*)** *See* BRACKET FUNGUS.

food begging Juvenile behaviour that may endure into adulthood (e.g. forming part of some *courtship rituals), in which young that are dependent on their parents for food stimulate adults to feed them.

food-chain The transfer of energy from the *primary producers (green plants) through a series of organisms that eat and are eaten, assuming that each organism feeds on only one other type of organism (e.g. earthworm → blackbird → sparrowhawk). At each stage much energy is lost as heat, a fact that usually limits the number of steps (*trophic levels) in the chain to four or five. Two basic types of food-chain are recognized: the *grazing and *detrital pathways. In practice these interact to give a complex food-web.

food-chain efficiency The ratio between the energy value (the nutritional value, discounting indigestible parts such as hair or feathers) of prey consumed by a *predator and the energy value of the food eaten by that prey. Maximum food-chain efficiency (gross ecological efficiency) occurs when the yield of prey to the predator is such that the surviving prey just consume all the available food: this implies that the food of the prey is being exploited to the best advantage by the predator.

food-web A diagram that represents the feeding relationships of organisms within an *ecosystem. It consists of a series of interconnecting *food-chains. Only some of the many possible relationships can be shown in such a diagram and it is usual to include only one or two carnivores at the highest levels. One complexity that is difficult to represent in a food-web diagram arises from the fact that even the highest carnivore may die and enter the detritus food chain and hence donate energy once more to lower trophic levels.

foolish seedling disease *See* BAKANAE DISEASE.

foraging **1.** All behaviour that is associated with the obtaining and consumption of food for which the animal must search or hunt. *Compare* FEEDING. **2.** Botanists apply the term to plants that extend over new areas of soil by means of runners, and periodically take root when conditions are appropriate. It has been shown in some plants that they extend their runners faster when crossing areas of low nutrient content and slow down (and take root) when the runners encounter richer soils.

forb A non-grassy, herbaceous species (e.g. legumes and composites).

Forbes, Edward **(1815–54)** A British naturalist who studied medicine but abandoned it for natural history following the publication in the *Magazine of Natural History* (1835–36) of the botanical results of a tour of Norway he made in 1833. He was appointed curator of the museum of the Geological Society of London (1842), professor of botany at King's College, London (1843), palaeontologist to the Geological Survey of Great Britain (1844), professor of natural history to the Royal School of Mines (1851), and professor of natural history at the University of Edinburgh (1854). It was in the *Memoirs* of the Geological Survey that, in 1846, he published his major contribution to the study of plant geography. In 'On the Connexion between the Distribution of the Existing Fauna and Flora of the British Isles, and the Geological Changes which have Affected their Area' he proposed that British plants may be considered as five distinguishable groups, most of which entered by migrating across land before, during, and after the Glacial (i.e. *Pleistocene) Epoch.

forced convection Mechanical turbulence, with the development of eddies, in air flowing over an uneven surface.

fore-dune In a coastal sand-dune system, the dune nearest to the sea; it is characterized by plants that can tolerate occasional flooding by sea water (e.g. the grass *Agropyron junceum*).

fore reef A *talus slope on the seaward side of a reef, which is constantly under attack by waves and currents.

foreset The slip-face of a *Gilbert-type delta.

foreshock A small *earthquake that precedes a major earthquake or volcanic eruption. Foreshocks sometimes occur in swarms.

foreshore The lower shore zone that lies between the normal high- and low-water marks. The foreshore may be either a plane slope dipping seawards at a low angle or one marked by the development of longshore bars (ridge and runnel topography), depending on the nature of the wave attack.

forest **1.** A plant formation that is composed of trees the crowns of which touch, so forming a continuous canopy (*compare* WOODLAND), or the trees that make up a forested area. **2.** In Britain, from Norman times, a district reserved for the hunting of deer, often belonging to the sovereign, to which special laws applied and which was administered by special officers. The word 'forest' is derived from the Latin *foris*, 'out of doors': the land lay beyond those areas enclosed for agriculture or parkland and was unfenced. The land set aside as forest was not necessarily tree-covered, especially in the uplands, and might include open *heathland, *grassland, and *bog, as well as wooded areas. Most of the land formerly under forest law has been disafforested (*see* DEFORESTATION), although a district may still bear the 'forest' designation (e.g. 'Dartmoor Forest'). Some Crown forests, never disafforested, came to be used for growing timber, especially for ship-building, and today they are managed by the Forestry Commission (e.g. New Forest, Forest of Dean). **3.** To plant with trees.

forest formation A major category of *forest, defined usually by features of its structure and *physiognomy.

forestry **1.** The practice of growing and managing forest trees for commercial timber production. This includes the management of specially planted forests, of native or *exotic species, as well as the commercial use of existing forest, and the genetic improvement of timber trees for selected purposes. **2.** The scientific study of tree growth and timber production systems.

formation **1. (plant formation)** In vegetation description and analysis, a classificatory unit which usually implies a distinctive *physiognomy rather than a distinctive species composition. In detail, the various *phytosociological traditions ascribe slightly different meanings and hierarchical status to the term. The early European schemes grouped associations into alliances and then formation groups (e.g. sclerophyllous scrub), and then formation classes (e.g. sclerophyllous woodland and scrub). The formation classes are roughly equivalent to the major world *biomes. **2.** *See* FORMATION TYPE. **3.** The fundamental unit used in lithostratigraphy. Specific features distinguish one rock formation from another. The thickness of the formation is unimportant in its definition, as a given formation may vary within different outcrops. Formations may be subdivided into members and together several formations constitute a group.

formation type A world vegetation type which has a relatively uniform appearance and life-forms. A formation is a geographically distinct component of a formation type. For example, the tropical rain forest is a formation type, which has a number of formations, each with a distinctive structure and *physiognomy. The term *'biome' is nearly synonymous with 'formation type', and 'formations' are also known as 'regions'. A formation type is generally regarded as climatic *climax vegetation.

form factor *See* DRAINAGE BASIN SHAPE INDEX.

form roughness *See* BED ROUGHNESS.

formula of vegetation A shorthand system for the rapid, precise description of vegetation. For example, in J. J. Christian's and T. O. Perry's (1953) scheme, A, B, and C respectively denote trees, shrubs, and herbs; 3, 2, and 1 indicate size as tall, medium, and short; density is recorded as x, y, and z, indicating respectively dense, average, and sparse. Average layer heights are also appended. There are many other such schemes, of varying complexity.

Fortin barometer A portable mercury barometer that was invented in 1800 by the French instrument maker Jean Nicholas Fortin (1750–1831). In other mercury barometers air pressure is calculated from the distance between the level of the mercury in the reservoir and in the tube. In the Fortin barometer the bottom of the reservoir is flexible (originally it was made from leather) and can be raised or lowered by means of a screw, allowing the surface of the mercury to be adjusted to a predetermined level. A vernier scale on the side of the barometer tube is then lowered until its base touches the top of the mercury in the reservoir and a reading is taken from the fixed scale marked on the tube. Before travelling the base of the reservoir is raised using the adjusting screw until both reservoir and tube are filled with mercury. The Fortin barometer makes no correction for changes in temperature, which affect the volume of mercury, or for capillarity, which introduces inaccuracies. The instrument is nevertheless sufficiently accurate for most purposes. *See also* FIDUCIAL POINT and KEW BAROMETER.

fossil **1.** Generally, anything that is ancient, especially if it is discovered buried below ground (e.g. *fossil fuel, fossil soil). In its original sense 'fossil' meant anything dug up from the earth, including mineral ores, precious stones, etc. **2.** In the modern sense, which dates from the late 17th century, a fossil is the remains or traces of a once-living organism. At one time the term was used only of material dating from before the end of the most recent glacial period, so fossils were more than 10 000 years old; remains and traces younger than 10 000 years were known as 'subfossils'. This restriction has been abandoned and now all ancient remains are called 'fossils' regardless of their age. Fossils include skeletons, tracks, impressions, trails, borings, and casts. Fossils are usually found in hard rocks, but not always; for example, insects and small plants or parts of plants (e.g. leaves and flowers) have been found preserved in *amber of *Tertiary age and woolly mammoths living 20 000 years ago have been recovered from the frozen *tundra of Siberia.

fossil fuel All deposits of organic material that are dug from the ground (from the Latin *fossilis*, meaning 'dug up') and are capable of being burnt for fuel; chiefly coal, oil, and gas. These are formed under pressure by alteration or decomposition of plant or animal remains.

fossilization The process by which a *fossil is formed. It is unusual for organisms to be preserved complete and unaltered; generally, the soft parts decay and the hard parts undergo various degrees of change. The completeness of a fossil indicates whether or not the organism was fossilized *in situ* (e.g. partly decayed crinoids are readily broken by being transported, even over a short distance). Solution and other chemical action may reduce the tissues to a thin film of carbon; this process is called 'carbonization'. Occasionally, rapid fires preserve flowers in three dimensions by carbonization and such fossils have been found that are 95 Ma old. The organism may be flattened by the compaction of sediments. If the organic content is preserved, the fossil is described as a 'compression'; if it is not preserved the fossil is an 'impression'. These terms do not necessarily imply that the material has been subjected to pressure. Porous structures (e.g. bones and shells) may be made more dense by the deposition of mineral matter by *groundwater. The internal physical structures of some shells may be changed as a result of solution and reprecipitation; in this process ('recrystallization') the original structure may be blurred or lost. Many shells which were originally composed of *aragonite are recrystallized into the more stable mineral *calcite. The solution of an original shell and the simultaneous deposition of another mineral material constitutes 'replacement'; this may occur molecule by molecule, in which case the microstructure is preserved, or *en masse*, when it is not. Common replacement minerals include silica or iron sulphide, but there are many others. If the original cell walls survive with some of their organic material, but supported by minerals (i.e. the cytoplasm is replaced), the process is called 'permineralization'. Because the organic material can be prepared on slides, such fossils yield more information than those produced by 'petrifaction', in which all the material has been converted to minerals, although the outlines of structures (e.g. organelles) can be studied on the polished surfaces. An organism may be fossilized in three dimensions as a nodule, formed if the decay process causes the nucleation of sediments around it. Organisms may also be preserved in almost perfect condition in *amber, but it is difficult to obtain information from them. The impression of skeletal remains in surrounding sediments constitutes a 'mould'. Where the external structures are preserved it is called an 'external mould' and where the internal features are preserved it is called an 'internal mould' or 'steinkern'. Filling of a mould cavity by mineral matter may produce a 'natural cast'. Tracks, trails, burrows, and other evidence of organic activity may also be preserved. These are called 'ichnofossils' or *trace fossils. *See also* DEATH ASSEMBLAGE and TAPHONOMY.

fossorial Burrowing.

founder effect The derivation of a new population (e.g. on an oceanic island) from a single individual or a limited number of immigrants. The founder(s) will represent a very small sample of the genetic pool to which it or they formerly belonged. *Natural selection operating on this more restricted genetic variety soon yields gene combinations quite different from those found in the ancestral population.

founder lineage In *phylogenetics, an ancestral lineage, often still extant, from which other lineages have risen. The term is usually applied to intraspecific studies of populations and used to describe *operational taxonomic units that occur at *internal nodes of a *phylogenetic tree.

^{14}C dating *See* RADIOCARBON DATING.

foveate Marked with small depressions or pits.

fowl cholera An acute, infectious, often fatal disease of fowl caused by the bacterium *Pasteurella multocida*.

fowl pest A disease of birds, which may be either *fowl plague or *Newcastle disease.

fowl plague A disease of birds (including fowl, ducks, and turkeys) which is of considerable commercial importance. Wild birds, particularly waterfowl, are naturally infected but do not normally show symptoms. In domestic fowl the disease affects the respiratory tract, but the central nervous system may also be involved. The disease is caused by an avian influenza virus.

fox-fire A light (bioluminescence) emitted by moist decaying wood or by certain types of fungal *fruit body.

fracto- *See* FRACTUS.

fractus (fracto-) The Latin *fractus*, meaning 'broken', used to describe a species of cloud that has an irregular or ragged form. The term is applied to *cumulus and *stratus. *See also* CLOUD CLASSIFICATION.

fragility The degree to which an *ecosystem is easily damaged. Fragile ecosystems include wetlands and coral reefs. Other ecosystems, such as short-turf grasslands, are generally more robust. *Compare* VULNERABILITY.

fragipan A subsoil *horizon, found deep in a soil *profile and having a high bulk density. It is a dense, brittle, and compact layer, apparently with little or no cementation horizon, associated with acid soil conditions.

Frasnian *See* DEVONIAN.

frame-shift mutation A *mutation in a transcribed gene (e.g. an *insertion or *deletion) that causes a new *open reading frame to be transcribed.

Fraser Darling effect Acceleration and synchronization of the breeding cycle that results from the visual and auditory stimulation experienced by individual members of large breeding colonies of birds. It tends to shorten the breeding season and to present predators with a sudden flush of eggs and young, thus increasing the likelihood that any given pair of adults will breed successfully. The effect is named after Sir Frank Fraser *Darling, who proposed its existence in 1938.

frass The fine powdery material to which phytophagous insects reduce plant parts; it is the faeces left by the insects.

fratricide The killing of *siblings.

frazil ice Flowing water ice that forms platelets rather than continuous sheets; it is often observed in Canadian rivers. The name is derived from the French *fraisil*, meaning 'cinder'.

free atmosphere The whole of the atmosphere above the *boundary layer, i.e. that part of the atmosphere in which air is not affected by direct contact with the surface. The upper margin of the boundary layer is usually taken to be at an average height of 500 m and the free atmosphere comprises about 95 per cent of the total mass of the atmosphere.

free-living **1.** Living independently (i.e. not parasitic on, or symbiotic with, any other organism). *See* PARASITISM and SYMBIOSIS. **2.** Not attached to a *substrate.

freezing nuclei Any nuclei, commonly ice crystals but sometimes the suitably shaped crystals of other substances, which, when present in clouds at temperatures below 0°C, will cause any supercooled droplets with which they collide to change to ice (in the form of a crystalline growth upon the nucleus).

frequency **1.** (f) The number of complete wavelengths which pass a given point in a specified time; units are hertz (Hz; one hertz is one cycle per second). The frequency of a periodic wave-form is given by $f = 1/T$, where T is the period; and by $f = v/\lambda$, where v is the velocity and λ the wavelength. **2.** In vegetation description, a term relating to the proportion of the samples taken that contains a particular species. It is expressed as a percentage.

fresh water Water containing little or no chlorine. According to the Venice system, which classifies brackish waters by their percentage chlorine content, fresh water contains 0.03 per cent or less of chlorine. *Compare* BRACKISH.

freshwater fish Fish that live only in fresh water. Many families and orders of fish are considered to include only primary freshwater species (i.e. species that evolved without contact with a marine environment). Some families have a worldwide distribution (e.g. the Cyprinidae are found in freshwater basins all over the world, except for South America and the Australian region). Others (e.g. the Cobitidae, loaches) are found only in the Eurasian region. A few families have a very restricted distribution (e.g. the Amiidae (bowfin) are found only in the eastern half of the USA).

friable Applied to the consistency or handling properties of soil, meaning that the soil crumbles easily.

fright substance *See* ALARM SUBSTANCE.

frigid *See* PERGELIC.

fringing forest **(gallery forest)** A *forest which often extends along river banks from the rain-forest belt into adjacent *savannah. The ribbon-like tracts generally resemble degenerate rain forest.

fringing reef *See* REEF.

Frisch, Karl von **(1886–1982)** An Austrian zoologist who was joint winner (with Konrad *Lorenz and Nikolaas *Tinbergen) of the 1973 Nobel Prize for Physiology or Medicine for their studies of animal behaviour. Von Frisch specialized in insect behaviour and in the 1940s showed that bees can navigate by the Sun and perform dances to communicate to other members of their hive the location of food sources (*see* DANCE LANGUAGE). He also discovered that fish have an acute sense of hearing.

front The boundary or boundary region that separates *air masses of different origins and characteristics. Temperature gradients in any horizontal surface are large through the front. Different types of front are distinguished according to the nature of the air masses separated by the front, the direction of the front's advance, and the stage of development. The term was first devised during World War I by the Norwegian school of meteorologists (headed by Professor V.

Bjerknes). *See also* ANAFRONT; COLD FRONT; KATAFRONT; OCCLUDED FRONT; POLAR FRONT; and WARM FRONT.

frontal wave A wave-like deformation of the line of a *front between two *air masses. The wave develops from the northward incursion of warm air and usually travels along the front, with colder air ahead and to the rear. Typically, frontal waves occur in sequences or 'families' of several waves, and develop into *depressions or storm centres travelling more or less eastward as 'secondaries' along the extended cold front to the rear of the original low. The secondaries tend to catch up and merge with the original depression as it slows up in its fully developed stage.

frontal zone The transition zone, sometimes amounting to a discontinuity, that separates adjacent *air masses. Some turbulent mixing takes place. The sloping zone separating a cold wedge under warm air typically extends about 1 km vertically and about 100 km horizontally.

frontogenesis The development and intensification of frontal boundaries between adjacent *air masses.

frontolysis The processes of dissolution or dissipation of a *front. Frontal decay results when different *air masses stagnate together, or move together or in succession along the same track at the same speed, or incorporate air of the same temperature.

frost The condition in which the prevailing temperature is below the freezing point of water (0 °C). This may lead to a deposit of ice crystals on objects (e.g. grass or trees). Such deposits result from condensation when the *dew-point temperature is below freezing. *See also* BLACK ICE.

frost heave (frost heaving) An upward movement of the ground surface or of individual particles, owing to the formation of lenses of ice up to 30 mm thick in the *regolith. It reaches its maximum in silt-dominated material, in which the greatest volume of ice may develop (more than 68 per cent ice by volume). When the total uplift of the surface is measured, it is found to be approximately equal to the sum of the thicknesses of the layers of ice. Surface stones may be heaved by the development of needle-ice columns ('pipkrakes').

frost heave test A laboratory test in which aggregate or soil is frozen under controlled conditions. A cylinder containing rock aggregate, 150 mm high and 100 mm in diameter, is placed in freezing conditions with its base in running water for 250 hours. The *frost heave must be less than 12 mm.

frost heaving *See* FROST HEAVE.

frost hollow An area (e.g. a valley bottom or a smaller hollow) that is very liable to severe and frequent *frosts as a result of dense, cold air moving down-slope (katabatic flow) and collecting there under conditions of radiation cooling (e.g. at night).

frost pull and frost push *Periglacial processes that bring about the upward migration of rock fragments (*clasts) through the *regolith. Frost push takes place when an ice lens forms beneath a clast and so pushes it upwards. Frost pull occurs when a clast adheres to ice within a freezing regolith and so is drawn upwards as the ground heaves.

frost shattering *See* FROST WEDGING.

frost smoke *See* ARCTIC SEA SMOKE.

frost table *See* TJAELE.

frost wedging (congelifraction, frost shattering, gelifraction, gelivation) The fracturing of rock by the expansionary pressure associated with the freezing of water in planes of weakness or in pore spaces.

Froude number (Fr) A dimensionless number equal to the ratio of water velocity to the speed of a gravity wave, used to assess whether flow in an open channel is critical, tranquil, or shooting. If the Froude number is less than 1, flow is said to be subcritical or slow; if Fr = 1 flow is critical; and if Fr is greater than 1 flow is fast or supercritical. It was calculated by the English engineer William Froude (1810–79).

frugivore An organism that feeds on fruit.

fruit Strictly, the ripened ovary of a plant and its contents. More loosely, the term is extended to the ripened ovary and *seeds together with any structure with which they are combined (e.g. the apple (a pome) in which the true fruit (core) is surrounded by flesh derived from the floral receptacle).

fruit body **(fruiting body)** A differentiated, *spore-bearing structure, particularly the ascocarps and basidiocarps of ascomycetes and basidiomycetes.

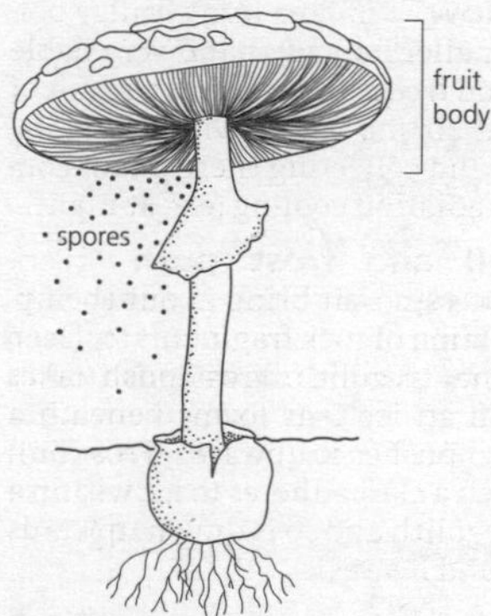

Fruit body

frustration The motivational state which results when the consequences of behaviour are less satisfying than previous experience had led an animal to expect. Often it leads to more determined attempts to obtain satisfaction, but it may lead to *irrelevant behaviour or *redirected behaviour (e.g. if a foe runs away, the remaining combatant may attack a tussock of grass or another inanimate object).

frustule The silica wall of a diatom.

fruticose Shrubby in habit (e.g. the thalli of certain lichens).

fugacious Short-lived; soon disappearing.

fugitive species **(opportunist species)** A species typical of unstable or periodically extreme environments, e.g. deserts or *ephemeral ponds, and characterized by a strong *dispersal ability. Such organisms are typically smaller than *equilibrium species and have shorter life cycles. Hence initially they can colonize an area rapidly during a favourable period, though in the longer term they may lose out in *competition with equilibrium species.

Fujita tornado intensity scale A six-point scale introduced in 1971 by Theodore Tetsuya Fujita (1920–98) and his colleagues to rank *tornadoes according their wind speeds and the damage they cause.

fuliginous Soot-coloured or dusky.

fuller's earth **1.** A clay consisting mainly of expanding clay minerals such as montmorillonite, which is used industrially for its absorptive properties. **2.** Capitalized, Fuller's Earth is the stratigraphic name of a *Jurassic clay formation outcropping in southern Britain.

fulvic acid A mixture of colourless organic acids that remains soluble in weak acid, alcohol, or water after its extraction from soil.

fulvous Reddish-brown or reddish-yellow in colour; tawny-coloured.

functional constraint The extent to which a region of DNA is intolerant of mutation, due to a reduction in its ability to carry out the function encoded.

functional type A system for classifying organisms that is based upon the role they play in an *ecosystem rather than their taxonomic status. It may refer to a structural feature, such as being deciduous or evergreen for plants (*compare* LIFE-FORM), or it may relate to physiological activity, such as nitrogen-fixing or cellulose-decomposing. This classification is proving valuable in studies of *biodiversity in relation to stability in ecosystems, because stability may be more closely related to the range of functional types present in an ecosystem than to the number of species found within it.

fundamental niche The potential *niche occupied by a viable population of a species in the absence of *competition from other species. *Compare* REALIZED NICHE.

fundatrix (pl. fundatrices) The parthenogenetic founder of a population (e.g. of aphids).

Fungi The taxonomic kingdom that comprises *eukaryotic, non-photosynthetic (*see* PHOTOSYNTHESIS) organisms, which obtain nutrients by absorbing organic compounds from their surroundings. Fungi may be unicellular, filamentous (*see* MYCELIUM), or plasmodial (i.e. forming an acellular, mobile feeding structure consisting of a mass of naked protoplasm with many nuclei), and most have cell walls containing *chitin. Many fungi live as *saprotophs and are important agents of organic decomposition. Others live as *symbionts or parasites (*see* PARASITISM). Some can cause disease in plants or animals. Fossils are rare, but fungi are believed to have left the sea about 400 Ma ago, when the first plants colonized dry land.

fungicide A chemical or biological agent that kills fungi.

fungicole An organism that grows on fungi.

fungus garden A 'garden' made by parasol ants (Attini) in which the ants cultivate fungi as a source of food.

funnel cloud A narrow, conical cloud that appears from the base of large *cumulonimbus the central part of which is rotating (forming a mesocyclone). The funnel cloud is a vortex formed by air spiralling inward to a region of low pressure, made visible by the condensation of water vapour due to the low pressure. If the funnel touches the surface it becomes a *tornado or *waterspout.

funnelling The constraining of an airflow by valleys, leading to higher wind speed, convergence, and uplift. Similar effects occur in the air between an advancing *front and the face of a mountain barrier.

furan Strictly C_4H_4O, but more commonly one of a range of polychlorinated dibenzofurans that are produced as contaminants from the incomplete incineration of chlorinated hydrocarbons, especially *polychlorinated biphenyls. Severe exposure to furans can cause chloracne in humans and they are suspected of causing liver damage and liver cancer.

furcipulate Pincer-like.

furfuraceous Scurfy; covered with small, bran-like scales.

fuscous Sombre in colour; dark-coloured or black.

fusiform Spindle-shaped; elongated with tapering ends.

fusinite *See* COAL MACERAL.

fusion **1.** Generally, the melting of a solid substance by heat. **2.** In nuclear fusion, the combining of two light atomic nuclei to form a heavier nucleus with the sudden release of energy (e.g. in the hydrogen bomb).

future-natural *See* NATURAL.

fynbos A South African name for the *sclerophyllous vegetation, physiognomically similar to *chaparral, which occurs in the Cape Province. The structure of the fynbos *community usually consists of an upper storey dominated by broad-leaved sclerophyllous *shrubs with small leaves. In higher latitudes mountain fynbos occurs. This consists of a low, open community, about 1 m in height, which is essentially a heath community (*see* HEATHLAND). Many *Erica* species and ericoid shrubs (e.g. *Phylica* and *Sympieza*) occur. These heaths, like those found in the northern hemisphere, have been greatly extended as a consequence of the anthropogenic influence of fire and grazing.

gaging station *See* GAUGING STATION.

Gaian hypothesis A hypothesis, formulated by James E. *Lovelock and Lynn Margulis and first advanced by Lovelock in 1968. It holds that the presence of living organisms on a planet leads to major modifications of the physical and chemical conditions pertaining on the planet, and that subsequent to the establishment of life the climate and major *biogeochemical cycles are mediated by living organisms themselves. Feedback mechanisms between the biotic and abiotic elements of the biosphere will tend to create stable conditions in which major environmental change is dampened. The resulting *homoeostasis resembles that found in the physiology of an individual organism (*see* GEOPHYSIOLOGY). The term Gaia was first proposed by the author William Golding. In Greek mythology, Gaia, or Ge, represented the Earth; she sprang from Chaos, gave birth to Uranus (the Heavens) and Pontus (the Sea), then became the mother, by Uranos, of the giants, Cyclopes, and Titans. This choice of nomenclature has led to the adoption and development of this hypothesis by New Age adherents, which was not the original intention of Lovelock.

gaining stream (influent stream) A stream that receives water emerging from a submerged spring or other *groundwater seepage which adds to its overall flow.

gait Manner of walking.

Galápagos Islands A group of oceanic islands, about 970 km from the west coast of South America, which *Darwin visited in 1835. He encountered a number of endemic (*see* ENDEMISM) species that were to prove influential in the development of his ideas on *evolution.

gale A wind blowing at between 14 m/s (32 mph; moderate gale on the *Beaufort scale) and 28 m/s (64 mph; storm on the Beaufort scale).

gall (cecidium) An abnormal growth or swelling in a plant. The formation of a gall may be induced by infection of the plant with bacteria or fungi, or by attack from certain mites, nematodes, or insects. Galls may be formed on roots, stems, or leaves. Some galls (e.g. clubroot and crown gall) are symptoms of disease; others appear to do little harm to their hosts, while some may actually be beneficial to the plant (e.g. nitrogen-fixing *root nodules of legumes). Galls are variously structured, ranging from a simple outgrowth to a large and histologically complex structure with up to five distinct tissue layers with nutritive zones. Most gall-forming species are members of the insect family Cynipidae (Hymenoptera), and often have complex, heterogynous life cycles, utilizing different parts of the same host or different hosts, which are generally *Quercus* species. The mechanism of gall initiation and development is little understood, and the study of galls offers an unparalleled opportunity for physiological and ecological research. The gall community supports a large number of parasitoids, hyperparasitoids, and inquilines (species which use the gall but do not kill its occupant).

gallery forest *See* FRINGING FOREST.

galvanotaxis A change in direction of locomotion in a motile organism or cell, made in response to an electrical stimulus.

game cropping The culling or husbanding of game animals for meat and other products, usually for human consumption. Game cropping is often advocated as an ecologically sound method of farming the African *savannah and similar environments: it is argued that the wide variety of game animals, with their different food preferences, utilize the primary production more efficiently than cattle, and are generally better adapted to the environment. This gives added reason for conservation of the wild ungulate population. Other social, economic, and political difficulties mean that game cropping is practised less widely than may seem desirable.

gamete A specialized *haploid cell (i.e. a sex cell) the nucleus and often the cytoplasm of which fuse with that of another gamete from the opposite sex or mating type in the process of *fertilization.

game theory The theory that relationships within a community (of organisms or of traits possessed by those organisms) can be regarded as a contest (i.e. a game) in which each participant seeks to secure some advantage. Numerical values can be attached to the gains and losses involved, allowing the contest to be simulated mathematically, usually by computer modelling. The application of game theory has produced many insights into ecological relationships and the significance of particular aspects of animal behaviour.

g

gametophyte A haploid phase of the life cycle of plants, during which *gametes are produced by mitosis. It arises from a haploid *spore produced by meiosis from a diploid *sporophyte. In lower plants (such as mosses), the gametophyte is the dominant and conspicuous generation. *See also* ALTERNATION OF GENERATIONS.

gamma rays Electromagnetic radiation, about 10^{-10} to 10^{-14} m in wavelength, similar to, but of shorter wavelength than X-rays, which is emitted by radioactive substances.

gamma-ray spectrometry An analytical method for the measurement of the intensities and energies of gamma radiation. Scintillation or semiconductor radiation detectors, coupled to various types of electronic circuitry, enable a spectrum to be accumulated. This may be used to identify the gamma-emitting radioisotopes, and their energy intensities can be used to determine the corresponding element concentrations.

gap In *alignment of *nucleic acid or *amino acid sequences, the introduction of spaces in one sequence to account for deletions in that sequence or insertions in another.

gap analysis A technique, first performed in 1978 by the American ecologist Michael Scott, for identifying ecosystems in need of conservation. Ranges are mapped for a variety of rare or endangered species. The maps are laid one above another and when all of them are overlaid on a map showing the location of reserves and protected areas gaps are revealed in which valuable ecosystems remain unprotected. Gap analysis identifies many more ecosystems deserving protection than the alternative *hot spot technique.

gap dynamics The process of colonization and the development of vegetation within an opening in a forest.

gap phase In forestry, the *pioneer phase during which trees begin to colonize a site.

garigue *See* GARRIGUE.

GARP *See* GLOBAL ATMOSPHERIC RESEARCH PROGRAMME.

garrigue **(garigue)** A low-growing, secondary vegetation which is widespread in the Mediterranean basin and is derived from the original *evergreen mixed forest. The dominant plants are aromatic herbs and prickly dwarf shrubs, with drought-resistant foliage, many belonging to the mint family (Lamiaceae) or Fabaceae. Garrigue is a degraded, fire-prone form of vegetation produced by intensive grazing and other human-based activities. *See also* MAQUIS and MEDITERRANEAN SCRUB.

GAS *See* GENERAL-ADAPTATION SYNDROME.

gas chromatography An analytical technique in which the components of a sample are separated by partitioning between either a mobile gas and a thin layer of non-volatile liquid held on a solid support (gas–liquid chromatography) or between the gas and a solid absorbent as the stationary phase (gas–solid chromatography). Partitioning occurs repeatedly throughout the column, and, as each solute travels at its own rate, a band corresponding to each solute will form. Solutes are eluted (washed out) in increasing order of partition ratio, and enter a detector attached to the column exit. The time of emergence of a peak on the display identifies a component, and the area under each peak is proportional to the component's concentration. Gas chromatography is used mainly in the analysis of *volatile organic compounds.

gaseous exchange The transfer of gases between an organism and the environment; it may occur in both *respiration and *photosynthesis.

gas-exchange method A method for measuring *primary productivity, based on rates of carbon-dioxide uptake and oxygen release, these being the easily monitored gaseous raw material and by-product of *photosynthesis. *See also* CARBON-DIOXIDE METHOD and OXYGEN METHOD.

gaster In Hymenoptera the abdomen, with the exception of the first abdominal segment, which has become included in the thorax from which it is separated by a narrow neck (petiole).

gastrolith A stone swallowed (e.g. by some reptiles and birds) to break up food and so assist digestion. Such stones acquire rounding and polish.

gastrozooid In polymorphic colonial coelenterates, a feeding *polyp.

gas vacuole A small, gas-filled vesicle, numbers of which are found in certain aquatic bacteria and *cyanobacteria. Their function appears to be that of giving buoyancy to the cells.

gauging station **(gaging station)** A point at which river flow or *groundwater levels are measured.

Gault Clay A glutinous marine deposit found in south-eastern England and France, which contains abundant fossil bivalve molluscs (Bivalvia), gastropods, ammonoids, and vertebrates. It is Lower *Cretaceous (Albian) in age.

Gause, Georgyi Frantsevich (1910–86) A Russian biologist and ecologist, best known for his *competitive exclusion principle, which he derived from his studies of competition among protists and described in a monograph, *The Struggle for Existence* (1934). Gause studied biology at Moscow University. In 1942 he and his wife, Maria Georgyevna Brazhnikova, isolated a strain of *Bacillus brevis* from which an antibiotic substance was produced and used later to treat infected wounds. For this Gause was awarded the Stalin Prize in 1946. From 1960 until his death he was director of the institute of antibiotics he and his wife had founded.

Gause principle *See* COMPETITIVE EXCLUSION PRINCIPLE.

gauss **(G)** The *c.g.s. unit of measurement of: (*a*) magnetic field; and (*b*) magnetic moment per unit volume. It has now been replaced by the SI units weber/m^2 (Wb/m^2) and tesla (T). $1\,\mathrm{G} = 10^{-4}\,\mathrm{T} = 10^{-4}\,\mathrm{Wb/m^2}$.

Gaussian curve *See* GAUSSIAN DISTRIBUTION.

Gaussian distribution **(Gaussian curve, normal distribution)** The values of a variable (x) are distributed systematically around their mean ($\bar{x}$) according to a frequency f, given by: $f = 1/(s(2\pi)^{-2}) \exp(-(x - \bar{x}^2)/2s^2)$, where s is the standard deviation. It is named after the German mathematician Karl Friedrich Gauss (1777–1855).

Geiger counter *See* GEIGER–MÜLLER COUNTER.

Geiger–Müller counter **(Geiger counter)** An instrument for detecting ionizing radiation which consists of a cylindrical metal cathode with a wire anode along its axis, the whole being enclosed in a thin-walled tube filled with low-pressure inert gas. In operation the cathode carries a charge of about 1 000 volts, which is just short of the level needed to produce an electrical discharge across the cathode–anode space. A charged particle or *gamma ray traversing this space collides with atoms of the inert gas, producing positive ions and negative electrons. Under the high voltage these are rapidly accelerated towards the cathode and anode, colliding on the way with other gas atoms and producing many more charged particles in a chain reaction. This avalanche arriving at the anode and cathode is registered as a pulse, which is amplified to produce a click in a headphone set or a succession of such pulses, which can be expressed as a meter reading in milliroentgens per hour or counts per second. It was invented in 1928 by the German physicists H. W. Geiger and W. Müller. For more accurate surveys (especially from the air) a scintillation counter is required, which is a more sensitive instrument.

Geikie, Sir Archibald **(1835–1924)** Director of the British Geological Survey from 1881 to 1901, Geikie made studies of glacial and fluvial *erosion, and attempted to calculate the age of the Earth from rates of denudation. This led to conflict with Lord Kelvin (William Thomson, 1824–1907), then the most eminent of British physicists and a pioneer in the study of thermodynamics, who had calculated a much shorter age for the Earth, based on its rate of cooling from an originally hot state. Geikie was also one of the first historians of geology, stressing in his *Founders of Geology* (1897, 1905) the importance of the work of his fellow Scot James Hutton.

gel A rigid or jelly-like material, in which molecules form a loosely linked network, formed by the coagulation of a colloid.

gel filtration A column chromatographic technique normally employing as a stationary phase polymeric carbohydrategel beads of controlled size and porosity. Mixture components are separated on the basis of their sizes and rates of diffusion into the beads. Smaller molecules tend to diffuse more rapidly into the beads, thereby leaving the mainstream of solvent and so becoming retarded with respect to larger molecules. This method can also be used to determine the molecular weight of an unknown substance.

gelifluction **(congelifluction)** The slow flow of water-lubricated, unsorted material and rock debris over perennially frozen ground, and on slopes as low as 1°. It is the cold-climate variety of *solifluction and occurs only in the *active layer (usually to a depth of 3 m).

gelifraction *See* FROST WEDGING.

geliturbate *See* GELITURBATION.

geliturbation **(congeliturbation, cryoturbation)** A general term that describes all frost-based movements of the *regolith, including *frost heaving and *gelifluction. The material disturbed by such movements is called 'geliturbate'.

gelivation *See* FROST WEDGING.

geminate **1. (adjective)** Combined in pairs. **2. (verb)** To combine in pairs. The word is from the Latin *geminus*, meaning 'twin'.

gendarme *See* ARÊTE.

gene The fundamental unit of inheritance, comprising a segment of DNA (or RNA in some *viruses) that codes for one or several related functions and occupies a fixed position (locus) on a *chromosome.

gene bank An establishment in which both somatic and hereditary genetic material are conserved. It stores, in a viable form, material from plants that are in danger of extinction in the wild and *cultivars which are not currently in popular use. The stored genetic information can be called upon when required. For example, a crop may be needed that possesses a quality (e.g. tolerance to adverse climatic conditions) which cannot be found in currently exploited cultivars but was present in more antiquated varieties. The normal method of storage is to reduce the water content of seed material to around 4 per cent and keep it at 0°C (pollen material may also be used but its longevity is considerably less). Stored this way, the material often remains viable for 10–20 years. When the desiccating process proves fatal, as is the case with tropical genera producing *recalcitrant seeds, where possible the material is maintained by growth. This may require considerable space, but in some cases the problem can be resolved using tissue-culture methods. All stored stock is periodically checked by germination. *See also* SEED BANK.

gene centre *See* CENTRE OF DIVERSITY.

gene conversion A nonsynonymous reciprocal *recombination process that results in one region of DNA becoming identical with another. *Compare* BIASED GENE CONVERSION.

gene diversity The mean expected *heterozygosity per *locus in a population.

gene flow The movement of genes within an interbreeding group which results from mating and the exchange of genes with immigrant individuals. The flow may occur in one direction or both.

gene-for-gene co-evolution (GFG co-evolution) A form of *co-evolution in which the virulence of a pathogen and the response of its host are governed by corresponding genes in the two organisms: resistance to a pathogen depends on the presence of both a resistance gene in the host and an avirulence gene in the pathogen. Evidence for the phenomenon has been found in associations between crop plants and pathogens, including grasses, herbs, shrubs, and trees, and fungi, viruses, and bacteria. It may also occur between plants and nematodes or insects.

gene library *See* LIBRARY.

gene pool The total number of genes or the amount of genetic information possessed by all the reproductive members of a population of sexually reproducing organisms.

general adaptation An adaptation that fits an organism for life in some broad environmental zone, as opposed to 'special adaptations' which are specializations for a particular way of life. Thus leaves are a general adaptation, while the particular kind of leaf is a special adaptation, and the wing of a bird is a general adaptation, while a particular kind of bill is a special adaptation. Major groups of organisms are differentiated very largely on the basis of general adaptations.

general-adaptation syndrome (GAS) A range of abnormal physiological systems reflecting pathological social *stress, and serving to regulate population growth, where no external resource limitation exists. Essentially it is a behavioural regulation of population growth. Mechanisms include the supression of the oestrus cycle, inadequate lactation, enlarged adrenal glands, increased aggression, etc. It has been demonstrated experimentally with overcrowded laboratory rats.

general circulation The term generally used to describe the large-scale circulation of the atmosphere over the globe, or over one hemisphere, with its more or less persistent features (which may be brought to prominence by considering long-term or even shorter-term averages) and all the transient features on various scales. Although in its nature it is a matter of the winds, the general circulation may be studied by means of barometric pressure maps because of the intimate relation between pressure and wind.

generalization The evocation of a learned pattern of behaviour by stimuli other than those to which an animal was trained to respond.

generation time The time that is required for a cell to complete one full growth cycle. If every cell in the population is capable of forming two daughter cells, has the same average generation time, and is not lost through lysis (*see* LYTIC RESPONSE), the doubling time of the cell number in a population will equal the generation time.

generic cycles, theory of A theory which envisages a life cycle for a species or genus that resembles that of an individual. The first stage is characterized by vigorous spread; the second by maximum phyletic activity giving rise to new forms; the third marks a phase of decline in area, owing to competition within the offspring species; and the fourth and final stage sees the extinction of the species. Although in many ways a useful analogy, there is no evidence that species become senile and die out spontaneously.

gene sharing The acquisition by a gene of a secondary function without the loss of its primary.

gene substitution The process whereby a new *allele becomes fixed in a population.

genet A genetic individual; the product of a *zygote.

genetic Pertaining to the origin or common ancestor or ancestral type. The term is widely used outside the life sciences. *See* GENETICS.

genetic code The set of correspondences between base (nucleotide pair) triplets (codons) in DNA and amino acids in protein. These base triplets carry the genetic information for protein synthesis. For example, the triplet CAA (cytosine, adenine, adenine) codes for valine.

genetic distance A measure of genetic similarity and evolutionary relationship (e.g. between two races of the same species), based on the frequency of a number of genes which can be easily detected and scored. A number of different indices of genetic distance are in use. *Compare* HOMOEOLOGY.

genetic drift The random fluctuations of gene frequencies in a population such that the genes among offspring are not a perfectly representative sampling of the parental genes. Although drift occurs in all populations, its effects are most marked in very small isolated populations, in which it gives rise to the random *fixation of alternative *alleles, so that the variation originally present within single (ancestral) populations comes to appear as variation between reproductively isolated populations.

genetic engineering The manipulation of DNA using restriction enzymes which can split the DNA molecule and then rejoin it to form a hybrid molecule: a new combination of non-homologous DNA (so-called recombinant DNA). The technique allows the by-passing of all the biological restraints to genetic exchange and mixing, and permits the combination of genes from widely differing species. Genetic engineering developed in the early 1970s, and is now one of the most fertile areas of genetics.

genetic equilibrium An equilibrium in which the frequencies of two alleles at a given locus are maintained at the same values generation after generation. A tendency for the population to equilibrate its genetic composition and resist sudden change is called genetic homoeostasis.

genetic erosion The loss of genetic information that occurs when highly adaptable *cultivars are developed and threaten the survival of their more locally adapted ancestors, which form the genetic base of the crop. For example, a new hybrid strain of maize (*Zea mays*), developed largely by Donald F. Jones from a variety discovered in 1917, had yields 25 per cent greater than standard maize. By the 1960s it was economically very favourable to use this single hybrid type, and more traditional varieties rapidly receded in distribution. The widespread adoption of this hybrid led to the narrowing of the genetic base (i.e. genetic erosion). In 1962 workers in the Philippines noticed that the fungus *Helminthosporium maydis* (southern corn leaf blight) was highly virulent on this hybrid. By 1970, 80 per cent of the US crop was vulnerable to *H. maydis*, because of the heavy dependence on one hybrid, and, during the wet summer of 1970, around 20 per cent of the US crop was lost to the blight. Fortunately, in this case the *genetic resources needed to produce a resistant strain had not become completely obsolete and so a recovery could be made.

genetic homoeostasis *See* GENETIC EQUILIBRIUM.

genetic load The average number of *lethal mutations per individual in a population. Such mutations result in the premature death of the organisms carrying them. Three main kinds of genetic load may be recognized: (*a*) input load, in which inferior alleles are introduced into the *gene pool of a population either by mutation or immigration; (*b*) balanced load, which is created by selection favouring allelic or genetic combinations which, by segregation and recombination, form inferior *genotypes every generation; and (*c*) substitutional load, which is generated by selection favouring the replacement of an existing allele by a new allele. Originally called the 'cost of natural selection' by the geneticist J. B. S. Haldane, substitutional load is the genetic load associated with transient *polymorphism. The term 'genetic load' was originally coined by H. J. Muller in 1950 to convey the burden that deleterious mutations provide, but it is probably better recognized as a measure of the amount of *natural selection associated with a certain amount of genetic variability, which provides the raw material for continued *adaptation and *evolution.

genetic polymorphism An occurrence in a population of two or more *genotypes in frequencies that cannot be accounted for by recurrent *mutation. Such occurrences are generally long-term. Genetic polymorphism may be balanced (such that *allele frequencies are in equilibrium with one another at a given *locus) or transient (such that a mutation is spreading through the population in a constant direction). In the former case, the different alleles may be maintained by different environmental conditions (in space or time), one being favoured under one set of circumstances and another under another set; or a heterozygous *genotype may be in some way superior to the genotypes that are homozygous at that locus; this is termed a 'heterozygous advantage'. It has been pointed out, however, that in many populations polymorphism is so high that to account for it all by *natural selection would entail an impossible *genetic load; thus, a good deal of polymorphism would be caused by chance increase or decrease of neutral alleles.

genetic resources The *gene pool in natural and cultivated stocks of organisms that are available for human exploitation. It is desirable to maintain as diverse a range of organisms as possible, particularly of domesticated *cultivars and their ancestors, in order to maintain a wide genetic base. The wider the genetic base, the greater the capacity for adaptation to particular environmental conditions (e.g. a pathogenic presence, *see* PATHOGEN). This has led to the establishment of *gene banks. *Compare* GENETIC EROSION.

genetics The scientific study of genes and heredity.

genetic system The organization of genetic material in a given species, and its method of transmission from the parental generation to its filial generations.

genetic variance A portion of *phenotypic variance that results from the varying *genotypes of the individuals in a population. Together with the environmental variance, it adds up to the total phenotypic variance observed among individuals in a population. When a particular phenotypic character is controlled by more than one gene, the genetic variance (V_G) is equal $V_D + V_I + V_A$, where V_D is the variance of components owing to the dominance of *alleles at a particular *locus, V_I is the sum of the epistatic (*see* EPISTASIS) interactions at different loci, and V_A is the additive genetic variance determined by the cumulative effects of alleles within and among loci (e.g. the resemblance of offspring to their parents is due to additive genetic variance). The phenotypic variance (V_P) is equal to the sum of the environmental variance (V_E) and V_G and the *heritability of a particular trait is given by V_A/V_P. (V is sometimes written as s^2.)

gene tree A *phylogenetic tree that has been constructed from information from *alleles of one or a few genes.

geniculate Resembling a knee or capable of bending as a knee bends. The word is derived from the Latin *geniculare*, meaning 'to bend' (like a knee).

genital lock The temporary inability of copulating partners to separate once intromission is achieved (e.g. in dogs the end of the penis is enlarged and a vaginal sphincter muscle closes around this enlargment).

genitus The growth of a new cloud from a mother-cloud, where only a limited part of the mother-cloud is affected by the change. *See also* CLOUD CLASSIFICATION and MUTATUS.

genome The total genetic information carried by a single set of *chromosomes (i.e. in a haploid nucleus). A single representative of each of all the chromosome pairs in a nucleus will therefore bear the genome of an individual.

genomic compartmentalization The presence within a cell of independent replication *genomes. In an animal cell the genomes include those of the cell nucleus and the mitochondria; in a plant cell those of the cell nucleus, mitochondria, and chloroplasts.

genotype The genetic constitution of an organism, as opposed to its physical appearance (*phenotype). Usually this refers to the specific allelic composition of a particular gene or set of genes in each cell of an organism, but it may also refer to the entire *genome.

genotypic adaptation *See* ADAPTATION.

geo- From the Greek *ge*, meaning 'Earth' (a version of 'Gaia'), a prefix meaning 'pertaining to the Earth'.

geo A narrow inlet of a cliffed coastline, which has developed along a major near-vertical joint or fault.

g

geobotanical anomaly The marked local concentration, above background levels, of one or more elements in an ecological assemblage, or a specific plant which may indicate the presence of an ore deposit or a concentration of hydrocarbons.

geobotanical exploration (biogeochemical exploration) Traditionally, the use of indicator plant species or assemblages to detect the possible presence of metal-rich deposits. It is based on the principle of *limits of tolerance, i.e. it assumes that only specialized species can withstand metal-contaminated soils. In practice, plant response may be confusingly more complex (e.g. plants may respond to low availability of essential nutrients rather than to high presence of toxic minerals) which makes such indicators unreliable. In modern use the concept includes the collection and chemical analysis of plant materials or soil layers, especially humus, in which metal ions may accumulate. It is a supplementary rather than a primary prospecting method. Geobotanical exploration can also be used to detect reserves of elements in the soil, because some species of plant accumulate particular elements. For example, certain *Astragalus* (Fabaceae) species accumulate selenium. *See also* LIMITING FACTOR.

geocarpic Applied to plants that fruit below ground (e.g. *Arachis hypogaea*, peanut).

geochronology The determination of time intervals on a geologic scale, through either absolute or relative dating methods. Absolute dating methods involve the use of radioactive elements and knowledge of their rates of decay: this yields an actual age in years for a given rock or fossil. Relative dating involves the use of fossils or sediments to place events and rock sequences in order, and does not provide absolute dates.

geochronometric scale (chronometric scale) A time-scale based on years BP (conventionally before 1950). Subdivisions on the scale are defined by particular units of duration (e.g. 10^6 years, 10^9 years) rather than reference points in actual rock successions. An example of such a subdivision is the placing of the boundary between the *Archaean and the *Proterozoic at 2 500 Ma (i.e. $2\,500 \times 10^6$ years) ago.

geochronometry The determination of the length of time intervals. Geochronometric resolutions for zonations based on different organisms may be calculated by dividing the time-span of a series by the number of zones and the intervals between zones. However, this will give only an approximate measure of time.

geo-electric section A diagrammatic section of stratified layers which is deduced from electrical (resistivity) depth probing or drilling, where layers are identified by their apparent resistivities. Such sections are useful in detecting *water-table levels and determining whether water is saline or fresh at the water-table.

geographical floral element A group of plant species that have marked geographical affinities. The flora of a given territory includes different geographical elements. For example, the Mediterranean element in the British flora comprises 38 species, concentrated in the extreme south and south-west of England. The majority are flowering *annuals or *biennials, categories of plants well represented in the sunny Mediterranean. When the full ranges of the 38 species are examined, they are all found to occur in the Mediterranean basin, with northerly extensions to Britain.

geographic information system (GIS) A computer database system that stores data organized by geographic coordinates, together with the sequences of operations needed to access and work with the data, and the personnel performing those operations and managing the system. Data are available on the World Wide Web. The user identifies a location, perhaps by pointing and clicking at a position on a map, and is able to download data relevant to that location. A wide range of geological, ecological, demographic, land-use, and other data is available, and can be displayed as maps, diagrams, satellite images, or photographs.

geologic time-scale A twofold scale that subdivides all the time since the Earth first came into being into named units of abstract time, and subdivides all the rocks formed since the Earth came into being into the successions of rock formed during each particular interval of time. The branch of geology that deals with the age relations of rocks is known as chronostratigraphy. The concept of a geologic time-scale has been evolving for the last century and a half, commencing with a relative time-scale (mainly achieved through biostratigraphy), to which it has gradually become possible to assign dates which are, nonetheless, subject to constant revision and refinement. Since the first International Geological Congress in Paris in 1878, one of the main objectives of stratigraphers has been the production of a complete and globally accepted stratigraphic scale to provide a historical framework into which all rocks, anywhere in the world, can be fitted. Such a standard scale is still a long way off, but the names for geologic-time units and chronostratigraphic units down to the rank of period/system are in common use; many epoch/series and age/stage names are still regionally variable. The International Union of Geological Sciences is the body responsible for agreeing the names and dates to be used in the time-scale. The most recent revision was published in 2004. See Appendix.

geomagnetic field The Earth's magnetic field; this shows variability on all time-scales, ranging from nanoseconds to millions of years. Most transient variations are of external origin, reflecting interactions between the solar wind and the Earth's atmosphere. Longer-term changes (secular variations) are of internal origin. The intensity of the field varies from about 30 μT near the equator to 60 μT near the observed geomagnetic poles at 73°N 100°W and 68°S 143°E.

geomagnetic reversal time-scale *See* MAGNETOSTRATIGRAPHIC TIME-SCALE.

geometric distribution *See* LOG-NORMAL DISTRIBUTION.

geometric series A numerical and graphical description of the relationship of species in a community according to their importance values. Curves approximating to a geometric series are common in pioneer or *ephemeral communities. In theory, the more successful *species restrict other species to the remaining *niche spaces (the pre-emption hypothesis) by occupying a larger proportion of the available niches. If the most successful species occupies 75 per cent of the available niche spaces, and the next species a similar proportion of the remaining spaces, and so on, then a geometric progression of importance values will result.

geomorphology The scientific study of the land-forms on the Earth's surface and of the processes that have fashioned them. Recently an extraterrestrial aspect has developed, resulting from studies of lunar and planetary surfaces.

geophysiology The scientific study of the Earth as a total system that comprises all its *biotic and *abiotic components (i.e. the *biota, atmosphere, hydrosphere, and lithosphere). The term was coined in about 1986 by James *Lovelock, whose *Gaia hypothesis proposes that a planet may be studied by methods analogous to those applied to the study of an individual organism.

geophyte A land plant that survives an unfavourable period by means of underground food-storage organs (e.g. rhizomes, tubers, and bulbs). Buds arise from these to produce new aerial shoots when favourable growth conditions return.

geostrophic current An ocean current that is the product of a balance between *pressure-gradient forces and the *Coriolis effect. This produces a current flow along the pressure gradient. Such a current does not flow directly from a region of higher pressure to one of lower pressure (i.e. 'down the slope' of the sea surface) but flows parallel to the gradient. All the major currents in the oceans are very nearly true geostrophic currents. The *Gulf Stream, for example, can be likened to a river that does not run down a hill but around the hill.

geostrophic wind The wind blowing above the *boundary layer that, by its strength and direction, represents the balance between the *pressure-gradient force, acting directly from the region of higher pressure towards the region of lower pressure, and the *Coriolis effect (force), deflecting moving air to the right (in the northern hemisphere and to the left in the southern hemisphere). When these are in balance the wind flows parallel to the isobars (*see* BUYS BALLOT'S LAW). Air is also subject to a centrifugal force, owing to the curvature of the air's path around a centre of low or high pressure. In the boundary layer, air experiences friction with the surface, causing it to flow across the isobars, at an angle of 10–20° over the sea and 25–35° over land (where friction is greater). *See also* GRADIENT WIND.

geotaxis A change in direction of locomotion in a motile organism or cell, made in response to the stimulus of gravity.

geothermal gradient The increase of temperature with depth below the ground surface. It usually refers to depths below 200 m. In the continents the gradient is usually between 20 and 40°C/km, although it can well exceed this in volcanic regions. In the oceans the depth of penetration of most core barrels is so short that the gradient can be determined over only a few metres and varies considerably. The average geothermal gradient at the surface of the Earth is about 24°C/km, but it is assumed to decrease with depth as widespread *mantle melting would otherwise occur. The observed gradients are therefore modified to result in an estimated temperature of about 1 200°C at the top of the seismic low-velocity zone in the upper mantle. Within the mantle, the increase of temperature with depth is considered to be less than 0.1°C/km greater than the adiabatic increase of 0.33°C/km.

geotropism A directional movement of a plant in response to the stimulus of gravity. Primary tap roots show positive geotropism; vertical primary shoots show negative geotropism; horizontal stems and leaves are diageotropic (*see* DIAGEOTROPISM); and branches and secondary roots at oblique angles are plagiogeotropic.

germination The beginning of growth of a seed, *spore, or other structure (e.g. *pollen), usually following a period of *dormancy, and generally in response to the return of favourable external conditions, most notably warmth, moisture, and oxygen. The internal biochemical status of the seed or spore must also be appropriate. In seeds germination may be epigeal (i.e. with cotyledons emerging above the ground) or hypogeal (i.e. with the cotyledons staying below ground).

germ plasm bank An establishment concerned primarily with the conservation of hereditary genetic material which may be lost through the process of *genetic erosion. Germ plasm loss is a major concern in Asia, parts of Africa, southern Europe, and countries bordering the Mediterranean, where antiquated *cultivars are rapidly replaced by new varieties. With the loss of older cultivars, qualities possessed by them may be lost permanently, and so cannot become incorporated in new varieties. *Gene banks are an important source of germ plasm.

gestalt The perception of a pattern or structure as a whole, not as a sum of its constituent parts. The German word *Gestalt* means 'form' or 'shape'. Ornithologists sometimes use the term 'jizz' as an alternative; this also incorporates an element of general behaviour of the organism.

gestation period The length of time from conception to birth in a *viviparous animal; it is usually a relatively fixed period for a particular species.

GFG co-evolution *See* GENE-FOR-GENE CO-EVOLUTION.

GHOST *See* GLOBAL HORIZONTAL SOUNDING TECHNIQUE.

ghost of competition past Term proposed by J. H. Connell in 1980 to describe one possible reason for observed differentiations in niches. Competing species may be less fit than a species which avoids competing because it occupies a *fundamental niche which does not overlap theirs. *Natural selection may then favour the non-competing species; its population increases and those of the competing species decrease. The observed differentiation is then the result of past competition and to explain it in this way is to invoke the ghost of competition past.

Ghyben–Herzberg relationship Beneath oceanic islands, the thickness (*d*) of a lens of fresh *groundwater (density ϱw) overlying sea water (density ϱm) can be determined if the height above sea level (*h*) of the top layer of the lens is known and the conditions are static. The relationship is $d = ah = \rho w/(\rho m - \rho w)$ where *a* is typically about 38. The relationship was discovered independently by W. Badon (sometimes spelled Baydon) Ghyben, who described it in 1888–9, and B. Herzberg, who described it in 1901.

giane *See* MEDITERRANEAN SCRUB.

gibber *See* GIBBER PLAIN.

gibber plain The Australian name for an extensive plain (normally a *pediplain) that is mantled by loose rock fragments (gibber). These fragments are typically the rubble left from the destruction of a silcrete *duricrust or from the breakdown of resistant conglomerates (in which case the gibber consists of quartz pebbles).

giga- (pronounced 'jigga') From the Greek *gigas* meaning 'giant', a prefix (symbol G) attached to *SI units, meaning the unit $\times 10^9$ (e.g. 2 Gm = 2 gigametres = 2×10^9 m).

Gilbert-type delta A river *delta which consists of a wedge-shaped body of sediment, comprising relatively thin, flat-lying, *topset sediments, long, steeply dipping *foresets which prograde (*see* PROGRADATION) from the river mouth, and thinner, flat-lying, *bottomset or *toeset deposits. Gilbert-type deltas are often developed in lakes, where river water and lake water are of the same density. It was first described by the American geologist Grove Karl Gilbert (1843–1918).

gilgai The undulating micro-relief of soils which contain large amounts of clay minerals (e.g. montmorillonite) that swell and shrink considerably on wetting and drying, to an extent that may be sufficient to fracture pipelines or move telegraph or fence poles from the vertical. The word is derived from a settlement in Queensland, Australia, where the soils of swelling clays are especially common. *See also* PATTERNED GROUND.

girdling Cutting transversely across the phloem in a stem so that downward transport of substances is unable to occur within the plant. For example, a tree is girdled by being cut right around its circumference to a depth that penetrates through the bark and into the wood; this kills the tree.

GIS *See* GEOGRAPHIC INFORMATION SYSTEM.

Givetian *See* DEVONIAN.

glabrous Smooth, lacking hairs.

glaci- (glacio-) Dominated by glacial ice. The prefix is followed by a term indicating the environment or process that is so dominated, e.g. glaciaquatic (of water derived from a *glacier), glacioeustasy (the theory that changes in sea level result from the growth and decay of ice sheets), glacifluvial (of sediments or land-forms produced by meltwater streams escaping from a glacier), glacioisostasy (the theory that local flexing of the Earth's crust occurs as a result of the loading and unloading that takes place as large ice sheets wax and wane), glacilacustrine (pertaining to a lake adjacent to a glacier), and glaciomarine (of sediments laid down in a sea environment near a glacier).

glacial breach A glacially eroded trough that cuts through a ridge and so breaches a former *divide. It is formed when the outflow of a *glacier (or ice sheet) is impeded, so that the thickness of the glacier consequently increases, and ultimately a higher escape route (the breach) is exploited. This process is called 'glacial diffluence' when a single glacier spills out of its valley, and 'glacial transfluence' when several breaches are formed, owing to the accumulation of a large ice sheet.

The many breaches through the western Highlands of Scotland were caused by the accumulation of transfluent ice east of the main watershed.

glacial diffluence *See* GLACIAL BREACH.

glacial diversion The displacement of a pre-glacial stream by the action of a *glacier. In upland areas, the new drainage often follows a *glacial breach. In lowlands, glacial *drift may block an existing valley and lead to stream diversion.

g

glacial drainage channel (meltwater channel) A channel cut by the action of glacial meltwater or by water from an ice-dammed lake (*overflow channel). Various types may be recognized, classified by the position of the channel with reference to the *glacier (e.g. icemarginal, *englacial, or *subglacial). Usually these channels are steep-sided and flat-floored, and are unrelated to the present drainage pattern.

glacial limit A line that marks the furthest extent of a former glacial advance. It may be identified on the ground through the recognition of features associated with glacial margins, including lateral and terminal *moraines, *outwash plains, marginal *glacial drainage channels, and *proglacial lakes. *See also* TRIM LINE.

glacial period A general term used to describe either a glacial stage (e.g. *Devensian) or an indeterminate period of glaciation.

glacial plucking (quarrying) The removal of relatively large fragments of bedrock by direct glacial action. The process involves several mechanisms, including the incorporation of rock fragments into the base of the *glacier when it freezes to weakened bedrock, and the removal of bedrock material when fragments already included in the ice are dragged over it.

glacial stairway The long profile of a *glacial trough: it is characterized by alternating rock bars (*riegels) and rock basins, giving the impression of a stairway. The structure is attributed to variations in the erosive power of ice, or to the influence of rock jointing.

glacial theory The theory first proposed in 1821 by Ignatz Venetz (1788–1859) and developed in the late 1830s and 1840s by Venetz, *Charpentier, and *Agassiz, that most of northern Europe, North America, and the north of Asia had been covered by ice sheets during a period later termed the *Pleistocene. The hypothesis was used to explain *erosion, and the subsequent deposition of *till or *boulder clay, and the extinction of species such as the mammoth. Since that time glacial theory has been developed to include multiple glaciation and evidence of much older ice ages.

glacial transfluence *See* GLACIAL BREACH.

glacial trough A relatively straight, steep-sided, U-shaped valley that results from glacial *erosion. Its cross profile approximates to a parabola, while its long profile is often irregular, with rock bars (*riegel) and over-deepened rock basins being typical features. The world's largest glacial trough is that of the Lambert Glacier, Antarctica, which is 50 km wide and about 3.4 km deep. *Fiords are glacial troughs flooded by the sea.

glaciaquatic *See* GLACI-.

glaciated rock knob *See* ROCHE MOUTONNÉE.

glaciation The covering of a landscape or larger region by ice; an *ice age.

glacier A large mass of ice that rests on a land surface (at coasts with a shallow shore slope this may include the seabed), is constrained by the topography (*compare* ICE SHEET), and that moves. Snow accumulates and changes into *firn in the *accumulation zone, where the climate is cold, and is lost from the *ablation zone lower down. The boundary between the accumulation and ablation zones is known as the *equilibrium line or firn line. The accumulation of ice in the accumulation zone and its loss from the ablation zone gradually alters the profile of the glacier, making it domed and increasing the gradient. This causes ice to move downwards and the degree of imbalance between the rate of accumulation and ablation determines the speed with which the glacier moves. This varies greatly. The Franz Josef Glacier,

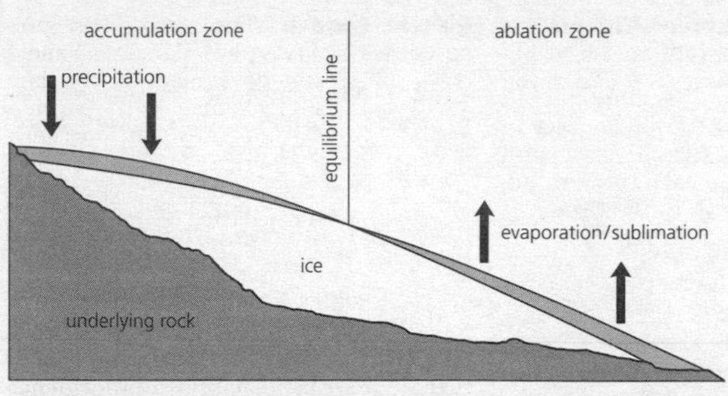

Glacier

New Zealand, moves about 300 m per year; the Meserve Glacier, Antarctica, moves about 3 m per year. Glaciers may be classified in several ways. The most useful division is based on temperature, and three categories are recognized. Temperate (or warm) glaciers are those where the ice is at *pressure melting point throughout the ice mass, except during winter when the top few metres may well be below 0°C and movement is largely by basal slip; glaciers in the European Alps are of this type. Polar (or cold) glaciers, such as those forming part of the Antarctic ice sheets, have temperatures well below the pressure melting point. Movement is slow and occurs largely by internal deformation. Subpolar glaciers, such as those in Spitzbergen, have temperate interiors and cold margins; they are therefore composite. Glaciers are also classed morphologically, based mainly on their size, shape, and the position of the ice mass. *Cirque glaciers, *valley glaciers, and *piedmont glaciers are among the types recognized.

glacier creep The deformation of *glacier ice in response to stress, by a process involving slippage within and between ice crystals. The rate of creep is dependent on both stress and temperature. When the shear stress is doubled the strain rate increases eight times, and a rise in temperature from −22 °C to 0 °C involves a tenfold increase in strain rate.

glacier ice *See* ICE.

glacier surge A relatively rapid movement of a *valley glacier or of an individual ice stream (*see* OUTLET GLACIER) within a major ice sheet. The movement may build up over a period ranging from a few months to several years and may be a hundred times faster than the 'normal' glacier movement velocity. Surging may result from an increase in ice thickness or from excessive basal water.

glacier wind *See* FIRN WIND.

glacifluvial *See* GLACI-.

glacilacustrine *See* GLACI-.

glacio- *See* GLACI-.

glacioeustasy (glacioeustasism) The theory that sea levels rise and fall in response to the melting of ice during *interglacials and the accumulation of ice during *glaciations. A major glacioeustatic oscillation has an amplitude of some 100 m.

glacioisostasy The adjustment of the *lithosphere following the melting of an ice sheet. The gradual rise of the land enables an estimation to be made of the rigidity of the lithosphere to flexing and the viscosity of the mantle. Adjustments need to be made to gravity readings in previously glaciated areas to allow for this effect, which occurs on a regional scale. Some 1 000 m of depression may have occurred in Scandinavia during the last ice age, where 520 m of recovery has subsequently been recorded. The process gives rise to warped shorelines.

glaciology The scientific study of ice in all its forms, including the study of ice in the atmosphere, in lakes, rivers, and oceans, and on and beneath the ground. Commonly, however, it is the study of *glaciers.

glaciomarine *See* GLACI-.

glaebules *See* CALICHE.

glaucous Sea-green or bluish-green in colour, or having a waxy, bluish-green bloom (e.g. as on a plum or the colour of the carpet formed on the surface of the open sea by swarms of sea 'lizards', *Glaucus eucharis*, members of the Nudibranchia). The original Greek word meant 'bright, sparkling, greyish', like the eyes of Pallas Athene.

glaze *See* CLEAR ICE.

Gleason, Henry Allan **(1882–1973)** An American ecologist, who worked at the New York Botanical Garden and, in 1917, challenged the *organismic *climax theory proposed by F. E. *Clements (and later by A. G. *Tansley in Britain), in favour of his own *individualistic hypothesis.

glenoid Pertaining to a socket. The word is derived from the Greek *glene*, meaning 'socket'.

gley The product of waterlogged soil conditions, and hence an anaerobic environment; it encourages the reduction of iron compounds by micro-organisms and often causes mottling of soil into a patchwork of grey and rust colours. The process is known as gleying or gleyzation (US usage).

Gleysols Soils with evidence of gleying (*see* GLEY) within 50 cm of the surface. Gleysols are a reference soil group in the *FAO *soil classification.

Glinka, Konstantin Dimitrievich **(1867–1927)** A Russian soil scientist who worked at the University of St Petersburg. Glinka was a student of *Dokuchaev and developed and organized his work. He was responsible for the soil surveys of most of European Russia and of Siberia. His book *Soil Science* was first published in 1908. *See* SOIL CLASSIFICATION.

Global Atmospheric Research Programme **(GARP)** An international project that aims to provide a comprehensive knowledge of atmospheric structure as an aid to prediction. The complex system is based on land and marine weather stations, together with upper-air soundings, satellite remote-sensing, and sensors carried by balloons.

Global Horizontal Sounding Technique **(GHOST)** A programme for the direct sensing of the atmosphere, using balloons designed to float at various constant-density levels. These are tracked and their various measurement readings are monitored by polar orbiting satellites. Sensors on the balloons record temperature, humidity, and pressure; tracking gives mean winds for each balloon. The project forms part of the World Weather Watch.

Global Positioning System **(GPS)** An aid to orientation and navigation that is based on transmissions from 24 satellites orbiting at a height of approximately 20 000 km. Each satellite takes 12 hours to complete one orbit and passes over the same point on the surface at intervals of 23 hours 56 minutes. Between five and eight satellites are above the horizon at any one time. A GPS receiver compares timing signals from several satellites with the time it receives them and measures the direction of the signal. The receiver automatically calculates the distance to the satellite from the time taken for the signal to reach it from the satellite. The distance and direction of several satellites allow the receiver to calculate its position and elevation, and it also records the time. Many GPS receivers are hand-held. The Standard Positioning Service (SPS), available to anyone free of charge, reports the position to within 100 m, the altitude to within 156 m, and the time with an accuracy of 340 nanoseconds.

global warming The modification of climates that would result from the retention of an increased proportion of *terrestrial radiation by certain atmospheric gases emitted mainly as by-products of human activities (i.e. an *anthropogenically induced *'greenhouse effect'). Computer models of the *general circulation predict that the warming would occur mainly in middle to high latitudes, but its precise extent and consequences (e.g. in terms of changes in sea level) are uncertain.

***Globigerina* ooze** Deep-sea ooze in which at least 30 per cent of the sediment consists of planktonic members of the order Foraminiferida, including chiefly *Globigerina*. It is the most widespread deposit to form from the settling out of material from overlying waters, covering almost 50 per cent of the deep-sea floor. *Globigerina* ooze covers most of the floor of the western Indian Ocean, the mid-Atlantic Ocean, and the equatorial and South Pacific. Species occurring in this deposit have been used to establish climatological and temperature criteria. *Globorotalia menardii* is supposed to indicate warmer conditions and *Globigerina pachyderma* to indicate colder temperatures. Another foraminiferan, *Globorotalia truncatulinoides* can coil in either a left- or right-handed manner and it is suggested that right coiling indicates warmer conditions and left coiling colder ones.

globose Spherical.

Gloger's rule Individuals of many species of insects, birds, and mammals are darkly pigmented in humid climates and lightly coloured in dry ones. This may well be a camouflage adaptation—moist habitats are usually well vegetated and tend to lack pale colours. There are many exceptions to this so-called rule. It was proposed by the German zoologist Constantin Wilhelm Lambert Gloger (1803–59). *See also* ALLEN'S RULE and BERGMANN'S RULE.

***Glossopteris* flora** The fossil flora that succeeds the *Permian glacial deposits of South Africa, Australia, South America, and Antarctica. It grew in a cold, wet climate, while the flora of North America and Europe existed under warm conditions. Plants with elongate, tongue-shaped leaves dominated the southern flora, with the genera *Glossopteris* and *Gangamopteris* being among the best known. *Glossopteris*, which gives its name to the flora, is characterized by a leaf with a fairly well defined midrib and a reticulate (net-like) venation. *G. indica* is the last species referred to the genus and to the family Glossopteridales. It is known from the *Triassic of India.

gnotobiotic Applied to a culture in which the exact composition of organisms is known, down to the presence or absence of bacteria. Such cultures usually develop during the formation of experimental laboratory ecosystems (*microcosms) from *axenic cultures. The word is derived from the Greek *gnosis*, meaning 'knowledge', and *bio*, meaning '(human) life'.

g

goal The state that is presumed to exist in the brain of an animal and which corresponds to a state of affairs the animal seeks to achieve. It is inferred from observation of the stimuli that terminate a behaviour.

golden spike *See* BOUNDARY STRATOTYPE.

Gondwana A former supercontinent of the southern hemisphere from which South America, Africa, India, Australasia, and Antarctica are derived. Their earlier connection explains why related groups of plants and animals are found in more than one of the now widely separated southern land masses; examples include the conifer *Araucaria* (monkey puzzle, Chile pine, hoop pine, etc.) common to South America and Australia. Throughout Gondwana there existed floristic assemblages represented by a few species of fossil plants that are thought to have grown in a cold climate. This view is supported by palaeomagnetic evidence and by indications of glaciation in South America, Africa, India, and Australia, which must then have been in a much higher latitude than they are today.

gonotheca In some colonial coelenterates, a cylindrical capsule containing a reproductive *polyp.

gonozooid In some colonial coelenterates, a reproductive *polyp.

goscojales *See* MEDITERRANEAN SCRUB.

gossan A near-surface, iron oxide-rich zone overlying a sulphide-bearing ore deposit, which is caused by the oxidation and *leaching of sulphides. It is useful in mineral exploration as a visible guide to sulphide mineralization by its yellow or red colour.

GPP (gross primary productivity) *See* PRIMARY PRODUCTIVITY.

GPS *See* GLOBAL POSITIONING SYSTEM.

grade **1.** A group of things all of which have the same value. **2.** A distinctive functional or structural improvement in the organization of an organism. Thus fish, amphibians, reptiles, and mammals represent successive vertebrate grades. Grades may occur within a single lineage or the same grade may be achieved independently in different ones (e.g. warm-bloodedness evolved independently in birds and mammals). **3.** The fraction of a sediment that falls within a particular size limit (e.g. sand grade, silt grade, and boulder grade). **4.** The quality of a mineral ore (i.e. the classification of an ore by the quantity or purity of the mineable metal in an ore body. **5.** In civil engineering, the gradient of a road. **6.** A balanced condition, especially of a river (river or stream grade) that has just sufficient energy to transport the load supplied from the drainage basin; a balance between erosion and deposition. The concept has also been applied to hillslopes ('graded slopes') that are stable dynamically and so maintain themselves in the most economical configuration. The term is no longer used widely as it oversimplifies the issues involved.

graded reach A length (reach) of a stream channel in which the gradient and cross-sectional form have become adjusted to carry just the discharge and sediment load that are normally supplied from upstream. Such a reach is said to be in equilibrium. Early definitions emphasized the smooth long profile of such a reach; it is now realized that a profile may be irregular and the reach still be graded.

graded slope *See* GRADE.

gradient analysis An ordination technique for the description of vegetation based on characteristics of the site rather than the composition of species. One or more environmental gradients are identified and *stands are arranged in order along them according to the characteristics of their sites. The vegetation is then examined to discover any related pattern.

gradient wind A wind that represents the balance among all the forces acting upon moving air. It may be expressed by V in the equation: $G = 2D\omega V \sin\varphi \pm DV^2/r + F$, where G is the pressure gradient, D the air's density, ω the angular velocity of the Earth's rotation about its axis, φ the latitude, r the radius of curvature of the air's path, F the friction, and V the air's velocity when the forces are in balance. The sign before the second term is + when the motion is around a cyclonic (low-pressure) centre,—when the motion is around an anticyclonic (high-pressure) centre. *See also* GEOSTROPHIC WIND.

grading curve (soil grading curve) A graph of grain size (horizontal logarithmic scale) plotted against percentage distribution (vertical arithmetic scale); a point on the curve indicates the percentage by weight of particles smaller in size than the grain size at the given point.

gradualism *See* PHYLETIC GRADUALISM.

graft **1. (noun)** A small piece of tissue implanted into an intact organism. **2. (verb)** To transfer a part of an organism from its normal position to another position on the same organism (autograft), or to a different organism of the same species (homograft), or an organism of a different species (heterograft). The source of the part that is grafted is called the donor (in animals) or the scion (in plants) and the organism to which it is united is called the host or recipient (in animals) or the stock (in plants). A graft hybrid is an organism made up of two genetically distinct tissues, owing to the fusion of host or scion and donor or stock after grafting. In cultivation a stem from one plant (the scion) is fused with a rooted portion of another (the stock) to form a single plant. Most fruit trees are produced by grafting, the type of fruit being determined by the scion, but the size of the tree by the stock.

grain roughness *See* BED ROUGHNESS.

Gram-negative bacteria *See* GRAM STAINING.

Gram-positive bacteria *See* GRAM STAINING.

Gram staining A test that distinguishes between two groups of similarly shaped bacteria by detecting differences in the structure of their cell walls. The test was devised in 1884 by the Danish physician and microbiologist Hans Christian Gram (1853–1939). A heat-fixed smear of bacteria is covered for 30 seconds with a solution of crystal violet and Gram's iodine solution. This is then rinsed off with acetone or *ethanol and the bacteria stained with a red dye. Bacteria with a large amount of peptidoglycan in their cell walls retain the violet dye; these are designated Gram-positive. The violet dye washes from bacteria with less peptidoglycan, but they retain the red dye; these are known as Gram-negative bacteria. Among bacteria capable of causing disease, Gram-negative species tend to be more harmful than Gram-positive species.

granulose Consisting of or covered with small granules or grains.

granulosis A disease of insects caused by certain DNA-containing viruses. The name of the disease derives from the granular inclusions seen in the cells of an infected insect. It affects mainly the larval stages of Lepidoptera (butterflies and moths).

Granville wilt (southern bacterial wilt) A disease of tobacco plants caused by the bacterium *Pseudomonas solanacearum*. Symptoms include wilting and yellowing of the leaves.

grass heath *See* ACIDIC GRASSLAND.

grassland Ground covered by vegetation that is dominated by grasses. Grassland constitutes a major world vegetation type and occurs where there is sufficient moisture for grass growth, but where the environmental conditions, both climatic and anthropogenic, prevent tree growth. Its occurrence therefore correlates with a rainfall intensity between that of desert and that of forest, and the range of grassland is extended by grazing and/or fire to form a *plagioclimax in many areas that were previously forested. The extensive mid-latitude grassland is known as *steppe or *prairie, whereas the corresponding tropical vegetation is called *savannah.

grass minimum temperature The minimum temperature recorded in open ground at night by a thermometer whose bulb is exposed over the tips of short grass.

graupel Soft hail, composed of particles resembling small snowballs, and formed by accretion when a snowflake falls through supercooled water droplets which freeze on contact and cover the flake.

graviportal Applied to limbs that are short and massive.

gravitational water Water that moves through soil under the influence of gravity and must be removed from the soil before this can attain *field capacity.

gravity corer A tube used to obtain samples of lake or marine sediments, which penetrates the material under its own weight. *See also* HYDRAULIC CORER.

gray (Gy) The *SI unit of absorbed radiation dose, being an energy of 1 joule imparted by ionizing radiation to 1 kilogram of matter. It is named after the British radiologist L. H. Gray (1905–65).

Gray, Asa (1810–88) An American botanist and taxonomist who did much to popularize the study of botany and to expound, but also criticize, *Darwin's evolutionary theory. In 1842 he was appointed Fisher Professor of Natural History at Harvard University and founded the Gray Herbarium and a library. He was an original member of the National Academy of Sciences and in 1872 was elected president of the American Association for the Advancement of Science. In 1859 he published a memoir on the relationships between the floras of Japan and North America, one of the earliest studies of discontinuous plant distribution. His *Manual of the Botany of the Northern United States*, first published in 1848, was probably his most enduring work, although it was his *Statistics of the Flora of the Northern United States* (1856–7) which established his academic reputation. He collaborated with John *Torrey to produce the two-volume *Flora of*

North America (1838–43) and published his *Synoptical Flora* in 1878.

grazing food-chain *See* GRAZING PATHWAY.

grazing pathway (grazing food-chain) A *food-chain in which the *primary producers (green plants) are eaten by grazing *herbivores, with subsequent energy transfer to various levels of *carnivore, e.g. plant (blackberry, *Rubus*) → herbivore (bank vole, *Clethrionomys glareolus*) → carnivore (tawny owl, *Strix aluco*). *Compare* DETRITAL PATHWAY.

g

great conveyor (Atlantic conveyor) A system of ocean currents forming a closed loop that transports cold water from the edge of the sea ice in the North Atlantic (*see* THERMOHALINE CIRCULATION). The water flows along the ocean floor to the Southern Ocean, then eastward and later northward, crossing the equator in the Indian and Pacific Oceans where it rises to about 1 000 m below the surface. Moving westward, it passes to the east of Madagascar, rounds the Cape of Good Hope, and flows northward through the Atlantic, completing the loop.

Great Interglacial *See* MINDEL/RISS INTERGLACIAL.

great soil groups A *soil classification that was devised in 1949, categorizing soils according to the climatic conditions in which they form. Soils were placed in three orders. Zonal soils are directly related to the climates in which they occur. Azonal soils are young soils and not related to climatic conditions. Intrazonal soils result from local conditions that are unrelated to climate; these include the soils of *marshes and *swamps, saline and alkaline soils, and *littoral deposits. The soil orders are then divided into 9 suborders, and the suborders into great soil groups. *See also* PEDALFER and PEDOCAL.

'greenhouse effect' The effect of heat retention in the lower atmosphere as a result of absorption and reradiation of long-wave (more than 4 µm) terrestrial radiation by clouds and gases (e.g. water vapour, carbon dioxide, methane, and chlorofluorocarbons), which make the atmosphere transparent to incoming short-wave radiation but partly opaque to reradiated long-wave radiation. The insu-

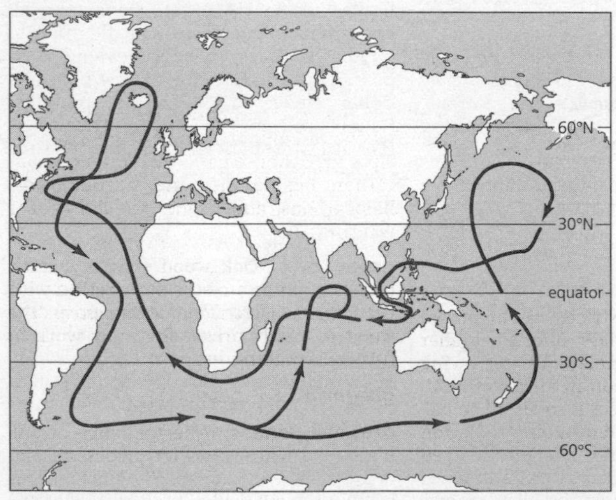

Great conveyor

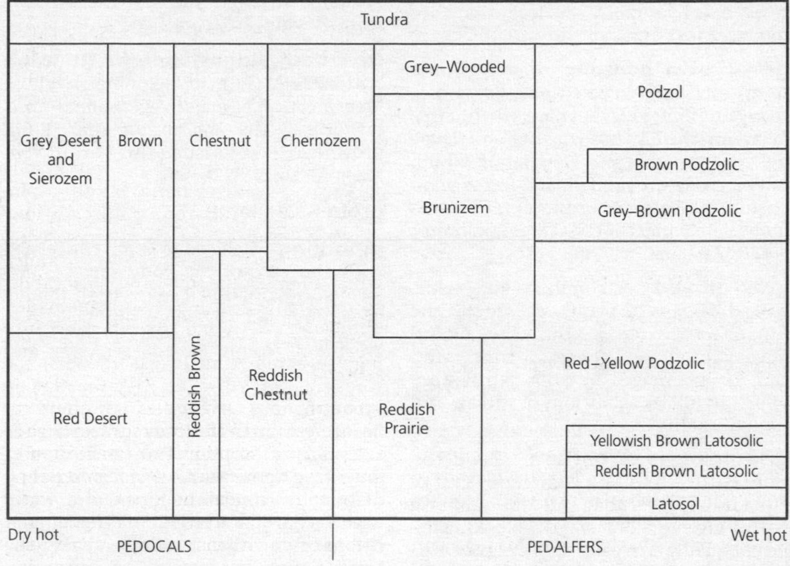

Great soil groups

Greenhouse effect

lating effect is not strictly analogous to that of greenhouse glass, since the higher temperature in a greenhouse is owing partly to the reduction in air movement. The atmospheric effect is to alter the balance of incoming and outgoing radiation in the Earth's energy budget. Marked increases in atmospheric carbon dioxide, generated, for example, by the combustion of fossil fuels, could result in a global increase of atmospheric temperatures if not offset by other (perhaps natural) changes.

greenhouse gas A gas that absorbs long-wave radiation and therefore contributes to *greenhouse-effect warming when present in the atmosphere. The principal greenhouse gases are water vapour, carbon dioxide, methane, nitrous oxide, halocarbons, and ozone. *See also* GLOBAL WARMING.

green oak Oak wood that is stained bluish-green as a result of infection with the fungus *Chlorociboria aeruginascens*. The wood is highly prized for inlay work in furniture-making, etc.

gregaria *See* LOCUST (1).

gregariousness The tendency of animals to form groups which possess a social organization (e.g. schools of fish, herds of mammals, flocks of birds). *Compare* AGGREGATION.

Grenz horizon *See* SUB-ATLANTIC.

gressorial Applied to legs that are modified for running.

grey-brown podzolic A soil-*profile term describing eluviated (*see* ELUVIATION), freely draining soils that have a distinctive *clay-enriched B *horizon. The soil develops under temperate woodlands which have a moderate rainfall, and forms good pasture-land after deforestation. The term is obsolete under the USDA Soil Taxonomy. *See also* ALFISOLS.

g

grey mould A common and widespread disease of plants, caused by the fungus *Botrytis cinerea*. A fluffy grey mould appears on flowers, leaves, or fruit. Infection is encouraged by poor ventilation, which produces conditions of still air and high humidity.

grid analysis of pattern The detection of pattern using a contiguous grid of *quadrats, rather than random sampling with a given quadrat size. By blocking adjacent quadrats in pairs, fours, eights, etc., the data may be analysed using increasingly larger quadrat sizes. This is important since the detection of pattern relates to the quadrat size used, with the most marked demonstration of contagion occurring when the quadrat size is approximately equal to the clump area (i.e. the size of the clumps). When the presence of contagion is not immediately obvious, detection of the scale of non-randomness (i.e. the quadrat size at which it is evident) provides very useful information, as it may suggest the likely cause of the clumping.

grike *See* LIMESTONE PAVEMENT.

Grime's habitat classification A scheme devised in the 1970s by J. P. Grime in which *habitats are categorized by the conditions they provide and the level of disturbance affecting them. This determines an appropriate life strategy for the plant species occupying them. Where resources are abundant and disturbance is rare, the habitat becomes crowded and species adopt a competitive strategy. Where disturbance is rare, but resources are scarce or conditions harsh, species adopt a tolerant strategy. Where conditions are benign, resources plentiful, but disturbance is common, species adopt a *ruderal strategy. Various intermediate strategies are also identified.

Grisebach, August Heinrich Rudolph (1814–79) A German taxonomist whose identification of major vegetational units introduced the concept of the *floral province. He expounded this idea in *Vegetation der Erde* (1872).

groin *See* GROYNE.

grooming Behaviour that is concerned with the care of the body surface, performed by an animal upon itself (*autogrooming), or by one animal upon another of the same species (*allogrooming). In some animals (e.g. rodents) the spreading of saliva on the fur during grooming may serve a role in thermoregulation. Autogrooming may serve to change the state of arousal of an animal; allogrooming strengthens social bonds and helps to provide all members of a group with a scent characteristic of the group, by means of which group members may identify one another.

groove mark A linear groove, cut in a muddy substrate by the dragging of an object through the sediment by flowing water, the orientation of which is parallel to the direction of the current. Subsequent infilling of the groove by sediment results in a groove cast being preserved on the base of the overlying bed.

gross ecological efficiency *See* FOOD-CHAIN EFFICIENCY and LINDEMAN'S EFFICIENCY.

gross primary productivity (GPP) *See* PRIMARY PRODUCTIVITY.

ground cover The area of ground that is covered by a plant when its *canopy edge is projected downwards perpendicularly.

ground flora A general term describing plants of the *field layer and *ground layer.

ground frost The condition in which the ground surface has a temperature below 0°C. *Compare* AIR FROST.

ground ice *See* ICE.

ground layer The lowest layer of a plant *community, comprising especially mosses, lichens, and fungi, together with low-growing herb species which often have trailing stems or *rosette forms.

ground moraine *See* MORAINE.

grounding line *See* ICE SHELF.

groundwater Water that occurs below the Earth's surface, contained in *pore spaces within *regolith and bedrock. It is either passing through or standing in the soil and underlying strata, and is free to move under the influence of gravity. Most groundwater derives from surface sources (*meteoric water); the remainder is either introduced by magmatic processes (*juvenile water) or is *connate water. *See also* VADOSE ZONE.

groundwater facies A *groundwater that has a particular character and chemistry. The concept of hydrochemical facies is based on the assumption that the chemical composition of groundwater at any point tends towards chemical equilibrium with the matrix rocks under the prevailing conditions.

group selection A type of selection affecting some kinds of behaviour (e.g. in certain species the decline in reproductive rates of a population that has passed the optimal density), which works to the advantage of populations rather than individuals. There is general scepticism about this theory.

group speed (of wave) In deep-water areas individual waves within a group move forward through the group at a faster speed than the group itself moves. The group form is produced by the interference of waves of different wavelengths (and therefore speeds) as they move forward from their source. New waves form at the rear of the group, lose amplitude as they reach the front of the group, and disappear. The group speed is half that of the individual waves that it comprises.

grouting Injecting liquid cement or chemical into the ground, where they set and subsequently impede or prevent the flow of water by sealing pore spaces and fractures in the subsurface rock.

grove **1.** A small wood, usually of less than eight hectares. **2.** An area that is composed entirely of timber trees (without *underwood) within a larger woodland which contains underwood.

growing season In seasonal climates, the period of rapid growth. Various definitions are used. In Britain, the growing season is the period when the mean temperature exceeds 6°C, though there is no agreement whether daily, weekly, or monthly means should be used, or whether these should be of ground or air temperatures. In the USA the growing season is defined as the period between the last *killing frost of spring and the first killing frost of autumn, a killing frost usually being regarded as one recorded on a thermometer exposed in a standard way in a *Stevenson screen.

growth The increase in size of a cell, organ, or organism. This may occur by cell enlargement or by cell division.

growth band (growth line) A band or line, found in many organisms, which marks growth. Growth lines in Brachiopoda, for example, record the former position of the commissure; in Gastropoda, growth may be intermittent and a band of rapid growth may be preserved in the shell.

growth form **1.** The *morphology of a plant, especially as it reflects physiological *adaptation to the environment. *See also* LIFE-FORM. **2.** The shape of population growth, as expressed by a growth curve (e.g. *J-shaped or *S-shaped).

growth line *See* GROWTH BAND.

growth regulator A synthetic compound that, when applied to a plant, promotes, inhibits, or otherwise modifies the growth of that plant.

growth retardant One of a group of synthetic compounds that reduce stem elongation in plants by inhibiting the activity of the subapical meristem. An important agricultural use is in the prevention of lodging in cereals. In horticulture retardants are used to produce more compact plants. Some of these compounds inhibit gibberellin synthesis.

growth substance A naturally occurring compound, other than a nutrient, which promotes, inhibits, or otherwise modifies the growth of a plant.

groyne (groin) A breakwater made from rock, concrete, wood, or metal, erected on a beach to inhibit the movement of sand and shingle and to protect against *longshore drift.

guan- A prefix derived from the Quechua *huanu*, meaning 'dung' (e.g. guanine can be extracted from *guano).

g

guanine A *purine base found in *nucleic acids. *See also* ADENINE, CYTOSINE, and THYMINE.

guano The accumulated droppings of birds, bats, or seals, found at sites where large colonies of these animals occur. Guano is rich in plant nutrients, especially calcium phosphate (bird guano is richer than bat or seal guano). Such deposits are found particularly on arid oceanic islands and in caves. Guano is worked industrially as a phosphate resource.

guerilla growth form The distribution that results when the rhizomes or *stolons by which a plant spreads by *clonal dispersal are long and often short-lived. The clonal shoots are widely spaced and *fugitive, constantly appearing and disappearing in different territories, like a guerilla army. *Compare* PHALANX GROWTH FORM.

guild A group of species all members of which exploit similar resources in a similar fashion.

Guinea zone The wettest and most wooded *savannah zone in West Africa in the classic threefold model of savannah vegetation. These are the Sahel zone, consisting of dry scrub with occasional grasses and patches of bare earth; typical grassy savannah; and the wetter, semi-wooded Guinea zone.

Gulf Stream The most important ocean-current system in the northern hemisphere, which stretches from Florida to north-western Europe. It incorporates several currents: the *Florida Current; the Gulf Stream itself; and an eastern extension, the *North Atlantic Drift. The Florida Current is fast, deep, and narrow, but after passing Cape Hatteras the Gulf Stream becomes less effective at depth and develops a series of large meanders which form, detach, and reform in a complicated manner. After passing the Grand Banks (off Newfoundland) the flow forms the diffuse, shallow, slower-moving North Atlantic Drift. The temperature (18–20°C) and salinity (36 parts per thousand) tend to be seasonally constant, unlike neighbouring coastal water masses.

gully A feature of surface *erosion that develops from the run-off of a violent torrent that bites deeply into topsoil and soft sediments. Gullies can develop on valley sides as valley-side gullies and also along valley floors as *arroyos. *See also* RILL-WASH.

Günz The first of four glacial episodes established by A. Penck and E. Bruckner in 1909. It is named after an Alpine river, so the term is really applicable only to its type area, but it has come to be used much more widely. The Günz may correlate with the *Menapian of northern Europe.

Günz/Mindel Interglacial An Alpine *interglacial stage, which is perhaps the equivalent of the *Cromerian stage of northern Europe.

gust A sharp increase in wind strength close to the ground, caused by mechanical disturbance in an air flow. Gusts may also be generated by temperature lapse rates and by wind shear (e.g. in *clear-air turbulence).

gustation The act of tasting.

gust front *See* SQUALL LINE.

guttation The extrusion of water and sometimes salts from the aerial parts of plants, particularly at night when *transpiration rates are low. Guttation may occur in plants that are less than about 10 m tall, where hydrostatic pressure is insufficient to prevent the flow of water into the xylem when the rate of transpiration is low. It also occurs in tropical plants, where high humidity inhibits transpiration.

Gy *See* GRAY.

gymnosperm A plant that produces seeds from ovules which are carried naked on the cone scales. The gymnosperms were formerly classified in the subdivision Gymnospermae, but this is now considered polyphyletic (*see* POLYPHYLY). The gymnosperms include the orders Ginkgoales (ginkgo), Cycadopsida (cycads), Coniferopsida (coniferous plants), and Gnetales (a possibly polyphyletic group containing *Welwitschia*, a plant found in the deserts of southern Africa, *Ephedra*, a shrub of American deserts, and *Gnetum*, tropical trees or shrubs). *Compare* ANGIOSPERM.

gynandromorph An organism in which part of the body is male and the other part female.

gynochory Dispersal of spores or seeds by motile females.

gynodioecious Applied to plants in which female and hermaphrodite flowers are borne on separate plants. *Compare* ANDRODIOECIOUS.

gynomonoecious Applied to plants in which female and hermaphrodite flowers are borne on the same plant. *Compare* ANDROMONOECIOUS.

gypcrete A gypsiferous (i.e. rich in gypsum) soil profile developed in arid regions. Gypcretes are formed by the precipitation of gypsum ($CaSO_4$) from saline waters drawn to the surface by capillary action.

gypsic Applied to a *soil horizon (a gypsic horizon) where secondary gypsum ($CaSO_4$) has accumulated through more than 150 mm of soil, so that this horizon contains at least 5 per cent more gypsum than the underlying horizon.

Gypsisols Soils that have a *gypsic horizon within 100 cm of the surface, or more than 15 per cent gypsum (calcium sulphate) below 100 cm. Gypsisols are a reference soil group in the *FAO *soil classification.

gyre **(ocean gyre)** A circular or spiral motion of water, the term usually being applied to a semi-closed current system. A major gyre exists in each of the main ocean basins, centred at about 30° from the equator and displaced towards the western sides of the ocean ('western intensification'). Gyres are generated mainly by surface winds and move clockwise in the northern hemisphere and anticlockwise in the southern hemisphere. Gyres form part of the global system of oceanic circulation (*see* GREAT CONVEYOR), which has an important function in the redistribution of energy around the world, especially from the tropics towards the poles. In the North Atlantic, the gyre was strongly affected around 13 000 years ago when fresh water draining from what is now the St Lawrence River formed a layer floating above the denser sea water. This suppressed the northward movement of the Gulf Stream (the North Atlantic Current or North Atlantic Drift). The outcome was a sudden cooling of northern Europe leading to the Younger *Dryas stadial.

gyrose Sinuous, curved; marked with curved lines or grooves.

gyttja **(nekron mud)** A rapidly accumulating, organic, muddy deposit, characteristic of *eutrophic lakes. The precise nature of gyttja varies with the producer organisms involved, which include small algae or macrophytes.

ha *See* HECTARE.

haar Especially in coastal areas of eastern Scotland and north-eastern England, a common name for a sea fog; these are frequent in early summer.

habit The typical growth form or occurrence of a plant (i.e. its form or shape).

habitat The living place of an organism or *community, characterized by its physical or *biotic properties.

habitat selection The choice by an organism of a particular habitat in preference to others (e.g. mayfly nymphs inhabit the underside of stones in fast-flowing streams and burrow in sediment in still water).

habituation A decrease in behavioural responsiveness which occurs when a stimulus is repeated frequently with neither reward nor punishment. The process involves learning to ignore insignificant stimuli and should not be confused with *accommodation.

haboob From the Arabic *habb* meaning 'to blow', a local name for a *dust storm in northern Sudan. Typically, the storm is experienced late in the day during summer.

hadal zone The part of the ocean that lies in very deep trenches below the general level of the deep-ocean floor (the *abyssal zone).

hadalpelagic zone The deepest part of the *pelagic zone of the ocean, extending from a depth of 6 000 m to the bottom of oceanic *trenches up to 11 000 m below the surface. Conditions in the hadalpelagic zone are extreme. No sunlight penetrates, the temperature is a constant 4°C, and the pressure is 60–110 MPa. *See* OCEAN DIVISIONS.

Hadean The interval of geologic time that lasted from 4 567.17 Ma until 3 800 Ma and preceded the *Archaean eon. Divisions of geologic time are required to have firmly dated events at their bases. This is not possible with the Hadean and consequently its name is used only informally.

Hadley cell One of the most fundamental divisions of the global wind circulation, comprising the net ascent of air over the lowest latitudes caused by convection, the compensatory downward motion in the subtropical anticyclone belt, and the resulting *trade winds which blow towards the meteorological equator (intertropical convergence). It is named after the English meteorologist George Hadley (1685–1768), who in 1735, seeking to account for the trade winds, proposed a single, large-scale convective cell representing a thermally driven, low-latitude atmospheric circulation.

Haeckel, Ernst Heinrich Philipp August (1834–1919) A German anatomist, zoologist, and field naturalist, who was appointed professor of zoology at the Zoological Institute, Jena, in 1865. He was an enthusiastic supporter of the Darwinian theory of evolution and *Darwin credited him with the success that theory enjoyed in Germany. Haeckel constructed genealogical trees for many living organisms and divided animals into Protozoa (single-celled) and Metazoa (multi-celled) groups. He was an accomplished field naturalist and is usually credited with having coined the word *'ecology' in his *Generelle Morphologie der Organismen*, published in two volumes in 1866 (as the German word *ökologie*, from the Greek *oikos*, meaning 'house' and *logos*, meaning 'discourse').

hag **1.** A parcel of wood that has been marked off for cutting. *See also* CANT and COUPE. **2.** An exposed face of peat that has been cut or eroded.

hail A form of *precipitation consisting of ice in the shape of balls or irregular particles (hailstones), whose concentric structure indicates a growth by *coalescence and freezing of supercooled water drops. Hail forms only in *cumulonimbus cloud and the larger the cloud the bigger the hailstones it generates.

hair hygrometer An instrument that indicates *relative humidity, based on the expansion and contraction of a treated human hair.

half-life *See* DECAY CONSTANT and RADIOMETRIC DATING.

halinity The extent to which particular water contains chloride. According to the Venice system, brackish waters (which are saline, but less so than sea water) are classified by the chloride they contain and divided into zones. The zones, with their percentage chlorinity (mean values at limits) are: euhaline 1.65–2.2; polyhaline 1.0–1.65; mesohaline 0.3–1.0; alphamesohaline 0.55–1.0; beta-mesohaline 0.3–0.55; oligohaline 0.03–0.3; fresh water 0.03 or less.

Hallian The last of two stages in the *Pleistocene of the west coast of North America, underlain by the *Wheelerian, overlain by the *Holocene, and roughly contemporaneous with the Upper Pleistocene Series of southern Europe.

halo- A prefix meaning 'pertaining to salt', derived from the Greek *hals*, *halos*, meaning 'salt'.

halo A rainbow-coloured, sometimes white, ring around the Sun or Moon, caused by the refraction of light by ice crystals when thin *cirrus cloud obscures the luminary. *Compare* CORONA.

halo blight A bacterial disease of dwarf and runner beans (*Phaseolus* species) caused by *Pseudomonas phaseolicola*. Symptoms include the appearance of brown spots on the leaves, each spot being surrounded by a yellow ring or halo, and infected plants may be stunted.

halocarbon An industrial compound containing carbon and a halogen (usually bromine, chlorine, or fluorine) which is commonly used in fire extinguishers. Halocarbons absorb long-wave electromagnetic radiation (i.e. they are *greenhouse gases) and, because of their chemical stability, they survive in the atmosphere long enough to enter the *stratosphere, where they are decomposed by ultraviolet radiation, releasing by-products that are implicated in the thinning of the *ozone layer. *See also* CHLOROFLUOROCARBON.

halocline A zone in which there are rapid, vertical changes in the *salinity. In low latitudes the halocline usually represents a decrease in salinity with increasing depth; in high latitudes it may represent the opposite. The halocline is usually well developed in coastal regions where there is much freshwater input from rivers producing surface waters of low salinity, a zone where salinity increases rapidly with depth (the halocline), and a deeper zone of more saline, denser waters.

halophile An *extremophile (domain *Archaea) that thrives in extremely saline environments.

halophilic Thriving in, or preferring to grow in, the presence of salt.

halophyte A terrestrial plant that is adapted morphologically and/or physiologically to grow in salt-rich soils and salt-laden air (e.g. *Salicornia* species, glassworts). *See also* SALT MARSH.

halosere A characteristic sequence of *communities associated with the developmental stages in plant *succession on *salt marshes or salt desert.

halo spot A disease of barley and other cereals, caused by the fungus *Septoria oxyspora* (*Selenophoma donacis*). Lesions on upper leaves have pale centres with dark margins. The disease is important in only a few regions of Britain, including the wetter western regions of England, Wales, and Ireland.

hamada **(hammada)** A rocky *desert, or desert region, which does not have surficial materials and which consists mainly of boulders and exposed bedrock. Two basic types occur: stony hamada, jaggedly developed across crystalline rocks; and pebbly hamada, cut across sedimentary material and mantled with bedrock fragments.

hammada *See* HAMADA.

handicap principle A controversial idea, advanced by A. Zahavi in 1975, which explains the existence of extravagant male traits (e.g. the tail of a peacock) by proposing that *sexual selection favours them because they indicate male prowess: the male demonstrates to females his ability to thrive despite the handicap (in this example of a heavy, cumbersome tail). Similarly, elaborate bird-song may demonstrate the success of the male in finding food quickly, allowing more time for song, and in avoiding predators, and elaborate plumage may demonstrate an ability to control skin and feather parasites. *Compare* RUNAWAY HYPOTHESIS.

hanging valley A tributary valley whose floor is well above that of the adjacent main valley and where there is therefore often a waterfall. It is typical of glaciated uplands, where it may result from glacial widening and/or deepening of the main valley.

haploid Applied to the number of chromosomes in a *gamete and conventionally symbolized by *n*. In somatic (i.e. non-sex) cells, the number of chromosomes is usually some multiple of this number (e.g. diploid 2*n*, triploid 3*n*, or tetraploid 4*n*, but sometimes polyploid many-*n*).

haplotype A set of closely linked genes that tend to be inherited together. The term is often applied to mitochondrial DNA in studies of animals, where the mitochondrial *genome is considered a haplotype due to the apparent lack of *recombination.

haptotropism *See* THIGMOTROPISM.

hardening *See* ACCLIMATIZATION.

hardness **1.** A measure of the ability of water to form a carbonate scale when boiled, or to prevent the sudsing of soap. Permanent hardness is due mainly to dissolved calcium and magnesium sulphate or chloride; the bicarbonate ion causes temporary hardness. Dissolved carbon dioxide and the weathering of carbonate rocks are the main sources of hardness in water. **2.** A physical property of minerals and one of the most useful tests for mineral identification. *Mohs's scale of hardness (H), which ranks minerals by their hardness and thus makes possible a diagnostic test in which one mineral is used to scratch another, was introduced in 1822 by the German-Austrian mineralogist Friedrich Mohs (1773–1839) and is still the standard used today. Useful tools for determining hardness are the fingernail (H about 2.5) and a penknife (H about 5.5). With a little practice, the hardness of a mineral may be determined by means of a scratch to within one or two points on Mohs's scale.

hardpan A hardened *soil horizon, usually found in the middle or lower parts of the *profile, which may be indurated (*see* INDURATION) or cemented by a variety of possible cementing materials. *See also* CALICHE; DURICRUST; and PAN.

hardwood The wood of angiospermous trees.

Hardy, Sir Alister Clavering (1896–1985) A British zoologist, who specialized in studies of *plankton, insects, and evolutionary processes. Following employment at the Ministry of Agriculture, Fisheries and Food Fisheries Laboratory, Lowestoft, in 1924 he was appointed chief zoologist to the *Discovery* expedition, which studied whales in Antarctic waters. During this voyage he invented the *continuous plankton recorder, a device that made possible the establishment of the most comprehensive monitoring system in the world. In 1928 he was appointed professor of zoology (and in 1931 also of oceanography) at University College Hull; in 1942 regius professor of natural history at the University of Aberdeen; and in 1945 Linacre professor of zoology at the University of Oxford, becoming an emeritus professor on his formal retirement in 1961. He was elected a Fellow of the Royal Society in 1940 and knighted in 1957.

Hardy–Weinberg law The law which states that in an infinitely large, interbreeding population, in which mating is random and in which there is no selection, migration, or mutation, gene and *genotype frequencies will remain the same between the sexes and constant from generation to generation, with no overlap between generations. In practice these conditions are rarely strictly present, but unless any departure is a marked one, there is no statistically significant movement away from equilibrium. Consider a single pair of *alleles, A and a, present in a diploid population with frequencies of p and q respectively. Three genotypes are possible, AA, Aa, and aa, and these will be present with frequencies of p^2, $2pq$, and q^2 respectively. The law was established in 1908 by G. H. Hardy and W. Weinberg.

harmattan wind (the doctor) A dry, dusty, north-easterly or easterly wind which occurs in West Africa north of the equator. Its effect extends from just north of the equator in January, almost to the northern tropic in July. In West Africa it is known as 'the doctor' because of its invigorating dryness compared with humid tropical air. The harmattan wind stream occasionally extends south of the equator during the northern winter as an upper air wind over the south-westerly *monsoon.

harvest method A productivity measuring technique, most commonly used for estimates of *primary productivity, especially in situations in which predation is low (e.g. among *annual crops, on certain *heathlands, in colonizing *grasslands, and sometimes in pond *ecosystems). Sample areas are harvested at intervals throughout the growing season, and the material is dried to estimate dry weight or *calorific value. The method may also be used for woodlands, although usually only one final felling and dry-weight estimation is feasible. In such situations it is generally more reliable and ecologically more desirable to use indirect, non-destructive estimates (e.g. by monitoring carbon dioxide profiles). *Compare* AERODYNAMIC METHOD. The harvest method is usually used only for above-ground *biomass and therefore neglects the large and important development of root biomass below ground level.

haustorium **1.** In certain parasitic *fungi, an outgrowth from a hypha that penetrates a host cell in order to absorb nutrients from it. **2.** In some parasitic flowering plants (angiosperms), outgrowths of the roots.

Hawaiian floral realm Part of R. Good's (1974, *The Geography of the Flowering Plants*) *Polynesian floral subkingdom, which falls within his *Palaeotropical floral kingdom. This is the most isolated floral region, which accounts for its great distinctiveness. About 20 per cent of the genera and over 90 per cent of the species are endemic (*see* ENDEMISM). No valuable economic or horticultural plants belong to this region. *See also* FLORAL PROVINCE and FLORISTIC REGION.

hay **1.** An enclosure or hedge. **2.** Part of a forest which has been fenced off for hunting. **3.** Grass that has been conserved by drying to be used later as feed or bedding for livestock.

haze An atmospheric condition in which visibility is reduced because of the dispersion of light by very small, dry particles (e.g. fine dust).

headcut *See* KNICK POINT.

headwall, glacial The steep rock slope at the head of a *cirque or *valley glacier. It is a site of active glacial *erosion, including *frost wedging.

heartwood The dead, woody centre of the trunk of a tree. The cells become impregnated with various organic compounds which cause a change in colour, so that this tissue is distinguished easily from the remainder of the wood.

heath forest A distinctive kind of *tropical rain forest, with *sclerophyllous, microphyllous leaves, found in south-eastern Asia, South America, and on a small area of the coast of Gabon, central Africa. It grows on siliceous, *podzolic soils. *See also* KERANGAS.

heathland Most typically, a lowland *community dominated by dwarf *shrubs belonging to the family Ericaceae. Sometimes, especially in Britain, 'heathland' is used more loosely to denote any shrub community developing on *acidic, *podzolized soils (e.g. grass heaths and lichen heaths). The term was first used in a narrow sense to describe shrubby vegetation dominated by the genus *Calluna* (heath or ling). *Compare* MOOR.

heavy-metal tolerance A biochemical and physiological adaptation to heavy metals (i.e. metals, e.g. copper and zinc, that have a density greater than 5 g/cm³) shown by plant species or *genotypes: such plants may therefore be found growing successfully on soils contaminated by metals, where other species or genotypes would fail. Many common grasses (e.g. *Agrostis tenuis* and *Festuca ovina*) have developed strains tolerant of heavy metals naturally on soils derived from geological strata rich in heavy metals. Seeds from these genotypes are used commercially in revegetating spoil contaminated by metals.

hect- *See* HECTO-.

hectare **(ha)** An *SI unit of area, equal to 100 ares (1 are = 100 m^2) or $10^4 m^2$ (=2.471 acres).

hecto- **(hect-)** From the Greek *hekaton*, meaning 'hundred', a prefix (symbol h) used with *SI units to denote the unit × 10^2.

Heinrich events The release into the ocean, during *glaciations, of large amounts of ice, as icebergs, at intervals of 8 000–10 000 years, with major consequences for the circulation of ocean water and the global climate. The phenomenon was first described in 1988 by Hartmut Heinrich.

hekistotherm A cold-tolerant plant of polar regions, according to A. L. P. de *Candolle's (1874) classic temperature-based scheme of world vegetation zones.

heliophyte A plant that is characteristic of, and shows adaptation to, bright, sunlit *habitats, as opposed to shade-tolerant or shade-preferring species (i.e. *sciophytes).

heliosis *See* SOLARIZATION.

heliotropic **(phototropic)** Applied to a plant or part of a plant that shows a directional growth movement in response to light.

heliotropism **(phototropism)** A *tropic response of a plant or plant organ to the stimulus of light.

helminth A worm, usually parasitic. The word is derived from the Greek *helmins, -inthos*, meaning 'intestinal worm'.

helm wind A local name for a type of lee wave that occurs frequently in winter and spring on the western (lee) slope of the Crossfell range in Cumbria, England. Associated with this cold, gusty, north-easterly wind, the helm itself is a thick bank of cloud along the mountain range, together with an outlier of narrow, almost motionless cloud away from the range.

helophyte A plant typical of marshy or lake-edge environments, in which the *perennating organ lies in soil or mud below the water level, but the aerial shoots protrude above the water (e.g. *Phragmites communis*, the common reed). *Compare* HYDROPHYTE.

helotism A relationship between two organisms (e.g. some groups of ants, *see* DULOSIS) in which one virtually 'enslaves' the other for its own benefit.

hemera A period of geological time determined by the maximum development of a *fossil plant or animal.

hemi- The Greek *hemi*, meaning 'half', used as a prefix meaning 'half' or 'affecting one half'.

hemicryptophyte One of *Raunkiaer's life-form categories, being a plant whose perennating buds are at ground level, the aerial shoots dying down at the onset of unfavourable conditions. Three sub-categories are recognized: (*a*) proto-hemicryptophytes, in which the lowest leaves on the stem are smaller than others, or scale-like, giving added protection to the bud (e.g. *Rubus idaeus*); (*b*) partial rosette plants, in which the best-developed leaves form a basal rosette, but some leaves are also present on aerial stems (e.g. *Ajuga reptans*, bugle); and (*c*) rosette plants, in which the leaves are confined to a rosette at the base of the aerial shoots (e.g. *Bellis perennis*, daisy). *Compare* CHAMAEPHYTE; CRYPTOPHYTE; PHANEROPHYTE; and THEROPHYTE.

hemiparasite **1.** **(meroparasite)** A plant parasite (*see* PARASITISM) that has chlorophyll and photosynthesizes, but which augments its nutrient supply by feeding on its host or uses its host for mechanical support. **2.** A plant parasite that grows from seed in the soil.

Hengelo **(Hoboken)** A Middle *Devensian *interstadial dating from about

40 000 years BP, when estimated July temperatures in the Netherlands were about 10°C. Together with the *Moershoofd and *Denekamp interstadials, the Hengelo is equivalent to the *Upton Warren interstadial of the British Isles.

Hennig, Willi (1913–76) A German zoologist who originated *phylogenetic systematics. He was awarded his PhD by the University of Leipzig in 1947 and conducted research on *Drosophila* larvae. His book, *Grundzüge einer Theorie der phylogenetischen Systematik* (1950), made no immediate impact, but when an English translation appeared in 1966, as *Phylogenetic Systematics* (with a second edition in 1979), it suddenly achieved authoritative status (comparable, perhaps, to *The Origin of Species!*). Hennig argued that, as *taxonomy aims to depict relationships and the only objective meaning of 'related' means sharing a common ancestor, taxonomy must be based on *phylogeny. This struck an immediate chord. Hennig coined such terms as *apomorphic, *plesiomorphic, and *sister groups, and offered a redefinition of *monophyly, which he insisted must be paramount in taxonomy.

Hennig's dilemma Phylogenetic trees constructed from an examination of two or more characters (e.g. two different genes) may result in two contradictory *phylogenies. In such cases no single tree can be constructed using all characters compatibly. (*Cladistics, in which the dilemma may arise, was founded by the German entomologist W. *Hennig.)

herb A small, non-woody, seed-bearing plant in which all the aerial parts die back at the end of each growing season. *Compare* SHRUB; SUBSHRUB; and TREE.

herbage **1.** The growing plants on which domestic animals feed. **2.** The ground vegetation (herbs), especially grass, when these are considered as an agricultural crop. **3.** The payment made to the owner of an area of land in return for permission to graze livestock on that land.

herbicide A chemical substance which suppresses, and is usually designed to eliminate, plant growth. It may be a non-selective weed-killer (e.g. paraquat); or selective, for example, eliminating dicotyledonous plants from among monocotyledonous stands (e.g. phenoxyacetic acids) or vice versa (e.g. dalapan).

herbivore A *heterotroph that obtains energy by feeding on primary producers, usually green plants. *Compare* CARNIVORE; DETRITIVORE; OMNIVORE. *See also* FOOD-CHAIN.

herding The formation by mammals of groups of individuals that have a social organization. Usually the term is restricted to the behaviour of large herbivores.

heritability A measure of the degree to which a *phenotype is genetically influenced and can be modified by selection. It is represented by the symbol h^2, which is equal to the additive *genetic variance divided by the *phenotypic variance. Parent–offspring correlations are estimates of familiality and not of heritability: they cannot account for environmental correlations between relatives. This definition of heritability is a narrow one: heritability in the broad sense (represented by H^2) is the fraction of total phenotypic variance that remains after exclusion of the variance due to environmental effects. Estimates of heritability are used widely by plant breeders to predict the likely effects of selection. If heritability estimates are low for a particular character, this indicates that the character is mainly influenced by the environment and suggests that the response to selection would not be rapid. *See also* FISHER'S FUNDAMENTAL THEOREM.

hermaphrodite An individual that possesses both male and female sex organs (i.e. it is bisexual).

hermaphroditic fish Fish that, when mature, possess both male (testes) and female (ovary) sex glands at the same time. In such species cross-fertilization can occur during spawning. In other species (e.g. Sparidae, sea bream) the testes develop first, these males turning into females as the fish grow larger and older. Most Serranidae (sea perches) are females first or have both sets of glands equally developed.

hermatypic Applied to corals that contain *zooxanthellae and are reef-forming. Modern scleractinian hermatypic corals are characterized by the presence of vast numbers of symbiotic zooxanthellae in their endodermal tissue. They live in waters of normal marine salinity, at depths of up to 90 m, in temperatures above 18°C, and grow vigorously in strong sunlight. *Compare* AHERMATYPIC.

hetero- A prefix meaning 'different from', derived from the Greek *heteros*, meaning 'other'. *Compare* HOMO-.

Heterobasidion annosum **(*Fomes annosus*)** *See* BRACKET FUNGUS.

h

heterochrony The dissociation in time of the development of factors such as shape, size, and maturity, so that organisms mature in these respects either at earlier stages of growth (e.g. in *acceleration, *predisplacement, and *hypermorphosis) or in later stages (e.g. in *neoteny, *progenesis, and *postdisplacement). This can lead in turn to *paedomorphosis or *peramorphosis (*recapitulation of phylogeny).

heteroecious **(heteroxenous)** Applied to a *parasitic organism (e.g. the *rust fungus *Puccinia graminis*) in which part of the life cycle occurs obligatorily in one host and the remaining part obligatorily in another. Heteroecism occurs in several groups of animal parasites and in the rust fungi, but in only a few plant species. *Compare* MONOECIOUS.

heterogametic sex The sex that has sex chromosomes that differ in morphology (e.g. XY), and therefore produces two different kinds of *gametes with respect to the sex chromosomes. The term may also refer to the possession of an unpaired sex chromosome (a single X-chromosome). In mammals and most other animals, the male is the heterogametic sex, usually with equal numbers of different gametes X and Y, although some men, for example, have XXY or even XXXY. In birds, reptiles, some fish and amphibians, and Lepidoptera, the female is the heterogametic sex.

heterogamy The alternation of reproduction by *parthenogenesis and bisexual reproduction. It is found in some aphids.

heterograft *See* GRAFT.

heterogynous Applied to species (e.g. many members of the Cynipidae) whose life cycle involves the production of alternating sexual and *agamic generations. Among insect plant parasites, the alternating generations may occur on different parts of the host plant, and sometimes on different host plants.

heterologous Applied to an *antigen and an *antibody that do not correspond, so that one is said to be heterologous with respect to the other. The term may also be used to refer to a *graft originating from a donor of a species different from that of the host.

heteromerous **1.** Composed of units (e.g. cells) of different types. **2.** Having unequal numbers of parts. **3.** Layered, as in a lichen thallus in which the algae are confined to a distinct layer.

heterophagous Applied to an organism that feeds on a wide variety of items (e.g. one that can parasitize many different hosts).

heterophylly The production of leaves of different shapes on the same plant.

heterophyte **1.** A plant that grows in a wide range of habitats. **2.** A *dioecious plant. **3.** A plant that lacks chlorophyll (i.e. it is parasitic).

heterosis **(hybrid vigour)** The increased vigour of growth, survival, and fertility of hybrids, as compared with the two homozygotes. It usually results from crosses between two genetically different, highly inbred lines. It is always associated with increased heterozygosis.

heterosphere The layers of the atmosphere, above 80 km altitude, through which the chemical composition of the air changes markedly with height, principally as a result of oxygen dissociation. *Compare* HOMOSPHERE.

heterospory *See* SPORE.

heterostyly A *polymorphism that occurs in some species of flowering plants, which produce flowers with anthers and styles of different lengths. This ensures cross-pollination (e.g. the pin-eyed (long style) and thrum-eyed (short style) forms of the primrose, *Primula vulgaris*).

heterosymbiosis *Symbiosis between organisms of different species. *Compare* HOMOSYMBIOSIS.

heterotopic Applied to an organism that occurs in a wide range of habitats.

heterotroph An organism that is unable to manufacture its own food from simple chemical compounds and therefore consumes other organisms, living or dead, as its main or sole source of carbon. Often, single-celled *autotrophs (e.g. *Euglena*) become heterotrophic in the absence of light.

heterotrophic succession *See* DEGRADATIVE SUCCESSION.

heteroxenous *See* HETEROECIOUS.

heterozygosis The presence of different *alleles at a particular gene *locus.

heterozygosity A measure of genetic variation, either in a population (the frequency of individuals *heterozygous at a particular *locus) or in an individual (the proportion of gene loci that are heterozygous).

heterozygote An organism or cell that is *heterozygous.

heterozygous Applied to an organism or cell possessing different *alleles at a particular gene *locus.

heterozygous advantage *See* GENETIC POLYMORPHISM.

hibernal In the winter. *See* AESTIVAL; PREVERNAL; SEROTINAL; and VERNAL.

hibernation A strategy for surviving winter cold that is characteristic of some mammals. Metabolic rate is reduced to a minimum and the animal enters a deep sleep, surviving on food reserves stored in the body during the favourable summer period. *Compare* AESTIVATION.

hierarchical and non-hierarchical classification methods The grouping of individuals by a series of subdivisions or agglomerations to form a characteristic 'family tree' (dendrogram) of groups. Alternatively, classification may be non-hierarchical, i.e. proceeding not by an organized series of progressive joinings or subdivisions, but instead achieving groupings by a series of simultaneous trial-and-error clusterings (successive approximation) until an optimum and stable pattern is found. A possible scheme, for example, is to select at random a number of starter individuals, equal to the number of groups required, and to which other individuals are added or from which they are removed according to their characteristics and the classificatory philosophy used (that of seeking to minimize internal group heterogeneity or of seeking to maximize differences between groups). The chief advantage of hierarchical over non-hierarchical methods is the clarity with which the routes to the final groupings may be followed, facilitating explanation of those groups. However, since most hierarchical classifications either dichotomize or pair at each division or join, natural clustering patterns may be distorted or poorly represented. Hierarchical classifications are also more likely to be unduly affected by irrelevant background information.

hierarchy **1.** A form of organization in which certain elements of a system regulate the activity of other elements. **2.** A form of social organization in which individuals, or groups of individuals, possess different degrees of status, affecting feeding, mating behaviour, etc.

high Arctic tundra The most northerly sector of the *tundra, distinguished by a lack of complete vegetation cover, except in the most favoured and usually very restricted *habitats. Such tundra vegetation as exists tends to be marshy, with little but *lichen in the more exposed situations.

higher categories In taxonomy, classifications above the level of *species, which are defined arbitrarily according to observed similarities among species, and which provide a useful hierarchical framework by which organisms may be described succinctly.

high forest A forester's term describing a natural forest, as opposed to a *plantation, usually comprising trees of all ages, as opposed to being even-aged. The term does not apply to *underwood or *coppice and applies equally whether the trees were self-sown or derived from coppice stools.

high-level waste *See* RADIOACTIVE WASTE.

Hiller peat-borer *See* PEAT-BORER.

hill fog Low cloud that covers high ground.

h

hillock tundra A poorly drained, marshy *tundra with numerous hummocks about 25 cm high, which give better drainage and may therefore support heathy plants and lichens. A range of microhabitats is provided by the hummocks and these are exploited by plants with different ecological requirements.

hispid Having short stiff hairs or bristles.

hist- A prefix, commonly with an added o (i.e. histo-) meaning 'web-like', derived from the Greek *histos*, meaning 'web'.

histic epipedon A surface *soil horizon, not less than 1 m in depth, high in organic carbon, and saturated with water for some part of the year. *See also* HUMUS (2). The name is from the Greek *histos*, meaning 'web'.

Histosols An order of organic soils composed of organic materials. Histosols must have a thickness of more than 40 cm when overlying unconsolidated mineral soil, but may be of any thickness when overlying rock. Histosols are a reference soil group in the *FAO *soil classification.

hitch-hiking effect The spread through a population of a selectively neutral *allele due to its close *linkage with a selectively advantageous allele which is undergoing positive selection.

hive An artificial nest provided by beekeepers for honey-bees (*Apis mellifera* in the West and in Africa; *A. cerana* in parts of Asia) and for some stingless bees (*Melipona* species) in South America.

Hjulstrom effect The contrast between the flow velocity at which a fine-grained, cohesive sediment may be deposited and that at which it will be eroded. Although fine-grained cohesive sediments (fine silt and clay) will be deposited only if flow velocities are very low, a very high velocity is required to erode the same sediment, once deposited. This is because of the cohesive nature of the sediment, which makes silt and clay more difficult to erode than pebbly sediment.

hoarding Storage by an animal of some commodity, usually food, in a central cache or in specific locations throughout a *home range.

hoar frost A thin layer of ice crystals, forming patterns ('feathers', 'needles', 'spines', etc.), seen on exposed surfaces that have chilled to below freezing temperature by radiation cooling, thereby reducing the temperature of the air in contact with the surfaces and raising its humidity to saturation. Atmospheric water vapour is deposited directly as ice. Hoar frost is seen particularly well on vegetation.

Hoboken *See* HENGELO.

Holarctica A formerly unified, circumpolar, biogeographic region, embracing what are now North America, Europe, and Asia (i.e. Laurasia). The legacy of this region is attested to by the great floral and faunal similarities between the three northern continents.

holdfast A differentiated structure in a seaweed or other algae, the function of which is to attach the thallus to a substrate (e.g. rocks, other plants, or shells). A holdfast may be superficially root-like or may be discoid and sucker-like.

holistic (holological) Relating to the whole. In *ecology the term is applied to studies which aim to understand *ecosystems as a whole (i.e. as entire systems), rather than examining their component parts. *Compare* MEROLOGICAL APPROACH.

Holling's disc equation A method for calculating the functional response of predators to increased prey density. The equation is based on laboratory experimental data simulating predation. A human predator gathers discs (prey) from boards with different 'prey' densities. In theory, the efficiency with which the predator consumes the prey should decline as the prey density increases, owing to extra time spent handling the prey. Thus the relationship between prey density and numbers consumed by predators is not a straight line but a curve. This relationship was first summarized mathematically by C. S. Holling in 1959 as $y = T_t ax/(1 - abx)$ where y is the number of discs removed, x is the disc density, T_t is the total experimental time, a is a constant describing the probability of finding a disc at a given density, and b is the time taken to pick up a disc. *Compare* DIVERSITY INDEX.

hollow phase A stage in the cyclical pattern of vegetation change in *grassland, at which a hollow forms from the erosion of an old grass hummock during the *degenerate phase. Colonization of the bare soil by grass seedlings triggers the development of a new hummock. *Compare* BUILDING PHASE; DEGENERATE PHASE; and MATURE PHASE.

holo- A prefix meaning 'whole', from the Greek *holos*, meaning 'whole' or 'entire'.

Holocene (Recent, Post-Glacial) The epoch that covers the last 10 000 years. It is often referred to as the Recent or Post-Glacial. *See also* FLANDRIAN. *See* ANTHROPOGENE.

holocoenotic Applied to a network of relationships (e.g. among the components of an *ecosystem) in which all factors act together, with no barriers separating them.

hological *See* HOLISTIC.

holomictic Applied to lakes in which the water turns over at least once a year. *Compare* DIMICTIC; MONOMICTIC; and POLYMICTIC.

holoparasite *See* PARASITISM.

holophyletic *See* HOLOPHYLY.

holophyly (adj. holophyletic) The condition of a group of taxa which not only are descended from a single ancestral species, but represent all the descendants of that ancestor. Holophyly represents a special case of *monophyly; some authors use the terms interchangeably.

holoplankton Zooplankton organisms that are planktonic throughout their life cycles. *Compare* MEROPLANKTON.

holotype *See* TYPE SPECIMEN. *Compare* LECTOTYPE; NEOTYPE; PARATYPE; and SYNTYPE.

holozoic Applied to the method of feeding in which nutrients are obtained by consuming other organisms (e.g. in most animals).

Holsteinian A north European *interglacial period, dating from about 0.3 Ma to 0.25 Ma, which is probably the equivalent of the *Mindel/Riss Interglacial of the Alpine areas and may also be equivalent to the *Hoxnian of East Anglia.

homeomorph An organism which, as a result of *convergent evolution, comes to resemble another to which it may not be closely related.

homeostasis *See* HOMOEOSTASIS.

homeotherm *See* HOMOIOTHERM.

home range The area within which an animal normally lives. The boundaries of the range may be marked (e.g. by *scent marking), and may or may not be defended, depending on species.

homing The return by an animal to a particular site which is used for breeding or sleeping. The term may apply to the return of an animal to its nest after foraging or to seasonal migrations between breeding and feeding grounds.

homo- A prefix meaning 'the same', derived from the Greek *homoios*, meaning 'like'. *Compare* HETERO-.

homoeology The study of *chromosomes which are related, but derived from different *genomes. Homoeologous chromosomes may occur within the same organism, e.g. in an *allopolyploid where the constitutive genomes are derived from different, but often closely related species.

homoeostasis (homeostasis) The tendency of a biological system to resist change and to maintain itself in a state of stable equilibrium.

homogametic sex The sex that has *sex chromosomes that are similar in morphology (e.g. XX); and hence the sex that produces only one kind of *gamete with respect to the sex chromosomes. *Compare* HETEROGAMETIC SEX.

homogeneous nucleation The spontaneous condensation or freezing of water in the atmosphere in the absence of substances to act as nuclei. This is most likely in supercooled air below –40 °C.

homograft *See* GRAFT.

homoiohydry The ability of an organism to maintain a constant internal water content by compensating for temporary fluctuations in the availability of water or evaporation. *Compare* POIKILOHYDRY.

homoiomerous Uniform in structure.

homoiotherm (homeotherm) An organism whose body temperature varies only within narrow limits. It may be regulated by internal mechanisms (i.e. in an *endotherm) or by behavioural means (i.e. in an *ectotherm), or by some combination of both (e.g. in humans, who light fires and wear thick clothes to keep warm in cold weather and wear light clothing to keep cool in warm weather, but who are also endothermic).

homologous chromosomes A pair of similar *chromosomes, one derived from each parent where both parents are of the same species. *Compare* HOMOEOLOGY.

homology **1.** The basic similarity of a particular structure in different organisms: it usually results from their descent from a common ancestor. Two organs sharing a similar position, a similar histological appearance, and a similar embryonic development are said to be homologous organs irrespective of their superficial appearance and function in the adult. *See also* HOMOLOGOUS CHROMOSOMES. **2.** The extent to which two DNA or *amino acid sequences resemble one another.

homonym In nomenclature, one of two or more separately published names for the same taxon or identical names for different taxa.

homoplasy In the course of *evolution, the appearance of similar structures in different lineages (i.e. not by inheritance from a common ancestor). The term includes *convergence and *parallelism.

homosphere The atmospheric layer from the Earth's surface to approximately 80 km altitude, in which the relative proportions of the various gaseous constituents, excluding water vapour, remain almost constant. *See also* HETEROSPHERE.

homospory *See* SPORE.

homosymbiosis *Symbiosis between organisms of the same species. *Compare* HETEROSYMBIOSIS.

homotaxis Literally, 'the same arrangement' (from the Greek *homos*, meaning 'same' and *tasso*, meaning 'arrange'), a term proposed by T. H. Huxley (1825–95), in an address to the Geological Society of London (published in the *Quarterly Journal of the Geological Society of London* in 1862), to describe strata from different areas which contain similar lithologic or fossil successions but are not necessarily of the same age.

homozygosis The presence of identical *alleles at a particular gene *locus.

homozygosity A measure of genetic uniformity, either in a population (the frequency of individuals *homozygous at a particular *locus) or in an individual (the proportion of gene loci that are homozygous).

homozygote An individual or cell that is *homozygous.

homozygous Applied to an individual organism or cell that possesses identical *alleles at a particular gene *locus.

honey Nectar collected by bees from flowers, and partly digested so that the complex sugars are broken down to simpler ones. It is concentrated by the evaporation of water, and also contains traces of gums, pollen, minerals, and enzymes. It is one of the principal foods of growing bee larvae, the other being protein-rich pollen. Honey is also eaten by adult bees, and is an energy-rich food which fuels their foraging trips.

honeycomb Double-sided sheets of hexagonal wax cells built by honey-bees (*Apis* species), in which pollen and honey are stored and the brood is reared.

honeydew Plant sap that has passed through the bodies of aphids (Aphididae).

honey fungus **(honey tuft)** *Armillaria mellea*, a fungus that is a serious parasite of many broad-leaved and coniferous trees and shrubs of temperate regions.

honey tuft *See* HONEY FUNGUS.

Hooker, Sir Joseph Dalton (1817–1911) A British botanist, who graduated in medicine at the University of Glasgow in 1839 and was then appointed assistant surgeon on board the *Erebus*, as a member of the Antarctic expedition led by Sir James Ross. On his return, in 1843, he published *Flora Antarctica* (1844–7), *Flora Novae Zelandiae* (1853–5), and *Flora Tasmanica* (1855–60). He explored the northern frontiers of India (1847–51), and published the *Flora of British India* (1855–97). The large number of rhododendrons he brought from India provided raw materials for many hybrids, which became popular ornamentals, transforming many British gardens. He prepared the fifth and sixth editions of *Bentham's *Handbook of the British Flora*, which then became known to generations of students as 'Bentham and Hooker'. He was appointed assistant director of the Royal Botanic Gardens at Kew in 1855 and succeeded his father, Sir William Jackson *Hooker, as director in 1865.

Hooker, Sir William Jackson (1785–1865) A British botanist and authority on cryptogamic botany, who became the first director (1841–65) of the Royal Botanic Gardens at Kew (and was succeeded by his son, Sir Joseph Dalton *Hooker). He studied the botany of Iceland (1809) and of France, Switzerland, and northern Italy (1814). He was appointed regius professor of botany at the University of Glasgow in 1820. He wrote prolifically, his works including *Tour of Iceland* (1811), two volumes of *Musci Exotica* (1818–20), *Flora Scotica* (1821), *Icones Filicum* (with R. K. Greville, 1829–31), *British Flora* (with G. A. W. Arnott et al., 1830), and *British Ferns* (1861–2).

hopeful monster An individual that carries a macromutation that is of no benefit to that individual (i.e. the individual is a monster) but which may prove beneficial to one of its descendants if it undergoes further, but relatively minor, evolutionary change (i.e. the monster is hopeful). For example, the loss of the tail in descendants of *Archaeopteryx* was followed by the development of the tail feathers which stabilize the flight of modern birds (assuming there was a direct evolutionary line from *Archaeopteryx* to the birds). The hypothesis is advanced in order to explain the evolution of structures that appear to confer no benefit, or even disadvantages, until they reach their completed form. Many of the structures involved are explicable without invoking 'hopeful monsters' (e.g. partially evolved feathers may be useless for flight but provide thermal insulation; and an eye that works inefficiently may nevertheless be better than no eye at all). Macromutations that involve a major reorganization of the genes may effectively sterilize the animal and in any case an individual whose appearance is markedly different from that of other members of its species may have difficulty finding a mate. The hypothesis is now applied mainly to macromutations that affect regulatory genes (the genes that activate or deactivate genes which cause the synthesis of proteins); these may have major effects but involve no major, and probably sterilizing, genetic alteration.

horizon **1.** In stratigraphy, an informal term that denotes a plane within a body of strata. It may be at a boundary of lithological change or is (commonly) a thin, distinctive bed within a lithological unit. **2.** An interface separating two media with different properties. **3.** *See* SOIL HORIZON.

horizontal gene transfer The transfer of genetic information from one organism to another by an infection-like process rather than by inheritance.

hormesis The benefit to health or stimulus to growth or reproduction which results from exposure to small doses of substances that are toxic in larger doses.

horotelic *See* HOROTELY.

horotely (adj. horotelic) A normal or average rate of *evolution per million years of genera within a given taxonomic group. Thus slowly or rapidly evolving lines may be horotelic during certain episodes in their history. *Compare* BRADYTELY and TACHYTELY.

horse latitudes Subtropical latitudes, coinciding with a major anticyclonic belt, which are characterized by generally settled weather and light or moderate winds. When sailing ships carrying cargoes of horses were becalmed in these latitudes, horses would sometimes be thrown overboard, mainly to reduce the demand for drinking water.

Horton analysis *See* DRAINAGE MORPHOMETRY.

Hortonian flow *See* SURFACE RUN-OFF.

host *See* GRAFT and PARASITISM.

hot spot An area which contains a large number of rare or endangered species and for that reason is designated for protection. Identifying hot spots is the traditional technique by which sites are selected for protection, but it tends to concentrate that protection on a small number of areas, leaving others, and the many species in them, unprotected. Many conservation biologists prefer to identify conservation areas by *gap analysis.

hot spot of mutation A region of DNA that exhibits an unusually high propensity to mutate.

Hoxnian An *interglacial period and a series of temperate-climate deposits named after Hoxne, Suffolk, England, with a characteristic vegetational sequence that occurs in *tills of the earliest glacial stages, sometimes filling deep channels. They may be equivalent to the *Holsteinian deposits of continental Europe. Sometimes during the Hoxnian Interglacial the sea rose to well above its present level. There is a correlation between this stage and the Boyn Hill terraces of the Thames Valley, and also with *raised beaches (30 m ordnance datum) found on the Sussex coast. Hand axes of this age (Acheulian culture) have been found and there is evidence for human modification of the environment by forest clearance and an increased incidence of fires associated with *Homo erectus* settlements in southern Britain.

hue *See* MUNSELL COLOUR.

Humboldt, Friedrich Heinrich Alexander, Freiherr von (1769–1859) A German naturalist, physical geographer, biogeographer, geologist, vulcanologist, and mining engineer, who began travelling extensively in 1796, after the death of his mother. Accompanied by the French surgeon and botanist Aimé Bonpland (1773–1858), he set out to join Napoleon in Egypt, but their plans changed and they went instead to Madrid. Their experiences in Spain made them decide to explore Spanish America. They embarked in 1799 and spent five years in the South American tropics, exploring the drainage basins of the Amazon and Orinoco, investigating the properties of guano, measuring the temperature of the ocean current that bears Humboldt's name, and studying the flora and fauna of the forests and savannah, returning with more than 30 cases of botanical specimens. He wrote on many subjects, his most important botanical work being the seven-volume *Nova genera et species plantarum* (1815–25), written in collaboration with A. J. A. Bonpland and C. S. Kunth as part six of the 30-volume *Voyage de Humboldt et Bonpland* (1805–34). In *Ideen zu einer Physiognomik der Gewächse* ('Ideas on a Physiognomy of Plants', 1806) he proposed the concept of biogeography.

Humboldt Current *See* PERU CURRENT.

humic acid A mixture of dark-brown organic substances, which can be extracted from soil with dilute alkali and precipitated by acidification to *pH 1–2 (in contrast with *fulvic acid, which remains soluble in acid solution). Its analysis in peat profiles provides a measure of the degree of *humification or decomposition, which in turn can provide an index of past wetness, possibly associated with climatic conditions.

humic coal *See* COAL.

humidity An expression of the moisture content of the atmosphere. Measures of humidity include statements of the total mass of water in 1 m^3 of air (absolute humidity), the mass of water vapour in a given mass of air (specific humidity), *relative humidity, vapour pressure, and the *mixing ratio.

humification The development of humus from dead organic material, by the action of *saprotrophic organisms which use this dead material as their food source. Humification is essentially an oxidation process in which complex organic molecules are broken down into simpler organic acids, which may subsequently be mineralized into simple, inorganic forms suitable for uptake by plants. Humification is therefore a vital stage in the cycling of nutrients. The degree of humification in a *peat *profile is often associated with the general colour of the peat, darker peats being better humified. The degree of humification relates to the wetness of the surface conditions at the time of peat formation, which can provide an index of climatic wetness at the time. Peat humification studies are now a major source of information about past climate change.

humilis The Latin *humilis*, meaning 'low', used to describe a species of shallow *cumulus cloud which typically has a flattened appearance. *See also* CLOUD CLASSIFICATION.

hummocky moraine A strongly undulating surface of ground *moraine, with a relative relief of up to 100 m, and showing steep slopes and deep, enclosed depressions. It results from the downwasting (i.e. thinning) of ice which is usually *stagnant. Blocks of ice may squeeze debris released from the ice into crevasses between the blocks.

humus **1.** Decomposed organic matter of soils that are aerobic for part of the year: it is dark brown and amorphous, having lost all trace of the structure and composition of the vegetable and animal matter from which it was derived. **2.** A term used by some horticulturists to describe any kind of organic matter in the soil. **3.** A surface organic *soil horizon that may be divided into types, e.g. *mor (acid and layered) or *mull (alkaline and decomposed). It is the *histic epipedon of the *USDA Soil Taxonomy. *See also* HUMIFICATION.

hunger The desire for food that, in human experience, is generally an unpleasant, even painful sensation, which may become so intense as to make the search for food dominate thought and action. Similar sensations in other animals are inferred from observation of their behaviour.

hunting wasp Any wasp species in which the females seek spider or insect prey that they sting and paralyse, then transport to the nest as food for their larvae. The term is applied to the following families (division Aculeata): Pompilidae (spider-hunting wasps), Vespidae subfamily Eumeninae (caterpillar and beetle-hunting wasps), and Ampulicidae and Sphecidae (digger wasps). In primitive pompilids and ampulicids, prey is caught before a nest is prepared, and the nest is often no more than a shallow scrape in the soil or a pre-existing cavity. In more advanced forms, a nest is excavated or built before the prey is caught.

hurricane **1.** The name given to a *tropical cyclone that develops over the North Atlantic and Caribbean. Hurricanes move westwards, then swing north, on tracks that often carry them across inhabited islands and coastal areas of Mexico and the United States. **2.** A wind blowing at more than 120 km/h (75 mph), which is Force 12 on the *Beaufort scale. *See* SAFFIR-SIMPSON HURRICANE SCALE.

Huygens' principle Each point of an advancing wave front can be thought of as a new source of secondary wavelets, so that the envelope tangent to all these wavelets forms a new wave front. The principle was discovered by the Dutch physicist Christiaan Huygens (1629–95).

hyaline Translucent or transparent; glass-like.

hybrid **1.** An individual plant or animal resulting from a cross between parents of differing *genotypes. Strictly, most individuals in an *outbreeding population are hybrids, but the term is more usually reserved for cases in which the parents are individuals whose *genomes are sufficiently distinct for them to be recognized as different species or subspecies. Good examples include the mule, produced by cross-breeding an ass and a horse (each of which can breed true as a species) and *Spartina townsendii*, produced by cross-breeding *Spartina maritima* (British cord grass) and the North American species *Spartina alterniflora* (each of which can breed true as a species). Hybrids may be fertile or sterile, depending on qualitative and/or quantitative differences in the genomes of the two parents. Hybrids like *Spartina townsendii*, whose parents are of different species, are sterile but generally reproduce vegetatively. **2.** By analogy with (1), any *heterozygote. Each heterozygote represents dissimilar alleles at a given locus, and this difference results from a cross between parental *gametes possessing differing alleles at that locus. **3.** (graft hybrid) *See* GRAFT.

hybrid dysgenesis A complex of genetic abnormalities, which occurs in certain *hybrids. The abnormalities may include sterility, enhanced rates of gene mutations, and chromosomal rearrangements. Hybrid dysgenesis occurs in the hybrid offspring of certain strains of *Drosophila*, in which it is thought to be owing to mutations induced by *transposon-like elements.

hybrid swarm A continuous series of *hybrids that are morphologically distinct from one another, resulting from the hybridization of two species followed by the crossing and *backcrossing of subsequent generations. The hybrids are very variable, owing to segregation of *alleles at each *locus.

hybrid vigour *See* HETEROSIS.

hybrid zone A geographical zone in which the *hybrids of two geographical races may be found.

hydr- From the Greek *hudor*, *hudatos*, meaning 'water', a prefix that means 'pertaining to water'.

hydrarch succession *See* HYDROSERE.

hydration A chemical reaction in which water combines with another substance.

hydraulic boundary Within a *groundwater system, the interface between regions of different hydraulic characteristics such as *porosity, storativity (*see* STORAGE COEFFICIENT), conductivity (*see* PERMEABILITY), or *transmissivity (e.g. where an *aquifer abuts an *aquiclude).

hydraulic conductivity *See* PERMEABILITY.

hydraulic conductivity The capacity of a material to allow fluids to pass through it, expressed as the volume flow rate of water through a cross-sectional area of a porous medium (*see* POROSITY) under the influence of a hydraulic gradient of unity at a specified temperature. This depends on properties of the fluid and the medium. It is measured in units of metres per second or metres per day and varies with temperature. Typical values range from 10^{-6} m/day for clay to 10^3 m/day for coarse gravel.

hydraulic corer **(piston corer)** A tube used to obtain samples from lake or marine sediments, which penetrates the material by hydraulic pressure. The *Mackereth corer is a hydraulic corer. *See also* GRAVITY CORER.

hydraulic geometry A description of the adjustments made by a stream in response to changes in discharge at a cross-section and in the downstream direction. Adjustments are made in width, mean depth, mean velocity, slope, frictional resistance, suspended-sediment load, and water-surface gradient. The relationship between discharge and adjustment is expressed as a power function: $y = aQ^b$, where y is the adjusting variable, Q is discharge, and a and b are coefficients.

hydraulic gradient A measure of the change in *groundwater head over a given distance. Maximum flow will normally be in the direction of the maximum fall in head per unit of horizontal distance, i.e. in the direction of the maximum hydraulic gradient. *See* HYDRAULIC HEAD.

hydraulic head In general, the elevation of a water body above a particular datum level. Specifically, the energy possessed by a unit weight of water at any particular point, and measured by the level of water in a *manometer at the laboratory scale, or by water level in a well, borehole, or *piezometer in the field. The hydraulic head consists of three parts: the elevation head (*see* ELEVATION POTENTIAL ENERGY), defined with reference to a standard level or datum; the *pressure head, defined with reference to atmospheric pressure; and the velocity head. Water invariably flows from points of larger hydraulic head to points of lower head down the *hydraulic gradient.

hydraulic radius The ratio between the cross-sectional area of a stream channel and the length of the water-channel contact at that cross-section (the wetted perimeter). It is a measure of channel efficiency: the higher the ratio, the more efficient is the channel in transmitting water.

hydric *See* MESIC.

hydrocarbon A naturally occurring compound that contains carbon and hydrogen. Hydrocarbons may be gaseous, solid, or liquid, and include natural gas, bitumens, and petroleum.

hydrochory Dispersal of spores or seeds by water.

hydrofuge hair Water-repellent, cuticular hair on the body of an insect, which serves a variety of functions. Mosquito larvae use hydrofuge hairs to seal their respiratory siphons when they dive; some beetles (Coleoptera) use them to create a plastron (i.e. a physical gill that traps air).

hydrogen bond A bond between molecules in one of which hydrogen atoms are bound to the electronegative atoms fluorine (F), nitrogen (N), or oxygen (O), producing a strongly polarized bond (e.g. O–H) that bonds weakly to another H (e.g. H–O–H, or H_2O). Hydrogen bonding gives water its peculiar properties and provides the link in *base pairing. *See also* COVALENT BOND, IONIC BOND, and METALLIC BOND.

hydrograph A graph on which the water flow in a water course, or the elevation of *groundwater in a borehole above a particular datum point, is plotted against time. The 'unit hydrograph' is the name given to a method of calculation which allows rainfall to be converted to stream flow and so facilitates the prediction of how particular river basins will respond to changing precipitation patterns. The discharge hydrograph shows the flow rate of water against time for a discharging water body. The stage hydrograph shows water level against time.

hydrologic cycle The flow of water in various states through the terrestrial and atmospheric environments. Storage points (stages) involve *groundwater and surface water, ice-caps, oceans, and the atmosphere. Exchanges between stages involve evaporation and transpiration from the Earth's surface, condensation to form clouds, and precipitation followed by run-off. *See also* RESIDENCE TIME.

hydrologic modelling (hydrologic simulation) The use of small-scale physical models, mathematical analogues, and computer simulations to characterize the likely behaviour of real hydrologic features and systems.

hydrologic network An integrated array of meteorological, *groundwater level, and stream-flow measuring stations which in combination give a complete measurement of the *hydrologic cycle for a particular area. In modern practice the various measuring and *gauging stations are sometimes linked by telemetry to a central monitoring unit for use in flood forecasting. *See* FLOOD PREDICTION.

hydrologic region One of the smaller units, with fixed boundaries, into which large tracts of country, sharing a common climate and geologic and topographical structure, are divided for the purpose of collecting hydrologic data on the same basis from year to year, thus facilitating historical analysis.

hydrologic simulation *See* HYDROLOGIC MODELLING.

hydrology The study of the *hydrologic cycle; this involves aspects of geology, oceanography, and meteorology, but emphasizes the study of bodies of surface water on land and how they change with time.

hydrolysis **1. (biochem.)** A reaction between a substance and water in which the substance is split into two or more products. At the points of cleavage the products react with the hydrogen or hydroxyl ions derived from water. **2. (pedol.)** The process of enriching the soil *adsorption complex with hydrogen after exchangeable metallic ions have been replaced by hydrogen ions. *Compare* WEATHERING.

h

hydronasty A *nastic movement induced in plant organs by changes in atmospheric humidity.

hydrophilic Applied to a molecule or surface that can become wetted or solvated by water. This ability is characteristic of polar compounds.

hydrophobic Applied to a molecule or surface that can resist wetting or solvation by water. The ability is characteristic of non-polar compounds.

hydrophyte A plant that is adapted morphologically and/or physiologically to grow in water or very wet environments. Adaptations include the development of finely divided submerged leaves, large floating leaves, the presence of aerenchyma, and the reduction of root systems. The *perennating bud lies at the bottom of fairly open water. With the leaves submerged or floating, only the inflorescence protrudes above the water surface (e.g. *Nuphar lutea*, yellow water-lily). *Compare* HELOPHYTE.

hydroponics Plant growth in a liquid culture solution rather than in soil. This technique is used commonly in experimental studies of mineral nutrient deficiencies or excesses and their effects. Hydroponics also has commercial applications, although to date it has not been used extensively in this way.

hydrosere **(hydrarch succession)** A sequence of *communities which reflects the developmental stages in a plant *succession starting on a soil submerged by fresh water.

hydrosphere The total body of water which exists on or close to the surface of the Earth.

hydrotaxis The locomotion of an organism in response to the stimulus of water.

hydrothermal vent A place on the ocean floor, on or adjacent to a mid-ocean ridge, from which there issues water that has been heated by contact with molten rock, commonly to about 300°C. The vent water often contains dissolved sulphides. These are oxidized by chemosynthetic bacteria, which fix carbon dioxide and synthesize organic compounds. Near the vents, at temperatures up to 40°C, there are highly productive communities comprising animals that utilize the organic compounds or live symbiotically (*see* SYMBIOSIS) with the chemosynthetic bacteria; these organisms support carnivores and detritivores. Vent fluids containing high concentrations of iron, manganese, and copper tend to be hot (about 350°C) and black. They are known as 'black smokers'. 'White smokers' flow more slowly, are cooler, and contain high concentrations of arsenic and zinc.

hydrotropism A directional growth of a plant organ towards wetter regions in response to the stimulus of water.

hydrous mica **(illite)** *See* CLAY MINERALS.

hygro- From the Greek *hugros*, meaning 'wet', a prefix meaning 'pertaining to moisture'.

hygrometer An instrument that is used to measure atmospheric *humidity. Types include the wet-bulb–dry-bulb, dew-point, and *hair hygrometers, and there is one type based on electrical resistance. *See also* PSYCHROMETER.

hygrophanous Translucent and watery in appearance.

hygrophilic *See* HYGROPHILOUS.

hygrophilous **(hygrophilic)** Growing in or preferring moist habitats.

hygroscopic nucleus A microscopic particle (e.g. of sulphur dioxide, salt, dust, or smoke) in the free air, on to which water vapour may condense to form droplets. *Aerosols that are soluble in water (e.g. salt or sulphuric acid) can induce condensation in unsaturated air (e.g. salt nuclei can induce it at a *relative humidity of less than 80 per cent. The size of nuclei may be from 0.001 μm to more than 10 μm (i.e. 'giant' nuclei such as particles of sea salt)). *See also* AITKEN NUCLEUS.

hygroscopic water Water absorbed from the atmosphere and held very tightly by the soil particles, so that it is unavailable to plants in amounts sufficient for them to survive. *Compare* CAPILLARY MOISTURE.

hygrotaxis The movement of an organism in response to the stimulus of humidity or moisture.

hygrothermograph (thermohygrograph) An instrument that records both the temperature and humidity of the air, on separate traces.

hyp- *See* HYPO-.

hyper- From the Greek *huper*, meaning 'beyond' or 'over', a prefix meaning 'exceeding' or 'greater than normal'.

hyper-arid Applied to a climate with an *aridity index of less than 0.05.

hypermorphosis A type of *heterochrony in which growth is prolonged, so that the adult morphology of the descendant is produced by a prolongation of the growth trajectory of its ancestor. The organism reaches its adult size and form well before the attainment of sexual maturity, and continues to develop into a 'super-adult'. This can facilitate evolution.

hyperparasite A parasite that exploits another parasite; a 'superparasite'.

hyperthermic *See* PERGELIC.

hyperthermophile An *extremophile (domain *Archaea) that thrives in environments where the temperature is extremely high, in some cases preferring a temperature of about 105°C, tolerating 113°C, and failing to multiply below 90°C. *Compare* THERMOPHILE.

hypertonic Applied to a cell in which the *osmotic pressure is higher than that in the surrounding medium. *Compare* HYPOTONIC and ISOTONIC.

hypervariable site A DNA site that shows high intraspecific variation.

hypervolume A region defined by more than three dimensions; an ecological *niche is often described as an *n*-dimensional hypervolume.

hypo- (hyp-) From the Greek *hupo*, meaning 'under', a prefix meaning 'below', 'slightly', or 'lower than normal'. 'Hypo-' is generally used before words beginning with a consonant, 'hyp-' before words beginning with a vowel.

hypobiosis *See* DORMANCY.

hypogeal Growing or occurring below ground, commonly with reference to a mode of seed *germination in which the cotyledons develop below the surface. *Compare* EPIGEAL.

hypogean Growing or occurring underground.

hypolimnion The lower, cooler, non-circulating water in a thermally stratified lake in summer. If, as often occurs, the *thermocline is below the *compensation level, the dissolved oxygen supply of the hypolimnion depletes gradually; replenishment by photosynthesis and by contact with the atmosphere is prevented. Reoxygenation is possible only when the thermal stratification breaks down in autumn.

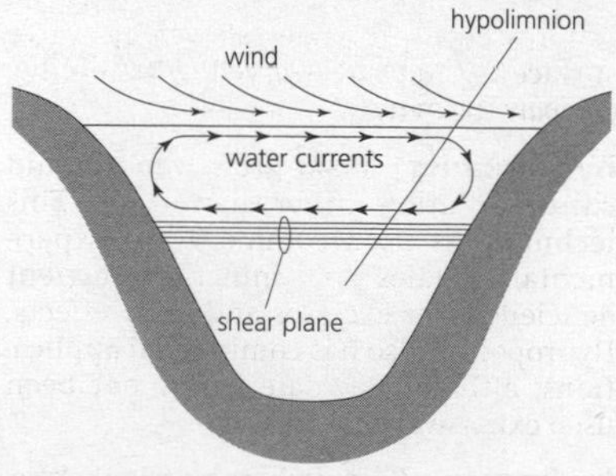

Hypolimnion

hypolith A photosynthetic organism that lives on the underside of a rock in a desert, where it is protected from scouring by wind-blown dust and sand and from *ultraviolet radiation, and where trapped moisture provides it with water. A community of hypoliths is known as a hypolithon.

hypolithon *See* HYPOLITH.

hyponasty A differential, *nastic growth of the lower part of a plant organ (e.g. the growth of the lower side of a leaf so that the leaf blade curves upwards).

hyponeustic *See* ZOONEUSTON.

hyponeuston The organisms living in the lower part of the surface film of water. *Compare* EPINEUSTON and NEUSTON.

hypophloeodal Growing or living beneath (or within) tree bark.

hypopycnal flow The flow developed at the mouth of a *delta where the fresh water from the river flows as a buoyant plume over the denser salt water. The salt water will ride under the fresh water and enter the river channel as an upstream-tapering wedge.

hypothesis An idea or concept that can be tested by experimentation. In inductive or inferential statistics, the hypothesis is usually stated as the converse of the expected results, i.e. as a null hypothesis (H_0). For example, if a specified feature were being compared in two samples, the null hypothesis would be that no difference existed in the populations from which the samples were taken. This helps workers to avoid reaching a wrong conclusion, since the original hypothesis H_1 will be accepted only if the experimental data depart significantly from the values predicted by the null hypothesis. Working in this negative way carries the risk of rejecting a valid research hypothesis even though it is true (a problem with small data samples); but this is generally considered preferable to the acceptance of a false hypothesis, which would tend to be favoured by working in the positive way.

hypothesis-generating method A term often applied to data-structuring techniques, such as classification and ordination methods which, by grouping and ranking data, suggest possible relationships with other factors. Appropriate data may then be collected to test these hypotheses statistically. For example, a classification or ordination of plant data from a heathland might suggest a relationship with soil-water status. If suitable soil data were then recorded, this hypothesized relationship could be tested.

hypothesis testing An evaluation of a hypothesis using an appropriate statistical method and significance test. *See also* STATISTICAL METHOD and NUMERICAL METHOD.

hypotonic Applied to a cell in which the *osmotic pressure is lower than that in the surrounding medium. *Compare* HYPERTONIC and ISOTONIC.

Hypsithermal *See* FLANDRIAN.

-ian The preferred suffix that is added to the geographical location of a type section or type area to form the name of a stage or age (e.g. Frasnian).

Iapetus Ocean (proto-Atlantic) The late *Precambrian and early *Palaeozoic ocean that lay between *Baltica and *Laurentia. The oceanic crust and upper mantle of the Iapetus Ocean floor is presumed to have been subducted during the early Palaeozoic, and the ocean to have disappeared completely by the latest *Silurian–early *Devonian (about 418 Ma ago). The ancient suture is thought to extend from WSW to ENE across the Solway Firth and Borders region, Scotland. The Caledonian orogenic belt, which removed all trace of the Iapetus Ocean, extends all along the border between the two ancient continents, affecting Norway, eastern Greenland, Scotland, northern England, Wales, Ireland, eastern Canada, and the eastern USA.

IBP *See* INTERNATIONAL BIOLOGICAL PROGRAMME.

ice Water which has frozen into a crystal lattice. Pure water freezes at 0°C at 1013.24 mb pressure. The presence of salts in solution depresses the freezing point of water. Liquid water has its maximum density at 4°C, in consequence of which ice floats on water. With increasing pressure, a series of denser polymorphs of ice forms, each designated by a Roman numeral, ordinary ice being ice I. Expansion on freezing (9.05 per cent in specific volume) generates very high pressures, which bring about *frost wedging. Such ice I converts into the denser ice III at lower temperatures, but the pressure exerted by it changes little. 'Ground ice' forms when *interstitial water freezes, and this may bring about heaving as well as frost wedging. 'Glacier ice' is a relatively opaque mass of interlocking crystals, and has a density of 0.85–0.91 g/cm. *'Regelation ice' is relatively clear and is formed by the freezing of meltwater beneath a temperate glacier.

ice ages Periods when ice has accumulated at the poles and the continents have been glaciated repeatedly. Exactly why glaciation occurred is not clear. There are suggestions of a middle *Precambrian glaciation about 2300 Ma ago in North America, South Africa, and Australia. More information exists to suggest that the Earth was glaciated between 950 and 615 Ma ago, and there are at least two glacial horizons in Africa, Australia, and Europe. There is good evidence for a glaciation at the end of the *Ordovician in North Africa, but glacial deposits described from elsewhere at this period are problematical, so the extent of the glaciation is not known. The Permo-Carboniferous glaciation of South America, South Africa, India, and Australia was widespread and is well documented. There is no evidence for further glaciation until the *Quaternary. Suggestions have been made for other ice ages during the *Palaeozoic but evidence for them is sparse. The *Pleistocene ice age is the best documented.

iceberg calving *See* CALVING.

ice cap An ice mass that is less than 50 000 km^2 in area, but is large enough to cover the underlying topography and whose flow behaviour is a consequence of its size and shape. It consists of a central *ice dome, where accumulation dominates, together with *outlet glaciers radiating from the periphery.

ice dome A main component (with the *outlet glacier) of an *ice sheet or *ice cap. It has a convex surface form, of parabolic shape, and tends to develop symmetrically over a land mass. Its thickness often exceeds 3 000 m.

ice field A nearly level field of ice, whose area may range from about 5 km^2 to near continental size, which is formed when the land surface is sufficiently high or uniform for ice to accumulate. It differs from an *ice cap in that it is not domed and its flow is controlled by the underlying relief.

Iceland low A low-pressure system in a region of the North Atlantic in which such systems occur so frequently that the average atmospheric pressure is low. Any one of these systems may be called 'an Iceland low', but the term is used mostly to describe the general statistical or climatic feature.

ice nucleus A crystalline, microscopic particle of ice or some other substance which can induce the growth of ice crystals upon it in clouds containing supercooled water droplets. The saturation vapour pressure is higher over a liquid water surface than over an ice surface. Consequently water evaporates from supercooled droplets and is deposited as ice on to existing ice crystals, especially at temperatures between −15°C and −25°C. During growth, splintering of crystals (e.g. as a result of updrafts) causes large numbers of new nuclei to be formed. Solid particles acting as ice nuclei are sometimes called sublimation nuclei. *See also* FREEZING NUCLEI.

ice prisms *See* DIAMOND DUST.

ice sheet A large mass of ice that lies on a land surface and expands laterally because it is not constrained topographically (*compare* GLACIER). Typically ice sheets cover areas greater than about 50 000 km^2. They are composed of *ice domes and *outlet glaciers. Where an ice sheet reaches a coast it may extend over the sea as an *ice shelf. The East and West Antarctic ice sheets are the world's largest; the East Antarctic sheet covers approximately 10 million km^2 and has a mean thickness of about 3 km and the West Antarctic sheet covers about 2 million km^2. The base of the East Antarctic ice sheet is above sea level over most of its area; the base of the West Antarctic ice sheet is below sea level, in places by as much as 2 km.

ice shelf The outer margin of an *ice sheet or *ice cap that extends beyond a coast. Its base rests on the seabed where the shore slope is shallow. Pressure from ice flowing downslope from the central *ice dome forces the sheet ice further into the sea, into deeper water where the ice floats. The grounding line marks the border between ice that rests on a solid surface and ice that floats. In Antarctica, 7 per cent of the total volume of ice is held in ice shelves. These are up to 200 m thick and typically end with cliffs up to 30 m high. The Ross Ice Shelf is the largest of all. During winter, the edges of ice shelves extend by accumulating sea ice driven onto them by the wind. Ice shelves lose mass by calving (the breaking away of ice blocks to become icebergs) and by melting at the base.

ice stream *See* OUTLET GLACIER.

ice wedge A tapering, vertically layered mass of ice, about a metre wide at the top and extending downward for some 3–7 m. It results from contraction cracking of the ground during extreme cold, followed by water penetration from the *active layer and subsequent freezing. It is characteristic of uniform sediment, such as river *alluvium, under *periglacial conditions. Fossil ice wedges found in currently temperate regions, such as eastern England, provide evidence of former *periglacial conditions.

ice-wedge polygon *See* PATTERNED GROUND.

ichnofossil *See* TRACE FOSSIL.

ichnology (palichnology) A sub-discipline of *palaeontology or, more specifically, of *palaeoecology, which is concerned with the study of *trace fossils and their value in the analysis of sedimentary sequences and the *morphology and behaviour of ancient organisms.

ICZN *See* INTERNATIONAL CODE OF ZOOLOGICAL NOMENCLATURE (INTERNATIONAL COMMISSION ON ZOOLOGICAL NOMENCLATURE).

-id *See* -IDE.

-idae A standardized suffix used to indicate a family of animals in the recognized codes of classification. For example, 'Felidae' is the cat family (including lions, tigers, lynxes, pumas, etc.), and 'Canidae' the dog family (including foxes, jackals, and wolves).

-ide (id, -ides) From the Greek *ides*, meaning 'son of', a suffix attached to the name of an element which is a member of a series (e.g. actinide, halide) or to the more electronegative element or radical in a binary compound (e.g. sodium chloride).

ideal free distribution The distribution of *foraging organisms that results if all the consumers are ideal in their assessment of the profitability of patches and free to move from one patch to another.

-ides *See* -IDE.

idiothetic Applied to information concerning its orientation in an environment that is obtained by an animal by reference to a previous orientation of its body, and without external spatial clues. *Compare* ALLOTHETIC.

-iformes A suffix often used, but not recommended by the *ICZN code, when a name refers to an order of animals (e.g. 'Strigiformes', the owls).

igapo *See* AMAZON FLORAL REGION.

igneous Applied to one of the three main groups of rock types (igneous, *metamorphic, and *sedimentary), to describe those rocks that have crystallized from a *magma. The work is derived from the Latin *ignis*, meaning 'fire'.

Illinoian The third (0.17–0.12 Ma) of four glacial stages recognized in North America. It is represented by deposits from ice moving from the north-east, and in formerly glaciated areas to the west pollen evidence suggests that mean annual temperatures were 2–3 °C cooler than they are now. At the end of this period the climate became warmer and drier. The Illinoian is approximately equivalent to the *Mindel and Riss glacials of the Alps.

illuviation A process of deposition (inwashing) of soil materials, either from suspension or solution, and usually into a lower *horizon, after removal from above or from a lateral source.

imago The fully developed adult among *pterygote insects.

imbibition The adsorption of liquid, usually water, into the ultramicroscopic spaces or pores found in materials such as cellulose, pectin, and cytoplasmic proteins in seeds.

imbricate With parts (e.g. plates) overlapping one another like tiles on a roof.

imitation The acquisition of patterns of behaviour by repeating similar behaviour observed in others, not necessarily of the same species.

immigration In genetics, the movement or flow of genes into a population, caused by immigrating individuals which interbreed with the residents. This is the usual source of new variation in a population, although the fundamental sources of all variation are gene mutation and recombination. *See also* MIGRATION.

immobilization The conversion of a chemical compound from an inorganic to an organic form as a result of biological activity; the compound is thereby removed from the reservoir of compounds available to plant roots.

immunity The natural or acquired resistance of an organism to a pathogenic micro-organism or its products. Immunity may be active (*see* ACTIVE IMMUNITY) or passive (*see* PASSIVE IMMUNITY).

importance value *See* DENSITY–FREQUENCY–DOMINANCE.

imprinting A general term applied to a form of *learning that only occurs early in a young animal's life. In it the young animal learns to direct some of its social responses to a particular object, usually a parent. The phenomenon was first described in detail by Konrad *Lorenz in 1937 in respect of *precocial birds, which learn to follow a moving object and in which it is most strongly developed. Young mammals are often imprinted to the scent or vocalizations of the adult and some birds are imprinted to the vocalizations of their parents (e.g. wood ducks (*Aix sponsa*) nest in restricted spaces in tree hollows; prior to hatching the mother makes a characteristic call; later, when the call is repeated, the chicks respond by leaving the nest, and when they are all assembled the mother moves away with the chicks following). In this case imprinting occurs only during a brief critical period, possibly of only a few hours, very early in the life cycle. Such imprinting is irreversible and influences subsequent behaviour patterns, most importantly sexual behaviour.

-ina A customary suffix, not specifically cited in the code of nomenclature, which indicates an animal subtribe.

-inae **1.** A standardized suffix used to indicate a subfamily of animals in the recognized code of classification. **2.** A standardized suffix used to indicate a ranking of plants at the subtribe level in the recognized code of classification.

inbreeding The interbreeding of closely related individuals (e.g. of *siblings, or cousins, or a parent and its offspring). The converse is *outbreeding. The inbreeding coefficient is the probability of *homozygosity owing to a zygote obtaining copies of the same gene from parents of common ancestry. Inbreeding increases homozygosity: unfavourable recessive genes are therefore more likely to be expressed in the offspring. In general, the genetic variability of an inbred population declines; this is usually disadvantageous for its long-term survival and success. Typically, vigour is reduced in inbred stock. Many organisms, including humans, have physical or social mechanisms that tend to discourage inbreeding and promote outbreeding.

inbreeding depression The decreased vigour in terms of growth, survival, or fecundity, which follows one or more generations of *inbreeding. This is the opposite outcome to that obtained by the crossing of two separate inbred lines. *Compare* HETEROSIS.

Inceptisols An order of *mineral soils that have one or more *horizons in which mineral materials have been weathered or removed. Inceptisols are in the early stages of forming visible horizons, and are only beginning the development of a distinctive soil *profile. The term embraces *brown earths.

incised meander *See* MEANDER.

incisor In mammals, one of the chisel-shaped teeth at the front of the mouth. In primitive forms there are three on each side of both jaws.

inclination **1.** The angle between a freely suspended compass needle and the horizontal. At the magnetic poles the needle will point directly downward, at an angle of 90°, conventionally taken to be +90° at the North Pole and −90° at the South Pole. The needle will be horizontal at the magnetic equator. *See also* DIP POLE. **2.** The angle between the orbital plane of a body orbiting the Sun and the plane of the Earth's *ecliptic.

inclusive fitness The *adaptive value (fitness) of an individual, taking account not only of that individual's own success, but also of the success of all its kin (i.e. those bearing some portion of the same

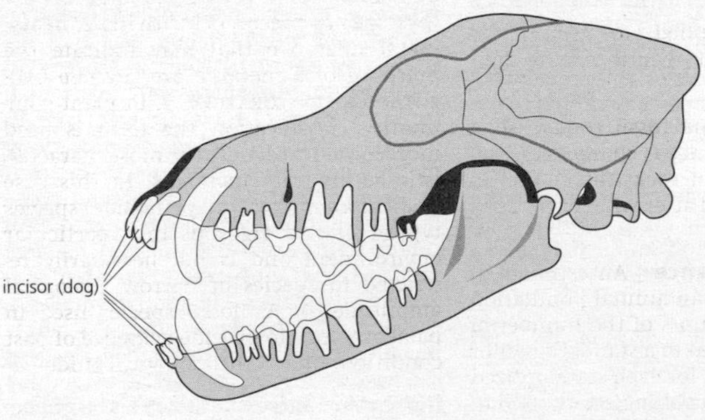

Incisor

*genotype). J. B. S. Haldane is reputed to have said that he would lay down his life for more than two brothers or sisters, eight cousins, 32 second cousins, etc., these numbers corresponding to the proportions of his own *genes shared by these relatives. *See also* KIN SELECTION.

incompatibility The relationship between patterns of behaviour that cannot be performed simultaneously because they are contradictory, because they require the simultaneous operation of reflexes that are mutually inhibitory, or because they are based on contradictory motivation.

inconsequent drainage (insequent drainage) A *drainage pattern which bears no apparent relationship to the underlying rock type or structure.

increment borer (tree borer, Pressler borer) A tool that is used to bore and extract a thin cylinder of wood (core) from a living tree without damaging the tree.

incus The Latin *incus*, meaning 'anvil', which is the name of the supplementary cloud feature that sometimes forms at the top of a *cumulonimbus cloud. It comprises a flattened, anvil-shaped feature composed of ice crystals that develops where the upward growth of the cloud is checked by the *tropopause and the top of the cloud is swept out by the wind. As crystals fall from the cloud they vaporize in the drier air below, giving the feature its characteristic shape. *See also* CLOUD CLASSIFICATION.

indehiscent Applied to fruits (e.g. berries) that do not open to release their *seeds.

index fossil (zonal fossil, zone fossil) A *fossil whose presence is characteristic of a particular unit of rock (the *zone) in which it occurs and after which that zone is named.

index of abundance An estimate of the relative size of an animal population calculated from counts of the number of individuals attracted to a standard bait or the number caught for each standardized unit of effort (e.g. of fishing effort, used in estimating the size of a fish stock).

Indian floral region Part of R. Good's (1974, *The Geography of the Flowering Plants*) Indo-Malaysian floral subkingdom within his *Palaeotropical floral kingdom, and more a region of floral mixing than a distinct entity, notwithstanding the extent of the subcontinent. There are only about 150 endemic genera (*see* ENDEMISM) and most of these are *monotypic. *See also* FLORAL REGION and FLORISTIC PROVINCE.

Indian Ocean One of the world's major oceans, lying between Africa, India, and Australia. It has a surface area of 77 million km^2 and an average depth of 3 872 m. The ocean receives a great deal of sediment from three of the world's major rivers (the Ganges, the Indus, and the Brahmaputra).

Indian summer *See* SINGULARITY.

indicator *See* ERRATIC.

indicator species **1.** A species that is of narrow ecological amplitude with respect to one or more environmental factors and which is, when present, therefore indicative of a particular environmental condition or set of conditions. For example, the lichen *Usnea articulata* occurs only where levels of atmospheric sulphur dioxide are low and the nettle (*Urtica dioica*) indicates high levels of phosphate. If species are long-lived their performance represents an integration of the influence of the factor with time and may give a better assessment of its importance than can a more precise physical measurement taken on a particular day. **2.** In geobotanical surveys, species or ecotypes with high *heavy-metal tolerance that may indicate the presence of a metallic ore. *See also* GEOBOTANICAL EXPLORATION. **3.** In plant community classification the term is used more loosely to denote the most characteristic community members. In this case 'indicator species' may include species typical of and vigorous in a particular environment and is not necessarily restricted to species of narrow ecological amplitude. **4.** A *fossil species used in palaeontology to provide evidence of past conditions or as a chronological guide.

indicator species analysis In general, a classificatory scheme in which the final groups are characterized by *indicator species derived from the data in the course of group definition. More specifically, the term refers to a *polythetic divisive classificatory scheme proposed by M. O. Hill in 1975. Sites are ranked by a reciprocal averaging ordination and divided into two groups at the mid-point (the 'centre of gravity') of all the weighted data values of the ordination. Indicator species (usually five) are then identified as those species exclusively, or most nearly, associated with one or other side of this division (positive and negative indicators). The site-indicator scores, effectively a rough secondary ordination, determine their final classification; and the process may then be repeated within the groups identified. The indicator species form a key, enabling new sites to be added easily into the classificatory framework without excessive recalculation.

indifferent species A species with no real affinity for any particular *community, but which is not rare (as an accidental species would be). It is fidelity class 2 in the *Braun-Blanquet phytosociological scheme (*see* FAITHFUL SPECIES). *Compare* ACCIDENTAL SPECIES; EXCLUSIVE SPECIES; PREFERENTIAL SPECIES; and SELECTIVE SPECIES.

indigenous *See* NATIVE.

indirect inhibition *See* MUTUAL INHIBITION COMPETITION TYPE.

individual distance The distance from an individual at which the presence of another individual of the same species incites aggression or avoidance.

individualistic hypothesis The view, first proposed by H. A. *Gleason in 1917, that vegetation is continuously variable in response to a continuously varying environment. Thus no two vegetation *communities are identical. It implies also that vegetation cannot be classified, and that recognition of particular individual communities will be difficult (the problem arising because of the difficulty of defining boundaries). This viewpoint underlies one of the two polarized approaches to the description and analysis of vegetation communities, which were much debated in the 1950s and 1960s. The individualistic hypothesis favours a continuum view of vegetation, for which ordination rather than classification methods are appropriate. *Compare* ORGANISMIC.

Indo-Malesian rain forest The rain forest found in western India and Sri Lanka, parts of continental south-east Asia, the Malay Peninsula, and the Malay archipelago, with outliers in Queensland, Australia. It is differentiated from the other rain forests by its floristic composition. Large areas of the Indo-Malesian rain forest have been cleared. *See* AFRICAN RAIN FOREST and AMERICAN RAIN FOREST.

induration The formation of hardened (i.e. indurated) *horizons or *hardpans that have a high *bulk density and are hard or brittle. Cementing materials may be present and responsible for the induration.

industrial melanism The development of melanic forms of organisms in response to soot and sulphur dioxide pollution. Industrial melanism is especially associated with various moth species and was first noted early in the nineteenth century. In the 1950s, studies by H. B. D. Kettlewell on the moth **Biston betularia* showed that melanism is a simple inherited trait favoured by natural selection in soot-polluted environments, where dark colouring gives better camouflage and so better protection from predators. In time the proportion of melanic forms increases in such polluted environments. The reverse process (i.e. selection in favour of pale forms) occurs in pollution-free environments where the paler, speckled form is better concealed on the tree lichens on which it settles. With appropriate calibration, the extent of melanism in a population may thus be used as an indicator of industrial pollution.

inertia **1.** The tendency of a body to resist forces that would set it in motion if it is at rest, or that would change its *velocity if it is in motion. **2.** The capacity of an *ecosystem to resist perturbation. It is one measure of the stability of an ecosystem. *See* STABILITY 2.

infauna Benthic organisms (*see* BENTHOS) that dig into the sea bed or construct tubes or burrows. They are most common in the subtidal and deeper zones (i.e. the area seaward of the low-water mark). *Compare* EPIFAUNA.

inferred tree A *phylogenetic tree based on empirical data pertaining to extant taxa.

infiltration The downward entry of water into soil.

infiltration capacity The maximum rate at which soils and rocks can absorb rainfall. The infiltration capacity tends to decrease as the soil moisture content of the surface layers increases. It also depends upon such factors as grain size and vegetation cover.

influent stream *See* GAINING STREAM.

information analysis An agglomerative classificatory scheme for describing vegetation *communities. It uses the *information statistic as the basis for the joining of groups.

information index *See* SHANNON–WIENER INDEX OF DIVERSITY.

information statistic A measure of the extent to which members of a group differ from one another (i.e. of disorder); it is zero when all individuals within the group are identical. In information analysis (an agglomerative, hierarchical classificatory technique, devised by W. L. T. Williams and others, and described in 1966) a hierarchy is constructed by repeatedly joining together those individuals or groups which exhibit the smallest increase in heterogeneity (disorder), and therefore the smallest change in information.

informative site In *phylogenetics, a *nucleotide or *amino acid site which is represented by at least two *character states, each represented by at least two sequences, from which a phylogenetic deduction can be reached.

infralittoral fringe In marine ecology, a term sometimes used to distinguish the intertidal region exposed only at the lowest (i.e. equinoctial) tides. It lies between the *medio- and *circalittoral zones. In the more widely used shore zonation scheme it corresponds to a transitional area between the *littoral and *sublittoral zones. In this zone *Zostera* species are very characteristic plants on soft substrates. The brown seaweeds (*Laminaria* species) start to become abundant in this zone.

infralittoral zone *See* SUBLITTORAL ZONE.

infusoria An old term for the teeming microscopic organisms, particularly protozoa, found in hay infusions (i.e. hay which has been left to soak in water).

ingrown meander *See* MEANDER.

inhibition **1.** The complete abolition of, or the decrease in the extent or rate of an action or process. **2.** During a succession, modification of the environment by a species in such a way as to reduce the suitability of that environment for a species that would otherwise become established in a later seral stage. It is the opposite of *facilitation.

-ini In classification by the nomenclature code, the recommended, but not mandatory suffix used to indicate a tribe of animals.

initial strontium ratio (common strontium) The ratio between the radioactively produced isotope ^{87}Sr and the 'ordinary', non-radiogenic isotope ^{86}Sr at the time when a rock crystallized. ^{87}Sr is produced by the decay of ^{87}Rb and measurements of the present proportions of rubidium and the two strontium isotopes can be used to calculate the initial strontium ratio and the age of the rock. This provides the basis for one of the most important geochronological methods.

inland sea An extensive body of water that is largely or wholly surrounded by land. Any connection to the open ocean is restricted to one or a few narrow sea passages (e.g. the Baltic and Mediterranean Seas).

innate **1.** Applied to behaviour that is not learned (i.e. behaviour that is acquired genetically). In the 1950s and 1960s there was much debate concerning whether particular behaviour is innate or learned; today most ethologists hold that most behaviour has both innate and learned components and studies concentrate on the relative contribution made by each. **2.** Applied to behaviour that may be learned but that has evolved by *natural selection. In this case parents behave in a particular way, and teach their offspring to behave likewise, and are more likely for this reason to reproduce successfully. The term is regarded by many ethologists as obsolete because it fails to allow for the modification of patterns of behaviour in response to environmental factors.

in-phase overlapping The translation of two or more proteins in the same *open reading frame.

inquiline A parasite that lives inside a plant *gall, but without parasitizing the other occupant of the gall.

inquilism An intimate association between two animals in which one partner lives within the host, obtaining shelter and, perhaps, a share of the host's food.

Insecta (insects) A class of *Arthropoda, members of which have three pairs of legs and usually two pairs of wings borne on the thorax. Typically there is a single pair of antennae and one pair of compound eyes. The oldest *fossil insects occur in *Devonian rocks and the first winged representatives are known from *Carboniferous rocks. The evolution of the flowering plants had a marked influence on insect development and many new forms appeared in the Cretaceous and *Tertiary Periods. More than 750 000 extant species of insects have been described. This is larger than the number of species belonging to all other animal classes combined.

insecticide A chemical substance that is toxic to insects and is used to control infestations of insect pests. Insecticides may be derived from substances that occur naturally in plants (e.g. pyrethrum, prepared from the flowers of *Chrysanthemum cinerariaefolium*, and derris (or rotenone), prepared from the roots of *Derris elliptica*) or are synthesized industrially.

inselberg A steep-sided, isolated hill that stands above adjacent, nearly flat plains (the word is German, from *Insel*, meaning 'island' and *Berg*, meaning 'mountain'). It may have a *pediment at its base. Locally, flared or steepened margins occur. It is best developed in a savannah climate. *See also* BORNHARDT.

insequent drainage *See* INCONSEQUENT DRAINAGE.

insertion A *mutation where one or more *nucleotides are inserted into a DNA sequence.

insertion sequence A transposable element that carries no genetic information beyond that which is required for *transposition.

insight learning The ability to respond correctly to a situation that is experienced for the first time and that is different from any experience encountered previously. This ability appears to involve some kind of mental reasoning. *See also* LATENT LEARNING and TRIAL-AND-ERROR LEARNING.

insolation The amount of incoming solar radiation that is received over a unit area of the Earth's surface. This varies according to season, latitude, transparency of the atmosphere, and aspect or ground slope. On average, equatorial areas receive approximately 2.4 times as much insolation as polar areas.

insolation weathering *See* THERMOCLASTIS.

instability The atmospheric condition in which displaced air tends to maintain its movement away from its original level. This occurs, for example, when rising air cools at the moist-adiabatic lapse rate while a greater *environmental lapse rate allows the air parcel to remain warmer than surrounding air (or even increases the temperature difference) so that continual buoyancy prevails. *See also* CONDITIONAL INSTABILITY; POTENTIAL INSTABILITY; and STABILITY.

instar An insect larva that is between one moult (ecdysis) of its exoskeleton and another, or between the final ecdysis and its emergence in the adult form. Instars are numbered and there are usually several during larval development.

instinct A genetically acquired force that impels animals to behave in certain fixed ways (i.e. *fixed-action patterns) in response to particular stimuli. The term is little used by modern ethologists because it is open to many of the same objections as the term *'drive', because it makes no allowance for environmental influences upon patterns of behaviour, and because behaviour formerly considered to be 'instinctive' is now known to result from several different categories of motivation.

instrumental conditioning *See* OPERANT CONDITIONING.

integrated pest control The control of agricultural and horticultural pests by using pesticides in such a way as to leave natural predators unharmed, thus integrating the positive features of chemical and biological control methods.

intensification In meteorology, the increase of the pressure gradient around a pressure system. *See also* WEAKENING.

intensity scale A sequence of thresholds such that progressively stronger stimuli are needed before the appropriate patterns of behaviour result. These stimuli may be increasingly intensive repetitions of a single stimulus (e.g. a spider that is only slightly hungry does not respond to prey; when rather more hungry it turns toward the prey; when still more hungry it initiates an attack).

intention movement The first stage in a recognized sequence of behaviour; by itself it indicates to an observer the intention of an animal to perform the full sequence. For example, the swimming of a stickleback low over a small area of a sandy river bed during the breeding season may be an intention movement that heralds the digging of a pit and the building of a nest. (This is thought to originate in *conflict between *aggression between male and female and nest-building, which has become ritualized.)

inter- The Latin *inter*, meaning 'between' or 'among', used as a prefix meaning 'between'.

interception **1.** The capture of rain water by vegetation from which the water evaporates and is thus prevented from reaching the *water-table and contributing to *surface run-off, *soil moisture, or *groundwater recharge. **2.** The abstraction of groundwater part of the way along its flow path, where otherwise the water might be lost (e.g. as coastal spring discharges).

interference *See* MUTUAL INTERFERENCE.

interference competition The *competition that occurs when two organisms demand the same resource and that resource is in short supply, and one of the organisms denies its competitor access to the resource. In essence, space is substituted for the resource as the prime object of competition and dominance of space provides an alternative to efficiency in resource exploitation. Territorial animals exhibit this type of competition, as do robust and aggressive plants (e.g. the mat grass, *Nardus stricta*). *Compare* EXPLOITATION COMPETITION.

interflow (throughflow) The lateral movement of water through the upper *soil horizons, most commonly during or following significant precipitation. If the interflow is shallow it may emerge at and flow for some time across the surface at the bottom of slopes (when it is known as 'return flow').

interfluve The elevated part of the landscape that extends between two adjacent valleys. It is normally seen as lying above the steeper slopes of each valley side.

interglacial A period of warmer climate that separates two glacial periods. Mid-latitude interglacials show a characteristic sequence of vegetation change. *Pollen of heathy *tundra is replaced in the pollen record by abundant herbaceous pollen, which in turn is replaced by *boreal and subsequently *deciduous forest, including pollen of *thermophilous species, e.g. *Tilia* (lime). From this peak the sequence reverses as the trend to colder conditions predominates.

intergrade A soil or *soil horizon that has the properties of two soils or horizons which do not share a common origin. An intergrade can be regarded as transitional between two distinctive soils or horizons.

intermediate disturbance hypothesis The proposition that the highest diversity of species in an *ecosystem is maintained by a level of disturbance intermediate between frequent and rare disturbance. If disturbance is frequent the succession may fail to develop beyond the *pioneer phase. If disturbance is rare, the climax will be established and diversity reduced according to the *competitive exclusion principle. At intermediate levels of disturbance, the arrival of new species will increase diversity in proportion to the interval between disturbances.

intermediate rock An *igneous rock with a chemical composition between those of *basic and *acid rocks. The limits are not fixed rigidly.

intermittent stream A stream which ceases to flow in very dry periods. The flow may occur when the *water-table is seasonally high, but there will not be flow when the water-table is significantly below the river-channel bed level. One view of the drainage network (*see* DRAINAGE PATTERN) is that it is made up of perennial streams, which flow all the time, intermittent streams, which flow when the water-table is seasonally high, and ephemeral streams, which flow only during storm conditions. Such streams tend to have permeable beds and discharge leaking through their beds (transmission losses) is added to the local *groundwater. The chalk bournes of southern England are special types of intermittent streams. *See also* LOSING STREAM.

intermontane **1.** Between mountains or mountain ranges. **2.** Applied to basins which are being infilled by sediment eroded from surrounding mountains.

interna A system of feeding 'roots' produced by certain parasitic barnacles (e.g. *Sacculina carcini*) which penetrates the entire body of the host.

internal node Within a *phylogenetic tree, a point where two branches join, representing an ancestral species or gene.

internal wave A wave that forms within a water mass at the boundary of two water layers that have different densities. The boundary may be abrupt or gradual, and the slow-moving waves can be detected only by instrumental observations of temperature or *salinity, and by acoustic scattering.

International Biological Programme (IBP) An international research programme conducted approximately from 1966 to 1975 with the aim of understanding the dynamics of whole *ecosystems from a range of world environments, e.g. *desert and *tropical rain forest. Objectives included the development of predictive mathematical models for ecosystem structure and function to help assess the effects of as yet unexperienced environmental impacts, as well as aiding the understanding of changes associated with past disruptions. Although not all the projects were equally successful and a complete reference set of models has not been produced and accepted, the models that do exist provide a rational basis for evaluating the likely consequences of proposed human interventions, for remedying the effects of previous human use of an area, and for adapting management systems to improve productivity and environmental quality.

International Code of Botanical Nomenclature A set of rules for the formal naming of plants, accepted by botanists, in which the underlying principle is the allocation of a single, unambiguous name to each *taxon. The Code comprises a set of six Principles to guide those who are selecting a new name, with 75 Articles and a number of Recommendations. Observance of the Articles is mandatory, but not of the Recommendations. This section is followed by translations in French and German and appendices listing conserved names. The starting-point for naming of plants is taken as *Species Plantarum*, published by *Linnaeus in 1753. Fungi, fossil plants, and bacteria start from different dates and authorities. The Code allows for the naming of *cultivars.

International Code of Zoological Nomenclature (International Commission on Zoological Nomenclature, ICZN) **1.** The regulations governing the scientific naming of animals. **2.** The international authority that draws up those regulations and supervises their application.

International Union for Conservation of Nature and Natural Resources (IUCN) An international, independent organization, founded in 1948, with headquarters in Switzerland, which promotes and initiates scientifically based conservation measures. Its members include more than 450 government agencies and conservation organizations in more than 100 countries, and it works closely with United Nations agencies. It publishes data books listing endangered species, and in 1980 it published the *World Conservation Strategy* in collaboration with the UN Environment Programme, the World Wide Fund for Nature, the Food and Agriculture Organization, and UNESCO.

interpluvial *See* PLUVIAL PERIOD.

intersex A class of individuals that belong to a species in which two sexes occur but which possess sexual characteristics that are intermediate between the two sexes. The condition may result from the failure of the sex-determining mechanism of the genes or through hormonal or other influences during development (e.g. the condition in cattle called free-martin, in which the female member of a pair of oppositely sexed twins has been influenced, probably hormonally, by her twin brother, through the fusing of their placental circulations).

interspecific competition *See* COMPETITION.

interstade *See* INTERSTADIAL.

interstadial (interstade) A phase of warmer climate within a glacial period, but of shorter duration (and thought to be less warm) that an *interglacial. Species demanding warmth, e.g. *Tilia* (lime), are not represented in the pollen record, which shows *boreal affinities. The absence of *thermophilous species may, however, be as much a consequence of the shorter time-span of an interstadial as of the lack of warmth.

interstitial Pertaining to the spaces (interstices) between sedimentary particles.

interstitial fauna Animals that inhabit the spaces between individual sand grains. The term is often used synonymously with *meiofauna, *mesofauna, and *microfauna.

intertidal zone An area between the highest and lowest tidal levels in a coastal region. *See* LITTORAL ZONE.

intertropical confluence An alternative term for the *intertropical convergence zone (ITCZ) which is preferred by some scientists because of the discontinuous occurrence of convergence within it.

Intertropical Convergence Zone (ITCZ) A belt surrounding the Earth close to the equator where the north-easterly *trade winds of the northern hemisphere and south-easterly trade winds of the southern hemisphere meet. Their convergence causes air to rise, producing low pressure at the surface (the *equatorial trough). When the rising air reaches the *tropopause it moves away from the equator, subsiding in the subtropics, some of the air rejoining the trade winds; this circulation comprises the *Hadley cells. Over the oceans (but not over land) the ITCZ coincides with the *thermal equator. The position of the ITCZ changes with the seasons and to a lesser extent from year to year due to changes in sea-surface temperatures. *See* INTERTROPICAL CONFLUENCE.

intertropical front The meeting-point of air brought by the *trade winds from the circulation of winds of the northern and southern hemispheres. It is not always a sharp *front, but is always a convergence zone. The equatorial rains are associated with this convergence.

intortus Twisted or entangled, the name of a species of *cirrus cloud. *See* CLOUD CLASSIFICATION.

intra- The Latin *intra*, meaning 'inside', used as a prefix meaning 'within' or 'on the inside'.

intraspecific competition *See* COMPETITION.

intraspecific parasitism A variety of *brood parasitism in birds, in which a female lays eggs in the nest of others of the same species.

intrinsic rate of natural increase *See* BIOTIC POTENTIAL.

introgression The incorporation of genes of one species into the gene pool of another. If the ranges of two species overlap and fertile *hybrids are produced, the hybrids tend to backcross with the more abundant species. This results in a population in which most individuals resemble the more abundant parents but also possess some of the characters of the other parent species.

intron A DNA segment of a transcribed gene which is removed during transcription and, therefore, does not appear in the mature *messenger-RNA.

i

inverse analysis The grouping of attributes based on an analysis of the individuals that possess or lack those attributes (e.g. in plant ecology the grouping of species according to their presence, absence, or relative abundance at different sample sites). The term is used particularly in numerical vegetation classification, and is sometimes referred to as a species classification. Ordination methods may similarly be described as either plot (individual) or species (attribute) ordinations. In plant ecology especially, inverse classifications are often used to complement normal analysis. Thus data will be analysed using both approaches and coincidence between the final groups examined. It is assumed that high coincidence implies the recognition of an important 'type' *community or *'nodum'. In its concept, nodal analysis is a method based on these principles. *See also* Q TECHNIQUE; *compare* NORMAL ANALYSIS and R TECHNIQUE.

inversion **1.** A *mutation that causes a segment of DNA to be turned back to front within the sequence. **2.** *See* TEMPERATURE INVERSION.

inverse stratification *See* THERMAL STRATIFICATION.

invertebrate drift The passive *dispersal of the larvae of invertebrates living in rivers, which are carried by the flow of water from the sites where they hatched to sites where they can develop further.

inverted relief An inverse relationship between the land surface and the underlying geologic structure, as when a hill is developed in a *syncline and a valley in an *anticline. It is a stage beyond the 'normal' (Jura-type) relief which is characterized by anticlinal hills and synclinal valleys. The term also denotes the more general case in which, through *erosion, a hill becomes a valley or vice versa.

in vitro Literally, 'in glass', but applied more generally to studies on living material which are performed outside the living organism from which the material is derived. Examples include the use of tissue cultures, cell homogenates, and subcellular fractions.

in vivo Applied to studies of whole, living organisms, on intact organ systems therein, or on populations of micro-organisms.

ion An atom that has acquired an electric charge by the loss (*cation; positive charge) or gain (*anion; negative charge) of one or more electrons.

-ion *See* ZURICH–MONTPELLIER SCHOOL OF PHYTOSOCIOLOGY.

ion exchange **(IX)** The reversible exchange of *ions for other ions in solution. Natural *zeolites are used to capture *anions and *cations from solution. Artificial ion-exchange resins with three-dimensional hydrocarbon networks are commonly used (e.g. in water softeners, for separating isotopes, in desalination, and in the chemical extraction of elements from ores).

ionic bond **(electrovalent bond)** A chemical bond in which electrons are transferred from one atom to other atoms and the atoms are held together by the resulting electrostatic attraction between them. *See also* COVALENT BOND, HYDROGEN BOND, and METALLIC BOND.

ionosphere The part of the atmosphere that lies above about 80 km altitude, with the highest concentrations of *ions and free electrons. The most intense concentration is at 100–300 km altitude.

Ipswichian The temperate, last *Pleistocene *interglacial, named after Ipswich, Suffolk, England, and the associated Late Pleistocene deposits that occur in river valleys, often associated with *terraces. Pollen diagrams compiled from them indicate that the climate was not much different from that of the present day. The deposits can perhaps be correlated with those of the *Eemian of north-western Europe.

iridescence A physical phenomenon in which fine colours are produced on a surface by the interference of light that is reflected from both the front and back of a thin film. Iridescence provides an alternative to pigmentation in many animals, as in beetle wing cases and peacock tail feathers. It also plays a part in *angiosperm petals, providing texture and sheen to the surfaces that may act as attractant and direction signal to the pollinating insect.

iridium anomaly The anomalously high (typically 50 p.p.b.) concentrations of iridium, relative to typical crustal abundances of less than 1 p.p.b., observed worldwide in sediments straddling the *Cretaceous–*Tertiary boundary. The anomaly is attributed to fallout resulting from a massive asteroidal or cometary impact, which is thought by many to have been responsible for the mass extinctions that define the Cretaceous–Tertiary boundary.

iron **(Fe)** An element required by plants. It is used in reactions in which rapid oxidation reductions occur by the transfer of electrons, as in photophosphorylation and oxidative phosphorylation. Other roles are not understood. Iron-deficient plants have chlorotic (*see* CHLOROSIS) young leaves; at first the veins remain green but later they too become chlorotic.

iron pan An indurated (*see* INDURATION) *soil horizon, found usually at the top of the B horizon, in which iron oxide is the main cementing material.

irrelevant behaviour Activity that interrupts a pattern of behaviour temporarily and does not contribute to the function of that pattern.

irrigation The artificial augmentation of the amount of water available to crops, either by spraying water directly on to the plants or making it available to their root systems through a series of surface channels or ditches.

irritability **(sensitivity)** The capacity of a cell, tissue, or organism to respond to a stimulus, usually in such a way as to enhance its survival.

irruption A sudden change or oscillation in the population density of an organism.

island biogeography The study of the distribution of plant and animal species on islands or in areas that are sufficiently isolated to resemble islands. Islands are numerous and their biotas (*see* ISLAND BIOTAS) are often small enough to be quantified. Accordingly it has been possible to determine a relationship between area and species number, postulated to be an equilibrium between immigration and extinction (*see* RESCUE EFFECT; and ISLAND BIOGEOGRAPHY, THEORY OF). This is the basis of island biogeography, and it may equally be applied on continents, where plant and animal communities are effectively reduced to islands in a sea of cultivation or urbanization.

island biogeography, theory of A theory, advanced in 1967 by R. H. MacArthur and E. O. Wilson and now shown to be too simplistic, so largely discredited, that the number of species on an island will reach a dynamic equilibrium between the continual immigration of species from a mainland source and the extinction of species already present. Once equilibrium is reached the species number will remain constant, but with a continually changing composition. The theory goes on to state that if the immigration rates and the extinction rates are known, then the species number at equilibrium can be calculated. They theory fails, however, to take account of species interaction or of habitat diversity which is, *de facto*, usually greater on big islands; it also makes a major and possibly erroneous assumption that immigration is independent of island size. The theory also compares topographically identical near and far islands in relation to the mainland source, maintaining that near islands would have a higher rate of immigration than far ones because the biogeographical barrier presented by the sea would be smaller. It also states that the extinction rate would be independent of island location, and therefore the resulting equilibrium point for the number of species would be higher for the nearer islands. The assumption that the extinction rate is independent of island location may be incorrect because of the *rescue effect. The concept has been extensively used as a basis for the selection of natural reserves, the assumption being that large areas with minimal boundaries located close to other similar areas will prove the richest and the most sustainable type of reserve. This use of the theory has been extensively tested with mixed results.

island biotas The plants and animals of oceanic islands. Because of their isolation, the biotas normally include numerous endemic (*see* ENDEMISM) taxa. Island biotas are generally fragile, unbalanced in that they lack plants and animals with poor transoceanic dispersal capacities, and vulnerable to disturbance by humans and introduced species.

island hopping The colonization of an island or islands by plants and animals from an adjacent island or islands. Birds are particularly likely to hop from one island to another. Over geological time, islands move away from their areas of origin by *continental drift. The descendant biotas may maintain themselves in the ancestral environment by island hopping on to successively younger islands as these emerge.

Island mammal region A concept proposed by the biogeographer Charles H. Smith, which combines the *Australian, *Madagascan, and West Indian faunal subregions. The development of the characteristic mammalian faunas of these insular faunal regions is at least as much the result of their isolation by the sea as of ecological and geographic factors.

iso- From the Greek *isos*, meaning 'equal', a prefix meaning 'equal'.

isobar A line on a weather map connecting points that are at the same atmospheric pressure. On surface charts the values are 'reduced' to sea level. Such isolines are drawn at a given interval in millibars. Contours of *isobaric surfaces may be drawn to represent surfaces in the upper atmosphere composed of points at the same pressure.

isobaric surface A surface on which any point experiences the same atmospheric pressure.

isochore A DNA segment that is homogeneous in base composition.

isocline A line joining points on a graph at which combinations of resource levels or population densities produce a similar rate of population growth for a particular species.

isofrigid *See* PERGELIC.

isogeneic (syngeneic) Applied to a *graft that involves a scion and stock that are genetically identical.

isohel A line on a climate map connecting points of equal average sunshine duration.

isohyet A line on a climate map connecting points of equal average rainfall.

isohyperthermic *See* PERGELIC.

isomer Either or any of two or more compounds that have the same molecular composition but different molecular structure. Isomers differ from each other in their physical and chemical properties.

isomesic *See* PERGELIC.

isoneph A line on a climate map joining points of uniform cloud cover.

isonome *See* TRANSECT.

isoprene C_5H_8, a volatile hydrocarbon compound that forms the structural base of many compounds synthesized by plants. Isoprene molecules form units that combine to make larger units, known as terpenes. Many plants, especially deciduous trees, emit both isoprenes and terpenes. These compounds are highly reactive and play a major role in the formation of ozone in the lower atmosphere. Isoprene contributes to the formation of photochemical smog and in rural areas it is the predominant hydrocarbon involved.

isopycnal A line on a map joining points of equal density within a water mass. A three-dimensional surface of equal density is called an isopycnal surface.

isotherm A line on a climate map connecting points of equal average temperature.

isothermic *See* PERGELIC.

isotonic Applied to a cell in which the *osmotic pressure is equal to that in the surrounding medium. *Compare* HYPERTONIC and HYPOTONIC.

isotope One of two or more varieties of a chemical element whose atoms have the same numbers of protons and electrons but different numbers of neutrons. There are 300 naturally occurring isotopes, but only 92 naturally occurring elements. Isotopes may be produced by various nuclear reactions. Frequently the products are radioactive. *See also* ISOTOPIC DATING.

isotope hydrology The use of naturally occurring and introduced *isotopes to date and identify water bodies. Among the most commonly used isotopes are tritium, deuterium, carbon-13, carbon-14, chlorine-36, and oxygen-18.

isotope tracer A radioactive *isotope, whose movement can be monitored, which is used to trace the pathways by which individual substances move through an organism, a living system, the *abiotic environment, etc. Non-radioactive chemical analogues of certain substances may be used for the same purpose if their movement can be monitored (e.g. caesium, which can be substituted for potassium).

isotopic dating A means of determining the age of certain materials by reference to the relative abundance of the parent *isotope (which is radioactive) and the daughter isotope (which may or may not be radioactive). If the decay constant (the 'half-life' or disintegration rate of the parent isotope) and the concentration of the daughter isotope are known, it is possible to calculate an age. *See also* DATING METHODS; RADIOCARBON DATING; and RADIOMETRIC DATING.

isotopic fractionation The analysis of the ratios of different stable *isotopes of an element that is present in a given material. The proportions of different stable isotopes of elements such as nitrogen and carbon can be used to trace the origin of the element within *ecosystems, depending upon the ratios of the different isotopes. Nitrogen isotopes, for example, have been used to detect the contribution of migrating salmon in inland streams to the influx and general nitrogen economy of surrounding terrestrial ecosystems.

isotropic Applied to substances whose optical or other physical properties are the same from whatever direction they are observed.

isozyme One of two or more *enzymes that have identical or similar functions, but are encoded by different loci (*see* LOCUS).

ITCZ *See* INTERTROPICAL CONVERGENCE ZONE.

iterative evolution A repeated *evolution of similar or parallel structures in the development of the same main line. There are many examples of iterative evolution in the fossil record, spanning a wide range of groups. This evolutionary conservatism is probably owing to the overriding morphogenetic control exerted by certain regulatory genes.

iteroparity The condition of an organism that has more than one reproductive cycle during its lifetime. *Compare* SEMELPARITY.

IUCN *See* INTERNATIONAL UNION FOR CONSERVATION OF NATURE and NATURAL RESOURCES.

IX *See* ION EXCHANGE.

J *See* JOULE.

Japan Trench The oceanic *trench that lies between the northern Japanese islands (an island arc) on the edge of the Eurasian Plate and the Pacific Plate.

japweed (strangleweed) The common name for the brown seaweed *Sargassum muticum*; the seaweed originated in Japan but has become established on the coasts of Europe and North America. It is regarded as a pest, fouling boat propellers and fishing nets and disturbing the natural balance of the indigenous marine flora and fauna.

Java Trench The oceanic *trench that marks the outer, deep edge of the East Indies subduction zone, where the Indo-Australian Plate is subducting beneath the Eurasian Plate. The trench is about 6 km deep along Java, but becomes shallower to the north-west because of progressive infilling by Bengal Fan turbidites (sediments deposited by *turbidity currents). There is an outer, non-volcanic arc, formed of an accretionary wedge, and a fore-arc basin.

jet stream A concentrated, high-speed air flow, generally in a broadly westerly (i.e. west to east) direction. The principal global jets are the *polar-front and the *subtropical jet streams, at heights of about 10–12 and 12–15 km respectively, and the polar-night or winter jet stream in the upper *stratosphere or *mesosphere at about 50–80 km. The intensity of the jets in narrow bands (the maximum velocity is commonly about 50–100 m/s, but greater speeds are sometimes observed) results from a large poleward increase in pressure gradient with altitude. This is a product of the pole–equator temperature gradient in the air beneath the jet. As pressure decreases more rapidly with height the lower the temperature, so pressure in the colder, polar air masses decreases more rapidly with height than over the regions with warmer air masses. Above the jet, wind speed diminishes as the pressure gradient declines with increasing height, owing to the effects of a different heating pattern in the stratosphere.

jizz *See* GESTALT.

jökulhlaup A sudden, violent, but short-lived increase in the discharge of a melt-water stream issuing from a *glacier or *ice cap, sometimes owing to volcanic activity beneath. A lake may develop above the heat source; this is subsequently breached to produce a torrent of melt-water (e.g. Lake Grimsvotn on Vatnajökull, Iceland). Flow velocity may reach 7–8 m/s and the discharge may attain 100 000 m^3/s (e.g. the Katlahlaups from Myrdalsjökull, Iceland), comparable to rates of flow of the Amazon river.

joule (J) The derived SI unit of energy, work, or quantity of heat. It is the work done or heat generated by a current of one ampere flowing for one second against a resistance of one ohm, or when the point of application of a force of one newton moves a distance of one metre in the direction of the force. 1 J = 0.239 calories = 10^7 ergs. The unit is named after the British physicist James Prescott Joule (1818–89). See Appendix.

J-shaped growth curve A curve on a graph that records the situation in which, in a new environment, the population density of an organism increases rapidly in an exponential (logarithmic) form, but then stops abruptly as environmental resistance (e.g. seasonality) or some other factor (e.g. the end of the breeding phase) suddenly becomes effective. It may be summarized mathematically as: $dN/dT = r$ (with a definite limit on N) where N is the number of individuals in the population, T is time, and r is a constant representing the *biotic potential of the organism concerned. Population numbers typically show great fluctuation, giving the characteristic 'boom and bust' cycles of some insects, or the pattern seen in algal blooms.

This type of population growth is termed 'density-independent' as the regulation of growth rate is not tied to the population density until the final crash. *Compare* S-SHAPED GROWTH CURVE.

Juan Fernandez floral region Part of R. Good's (1974, *The Geography of the Flowering Plants*) *Neotropical floral kingdom, which corresponds geographically to the islands of Juan Fernandez and the Desventuradas Islands, 650–1 000 km west of the coast of Chile. There is a remarkable degree of *endemism, including one family, 18 genera, and nearly 160 species. *See also* FLORAL REGION and FLORISTIC PROVINCE.

jungle A popular term for *tropical rain forest, derived from the Sanskrit *jangala*, meaning 'desert', 'forest', or 'an impenetrable tangle'.

junk DNA Genomic DNA that has no known function.

Jurassic One of the three *Mesozoic periods, about 199.6–145.5 Ma ago, which followed the *Triassic and preceded the *Cretaceous. The Jurassic Period is subdivided into 11 stages, with clays, calcareous sandstones, and limestones being the most common rock types. Brachiopoda, Bivalvia, and Ammonoidea were abundant fossils, along with many other invertebrate stocks. Reptiles flourished on land and in the sea, but mammals were relatively insignificant and are presumed to have been predominantly nocturnal. The first birds, including *Archaeopteryx*, appeared in the Late Jurassic.

juvenile water Original water, formed as a result of magmatic processes, which has never been in the atmosphere. Magmatic water can form in very large quantities. A magma body with a density of 2.5, an assumed water content of 5 per cent by weight, a thickness of 1 km, and an area of 10 km^2 contains some $1.25 \times 10^9 m^3$ of water. *See also* GROUNDWATER.

K *See* POTASSIUM.

Kainozoic *See* CENOZOIC.

kame A steep-sided mound, composed of bedded sand and gravel, which often shows signs of marginal slumping. It is a land-form of glacial deposition, associated with *stagnant ice whose removal by melting causes the collapse.

kame delta A flat-topped mound of stratified sand and gravel laid down in standing water at an ice margin. Subsequent melt of the ice margin leads to loss of support and brings about collapse on the former ice-contact side.

kame terrace A continuous valley-side land-form, consisting of stratified sand and gravel, the outer edge of which typically shows collapse features. It is laid down by meltwater at the junction between an ice mass and the valley wall. It may be connected to *eskers.

kampfzone An altitudinal belt of stunted and often prostrate trees, found between the upper limit of tall, erect trees growing in forest densities (waldgrenze) and the extreme upper limit of tree growth (the species limit or *tree line). *See also* ELFIN WOODLAND.

Kansan The second (0.48–0.23 Ma) of four glacial stages occurring in North America, during which isotope evidence suggests that the climate was less extreme than during the *Nebraskan episode, although the Kansan glaciation extended further south. The Kansan is approximately equivalent to the *Günz glacial of the Alps.

K–Ar method *See* POTASSIUM–ARGON DATING.

karren The German name for a group of solutional features developed at the surface of an outcrop of hard limestone. Forms range from runnels (shallow troughs) a few millimetres deep to fissures (grikes) extending several metres into the limestone.

karst Specifically, an area of the Dinaric Alps of Bosnia Hercegovina that is underlain by limestone. By extension, any region that is underlain by limestone and characterized by a set of land-forms resulting largely from the action of carbonation or of other processes produces similar types of land-forms (e.g. *thermokarst in the *periglacial environment).

karstic aquifer An *aquifer within a *karst limestone rock matrix. Such aquifers are normally characterized by large void spaces, relatively high values for hydraulic conductivity (*see* PERMEABILITY), flat *water-tables, and extensive networks of solution channels within which *Darcy's law is not obeyed and flow may be turbulent.

karyotype The entire chromosomal complement of an individual or cell, which may be observed during mitotic metaphase.

Kastanozems Soils that have a *mollic horizon more than 20 cm deep and also concentrations of calcium compounds within 100 cm of the surface. Kastanozems are a reference soil group in the *FAO *soil classification.

katabatic wind (drainage wind, mountain breeze) A generic term for the wind that occurs when cold, dense air, chilled by radiation cooling, usually at night, moves downslope gravitationally beneath warmer, less dense air. The occurrence is frequent and widespread in, for example, the *fiords of Norway, and as an outblowing wind over ice-covered surfaces in Antarctica and Greenland, where the wind may be extremely strong near the coasts and less severe in many mountain regions. *Compare* ANABATIC WIND.

katafront A weak frontal condition in which warm-sector air sinks relative to colder air. The term was coined by T. Bergeron. *Compare* ANAFRONT.

kb *See* KILOBAR.

kelp **1.** Brown seaweeds that grow below the low-tide level. Large brown algae (e.g. *Laminaria* species) which anchor themselves firmly to the sea-bed are typical. In some places they are harvested for use as fertilizer, either directly or after burning or processing into a liquid manure. **2.** The ash obtained by burning various large, brown seaweeds, used as an agricultural fertilizer and as a source of iodine, potash, and soda.

kennarten species **(characteristic species)** Like 'faithful species', a former collective term for species of fidelity classes 3–5 (i.e. preferential, selective, and exclusive species). As *phytosociological work progressed, the need became evident for finer distinctions within the kennarten species. *See also* FAITHFUL SPECIES.

kentron A hollow, dart-like structure produced by certain parasitic barnacles (e.g. *Sacculina carcini*) which pierces the *cuticle of the host.

kerangas In Borneo, the name given to *heath and forest, derived from an Iban word meaning 'land which will not grow rice'.

kettle hole **(kettle)** A depression in the surface of glacial *drift (especially ablation or kettle moraine), which results from the melting of an included *stagnant ice mass. It may be filled with water to form a small lake ('kettle lake').

kettle lake *See* KETTLE HOLE.

Kew barometer A *barometer in which the scale markings are adjusted to take account of the changes of the mercury level in the cistern, so eliminating the need to adjust the cistern to the *fiducial point, as is required in the *Fortin barometer. Kew barometers are often used at sea, because they can be made to minimize the oscillations in the height of the mercury caused by movements of the ship, and they can be transported horizontally or inverted, with the cistern at the top and the tube filled with mercury.

keystone species A species that has a disproportionately strong influence within a particular *ecosystem, such that its removal results in severe destabilization of the ecosystem and can lead to further species losses.

khamsin A hot, dry, dusty wind that originates in North Africa and blows across Egypt, usually between April and June, ahead of *depressions which move eastward or north-eastward in the Mediterranean Sea or across North Africa, with high pressure to the east. The term is also applied to very strong southerly or south-westerly winds in the Red Sea. *See* DUST STORM.

kieselguhr *See* DIATOMACEOUS EARTH.

killing frost A sharp fall in temperature which damages a plant so severely as to cause its death, or prevents the reproduction of an *annual, *biennial, or *ephemeral plant. As the falling temperature approaches freezing point, some water is lost from the vacuole into the intercellular spaces, where a further drop in temperature causes it to form small crystals of ice. If the temperature then rises slowly, the water will be reabsorbed by the cell as the ice melts and the cell will recover, but if the thaw is rapid the water will be lost and the cell will die from dehydration. If the freezing temperature is prolonged, ice may be lost by sublimation (i.e. the direct change from the solid to the gaseous phase). This will also cause dehydration. Reproduction will be prevented if freezing causes such damage to flowers or to developing fruit that has not yet produced viable seed. Seeds themselves contain little water and are seldom damaged by frost. *See also* GROWING SEASON.

kilo- From the Greek *khilioi*, meaning 'thousand', a prefix (symbol k) used with *SI units to denote the unit $\times 10^3$.

kilobar **(kb)** A unit of pressure, equal to 1 000 bars (986.923 atmospheres, 10^8 N/m^2, or 10^8 Pa).

Kimura, Motoo **(1924–1994)** A Japanese population geneticist, working at the National Institute of Genetics in Mishima, who made important contributions to evolutionary theory by his mathematical modelling of evolutionary change and his proposal of the *neutrality theory of evolution.

kinaesthetic Applied to sensory receptors or organs that detect movement or changes in their own position.

kinaesthetic orientation The behaviour of an animal that moves through familiar terrain in the absence of sensory information (e.g. in total darkness) by the repetition of actions remembered from past experience of the terrain.

kinesis The phenomenon in which a *motile organism or cell changes its rate of locomotion (or frequency of turning) in response to the intensity of a particular stimulus. The direction of locomotion remains random and is unrelated to the direction of the stimulus. *Compare* TAXIS.

k

kingdom In *taxonomy, one of the major groups into which organisms are placed. In the widely used five-kingdom system of classification the kingdoms are: *Bacteria; *Protoctista; *Animalia; *Fungi; and *Plantae. The kingdoms are grouped into two superkingdoms: *Prokarya, containing the kingdom Bacteria; and *Eukarya, containing the remaining four kingdoms. In the three-domain classification system, kingdoms are ranked below the *domains.

kin selection A form of *natural selection in which the *altruism of an individual benefits its close relatives, and thereby helps to ensure the survival of at least some of its own genes. *See also* INCLUSIVE FITNESS.

kipuka A Hawaiian name for an 'island' of land that is completely surrounded and cut off by lava. Kipukas are also known by the term 'dagalas' in Italy.

kissing bug A name given to several American genera of Reduviidae which feed on the blood of mammals and birds. Some species live in and around human dwellings, where they can transmit the blood parasite *Trypanosoma cruzi*, responsible for the debilitating *Chagas's disease.

kleptoparasitism A *parasitism that is based upon the theft of food from other organisms. Skuas practise this form of parasitism upon gulls and terns. Ants often attempt this parasitism upon carnivorous plants, robbing them of their insect prey. One estimate suggests that carnivorous plants may lose as much as 50 per cent of their prey to kleptoparasites.

klino-kinesis A change of direction of movement of an animal in response to a stimulus such that the rate at which the direction changes is proportional to the strength of the stimulus.

klino-taxis The movement of an animal in response to a stimulus; the animal compares the intensity of the stimulus to either side of its body and moves either towards or away from the stimulus, typically along a sinuous path with constant turning of the head from side to side.

knick point (headcut) An abrupt change of gradient in the generally smooth long-profile of a stream, typically separating two concave-up segments. It is often attributed to a fall in base level: this, it is said, initiates a knick point which then slowly travels upstream. It may alternatively be owing to a change in rock type or load size, or to tributary entry.

knob and kettle (sag and swell topography) The landscape sometimes found on a recent terminal *moraine complex, consisting of a hummocky mound (the 'knob') alternating with a depression (the 'kettle'). The 'kettle' results from the melting of a block of ice enclosed in the *drift.

knock and lochan A glaciated landscape of low relief which is made up of ice-moulded hillocks and intervening lochans (small lakes) eroded along zones of rock weakness. It is especially well developed in the Lewisian gneiss area of the coastal lowlands of north-western Scotland.

knot A unit of speed equal to one nautical mile per hour (0.515 m/s). It is still used in many countries as a measure of wind speed and current velocity (as well as for the speed of ships and aircraft).

Koenigia islandica *See* BIPOLAR DISTRIBUTION.

kona storm A type of storm that is associated with very strong southerly winds over the Hawaiian islands, bringing heavy rainfall. The winds blow in conjunction with the passage of a *depression to the north of the islands.

kopje *See* KOPPIE.

Köppen climate classification A system devised in 1918 by Wladimir Peter Köppen (1846–1940), with modifications that were completed in 1936, by which climates are divided into six broad groups according to the major vegetation types associated with them, broadly determined by critical temperatures and the seasonality of precipitation. For example, a summer temperature of 10°C defines the poleward limit of tree growth; a winter temperature of 18°C is critical for certain tropical plants; a temperature of –3°C indicates some period of regular snow cover. The groups are: (A) tropical rainy climates with temperatures in the coldest month higher than 18°C; (B) arid climates; (C) warm, temperate, rainy climates in which temperatures in the coldest month are between 3°C and 18°C, and in the warmest month higher than 10°C; (D) rainy climates typical of boreal forest, in which temperatures in the coldest month are lower than –3°C (in US usage modified to 0°C), and in the warmest month higher than 10°C; (E) tundra, in which temperatures in the warmest month are 0–10°C; (F) permanent frost and ice caps, in which temperatures in the warmest month are below 0°C. H denotes highland climates, where temperatures are low due to altitude. Subsets of the main classes (written as capital and lower case letters, e.g. Cs) are: absence of a dry season (f); a dry summer season (s); a dry winter season (w); a monsoon climate, with a dry season and rains at other times (m). Arid climates (B) are subdivided into semi-arid steppe-type (S) and arid desert (W). The temperatures within class (B) are indicated as: mean annual temperature higher than 18°C (h); mean annual temperature lower than 18°C with the warmest month higher than 18°C (k); mean annual temperature and warmest month both lower than 18°C (k′). Some criticism of the system centres on its arbitrary criteria of temperature associated with fixed boundaries, and the failure of data on temperature and precipitation to account fully for the effectiveness of the precipitation. *See also* STRAHLER CLIMATE CLASSIFICATION and THORNTHWAITE CLIMATE CLASSIFICATION.

koppie **(kopje)** An Afrikaans word (the literal meaning is 'little head'), applied to a land-form widely described in Africa. It is similar to a *tor, being a steep-sided, isolated land-form, usually at least the size of a house, and best developed on granitic outcrops. It often shows an angular, castellated outline, when it is called a 'castle koppie'. A koppie may be a late stage in the destruction of a *bornhardt.

kosava A local wind in the Danube valley. The term is also used in a generic sense to refer to a type of *ravine wind.

krasnozem *See* RED PODZOLIC SOIL.

kronism The killing and eating of offspring. The word is derived from the name Kronos, in Greek mythology the Titan who swallowed his children (apart from Zeus, who escaped because his mother substituted a stone). The analogy is not perfect, as the victims did not die and eventually were vomited by Kronos, and subsequently killed him.

krotovina **(crotovina)** An animal burrow that has been filled with organic or mineral material from another *soil horizon.

krummholz Gnarled, stunted, and usually bush forms of trees, typically conifers, which grow in the *kampfzone between waldgrenze and *tree line. Krummholz trees can also be found at relatively low altitudes, whenever there is marked exposure to strong winds.

***K*-selection** Selection for maximizing competitive ability, the strategy of *equilibrium species. Most typically it is a response to stable environmental resources. This implies selection for low birth rates, high survival rates among offspring, and prolonged development. *K* represents the carrying capacity of the environment for species populations showing an *S-shaped population-growth curve. *See also* BET-HEDGING; *compare* *R*-SELECTION.

kunkar *See* CALCIFICATION.

Kuril Trench The oceanic *trench that marks part of the destructive plate margin between the Pacific and North American Plates. The Kuril Trench is backed by the Kuril Island Arc, running from Kamchatka to northernmost Japan.

Kuroshio Current (Kuro Shiwo Current) The oceanic surface current that flows northwards from the Philippines, along the Japanese coast, and then out into the North Pacific. It is an example of a western *boundary current: fast-flowing (up to 3 m/s), narrow (less than 80 km), and relatively deep. It is second in strength only to the *Gulf Stream. The warm water transports heat polewards. The volume transport is variable but is normally about $4 \times 10^7 \, m^3/s$.

Kuro Shiwo Current *See* KUROSHIO CURRENT.

***k*-value** The 'killing power' of an environment, calculated as the number of individuals lost from a specified stage in the life cycle of a species and expressed logarithmically ($k = \log_{10} a_x - \log_{10} a_{x+1}$, where a_x is the number of individuals at the start of the stage, time x, and a_{x+1} is the number at the end of the stage, time $x + 1$).

k

Labrador Current An oceanic current that brings cold Arctic waters southwards into the North Atlantic along the western margin of Greenland. Frequently it carries icebergs southwards, concentrating them in the area to the east of the Grand Banks in late spring to early summer. Fog banks often occur off the coast of Newfoundland where the Labrador Current meets the *Gulf Stream.

lacunosus The Latin *lacunosus*, meaning 'with holes', used to describe a variety of cloud usually associated with the genera *altocumulus and *cirrocumulus. Thin layers or sheets of cloud display a fairly regular set of holes with frilled edges, so forming a net. *See also* CLOUD CLASSIFICATION.

lacustrine Pertaining to lakes.

labyrinth fish Fish that have accessory respiratory organs in the gill chambers, enabling them to utilize atmospheric oxygen when necessary.

laciniate Deeply cut into irregular, narrow segments or lobes.

lacunose Having a surface pitted with cavities or indentations.

LAD *See* LAST-APPEARANCE DATUM.

LAI *See* LEAF-AREA INDEX.

lagg The *rheotrophic *fens that surround a *raised bog.

lagoon A coastal body of shallow water, characterized by a restricted connection with the sea. The water body is retained behind a reef or islands.

Lagrangian current measurement A technique for measuring water movements by tracing the path of a water particle over a long time interval. A device is released into the water and allowed to drift passively with it. Measurements can be made, e.g. by following and plotting the progressive position of a neutrally buoyant float for subsurface currents, or a drift pole or buoy for surface-water movements.

lahar (mudflow) A catastrophic mudflow on the flank of a volcano which is a notable feature of the volcanic areas of Indonesia, especially Java. Lahars are the cause of most volcanic fatalities. They may reach more than 100 km from the source volcano, when confined along pre-existing valleys.

lake forest A forest dominated by conifers, which occurs in the eastern half of North America, in Minnesota, Michigan, northern Pennsylvania, southern Ontario, and northern New England. Although a distinct formation it is in many respects transitional between *boreal coniferous forest and southern *deciduous forest. Very little lake forest survives, having been cut in the latter part of the nineteenth century.

Lamarck, Jean Baptiste Pierre Antoine de Monet, chevalier de (1744–1829) A French naturalist who, in 1809, advanced the theory that evolutionary change may occur by the inheritance of characteristics acquired during the lifetime of the individual. This theory was also the basis for *Lysenko's arguments on the inheritance of acquired plant characteristics. It is interesting to note that the theory of the inheritance of acquired characteristics did not hold a central position in Lamarck's own writings. His cardinal point was that evolution is a directional, creative process in which life climbs a ladder from simple to complex organisms. He believed the inheritance of acquired characteristics provided a mechanism for this evolution. Lamarck explained that this progress of life up the ladder of complexity is complicated by organisms being diverted by the requirements of local environments; thus cacti have reduced leaves (and giraffes have long necks). *Compare* DARWIN, CHARLES ROBERT.

Lamarckism The theory of *evolution propounded by Lamarck. *See* ACQUIRED CHARACTERISTICS.

Lamb's dust-veil index An index of the amount of finely divided material suspended in the atmosphere after great volcanic eruptions, and of the duration of an effective veil intercepting the Sun's radiation. It can be calculated from estimates of the amount of solid matter thrown up, from the reduction of intensity of the solar beam, or from the reduction of temperatures prevailing at the surface of the Earth. The latitude of the volcano also affects the index values, because the maximum extent of the veil over the Earth varies, being greatest after eruptions in low latitudes. (The great eruption of Krakatoa in Indonesia in 1883, which ejected about 17 km^3 of particulate matter into the atmosphere, where it remained for three years, gave an index value of 1000.) The index was devised by Professor H. Lamb.

laminar flow A type of flow (normally in water) that is characterized by the movement of fluid particles parallel to each other, with no transverse movement or mixing. Velocity increases steadily away from the bed. Laminar flow in channels is found only at low velocities and adjacent to smooth surfaces. It is almost ubiquitous in soil moisture and *groundwater (except in *karstic aquifers).

land and sea breezes Circulations of air, common along coasts, which are caused by a low-level pressure gradient owing to the differential heating of land and sea. On summer days solar radiation warms the land surface more strongly than the adjacent sea: a pressure gradient from sea to land results in a gentle, cooling, landward 'sea breeze' whose maximum strength is usually developed by late afternoon. Upward movement of warm air over the land and movement towards the sea at greater height, followed by subsidence, produces a shallow convection cell. At night and in early morning cooler land and relatively warmer sea produce a reverse-flow convection cell, with a seaward 'land breeze'. The horizontal extent of well-developed land and sea breezes is typically limited to about 40 km from the coast, but associated air movements can often be detected over a much wider coastal belt.

land bridge A connection between two land masses, especially continents (e.g. the Bering land bridge linking Alaska and Siberia across the Bering Strait), which allows migration of plants and animals from one land mass to the other. Before the widespread acceptance of continental

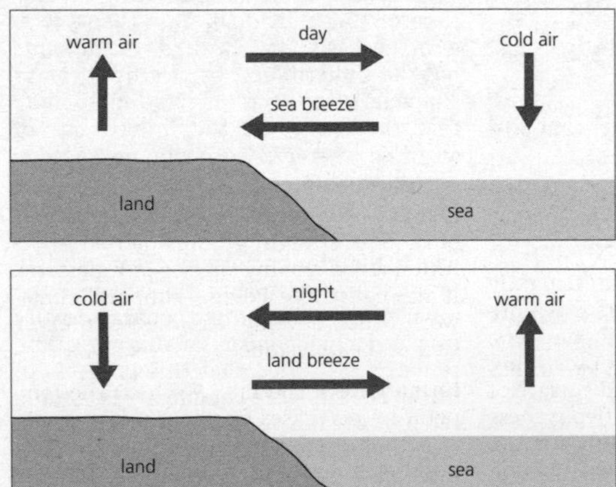

Land and sea breezes

drift, the existence of former land bridges was often invoked to explain faunal and floral similarities between continents now widely separated. On a smaller scale, the term may be applied to land connections that have now been removed by recent tectonics or the *Flandrian rise in sea level (e.g. between northern France and south-eastern England).

landfill **(sanitary landfill)** A method for the hygienic disposal of bulky wastes, in which the waste is deposited in a hollow that is excavated and sealed to prevent the contamination of adjacent land or water by *leaching. At intervals, usually of one day, the waste is levelled, then buried beneath soil. *See also* MADE GROUND.

land-locked fish Marine fish which no longer migrate back to the sea, but have become established permanently in freshwater lakes and drainage basins. There are, for example, land-locked populations of *Salmo salar* (salmon) in parts of North America and Sweden.

landnam A Danish word meaning a primary forest clearance, producing a characteristic horizon in the *pollen chronologies of Britain and western Europe, and also found elsewhere though at different stratigraphic dates. The absolute quantity of tree pollen in relation to that of other plants is greatly reduced. Sometimes a charcoal layer is found at a similar stratigraphic level. The non-tree pollen shows an increase in arable weed species, e.g. plantain and nettles, and cereal pollens are first recorded in some abundance at this level. Together with archaeological evidence, the horizon is strongly suggestive of a first phase of deliberate forest clearance by Neolithic agriculturists. The landnam phase is dated to about 5 000 BP throughout north-western Europe.

Landsat A series of US satellites, mainly carrying multispectral scanners and more recently also carrying thematic mappers, which are primarily designed for scanning the vegetation cover of the land surface of the Earth and evaluating the effectiveness of satellite-based scanning systems for routine monitoring. The recordings are available in digital form on magnetic tape for computer analysis or in photographic form (usually known as Landsat images). Landsat is complemented by Seasat. *See* SPOT.

landscape architecture A branch of architectural studies which is particularly concerned with design in relation to the scenic environment, e.g. the use of appropriate tree species to blend with buildings or landscape morphology, the harmonious design of way-marking, shelters, etc. The subject also embraces the engineered modification of landscapes to provide appropriate settings or screening for buildings. The term is often used to include the design of gardens (i.e. synonymously with landscape gardening).

landscape ecology The study of landscapes, taking account of the ecology of their biological populations. The subject thus embraces geomorphology and ecology and is applied to the design and architecture of landscapes (*see* LANDSCAPE ARCHITECTURE).

landscape evaluation The assessment of landscape as a scenic resource. It is controversial, since such concepts as 'scenic beauty' are neither generally agreed nor readily quantifiable. Nevertheless evaluations are vital for the conservation of high-quality and/or traditional landscapes, especially in densely populated, developed regions.

landslide *See* MASS-WASTING.

lanugo The soft, downy hair that covers the body of some juvenile *mammals and is replaced by hair of the adult type as the animal develops. The retention of lanugo in humans is regarded as evidence of *neoteny.

lapiés A variety of *karren that consists of shallow, straight grooves incised by solution into a sloping surface of limestone. It may constitute a dense, subparallel network which develops rapidly upon exposure of the limestone.

lapse rate The rate at which the temperature decreases for each unit increase of height in the atmosphere. In the *troposphere the average rate is approximately 6.5°C per 1 000 m. *See also* ADIABATIC and ENVIRONMENTAL LAPSE RATE.

larviparous Applied to animals (e.g. flesh flies, *Sarcophaga carnaria*) that reproduce by depositing larvae, rather than eggs.

last appearance datum (LAD) The last recorded occurrnce of a key taxon in biological history.

late blight of potato (potato blight) A widespread and serious disease affecting the potato and related plants. Symptoms include the appearance of brown patches on the leaves, often with white mould on the undersides. Under damp conditions the entire foliage may collapse. Brown lesions also develop on tubers, spreading to the entire tuber in a dry *brown rot. The disease is favoured by wet weather. It is caused by the oomycete *Phytophthora infestans* and was responsible for the Irish potato famine in the 1840s.

Late Devensian Interstadial *See* WINDERMERE INTERSTADIAL.

late glacial A term usually applied to the time between the first rise of the temperature curve after the last minimum of the *Devensian glaciation, and the very rapid rise of temperature that marks the beginning of the post-glacial, or *Flandrian period. The late Devensian extends from about 15 000–10 000 BP and shows a characteristic threefold climatic and hence depositional sequence, from cold, older *Dryas deposits, to warmer *Allerød, to colder, younger Dryas. In Europe there is some evidence for an additional warmer phase, the *Bølling interstadial, dating about 13 000 BP (i.e. during older Dryas times).

latent heat of transition The heat required to activate a phase change from a solid to a liquid or from a liquid to a gas, i.e. to a higher energy state (e.g. latent heat of melting), measured in joules per mole (J/mol). Latent heat is released by reactions in the reverse direction (e.g. latent heat of crystalization).

latent learning The formation of associations that appear to bring the animal neither reward nor punishment (although novel experiences may themselves constitute a reward). There is a delay between exposure to the learning situation and the performance of a behaviour pattern that demonstrates the effect of learning. For example, rats will explore a maze without apparent reward or punishment and will remember the information they obtain; later, when rewards are presented, those rats will perform better at finding their way through the maze than rats with no exploratory experience.

latent root *See* EIGEN VALUE.

lateral moraine *See* MORAINE.

laterite A weathering product of rock, composed mainly of hydrated iron and aluminium oxides, hydroxides, and *clay minerals, but also containing some silica. It is related to bauxites and is formed in humid, tropical settings by the weathering of such rocks as basalts. After exposure at the surface it can become a hard, impermeable layer and it has been used as a building material.

laterization The *weathering process by which *laterite is formed.

Latin American mammal region A concept that is considered by the biogeographer Charles H. Smith to be more appropriate than the *Neotropical faunal region in respect of mammals. It extends northwards from Tierra del Fuego to just across the southern borders of the United States.

latitudinal diversity gradient Most groups of plants and animals have more species in the low latitudes than in the high latitudes. There are many possible explanations for this and many different opinions on the relative merits of those explanations. Solar energy is more abundant in the tropics and overall *primary productivity is higher. Greater *biomass in the tropical *ecosystems (especially rain forest) leads to higher structural complexity and consequently greater opportunities for *niche diversification. The climatic changes affecting the higher latitudes during the *Pleistocene glaciations may not have caused such great disturbance in the tropics, though this is debatable. Evolutionary rates may be faster in the tropics, hence this may be the region in which many new taxa are generated, subsequently spreading to the higher latitudes.

latitudinal vegetation zone A major vegetation belt that coincides with a broad, global, climatic regime. Thus rain forest relates to the equatorial climate, and *savannah to the seasonal tropics roughly 10–20° on either side of the equator. The simple pattern is complicated by the effects of altitude, so that in each latitudinal vegetation zone there are found vegetation types more characteristic of latitudes nearer the poles.

Laurasia The northern continental mass produced in the early *Mesozoic by the initial rifting of *Pangaea along the line of the northern Atlantic Ocean and the *Tethys sea. Laurasia included what was to become North America, Greenland, Europe, Asia, and Malesia east to Sulawesi, while the large, southern continental mass (called *Gondwana) was later to divide into South America, Africa, India, Australasia, Malesia east of Sulawesi, and Antarctica. Fossil evidence indicates that the Laurasian floral assemblage included many species of tropical plants that were incorporated into sediments to form the extensive *coal measures that are mined throughout Europe and the eastern USA.

Laurentia **(Laurentian Shield)** The *Precambrian shield of central eastern Canada which forms the ancient 'core' of Canada, around which younger mountain belts have been accreted. The name, derived from the St Lawrence River, has been applied to a series of granites, gneisses, and metasediments that are older than 2 500 Ma.

Laurentian Shield *See* LAURENTIA.

Laurentide ice sheet An area of continental ice that lay over the eastern part of Canada during the *Pleistocene glaciations. The centre of the ice mass may have originated in or near northern Quebec, Labrador, and Newfoundland, and spread out to the south and west. At its maximum spread it may have covered an area of 13×10^5 km^2.

lava Molten rock, normally a silicate, which is expelled by a volcano. Its behaviour on extrusion and its relief-forming capacity depend largely on its viscosity, which is affected by its silica content, temperature, and the amount of dissolved gases and solids it contains. Generally, the less viscous the lava the faster the flow, and the more viscous the lava the greater the tendency towards explosive eruption.

law of faunal succession *See* FAUNAL SUCCESSION.

law of superposition of strata *See* FAUNAL SUCCESSION.

law of the minimum *See* LIEBIG'S LAW OF THE MINIMUM.

layer cloud One of the principal forms of cloud, with flattened, sheet-like appearance and of limited vertical extent. Common types of cloud exhibiting this form are: (*a*) low-level layer clouds (e.g. *fog and *stratus); and (*b*) multi-layered clouds (e.g. *altostratus, *cirrostratus, and *nimbostratus). *See also* CLOUD CLASSIFICATION.

leachate The solution formed when water percolates through a permeable medium. When derived from solid waste, in some cases the leachate may be toxic or carry bacteria. In mining, leaching of waste tips can produce a mineral rich leachate which is collected for further processing.

leaching The removal of soil materials in solution. Water may percolate downwards through a soil, removing humus and mineral bases in solution before depositing them in underlying layers by *illuviation. The upper layer of leached soil becomes increasingly acidic and deficient in plant nutrients.

leading dominant According to the *Wisconsin School ordination scheme, the species in any given stand or *quadrat which has the highest importance value.

leaf-area index **(LAI)** Of a plant, the total leaf surface area exposed to incoming light energy, expressed in relation to the ground surface area beneath the plant (e.g. an LAI of 4 means that the leaf area exposed to light is 4 times the ground surface area).

leaf-area ratio The photosynthetic surface area per unit dry weight of a plant. It is a measure of the efficiency with which a plant deploys its photosynthetic resources. Typically, it is increased by low light intensities.

leaf-cutter bee Any species of the cosmopolitan family Megachilidae, the female of which cuts leaf pieces for use as material for lining or closing the nest. Sometimes the leaf fragments are augmented with mud or resin.

leaf physiognomy The form of leaves, variations in which are largely influenced by rainfall and temperature and can be used in palaeoclimatological studies. In simple studies the percentage of leaves with entire (i.e. not serrated) margins indicates temperature (e.g. if 68–70 per cent of leaf margins are entire the mean annual temperature (MAT) is 20 °C, a deviation of 4 per cent in the proportion of entire leaf margins indicating a temperature deviation of 1 °C) and the length of leaves is related to the availability of water. For more reliable interpretation, these measurements are used in conjunction with other factors (e.g. the percentage of leaves with apices that are attenuated (drip tips) rather than rounded; the size of mesophylls, notophylls, and nanophylls; and the ratio of leaf length to width). Studies of leaf physiognomy have indicated that both temperature and rainfall increased substantially at the commencement of the *Tertiary.

learning The acquisition of information or patterns of behaviour other than by genetic inheritance, or the modification of genetically acquired information or behaviour as a result of experience. *See also* CONDITIONING; CULTURAL; EXPLORATORY LEARNING; LATENT LEARNING; OPERANT CONDITIONING; and TRIAL-AND-ERROR LEARNING.

least-work principle The theory that geomorphological processes always operate in such a way as to achieve the work that has to be done with a minimum expenditure of energy (and maximum entropy). This is typically achieved by the adoption of a certain profile or shape (e.g. a river *meander may be that shape best suited for carrying the discharge and sediment with the least loss of energy). *See* LEAST-WORK PROFILE.

least-work profile That profile whose gradient is just sufficient for the associated geomorphological process to occur with the minimum possible expenditure of energy. An example is a long river profile, whose concave-up form is the shape best suited for the transfer of increasing quantities of water and sediment in accordance with the *least-work principle. Such a profile expresses a state of high entropy.

lectotype One of a collection of *syntypes which, subsequent to publication of the original description, is chosen and designated through published papers to serve as the *type specimen.

lee depression A *non-frontal depression that develops on the lee side of an upland barrier across the airflow as a result of contraction leading to cyclonic curvature. Dynamic processes are responsible for the low-pressure system rather than wave development along a *front. Such depressions are common, for example, in winter on the southern lee side of the Alps.

lee waves Air waves in the lee of a mountain barrier, where a stable layer of air, after displacement by movement over the barrier, returns to its original level. This process results in a series of stationary ('standing') waves extending downwind from the lee side of the barrier. Clouds often form along the wave crests in lenticular form: they may appear stationary, owing to condensation of water vapour at the upward side, caused by the upward air movement, and evaporation on the downward side of the wave. The wavelength can be up to 40 km and the wave amplitude is most pronounced in the intermediate levels of the airstream. Circular air motion (in the vertical plane) beneath the wave crests may reverse wind direction locally within the general air flow. This phenomenon is termed a 'rotor'. In addition to stable air at an intermediate level, lee-wave formation requires a constant wind of at least 15 knots. Well-known wave clouds on the lee side of barriers include the Sierra wave of the Sierra Nevada, California, the helm wave of Cumbria, England, and the moazagotl of Silesia.

legionary In ants, applied to the type

of foraging in which groups of workers seek food together.

lek An area within which a number of males occupy very small territories where they display in order to attract the attention of females for the purpose of mating. The area occupied by each male is thus a mating station and it is defended against rivals. The lek contains no food or other resource of use to females. The males remain at the lek throughout the breeding season. Females, whose breeding activity is not synchronized, visit the lek when they are ready to mate, select a male for the quality of his display, and leave immediately after mating. Since there are always more males than females present at the lek, competition among the males is intense and as well as display performance this translates into competition for the best stations. In many species females find the territories near the centre of the lek especially attractive and these are occupied by the most dominant and successful males. Lekking has been observed among insects, fishes, and mammals, but has been studied in most detail in birds.

length abridgement The shortening of *pseudogenes, due to the accumulation of *insertions and *deletions over time.

lentic Applied to a freshwater habitat characterized by calm or standing water (e.g. lakes, ponds, swamps, and bogs).

lenticular Shaped like a biconvex lens.

lenticularis The Latin *lenticularis*, meaning 'biconvex' or 'lens-shaped', used to describe a form of cloud consisting of clearly defined, elongated lenses. The form is typical of *lee-wave clouds and may affect such clouds as *stratocumulus, *altocumulus, and *cirrocumulus. *See also* CLOUD CLASSIFICATION.

Leopold, Aldo (1887–1948) An American ecologist and author of *Sand County Almanac* (1949), in which he expounded his view that all living organisms are related and the biosphere they inhabit is a partnership to which humans belong. His book became very influential in the 1960s, in the early days of the environmental movement. Leopold was born in Burlington, Iowa, and educated in New Jersey and then Yale University, where he obtained a master's degree in forestry. He worked for the US Forest Service, then for the Sporting Arms and Ammunition Manufacturing Association, promoting game management. He was a keen hunter throughout his life. He became president of the Ecological Society of America and the Wildlife Society and was one of the founders, in 1935, of the Wilderness Society. From 1943 until his death he was a member of the Wisconsin Conservation Commission.

lepidote Covered in small scales (e.g. the leaves of Aetoxicaceae).

leprose Consisting of or bearing powdery or scurfy granules.

leptodermous Thin-skinned or thin-walled.

leptokurtic Applied to a distribution that is more peaked than a *Gaussian distribution (i.e. a few points occur far from the origin, but most are very close to it). This is typical of wind-dispersed *propagules, such as seeds or pollen grains.

leptophyll In the classification of leaf sizes, a leaf up to 25 mm^2 in area.

Leptosols Soils in which there is hard rock within 25 cm of the surface, or that overlie material containing more than 40 per cent calcium carbonate within 25 cm of the surface, or that have less than 10 per cent fine-grained material to a depth of 75 cm. These are weakly developed soils. Leptosols are a reference soil group in the *FAO *soil classification.

lessivage The *eluviation of insoluble particles to a deeper level within the soil. This process often produces *cutan.

leste A regional wind, affecting North Africa and Madeira, which blows ahead of a low-pressure area and brings hot, dry conditions.

lethal mutation A gene mutation whose expression results in the premature death of the organism carrying it. Dominant lethals kill both *homozygotes and *heterozygotes, recessive lethals kill homozygotes only.

leucon Applied to the body structure of a sponge if it is complex and consists of many chambers.

levanter A local wind from the east which occurs in the Straits of Gibraltar and is associated with standing waves in the lee of the Rock of Gibraltar. The wind is especially prevalent in late summer and autumn and brings high humidity.

leveche A local wind that affects south-eastern Spain, especially in summer. It is similar to such other hot, dry, dusty winds of tropical continental origin as the *sirocco and *khamsin, which blow in the Mediterranean region.

levee A raised embankment along the edge of a river channel, showing a gentle slope away from the channel. It results from periodic overbank flooding, when coarser sediment is immediately deposited, owing to a reduction in velocity. This may lead to a situation in which the river flows well above the level of its outer *flood-plain.

l

liana *See* LIANE.

liane **(liana)** Any wiry or woody, free-hanging, climbing plant.

libeccio A local south-westerly wind which brings stormy conditions, especially in winter, to the central Mediterranean.

library **(gene library)** A random collection of cloned (*see* CLONE) DNA fragments in a number of vectors, which ideally includes all the genetic information of that species.

lichen A type of composite organism, which consists of a fungus (the mycobiont) and an alga or cyanobacterium (the phycobiont) living in *symbiotic association. A lichen thallus may be crustlike (crustose), scaly or leafy (foliose), or shrubby (fruticose), according to the species. Lichens are classified on the basis of the fungal partner; most belong to the Ascomycotina. Specialized asexual reproductive structures may be produced. Many lichens are extremely sensitive to atmospheric pollution and have been used as pollution indicators.

lichenicolous Growing on *lichens.

lichen woodland The *facies of the boreal conifer zone, situated south of the forest *tundra *ecotone and north of the closed forest proper. The characteristic is that of open woodland, with a sparse stand of conifers set in a ground layer dominated by *lichens (e.g. *Cladonia*).

lichen zone An area with a characteristic *lichen flora ranging from complete absence to a full complement of foliose and fruticose species, depending on the level and type of air pollution present in the area. The standardized lichen zones for England and Wales range from zone 0 (complete absence) typical of heavily polluted industrial areas, to zone 10 in which bearded lichens (e.g *Usnea* and *Ramalina* species) are present in association with many other species. Zone 10 is typical of unpolluted areas in which virtually no sulphur dioxide, fluorine compounds, or other air pollutants are present.

Liebig, Baron Justus von **(1803–73)** A German chemist who contributed much to the systematization of organic chemistry, the early development of biochemistry, and agricultural chemistry. In 1840 he published *Die organische Chemie in ihrer Anwendung auf Agrikulturchemie und Physiologie* ('Organic Chemistry in its Application to Agricultural Chemistry and Physiology'), in which he showed that plants take up nutrients in simple chemical form and that nutrient deficiencies in soils may be remedied by the application of mineral fertilizers. He maintained that plant growth is limited by the availability of the scarcest essential nutrient (*see* LIEBIG'S LAW OF THE MINIMUM). Others expanded on this work later to produce a broader appreciation of the 'limits of tolerance', recognizing that maximum as well as minimum thresholds exist for all commodities (not only chemicals) that are essential to plant and animal growth. *See also* SHELFORD'S LAW OF TOLERANCE.

Liebig's law of the minimum The concept first stated by J. von *Liebig in 1840, that the rate of growth of a plant, the size to which it grows, and its overall health depend on the amount of the scarcest of its essential nutrients that is available to it. This concept is now broadened into a general model of *limiting factors for all organisms, including the limiting effects of excesses of chemical nutrients and other environmental factors. *See also* SHELFORD'S LAW OF TOLERANCE.

life A state of physical entities that utilize substances derived from outside themselves for the purposes of growth, the repair of their own structure, and the maintenance of their functional systems, and that also reproduce. The oldest known fossils are from rocks about 3 300 Ma old and the Earth is estimated to be about 4 600 Ma old. If it is accepted that life began on Earth (rather than being introduced from elsewhere), then it must have done so during the intervening 1 300 Ma. There is disagreement on the mechanism by which this happened. The classic view is that lightning discharges in a reducing atmosphere produced prebiotic substances which became dissolved in the oceans where, protected from ultra-violet light, they were somehow assembled into the first organisms. Recently, evidence has been advanced to suggest that life may have originated in small freshwater pools in an atmosphere of carbon dioxide with traces of ammonia.

life assemblage (biocoenosis) A *fossil community that is interpreted as representing a former living community. Most assemblages interpreted as life assemblages represent only a small fraction of a former community.

life cycle A series of developmental changes undergone by the individuals comprising a population, including *fertilization, reproduction, and the death of those individuals, and their replacement by a new generation. The life cycle in fact is linear in respect of individuals but cyclical in respect of populations. In many plants there is a succession of individuals in the entire cycle, with *sexual or *asexual reproduction linking them.

life-form The structure, form, habits, and life history of an organism. In plants especially characteristic life-forms, in particular morphological features, are associated with different environments. This observation has formed the basis of several attempts at life-form classifications of vegetation. *Raunkier's scheme (1934) is the best-known and most widely applied life-form scheme.

life span **1.** The length of time between the birth and death of an individual. **2.** The mean or maximum duration of the lives of members of a group of organisms. **3.** The duration of the existence of a taxon.

life table A table showing the mortality within a population arranged by age groups (cohorts). *See* STATIC LIFE TABLE.

life-zone An original concept of C. H. Merriam (1894) describing the way in which changing vegetation forms give a series of life-zones in relation to temperature gradients. In modern ecology these life-zones are defined by reference to a range of interacting environment gradients, and reflect animal as well as plant characteristics. On a world scale life-zones are thus synonymous with the major *biomes. However, life-zone is more frequently used for more local changes (e.g. the altitudinal zonation of *communities on mountains).

lifting condensation level The level at which air becomes saturated when it is forced to rise.

light-and-dark bottle technique *See* OXYGEN METHOD.

light-dependent stage *See* LIGHT REACTIONS.

light reactions (light-dependent stage) During *photosynthesis, those reactions which require the presence of light (e.g. photophosphorylation and photolysis). During the light reactions, light strikes chlorophyll molecules, which absorb its energy. This allows one electron to escape for each photon of light absorbed. The electron attaches to a neighbouring molecule, thereby ejecting another electron and causing electrons to move along an electron-transport chain of molecules. Some of the transported energy is used to attach phosphate groups to molecules of adenosine diphosphate (ADP) converting them to adenosine triphosphate (ATP); this reaction is called photophosphorylation. ATP is used to carry energy to

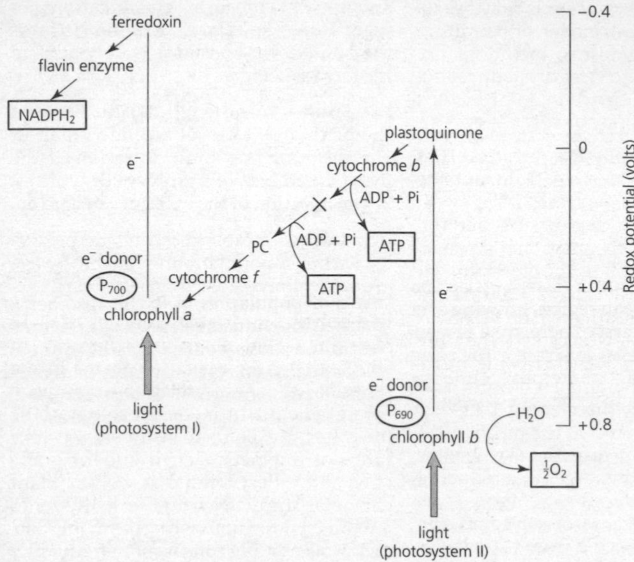

Light reactions

wherever it is needed in the organism. Energy not used in the ADP→ATP reaction is used to split a water molecule into H^+ and OH^- ions; the reaction is called photolysis. The H^+ attaches to nicotinamide adenine dinucleotide phosphate (NADP), converting it to NADPH. The OH^- passes one electron to the chlorophyll molecule, restoring the neutrality of both chlorophyll and hydroxyl. Hydroxyls then combine to form water ($4OH \rightarrow 2H_2O + O_2\uparrow$).

lignicolous Growing on *decorticate wood.

lignified Applied to cells that have a large amount of *lignin deposited in their *cell walls, giving them a rigid, woody structure.

lignin A carbohydrate polymer formed by the condensation of coniferyl alcohol that is found in wood, where it comprises about 25 per cent of the material, and in the cell walls of strengthening tissue (sclerenchyma) in plants. Lignin makes tissues more resistant to compression and tension.

lime **1.** The fruit of *Citrus aurantifolia*. **2. (linden)** The common name for trees of the genus *Tilia*. **3.** Compounds mostly of calcium carbonates, but also other basic (alkaline) substances, used to correct soil acidity and occasionally as fertilizers to supply magnesium.

limes convergens The Latin name (*limen* means 'threshold') given to a well-defined boundary zone between two fairly uniform major *habitat types. An example is a flood meadow zone, which is subject to periodic catastrophic flooding: this includes species characteristic of wet conditions, as well as those typical of trampled and grazed areas (which reflect the normal use when the area is not flooded).

limes divergens The Latin name (*limen* means 'threshold') given to a diffuse boundary zone in which one major *habitat type changes gradually into another, typically with both types showing internal variation. It is more stable than the *limes convergens, since the change in environment is spatially and temporarily gradual and persistent, rather than catastrophic and/or intermittent.

limestone A sedimentary type of rock, composed mainly of *calcite and/or dolostone, with which the calcite is often interbedded. Dolostone comprises the mineral dolomite ($CaMg(CO_3)_2$), produced by reactions between calcite and magnesium-bearing solutions.

limestone forest A distinctive forest formation, found within *tropical rainforest regions of south-eastern Asia, and also in the Caribbean region, growing over limestone hills (karst). There are a few endemic (*see* ENDEMISM) genera (e.g. the palm *Maxburretia*) and numerous endemic species restricted to limestone forests in Malesia.

limestone pavement An extensive bedding-plane surface, exposed by *erosion, which consists of hard, almost horizontally bedded *limestone dissected into blocks (called clints) by smooth gutters or runnels (called grikes). The grikes support a flora that often includes rare species.

limiting factor (ecological factor) Defined originally as whichever essential material is available in an amount most closely approaching the critical minimum needed, but now used more generally to describe any environmental condition or set of conditions that approaches most nearly the limits (maximum or minimum) of tolerance for a given organism.

limits of tolerance The upper and lower limits to the range of particular environmental factors (e.g. light, temperature, availability of water) within which an organism can survive. Organisms with a wide range of tolerance are usually distributed widely, while those with a narrow range have a more restricted distribution. *See also* SHELFORD'S LAW OF TOLERANCE.

limnetic zone (sublittoral zone) The area in more extensive and deeper freshwater ecosystems which lies above the *compensation level and beyond the *littoral (lake-edge) zone. This zone is mainly inhabited by plankton and *nekton with occasional *neuston species. The limnetic and littoral zones together comprise the *euphotic or well-illuminated zone. In very small and shallow lakes or ponds the limnetic zone may be absent.

limnology The study of freshwater ecosystems, especially lakes.

limnophilous Applied to organisms that thrive in ponds or lakes.

Lincoln index A simple estimate of animal population density, based on data obtained by *mark-recapture techniques. It is written as: $P = an/r$, where P is the estimate of population, a is the number of marked individuals released, n is the number in the subsequent sample of captured individuals, and r is the number of marked individuals recaptured. It was devised in 1930 by the American ornithologist Frederick C. Lincoln.

Lindeman's efficiency The ratio of energy assimilated at one *trophic level to that assimilated at the preceding trophic level; the ratio of energy intake at successive trophic levels. Described by R. L. Lindeman in a paper published in *Ecology* in 1942, it is one of the earliest and most widely applied measures of *ecological efficiency.

LINE *See* LONG INTERDISPERSED ELEMENT.

lineage *See* EVOLUTIONARY LINEAGE.

lineage zone (evolutionary zone, morphogenetic zone, phylogenetic zone) A unit of rock strata which contains a clearly defined portion of an *evolutionary lineage, marked above and below by some distinct and specified change in form.

linear sand ridge A submarine sand mound, typical of shallow sea and wide *continental-shelf areas, 3–10 m high, 1–2 km wide, which may extend for tens of kilometres across the shelf. Individual ridges have an average spacing of about 3 km. Such ridges have been described from the North Sea (off Norfolk, England) and the eastern seaboard of the USA. They are the product of storm and tidal action.

line transect A tape or string laid along the ground in a straight line as a guide to a sampling method used to measure the distribution of organisms. Sampling is rigorously confined to organisms that are actually touching the line. *Compare* BELT TRANSECT.

linkage The association of genes that results from their being on the same *chromosome. The nearer such genes are to each other on a chromosome, the more closely linked they are, and the less often they are likely to be separated in future generations by crossing-over. All the genes in one chromosome form one linkage group.

linkage disequilibrium A deviation from the *Hardy–Weinberg equilibrium due to a higher probability than would be predicted from Mendel's second law (the law of random segregation, *see* MENDEL'S LAWS) that two or more genes being studied will be inherited together, due to their close *linkage.

linkage group *See* LINKAGE.

Linnaeus, Carolus (Carl von Linné) (1707–78) A Swedish naturalist remembered for his large contributions to plant classification and his introduction of the binomial system of nomenclature. Trained as a botanist and physician in Uppsala, he went to Holland to continue his studies. While there he published the *Systema Naturae* (1735), a classified list of plants, animals, and minerals. The list grew in later editions and the tenth edition (1758) is the starting-point of zoological nomenclature. In the first edition of his *Genera Plantarum* (1737) he gave more details of his *artificial classification of plants, based largely on the number of stamens and pistils in a flower and the manner in which they occurred. The system was very popular, because it allowed students to catalogue and recognize vegetation quickly; its popularity caused some difficulties when it had to be replaced by *natural classifications. In 1741 he was appointed to the chair of medicine at Uppsala, but exchanged it within a year for the chair of botany. He then began listing species, grouping them into genera, genera into classes, and classes into orders. In 1749 he introduced binomial classification, a system which used a Latin generic noun followed by a specific adjective. Until this time, polynomial plant names comprised concise Latin descriptions, which restricted the growth of classification. Ironically, it was largely this new system that allowed the development of ideas about the evolution of species, a concept to which Linnaeus was opposed. Linnaeus published the specific names in his most important botanical work, *Species Plantarum* (1753), still the official starting-point of current botanical nomenclature. Altogether he completed around 180 works and was considered to be an excellent tutor. After his death, his collections and library were sold in 1784 to Sir James Edward Smith, the first president of the Linnean Society of London (founded in 1788) and in 1828 the Society purchased the collection, which it still holds. *See also* INTERNATIONAL CODE OF BOTANICAL NOMENCLATURE and INTERNATIONAL CODE OF ZOOLOGICAL NOMENCLATURE.

liquid limit *See* ATTERBERG LIMITS.

Liriodendron *See* BICENTRIC DISTRIBUTION.

litho- A prefix meaning 'pertaining to rock or stone', from the Greek *lithos*, meaning 'stone'.

lithologic trap *See* STRATIGRAPHIC TRAP.

lithosere The sequence of plant *communities, proceeding through all the stages of a *succession to a *climax vegetation, which begins on a bare rock surface.

lithosphere The upper (oceanic and continental) layer of the solid Earth, comprising all crustal rocks and the brittle part of the uppermost *mantle. Its thickness is variable, from 1–2 km at mid-oceanic ridge crests, but generally increasing from 60 km near the ridge to 120–140 km beneath older oceanic crust. The thickness beneath continental crust is uncertain, but is probably some 300 km in places.

lithotroph An organism that obtains energy from the oxidation of inorganic compounds or elements (*compare* ORGANOTROPH). Sometimes the term is inaccurately used as a synonym of *autotroph.

litter **(L-layer)** An accumulation of dead plant remains on the soil surface.

Little Ice Age A period between about 1450 and 1860 during which the climate of the middle latitudes in both hemispheres became generally harsher and there was a worldwide expansion of *glaciers. The effects have been recorded in the Alps, Norway, and Iceland, where farm land and buildings were destroyed. Evidence of the Little Ice Age has also been found in New Zealand. There were times of especial severity during this period (e.g. the early 1600s, when glaciers were particularly active in the Chamonix valley, in the French Alps).

littoral Pertaining to the shore, from the Latin *litus*, meaning 'shore' and *litoralis*, 'of the shore'.

littoral drift *See* BEACH DRIFT and LONGSHORE DRIFT.

littoral fish **1. (freshwater)** Those fish that are found along the shores of a lake from the edge of the water down to the limits of rooted vegetation. **2. (marine)** Those fish that are found in the intertidal zone of the seashore, most of the fish moving in and out with the tide.

littoral fringe The uppermost reaches of the *littoral zone of marine ecosystems. On rocky shores it is the main area for *Littorina* species (periwinkles), and the black lichen, *Verrucaria*, and is distinguished from the lower, *eulittoral zone by the absence of barnacles (e.g. *Balanus* species and *Chthamalus* species). On sandy shores crabs, beach amphipods, and some land insects are characteristic.

littoral zone **1.** The area in shallow, fresh water and around lake shores, where light penetration extends to the bottom sediments, giving a zone colonized by rooted plants. **2.** In marine ecosystems the shore area or intertidal zone, where periodic exposure and submersion by tides is normal. Since the precise physical limits of tidal range vary constantly, a biological definition of this zone, which essentially reflects typical physical conditions rather than rarely experienced events, is generally more useful. Thus in Britain the littoral zone is defined as the region between the upper limit of species of the seaweed *Laminaria* and the upper limit of *Littorina* species (periwinkles) or of the *lichen *Verrucaria*.

live oak An American term covering all evergreen species of oak, especially *Quercus virginiana* (despite its name, the state tree of Georgia), and *Q. geminata* (sand live oak).

living fossil A member of a living animal or plant species that is almost identical to species known from the fossil record (not the recent fossil record), i.e. they have changed very little over a long period. For example, *Latimeria chalumnae* (coelacanth) is close to Late *Palaeozoic genera; *Xiphosura* species (horseshoe crabs) are close to Middle Palaeozoic genera; *Tapirus pinchaque* (Andean tapir) is close to a *Miocene species; and *Sphenodon* (tuatara) has changed little since the early *Mesozoic. The dawn redwood (*Metasequoia glyptostroboides*), a deciduous coniferous tree found growing in China in 1948, had been known previously only as fossils in America, from which it differs little; it is now widely cultivated as an ornamental. Note, however, that none of these species is identical to any fossil taxon; evolution has been slow, but not absent.

Lixisols All soils that have an *argic horizon within 100–200 cm of the surface, apart from *Acrisols, *Albeluvisols, *Alisols, and *Luvisols. Lixisols are a reference soil group in the *FAO *soil classification.

llanos The *savannah grasslands of the Orinoco basin of Venezuela. Some authorities have suggested that the paucity of trees is owing to the impermeable subsoil of the region, which produces very wet conditions during the rainy season. It is likely, however, that fire and grazing are also responsible for preventing the development of woodland.

L-layer *See* LITTER.

load The total amount of material carried by a stream or river, or the mass of rock overlying a geologic structure.

loam A class of soil texture that is composed of *sand, *silt, and *clay, which produces a physical property intermediate between the extremes of the three components. It is an easily worked soil, much prized by farmers.

localized repeat sequences Repetitive DNA that occurs in a *tandem array, often with a short repeating unit (e.g. *satellite DNA).

lochan *See* KNOCK and LOCHAN.

Lochkovian *See* DEVONIAN.

Loch Lomond Stadial A relatively cold period, equivalent to the Younger *Dryas stadial (*see* STADE), that occurred towards the end of the last (*Devensian) glaciation in Scotland. The event took place about 11 000–10 000 radiocarbon years BP. It is characterized by the development of small ice caps and *cirque glaciers in the Highlands.

locus (pl. loci) The specific place on a *chromosome where a gene is located. In *diploids, loci pair during *meiosis and unless there have been translocations, inversions, etc., the *homologous chromosomes contain identical sets of loci in the same linear order. At each locus is one gene; if that gene can take several forms (*alleles), only one of these will be present at a given locus.

locust 1. One of several species of acridids (order Orthoptera, family Acrididae) which show density-related changes in their morphology and behaviour. At low population densities the insects develop as solitary, cryptically coloured grasshoppers (phase *solitaria*). At higher densities, such as may result from an abundance of food after rain, the insects develop into gregarious, brightly coloured individuals, which swarm and migrate, often causing great destruction to vegetation (phase *gregaria*). Major species include *Locusta migratoria* (migratory locust), *Schistocerca gregaria* (desert locust), and *Nomadacris septemfasciata* (red locust). **2. (carob)** The pod and seeds of trees belonging to the leguminous tree *Ceratonia siliqua*. **3.** A common name for leguminous trees and shrubs of the genus *Robinia*, also known as false acacia.

lodgement till *See* TILL.

lodging In plants, a state of permanent displacement of a stem-crop stem from its upright position. This can cause considerable reduction in yield. Normally it is caused by storm damage, but it may be produced by rots, insects, or excess nitrogen.

loess Unconsolidated, wind-deposited sediment composed largely of *silt-sized quartz particles (0.015–0.05 mm diameter) and showing little or no stratification. It occurs widely in the central USA, northern Europe, Russia, China, and Argentina. It can give rise to a rugged topography with steep slopes (up to 70°). The soils derived from loess are of a very high quality and support excellent crop yields.

logistic equation (logistic model) A mathematical description of growth rates for a simple population in a confined space with limited resources. The equation summarizes the interaction of *biotic potential with environmental resources, as seen in populations showing the *S-shaped growth curve, as: $dN/dt = rN(K - N)/K$ where N is the number of individuals in the population, t is time, r is the biotic potential of the organism concerned, and K is the saturation value or *carrying capacity for that organism in that environment. The expression $(K - N)/K$ is the term that ensures the slowing down of the growth rate as N approaches K. When $N = K$ the term is zero and population growth ceases. The resulting growth rate or logistic curve is a parabola, while the graph for organism numbers over time is sigmoidal. *Compare* J-SHAPED GROWTH CURVE.

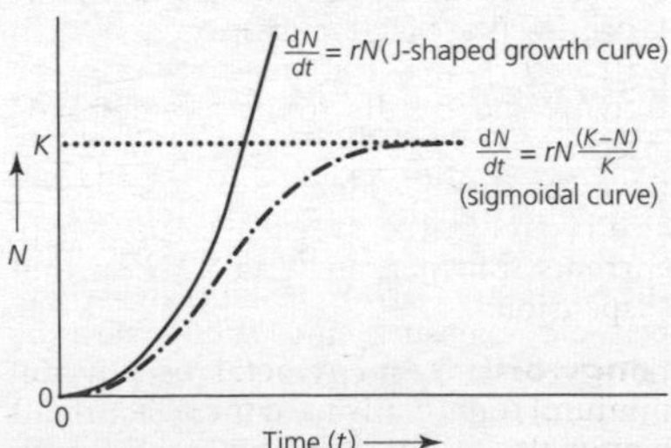

Logistic equation

logistic model *See* LOGISTIC EQUATION.

log-normal distribution (geometric distribution) A distribution in which the logarithms of the values have a *Gaussian (i.e. normal) distribution.

long-day plant A plant in which flowering is favoured by long days (i.e. days when there are more than 14 hours of daylight) and correspondingly short dark periods. There are two groups of such plants, species in which there is an absolute requirement for these conditions (such that flowering will not begin without them) and others in which flowering is merely hastened by them. Spinach, lettuce, and grasses are long-day plants; in Britain all of them flower in summer.

longevity The persistence of an individual for longer than most members of its species, or of a genus or species over a prolonged period of geological time.

long interdispersed element (LINE) A DNA sequence of more than 5 000 *base pairs with a *copy number greater than 10 000, that occurs throughout the *genome rather than in a *tandem array.

longshore bar A linear ridge of sand whose long axis is parallel to the shore and which is in, or immediately seaward of, the intertidal zone. *See also* RIDGE and RUNNEL.

longshore current A current that flows parallel to the shore within the zone of breaking waves: it is generated by the oblique approach of waves.

longshore drift (littoral drift) The movement of sand and shingle along the shore. It takes place in two zones. *Beach drift occurs at the upper limit of wave activity, and results from the combined effect of *swash and *backwash when waves approach at an angle. Movement also occurs in the *breaker zone, where currents transport material thrown into suspension.

Longworth trap A metal (usually aluminium) trap that is used to collect small mammals without injuring them. It consists of a nest-box, in which appropriate nesting material and food are placed, and a tunnel leading into it. As an animal enters it trips a door which falls shut, so it cannot escape and must remain in the nest-box.

Loopstedt *See* BRØRUP.

loose smut A disease of plants, caused by a fungus of the Ustilaginales (e.g. species of *Ustilago*) in which the masses of spores are exposed at maturity and can be dispersed freely by wind, etc.

lop and top The branches and top cut from a tree that has fallen or been felled or, more rarely, from a standing tree.

lordosis A curvature of the spine. In some female mammals, a sexual posture in which the back is arched.

Lorenz, Konrad Zacharias (1903–89) An Austrian zoologist who was joint winner (with Karl von *Frisch and Nikolaas *Tinbergen) of the 1973 Nobel Prize for Physiology or Medicine for their studies of animal behaviour. Lorenz worked for some years at a centre for behavioural physiology built for him by the Max Planck Institute at Seewiesen, Germany, and it was there he conducted the studies of *imprinting, especially with geese, for which he became widely known.

losing stream A stream that has a permeable bed through which water can seep (transmission losses) to the *water-table. *See also* INTERMITTENT STREAM.

loss on ignition A method for determining the organic matter content of a soil. A dried sample of soil is subjected to high temperature in a *muffle furnace in order to oxidize all of the organic matter. The loss in weight after cooling then gives the organic matter content. Since calcium carbonate also decays at high temperature to release carbon dioxide, any lime present in the soil must be removed first.

lotic Applied to a freshwater *habitat characterized by running water (e.g. springs, rivers, and streams).

Lotka–Volterra equations Mathematical models of competition, devised in the 1920s by A. J. Lotka and V. Volterra, between resource-limited species living in the same space with the same environmental requirements. They have been modified subsequently to simulate simple predator–prey interactions. The competition model predicts that coexistence of such species populations is impossible;

one is always eliminated, as was verified experimentally by G. F. Gause. The predation model predicts cyclic fluctuations of predator and prey populations. Reduction of predator numbers allows prey to recuperate, which in turn stimulates the population growth of the predator. Increasing predator numbers depress the prey population, leading eventually to a reduction in the predator population. This was also tested experimentally by Gause in 1934 and more recently and more convincingly by S. Utida in 1950 and 1957. *See also* COMPETITIVE EXCLUSION PRINCIPLE.

Lovelock, James Ephraim (b. 1919) A British chemist, biophysicist, inventor, and principal author of the *Gaia hypothesis, the development of which began in the 1960s, while he was working at the Jet Propulsion Laboratory in Pasadena, California, designing instruments in connection with the NASA lunar and Mars programmes. Earlier he invented the electron capture detector, an instrument capable of measuring extremely small amounts of substances (e.g. pesticide residues, methyl mercury, tetraethyl lead, *polychlorinated biphenyls (PCBs), *chlorofluorocarbons (CFCs), and nitrous oxide) in the natural environment. During the 1972–3 voyage of the research ship *Shackleton*, Lovelock used the electron capture detector to discover the global distribution of CFCs, dimethyl sulphide, methyl iodide, and carbon disulphide. In collaboration with R. J. Charlson, M. O. Andreae, and S. G. Warren, he hypothesized a link between marine algae and cloud formation that is now accepted by meteorologists and for which the four were awarded the Norbert Gerbier Prize and Medal of the World Meteorological Organization in 1988. He was elected a Fellow of the Royal Society in 1974 and received a CBE in 1989. For his contributions to environmental science Lovelock was awarded the Amsterdam Prize of the Netherlands Royal Academy in 1990. From 1987 to 1991 he was president of the Marine Biological Association.

low The common name for a low-pressure system (e.g. a *depression).

low Arctic tundra The southernmost latitudinal sector of the Arctic *tundra, distinguished from the more northerly belts by the presence of continuous vegetation cover in most areas. The plant *communities form a vegetation mosaic, which reflects the importance of microhabitats in this harsh, exposed environment.

low-level waste *See* RADIOACTIVE WASTE.

luminous night clouds *See* NOCTILUCENT CLOUDS.

lumpiness In *ecosystems, the hierarchical organization of controlling factors, such that vertebrate body sizes tend to form groups separated by gaps and changes to the ecosystem as a whole tend to occur in widely separated cycles.

lunate Shaped like a half-moon.

Lusitanian floral element A geographical element of the British flora, including plants such as *Erica mackaiana* (Mackay's heath) which are found in western Ireland and reappear again only far to the south, in the Iberian peninsula. These plants are generally thought to have migrated north to Ireland early in the post-glacial period, before sea level was fully restored.

Lutz phytograph In the *phytosociological assessment of woodland *communities (especially tropical woodlands), a polygonal figure representing four structural characteristics of each major tree species present, usually (*a*) the percentage of the total number of trees that are larger than 25 cm diameter at breast height (dbh); (*b*) the percentage frequency of a particular species in the total number of trees larger than 25 cm dbh; (*c*) the occurrence of the species in each of five size classes reflecting maturity; and (*d*) the dominance of the species as reflected by its percentage of the total tree basal area.

Luvisols Soils that have an *argic horizon with a *cation-exchange capacity greater than 24 $cmol_c$/kg, and with illuvial (*see* ILLUVIATION) accumulations of *clay. Luvisols are a reference soil group in the *FAO *soil classification.

Lydekker's line A line that defines the easternmost extension of oriental animals into the zone of mixing between the *Oriental and *Australian *faunal regions. The corresponding western limit of the zone is known as *Wallace's line, and marks the maximum extent of marsupials in that direction. It is named after Richard Lydekker (1849–1915). *See map accompanying Wallace's line.*

Lysenko, Trofim Denisovich (1898–1976) A fanatic anti-geneticist, who attained extraordinary power in the USSR under Stalin, and retained it under Krushchev. He sought successfully the suppression of research in genetics and was responsible for the imprisonment and execution of a number of noted Soviet geneticists. The failure of his own attempts, based on a kind of *Lamarckism, to increase grain production led to his own downfall and that of Krushchev.

lysimeter A device for the direct estimation of *evapotranspiration. Typically it comprises a vegetated block of soil 0.5–1 m^3, to which the amount of water added is known, and from which the amount lost as run-off or *percolation may be measured. Recording the changing weight of the soil vegetation system (keeping vegetation change due to growth static or monitored) reveals the amount of water retained by the system, and thus by difference the amount lost as evapotranspiration. For geographic comparisons, easily standardized, short, grass vegetation cover is used. For water-budget experiments, vegetation cover may be varied to simulate different crop types or *semi-natural communities.

lysis *See* LYTIC RESPONSE.

lysogeny A stable, non-destructive relationship between a bacteriophage and its host bacterium; the genome of the bacteriophage may become integrated with that of its host. Under certain conditions the stable relationship breaks down and the phage may then destroy the bacterium.

lytic response The rupture and death (lysis) of a bacterial cell following its infection by a *bacteriophage which then reproduces inside the cell, as opposed to a lysogenic response (*see* LYSOGENY) in which the infecting bacteriophage does not multiply but instead behaves as a *prophage.

m- *See* MILLI-.

M- *See* MEGA-.

Ma An abbreviation meaning 'million years'.

maar A crater, often occupied by a shallow lake, which is produced by an explosive volcanic eruption. Normally it is surrounded by a low rampart or ring of ejected material. Typically, maars are formed by the explosive interaction of volcanic magma with *groundwater. The ejected material is a mixture of country rock and highly fragmented ash.

Macaronesian floral region A region comprising the Canaries, Madeira, and the Azores, in which few genera are endemic, perhaps less than 30 (*see* ENDEMISM); surprisingly, the most remote island group, the Azores, has no endemics. The flora of the Canaries displays a notably high frequency of succulents belonging to the Crassulaceae, in particular the genus *Sempervivum*. *See also* FLORAL PROVINCE and FLORISTIC REGION.

macchia *See* MAQUIS.

machair An area of low, undulating tracts, supporting stable, herb-rich grassland growing on shell sand, which has developed over a long period by the accumulation of blown sand behind coastal sand-dunes, occurring most typically in the Hebrides and along the north-west coast of Scotland.

mackerel sky A pattern of wavy *cirrocumulus (or *altocumulus) cloud with holes which produces an overall resemblance to the body markings of mackerel. *See also* VERTEBRATUS.

Mackereth corer A form of hydraulic corer that is commonly used to obtain lake-sediment cores.

macro- A prefix meaning 'large' or 'long', from the Greek *makros*, meaning 'large' or 'long'.

macrobiota (macrofauna, macroflora) A general term for the larger soil organisms which may be hand-sorted from a soil sample. *'Macrofauna' in particular refers to burrowing vertebrate animals (e.g. rabbits and moles), while 'macrobiota' generally also includes larger plant material (e.g. tree roots). Some workers also include larger insects and earthworms in this category, but others consider them part of the mesobiota. *Compare* MESOBIOTA and MICROBIOTA.

macroclimate The climate character of a large region.

macroconsumer *See* CONSUMER ORGANISM.

macroecology A term coined by James H. Brown of the University of New Mexico that is used in relation to biogeographic studies (*see* BIOGEOGRAPHY) of population and species interactions on a large rather than a local scale. It brings to bear both geographic and historical considerations in understanding the local abundance, distribution, and diversity of species.

macro-evolution Evolution above the species level (i.e. the development of new species, genera, families, orders, etc.). There is no agreement as to whether macro-evolution results from the accumulation of small changes due to *micro-evolution, or whether macro-evolution is uncoupled from micro-evolution.

macrofauna The larger soil animals; the term is sometimes used to include larger insects and earthworms in this category, but otherwise these form part of the mesofauna. *Compare* MACROBIOTA; MEIOFAUNA; MESOFAUNA; and MICROFAUNA.

macroflora *See* MACROBIOTA.

macrofossil *See* MEGAFOSSIL.

macromutation A *mutation that has very large phenotypic (*see* PHENOTYPE) effects (e.g. a mutation affecting early *ontogeny). Macromutations have been proposed as the leading mechanism of evolution, as in the *'hopeful monster' hypothesis.

macronutrient An organic or inorganic element or compound, that is needed in relatively large amounts by living organisms. Organic macronutrient groups include *amino acids, *carbohydrates, and fats. *Compare* MICRONUTRIENT; *see also* ESSENTIAL ELEMENT.

macrophyll In the classification of leaf sizes, a leaf that is more than 250 mm long, or 18 225–164 025 mm^2 in area.

macrotidal Applied to coastal areas where the tidal range is in excess of 4 m. Tidal currents dominate the processes active in macrotidal areas (e.g. the coast of the British Isles).

Madagascan faunal subregion The strongly endemic (*see* ENDEMISM) and insular fauna of Madagascar, which includes five orders found nowhere else. For example, four of the five families of lemurs survive only in Madagascar; and this is the only part of the African continent from which the more advanced apes and monkeys are entirely absent.

Madagascan floral region Part of R. Good's (1974, *The Geography of the Flowering Plants*) *Palaeotropical kingdom, which geographically includes Madagascar, the Seychelles, and the Mascarenes. A high proportion of the genera, perhaps exceeding 200, is endemic (*see* ENDEMISM); and depending on the taxonomic system adopted, there are two or three endemic families. *See also* FLORAL PROVINCE and FLORISTIC REGION.

made ground **(made land)** An area of dry land that has been made by people, generally through the reclamation of marshes, lakes, or shorelines. An artificial fill (*landfill) is used, consisting of natural materials, refuse, etc.

made land *See* MADE GROUND.

maerl The Breton name for a mixture of carbonate-rich (skeletal) sand and seaweed used as an agricultural dressing. *See also* MARL.

maestro A local north-westerly wind of the Adriatic Sea which affects the western coasts, especially in summer. The term is also applied to north-westerly winds in the Ionian Sea and to winds off the coasts of Corsica and Sardinia.

magma Molten rock, silicate, carbonate, or sulphide in composition and containing dissolved *volatiles and suspended crystals, which is generated by partial melting of the Earth's crust or *mantle and is the raw material for all *igneous processes.

magma chamber A region, postulated to exist below the Earth's surface, in which *magma is received from a source region in the deep crust or upper *mantle, stored, and from which it moves to the Earth's surface at the site of a volcano. When magma moves rapidly from the chamber, the unsupported chamber roof may collapse to produce a caldera at the surface.

magnesium **(Mg)** An element that is found in high concentrations in plants. It plays an important role in the chemical structure of *chlorophyll and of membranes and is involved in many *enzyme reactions, especially those catalysing the transfer of phosphate compounds. Deficiency can produce various symptoms, including *chlorosis and the development of other pigments in leaves. Magnesium is a relatively abundant component of sea water and is thus quite plentiful in the rainfall of oceanic regions.

magnetic dating The use of magnetic properties for assessing the age of archaeological and geologic materials, based on the *natural remanent magnetization acquired at a specific time. *See* ARCHAEOMAGNETISM.

magnetic orientation The sensing by certain organisms of the direction of the geomagnetic field and their use of it for purposes of orientation. *See* BIOMAGNETISM.

magnetic storm A major disturbance of the Earth's magnetic field resulting from the passage of high-speed, charged solar particles, following a solar flare. The particles are deflected towards the regions of the Earth's magnetic poles.

magnetic susceptibility The degree to which a material is able to retain induced magnetic properties. It is used in the study of lake sediment stratigraphy, where high magnetic susceptibility indicates the presence of ferrous metals, usually associated with soil erosion into the lake basin. Soil erosion can in turn indicate periods of human clearance of forest in catchments.

magnetosphere The space around a planet in which ionized particles are affected by the planet's magnetic field. The Earth's magnetosphere reaches far beyond the atmosphere. In the magnetosphere, charged particles are concentrated at altitudes of about 3 000 km and 1 600 km. The charged particles oscillate between the northern and southern hemispheres. The outer boundary of the magnetosphere is sharp and well defined, extending to about 10 Earth radii on the sunlit side of the Earth and to perhaps 40 Earth radii on the dark side; but the boundary changes its position in response to solar activity, being depressed by the solar wind. *See also* EXOSPHERE and IONOSPHERE.

magnetostratigraphic time-scale (polarity time-scale, geomagnetic reversal time-scale, reversal time-scale) A time-scale based on the periodic polarity reversals in the Earth's geomagnetic field. Magnetic minerals within a rock retain an orientation induced by the field at the time the rock was formed (*see* NATURAL REMANENT MAGNETISM). Provided they include suitable minerals, strata from all over the world thus contain a record of the normal (as at present) or reversed state of the geomagnetic field at the time of their formation. This reversal pattern has been correlated between different successions of rocks to produce a sequence which, when combined with an appropriate dating method, has given a time-scale measured in units of normal or reversed polarity. The scale was first established in detail for the last 4.5 Ma using data from terrestrial, mainly extrusive, rocks; it has now been extended back to the Upper *Jurassic by means of the magnetic-anomaly patterns in oceanic crust.

Mahalanobis's D^2 A measure of generalized distance between samples based on the means, variances (*see* MEAN SQUARE), and covariances of various properties of replicate samples in multivariate analysis. The larger is D^2, the greater is the difference between the samples (or the properties measured).

maiden A tree arising from a seed, or more rarely a sucker, which has not been *coppiced or *pollarded.

maintenance evolution *See* STABILIZING SELECTION.

Malaysian floral region Part of R. Good's (1974, *The Geography of the Flowering Plants*) Indo-Malaysian floral subkingdom within his *Palaeotropical floral kingdom, and including the Malay peninsula and archipelago of south-eastern Asia. The flora is one of the richest, if not the richest, in the world. It has yielded many plants of value to humans, of which bananas, nutmeg, cinnamon, and cloves are perhaps the best known. *See also* FLORAL PROVINCE and FLORISTIC REGION.

Malesian flora A name that was introduced to replace 'Malaysia' which was used formerly, and so to distinguish a *floral region from the state of Malaysia. It should be restricted to the flora of the islands on and between the Sunda and Sahul shelves, an area comprising Malaysia, Papua New Guinea, and the islands of Indonesia, Brunei, and the Philippines. It is in this sense that the term is used in this dictionary. Malesia is sharply delimited by major floristic changes at about the latitude of the Kra isthmus in the west and the Torres Strait in the south-east. *See also* FLORAL PROVINCE and FLORISTIC REGION.

mallee A southern Australian *sclerophyllous scrub community, about 2–3 m high, in which most of the species belong to the genus *Eucalyptus*. The scrub is very similar to that found in other continents where *Mediterranean-type climates occur (e.g. the *chaparral of California and the *maquis of Europe).

Malthus, Thomas Robert (1766–1834) An English mathematician and economist, who took holy orders in 1788 and was appointed in 1805 to the first professorship of political economy in Britain, at Haileybury College, founded by the East India Company, a position he occupied until his death. He was elected a Fellow of the Royal Society and a member of the French Institute and Berlin Royal Academy. He is best remembered for *An Essay on the Principle of Population, as it affects the future improvement of society*, written to refute the ideas on social and economic reform of William Godwin, the Marquis de Condorcet, and others, first published anonymously in 1798, with a much revised second edition in 1803 and a summary in 1830. Malthus maintained that human populations have the capacity to increase more rapidly than resources can be mobilized to sustain them, and their reproductive capacity is constrained by disease, hunger, and other forms of suffering produced by poverty. Attempts to alleviate poverty, therefore, can lead only to additional births and further suffering. *See* MALTHUSIANISM.

Malthusianism The view, derived from the theory propounded by *Malthus, that poverty (in its modern interpretation linked to environmental degradation) is caused primarily by the pressure on resources exerted by the increasing size of the human population and that no remedy is practicable which does not include measures to reduce the birth-rate, especially among the poor.

mamma (mammatus) The Latin *mamma*, meaning 'udder', used to describe a cloud feature consisting of projections from the basal surface of *altocumulus, *altostratus, *stratocumulus, *cumulonimbus, *cirrus, and *cirrocumulus. *See also* CLOUD CLASSIFICATION.

mammal A member of a class (Mammalia) of *homoiothermic, chordate animals in which the young are fed milk secreted by mammae (which give the class its name). The head is supported by a flexible neck typically with seven vertebrae, articulating through two occipital condyles, the side wall of the skull is formed by the alisphenoid bone, the lower jaw is formed from the dentary bone and articulates with the squamosal, the quadrate and articular bones form auditory ossicles, and the angular bone forms the tympanic bone. Typically teeth are present and the mouth cavity is separated by a hard palate from the nasal cavity. The thorax and abdomen are separated by a diaphragm, the heart is four-chambered, the right aortic arch is absent, except in Monotremata (platypus and echidnas), the egg is small and develops in the uterus, and the skin has at least a few hairs. Many mammalian features were present in therapsids (mammal-like reptiles) during the *Triassic. Mammals are believed to have appeared first towards the end of the Triassic and to have diversified rapidly from the end of the *Mesozoic, 100 Ma later, following the mass extinction which marks the Mesozoic-Tertiary boundary 65 Ma ago.

mammal regions A biogeographic scheme that has been proposed by the biogeographer Charles H. Smith on the basis of statistical analyses. Four regions are recognized: *Holarctica (the only region consistent with traditional faunal regions); *Afro-Tethyan; *Island; and *Latin American.

mammatus *See* MAMMA.

mammillated topography Hill relief with a streamlined, rounded, and smoothed appearance, normally resulting from the scouring action of an *ice sheet, as in the Adirondack Mountains, eastern USA. However, some crystalline rocks may support a pseudo-mammillated surface formed by non-glacial processes, as on Ben Lomond, Tasmania.

management *See* ECOSYSTEM MANAGEMENT.

mangal *See* MANGROVE FOREST.

manganese (Mn) An element that is required in small amounts by plants. It is involved in the light reactions of photosynthesis and also binds to proteins. The leaves of plants deficient in manganese show interveinal *chlorosis and may become malformed.

manganese nodule A concretion of iron and manganese oxides which also contains copper, nickel, and cobalt. Nodules are variable in size, shape, and composition, and are layered internally. The average composition is: manganese 30 per cent; iron 24 per cent; nickel 1 per cent; copper 0.5 per cent; cobalt 0.5 per cent. Manganese nodules are widely distributed on the sea floor of every ocean and in some temperate lakes. They are found in areas of negligible sedimentation and/or strong bottom currents, e.g. on the North Pacific *abyssal plain at depths of 3 500–4 500 m, and on the shallow, current-swept Blake Plateau off the east coast of the USA. Submarine mining of these deposits is now considered viable to recover copper and other scarce metals. Growth of manganese nodules apparently reached a peak in the early *Tertiary Period.

mangrove forest **(mangal)** A swamp forest of saline or brackish water, which develops on tropical and subtropical coasts (*see* TIDAL FLAT). Characteristically, mangrove forest has a dense tangle of aerating roots projecting above the mud. Virgin mangrove can reach 30 m tall.

man-induced turnover The additional flow of an element through the active part of a *biogeochemical cycle, which results from human activity. For example, by burning fossil fuels humans add an extra 5 billion tonnes per year of carbon to the turnover of the carbon cycle, which is naturally about 75 billion tonnes per year. (One billion is equal to one thousand million, 10^9.)

manometer An instrument used to measure differences in pressure, usually by comparing the heights of two liquid columns. The simplest version comprises a U-shaped tube containing liquid. One end of the tube is open, the other is connected to the container holding the pressure to be measured, and the pressure is registered as the difference in the level of the liquid in the two sides of the tube.

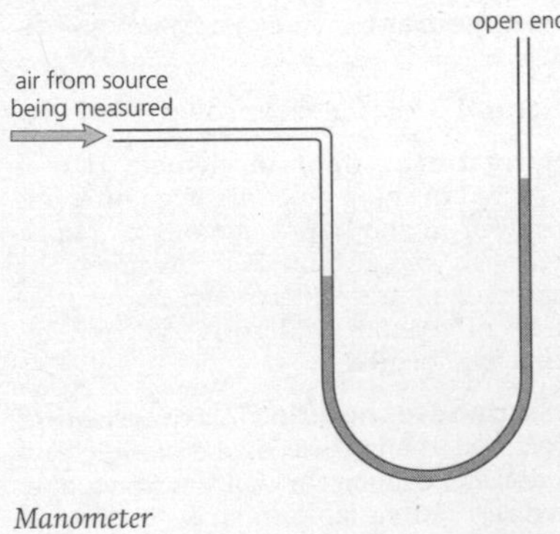

Manometer

mantle **1. (pallium)** In *molluscs and some brachiopods (in which it is known as the mantle lobe), a fold of skin on the dorsal surface that encloses a space (the mantle cavity) containing the gills. The mantle is responsible for the secretion of the shell. **2.** In Cirripedia (barnacles), the name often given to the carapace. **3.** The zone lying between the Earth's crust and core, approximately 2 300 km thick, and representing about 84 per cent of the Earth's volume and 68 per cent of its mass.

maquis **(Italian *macchia*; Spanish *matorral*)**. A French term for drought-resistant *Mediterranean scrub, taller than *garrigue, and composed of evergreen shrubs and small trees with thick, leathery leaves (sclerophylls) or spiny foliage, e.g. *Olea europaea* (wild olive), *Cistus* species (cistus), *Erica* species (heather), and *Genista* species (broom). For the most part this *sclerophyllous formation has been derived by a combination of burning and grazing from the original mixed evergreen *Mediterranean forest.

mares' tails The popular name for tufted *cirrus clouds with *virga (precipitation trails) seen below each cloud.

marginal sea A semi-enclosed body of water that is adjacent to, and widely open to, the ocean (e.g. the Gulf of Mexico, the Caribbean Sea, and the Gulf of California).

marginal value theorem A mathematical rule, proposed by E. L. Charnov in 1976, according to which the optimum time a *foraging animal remains in a patch is defined in terms of the rate at which the forager is extracting energy at the time it leaves (the marginal value of the patch). The optimum foraging strategy is to abandon each patch when the rate of energy extraction from it falls to a certain level, this level being the same for all patches. The theorem predicts that foragers will remain a shorter time in patches with little food than in patches with more food, patches will be abandoned more quickly when they are close together than when they are scattered, and patches will be abandoned more quickly in an area of abundance than in a poorer area.

Marianas Trench The oceanic *trench, about 11 km deep, which marks the destructive margin between the Pacific and Philippine Plates.

marine platform *See* SHORE PLATFORM.

maritime air An *air mass having properties of temperature and humidity derived either from a *source region in, or from long passage over the oceans.

maritime climate A climate that is much modified by oceanic influences. Typical characteristics include relatively small diurnal and seasonal temperature variation and increased precipitation owing to more moist air.

mark-recapture technique A technique for estimating the population density of more elusive or mobile animals. A sample of the population is captured, marked, and released. Assuming that these marked individuals become randomly distributed through the wild population, and that subsequent trapping is random, any new sample should contain a representative proportion of marked to unmarked individuals. From this, the size of the population may be estimated, most simply by multiplying the number in the first sample by the number in the second sample, and dividing the product by the number of marked individuals in the recaptured sample. This calculation is appropriate only when the population is fairly static or where changes (owing to migration, natality, or mortality) are known. It may also be distorted if marked individuals become more vulnerable to predators and if particular individuals become 'trap-addicted' or 'trap-shy'. When a population is fluctuating rapidly, as is common in insect populations, more sophisticated indices, which allow for the probabilities of change, are preferable.

marl Lime-rich *clay, usually found as an *alluvial deposit and containing a high proportion of soft calcium carbonate. *See also* MAERL.

marlstone A semi-lithified *marl; a fully lithified marl is called an *argillaceous *limestone.

marsh A more or less permanently wet area of *mineral soil, as opposed to a *peaty area. Marsh often occurs around the edges of a lake or on the flood-plain of a river. In North American usage, a marsh is a herbaceous wetland in which the *water-table is permanently above the soil surface, equivalent to the British 'swamp'.

marsh gas *See* METHANOGEN.

marshy tundra Marshy tracts (*see* MARSH) that are found as important components in the low-, middle-, and higharctic *tundra belts, because drainage is frequently poor as a result of the widespread presence of permanently frozen ground at no great depth. Grasses, sedges, and dwarf willows are the most prolific vascular plants of marshy tundra, often accompanied by luxuriant growths of moss.

mason bee Any member of the genera *Anthocopa*, *Chalicodoma*, *Dianthidium*, *Heriades*, *Hoplitis*, or *Osmia*, of the cosmopolitan family Megachilidae. The female collects soft, malleable building materials which are shaped at the nest site into durable structures. The nests may be inside existing cavities in timber, under stones, or built on exposed surfaces (e.g. rocks, or leaves and branches of woody plants). According to the bee species, the collected materials may be mud, pebbles, quartz chips, resin, a mastic of chewed leaves or petals, or a combination of two or three of these substances.

mass flow A down-slope slide of sediment which moves under the force of gravity. Mass flows include rockfalls (accumulations of *scree), slumps and slides (where masses of sediment move down-slope along discrete shear planes), debris flows (in which ill-sorted masses of sediment move down-slope following the loss of internal strength of the sediment mass), liquefied sediment flows, grain flows, and turbidity flows.

massif A very large topographic or structural feature, which is usually more rigid than the surrounding rock.

mass movement *See* MASS-WASTING.

mass provisioning The provision of all the food required for larval development by wasps and bees. The food, either insect or spider prey (for wasps) or honey and pollen (for bees), is stored in a specially prepared space (cell) within the nest. An egg is laid on or in the food, and the cell is sealed. Mass provisioning is very largely found in solitary (non-social) species, but is also practised by the highly social stingless bees.

mass spectrometry A technique that allows the measurement of atomic and molecular masses. Material is vaporized in a vacuum, ionized, then passed first through a strongly accelerating electric potential, and then through a powerful magnetic field. This serves to separate the *ions in order of their charge: mass ratio; detection is commonly made using an electrometer, which measures the force between charges and hence the electrical potential.

mass-wasting **(mass movement)** The transfer of material down hill-slopes. It is the result of gravity acting on material that has lost cohesion, typically as a result of an increase in water content. It includes four main categories: flow, slide, fall, and creep, of which creep is the most important if least spectacular. An avalanche is a rapid and often destructive flow of rock or snow. A slide (or landslide) is a comparatively rapid displacement of material over one or more failure surfaces which may be curved or planar. Failure on an arcuate surface is typical of clays, and gives rise to rotational slides such as those of Folkestone Warren, England.

mast A fruit, especially of beech but also of oak, elm, and other forest trees, formerly often used as food for pigs.

master chronology An average tree-ring chronology for a particular region, or one derived locally from a number of closely matching individual tree-ring chronologies. The master chronology forms the reference against which new ring series may be compared and dated.

Mastigophora *See* PROTOZOA.

mast year A year in which there is a particularly high production of *mast. In oak, for example, there are often several years of poor acorn production between mast years.

match In *alignment, the existence of the same base at a homologous position in two sequences.

mating The union of two individuals of opposite sexual type to accomplish sexual reproduction.

mating system In *artificial selection, a procedure that is used to control the genetic constitution of offspring (usually the degree of *homozygosity due to inbreeding). Mating may be assortative (i.e. between similar *phenotypes), resulting in increased homozygosity; or disassortative (i.e. between unlike phenotypes), tending to maintain or increase *heterozygosity.

mating type The equivalent in lower organisms (especially micro-organisms) of *sexes in higher organisms. Microorganisms may be subdivided into mating types on the basis of their physiology and mating behaviour. Different mating types are usually identical in physical form, although individuals of one mating type possess on their surfaces proteins that will bind to complementary proteins or polysaccharides found only on the coats of individuals of the opposite mating type. In this way, only individuals of different mating types will undergo conjugation.

matorral *See* MAQUIS.

mature phase The stage in the cyclical pattern of *community change in *grasslands, *heathlands, and forests. In grasslands, the mature phase is the stage in which the grass hummocks attain their maximum height (about 4 cm) and begin to be colonized by *lichens. In heathlands, it is the stage at which heaths (e.g. *Calluna vulgaris* bushes) about 12–18 years old start to become less dense, allowing other species, especially bryophytes, to grow. *Compare* BUILDING PHASE; DEGENERATE PHASE; HOLLOW PHASE; and PIONEER PHASE.

maximum-likelihood tree In *phylogenetics, a tree-building method that uses the maximum-likelihood statistical estimator to calculate the *topology with the highest probability of being correct under assumed rates of *character change.

maximum-parsimony tree The method for selecting a *phylogenetic tree from all possible tree *topologies that requires the smallest number of substitutions.

maximum sustained yield (MSY) *See* OPTIMUM YIELD.

maximum thermometer A thermometer that records the highest temperature to which it has been exposed (e.g. by allowing a rise of mercury past a restriction in the tube, but preventing the mercury's return on contraction). It is most commonly used to record maximum daily temperatures.

MCR (Mutual Climatic Range) *See* BEETLE ANALYSIS.

meadow steppe A variant of Eurasian *steppe adjacent to the forest to the north, and comprised primarily of various sod-forming grasses and subordinate tussock grasses. The grasses grow to over a metre, and associated with them are broad-leaved flowering herbs, the whole giving a meadow-like aspect in summer. Meadow steppe has long been greatly reduced in extent by agricultural practices.

meander The sinuous trace of a stream channel whose length is normally equal to or greater than 1.5 times the down-valley (or straight-line) distance. It is best developed in cohesive *flood-plain *alluvium. The relationships between its geometric properties vary little with size: e.g. meander wavelength (the straight-line distance between two points at similar positions (e.g. outward extremities of curve) on two successive curves along the trace) is normally 10–14 times the channel width, irrespective of size. Meander origin is uncertain, but the sinuous curve may be that shape best fitted for the transfer of channelled flow in accordance with the *least-work principle. Over time, a meander may move laterally and/or vertically. The process of sideways movement is known as 'meander migration'; it involves the deposition of point *bars on the inner sides of bends and erosion on the outer, and is limited to a tract of flood-plain called the 'meander belt'. The migration of two adjacent, concave bands may narrow the flood-plain between them, and the restriction is a 'meander neck'. This widens out to form a bulbous feature, the 'meander core', around which the river swings. The surface of a core may show 'meander scrolls', which are low, curved ridges of relatively coarse material lying parallel to the main channel and deposited by the stream. An 'incised meander' results from down-cutting, and two types are found. (*a*) If incision is fairly slow, and sideways movement occurs, the result is an 'ingrown meander'; the slope down which the stream has migrated during incision is called a 'slip-off slope'. (*b*) When incision is rapid, with mainly vertical erosion, the consequence is an 'entrenched meander'.

meander belt *See* MEANDER.

meander core *See* MEANDER.

meander migration *See* MEANDER.

meander neck *See* MEANDER.

meander scroll *See* MEANDER.

meander wavelength *See* MEANDER.

mean sea level The average height of the surface of the sea, for all stages of the tide, over a long (usually 19-year) period, being determined from hourly readings of tidal height.

mean square (variance) The square of the mean variation of a set of observations around the sample mean.

mechanical senses Senses that detect mechanical stimuli (e.g. touch, hearing (detecting air vibration), and balance).

mechanical weathering The *in situ* breakdown of rocks and minerals by a set of disintegration processes that do not involve chemical alteration. The chief mechanisms are: crystal growth, including *frost wedging and salt weathering; *hydration shattering; insolation weathering (*thermoclastis); and pressure release.

medial moraine *See* MORAINE.

median network A specialist type of *maximum-parsimony tree method for phylogenetic reconstruction in which all the most parsimonious trees are represented in a reticulate grid in a three-dimensional perspective. Median networks are applied to data where the degree of *homoplasy is very high and the number of informative sites very low (e.g. with data derived from populations).

median valley (axial rift, axial trough) The valley which lies along the axis of some oceanic ridges. Median valleys develop on slower-spreading ridges (e.g. *Mid-Atlantic and Carlsberg) and are up to 3 km deep.

medieval woodland A woodland that is known to have existed prior to the seventeenth century. Some authorities use 1600, others 1650, as the base date. *See also* ANCIENT WOODLAND.

mediocris The Latin *mediocris*, meaning 'medium', used to describe a species of *cumulus with limited vertical development and characterized by very slight projections on the upper surface. *See also* CLOUD CLASSIFICATION.

m

mediolittoral Applied to that area of a shore approximately equivalent to the intertidal or *littoral zone, but excluding the lowest reaches, which are uncovered for only very limited periods. The term is used in an alternative system for subdividing the near-shore area of marine ecosystems. *Compare* INFRALITTORAL and CIRCALITTORAL; *see also* LITTORAL ZONE.

Mediterranean climate The distinctive climatic type which occurs around latitude 35°N and 35°S, and is associated with warm-temperate west coasts, Summers generally are hot and dry, winters mild to cool and rainy. The climate is strongly influenced by westerly airstreams in winter, and subtropical high pressure in summer. In the type area, the Mediterranean basin, there is a variety of climatic regimes owing to the complex configuration of seas and mountainous peninsulas in the 3 000 km incursion into Eurasia. Annual rainfall is broadly 500–900 mm, but less in more continental locations. Other areas of this type include the coasts of Chile in corresponding latitudes, southern California, south-western Africa, and south-western Australia. Cool ocean currents offshore bring lower temperatures to parts of the Chilean and Californian coasts, but rainfall and temperatures are much affected by the differences of slope and elevation inland.

Mediterranean faunal subregion *See* AFRO-TETHYAN MAMMAL REGION.

Mediterranean floral region Part of R. Good's (1974, *The Geography of the Flowering Plants*) *Boreal Realm, and geographically a relatively small region, being restricted to Iberia and the Mediterranean coast, but with an important extension northward along the western coast of Europe. The vegetation is specialized, reflecting the distinctive climatic setting, and consequently contains a high proportion of endemics (*see* ENDEMISM). The Mediterranean has been called 'the cradle of civilization' and its long association with mankind has yielded many garden, horticultural, and agricultural plants, and has led to extensive and intensive human influence on the natural vegetation. *See also* FLORAL PROVINCE and FLORISTIC REGION.

Mediterranean forest The *climatic climax of most of the Mediterranean basin, now virtually destroyed and over vast areas replaced by *garrigue and *maquis. The trees were *evergreen, both conifers and broad-leaved hardwoods. They were probably mixed, but there may well also have been segregation of the two types.

Mediterranean scrub Two principal varieties are recognized on the basis of the stature of the constituent *shrubs. The tall variety is called *maquis, the short variety *garrigue; but the distinction is somewhat arbitrary and there are many intergrade types. Both occupy vast areas and were derived by burning and grazing from an earlier forest cover. Indeed, constant use of fire has created distinctive plant associations within the scrub, such as the *giane* (rosemary and *Genista* species) and *ericeto* (tree heath) in southern Italy. In Iberia such associations are even more extensive and include the *brezales* (heathers), *jorales* (*Cistus* species), *goscojales* (*Quercus coccifera*, kermes oak) and *bujedales* (*Buxus sempervirens*, box).

Mediterranean water A water mass, formed in the arid eastern Mediterranean, which flows westward, sinking in the Algero-Ligurian and Alboran basins to a depth of approximately 500 m, owing to its high salinity (36.5–39.1 parts per thousand). This dense water flows into the Atlantic Ocean through the relatively shallow Straits of Gibraltar at a depth below 150 m, while above it lighter, Atlantic water flows eastward into the Mediterranean Sea. The Mediterranean water in the Atlantic then sinks to about 1 000 m, where it forms a clearly identifiable water mass.

mega- (M-) From the Greek *megas*, meaning 'great', a prefix meaning 'very large'. Attached to SI units, it denotes the unit $\times 10^6$.

megafauna **1.** Animals that are large enough to be seen with the naked eye. **2.** The term is often used of very large mammals and birds, i.e. those exceeding 50 kg in body weight. Much controversy concerns the extinction of many members of the megafauna in the early part of the *Holocene, especially in Europe and North America. Climatic change is a possible factor, but most of the species involved had already survived previous *interglacials. The distinctive aspect of our present interglacial is the rise to prominence of humans, and many megafaunal species became extinct at a time when human hunting pressures were gaining strength.

megafossil (macrofossil) A fossil that is large enough to examine without the aid of a microscope. *Compare* MICROFOSSIL.

megaphyll In the classification of leaf sizes, a leaf that is more than 164 025 mm^2 in area.

megaripple *See* DUNE BEDFORM.

megatherm In Alphonse de *Candolle's (1874) classic temperature-based scheme of world vegetation zones, a plant of the warmest (i.e. tropical) environments, where each month has a temperature mean of no less than 18°C and moisture supply is not limited. *Compare* XEROPHILE.

megathermal Applied to the seasonal patterns of river flow in certain equatorial regions, where the main characteristics are high temperatures and evaporation rates all year round, so that river flow maxima reflect the months of maximum precipitation.

megathermal climate A high-temperature climate type, known more commonly in Europe (e.g. in the *Köppen classification) as humid subtropical or tropical, with the coldest monthly mean temperature above 18°C. The term is also defined in the *Thornthwaite classification by potential evapotranspiration and moisture-budget criteria.

meiofauna That part of the *microfauna which inhabits algae, rock fissures, and the superficial layers of the muddy sea-bottom. The term is often used synonymously with *interstitial fauna. *Compare* MACROFAUNA; MESOFAUNA; and MICROFAUNA.

meiosis (reduction division) A form of nuclear division whereby each *gamete receives only one member of a *chromosome pair (this forms one of the bases of *Mendel's first law of genetic segregation) and genetic material can be exchanged between *homologous chromosomes. Two successive divisions of the *nucleus occur, with corresponding cell divisions, following a single chromosomal duplication. Thus a single *diploid cell gives rise to four *haploid cells. This produces gametes (in animals) or sexual spores (in plants and some protozoa) that have one half of the genetic material or chromosome number of the original cell. This halving of the chromosome number ($2n$ to n) compensates for its doubling when the gametes ($n + n$) unite ($2n$) during sexual reproduction. *Compare* MITOSIS.

meiotic drive *Meiosis, where there is an unequal production of allelic alternatives in the *gametes of a heterozygous parent.

Melanesian and Micronesian floral region Part of R. Good's (1974, *The Geography of the Flowering Plants*) *Polynesian floral subkingdom, within his *Palaeotropical floral kingdom, and geographically coinciding with Micronesia (Marianas, Carolines, Marshall, Kiribati, and Tuvalu) and Melanesia (New Guinea, Solomon Islands, Vanuatu, and Fiji, but not New Caledonia). The flora is not rich and although there is an appreciable endemic (*see* ENDEMISM) component at the species level, most of the species belong to large, widespread genera. The status of the endemism is not high, therefore, and reflects the fact that the flora is largely derived from *Malesia to the west. *See also* FLORAL PROVINCE and FLORISTIC REGION.

melanism In an animal population, the occurrence of individuals that are black owing to the presence in excess of the pigment melanin. *See* INDUSTRIAL MELANISM.

Mellanby, Kenneth (1908–93) A British ecologist who was one of the first scientists to warn of the harmful effects on wildlife of certain insecticides and who became a leading world authority on all forms of environmental pollution. Mellanby studied zoology and entomology at the University of Cambridge, conducted research in medical entomology, which he continued during the Second World War as a major in the Royal Army Medical Corps, and after the war was appointed principal, and professor of parasitology, at the University of Ibadan, Nigeria. In 1955 he became head of entomology at the Rothamsted Experimental Station and from 1961 to 1974 he was director of the Monks Wood Experimental Station (Nature Conservancy, later Institute of Terrestrial Ecology). It was there that major studies were conducted on the effect of toxic chemicals on wildlife. Despite his strong commitment to environmental improvement, he was scathing about the pessimism of many environmentalists. He was the author of several books, including *Pesticides and Pollution* (1967), *Can Britain Feed Itself?* (1975), *Waste and Pollution* (1992), and *The DDT Story* (1992).

melliferous Producing honey.

meltemi *See* ETESIAN WINDS.

meltout till *See* TILL.

meltwater channel *See* GLACIAL DRAINAGE CHANNEL.

MEM *See* MICRO-EROSION METER.

memory The ability to store and recall the effects of experience. In non-human animals, memory may be studied by observing the persistence of learned responses.

Menapian A glacial stage (0.9–0.8 Ma) of the northern European plain, which is perhaps equivalent to the *Günz glacial of the Alps and the *Kansan of North America.

Mendel, Gregor Johann (1822–84) An Austrian Augustinian monk in the monastery at Brünn (now Brno), whose interest in natural science led him to conduct the experiments which resulted in his discovery of the underlying principles of heredity. His experiments began in 1856 and were conducted in the monastery garden. They involved crossing varieties of garden peas, each of which had certain distinctive features, and recording the appearance of those features in the progeny. Mendel took part in the meetings of the Naturforscher Verein (natural science society) in Brünn and it was there that he reported his results on 8 March 1865. In 1900, while searching the literature in connection with their own research, K. E. Correns, E. Tschermak von Seysenegg, and H. de Vries came across Mendel's paper describing results similar to their own. Mendel was elected abbot of his monastery in 1854.

Mendelian character A character that follows the laws of inheritance formulated by Gregor *Mendel.

Mendelian population An interbreeding group of organisms that share a common *gene pool.

Mendel's laws Two general laws of inheritance formulated by the Austrian monk Gregor *Mendel. Re-expressed in modern terms, the first law, of segregation, is that the two members of a gene pair segregate from each other during meiosis, each *gamete having an equal probability of obtaining either member of the gene pair. Re-expressed in modern terms, the second law, of independent assortment, is that different segregating gene pairs behave independently. This second law is not universal, as was originally thought, but applies only to unlinked or distantly linked pairs (*see* LINKAGE). At the time of Mendel, genes had not been identified as the units of inheritance: he considered factors of a pair of characters segregating and members of different pairs of factors assorting independently.

mendip A hill or ridge that has been buried by younger rocks and subsequently exposed by *erosion. The type example is provided by the Mendip Hills, England, described by W. M. Davis in 1912, but today the term is used only very rarely.

Mercalli, Giuseppe (1815–1914) An Italian professor of natural sciences who studied the major earthquake districts of Italy, and is best known for his scale of earthquake intensity, first formulated in 1897 and refined several times. *See* MERCALLI SCALE.

Mercalli scale An earthquake-intensity scale based on direct observation. Devised by G. *Mercalli, it ranges from scales I (not felt) and II (felt by most people at rest) to VII (difficult for a person to remain standing), X (most structures destroyed), and XII (total devastation). It is not a reliable scale for energy release because the extent of destruction depends on the local geology, type of buildings, etc. *See also* RICHTER SCALE.

meridional circulation Air circulation in a cell, or general flow to or from different latitudes, with a marked south-to-north or north-to-south component. *See also* HADLEY CELL and ZONAL FLOW.

mero- From the Greek *meros*, meaning 'part', a prefix meaning 'partial' or 'part of'.

merocenosis *See* MEROTOPE.

merological approach An approach to *ecosystem studies in which the component parts of an ecosystem are studied in detail in an attempt to compose a picture of the whole system. *Compare* HOLISTIC.

meromictic Applied to lakes in which part of the water column is stratified permanently, usually because of some chemical difference (e.g. contrasting salinities, and hence densities) between *epilimnion and *hypolimnion waters. Meromictic lakes may be *amictic or *oligomictic.

meroplankton Temporary zooplankton (i.e. the larval stages of other organisms). *See* PLANKTON; *compare* HOLOPLANKTON.

merotope A *microhabitat that forms part of a larger unit (e.g. a fruit or a pebble). Organisms colonizing the merotope form a merocenosis.

mesa A flat-topped hill of limited extent, but wider than a *butte and normally underlain by near-horizontally bedded sediments. A land-form similar to a mesa but larger is called a 'plateau'.

mesaxonic Applied to vertebrate limbs in which the weight of the animal passes through the third digit.

mesic Applied to an environment that is neither extremely wet (hydric) nor extremely dry (xeric). *See also* PERGELIC.

meso- A prefix, derived from the Greek *mesos*, meaning 'middle' (e.g. *mesotrophic, neither nutrient-poor nor nutrient-rich; and *mesobiota, the medium-sized soil organisms).

mesobiota A general term for soil organisms of intermediate size. *Compare* MACROBIOTA and MICROBIOTA.

mesoclimate A general term applied to the characteristics of a relatively small region (e.g. a valley or an urban area).

mesocyclone *See* FUNNEL CLOUD.

mesofauna Small, invertebrate animals found in the soil, characteristically annelids (worms), arthropods (*Arthropoda), nematodes, and *molluscs. These organisms are readily removed from a soil sample by using a *Tullgren funnel or a similar device. *Compare* MACROFAUNA and MICROFAUNA.

mesoflora *See* MESOBIOTA.

mesohaline water *See* BRACKISH and HALINITY.

mesopause In the atmosphere, the inversion at about 80 km height, which separates the *mesosphere from the *thermosphere above. The first 10 km of the mesopause are almost isothermal. The mesopause is liable to be marked by *noctilucent cloud composed of ice crystals on meteoric dust.

mesopelagic zone The part of the *pelagic zone of the ocean that extends from the base of the *epipelagic zone, at a depth of 200 m, to the upper boundary of the *bathypelagic zone, 700–1 000 m below the surface, and forms the upper region of the *aphotic zone. The temperature at the bottom of the mesopelagic zone is 10°C and the pressure through the zone ranges from 2 MPa at the upper margin to 7–10 MPa at the base. *See* OCEAN DIVISIONS.

mesophile An organism that grows best at moderate temperatures, often quoted as 20–45°C.

mesophyll **1.** In the classification of leaf sizes, a leaf that is 125–250 mm long, or 2 025–18 225 mm^2 in area. **2.** Plant tissue composed of unspecialized cells that lies immediately beneath the outer layer of leaf cells. *Photosynthesis takes place in mesophyll cells and starch is stored in them.

mesophyte A plant adapted to environments that are neither extremely wet nor extremely dry. *Compare* HYDROPHYTE and XEROPHYTE.

mesosphere **1.** An upper atmospheric layer above the *stratopause (at 50 km) through which temperature decreases with height up to about 80 km, where temperatures reach a minimum of about −90°. This level is the *mesopause, an inversion above which temperatures rise again. **2.** That part of the Earth underlying the asthenosphere. The term is no longer in current use in this sense.

mesotherm A plant of warm-temperate areas, where the hottest month has a mean temperature of more than 22°C and the coldest month a mean of not less than 6°C, according to Alphonse de *Candolle's (1874) classic temperature-based scheme of world vegetation zones.

mesothermal Applied to the seasonal patterns of river discharge of certain warm, subtropical, and temperate areas.

mesothermal climate A climatic type with moderate temperatures, known most commonly in Europe (e.g. in the *Köppen classification) as a warm-temperate rainy climate having a coldest month with temperatures of −3°C to +18°C and a warmest month above +10°C. Such climates are found typically in latitudes 30–45°C. The *Thornthwaite classification defines this type according to *evapotranspiration and moisture budget.

mesotidal Applied to coastal areas where the *tidal range is 2–4 m. Tidal action and wave activity both tend to be important in such areas.

mesotrophic Applied to waters having levels of plant nutrients intermediate between those of *oligotrophic and *eutrophic waters.

Mesozoic The middle of three eras that constitute the *Phanerozoic period of time. The Mesozoic (literally 'middle life') was preceded by the *Palaeozoic Era and followed by the *Cenozoic Era. It began with the *Triassic approximately 251 Ma ago and ended around 65.5 Ma at the start of the *Tertiary. The Mesozoic comprises the Triassic, *Jurassic, and *Cretaceous Periods.

messenger-RNA (m-RNA) A single-stranded *RNA molecule that is responsible for the transmission to the *ribosomes of the genetic information contained in the nuclear *DNA. It is synthesized during *transcription and its base sequencing exactly matches that of one of the strands of the double-stranded DNA molecule.

meta- The Greek *meta*, meaning 'with' or 'after', used as a prefix implying change and meaning 'behind', 'after', or 'beyond'.

metabolic pathway A sequential series of enzymatic reactions involving the synthesis, degradation, or transformation of a *metabolite. Such a pathway may be linear, branched, or cyclic, and directly or indirectly reversible.

metabolic rate The rate at which the chemical reactions constituting the *metabolism of a living organism take place. This varies according to the species of the organism, its gender, age, and level of physical activity. Metabolic rate is usually expressed as energy per unit surface area per unit time (e.g. $kJ\,m^{-2}\,h^{-1}$).

metabolism The total of all the chemical reactions that occur within a living organism (e.g. those involved in the digestion of food and the synthesis of compounds from *metabolites so obtained).

metabolite Any compound that takes part in or is produced by a chemical reaction within a living organism.

metachromism The widely accepted evolutionary theory which holds that the primitive colour of mammalian hair is *agouti. Subsequent evolution leads to saturation by one or other colour (i.e. *eumelanic or *phaeomelanic) owing to the elimination of the other colour, followed by bleaching, eventually to white. In the course of these changes the two primitive colours may be distributed about the body, producing distinctive patterns.

metagenesis The occurrence, during the life cycle of an organism, of two types of individual, both *diploid, both of which reproduce, one sexually and the other asexually. The term is sometimes used loosely (but incorrectly) as a synonym for the *alternation of generations. For example, the life cycle of hydrozoans (e.g. *Obelia geniculata*) comprises an asexual sedentary generation followed by a sexual medusoid generation, both of which are diploid. The medusoid individuals then become sedentary (and asexual).

metalimnion That part of a water column where the *thermocline and *pycnocline are steepest.

metallic bond The chemical bond that links the atoms in a solid metal. The atoms are ionized and electrons move fairly freely among them as an 'electron gas', the bond being between the electropositive atoms and the electrons. It is the free electrons that give metals their high electrical and thermal conductivities. *See also* COVALENT BOND, HYDROGEN BOND, and IONIC BOND.

metamorphic rock An aggregate of minerals formed by the recrystallization of pre-existing rocks in response to a change of pressure, temperature, or *volatile content.

metamorphism The process of changing the characteristics of a rock, usually by the crystallization of new mineral phases. *See* METAMORPHIC ROCK.

metamorphosis Abrupt physical change. The word derives from the Greek **meta* and *morphe* ('form'). In a restricted sense, the transformation from the larval to the adult condition, as in the classes *Insecta (insects), *Amphibia (amphibians), etc.

metapopulation A group of *conspecific populations that exist at the same time, but in different places. In other words, it is a population that is fragmented into numerous sub-populations.

metastable Applied to a situation or condition which is apparently stable but is capable of reaction if disturbed, usually because of the slowness of a system to attain equilibrium (e.g. a supersaturated solution).

meteoric water Water of atmospheric origin which reaches the Earth's surface crust from above, either as rainfall or as seepage from surface-water bodies. *See also* GROUNDWATER.

methane (CH_4) The simplest *hydrocarbon compound, which is released as a gaseous by-product of the metabolic activity of certain bacteria. The principal sources of atmospheric methane are swamps, marshes, and natural wetlands (which may also be *nature reserves), paddy-rice fields, and cellulose-digesting bacteria in the guts of termites and ruminant cattle. Methane is an important *greenhouse gas, absorbing long-wave radiation at wavelengths of about 10 μm.

methanogen A single-celled organism that derives metabolic energy by using hydrogen to reduce carbon dioxide to

m

methane, emitting the methane as a by-product. Methanogens are obligatory anaerobes (*see* ANAEROBIC) that inhabit *swamps and *marshes where other organisms have consumed all the oxygen. The methane bubbling to the surface is known as marsh gas. Methanogens belong to the *Archaea, ranked as a subkingdom in the five-kingdom taxonomic system and as a *domain in the three-domain system.

methanotroph A bacterium that can use *methane as a nutrient.

methylmercury compounds *See* ORGANOMERCURY COMPOUNDS.

methylotroph An organism that can use (as its sole source of carbon and energy) organic compounds that contain only one carbon atom (i.e. compounds such as methane and methanol).

Mg *See* MAGNESIUM.

micrinite *See* COAL MACERAL.

micro- A prefix meaning 'small', from the Greek *mikros*, meaning 'small'. Attached to *SI units it means the unit × 10^{-6} (abbreviated as μ).

micro-aerobic Applied to an environment in which the concentration of oxygen is less than that in air.

micro-aerophilic Applied to an organism that grows best under *micro-aerobic conditions.

microbe An organism that can be seen only with the help of a microscope. The term covers *bacteria, *fungi, and *viruses, but does not include *protozoa and *algae. *See also* MICRO-ORGANISM.

microbial (microbic) Of or pertaining to *micro-organisms.

microbic *See* MICROBIAL.

microbiocoenosis *See* BIOCOENOSIS.

microbiota (microfauna, microflora) The smallest soil organisms, comprising bacteria, fungi, algae, and protozoa.

microbivore An animal which feeds on micro-organisms.

microclimate The atmospheric characteristics prevailing within a small space, usually in the layer near the ground that is affected by the ground surface. Special influences include the impact of vegetation cover on humidity (by *evapotranspiration) and on temperature and winds. *See also* URBAN CLIMATES.

microconsumer *See* CONSUMER ORGANISM.

microcosm 1. A late nineteenth-century American term encompassing essentially the same ideas as the word *'ecosystem'. **2. (ecotron, micro-ecosystem)** A small-scale, simplified, experimental ecosystem, laboratory- or field-based, which may be: (*a*) derived directly from nature (e.g. when samples of pond water are maintained subsequently by the input of artificial light and gas-exchange); or (*b*) built up from *axenic cultures until the required conditions of organisms and environment are achieved.

micro-ecosystem *See* MICROCOSM.

micro-erosion meter (MEM) A device for measuring the rate at which an exposed rock surface is lowered, perhaps by *weathering, consisting of a gauge that records the extension of a probe. The unit is placed on three studs, set in the rock surface to provide a reference level, and the extension of the probe provides a measure of *erosion.

micro-evolution Evolutionary change within species, which results from the differential survival of the constituent individuals in response to *natural selection. The genetic variability on which selection operates arises from *mutation and sexual reshuffling of gene combinations in each generation.

microfauna The smallest soil animals (i.e. micro-organisms). *Compare* MACROFAUNA; MEIOFAUNA; and MESOFAUNA.

microflora *See* MICROBIOTA.

microfossil Any *fossil that is best studied by means of a microscope. Material may include dissociated fragments of larger organisms, whole organisms of microscopic size, or embryonic forms of larger fossil organisms.

microhabitat A precise location within a *habitat where an individual species is normally found (e.g. within a deciduous oak woodland habitat woodlice may be found in the microhabitat beneath the bark of rotting wood).

micronutrient An organic or inorganic element or compound that is needed only in small amounts by living organisms. Vitamins are the main group of organic micronutrients. *Compare* MACRONUTRIENT; *see also* ESSENTIAL ELEMENT.

micro-organism Literally, a microscopic organism. The term is usually taken to include only those organisms studied in microbiology (i.e. bacteria, fungi, microscopic algae, protozoa, and viruses), thus excluding other microscopic organisms such as eelworms and rotifers.

micropalaeontology The study of *microfossils, a discipline A. D. d'Orbigny (1802–57) is credited with founding. Commercial or applied micropalaeontology began in 1877, since when it has become a powerful tool in geologic investigations. Many oil companies either have their own laboratories devoted to micropalaeontology or employ consultants.

microphyll In the classification of leaf sizes, 25–75 mm long, or 225–2 025 mm^2 in area.

microsatellite DNA *See* SATELLITE DNA.

microspecies A population of uniparental plants that is genotypically uniform and has recognizable phenotypic expression; in both respects it is distinctive within its own uniparental group.

microtherm A plant of cool temperate environments, where temperatures in the warmest month lie between 10°C and 22°C and the coldest monthly mean temperature does not fall below 6°C, in Alphonse de *Candolle's (1874) classic temperature-based scheme of world vegetation zones.

microthermal Applied to the seasonal patterns of river discharge in areas where at least one month has a mean temperature below –3°C.

microthermal climate A low-temperature climate of short summers, defined in the *Köppen classification as having mean winter temperatures of less than –3°C. Examples include the cold *boreal forest climate types in continental interiors, and along some eastern seaboards in latitudes 40–65°. The term is also applied in the *Thornthwaite classification according to potential evapotranspiration and moisture-budget criteria.

microtidal Applied to coastal areas in which the tidal range is less than 2 m. Wave action dominates the processes active in microtidal areas (e.g. the Mediterranean Sea and the Gulf of Mexico).

microtine cycles The periodic, *density-dependent fluctuations in population size, involving mass migrations, of certain animal species, the best known of which is the Scandinavian lemming (*Myodes lemmus*), of the rodent subfamily Microtinae, after which the phenomenon is named. An abundance of nutritious food (phosphate being the limiting nutrient for lemmings) causes a rapid increase in numbers; their feeding reduces plant productivity until overcrowding triggers the migration. Cyclical *locust migrations can cause catastrophic damage to agricultural crops over large areas.

Mid-Atlantic Ridge The oceanic *ridge which separates the North and South American Plates from the African and Eurasian Plates. It is a slow-spreading ridge, with rugged topography and a well-developed *median valley.

middle Arctic tundra The middle Arctic belt of the *tundra. The tundra vegetation at this latitude is intermediate in floristic composition and luxuriance of development between that of the low Arctic and the high Arctic.

mid-latitude mixed forest A forest comprising coniferous and broad-leaved trees, and belonging to one of two broad categories: (*a*) *ecotone mixed forest, which has characteristics transitional between those of the two great belts of *boreal coniferous forest and mid-latitude broad-leaved *deciduous forest; and (*b*) a second type that seems to have the status of true *climatic climax, in which both the conifers and broad-leaved trees are *evergreens. These evergreen mixed forests were once extensive in the Mediterranean basin, and in the southern hemisphere are found in Chile, southern Brazil, Tasmania, northern New Zealand, and the Cape Province (South Africa).

mid-oceanic ridge *See* RIDGE.

mid-water Applied to fish that swim clear of the sea bottom.

Mie scattering The scattering of electromagnetic radiation, mainly in a forward direction, by spherical particles. The theory was proposed by Gustav Mie (1868–1957) in 1908. *See also* RAYLEIGH SCATTERING.

migration **1.** The movement of individuals or their propagules (seeds, spores, larvae, etc.) from one area to another. Three cases may be distinguished: (*a*) emigration, which is outward only; (*b*) immigration, which is inward only; and (*c*) migration, which in this stricter sense implies periodic movements to and from a given area and usually along well-defined routes. Such migratory movement is triggered by seasonal or other periodic factors (e.g. changing day-length), and occurs in many animal groups. **2.** In plant *succession, specifically the arrival of migrating propagules (migrules) at a newly denuded area.

migration route A link between two biogeographical regions which permits the interchange of plants and/or animals. Various types are recognized in the literature; for example. G. G. *Simpson's *'corridors', *'filters', and *'sweepstakes routes' are widely referred to in connection with mammalian, and more recently reptilian migrations.

migratory locust **(*Locusta migratoria*)** *See* LOCUST.

migrule *See* MIGRATION.

Milankovitch solar radiation curve A graphical representation of the combined effects of the precession of the equinoxes, the obliquity of the ecliptic, and the eccentricity of the Earth's orbit, which was calculated early in this century by Milutin Milankovitch (1879–1958) and used to account for the variations of climate during the ice ages. Theoretically, each period of radiation minimum caused an ice age. There were nine minima in all, with an irregular spacing pattern, and each of them caused an ice advance. The three factors coincided about 13 000 years BP, at the start of the Older *Dryas.

Milazzian *See* QUATERNARY.

mildew Any fungal disease of a plant in which the mycelium of the causal agent is visible as white or pale-coloured, cottony or powdery patches on leaves, etc. *See also* DOWNY MILDEW and POWDERY MILDEW.

milli- **(m-)** From the Latin *mille*, meaning 'one thousand', a prefix meaning 'one-thousandth' (e.g. a 'milli-equivalent' is one-thousandth of an equivalent weight). Attached to *SI units it denotes the unit 10^{-3}.

millibar *See* BAR.

milt A mass of spermatozoa and other secretions of the sperm ducts exuded by male fish during spawning, or during copulation in the case of the live-bearing species.

mimicry *See* AUTOMIMICRY; BATESIAN MIMICRY; and MÜLLERIAN MIMICRY.

Mindel The second of four glacial episodes, taking its name from an Alpine river, which may be equivalent to the *Elsterian of northern Europe.

Mindel/Riss Interglacial **(Great Interglacial)** An Alpine *interglacial stage that is possibly the equivalent of the *Hoxnian of East Anglia, England, or the *Holsteinian of northern Europe.

mineral A usually inorganic substance which occurs naturally, and typically has a crystalline structure whose characteristics of hardness, lustre, colour, cleavage, fracture, and relative density can be used to identify it. Each mineral has a characteristic chemical composition. Rocks are composed of minerals. More loosely, certain organic substances obtained by mining are sometimes termed 'minerals'. A further loose use of the term is the expression 'mineral nutrition' when applied to the uptake from the soil of elements required for plant growth.

mineral cycle *See* NUTRIENT CYCLE.

mineralization The conversion of organic tissues to an inorganic state as a result of decomposition by soil microorganisms.

mineral nutrition *See* MINERAL.

mineral soil A soil, composed principally of mineral matter, the characteristics of which are determined more by the mineral than by the organic content.

minimal area The smallest area that contains the species representative of a particular plant *community. When the number of species recorded in increasingly larger sample units is plotted graphically (to give a species–area curve) the minimal area is the point at which the curve becomes horizontal. In practice, the curve rarely becomes truly horizontal because of natural heterogeneity, and some subjective assessment of the minimal area is made from the curve. Alternatively, the 'minimum quadrat number' may be defined as the 'equivalence point' (i.e. the point at which the number of species and the number of *quadrats are equal).

minimum quadrat number *See* MINIMAL AREA.

minimum temperature The lowest temperature recorded diurnally, monthly, seasonally, or annually, or the lowest temperature of the entire record. Daily air-temperature minima are recorded by the screen minimum thermometer. *See also* GRASS MINIMUM TEMPERATURE.

minimum thermometer A thermometer that records the lowest temperature to which it has been exposed (e.g. by allowing a fall of mercury past a restriction in the tube, but preventing the mercury's return on expansion). It is most commonly used to record minimum daily temperatures.

minisatellite DNA *See* SATELLITE DNA.

mio- From the Greek *meion*, meaning 'less', a prefix meaning 'less' (e.g. *Miocene, 'mio' plus 'cene' (from the Greek *kainos*, 'new'), meaning 'less new').

Miocene The fourth of the five epochs of the *Tertiary Period, extending from the end of the *Oligocene, 23.03 Ma ago, to the beginning of the *Pliocene, 5.332 Ma ago. Many mammals with a more modern appearance evolved during this epoch, including deer, pigs, and several elephant stocks.

mirage An optical effect in which major vertical variation in temperature of the lower atmosphere produces differential refraction of light, resulting, for example, in raised images and in gaps, which may give the appearance of a water surface.

mire A *peat-producing *ecosystem.

misfit stream (underfit stream) A stream that is too small to have cut the valley it currently occupies. The term applies particularly to a meandering stream whose dimensions are much smaller than those of the meandering valley through which it flows.

mismatch In *alignment, the occurrence of different bases in homologous positions in two different sequences.

missing-plot technique A standard formula for the estimation of a missing datum observation (or, with suitable modification, the estimation of several missing values) in the analysis of the variance of data collected according to a recognized experimental design. The missing observation (x'_{ij}) may be estimated as: $x'_{ij} = tT'_i + bB'_j - G'/(t-1)(b-1)$, where T'_i, B'_j, and G' are the treatment, block, and grand totals for the available observations, i is the ith treatment, j is the jth block, t is the number of treatments, and b is the number of blocks.

Mississippian **1.** The Early *Carboniferous subperiod which is followed by the *Pennsylvanian and comprises the Tournaisean, Visean, and Serpukhovian *Epochs. It is dated at 359.2–318.1 Ma. **2.** The name of the corresponding North American subsystem which comprises the Kinderhookian, Osagean, Meramecian, and Chesterian Series. It is roughly contemporaneous with the Dinantian plus the Namurian A of western Europe.

mist A surface-layer atmospheric condition in which visibility is reduced by very fine, suspended water droplets. In *synoptic meteorology, the *relative humidity in a mist condition is more than 95 per cent and overall visibility is at least 1 km. *See also* HAZE and FOG.

mistral A strong, cold, northerly wind that blows offshore with great frequency along the Mediterranean coast from northern Spain to northern Italy, and which is particularly frequent in the lower Rhône valley. The wind may persist for several days, and is best developed when a *depression is forming in the Gulf of Genoa to the east of a ridge of high pressure. The airstream that feeds the mistral is commonly derived from polar air of maritime origin. In the Rhône valley and similar areas of occurrence, the airflow is strengthened by *katabatic and funnelling effects producing speeds of up to 75 knots (139 km/h), compared with the typical 40 knots (74 km/h) experienced along the coast.

mitochondrial-DNA (mt-DNA) Circular *DNA that is found in mitochondria. It is entirely independent of nuclear DNA and, with very few exceptions, is transmitted from females to their offspring with no contribution from the male parent. Mitochondrial-DNA codes for specific *RNA components of *ribosomes that are unique to those *organelles. It also codes for some of the respiratory enzymes found in mitochondria.

mitosis The normal process of nuclear division by which two daughter nuclei are produced, each identical to the parent nucleus, resulting in two daughter cells with identical nuclear contents. *Compare* MEIOSIS.

mixed cloud Cloud that contains unfrozen water droplets as well as ice crystals. Typically, the condition is found in *cumulonimbus, *nimbostratus, and *altostratus.

mixed woodland A woodland in which the minority of trees, but amounting to not less than about 20 per cent, are either coniferous or broad-leaved.

mixing condensation level The lowest level at which condensation can occur in air that is mixing vertically (e.g. through the amalgamation of *air masses with different temperatures), leading to conditions of oversaturation and condensation.

mixing depth Measured from the surface of the Earth, the extent of an atmospheric layer (usually a sub-inversion layer) in which convection and turbulence lead to mixing of the air and of any pollutants in it.

mixing ratio The ratio of the mass of a given gas (e.g. water vapour) to that of the remaining gas (e.g. dry air) in the mixture. (The examples given yield the 'humidity mixing ratio', expressed most conveniently in grams per kilogram of dry air.)

mixotroph An organism in whose mode of nutrition both organic and inorganic compounds are used as sources of carbon and/or energy.

Mn *See* MANGANESE.

mnemon A *memory unit within the brain which is associated with a particular sensory system and with a particular set of behavioural responses to stimuli received by that system.

Mo *See* MOLYBDENUM.

mobbing The harassment of a predator by members of a prey species.

model A representation of reality in which the main features of some aspect of the real world are presented in simplified terms in order to make that aspect easier to comprehend, and often to facilitate the making of predictions.

modern synthesis (neo-Darwinism) The fusion of Mendelian genetics and Darwin's *natural selection. A further synthesis has been achieved in recent years with the incorporation of knowledge of evolution at the molecular level.

modular organism An organism in which the *zygote develops into a discrete unit which then produces more units like itself, rather than developing into a complete organism. Modular organisms (e.g. plants, fungi, sponges, etc.) usually have a branching structure and an overall shape that is highly variable and determined mainly by environmental influences. Mutations of cells along branches may lead to heterogeneity between different parts of the organism.

Moershoofd (Poperinge) A Middle *Devensian *interstadial from the Netherlands during which the climate was relatively cool, with average July temperatures of 6–7°C. This division together with the *Hengelo and *Denekamp interstadials, is perhaps equivalent to the *Upton Warren interstadial of the British Isles.

moisture balance *See* MOISTURE BUDGET.

moisture budget (moisture balance) The balance of water, as represented broadly by the equation: balance = precipitation − (runoff + evapotranspiration + the change in soil-moisture). Over the year, for example in mid-latitudes, the budget is balanced by a high level of potential evapotranspiration and utilization of soil moisture in summer, compensated by a water surplus and recharge of soil moisture in winter, when evaporation is less and precipitation is sometimes greater.

moisture index The term used instead of 'moisture budget' (e.g. by C. W. Thornthwaite (1955)), and calculated from the aridity and humidity indices, as $I_m = 100 \times (S - D)/PE$, where I_m is the moisture index, S is the water surplus, D is the water deficit, and PE is the potential evapotranspiration.

molar In mammals, one of the posterior teeth, commonly used for crushing, which are not preceded by milk teeth. Usually molars have several roots and their biting surface is formed from patterns of projections and ridges.

mold *See* MOULD.

molecular clock The idea that molecular *evolution occurs at a constant rate, so that the degree of molecular difference between two species can be used as a measure of the time elapsed since they diverged. Its accuracy depends on the validity of the *neutrality theory of evolution.

molecular drive A hypothesis advanced principally to explain the occurrence of *concerted evolution. The hypothesis proposes that cohesive genetic change can occur in a population through a process of constant homogenization taking place through the genetic processes of *transposition, *unequal crossing over, gene conversion, and *biased gene conversion, in conjunction with a high rate of sexual randomization of *chromosomes relative to the genetic processes. The evolution of such loci (*see* LOCUS) may be directed (driven) by biases within the genetic processes which are independent of *natural selection.

mole drain A drain that can be made in soils by pulling a bullet-shaped device through the soil so that the compacted sides of the tunnel maintain that form for several years.

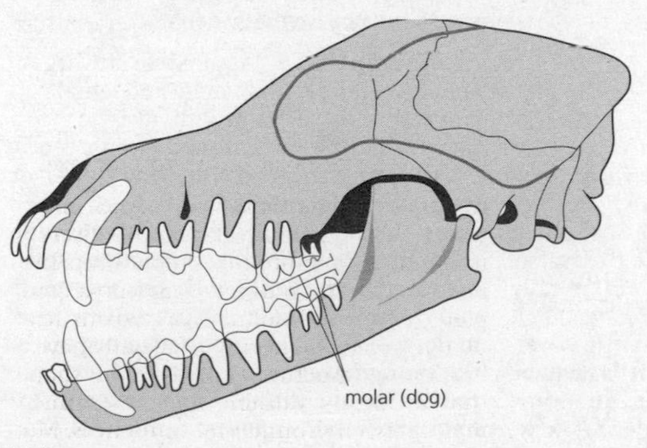

Molar

mollic horizon A surface *horizon of *mineral soil that is dark in colour, and relatively deep, and contains (dry weight) at least 1 per cent organic matter or 0.6 per cent organic carbon, the determination of either being acceptable. It is the *diagnostic horizon of *Mollisols and is associated with base-rich materials and grassland vegetation. The name is from the Latin *mollis*, meaning soft.

Mollisols An order of *mineral soils, which are identified by a deep *mollic surface *horizon (well decomposed and finely distributed organic matter) and base-rich mineral soil below. Mollisols form mainly in grasslands areas where moisture may be seasonally deficient (e.g. the Great Plains of North America and the pampas of South America). They are among the most fertile soils in the world and now produce most of the world's cereals.

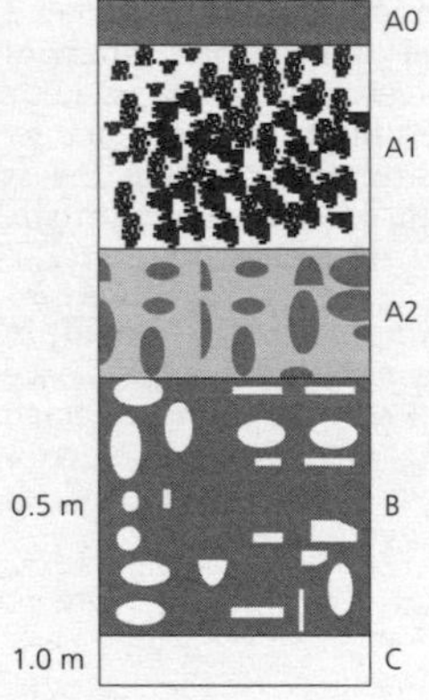

Mollisols

mollusc A member of a phylum (Mollusca) of invertebrate animals, most of them aquatic, comprising classes which are morphologically quite diverse. Molluscs are fundamentally *bilaterally symmetrical. Calcium carbonate shell material is secreted by the mantle covering the visceral hump and shells may be univalve, bivalve, plated, or, in some groups, modified to serve as internal skeletons. Most have a well differentiated head, with radula and salivary glands. A ventral muscular foot is very common. Molluscs first appeared in the Lower *Cambrian. There are six extant and three extinct classes, with more than 80 000 species.

molybdenum (Mo) An element that is required in small amounts by plants and is found largely in the enzyme nitrate reductase. A symptom of deficiency is interveinal *chlorosis.

monadnock An isolated hill or range of hills standing above the general level of a *peneplain, which results from the *erosion of the surrounding terrain. It may be located on relatively resistant rock or in a *watershed position where erosion is least. It is named after Mount Monadnock, New Hampshire, USA.

monilliform Resembling a string of beads.

monkey puzzle tree *See* DISJUNCT DISTRIBUTION.

mono- From the Greek *monos*, meaning 'alone', a prefix meaning 'single' or 'one'.

monoclimax theory *See* CLIMAX THEORY.

monoculture The growing over a large area of a single crop species (e.g. *Triticum aestivum*, bread wheat) or of a single variety of a particular species. Monocultures are especially vulnerable to pest and disease infestation, but uniformity of height, development, etc. in a crop facilitates management, especially harvesting. The economic and ecological wisdom of monoculture is widely debated.

monoecious **1.** Applied to an organism in which separate male and female organs occur on the same individual (e.g. to a plant which bears male and female reproductive structures in the same flower or separate male and female flowers on the same plant, or to a *hermaphrodite animal). Some authors restrict the term botanically to plants with separate male and female flowers; plants which bear male and female reproductive organs in the same flower are then called hermaphrodite. *See also* ANDROECIOUS and UNISEXUAL FLOWER; *compare* DIOECIOUS. **2.** Applied to a parasite (*see* PARASITISM) that utilizes only one host. *Compare* HETEROECIOUS.

monogamy The exclusive pairing of a male with a female for the purpose of *mating (i.e. neither mates with any individual other than the partner). Usually the bond operates through the breeding season and in some cases it may extend through the adult life of the two individuals.

monohybrid A cross between two individuals which are identically *heterozygous for the *alleles of one particular gene (e.g. *Aa* × *Aa*).

monomictic Applied to lakes in which only one seasonal period of free circulation occurs. In cold monomictic lakes, typical of polar latitudes, the seasonal overturn occurs briefly in summer and the water temperature never rises above 4°C, so inducing density stratification. In warm monomictic lakes, typical of warm temperate or subtropical regions, the seasonal overturn occurs in winter. At other times thermal stratification, with the formation of a distinct epilimnion, prevents free circulation through the depth of the lake.

monomorphic Applied to a population in which all individuals have the same *allele at a particular *locus. *Compare* POLYMORPHIC.

monophagous Applied to an organism (e.g. many insect larvae) that feeds on only one type of plant or prey, or on species that are closely related. *Compare* OLIGOPHAGOUS and POLYPHAGOUS.

monophyletic Applied to a group of species that share a common ancestry, being derived from a single interbreeding (or *Mendelian) population, as opposed to a polyphyletic group, which is derived from many such populations. If the members of a given taxon are descended from a common ancestor they are said to be monophyletic (e.g. the families within a class would be monophyletic if they were all descended from the same family or a lower taxonomic unit). Under the strictest definition they would all have to be descended from a single species. Monophyly is sometimes used interchangeably with *holophyly.

monopodial A type of branching in which branches arise laterally from a main, central stem, not from its apex. *Compare* SYMPODIAL.

monothetic In numerical classification schemes, the use of a single criterion or attribute as the basis for each subdivision of a sample population, as for example in association analysis. *Compare* POLYTHETIC.

monotypic Applied to any taxon that has only one immediately subordinate taxon. For example, a genus that contains only one species would be described as monotypic, as would a family containing only one genus. *Compare* POLYTYPIC.

monoxenous *See* AUTOECIOUS.

monsoon From the Arabic *mausim*, meaning 'season', a seasonal change of wind direction and properties associated with widespread temperature changes over land and water in the subtropics. Seasonal alternations of pressure systems together with shifting upper wind patterns and jet streams produce seasonal winds, called 'monsoon winds'. The climate of the Indian subcontinent is especially characterized by the monsoon, where a distinct rainy season occurs in the south-westerly monsoon. Other major areas of monsoons are eastern and south-eastern Asia, the west African coast (latitude 5–15°N) and northern Australia.

monsoon forest A term for tropical seasonal forest in Asia. There is a marked seasonal rhythm induced by the alternating wet and dry seasons produced by the two monsoons. Most of the canopy top trees lose their leaves in the dry season. Monsoon forests contain valuable hardwood species (e.g. the teak forests of Burma).

montane Pertaining to a mountain or mountains.

montane forest A forest in the *montane zone. It differs in floristic composition and ecological character from those found at lower elevations in the same latitude, and in both respects often has strong affinities with forest found in the lowlands of adjacent higher latitudes.

month degrees The excess of mean monthly temperatures above 6°C (43°F), added together and used as accumulated temperature (indicative of conditions for vegetation growth) in some climate classifications (e.g. that of A. Miller, 1951).

Montpellier school of phytosociology *See* ZURICH–MONTPELLIER SCHOOL OF PHYTOSOCIOLOGY. *Compare* UPPSALA SCHOOL OF PHYTOSOCIOLOGY.

moor An acidic area, usually high-lying and with *peat development, and most typically dominated by low-growing ericaceous shrubs (especially *Vaccinium myrtillus*, bilberry), though including grass and sedge-dominated areas. A. G. *Tansley (1939) maintained a traditional distinction between upland and lowland heaths, as opposed to heather moors, which have deeper peat development rather than any distinctive floral characteristics. *Compare* HEATHLAND.

moot *See* COPPICE STUMP.

mor A type of surface *humus *horizon that is acid in reaction, low in microbial activity except that of fungi, and composed of several layers of organic matter in different degrees of decomposition. It forms beneath conifer forest and on open heath and moorland in cool, moist climates, and is very acidic.

moraine The term originally applied to the ridges of rock debris around the margins of Alpine *glaciers. Subsequently its meaning has been widened to include *till deposits. For example, 'ground' moraine may denote an irregularly undulating surface of till, glacial *drift, or *boulder clay, or it may describe the deposit itself. An 'end' or 'terminal' moraine is a ridge of glacially deposited material laid down at the leading edge of an active glacier. Its height is in the range 1–100 m, and it is accumulated by a combination of glacial dumping and pushing. A 'recessional' moraine is morphologically similar and is laid down at the terminus of a glacier during a period of stillstand that interrupts a sustained period of retreat of the ice margin. A 'lateral' moraine is a ridge of debris at the margin of a valley glacier and largely derived from rock fall. It is a prominent feature of many contemporary Alpine glaciers. A 'medial' moraine results from the merging of lateral moraines when two glaciers converge. A 'washboard' moraine is a single ridge in a closely spaced pattern (perhaps 9–12 per kilometre) and stands some 1–3 m above the adjacent depressions. It is found in the 'end' moraine belt. A 'push' moraine is a morainic ridge made of unconsolidated rock debris and pushed up by the *snout of an advancing glacier. *See also* DE GEER MORAINE; FLUTED MORAINE; and HUMMOCKY MORAINE.

morbidity **1.** The proportion of individuals in a population suffering from a particular disease. **2.** The state or condition of being diseased.

Morgan's canon The injunction, first stated by Lloyd Morgan in 1894, that animal behaviour should never be attributed to a high mental function if the behaviour can be explained in terms of a simpler process.

morpho- From the Greek *morphe*, meaning 'form', a prefix meaning 'pertaining to form or shape'.

morphogenetic zone **1.** An area that is characterized by a distinctive assemblage of land-forms and coincides with a major climatic type. It is believed that the land-forms largely result from the action of a unique combination of surface processes controlled by climate. **2.** *See* LINEAGE ZONE.

morphological mapping A method of mapping the form of a land surface. It is based on the assumption that land surfaces can be divided into a number of components, each of which has a uniform gradient or curvature, and which are separated by abrupt or gentle changes of slope. The nature of the change of slope is shown on the map by a standard set of symbols. The angle of each component may be measured instrumentally.

morphological system In *geomorphology, a theoretical construct consisting of the relationship between the physical properties of a natural (geomorphological) system. For example, the physical dimensions of a beach (angle of slope seaward, average grain size, and porosity) may be related to each other in an orderly manner, and so constitute a morphological system, and the geometric properties of a valley-side slope are typically correlated with certain characteristics of soil and vegetation.

morphology The form and structure of individual organisms, as distinct from their anatomy (which involves dissection). *Compare* PHYSIOGNOMY.

morphometric analysis *See* DRAINAGE MORPHOMETRY.

morphospecies A group of biological organisms whose members differ from all other groups in some aspect of their form and structure (*see* MORPHOLOGY), but are so similar among themselves that they are lumped together for the purposes of analysis.

mosaic A general term for a virus disease of plants in which the symptoms include the appearance of angular areas of yellow colour on the leaves, forming a mosaic pattern.

mosaic evolution The differential rates of development of various adaptive attributes that occur within the same evolutionary lineage. For example, a particular plant taxon might show greatly different rates of change with respect to the leaves, shoots, and roots and an animal taxon with respect to the head, body, and limbs. This is a common phenomenon and makes the reconstruction of transitional fossil types very difficult.

moss animals *See* BRYOZOA.

mossy forest A tropical *montane forest, typically with contorted trees no more than 10–15 m high with dense crowns, their trunks, boughs, and twigs being festooned with mosses, *lichens, and liverworts. The growths of 'moss' are particularly luxuriant where mists prevail.

mother cloud An incipient cloud from which a well-defined cloud type can form and develop. *See also* CLOUD CLASSIFICATION.

motile Capable of independent locomotion.

motivation The cause for a spontaneous change in the behaviour of an animal which occurs independently of any outside stimulus, or of a change in the threshold of responsiveness of an animal to a stimulus, and which is not due to fatigue, *learning, or the maturation of the animal.

motivational conflict The simultaneous existence of two or more *motivations that lead to contradictory patterns of behaviour (e.g. to approach a human offering food in order to obtain the food and to flee from the human). *See* CONFLICT.

mottle A general term for a virus disease of plants in which the symptoms include the appearance of rounded or diffuse areas of yellow colour on the leaves.

mottling A patchwork of different colours in *mineral soil (usually orange or rust against a background of grey or blue) which indicates periods of anaerobic conditions.

mould **(mold)** **1.** Any fungus. **2.** Any fungus of 'mouldy' appearance, i.e. one with abundant, visible, woolly mycelium upon which dusty or powdery conidia can be seen (e.g. *Penicillium* species). **3.** *See* FOSSILIZATION.

moulting The periodic, often seasonal, shedding of hair or feathers by animals. In birds it is the process by which feathers are periodically renewed, at least once a year, sometimes twice or (rarely) three times. Feather shedding is usually a gradual process and does not affect flight or other functions, but Anatidae (ducks, geese, and swans) lose their flight feathers simultaneously, becoming temporarily flightless. The usual sequence of moult begins with the loss of primaries, followed by secondaries, tail feathers, and then body feathers. *See also* ECDYSIS.

mountain breeze *See* MOUNTAIN WIND.

mountain fynbos *See* FYNBOS.

mountain wind **(mountain breeze)** A *katabatic wind or breeze and its counterpart, the upslope anabatic wind or breeze, which occurs on warm days over heated mountain slopes.

m-RNA *See* MESSENGER-RNA.

MSY *See* OPTIMUM YIELD.

mt-DNA *See* MITOCHONDRIAL-DNA.

muck **1.** Highly decomposed organic matter in which original plant material cannot be recognized. **2.** Farmyard manure (FYM) composed of animal faeces and urine mixed with straw and highly decomposed. **3.** *See* MUCK SOIL.

muck soil An organic soil (i.e. a silty *peat) from which 35–65 per cent of its material by weight is lost on combustion. *See* HISTOSOLS.

mud-cracks *See* DESICCATION CRACKS.

mud drape A layer of mud deposited over a pre-existing morphological feature (e.g. a *bar, *ripple, or *dune).

m

mudflat An area of a coastline where fine-grained *silt or sediment and *clay is accumulating. Its development is favoured by ample sediment, by sheltered conditions, and by the trapping effect of vegetation. It is an early stage in the development of a *salt-marsh or *mangrove forest.

mudflow **1.** A heavily loaded *ephemeral stream whose viscosity increases with evaporation as it flows over a desert fan. **2.** A rapidly moving variety of *earthflow. This is a typical phenomenon of areas underlain by sensitive clays, which may liquefy and flow following a shock, perhaps initiated by sliding. **3.** *See* LAHAR.

mud mound A build-up of carbonate sediment in the form of a bank or mound, dominated by mud. The accumulation of mud occurs by its deposition in the lee of *in situ* organisms (e.g. corals or crinoids), by the sweeping of mud into banks by currents, or by the entrapment and precipitation of carbonate mud by algae and other organisms acting as baffles.

mudrock A lithified mud.

mudstone An *argillaceous or clay-bearing sedimentary rock which is not plastic and occurs as a mass, lacking any form or structure.

muffle furnace A furnace, usually heated electrically, in which heat is applied to the charge from the outside of a refractory (heat-resistant) chamber (the muffle).

mulch A loose surface *soil horizon, either natural or man-made, composed of organic or mineral materials. It protects soil and plant roots from the impact of rain, temperature change, or evaporation.

mull A type of surface *humus *horizon that is chemically neutral or alkaline in reaction, well aerated, and provides generally favourable conditions for the decomposition of organic matter. Mull humus is well decomposed and intimately mixed with mineral matter. It forms a ground surface layer in deciduous forest and is typical of *brown earths.

Müller, Fritz Johann Friedrich (1821–97) The German-born Brazilian zoologist who first recognized, in 1879, the type of mimicry named after him.

Muller, Hermann Joseph (1890–1967) An American geneticist, who was awarded the 1927 American Association for the Advancement of Science Prize and the 1946 Nobel Prize for Medicine for his discovery in 1927 that X-rays increase the rate of gene *mutation. His research into mutations and studies of genetic theory contributed greatly to the understanding of heredity and evolution. He taught at the University of Texas, moved to Germany in 1932, and from 1933–7 was senior geneticist at the Institute of Genetics of the Academy of Sciences of the USSR. Following his return to the United States, he held several posts before being appointed professor of zoology at Indiana University in 1945.

Müllerian mimicry The similarity in appearance of one species of animal to that of another, where both are distasteful to predators. Both gain from having the same warning coloration, since predators learn to avoid both species after tasting either one or the other. The phenomenon is named after Fritz *Müller, who described it in relation to insects in South America. *Compare* BATESIAN MIMICRY.

Muller's ratchet A model, proposed by H. J. *Muller in 1964, which explains the advantage enjoyed by sexually reproducing organisms over those reproducing parthenogenetically in terms of the ability to eliminate deleterious *mutations. The best-adapted *genotypes are occasionally lost from a population. If the members of that population reproduce sexually, the genotype can be recreated by crossing-over, but if reproduction is parthenogenetic it can be recreated only by favourable mutation. Favourable mutations being less common than deleterious mutations, each loss of a best-adapted genotype acts like a ratchet to increase the load of deleterious mutations until it becomes intolerable and the *clone dies. The ratchet prevents any line in an asexual population from reducing its deleterious mutation load to a level lower than that of the line with the lowest load.

multifurcation In a *phylogenetic tree, the occurrence of a split in an ancestral branch into more than two branches at an *internal node, because the order in which the progenic branches occur cannot be determined.

multigene family A group of genes derived by one or more *duplication events, originally from an ancestral gene, which show more than 50 per cent similarity.

multiple land-use strategy The designed use of an area so that a range of compatible uses, or activities that can be rendered compatible by careful management, may be practised in a single locality (for example, forestry with rambling, camping, nature conservation, flood control, etc.).

multiple substitutions The occurrence at a base site in a DNA sequence of more than one *substitution relative to another sequence in an *alignment. In such cases the true genetic distance between two sequences is greater than that which is directly observed in an alignment.

multiseriate Arranged in many rows.

multistorey sandbody A series of sandstone beds, each deposited by the infilling of a river channel, stacked one above the other with little or no intervening *mudstone. The multistorey body is formed by the repeated and rapid migration of the channel network over the *alluvial plain, so allowing little chance for the preservation of fine *flood-plain sediment.

Munsell colour A soil-colour system, devised originally in the USA, which is based on the three variables of colour—hue, value, and chroma—and given notation such as '10YR 6/4'. It is now used as an international method for reporting soil colour.

muriform Resembling a brick wall; having both transverse and longitudinal cross-walls.

murine Of or like a mouse; mouse-coloured.

Muschelkalk **1.** The seaway (Muschelkalk Sea) that extended across north-western Europe during the *Triassic, from the present-day British Isles in the west to northern Germany and Poland in the east. It was bordered to the south by the Bohemian Massif and in the north by the Baltic Shield (*see* BALTICA). **2.** The mid-Triassic, when using the European tripartite divisions of the Triassic System (Bunter, Muschelkalk, and Keuper). The German word *Muschelkalk* means 'shelly limestone'.

muscicolous Growing on or among mosses (Musci).

mushroom *Agaricus bisporus* (the cultivated mushroom) or any edible fungus similar to it in appearance. *Compare* TOADSTOOL.

mushroom rock *See* PEDESTAL ROCK.

muskeg An ill-drained, often extensive, boggy tract, with at best sparse, stunted trees, which occurs within the *boreal forest. The vegetation is dominated by **Sphagnum* species (bog moss) and *Eriophorum* species (cotton sedge).

musth In sexually mature male elephants, a physiological and behavioural condition associated with sexual activity. It lasts 2–3 months, during which testosterone levels are high; the temporal glands become enlarged and temporin secretion becomes copious; urine is discharged continually, and behaviour becomes more aggressive. Musth occurs annually or biennially, but at any time of year rather than within a particular season. Its occurrence is not synchronized among groups of males.

mutagen An agent that increases the *mutation rate within an organism. Examples of mutagens are X-rays, gamma rays, neutrons, and certain chemicals, such as carcinogens.

mutagenic Causing a *mutation.

mutant **1.** A cell or organism that carries a gene *mutation. **2.** A gene that has undergone mutation.

mutation **1.** A change in the structure or amount of the genetic material of an organism. **2.** A gene or chromosome set that has undergone a structural change. The majority of mutations are changes within individual genes, but some are gross structural changes of chromosomes or changes in the number of chromosomes per nucleus. A mutation that occurs in a body cell is called a somatic mutation: it is transmitted to all cells derived by mitosis from that cell. Mutations that occur in *gametes or in cells destined to be gametes and therefore are inherited form the raw material for *evolution by providing the source of all variation. Most mutations are deleterious; evolution progresses through the few that are favourable. Some mutations are recurrent: they occur repeatedly within a population or over long periods of time (as does haemophilia, for example).

mutational bias A pattern of *mutation in DNA that is disproportional between the four bases, such that there is a tendency for certain bases to accumulate.

mutation rate In *phylogenetics, the number of *mutations arising at a single *nucleotide site per gene per unit time.

mutatus A form of cloud evolution in which the *mother cloud is greatly affected by the development of cloud from it.

mutual antagonism A relationship in which the effect of competition between two or more species (interspecific competition) exceeds that of competition within each species (intraspecific competition).

Mutual Climatic Range **(MCR)** *See* BEETLE ANALYSIS.

mutual inhibition competition type A form of direct competition between populations of two species in which both species actively inhibit one another. This is quite distinct from indirect competition, which arises from competition for a common resource that is in short supply.

mutual interference Behavioural interactions among feeding organisms that reduce the time each individual spends obtaining food or the amount of food each individual consumes. It occurs most commonly where the amount of food is small or the number of animals feeding is high. When the searching efficiency of an individual consumer is calculated (*see* HOLLING'S DISC EQUATION) and plotted against consumer density on a graph, both logarithmically, the resulting negative slope is given a value, $-m$, and m is known as the coefficient of interference.

mutualism An interaction between members of two species which benefits both. Strictly, the term may be confined to obligatory mutualism, in which neither species can survive under natural conditions without the other. Sometimes the term is used more generally to include *facultative mutualism (*protocooperation). *See* SYMBIOSIS.

mycelium A mass of thread-like filaments (hyphae) which makes up the vegetative state of many *fungi and *actinomycetes.

mycetocyte In certain insects, a cell that contains symbiotic fungi or bacteria.

mycobiont The fungal *symbiont in a *lichen.

mycocecidium A *gall formed on a plant as a result of infection by a fungus.

mycology The study of *fungi.

mycorrhiza A close physical association between a fungus and the roots of a plant, from which both fungus and plant appear to benefit; a mycorrhizal root takes up nutrients more efficiently than does an uninfected root. A very wide range of plants can form mycorrhizas of one form or another and some plants (e.g. some orchids and some species of *Pinus*) appear incapable of normal development in the absence of their mycorrhizal fungi. *See also* ECTOMYCORRHIZA and ENDOMYCORRHIZA.

mycosis Any disease of humans or other animals in which the causal agent is a fungus.

mycotoxin A toxin produced by a fungus.

mycotrophic Applied to a plant that is associated with a fungus in a *mycorrhiza.

mycovirus A *virus that infects and replicates in fungi.

myrmecochory Dispersal of spores or seeds by ants.

myrmecophile Applied to a species which relies on ants for food or protection in order to complete its life cycle. Food is obtained from the ants by direct parasitism, predation, stealing, or scavenging.

myrmecophyte A plant that is associated with ants in a mutualistic relationship (*see* MUTUALISM).

m

N **1.** See NEWTON. **2.** *See* NITROGEN.

n *See* NANO-.

Na *See* SODIUM.

nacreous Applied to the iridescent appearance of the pearly inner surface of some molluscan shells (mother-of-pearl) in which the mineral is laid down in thin, lustrous sheets, each deposited over a thin, organic matrix.

nacreous clouds (mother-of-pearl clouds) A type of cloud seen occasionally at great altitude (approximately 22–24 km), just before sunrise or after sunset, characterized by iridescent colouring. The cloud is fine and rather lenticular.

NADW *See* NORTH ATLANTIC DEEP WATER.

nannofossil A *fossil of a member of the nanoplankton, the smallest of all marine organisms. Calcareous nannofossils, including *coccoliths and an extinct (and probably artificial) group called nannoliths, are typically smaller than about 30 μm and may consist of *spicules, juvenile foraminiferans, or *calcispheres. Silicoflagellates, which are flagellates (*see* PROTOZOA) that produce silica skeletons, also leave nannofossils. Nannofossils are used as stratigraphic indicators.

nanoplankton *See* PLANKTON.

nano- (n) From the Greek *nanos*, meaning 'dwarf', a prefix meaning 'extremely small' (e.g. *nanophyll). Attached to *SI units it denotes the unit $\times 10^{-9}$.

nanophyll In the classification of leaf sizes, a leaf that is less than 25 mm long, or 25–225 mm^2 in area.

nastic *See* NASTY.

nasty (adj. nastic) The response of a plant organ to a non-directional stimulus (e.g. the opening or closing of flowers in response to changes in light intensity or temperature). The plant may respond by changes in cell growth or changes in *turgor.

nasus *See* NASUTE.

nasute A termite soldier that is equipped with a projection on the head (nasus). It squirts at enemies an adhesive or toxic fluid produced by the frontal gland.

natal dispersal *See* DISPERSAL.

national park As defined by the *IUCN (1975), a large area of land containing ecosystems that have not been materially altered by human activities, and including plant and animal species, landscape features, and *habitats of great scientific interest, or of beauty, or recreational or educational interest. It is under the direct control of the state, and the public is allowed to visit it for inspirational, cultural, and recreational purposes. Implicitly, a national park is an area set aside in perpetuity for *conservation, within which such public recreational activity is permitted as is compatible with the primary conservation objectives. British national parks do not comply strictly with the IUCN definition, since they comprise areas that have been altered significantly during a long period of human occupation and since they are still actively farmed and occupied, much of the land being owned privately rather than publicly. Land is so designated in order to maintain long-established, cultural landscapes. Similar schemes have been developed elsewhere, but they are often given distinctive and less confusing titles, e.g. *parcs naturels et régionaux* in France, and 'greenline parks' in the USA.

native (indigenous) Applied to a species that occurs naturally in an area, and therefore one that has not been introduced by humans either accidentally or intentionally. Of plants found in a particular place, the term is applied to those species that occur naturally in (i.e. are indigenous to) the region and at the site.

native element An element which occurs in a free state as a *mineral (e.g. gold, copper, and carbon).

natric horizon A *mineral *soil horizon that is developed in a subsurface position in the *profile, that satisfies the definition of an *argillic horizon, and that also has a columnar structure and more than 15 per cent saturation of the exchangeable *cation sites by sodium. The name is from the Latin *natrium*, meaning sodium (chemical symbol Na).

natron lake A saline lake, rich in the sodium carbonate salt natron ($Na_2CO_3.10H_2O$).

natural A term that is applied to a *community of *native plants and animals. 'Future-natural' describes the community that would develop were human influences to be removed completely and permanently, but allows for possible changes in climate or site. 'Original-natural' describes a community as it existed in the past, with no modification by humans. 'Past-natural' describes the condition in which the present features are derived directly from those existing originally, with relatively little modification by humans. 'Potential-natural' describes the community that would develop were human influence removed, but the consequent *succession completed instantly (future changes in climate or site are not taken into account). 'Present-natural' describes the community that would exist now had there been no human modification. Because of the dynamic nature of any ecosystem, this condition may not be identical to the last original-natural state before human intervention began.

natural cast *See* FOSSILIZATION.

natural classification The ordering of organisms into groups on the basis of their evolutionary relationships. *Compare* ARTIFICIAL CLASSIFICATION.

natural gas Gaseous hydrocarbons, chiefly methane (CH_4), ethane (C_2H_6), propane (C_3H_8), and butane (C_4H_{10}), trapped in pore spaces in rocks with or without liquid petroleum. It has a high heat value, burns without smoke or soot, and provides raw material for the chemical industry for making plastics, detergents, fertilizers, etc. Gas of this composition is also termed 'natural gas' if it occurs as an industrial by-product.

natural immunity Immunity that has been inherited, in contrast to one that has been acquired. If an individual acquires immunity by having the disease in question, as opposed to acquiring it as the result of vaccination, then that individual is said to have natural acquired immunity.

naturalized Applied to a species that was originally imported from another country but now behaves like a *native in that it maintains itself without further human intervention and has invaded native communities.

natural remanent magnetism (NRM) The *remanent magnetization of rocks and naturally occurring objects which has been acquired by normal processes. In *igneous rocks and fired archaeological materials, NRM is normally of thermal or chemical origin; in sediments it is usually of depositional origin; and in sedimentary rocks it is usually of chemical origin.

natural selection ('survival of the fittest') A complex process in which the total environment determines which members of a species survive to reproduce and so pass on their *genes to the next generation. This need not involve a struggle between organisms. Natural selection is not necessarily the only mechanism for evolution (*see* NEUTRALITY THEORY OF EVOLUTION).

natural turnover rate The normal rate of transfer of an element through the active part of a biogeochemical cycle. *Compare* MAN-INDUCED TURNOVER. *See also* TURNOVER.

natural woodland A woodland comprising trees that have not been planted by humans, and where no human interference has occurred.

nature and nurture Synonyms for heredity and environment as they affect a character. Both may affect observed variation among individuals; only variations due to 'nature' are inherited, and it is these that form the subject of quantitative genetics. *See* ENVIRONMENTALIST.

nature conservation *See* BIOLOGICAL CONSERVATION and CONSERVATION.

nature reserve An area of land set aside for nature *conservation and associated scientific research, usually with strong legal protection against other uses. Public access may be restricted partially or completely. The precise status and definition of reserves varies from one country to another. The relative merits of the many criteria that govern the selection of nature reserves have been discussed extensively in recent ecological literature. In Britain, a key objective is to maintain a fully representative range of *habitats and their flora and fauna; this is done by means of national nature reserves (NNR), controlled by statutory bodies, complemented by local nature reserves (LNR), of local rather than national or international interest and importance, managed by various agencies, especially by county naturalists' trusts, on behalf of local government. The Royal Society for the Protection of Birds (RSPB) also manages many reserves.

nautical mile The unit of length that is used in ocean and air navigation, equivalent to one minute (1/60°) change in latitude. It is internationally defined as being equal to 1852 metres.

navigation The *orientation of itself by an animal towards a destination, regardless of its direction, by means other than the recognition of landmarks. *Compare* COMPASS ORIENTATION and PILOTAGE.

neap tide The tide of small range that occurs every 14 days, near the times of the first and last quarter of the Moon, when the Moon, Earth, and Sun are at right angles. The neap tidal range is 10–30 per cent less than the mean tidal range.

Nearctic faunal region The fauna of North America, south to Mexico. At the order and family level the fauna is essentially the same as that of the *Palaearctic faunal region, but some genera and more especially species are distinctive to the Nearctic. Because of their strong affinities, the Nearctic and Palaearctic regions are often grouped into one unit, *Holarctica, reflecting their former connection via the Bering *land bridge.

nearest-neighbour analysis (nearest-neighbour measure) A method for testing the pattern of distribution of individuals. The mean nearest-neighbour distance for all, or for a *random sample of, individuals in a given area is compared with the expected mean distance if the same individuals were randomly distributed throughout the area. The ratio of the two mean values (R) is a measure of the departure from randomness, and may be tested for statistical significance. The value of $R = 1$ indicates randomness; $R = 0$ indicates maximum aggregation; and $R = 2.149$ indicates maximum possible spacing (i.e. a hexagonal pattern).

nearest-neighbour measure *See* NEAREST-NEIGHBOUR ANALYSIS.

nearest-neighbour sampling method A method of plotless sampling in which the distance is measured from the first individual (the nearest to the *random sampling point) to its nearest neighbour. This permits the calculation of the density of individuals, or of its reciprocal, the mean area per individual.

near-natural community A vegetation *community consisting wholly or predominantly of native species which, though modified by human use, has not been deflected from its natural course of *succession. Examples include traditional wood-pasture (i.e. an area in which some trees have been removed, but none planted) and *coppiced woodland. *Compare* SEMI-NATURAL COMMUNITY.

near-shore current system A system of currents that is caused by wave activity within and adjacent to the *breaker zone. The current system includes: the shoreward mass transport of water; *longshore currents; and seaward-moving *rip currents. This wave-induced current system often has a reversing tidal-current system superimposed on it.

nebulosus The Latin *nebulosus*, meaning 'fog-covered', used to describe a form of *stratus or *cirrostratus cloud or cloud-sheet which is rather indistinct and seen as a nebulous layer. *See also* CLOUD CLASSIFICATION.

necrology The scientific study of all

the processes affecting dead animal and plant material, including decomposition and *fossilization.

necromass The mass of dead plant material lying as *litter on the ground surface.

necrophoresis The removal of dead individuals by carrying them away from places inhabited by living individuals.

necrosis The death of a circumscribed piece of tissue. Necrotic wounds in mammals are produced, for example, by the bite of *Loxosceles reclusa* (brown recluse spider).

necrotrophic Applied to a parasitic organism (*see* PARASITISM) that obtains its nutrients from dead cells and tissues of its host organism.

needle ice *See* PIPKRAKE.

negative feedback In a system, the mechanism by which a process is limited internally, i.e. without reference to factors outside the system. Ecologically such mechanisms favour the maintenance of equilibrium in organisms, populations, and ecosystems. *Compare* POSITIVE FEEDBACK.

nekron mud *See* GYTTJA.

nekton Free-swimming organisms in aquatic *ecosystems (e.g. fish, amphibians, and large swimming insects). Unlike *plankton, they are able to navigate at will.

nematocyst In Cnidaria (sea anemones, jellyfish, and corals), part of an epidermal cell (cnidoblast), comprising a sac with a lid, the open end of the sac being extended into a long, hollow thread. The thread is stored inside the sac, coiled and immersed in a fluid. When a sensory bristle (cnidocil) is stimulated, the thread is expelled to capture or hold prey or as a weapon, depending on the type of nematocyst. Some nematocysts are barbed and some inject venom.

nematophagous fungi Fungi that are *parasitic or predatory on nematode worms; most can also grow *saprotrophically. Nematophagous fungi belong to various taxonomic groups and are found, for example, in soil and in decaying plant matter.

neo- From the Greek *neos*, meaning 'new', a prefix meaning 'new', 'modern', or 'revived' (e.g. neo-Darwinism, a modern version of Darwinism).

neocatastrophist *See* CATASTROPHISM.

neo-Darwinism *See* DARWIN and MODERN SYNTHESIS.

neoendemic *See* ENDEMISM.

Neogea A neotropical zoogeographical region that comprises Central and South America. *See also* FAUNAL REGION. *Compare* ARCTOGEA and NOTOGEA.

Neogene The later of the two periods comprising the *Tertiary sub-Era, preceded by the *Palaeogene, followed by the *Pleistogene, and dated at 23.03–1.806 Ma. The Neogene Period is subdivided into the *Miocene and *Pliocene Epochs.

neo-Lamarckism Any modern variant of the theory of evolution by the inheritance of acquired characteristics which was proposed by *Lamarck. For example, in 1980 E. J. Steele proposed that what in effect is Lamarckian evolution may occur by the insertion of new genetic material into the host genome by a *retrovirus.

neoteny A form of *heterochrony that involves the slowing down in a descendant of part or all of its ancestor's rate of development, so that at least some aspects of the descendant resemble a (generally large-sized) juvenile stage of the ancestor. This may lead to *paedomorphosis. Since the juvenile stages of many organisms are less specialized than the corresponding adult stages, such shifts allow the organisms concerned to switch to new evolutionary pathways (*see* EVOLUTION). The word comes from the Greek *neos* (meaning 'youthful'). Neoteny is common among Urodela (newts and salamanders) and some features of human evolution (e.g. lack of body hair) have been ascribed to it.

Neotethys *See* TETHYS SEA.

Neotropical faunal region The region which includes South and Central America, including southern Mexico, the West Indies, and the Galápagos Islands.

Much of the area was isolated for the greater part of the *Tertiary Period, which explains the distinctiveness of the fauna and the survival of ancient forms of mammals (e.g. the marsupials and edentates).

Neotropical floral kingdom (Neotropical floral realm) One of the major floral subdivisions, covering Central and South America, with the exception of the southern tip, which is included in the Antarctic kingdom. The floristic distinctiveness results from the isolation of South America from other land masses for much of *Cenozoic time. *See also* FLORAL PROVINCE and FLORISTIC REGION.

Neotropical floral realm *See* NEOTROPICAL FLORAL KINGDOM.

neotype In taxonomy, the specimen that is chosen to act as the 'type' material subsequent to a published original description. This occurs in cases where the original types have been lost, or where they have been suppressed by the *ICZN.

NEP *See* PRIMARY PRODUCTIVITY.

nephanalysis The interpretation of cloud type and amount from satellite pictures in facsimile or digitized form.

nepheloid layer A body of water with a high concentration of suspended sediment, which occurs near the deep ocean bottom, close to the base of the *continental slope. Nepheloid layers are usually 1–300 m thick, and carry sediment up to 12 μm in size, in concentrations of 0.3–0.01 mg/l. The sediment is suspended and transported by the movement of the oceanic *thermohaline circulation. *See* CONTOUR CURRENT.

neritic province *See* NERITIC ZONE.

neritic zone (neritic province) The shallow-water, or near-shore, marine zone extending from the low-tide level to a depth of 200 m. This zone covers about 8 per cent of the total ocean floor and is the area most populated by benthic organisms (*see* BENTHOS), owing to the penetration of sunlight to these shallow depths.

nestling 1. A young bird, before it leaves the nest. **2.** A mode of life adopted by certain bivalve *molluscs (Bivalvia) which live in crevices and depressions in hard substrates, which have not been excavated by the bivalve itself.

nest parasite 1. A parasite (*see* PARASITISM) that lives in the nest of its host. **2.** *See* BROOD PARASITE.

net ecosystem productivity *See* PRIMARY PRODUCTIVITY.

net plankton *See* PLANKTON.

net primary productivity (NPP or P_n) *See* PRIMARY PRODUCTIVITY.

neurotoxin A *toxin that affects the functioning of the nervous system.

neuston A collective term for organisms that are resting or swimming on the surface of an aquatic *ecosystem. *Compare* EPINEUSTON and HYPONEUSTON.

neutralism A situation in which two species populations coexist, with neither population being affected by association with the other.

neutrality theory of evolution (neutral mutation theory) A theory proposed in 1983 by the Japanese geneticist Motoo *Kimura, which asserts that many genetic *mutations are adaptively equivalent (effectively neutral), and do not affect significantly the fitness of the carrier. Thus they can become fixed in the *genome at a random rate. Changes in their frequencies are more often the result of chance than of *natural selection. The theory applies only to protein evolution and does not deny the role of natural selection in shaping morphological and behavioural attributes, but it is critical of the role of selection in maintaining *polymorphism, which it regards as a transient phase of *molecular evolution resulting from a balance between mutational input and random extinction by drift. Kimura made important contributions to the understanding of evolutionary change by constructing mathematical models based on it.

neutral mutation A mutation which has no effect on the organism's fitness.

neutral mutation theory *See* NEUTRALITY THEORY OF EVOLUTION.

neutral soil Soil with a *pH value of 6.6–7.3.

neutral theory of biogeography An approach to *biogeography, developed by Stephen Hubbell, a staff scientist at the Smithsonian Tropical Research Institute, that regards *communities as random collections of species cast together by the vagaries of history, dispersal, and chance. The theory of *island biogeography is an example of this approach, since it is based upon the assumption that all species within the system are equivalent to one another.

neutrophilous Preferring a habitat that is neither acid nor alkaline.

névé *See* FIRN.

New Caledonian floral region Part of R. Good's (1974, *The Geography of the Flowering Plants*) *Polynesian floral subkingdom, within his *Palaeotropical floral kingdom, which also includes Lord Howe and Norfolk Islands. It has a remarkably rich flora, with a high degree of *endemism. Some of the endemic genera, of which there are well over 100, constitute separate families (or nearly so). *See also* FLORAL PROVINCE and FLORISTIC REGION.

Newcastle disease An acute, infectious disease which affects birds, including domestic fowl. Although not necessarily fatal, the disease is economically important since affected birds are less productive. The disease affects mainly the respiratory tract, but the nervous system may become involved. Mortality rates vary. It is caused by a paramyxovirus.

Newer Drift The deposits that mark the maximum extent of the last (*Devensian) glaciation. In the British Isles, a morphological distinction occurs between the more weathered *drift to the south and the less weathered drift to the north. The boundaries between the less weathered Newer Drift and the *Older Drift have been shown to be less distinct than was once supposed.

newton **(N)** The derived *SI unit of force, named after Sir Isaac Newton (1642–1727), being the force required to produce an acceleration of 1 m/s^2 in a mass of 1 kg; 1 N = 1 J/m.

New Zealand floral region Part of R. Good's (1974, *The Geography of the Flowering Plants*) Antarctic kingdom. Approximately 30 genera are known to be endemic (*see* ENDEMISM) to it, but the surprisingly large number of endemic species (about 75 per cent of the total) belong very largely to non-endemic genera. *See also* FLORAL PROVINCE and FLORISTIC REGION.

niche **(ecological niche)** The functional position of an organism in its environment, comprising the *habitat in which the organism lives, the periods of time during which it occurs and is active there, and the resources it obtains there. In other words, its niche is the role that a species plays in a community.

nictonasty **(sleep movement)** A diurnal, nastic (*see* NASTY) movement (e.g. the opening of certain flowers in the day and their closing at night).

nidicolous Applied to a young bird that stays in the nest until it is able to fly. Nidicolous young are usually naked and incapable of locomotion.

nidifugous Applied to a young bird that leaves the nest on hatching or soon after. Nidifugous young are down-covered and capable of locomotion.

nimbostratus From the Latin *nimbus*, meaning 'cloud', and *stratus*, 'spread out', a dark or grey cloud that obscures the Sun, associated with more or less continuous rainfall, which makes the cloud base diffuse, *See also* CLOUD CLASSIFICATION.

Nitisols Soils that have a *nitric horizon more than 39 cm deep, with a *cation-exchange capacity of less than 36 $cmol_c/kg$ and no evidence of clay *lessivage within 100 cm of the surface.

nitrification The oxidation of ammonia to nitrite, and/or nitrite to nitrate, by *chemo-lithotrophic bacteria of the family Nitrobacteraceae.

nitric horizon A clay-rich *soil horizon in which more than 30 per cent of the clay consists of 1 : 1 clay minerals. These are sheet silicates in which one layer of SiO is coupled to a layer of brucite or gibbsite; the group includes serpentine, kaolinite, and kandite clays.

nitrogen **(N)** An element that is essential to all plant and animal life. It is found reduced and covalently bound in many organic compounds, and its chemical properties are especially important in the structures of proteins and nucleic acids. Nitrogen-deficient plants are chlorotic (*see* CHLOROSIS) and may be etiolated (*see* ETIOLATION), with the older parts becoming affected first.

nitrogen cycle A description of the balance, changes, and nature of the nitrogen-containing compounds circulating between the atmosphere, the soil, and living matter. For plants, *nitrogen fixation by soil bacteria, which renders nitrogen readily available for assimilation by the plants, is an essential and crucial part of the nitrogen cycle.

nitrogen fixation The reduction of gaseous molecular nitrogen and its incorporation into nitrogenous compounds. In nature this occurs during thunderstorms by means of the electrical energy released as lightning, by photochemical fixation in the atmosphere, and by the action of nitrogen-fixing micro-organisms. Free-living nitrogen-fixing soil and aquatic bacteria include *Azotobacter* species, *Bacillus* species, *Clostridium* species, and cyanobacteria (e.g. *Nostoc*). *Symbiotic nitrogen-fixing bacteria include *Rhizobium* and *Bradyrhizobium* species, which form the characteristic *root nodules of leguminous plants. The bacteria supply the legume with ammonia and receive carbohydrate from the legume. Certain non-leguminous plants, e.g. *Alnus* species (alder), *Myrica* species (bog myrtle), and *Casuarina*, typically plants of poorly drained or nutrient-depleted habitats, form symbiotic associations with nitrogen-fixing Actinobacteria (e.g. *Frankia*). The other main type of symbiotic nitrogen-fixing association occurs in certain *lichens, in which a nitrogen-fixing cyanobacterium may be the main phycobiont or may occur in specialized cephalodia.

nitrosamine One of a range of compounds formed by the reaction of nitrous acid (HNO_2) with a secondary or tertiary *amine. Most nitrosamines are carcinogenic in experimental mammals and fears have been expressed that nitrosamines forming in the human stomach from nitrous acid derived from water contaminated by fertilizer nitrate may cause stomach cancer. This effect has been found only where nitrate intake is extremely high.

nival Applied to those geomorphological processes that result from the action of snow.

nivation The complex of surface erosional processes that act beneath a cover of snow, including *frost wedging and the removal of shattered debris by *solifluction and the movement of melted snow. Nivation is an initial process in *cirque development.

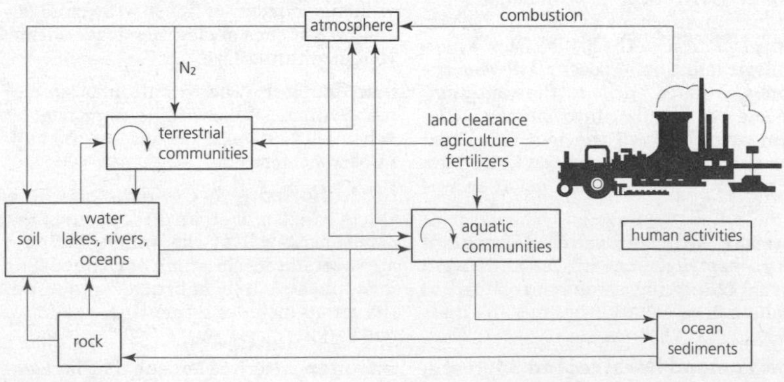

Nitrogen cycle

noble rot The rot of white grapes which is caused by the fungus *Botrytis cinerea*; infected fruit is used to make high-quality, sweet wines.

noctilucent clouds (luminous night clouds) Clouds, occurring at 80–85 km altitude, near the upper limit of the *stratosphere, which are blue to yellow in colour and similar in appearance to *cirrostratus. The clouds are seen on summer nights at latitudes between about 50° and 65° in both hemispheres; they move rapidly, at speeds up to 300 knots (556 km/h), often in a wave formation.

nocturnal radiation Long-wave radiation from the surface of the Earth at night, which exceeds the amount of incoming radiation from the atmosphere. *See also* ATMOSPHERIC 'WINDOW'; RADIATION BUDGET; and TERRESTRIAL RADIATION.

nodal analysis *See* INVERSE ANALYSIS.

node In a *phylogenetic tree, a representation of an extant (terminal node) or ancestral (internal node) operational taxonomic unit.

nodum In plant ecology, a characteristic vegetation unit. The term was first applied by M. E. D. Poore (1956) to similar sites grouped together, using modified, traditional, *phytosociological techniques.

nomen abortivum In taxonomy, a name which contravened the Code in operation at the time.

nomen ambiguum In taxonomy, a name which is ambiguous, because different authors apply it to different taxa.

nomen conservandum In taxonomy, a name, otherwise unacceptable under the rules of nomenclature, which is made valid using specified procedures, with either the original or altered spelling.

nomen correctum In taxonomy, a name whose spelling is required or allowed to be intentionally altered under the rules of nomenclature but which does not have to be transferred from one *taxon to another.

nomen dubium In taxonomy, a name which cannot be attached certainly to any particular taxon, and is therefore dubious.

nomen illegitimum In taxonomy, a name which must be rejected under the rules of the Code and is therefore illegitimate.

nomen imperfectum In taxonomy, a name that, as originally published, meets all the mandatory requirements of the rules of nomenclature, but which contains a defect needing correction (e.g. incorrect original spelling).

nomen invalidum In taxonomy, a name that has not been published properly, or is unavailable, and which is therefore invalid.

nomen inviolatum In taxonomy, a name that, as originally published, meets all the mandatory requirements of the rules of nomenclature and is not subject to any sort of alteration.

nomen neglectum In taxonomy, a name that was published at some time in the past, but which has subsequently been overlooked.

nomen novem In taxonomy, a new name that is proposed as a replacement or substitute for an existing name.

nomen nudum In taxonomy, a name that, as originally published, fails to meet all of the mandatory requirements of the rules of nomenclature and thus has no status in the nomenclature, even if corrected.

nomen oblitum In taxonomy, a forgotten name; i.e. the name of a senior *synonym that has not been used in the zoological literature for at least 50 years. Such names are not to be used unless permission is sought first from the *ICZN.

nomen perfectum In taxonomy, a name that, as originally published, meets all of the requirements of the rules of nomenclature, needing no correction of any kind, but which nevertheless is validly alterable by a change of ending.

nomen substitutum In taxonomy, a new, replacement name, published as a substitute for an invalid one (e.g. a junior *synonym).

nomen translatum In taxonomy, a name that is derived by the valid change of a previously published name as a result of a transfer from one taxonomic level to another within the group to which it belongs.

nomen triviale In taxonomy, a species (or trivial) name.

nomogenesis An evolutionary model holding that the direction of evolution operates to some degree by sets of rules or laws, independently of natural selection. For a long time it was regarded as an outmoded hypothesis but it has recently been maintained that it corresponds rather well with observations of evolution in the fossil record and that such mechanisms as *heterochrony and *molecular drive would produce nomogenetic effects. *See also* ARISTOGENESIS; ENTELECHY; and ORTHOGENESIS.

nondegenerate site A *nucleotide site within an *open reading frame of a gene at which any *substitution will result in an *amino acid change.

non-frontal depression A low-pressure system that does not develop from a frontal wave (as do typical mid-latitude frontal cyclones or *depressions). Most tropical cyclones are non-frontal. Various conditions can lead to the formation of depressions without frontal characteristics. *See also* COLD LOW; LEE DEPRESSION; POLAR-AIR DEPRESSION; and THERMAL LOW.

nonfunctionalization The production of a *pseudogene from a functional gene due to a *mutation event.

non-hierarchical classification method *See* HIERARCHICAL and NON-HIERARCHICAL CLASSIFICATION METHODS.

non-parametric test *See* STATISTICAL METHOD.

non-renewable resource (finite resource) A resource that is concentrated or formed at a rate very much slower than its rate of consumption. *Compare* RENEWABLE RESOURCE.

nonsense codon A *codon which causes the termination of translation.

nonsense mutation A *mutation which results in a *nonsense codon.

nonsynonymous substitution A nucleotide *substitution in a *codon which results in a change of the *amino acid encoded.

normal analysis Groupings based on analysis of the attributes describing individuals (e.g. in plant ecology, the grouping of sample sites by analysis of their species composition). The term is particularly applied to numerical vegetation classification, and sometimes implies a plot, as distinct from species, classification. *See also* R TECHNIQUE. *Compare* INVERSE ANALYSIS and Q TECHNIQUE.

normal distribution *See* GAUSSIAN DISTRIBUTION.

normalizing selection *See* STABILIZING SELECTION.

norte (norther) A cold, northerly, local wind that affects the coasts of the Gulf of Mexico, most commonly in winter. Sometimes it brings rainfall.

North American lowland coniferous forest In contrast to the situation in Eurasia, North America once possessed great tracts of lowland coniferous forest to the south of the *boreal conifer zone. They comprised several formations, namely the lake forest, Pacific Coast forest, and the pine barrens of the north-eastern and south-eastern USA. The lake forest has been all but destroyed.

North Atlantic deep water (NADW) A water mass (salinity 34.9–35.03 parts per thousand, temperature 1.0–2.5°C) that was originally believed to form in an area off the southern tip of Greenland, where winter cooling of saline waters was thought to cause a body of water to sink and spread south. It is now recognized that the main source is in the Norwegian Sea, from which deep water flows over the sills between Scotland, Iceland, and Greenland and cascades into the depths of the Atlantic.

North Atlantic Drift An oceanic surface current in the North Atlantic, which flows from the Grand Banks off Newfoundland eastwards to north-western Europe, forming a northerly extension of the *Gulf Stream. This diffuse, shallow and relatively warm current has an ameliorating influence on the climate of the coastal regions of north-western Europe.

north-east African highland-and-steppe floral region Part of R. Good's (1974, *The Geography of the Flowering Plants*) African subkingdom, within his *Palaeotropical floral kingdom. It contains about 50 endemic (*see* ENDEMISM) genera, although the island of Socotra accounts for nearly half of these despite the fact that it is not notably isolated. *Coffea arabica* is native to this region. *See also* FLORAL PROVINCE and FLORISTIC REGION.

norther *See* NORTE.

North Pacific Current An oceanic surface current in the North Pacific which flows eastwards, as an extension of the *Kuroshio Current, towards California. It occupies a position in the North Pacific similar to that in the Atlantic of the system comprising the *Gulf Stream and *North Atlantic Drift.

nor'wester A convective storm, usually from the north-west, which affects Assam and Bengal between March and May. It is characterized by violent conditions, including a *line squall.

Nothofagus *See* BICENTRIC DISTRIBUTION.

Notogea The Australasian *faunal region, which possesses a very distinctive fauna. It comprises Australia, New Guinea, Tasmania, New Zealand, and the islands to the south and east of *Wallace's line. It can be subdivided into the Australian, Polynesian, and Hawaiian regions. *Compare* ARCTOGEA and NEOGEA.

notophyll In the classification of leaf sizes, a leaf that is 75–125 mm long, or 2 025–4 500 mm^2 in area.

NPP **(P_n)** Net primary production. *See* PRIMARY PRODUCTIVITY.

NRM *See* NATURAL REMANENT MAGNETISM.

nuchal Of, or pertaining to, the nape of the neck.

nuclear waste *See* RADIOACTIVE WASTE.

nucleic acid *Nucleotide polymers, with high relative molecular mass, produced by living cells and found in both the nucleus and cytoplasm of cells. They occur in two forms, designated DNA and RNA, and may be double- or single-stranded. DNA embodies the genetic code of a cell or organelle, while various forms of RNA function in the transcriptional and translational aspects of protein synthesis.

nuclide A widely used alternative name for an atom. The composition of any nuclide can be given by means of the chemical symbol of the element, with the mass number written as a superscript, and the atomic number written as a subscript (e.g. ${}^{14}_{6}C$ is an atom of carbon (C) having six protons (the atomic number) and 14 nucleons (the mass number)).

nucleotide The basic monomer that makes up a DNA molecule, which is composed of a nitrogen base, a pentose sugar, and a phosphate group.

nucleotide diversity A measure of *polymorphism, expressed as the mean number of nucleotide *substitutions per site between any two randomly selected DNA sequences in a population.

nucleotide substitution *See* SUBSTITUTION.

nudation The initiation of a new plant *succession by a major environmental disturbance (e.g. a volcanic eruption).

Nukumaruan *See* CASTLECLIFFIAN.

null hypothesis *See* HYPOTHESIS.

null point **1.** In near-shore, shoaling waters, the (hypothetical) place at which there is no net movement of a particle of sediment landward or seaward, because there is a balance between the component of gravity acting down the slope in a seaward direction and the force on the particle resulting from the difference between the crest and trough velocities of the waves which tends to move the particle landwards. **2.** In an *estuary, the point at which the residual landward flow of sea water is balanced by the residual seaward flow of river water.

numerical method **(numerical taxonomy)** Any numerical description of an individual or community, or comparisons of these (such as similarity indices), or data structuring methods (such as principal components analysis and classification). The essential point is that numerical methods make no assumptions about the data (e.g. the sampling approach used, the distribution of the data, or the probability of particular patterns or frequencies). In contrast, statistical methods do make certain assumptions about the form of the data, on which the validity of significance testing depends. *Compare* STATISTICAL METHOD.

numerical taxonomy *See* NUMERICAL METHOD.

nunatak A rocky summit or mountain range that stands above a surrounding *ice sheet in an area that is currently being glaciated. It has been proposed that the presence of nunataks during a glacial episode provides a means of periglacial survival by plants and animals.

nurse tree A fast-growing tree (e.g. birch or pine), which is planted at the same time as a slower-growing and ultimately more valuable tree (e.g. oak). By growing faster, the nurse tree shelters and protects the other seedling while it is young and is removed when the young tree is established.

nutrient cycle **(mineral cycle)** A *biogeochemical cycle, in which inorganic nutrients move through the soil, living organisms, air, and water, or through some of these (e.g. nitrogen as NO_3 and NH_4, etc.). The use of the term 'mineral' can cause confusion, because, although some nutrients (e.g. potassium, magnesium, etc.) are derived originally from *minerals in the strict sense, minerals are involved in the cycle only as sources of replenishment.

$^{18}O:^{16}O$ *See* OXYGEN-ISOTOPE RATIO.

oak-apple gall **(King Charles's apple)** A *gall formed in *Quercus robur* by the unisexual generation of the wasp *Biorrhiza pallida*. Wingless females arise from root galls and climb the trunk of the tree in spring to lay many eggs in the bases of leaf buds. The eggs cause the formation of multilocular, pale pink, spongy galls that resemble small apples. The galls mature by midsummer and give rise to the sexual insect generation whose members emerge and mate in June and July. Mated females penetrate the soil around the tree and lay eggs on rootlets, giving rise to small, brown, rounded galls whose occupants emerge at the end of their second winter. May 29 is Oak Apple Day and commemorates the restoration of the English monarchy in 1660 and the birthday of Charles II; according to legend, the king once hid from his pursuers by climbing an oak tree.

oak-marble gall A dark brown, hard, spherical *gall on *Quercus robur* formed by the sexual generation of the gall wasp *Andricus kollari*. The galls contain members of the unisexual generation (all females) which emerge in spring to form 'ant-pupae' galls on the axillary buds of *Quercus cerris* (Turkey oak). The insect was introduced to Britain in 1830 for use in the dyeing and ink industry, the galls containing up to 17 per cent dry weight of tannic acid.

oak-nut A general common name given to any *gall that forms on an oak tree.

oak-spangle gall A generic name for the unisexual stage *galls found on the leaves of *Quercus robur* and produced by four species of wasps of the genus *Neuroterus*. The eggs laid by mated females give rise to generally small (1 mm diameter), flat, red galls. Gall densities may be high but competition seems to be avoided by niche separation within a single leaf and host tree, some species preferring leaf margins or apices, others the peripheral foliage. The galls overwinter in the leaf litter, often having become detached from the leaves.

oak wilt A disease of oak trees (*Quercus* species) that is caused by the fungus *Ceratocystis fagacearum*. The fungus is carried by insects, particularly sap-sucking beetles.

oasis **1.** In an arid region, an isolated area that supports water-loving plants throughout the year. Oases most commonly occur in depressions, where the water-table lies close enough to the surface to be within the reach of plant roots. The *groundwater supplying the vegetation is often enriched with salts and the vegetation can be zoned according to the concentration and type of salts present. **2.** By analogy, a small area supporting vegetation of one type (e.g. woodland) which is surrounded on all sides by a much larger area supporting vegetation of a quite different type (e.g. an arable field).

Obik Sea *See* URAL SEA.

obligate parasite *See* PARASITISM.

obsequent Applied to a land-form whose orientation is opposite to that which might have been expected (e.g. an obsequent stream flows against the dip of underlying strata and an obsequent fault-line *scarp faces the opposite way to the original fault scarp).

observation well A well that is used to observe changes in *groundwater levels over a period, or more specifically during a pumping test. Pumping does not normally take place from observation wells, which are often relatively small in diameter.

Occam's razor **(Ockham's razor)** The axiom, proposed by William of Occam (William Ockham, *c*.1280–1349), that *pluralitas non est ponenda sine necessitate* ('multiplicity ought not to be posited without necessity'); i.e. when alternative hypotheses exist, the one requiring the fewest assumptions should be preferred.

occluded front The composite front formed when a *warm front and the sector of warm air behind it are lifted and overtaken by a *cold front. *See also* OCCLUSION.

occlusion The stage in the development of a frontal *depression during which the warm-sector air is gradually lifted from the ground and above the colder air. The word is used to describe an *occluded front. An occlusion may be a 'warm occlusion', such that warmer air follows the frontal system, or a 'cold occlusion', such that colder air follows in the rear. A 'warm occlusion' has characteristics similar to those of a warm front, but there is an upper cold front ahead and aloft, which brings showers from *cumulus and *cumulonimbus cloud. A 'cold occlusion' exhibits cold-front features, but is preceded by warm-front cloud types. Occlusions formed in the later stages in the development of travelling depressions commonly cross north-western Europe, especially the 'warm' type in winter and the 'cold' type in summer.

occult precipitation *Precipitation arriving at a location by processes that would normally go unrecorded by a standard rain gauge, e.g. the condensation of *mist and *fog on foliage.

ocean The salt-water mass that occupies more than two-thirds of the surface of the Earth (70.8 per cent). The oceans contain 1 370 × 10^6 km^3 of water; the average depth is 3 730 m.

ocean-basin floor The ocean floor in those parts of the oceans that are more than 2 000 m deep. It occupies approximately one-third of the Atlantic and Indian Ocean floors, and three-quarters of the Pacific Ocean floor.

ocean current A large-scale water movement in an ocean, arising from three main causes: (*a*) wind stresses acting on the surface of the sea; (*b*) tidal motion caused by the variable attractions of the Sun and Moon; and (*c*) density differences in sea water, caused by differential heating and cooling, *salinity differences, or variations in the suspended-sediment concentration of sea-water masses.

ocean divisions The ecological regions into which the oceans are divided horizontally and vertically. Open water comprises the *pelagic zone, the water adjacent to the ocean floor the *benthic zone. Shallow water close to the shore comprises the *littoral zone and open water overlying the *continental shelf the *neritic zone. The *oceanic zone includes all other open water. Pelagic waters are divided according to depth into the *epipelagic, *mesopelagic, *bathypelagic, *abyssalpelagic, and *hadalpelagic zones. The benthic regions are divided by depth into the *bathyal, *abyssal, and *hadal zones.

ocean gyre *See* GYRE.

oceanic Applied to the regions of the sea that lie beyond the *continental shelf, with depths greater than 200 m.

oceanicity The effects of maritime influences on a climate. *Compare* CONTINENTALITY.

oceanic plateau An extensive, topographically high area of an ocean floor that rises to within 2–3 km of the sea surface above the abyssal floor (e.g. the plateau on which Iceland stands, the Galápagos Islands platform, and the Azores platform). Many Pacific plateaux (e.g. the Magellan Rise and Ontong Java Plateau) have a thick covering of carbonate oozes overlying volcanic rocks. The origin of such plateaux is volcanic, but many are now inactive.

oceanic trench *See* TRENCH.

oceanic zone The waters of the open ocean, extending from the edge of the *continental shelf. *See* OCEAN DIVISIONS.

ocean wave A disturbance of the ocean's surface, seen as an alternate rise and fall of the surface. Ocean waves are of several types: (*a*) wind-generated waves (e.g. sea waves with a chaotic wave pattern) and swell (long-period waves); (*b*) catastrophic waves (e.g. *tsunamis, landslide surges, and storm surges); and (*c*) internal waves (subsurface waves at the boundary between two water layers).

ochric horizon A light-coloured, mineral *soil horizon, usually at the soil

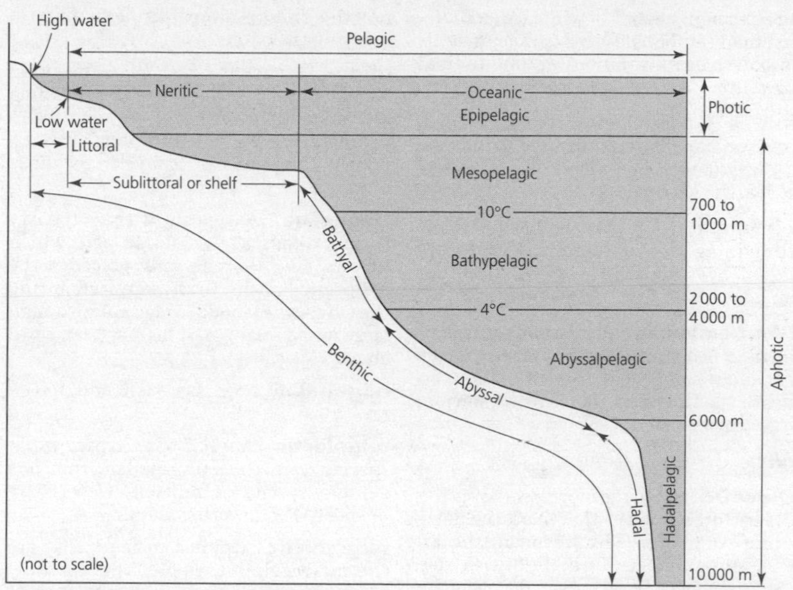

Ocean divisions

surface, and characteristic of arid- environment soils. The name is from the Greek *ochros*, meaning pale.

Ockham's razor *See* OCCAM'S RAZOR.

Odderade The third *interstadial within the *Weichselian, which occurred between 70 000 years BP and 60 000 years BP, between the *Brørup and *Moershoofd interstadials. The vegetational evidence indicates *tundra conditions with some temperate boreal elements having been present. After the Odderade interstadial polar-desert conditions prevailed until the Moershoofd interstadial.

Odintsovo An *interstadial in the Saale Drift of European Russia, marked by a temperate, broad-leaved, forest flora.

Odum, Eugene P. (1913–2002) An American ecologist who worked mainly at the University of Georgia. He pioneered the study of *ecosystems, population dynamics, and ecological energetics, and has a particular interest in wetland ecology, resource use, and ornithology. His *Fundamentals of Ecology* (first edition published in 1953) has been widely used for many years.

oestrus cycle **(estrus cycle)** In female mammals (other than most primates) the hormonally controlled, regularly repeated stages by which the body is prepared for reproduction. In anoestrus the female reproductive apparatus is inactive; in prooestrus it becomes active; and in oestrus ovulation usually occurs (in some species ovulation is triggered by copulation) and the female becomes receptive to males. Unless fertilization occurs, oestrus gives way to anoestrus as the cycle repeats.

offshore bar *See* BAR.

offshore zone The zone extending seaward from the point of low tide to the depth of wave-base level or to the outer edge of the *continental shelf.

ogive A banded pattern on the surface of a *glacier, normally convex down-glacier due to relatively high velocities at the centre. The banding effect may be due to alternating bands of white ice (containing many air bubbles) and dark ice (few air bubbles), or to variations in longitudinal pressure.

-oidea In taxonomy, a recommended, but not mandatory, suffix used to indicate a superfamily of animals (e.g. Chelonoidea, the sea turtles).

oil shale Dark grey or black *shale, containing organic substances that yield liquid hydrocarbons on distillation, but do not contain free *petroleum.

Oka/Demyanka A Russian equivalent of the *Elsterian (*Mindel) glacial, or its *Anglian equivalent, which passes beneath the Dnepr *drift. Little is known of its occurrence.

okta *See* CLOUD AMOUNT.

Older Drift All the more weathered (i.e. older) *drift deposits that mark the maximum extent of ice advance during the last (*Devensian) glaciation in the British Isles. *Compare* NEWER DRIFT.

Older Dryas *See* DRYAS.

Oldest Dryas *See* DRYAS.

old-field ecosystem An ecosystem that develops on abandoned farmland by old-field succession. The terms are most commonly used in the USA.

old-growth forest In N. America, a forest at a late seral stage and, by implication, *primary forest that existed prior to European settlement. *Compare* ANCIENT WOODLAND.

Old Red Sandstone Continent The continental *facies of the *Devonian in the British Isles. It is characterized by red sandstones and conglomerates which were deposited under terrestrial conditions.

olfaction Detecting by means of smell.

olfactometer A device used to determine the response of an animal to odours. The animal is placed inside the stem of a Y-shaped tube; equal air currents flow through each arm of the Y, one of them after crossing a source of the odour being tested, and a record is made of whether or not the animal moves through the arm carrying the odour.

oligo- From the Greek *oligos*, meaning 'small' or (*oligoi*) 'few', a prefix meaning few or small; in ecology it is often used to denote a lack (e.g. *oligotrophic, nutrient-poor; *oligomictic, subject to little mixing).

Oligocene An epoch of the *Tertiary Period, about 33.9–23.03 Ma ago, which follows the *Eocene and precedes the *Miocene Epochs. Grasses evolved during the Oligocene (about 26 Ma ago), although large grass plains did not appear until about 5 Ma ago.

oligohaline *See* BRACKISH and HALINITY.

oligolectic Applied to bee species that specialize in collecting pollen from one genus or species (or from only a few genera or species) of flowering plants.

oligomictic Applied to lakes that are thermally almost stable, mixing only rarely. This condition is characteristic of tropical lakes with very high (20–30°C) surface temperatures.

oligophagous Applied to an organism that feeds on a few types of plants or prey. *Compare* MONOPHAGOUS and POLYPHAGOUS.

oligophotic zone *See* PHOTIC ZONE.

oligotrophic Applied to waters or soils that are poor in nutrients and with low *primary productivity. Typically, oligotrophic lakes are deep, with the *hypolimnion much more extensive than the *epilimnion. The low nutrient content means that plankton blooms are rare and littoral plants are scarce. The low organic content means that dissolved oxygen levels are high. By comparison with *eutrophic lakes, oligotrophic lakes are considered geologically young, or little modified by weathering and erosion products.

oligotrophication The process of nutrient depletion, or reduction in rates of nutrient cycling, in aquatic ecosystems. It arises as a consequence of acidification, typically the result of *pollution and most notably associated with air pollution and *acid precipitation.

olive knot A type of *gall that develops on olive trees (*Olea*) infected with the bacterium *Pseudomonas savastanoi*.

ombrogenous bog A peat-forming vegetation community lying above *groundwater level: it is separated from the ground flora and the *mineral soil, and is thus dependent on rain water for mineral nutrients. The resulting lack of dissolved bases gives strongly acidic conditions and only specialized vegetation, predominantly *Sphagnum* species (bog mosses), will grow. Two types of ombrogenous bogs are commonly distinguished: raised bogs and blanket bogs. *Compare* VALLEY BOG. The elevated bog forest found in the coastal regions of some of the islands of south-east Asia comprises a third type of ombrogenouos bog. These bog forests form extensive domes, often more than 10 km in diameter and their surfaces are elevated 5–10 m above the groundwater table.

ombrotrophic Applied to a *mire system that is fed by rain water. *Compare* RHEOTROPHIC.

omnivore **(diversivore)** A heterotroph that feeds on both plants and animals, and thus operates at a range of *trophic levels.

oncogenic Capable of causing the formation of tumours.

ontogenetic allometry *See* ALLOMETRY.

ontogeny The development of an individual from fertilization of the egg to adulthood.

ooze A mud that consists of the calcareous or siliceous remains of *pelagic organisms (e.g. coccoliths and tests of foraminiferids and diatoms) and *hemipelagic clay minerals. Calcareous oozes accumulate in the deep oceans at depths shallower than the *carbonate compensation depth. *See also* DIATOM OOZE; *GLOBIGERINA* OOZE; PTEROPOD OOZE; and RADIOLARIAN OOZE.

opacus The Latin *opacus*, meaning 'shady', used to describe a variety of extensive cloud that obscures the Sun or Moon. The term is used to describe *stratus, *stratocumulus, *altostratus, and *altocumulus. *See also* CLOUD CLASSIFICATION.

open canopy Tree communities in which the individual tree crowns do not overlap to form a continuous canopy layer but are more widely spaced, leaving areas of the ground open to the sky.

open-cast mining **(strip mining)** Mining in which large strips of land are excavated in order to extract materials without sub-surface tunnelling. It may result in short- and long-term environmental damage.

open-pit mining A method of extraction that is used where the *overburden is limited and easily stripped, but where waste has to be transported to external dumps. It is generally used where deposits are limited laterally but are thicker than in *open-cast mining.

open population A population that is freely exposed to *gene flow, as opposed to a closed one in which there is a barrier to gene flow. *Compare* CLOSED POPULATION.

open reading frame **(ORF)** A DNA sequence that is potentially translatable into a protein. In protein-coding regions of genes there are three potential open reading frames, because there are three ways in which the DNA sequence can be broken into *codon triplets.

open-space area Any land area to which the public has unrestricted access (although access may extend only to permission to move freely over the area on foot).

open system A system in which energy and matter are exchanged between the system and its environment (e.g. a living organism, or an *ecosystem).

operant conditioning **(instrumental conditioning)** *Conditioning in which an animal forms an association between a particular behaviour and a result that *reinforces the behaviour, its behaviour being operant (or instrumental) in producing the result. For example, a bird that turns over dead leaves may find food beneath them, so it may come to associate turning over dead leaves with finding food. The process may be negative, as when an animal learns to associate a particular activity with an unpleasant result.

operational taxonomic unit (OTU) In *phylogenetics, any of the data sets under examination. During the tree-building process, OTUs may become combined into a composite which is then treated as a single OTU.

Oppel zone A *concurrent range zone, originally described by the German geologist and palaeontologist Albert Oppel (1831–65), in which the lower boundary is usually marked by the first appearance of one diagnostic taxon and the upper boundary by the last appearance of another (either may extend above or below into a region of overlap with the other taxon at the upper or lower end of its range). Other characteristic taxa may be contained within the zone or extend to either side, although long-ranging and slowly evolving lineages are not usually included in the diagnostic assemblage. Not all the significant taxa are required to be present at all levels and in all places. The zone is named after one of the diagnostic species. Oppel zones are more flexible and subjective than concurrent range zones in the rigorously applied sense of that term, although 'concurrent range zone' is commonly used for what is in fact an Oppel zone.

opportunist pathogen A usually non-aggressive micro-organism, commonly a normal member of the population present in the body, which under certain conditions takes on the role of a *pathogen. Predisposing factors include surgery, antibiotic therapy, and suppression of the normal immune responses.

opportunist species *See* FUGITIVE SPECIES.

optimality theory The theory that behavioural strategies involve decisions which tend to maximize the efficiency of that behaviour.

optimum foraging theory The theory that *foraging strategies may involve decisions which maximize the net rate of food intake or some other measure of foraging efficiency.

optimum yield (maximum sustained yield, MSY) The theoretical point at which the size of a population is such as to produce a maximum rate of increase. If the population has a symmetrical, *S-shaped growth curve this is equal to half the *carrying capacity. The concept is of practical use in farming and has been applied widely to commercial fisheries. It forms the basis for models that predict the stocking density required to maintain optimum fish production, and the harvesting methods and food supply needed to maintain production at that level.

oral Pertaining to the mouth.

orbicular Disc-shaped, circular, or globular.

ordination method A method for arranging individuals (or sometimes attributes) in order along one or more lines. The method is used, with many techniques, in the biological and earth sciences, and especially in biology. There is an extensive ecological literature that discusses ordination methods, their applicability to different situations, and the relative merits of ordination as opposed to classification schemes. *See, for example*, CONSTELLATION DIAGRAM; GRADIENT ANALYSIS; HYPOTHESIS-GENERATING METHOD; INVERSE ANALYSIS; and RECIPROCAL AVERAGING.

Ordovician The second of six periods that constitute the *Palaeozoic Era, named after an ancient Celtic tribe, the Ordovices. It lasted from about 488.3 to 443.7 Ma ago. The Ordovician follows the *Cambrian and precedes the *Silurian. It is noted for the presence of various rapidly evolving graptolite genera (Graptolithina) and of the earliest jawless fish. Algae were the predominant plants. Some may have been terrestrial, forming thick, moss-like mats on wet ground.

ore A *mineral or rock that can be worked economically.

orebody An accumulation of *minerals, distinct from the host rock, which is rich enough in a metal to be worth commercial exploitation.

ore mineral A metalliferous *mineral that may be extracted profitably from an *orebody.

ORF *See* OPEN READING FRAME.

organic evolution The evolution of life.

organic matter In particular, the organic material present in soils; more generally, the organic component of any ecosystem. It is usually measured by determining the weight *loss on ignition, or more gently by oxidation using 30 per cent hydrogen peroxide.

organic selection The selection of *traits that are best suited to the way of life of an organism. If an organism responds to novel environmental conditions by appropriately modifying its behaviour or habit, such modification will tend to confine it to that environment, which may favour the selection of mutations tending in the same direction.

organic soil Soil with a high content of organic matter and water. The term usually refers to peat. The *USDA defines an organic soil as one with a minimum of 20–30 per cent organic matter, depending on the *clay content.

organismic Applied to groups of organisms, or *communities, that are thought to have properties (e.g. homoeostasis or reproduction) similar to those of a single living organism, making them 'superorganisms'. The term is most used to describe plant communities by those who consider that discrete *climax vegetation entities, e.g. beech–oak woodlands, may be identified. According to this concept, these units will necessarily show a high degree of internal interdependence of species, and on the maturity and death of a community another identical plant association will replace it. This organismic concept forms the theoretical basis for a classificatory approach (e.g. those of F. E. *Clements and A. G. *Tansley) to the description of vegetation communities and their analysis. *Compare* INDIVIDUALISTIC HYPOTHESIS.

organochlorine A compound in the molecules of which chlorine is bound to the carbon of hydrocarbon groups. The organochlorines include the *polychlorinated biphenyls and the class of insecticides that includes *DDT, *aldrin, and *dieldrin. These vary in their toxicity, but tend to accumulate along food-chains, sometimes reaching harmful concentrations.

organomercury compounds A range of substances in which mercury is combined with carbon and hydrogen (e.g. a phenyl group, C_6H_5, or methyl group, CH_3) that are used as seed dressings, and sometimes applied to timber, to prevent fungal infestation. Dressed seed is dispersed widely, distributing the mercury present in the fungicide at concentrations well below those present naturally in most soils, but birds have been harmed by ingesting seed dressed with methylmercury compounds. Phenylmercury compounds are of low toxicity to vertebrates; methylmercury compounds are much more toxic.

organophosphorus compounds Chemical compounds in the molecules of which an atom of phosphorus is linked to the carbon atom of a hydrocarbon group. These compounds have a number of industrial uses, but their principal application is as insecticides. Early organophosphorus insecticides (e.g. parathion) were highly toxic to mammals and birds. Later compounds were less toxic and dangers to wildlife were reduced, although careless or improper use can cause serious harm.

organotroph An organism that obtains energy from the metabolism of organic compounds, sometimes inaccurately used as a synonym of *heterotroph. *Compare* LITHOTROPH.

Oriental faunal region The area that encompasses India and Asia south of the Himalayan–Tibetan mountain barrier, and the Australasian archipelago, excluding New Guinea and the Sulawesi. There are marked similarities with the *Ethiopian faunal region (e.g. both have elephants and rhinoceroses) but there are endemic (*see* ENDEMISM) groups (e.g. pandas and gibbons).

orientating response A spontaneous reaction of an animal to a stimulus in which the head and/or body are moved so that the source of the stimulus may be examined more thoroughly.

orientation A change of position by an animal or plant in response to an external stimulus.

original-natural *See* NATURAL.

ornithocoprophilous Preferring habitats rich in bird manure.

ornithosis A disease of birds caused by the bacterium *Chlamydia psittaci*. Humans can also be infected. Sometimes the term 'ornithosis' is used synonymously with 'psittacosis', and sometimes it is reserved for the infection in birds only, 'psittacosis' being reserved for the disease in humans.

orogeny A mountain-building episode, especially one that is caused by the compression of a part of the Earth's crust to form a chain of mountains.

orographic Applied to the rain or cloud caused by the effects of mountains on air streams that cross them. Orographic cloud and rain are produced by the forced uplift of moist air and the consequent condensation when this is cooled to saturation point.

ortho- From the Greek *orthos*, meaning 'straight', a prefix meaning 'straight', 'rectangular', or 'correct'.

orthodox seed A seed that can be desiccated and then stored. Storage life is increased by low temperature. Most temperate plant species and tropical *pioneer species have orthodox seeds. *Compare* RECALCITRANT SEED.

orthogenesis Evolutionary trends that remain fairly constant over long periods of time and so appear to lead directly from ancestor organisms to their descendants. This was once explained as the result of some internal directing force or 'need' within the organisms themselves. Such metaphysical interpretations have been displaced by the concepts of *orthoselection and species selection. *See also* ARISTOGENESIS; ENTELECHY; and NOMOGENESIS.

orthokinesis The movement of an animal in response to a stimulus, such that the speed of movement is proportional to the strength of the stimulus.

orthology In *phylogenetics, pertaining to genes which have been separated by a speciation event. *Compare* PARALOGY and XENOLOGY.

orthoselection A primary selective pressure of a directional kind, which results in a self-perpetuating evolutionary trend. Species selection, via the *effect hypothesis, has been advanced as an alternative explanation for such trends. *See also* DOLLO'S LAW.

ortstein An indurated *horizon in the B horizon of *podzols (*Spodosols), in which the cementing materials are mainly iron oxide and organic matter.

Osborn, Henry Fairfield (1857–1935) An American palaeontologist and evolutionary theorist, who taught at Princeton and Columbia Universities. He developed the concept of *adaptive radiation. He also arranged the mammalian palaeontology exhibits at the American Museum of Natural History, New York.

oscillation ripple (wave ripple mark) A small ridge of sand formed by wave action on the floor of a sea or lake. Such ripples, which are usually less than 10 cm in height, commonly display rounded symmetrical troughs and rounded, or sometimes sharp, symmetrical peaks, although many wave-generated ripples are not symmetrical owing to an inequality in forward and backward motion associated with the wave action. The *ripple index of oscillation ripples is usually between 6 and 10.

oscillatory wave A wave that causes a mass of water to move to and fro about a point but to undergo no appreciable net displacement in the direction of wave advance. The wave-form advances, but the individual water particles move in closed or nearly closed orbits.

osmophilic Applied to organisms that grow best in *habitats containing relatively high concentrations of salts or sugars (i.e. those having a relatively high *osmotic pressure).

osmoregulation The process whereby an organism maintains control over its internal *osmotic pressure irrespective of variations in the environment.

osmosis The net movement of water or of another solvent from a region of low solute concentration to one of higher concentration through a *semi-permeable membrane.

osmotic potential **(solute potential)** The part of the water potential of a tissue that results from the presence of solute particles. It is equivalent to *osmotic pressure in concept but opposite in sign. The concept has been largely replaced by that of *water potential.

osmotic pressure The pressure that is needed to prevent the passage of water or another pure solvent through a *semi-permeable membrane separating the solvent from the solution. Osmotic pressure rises with an increase in concentration of the solution. Where two solutions of different substances or concentrations are separated by a semi-permeable membrane, the solvent will move to equalize osmotic pressure within the system. The concept has been largely replaced by that of *water potential.

osmotrophic Applied to an organism that absorbs nutrients from solution, as opposed to ingesting particulate matter.

otic Pertaining to the ear.

OTU *See* OPERATIONAL TAXONOMIC UNIT.

ouadi *See* WADI.

outbreeding The crossing of plants or animals within a species which are not closely related genetically, in contrast to inbreeding, in which the individuals are closely related. *See also* CROSS-BREEDING.

outgroup In *phylogenetics, a species which is the least related to the species under analysis. The inclusion of a known outgroup allows the identification of *plesiomorphic and *apomorphic *character states, which might otherwise remain unclear, a situation that can give rise to *topological errors. A tree that includes an outgroup is said to be rooted.

outlet glacier A tongue of ice that extends radially from an *ice dome. It may be identified within the dome as a rapidly moving ribbon of ice (an 'ice stream'), while beyond the dome it typically occupies a shallow, irregular depression. The 700 km long Lambert Glacier, Antarctica, is one of the world's largest outlet glaciers.

outlier **1.** An organism that occurs naturally some distance away from the principal area in which its species is found. **2.** An area in which younger rocks are completely surrounded by older rocks.

out-of-phase overlapping The occurrence of two or more proteins that are encoded in the same DNA sequence, but in different *open reading frames.

outwash **1.** The stratified sands and gravels deposited at or near ice margins. **2.** Meltwater escaping from the terminal zone of a *glacier. The resulting streams are typically *braided, and show marked seasonal variations in discharge.

outwash plain **(sandur, pl. sandar)** An extensive accumulation of rock debris built up by *outwash in front of a *glacier. Its constituents are very coarse close to the ice, and diminish in size further away. Its surface may be dissected by *braided channels. Fossil outwash plains are found at the margins of many *Pleistocene glaciers. The name 'sandur' is Icelandic.

outwelling The enrichment of coastal seas by nutrient-rich estuarine waters.

overburden **1.** Any loose material which overlies bedrock. **2.** In a sedimentary deposit, the upper strata which cover, compress, and consolidate those beneath. **3.** Any barren material, consolidated or loose, which overlies an *ore deposit. The depth and type of overburden may control whether an ore deposit is worked by underground or *open-cast methods. The proportion of overburden thickness to mineral deposit is called the overburden ratio.

overburden ratio *See* OVERBURDEN.

overcrowding effect The regulation of population growth by behavioural abnormalities caused by malfunction of the endocrine glands, which themselves occur in response to overcrowding.

overdispersion **(contagious distribution)** In plant ecology, a situation in which the pattern formed by the distribution of individuals of a given plant species within a *community is not random but shows clumping, so that both empty and heavily populated *quadrats are recorded. *See also* PATTERN ANALYSIS.

overflow channel (spillway) A channel cut by meltwater escaping from a *proglacial lake. Generally it is trough-shaped in form, lacks tributaries, and is not integrated with the local drainage pattern. It is often difficult to distinguish from other varieties of *glacial drainage channel.

overflowing well *See* ARTESIAN WELL.

overland flow *See* SURFACE RUN-OFF.

over-representation In *palynology, the presence in considerable abundance of immediately local *pollen which tends to obscure the general pattern of changes in the regional *pollen-rain record (e.g. the predominance of pollen from copious pollen producers, such as *Pinus* species, and from bog plants at a peat bog site).

overspecialization An old theory which held that straight-line evolution or orthogenetic trends (*see* ORTHOGENESIS) might proceed to the point at which the lineage was at an adaptive disadvantage. Overspecialization was therefore considered as one of the causes of extinction. There is no reason to believe, however, that *natural selection would permit evolution to proceed beyond maximum adaptation (*see* RUNAWAY HYPOTHESIS). More recently, the term has been applied to highly specialized organisms which have proved incapable of responding to environmental change and so have become extinct.

overturn *See* THERMAL STRATIFICATION.

ovipary The method of reproduction in which eggs are laid and embryos develop outside the mother's body, each egg eventually hatching into a young animal. Little or no development occurs within the mother's body. Most invertebrates and many vertebrates reproduce in this way. *Compare* OVOVIVIPARY and VIVIPARY.

oviposition The act of depositing *eggs (i.e. ovipositing).

ovovivipary The method of reproduction in which young develop from eggs retained within the mother's body but separated from it by the egg membranes. The eggs contain considerable yolk, which provides nourishment for the developing embryo. Many insect groups, fish, and reptiles reproduce in this way. *Compare* OVIPARY and VIVIPARY.

Owen, Sir Richard (1804–92) An English anatomist and palaeontologist, who worked on fossil mammals and reptiles, including those brought back from South America by *Darwin. He coined the name 'dinosaur', and reconstructed fossil reptiles such as Iguanodon. He believed that animals within a major group (e.g. vertebrates) were variations on a single theme, or 'archetype'. His crowning achievement was the founding of the Natural History Museum in South Kensington, London, in 1881.

ox bow *See* CUT-OFF.

oxic horizon A mineral subsoil *horizon that is at least 30 cm thick and is identified by the almost complete absence of weatherable primary minerals, by the presence of kaolinite clay, insoluble minerals such as quartz, hydrated oxides of iron and aluminium, and small amounts of exchangeable bases, and by low *cation-exchange capacity. It is the distinguishing subsoil horizon (B horizon) of an *Oxisol. The name is from the French *oxide*, meaning oxide.

oxidation A reaction in which atoms or molecules gain oxygen or lose hydrogen or electrons.

oxidative potential The potential to cause a *redox reaction to occur through loss of electrons.

Oxisols An order of *mineral soils that is distinguished by the presence of an *oxic horizon within 2 m of the soil surface, or with *plinthite close to the soil surface, and without a *spodic or *argillic horizon above the oxic horizon. Oxisols occur over a large part of the humid tropics, where they are usually heavily leached and exhausted by hand tillage, resulting in low fertility.

oxygen demand A measure of the amount of oxidizable material present in an effluent, stream, etc. (i.e. a measure of organic pollution). *See* BIOLOGICAL OXYGEN DEMAND.

oxygen-isotope analysis A method for estimating past ocean temperatures. The ratio of the stable oxygen isotopes, ^{18}O and ^{16}O, is temperature-dependent in water, ^{18}O increasing as temperature falls. Oxygen incorporated in the calcium-carbonate shells of marine organisms reflects the prevailing $^{18}O:^{16}O$ ratio.

oxygen-isotope ratio **($^{18}O:^{16}O$ ratio)** The abundance ratio between two of the three isotopes of oxygen. They have similar chemical properties because they have the same electronic structure, but because of the differences in mass between their nuclei they have different vibrational frequencies which cause them to behave slightly differently in physicochemical reactions. These differences can provide information (e.g. regarding the source of water in a past environment or the temperature at which various interactions have taken place). For example, surface waters vary in their oxygen isotopes; light water ($H_2{}^{16}O$) has a higher vapour pressure than $H_2{}^{18}O$ and therefore is concentrated by evaporation so that fresh water and polar ice are light but sea water is heavy. $CaCO_3$ or SiO_2 are richer in the heavier isotope when precipitated from sea water than when precipitated from fresh water. Moreover, because of meteorological cycles of evaporation/condensation, there is a steady depletion of ^{18}O in sea water towards the poles. *See also* OXYGEN-ISOTOPE ANALYSIS.

oxygen-isotope stage One of the glacial or *interglacial stages revealed by oxygen-isotope analysis. Graphical plots from Atlantic and Pacific deep-sea cores were divided by C. Emiliani into 16 stages, their fluctuations being correlated with ice-sheet growth and decay. N. J. Shackleton and N. D. Opdyke extended these subdivisions in 1973, from their studies of a core from the western Pacific in which 23 stages were recognized. These are presumed to represent a continuous record from about 870 000 years BP. In 1976 J. Van Donk obtained curves from a core in the equatorial Atlantic which yielded 42 isotope stages, representing 21 glacial and 21 interglacial stages.

oxygen method An estimation of gross *primary productivity by monitoring rates of production of oxygen, the by-product of *photosynthesis. In field practice this principle is most often used for productivity measurement in aquatic ecosystems. Paired 'light and dark' bottles are filled with water from the required sample depth and suspended at this level for several hours. The dark bottle records oxygen uptake by respiration of phytoplankton and other micro-organisms during the measurement period. Changes in oxygen levels in the light bottle reflect production in photosynthesis as well as loss in respiration. Assuming that respiration rates in the paired bottles are similar, gross primary productivity may be estimated by adding oxygen loss in the dark bottle to the change recorded in the light bottle. The light bottle alone records net (*ecosystem) productivity. Net primary productivity estimates are only possible in dense phytoplanktonic communities (e.g. algal blooms in which the oxygen loss caused by respiration of other micro-organisms is proportionately very small). An important limitation and source of error in the application of the oxygen method to macrophytes arises from the internal storage and utilization of oxygen produced by photosynthesis.

oxygen quotient **(QO_2)** The volume, in microlitres, of oxygen taken up at normal temperature and pressure in 1 hour by 1 milligram of plant tissue.

oxygen sag The characteristic fall and recovery in the curve of dissolved oxygen percentage saturation levels in streams, rivers, etc. downstream from a major *pollution source such as a sewage works, a brewery, or a paper-mill.

Oyashio Current The western *boundary current in the subpolar *gyre of the North Pacific. It originates in the Bering Sea and flows south-west off the Kuril Islands to meet the *Kuroshio Current east of northern Japan. The current flows at less than 0.5 m/s, it is cold (4–5 °C at 200 m depth), and has a low *salinity (33.7–34.0 parts per thousand).

ozone 'hole' An increase in the natural seasonal decrease in the concentration of stratospheric ozone that was first observed over Antarctica in 1985. The decrease occurs naturally during later winter and early spring (September to December) in latitudes higher than 60°S, because the reactions causing the formation of ozone depend on sunlight, but the reactions destroying ozone do not. The increase in the depletion observed in 1985 was found to be due to reactions involving chlorofluorocarbon (CFC) compounds, and to a lesser degree to halons (compounds containing bromine). CFCs and halons are chemically stable, but they break down when exposed to *ultraviolet radiation in the stratosphere, releasing free chlorine and bromine respectively. These destroy ozone in a series of reactions. For chlorine, these are:

$$Cl + O_3 \rightarrow ClO + O_2$$
$$ClO + O \rightarrow Cl + O_2$$

Each chlorine atom is thus able to destroy many molecules of ozone. Ozone did not disappear entirely, but in some years the concentration fell from the usual 220–460 *Dobson units (DU) to little more than 100 DU.

ozone layer The stratospheric layer at 20–30 km altitude, in which ozone (O_3) is concentrated at 1–10 parts per million (220–460 DU). Ozone also occurs in very low concentrations at altitudes of 10–15 km and 30–50 km. The highest ozone concentration is found at about 30 km over the equator and about 18 km over the poles. Atmospheric ozone is produced by the photochemical dissociation of oxygen (O_2), resulting from the absorption by oxygen molecules of *ultraviolet radiation (UV) at a wave-length of 240 nanometres and an energy of 5.16 electron volts (known as hard UV or UVC). The reaction occurs in two steps:

$$O_2 + \mathit{photon} \rightarrow O + O$$
$$O_2 + O + M \rightarrow O_3 + M$$

M is a molecule of any other substance, but usually nitrogen. The first reaction is strongest at altitudes above 80 km. The natural mixing of air then carries atomic oxygen to a lower altitude, where the second reaction takes place. Ozone then absorbs more UV radiation (mostly UVB with some UVC), which causes it to break down:

$$O_3 + \mathit{photon} \rightarrow O_2 + O$$

It also breaks down by collision with atomic oxygen:

$$O_3 + O \rightarrow 2O_2$$

The formation and destruction of ozone in the upper atmosphere shields the surface from UVC and most UVB radiation.

P 1. *See* PETA-. 2. *See* PHOSPHORUS.

P_n **(NPP)** Net primary productivity. *See* PRIMARY PRODUCTIVITY.

Pa *See* PASCAL.

Pacific and Indian Ocean common water **(PIOCW)** The deep waters of the Pacific and Indian Oceans, classed as one mass because they are so similar in character. The mean temperature is 1.5°C and *salinity is 34.70 parts per thousand.

Pacific–Antarctic Ridge The *ridge which lies between the Pacific and Antarctic Plates and joins the South-East Indian and East Pacific Rises.

Pacific Coast forest The densest and most majestic coniferous forest in the world, extending from the north of California to the south of British Columbia. It is famed for its giant trees, the best known being *Sequoia sempervirens* (coastal redwood), in California, and *Pseudotsuga taxifolia* (Douglas fir) further north. All may exceed 100 m in height and even 110 m in the case of the big tree (*Sequoiadendron giganteum*), and the girths of certain redwoods are often greater than 20 m.

Pacific North American floral region Part of R. Good's (1974, *The Geography of the Flowering Plants*) *Boreal Realm, distinguished by about 300 endemic genera (*see* ENDEMISM). Although the flora as a whole has contributed few economic plants, it has furnished numerous garden species. Like all floras of *Mediterranean-type climates, the Californian part of this region shows a marked degree of endemism, with 30–50 per cent of its species restricted to the state. *See also* FLORAL PROVINCE and FLORISTIC REGION.

Pacific Ocean The largest of the world's major oceans (179.7×10^6 km^2) and also the coldest (average 3.36°C), deepest (4 028 m average), and least saline (34.62 parts per thousand).

paedogenesis Reproduction by larval or other immature forms.

paedomorphosis Evolutionary change that results in the retention of juvenile characters into adult life. It may be the result of *neoteny, of *progenesis, or of *postdisplacement. It permits an 'escape' from specialization, and has been invoked to account for the origin of many taxa, from subspecies to phyla.

PAH *See* POLYAROMATIC HYDROCARBON.

Palaearctic faunal region The region that includes Europe, Asia north of the Himalayan–Tibetan physical barrier, North Africa, and much of Arabia. Despite its size and great range of habitats, the faunal variety in it is far less than that of the *Ethiopian and *Oriental faunal regions to the south of it (e.g. monkeys are nearly absent and reptiles few in number). The region is similar at the family level, and rather less so at the generic level, to the *Nearctic region; the Bering *land bridge connected the two for much of the *Cenozoic Era.

palaeo- **(paleo-)** From the Greek *palaios*, meaning 'ancient', a prefix meaning 'of ancient times'.

palaeobiology The scientific discipline that attempts to interpret the biology of *fossil organisms. It is sometimes combined with functional morphological studies in the more general study of *palaeoecology.

palaeobotany The study of fossil plants.

Palaeocene **(Paleocene)** The lowest epoch of the *Tertiary Period, about 65.5–56.5 Ma ago. The name is derived from the Greek *palaios* 'ancient', *eos* 'dawn', and *kainos* 'new', and means 'the old part of the *Eocene' (the subsequent epoch).

palaeoclimatic indicator One of the sources from which evidence concerning past climates can be obtained. Such indicators include glacial, *periglacial, and pluvial deposits, which provide morphological information related to climate and cave deposits, dunes, and dunefields, which yield lithologic information. Plants (including *pollen; *see also* LEAF PHYSIOGNOMY), molluscs, foraminifera, beetles (*see* BEETLE ANALYSIS), and ostracods are among the organisms that have been used to derive biotic information.

palaeoclimatology The study of past climates from *fossils and traces left in the geologic record. *See* BEETLE ANALYSIS; LEAF PHYSIOGNOMY; PALAEOCLIMATIC INDICATOR; POLLEN ANALYSIS; and VARVE.

palaeocurrent analysis The collection, presentation, and interpretation of directional data measured from sedimentary structures and textures which were formed under flowing water, the wind, or moving ice. Ranging from the study of a single structure to data collected across a whole sedimentary basin, palaeocurrents can yield a hierarchy of information, from the local direction of flow which yielded a single ripple train, through the direction of migration of a *bar or channel, to the movement of sediment through a river or deltaic network (*see* DELTA), and the regional pattern of sediment provenance and dispersal through a basin.

palaeoecology The application of ecological concepts to fossil and sedimentary evidence to study the interactions of Earth surface, atmosphere, and *biosphere in former (prehistoric) times.

palaeoendemic *See* ENDEMISM.

palaeoethnobotany The study of plants in relation to early human cultures.

Palaeogene The period that comprises the first three epochs of the *Tertiary Sub-Era: the *Paleocene, *Eocene, and *Oligocene, preceded by the *Cretaceous, followed by the *Neogene, and dated at 65.5–23.03 Ma.

palaeogeography The reconstruction of the physical geography of past geologic ages. A palaeogeographical map usually shows the *palaeolatitude of the area it covers, together with the location of inferred shorelines, drainage areas, *continental shelves and depositional environments. Today, the base map is based on palaeomagnetic data, but many maps in earlier publications used the present geographical positions of the continents as a foundation. With the advent of balanced sections in structural geology, many palaeogeographical maps are now being produced which take account of the physical shortening involved and attempt to restore areas to their 'depositional' condition.

palaeohydraulics The analysis of the geometry of ancient *fluviatile features (e.g. preserved channel forms, lateral accretion surfaces, and channel-fill sequences) to estimate the hydraulic characteristics at the time a channel system was formed or when a sediment or large boulders were deposited.

palaeolatitude The position, relative to the equator, of a geographic or geologic feature at some time in the past. The evidence for the position may come from palaeoclimatic data (*see* PALAEOCLIMATIC INDICATOR and PALAEOCLIMATOLOGY) or, more normally, from palaeomagnetic evidence.

palaeolimnology The study of the history and development of freshwater *ecosystems, especially lakes.

palaeontology The study of *fossil flora and fauna. Information thus gained may be used to establish the characteristics of ancient *environments.

palaeosol (paleosol, relict soil) A soil formed during an earlier period of *pedogenesis, and which may have been buried, buried and exhumed, or continuously present on the landscape until the current period of pedogenesis.

palaeospecies (chronospecies) A group of biological organisms, known only from *fossils, which differs in some respect from all other groups.

Palaeotethys In the view of some authors, a late *Palaeozoic gulf at the margin of *Panthalassa, between the parts of *Pangaea that were later to separate as *Laurasia and *Gondwana. This ocean, or large parts of it, were consumed in a Palaeocimmerian–Indosinian subduction zone. In the *Jurassic a new ocean, or oceans, opened along the northern margin of Palaeotethys. This Jurassic–*Cretaceous ocean has been called Neotethys or Tethys (*sensu stricto*).

Palaeotropical floral kingdom (realm) A region whose geographical extent varies with the authority consulted. R. Good's (1974, *The Geography of the Flowering Plants*) definition embraces the African (excluding the Cape), Indo-Malaysian, and *Polynesian floral subkingdoms. Other authorities adopt a more restricted definition, embracing most of Africa, Arabia, and the north-west part of the Indian subcontinent. *See also* FLORAL PROVINCE and FLORISTIC REGION.

Palaeozoic (Paleozoic) The first of the three eras of the *Phanerozoic, about 542–251 Ma ago. The *Cambrian, *Ordovician, and *Silurian Periods together form the Lower Palaeozoic Sub-Era; the *Devonian, *Carboniferous, and *Permian the Upper Palaeozoic Sub-Era. It was an era of great evolutionary change among plants, which began to invade the land at its beginning. By the end of the era, amphibians and reptiles were major components of various communities and giant tree-ferns, horsetails (Calamitaceae), and cycads gave rise to extensive forests. The faunas of the Palaeozoic are noted for the presence of many invertebrate organisms including trilobites, graptolites, brachiopods, cephalopods, and corals. The name is derived from *palaeo- and the Greek *zoe*, meaning 'life'.

pale A boundary. Originally, a deer-proof fence erected around the perimeter of a park.

paleo- *See* PALAEO-.

palichnology *See* ICHNOLOGY.

palimpsest **1.** A word derived from the Greek *palimpsestos*, meaning 'to rub smooth again', and used to describe a re-used parchment, paper, or ornamental brass from which the original writing or engraving has been only partially erased. **2.** A landscape that bears the imprint of two or more sets of earlier geomorphological processes. For example, much of the Sahel region of Africa shows landforms resulting from former wet and dry episodes.

palindromic DNA A sequence of DNA which reads the same on both strands (like a palindrome). Such sequences often form the recognition sites for enzymes such as *restriction endonucleases.

Palouse prairie (bunch grass prairie) A distinctive tract of *prairie in the Palouse region of south-eastern Washington (USA), but similar prairie also extends into Oregon, southern Idaho, and northern Utah. A rich growth of bunch grasses, especially *Agropyron spicatum* (blue-bunch wheat-grass), occurs on the deep, highly fertile soils formed in *aeolian parent materials. Much of the Palouse prairie is now cultivated for wheat.

palsa A mound or ridge, largely made from peat, which contains a perennial ice lens and is found in the damper sites of mires in periglacial areas. Widths are in the range 10–30 m, lengths 15–150 m, and heights 1–7 m. Probably palsas are a result of heaving associated with the growth of segregated ice. *See also* PINGO.

palsa mire Peaty *tundra, in which ground ice is partly responsible for the formation of peat hummocks, called palsas or palsa mounds, up to 35 m long, 15 m wide, and 7 m high. Palsas have a core of frozen peat and develop cyclically, but the remainder of the mire remains clear of ice in summer. The palsa surface develops a cover of *lichen. This gives it a high *albedo, preventing the interior from warming in summer and eventually allowing a block of *permafrost to form. As the surface becomes unstable it erodes, removing the lichen and revealing the underlying peat, which is dark and reduces the albedo, causing the peat to absorb more heat. Pools form as the palsa warms and its core of ice melts, and the sequence begins again. Palsa mires are circumpolar in distribution. They are forming and decaying constantly, but there is some evidence that they have formed more rapidly at certain periods in the past, possibly due to climatic differences.

paludal Applied to organisms, soils, etc., that are of or associated with a *marsh.

palynology The study of living and *fossil *pollen grains, *spores, and certain other *microfossils (e.g. dinoflagellates and coccoliths). Palynology was developed from *pollen analysis and deals principally with structure, classification, and distribution. It has many applications (e.g. in medicine, archaeology, petroleum exploration, and *palaeoclimatology).

pamirs *See* PUNA.

pampas A temperate South American grassland, the largest continuous area being in the eastern Argentinian province of Buenos Aires, together with parts of the adjacent provinces. Much of the moister sector of the pampas is now cultivated, but formerly it was covered by bunch grasses, in between the individual tussocks of which the soil lay bare. To the west and south-west drier conditions prevail, and here the pampas comprises a mixture of short grasses and *xerophytic shrubs.

Pampas floral region Part of R. Good's (1974, *The Geography of the Flowering Plants*) *Neotropical floral kingdom: it lies east of the Andes, and between the tropical flora of Brazil and the temperate flora of Patagonia. The flora is relatively poor, with only about 50 endemic genera (*see* ENDEMISM), all of them small. In terms of vegetation the region is dominated by the pampas. Few garden plants or important economic plants come from this region, although the pineapple (*Ananas comosus*) comes from Paraguay. *See also* FLORAL PROVINCE and FLORISTIC REGION.

pampero A regional storm of the *line-squall type which affects Argentina and Uruguay. Sometimes the storm brings rain, thunder, and lightning, and its passage is marked by a fall in temperature. The storm moves ahead of a south-westerly wind, which follows a *depression.

PAN *See* PHOTOCHEMICAL SMOG.

pan- From the Greek *pan, pas*, meaning 'all', a prefix meaning 'of universal application'.

pan A *soil horizon, usually in the subsoil, that is strongly compacted, indurated (*see* INDURATION), cemented, or very high in clay content, when it is called a 'clay pan'.

Panama Isthmus The narrow neck of land that connects North America and Mexico with South America. During the *Mesozoic there was an open marine connection between the Atlantic and Pacific Oceans. At the culmination of the Andean–Laramide *orogeny (Late *Cretaceous) a temporary connection between North and South America was formed, which explains the similarity between the floras of California and northern South America at that time. Some early *mammals also migrated southwards across the isthmus. During the early *Cenozoic the land connection was broken but it became re-established in the *Pliocene, thereby allowing another 14 mammal families to migrate from north to south. Some families, notably the elephants, appeared to be unable to migrate southwards, perhaps being constrained by the climatic zones that had to be traversed in such a migration.

panbiogeography A term coined in 1962 by L. Croizat to describe a synthesis of the sciences of plant and animal distribution. The main features are that consistencies recur in distribution patterns and that analysis of these produces 'tracks' (joining areas of common floras and faunas) and 'nodes' (where different tracks meet). These ideas form the basis of the new school of *vicariance biogeography.

panendemic distribution *See* COSMOPOLITAN DISTRIBUTION.

Pangaea A single supercontinent which came into being in late *Permian times and persisted for about 40 Ma before it began to break up at the end of the *Triassic Period; or, in the view of some people, which existed throughout most of the Earth's history prior to the Triassic (the matter remains unresolved). It was surrounded by the universal ocean of *Panthalassa.

panmictic unit A local population in which mating is completely random.

panmixia The process of *panmixis.

panmixis The random mating of individuals. *Compare* ASSORTATIVE MATING.

pannage **1. (commonly used of *mast)** The right to allow pigs to feed on common land where it comprises oak or beech woodland. **2.** Money paid for permission to allow pigs so to feed.

pannus The Latin *pannus*, meaning 'shred', used as an accessory cloud term applied to ragged cloud either beneath or attached to another cloud, usually *cumulonimbus, *cumulus, *nimbostratus, or *altostratus. *See also* CLOUD CLASSIFICATION.

panplain **(panplane, planplain)** An area of very subdued relief that consists of coalesced *flood-plains. It is, therefore, owing to lateral stream migration and is a component of a *peneplain. Good examples are found in the Carpentaria region of Australia.

panplane *See* PANPLAIN.

Panthalassa The vast oceanic area that surrounded *Pangaea when that supercontinent was in existence. *Tethys was a minor arm of this ocean. Once Pangaea began to split in the *Triassic, the names of all the modern oceans are normally applied to the developing ocean basins even though they were, at that time, still very small.

pantropical distribution The distribution pattern of organisms that occur more or less throughout the tropics. The plants in this category are mainly weedy, herbaceous species, but there are many families (e.g. Arecaceae, palms).

papaya *See* DISJUNCT DISTRIBUTION.

paper wasps Social wasps (superfamily Vespoidea, family Vespidae) which build their nests from chewed wood pulp. Most belong to the subfamilies Vespinae, Polistinae, and Polybiinae, and the name is sometimes restricted to the Polistinae. The nest structure is variable, with or without outer coverings, but is usually spherical and contains a large number of cells. Nest construction is started by the queen in spring and the nest is subsequently enlarged by many generations of workers. Nests sometimes attain 35 cm or more in diameter.

para- The Greek *para*, meaning 'beside' or 'beyond', used as a prefix meaning 'beside' or 'beyond'.

parabolic dune *See* DUNE.

paradigm Essentially, a large-scale and generalized *model that provides a viewpoint from which the real world may be investigated. It differs from most other models, which are abstractions based on data derived from the real world.

parallel evolution Similar evolutionary development that occurs in lineages of common ancestry. Thus the descendants are as alike as were their ancestors. The nature of the ancestry imposes or directly influences the development of the parallelism.

Parallel evolution

paralogy In phylogenetics, pertaining to genes which have been separated by a gene duplication event. *Compare* ORTHOLOGY and XENOLOGY.

parametric test *See* STATISTICAL METHOD.

paramo A name for the humid Arctic-alpine meadows and *scrub vegetation in which mosses and *lichens are common. It occurs in the Andes, situated between the tree line and the snow line, and deriving much moisture from cloud or mist.

parapatric Applied to *species whose *habitats are separate but adjoining. *Compare* ALLOPATRIC and SYMPATRIC.

parapatric speciation Speciation that occurs regardless of minor gene flow between *demes. In many species selective pressures are sufficiently strong on the whole to prevent homogenization of the immigrant genes by interbreeding.

paraphyletic Of a taxon, including some but not all descendants of the common ancestor (i.e. not *holophyletic).

pararetrovirus A *virus that contains a gene for reverse transcriptase, but cannot insert itself into a host *genome.

parasite *See* PARASITISM.

parasitism An interaction of *species populations in which one (typically small) organism (the parasite) lives in or on another (the host), from which it obtains food (when the parasite may be called a biotroph), shelter, or other requirements. Whereas a predator kills its host (i.e. it lives on the capital of its food resource) a parasite does not (i.e. it lives on the income). Parasitism usually implies that some harm is done to the host, but this interpretation must be qualified. Effects on the host range from almost none to severe illness and eventual death, but even where such obvious immediate harm accrues to the individual host, it does not follow that the relationship is harmful to the host species in the long term or in an evolutionary context (e.g. it might favour beneficial adaptation in the host species population). Obligate parasites (holoparasites) can live only parasitically. Facultative parasites may live as parasites or as independent *saprotrophs. Partial parasites (semiparasites) are facultative parasites that live more successfully as parasites than they do independently. Ectoparasites live externally on the host. Endoparasites live inside the body of the host. *Compare* COMMENSALISM; MUTUALISM; and NEUTRALISM; *see also* HEMIPARASITE; HYPERPARASITE; and NECROTROPHIC.

parasitoid An organism that spends part of its life as a parasite and part as a predator (e.g. many wasps that are parasites during their larval stages and predators when mature).

parasitology The study of small organisms (parasites) living on or in other organisms (hosts), regardless of whether the effect on the hosts is beneficial, neutral, or harmful. The study uses the term 'parasite' in a wider sense than is usually associated with *parasitism.

parataxon In taxonomy, an artificial classification that is suggested for certain common organisms of doubtful affinities, or as yet unknown origins (e.g. fossil spores, dinosaur footprints).

Paratethys (Central European Sea) A large, arcuate seaway that, at its maximum development, extended a distance of 4 500 km from just north of the Alps to just east of the Aral Sea. By the end of the *Oligocene it had been separated from the Boreal Sea by closure of the Ural, Polish, and Alsace Straits. Towards the end of the *Miocene it became more lagoonal in character and by the *Pliocene was represented only by a series of land-locked lakes: Lake Balaton in Hungary, the Black Sea (rejoined to the Mediterranean Sea by *Quaternary faulting), the Caspian Sea, and the Aral Sea.

paratype In taxonomy, a specimen, other than a *type specimen, which is used by an author at the time of the original description, and designated as such by the author.

parcel of air A quantity of air that has more or less uniform characteristics throughout.

parental care All activities that are directed by an animal towards the protection and maintenance of its own offspring or those of a near relative.

parent material The original material from which the soil profile has developed through *pedogenesis, usually to be found at the base of the profile as weathered but otherwise unaltered mineral or organic material.

park **1.** In ancient farming systems, the enclosed fields lying between the inner fields next to the farm buildings and the larger, outer fields used only seasonally for pasture. **2.** Enclosed land on which deer are or were kept. **3.** Land, usually wood-pasture, enclosed by a *pale, and intended for the keeping of deer. **4.** In modern use: (*a*) an area of land set aside for public enjoyment and designed to resemble semi-natural land; (*b*) an enclosure for semi-wild animals. **5.** *See* NATIONAL PARK.

park woodland A woodland in which there is an open canopy of mature trees over pasture.

parr The juvenile, freshwater stage of salmon. Parr bearing dark spots or bars tend to stay in the rivers for periods of at least two years before turning into *smolt and migrating to the sea.

parthenogenesis The development of an individual from an egg without that egg undergoing fertilization. It occurs in some groups of animals (e.g. flatworms, leeches, aphids, rotifers), in which males may be absent, and in some plants (e.g. dandelion). The eggs (ova) that develop in this way are usually diploid, so all offspring are genetically identical with the parent. Commonly, parthenogenesis with only females in the population alternates with ordinary sexual reproduction, which allows the recombination of genetic material and presents a need for males. This alternation is called heterogamy. *See* APOMIXIS.

partially permeable *See* SEMI-PERMEABLE.

partial parasite *See* PARASITISM.

partial pressure In a mixture of gases, the pressure one of those gases would exert were it alone in the same space as that occupied by the whole. The interaction of the partial pressures of each of the gases equals the total pressure of the mixture.

partial random sample *See* STRATIFIED RANDOM SAMPLE.

partial refuge A feeding patch in which the density of prey is low and, therefore, prey individuals have a relatively low risk of being attacked. *See also* AGGREGATIVE RESPONSE.

partial rosette plant *See* HEMICRYPTOPHYTE.

particle density The mass per unit volume of soil particles, usually expressed in grams per cubic centimetre. *Compare* BULK DENSITY.

particulate matter Fine particles that are suspended in the atmosphere. Many are of natural origin, e.g. *pollen grains, fungal *spores, and volcanic dust. Others are injected into the air by human activity, e.g. ploughing of dry soil, burning of vegetation, and the burning of fossil fuels. Small particles can be harmful to health when inhaled, because they are able to penetrate deep into the lungs. Particles smaller than 25 μm in diameter are known as PM-25; these are believed to cause a number of respiratory illnesses. Particles 10 μm in size, known as PM-10, are possibly even more injurious.

partridge wood Oak wood that has been subjected to a *pocket rot caused by the fungus *Stereum frustulatum*, which gives the wood a speckled appearance.

pascal **1.** **(Pa)** The derived SI unit of pressure, equal to 1 N/m^2. **2.** A high-level computer programming language. Both are named after the French mathematician Blaise Pascal (1623–62).

passive chamaephyte *See* CHAMAEPHYTE.

passive dispersal *See* DISPERSAL.

passive immunity Immunity against a given disease which is acquired by the injection into the host of serum containing *antibodies that have been formed by a donor organism itself possessing active immunity to the disease.

Pasteur effect The transition from an anaerobic to an aerobic lifestyle, which occurs among certain organisms when the oxygen content of the atmosphere is 1 per cent of that obtaining now. The critical point of transition is the 'Pasteur point'. The gradual oxygen enrichment of the Earth's atmosphere during the *Precambrian passed through the Pasteur point approximately 700 Ma ago, resulting in a general transition to an aerobic lifestyle.

Pasteur point *See* PASTEUR EFFECT.

past-natural *See* NATURAL.

Pastonian A Middle *Pleistocene stage represented by estuarine silts and freshwater *peat, revealed as marine clays in the borehole at Ludham, Norfolk, England.

Patagonian floral region Part of R. Good's (1974, *The Geography of the Flowering*

Plants) Antarctic floral kingdom: it has a small flora with a modest number of endemic genera (*see* ENDEMISM). It comprises the lowland regions of the southern extremities of South America. The Falkland Islands (Malvinas) are also included, but despite their isolation they have no well-known endemic genera. There are clear relationships with the flora of New Zealand. *See also* FLORAL PROVINCE and FLORISTIC REGION.

patch dynamics In *metapopulation studies, the study and mathematical modelling of the proportion of the feeding areas (patches) in a habitat that are occupied by the species under study at any given time. A patch may be empty, may contain just the one species (resulting in intraspecific competition), may contain a different species, or may contain both species (resulting in interspecific competition).

patch reef *See* REEF.

patch residence time The length of time an animal spends foraging in a particular area (i.e. a patch) of its habitat before diminishing returns compel it to move elsewhere.

'patella' beach An old name for a raised beach standing about 3 m above the present high-water mark and found locally on the south coast of England, in southern and western Ireland, and along the Channel coast of France. Traditionally it was given an *interglacial age, but the term is now little used, owing to problems of correlation. It is named after the large number of limpets (*Patella vulgata* is the common limpet) it typically contains.

paternoster lake A body of water in a formerly glaciated environment which is aligned with neighbouring lakes, so that it looks like a paternoster in a rosary. It is caused by irregular glacial scouring along a zone of weakness.

pathogen Any *micro-organism that causes disease. Pathogens may be ecologically important in controlling the distribution of *species and interspecific and intraspecific *competition.

pathogenesis The process of disease development.

pathognomonic Applied to symptoms that are characteristic of a particular disease and may therefore be useful in diagnosis.

pattern analysis Any analysis for the detection of a non-random distribution of organisms (e.g. *nearest-neighbour analysis or classification techniques). However, the term is applied more particularly to the detection of patterns in the distribution of individuals of a given plant species in a community, by comparing the observed number of individuals per *quadrat with the expected number derived from the Poisson series (based on a population that is dispersed at random). The correspondence is tested for statistical significance either by comparing observed numbers with those predicted using the χ^2 test, or by comparing the variance:mean ratio for the data with the value for the Poisson series of unity. Where this ratio is significantly greater or less than unity, the population is respectively overdispersed (contagious) or underdispersed (regular). Though first applied to studies of plant ecology, the technique is now widely used in other fields.

patterned ground An assemblage of small-scale, geometric features typically found at the surface of a *regolith that has been disturbed by frost action. The group includes circles, polygons, and nets, which normally occur on level or gently sloping surfaces, and steps and stripes, which are found on steeper gradients. Both sorted and non-sorted varieties are recognized. The sorted varieties are typically outlined by coarse, stony material, and so are termed 'stone circles', 'stone polygons', 'stone nets', 'stone steps', and 'stone stripes'. The origin of patterned ground involves a complex interaction of several geomorphological processes, including ground cracking, frost sorting, *frost heaving, and *mass movement. The 'ice-wedge polygon' is an important member. It is usually 15–30 m in diameter and bounded by *ice wedges up to 3 m wide and about 10 m deep which occupy contraction cracks that form under very low temperatures. The wedges define raised zones (when freezing is active) or depressions (owing to thaw). The 'stone garland' is a variety of sorted step, which ends in a stony riser (less than 1 m high) supporting a relatively bare tread (less than 8 m long) upslope. It is found on gradients of 5–15° (in Alaska) and may be caused by a combination of *frost pull and frost push, which heave stones to the surface, and mass movement. Patterned ground may also be found in areas underlain by montmorillonitic soils experiencing markedly seasonal rainfall, where the microrelief forms are called *gilgai.

Pavlov, Ivan Petrovich **(1849–1936)** A Russian physiologist who is renowned for his research into blood circulation, digestion (for which he was awarded the 1904 Nobel Prize), and conditioned behaviour in animals. From 1928 until his death he concentrated on the application of the principles of conditioning to psychiatric therapy. His approach to his behavioural work was based on the application of objective experiment to mental processes and the rejection of the mind–body dualism characteristic of earlier psychological studies.

PAW *See* PLANT AVAILABLE WATER.

pawpaw *See* DISJUNCT DISTRIBUTION.

PCB *See* POLYCHLORINATED BIPHENYL.

PCR *See* POLYMERASE CHAIN REACTION.

PE *See* POTENTIAL EVAPOTRANSPIRATION.

peach leaf curl A disease which affects *Prunus* species, including peach, nectarine, almond, and (less often) apricot trees. The characteristic symptoms are the curling, thickening, and distortion of the leaves, which typically become bright red. The disease is caused by a fungus, *Taphrina deformans*.

peak zone *See* ACME ZONE.

peat An organic soil (*Histosol) in which the O *Soil horizon is at least 40 cm thick (and often much deeper) and the dry weight contains a minimum of 65 per cent organic matter, the remainder being of mineral material. *Ombrotrophic peat, which has developed in situations where the sole water supply is from rainfall, generally has a much higher proportion of organic matter, often as high as 99 per cent dry weight. Peat formation occurs when decomposition of plant material is slow owing to the *anaerobic conditions associated with waterlogging. Decomposition of cellulose and hemicellulose is particularly slow for *Sphagnum* plants (bog mosses), which are characteristic of such sites and hence among the principal peat-forming plants. In addition to *Sphagnum* the plant material consists mainly of moor grasses and heather. The extent of decomposition increases with depth, so there is a progressive transition from fibrous and often identifiable plant residues near the surface to highly humified (*see* HUMIFICATION) material lower down. *Fen and *bog peats differ considerably. In fen peats the presence of calcium in the *groundwater neutralizes the acidity, often leading to the disappearance of plant structure, giving a black, structureless peat. Bog peats, formed in much more acidic waters, vary according to the main plants involved. It often remains possible to identify animal and plant species for a considerable time. Recent *Sphagnum* peat is light in colour, with the structure of the mosses perfectly preserved.

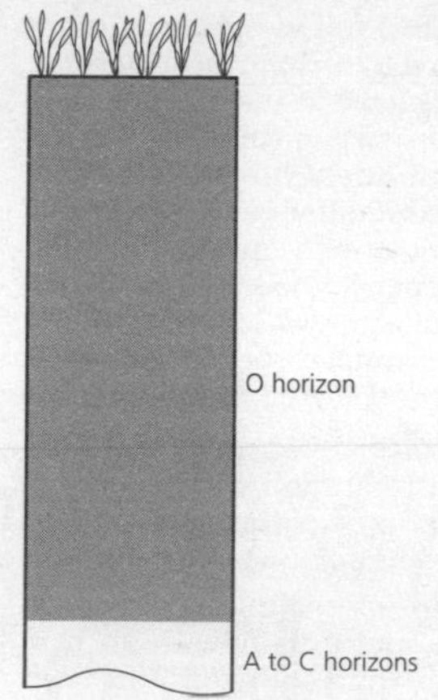

Peat

peat-borer An implement designed to extract peat cores with the minimum of disturbance. The most familiar is the Hiller peat-borer, which consists of a short screw auger head to ease penetration of the peat, backed by a chamber which can be opened and closed at the required sample depth, the sharp cutting edge of the chamber assisting detachment of the sample in more consolidated peats. The principal alternatives are the piston sampler, which is particularly good for loose peats, and the Russian borer, which allows easier removal of the complete peat core than is possible with the Hiller borer, but which is more difficult to use in compacted peats since it has no screw auger head.

peat podzol A *podzol soil *profile distinguished by having a surface *mor (peaty) *humus up to a maximum thickness of 30 cm, and usually with an iron pan at the top of the B *horizons. The term occurs in most of the classification systems derived originally from the work of V. V. *Dokuchaev, published in 1886. It has been superseded by the *USDA Soil Taxonomy, where podzols fall within the order *Spodosols. *See* SOIL CLASSIFICATION.

pecking order The name given to the hierarchical social organization found in some insect species (e.g. ants and bees) and in many vertebrate species. It is so called because the phenomenon was first described for chickens. Based on *dominance, it allows each member of the group to threaten (or actually peck) the individual immediately subordinate to it and so gain prior access to food or other resources.

pectinate Resembling a comb in arrangement or shape.

ped A unit of soil structure (e.g. an aggregate, crumb, granule, or prism) that is formed naturally. *Compare* CLOD.

pedalfer A freely draining *acid soil that develops in regions with a wet climate. The soil is wetted to its *field capacity and water drains all the way to the *groundwater, leaching out the soluble constituents of the soil. Water moving downwards removes aluminium (*al*) silicates (e.g. clays) and iron (*fer*) oxides from the soil (*ped*) and precipitates them 30–60 cm below ground level. The acidity of the water decreases as it moves downward, and alkaline earths and alkalis

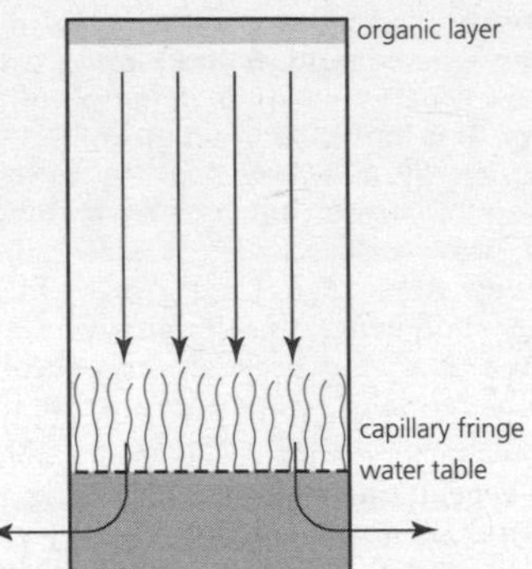

Pedalfer

drain away from the soil completely. The process is known as *podzolization and the resulting soil is a *podzol. C. F. Marbut introduced the term in 1928 as part of a twofold division in his system of *soil classification based on *pedogenesis, with the intention of using the basic soil type used in mapping as the lowest category in classification. Whether this is possible is controversial, but the Marbut classification was adopted by the US Soil Survey, eventually to be replaced by the *USDA Soil Taxonomy, and the term is still widely used. *Compare* PEDOCAL.

pedestal rock (mushroom rock) An unstable, mushroom-shaped land-form found typically in arid and semi-arid regions. The undercut base was formerly attributed to wind abrasion, but is now believed to result from enhanced chemical weathering at a site where moisture would be retained longest. A famous example is Pedestal Rock, Utah, USA.

pediment (concave slope, waning slope) A surface of low relief, partly covered by a skin of rock debris, which is concave-upward and slopes at a low angle (normally less than 5°) from the base of a mountain zone or *scarp. Classically it is developed and has been investigated in the arid and semi-arid regions of the western USA. Pediments may coalesce to form a *pediplain.

pediplain An extensive plain, best developed in arid and semi-arid regions, showing gently concave or straight-slope profiles and terminated abruptly by uplands. A result of scarp recession rather than of surface lowering, it consists of coalesced *pediments.

pedocal An *alkaline soil that develops in regions with a dry climate, where the rainfall is sufficient only to wet the upper layer of soil. Although the soil drains freely, water does not reach the *water-table. Vegetation is sparse, *leaching is light, and the soil air is only slightly enriched in carbon dioxide resulting from decomposition. These conditions favour the precipitation of carbonates, especially calcium carbonate. All the transported material accumulates in a layer at the depth to which water penetrates. The process is called *calcification (*cal*). C. F. Marbut introduced the term in 1928 as part of a twofold division in his system of *soil classification based on *pedogenesis, with the intention of using the basic soil type used in mapping as the lowest category in classification. Whether this is possible is controversial, but the Marbut classification was adopted by the US Soil Survey, eventually to be replaced by the *USDA Soil Taxonomy, and the term is still widely used. *Compare* PEDALFER.

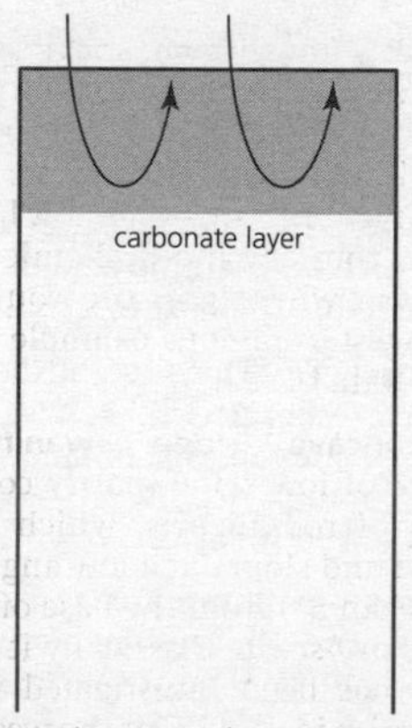

Pedocal

pedogenesis The natural process of soil formation, including a variety of subsidiary processes such as humification, *weathering, *leaching, and *calcification.

pedology The scientific study of the composition, distribution, and formation of soils, as they occur naturally.

pedon A three-dimensional sampling unit of soil, with depth to the parent material and lateral dimensions great enough to allow the study of all *horizon shapes and *intergrades below the surface.

peel technique A technique originally developed for palaeobotanical work, but then refined and now used extensively in carbonate sedimentology and palaeontological work. Calcareous material is etched in a weak solution of hydrochloric acid and differences in relief are produced. After washing, the surface is flooded with acetone, and polyvinylacetate (PVA) sheeting is rolled on to the surface. The acetone softens the sheeting and moulds it to the etched rock surface. After drying the sheeting is peeled from the surface, bringing a thin layer of the surface with it. This 'peel' can then be examined in transmitted light. A series of peels can be taken to reveal and reconstruct buried structures, and staining with various chemicals may reveal additional details.

pelage The hair covering the body of a mammal.

pelagic **1.** In marine ecology, applied to the organisms that inhabit open water, i.e. *plankton, *nekton, and *neuston (although neuston are fairly unimportant in such environments). **2.** In ornithology, applied to sea-birds that come to land only to breed, and spend the major part of their lives far out to sea.

pelagic ooze Deep-ocean sediments that accumulate by the settling out of materials from the overlying ocean waters. The dominant constituents are microscopic *pelagic organisms (e.g. the calcareous globigerina and pteropods and the siliceous diatoms and radiolaria). Minor amounts of fine volcanic, terrige-

p

nous, and extraterrestrial debris also contribute to pelagic ooze.

pelagic zone Waters of the open ocean, extending from the edge of the *continental shelf. It is subdivided horizontally on the basis of conditions obtaining at different depths into the *epipelagic, *mesopelagic, *bathypelagic, *abyssalpelagic, and *hadalpelagic zones. *See* OCEAN DIVISIONS.

pellicular water (film water) Thin films of water that cling to soil and rock particles above the *water-table.

pene- From the Latin *paene*, meaning 'almost', a prefix meaning 'almost' or 'nearly' (e.g. a *peneplain is almost a plain).

peneplain (peneplane) Literally, almost a plain: an extensive area of low relief, dominated by convex-up hill-slopes mantled by a continuous *regolith, and by wide, shallow river valleys. Locally, *monadnocks may occur. A peneplain is the end-product of a cycle of erosion, produced by the action of down-wearing over a long period of time, and it is the end-product of the *Davisian cycle.

peneplane *See* PENEPLAIN.

penesaline A level of *salinity intermediate between normal marine and hypersaline, ranging from 72 parts per thousand to 352 parts per thousand. These salinity levels are high enough to be toxic to normal marine organisms and can be tolerated only by a restricted range of fauna and flora. The characteristic sediments of penesaline zones are evaporitic carbonates interbedded with anhydrite or gypsum. Penesaline environments are often encountered in the back-barrier and backreef zones.

penetrance The proportion of individuals of a specified *genotype who manifest that genotype in the *phenotype under a defined set of environmental conditions. If all individuals carrying a lethal *mutation die prematurely, then the mutant gene is said to show complete penetrance. An organism may not express the phenotype normally associated with its genotype because of the presence of modifiers, epistatic genes (*see* EPISTASIS), or suppressors in the rest of the *genome; or because of a modifying effect of the environment. Expressivity describes the degree or extent to which a given genotype is expressed phenotypically in an individual.

penicillin A type of antibiotic produced, for example, by fungi of the genus *Penicillium*. Penicillins are active against certain types of bacteria (mainly Gram-positive species, *see* GRAM STAINING) and are widely used in the treatment of diseases in animals caused by those bacteria. (Pencillin G was one of the first antibiotics to be used for the treatment of disease.) There is now a wide range of chemically modified penicillins, each with slightly different properties. They function by inhibiting the synthesis of bacterial cell walls.

Pennsylvanian **1.** The late *Carboniferous sub-period, preceded by the *Mississippian, comprising the Bashkirian, Moscovian, Kasimovian, and Gzelian Epochs, and dated at 318.1–299.0 Ma. **2.** The name of the corresponding North American subsystem, comprising the Morrowan, Atokan (Derryan), Desmoinesian, Missourian, and Virgilian Series, and roughly contemporaneous with most of the Silesian subsystem (i.e. above Namurian A).

peppered moth *See* *BISTON BETULARIA*.

peptide A linear molecule comprising two or more *amino acids linked by *peptide bonds. The simplest peptide is $H_2N.CH_2.CO.NH.CH_2.CO_2H$ (glycylglycine).

peptide bond A chemical bond (–CO.NH–) by which one *amino acid molecule may be linked to the next in a chain.

perambulation **1.** The established boundaries of a forest or parish. **2.** A legal document that defines an area of land by describing its boundaries. **3.** A walk around established boundaries.

peramorphosis Evolutionary change that results in the descendant incorporating all the ontogenetic stages of its ancestor, including the adult stage, in its *ontogeny, so that the adult descendant 'goes beyond' its ancestor. It may occur by *acceleration, *hypermorphosis, or *predisplacement. Such progeny will show *recapitulation of phylogeny.

percentage cover *See* COVER.

perception The appreciation of the external environment by means of the senses.

perched aquifer *See* AQUIFER.

perched block A glacially transported boulder that rests on bedrock where it was deposited by melting ice. Some examples are spectacular and give rise to local legends.

perched water-table The upper boundary of water that is held above the general *water-table by a feature such as a layer of impermeable material beneath the material in which the water is held. For example, water in the *acrotelm of a *raised bog is held in an elevated position by the impermeable catotelm peat beneath. *See* AQUIFER.

percolation The downward movement of water through soil, especially through soil that is saturated or close to saturation.

perennating bud **(perennating organ)** The vegetative means whereby *biennial and *perennial plants survive periods of unfavourable conditions. The aerial parts die back to a minimum at the onset of unfavourable conditions, and food for the new shoots of the next growing season is stored in underground organs (e.g. tubers, bulbs, rhizomes), or in buds on the stems of woody plants. Seeds may also be considered perennating organs.

perennating organ *See* PERENNATING BUD.

perennial A plant that normally lives for more than two seasons and which, after an initial period, produces flowers annually.

perennial stream A stream that normally flows throughout the year, albeit with low dry-weather flows in occasional drought years.

pergelic The lowest of the soil-temperature classes for family groupings of soils in the *USDA Soil Taxonomy system, applied to soils in temperate regions. The assessment of soil temperature is based on mean annual temperatures, and on differences between mean summer and mean winter temperatures, measured at a depth of 50 cm or at the surface of the underlying rock, whichever is shallower. Higher temperature classes in temperate region soils are called cryic, frigid, mesic, thermic, and hyperthermic, and in tropical regions the scale from cold to hot is isofrigid, isomesic, isothermic, and isohyperthermic.

pergelisol *See* PERMAFROST.

peri- The Greek *peri*, meaning 'around', used as a prefix meaning 'around' or 'enveloping'.

periglacial Applied strictly to an area adjacent to a contemporary or *Pleistocene *glacier or *ice sheet, but more generally to any environment where the action of freezing and thawing is currently, or was during the Pleistocene, the dominant surface process.

periphyton **(aufwuchs)** Organisms attached to or clinging to the stems and leaves of plants or other objects projecting above the bottom sediments of freshwater *ecosystems.

peritidal Applied to the zone extending from above the level of the highest tide to below that exposed at the lowest tide; a belt somewhat wider than the *intertidal zone.

perlucidus The Latin *perlucidus*, meaning 'allowing the passage of light', used to describe a variety of cloud comprising extensive layers or sheets with holes which allow a view beyond the cloud. The term is applied to *stratocumulus and *altocumulus. *See also* CLOUD CLASSIFICATION.

permafrost **(pergelisol)** The permanently frozen ground which occupies some 26 per cent of the Earth's land surface under thermal conditions where temperatures below 0°C have persisted for at least two consecutive winters and the intervening summer. Considerable thicknesses may develop (e.g. 600 m on the North Slope of Alaska and 1 400 m in Siberia, but these are partly relicts of the last glaciation). Permafrost may contain an unfrozen unit, called 'talik', and may be overlain by an *active layer. The permafrost may be continuous, discontinuous, intermittent, or sporadic.

permafrost table The upper limit of *permafrost. *Compare* TJAELE.

permanent quadrat A *quadrat repeatedly recorded in detail over a number of years in order that plant relationships and changes in them over time may be studied in detail.

permanent wilting percentage *See* PERMANENT WILTING POINT.

permanent wilting point (PWP, permanent wilting percentage, wilting coefficient, wilting point) As moisture is lost from the soil, the point at which the force with which the remaining moisture adheres to soil particles exceeds that exerted by plant roots. Plants are therefore unable to absorb moisture and *wilting results. Since this condition arises from the amount of water present in the soil, plants will not recover unless water is added to the soil, i.e. the wilting is permanent. The permanent wilting point usually occurs when soil moisture is held with a force of about 15 MPa (15 bar). It is also measured as the percentage of moisture remaining in the soil after a specified test plant has wilted under defined conditions and will not recover unless water is added to the soil. *Compare* TEMPORARY WILTING.

permeability **1.** A property of a membrane or other barrier, being the ease with which a substance will diffuse or pass across it. **2.** The ease with which gases, liquids, or plant roots penetrate into or pass through a layer of soil.

permeability coefficient A quantitative estimate of the rate of passage of a solute across a membrane. In a concentration of 10 moles/litre it represents the net number of solute molecules crossing 1 cm^2 of membrane per unit time.

permeameter A laboratory device for measuring the hydraulic conductivity of rock and soil samples or the *permeability coefficient of soil. Two main types are commonly used: those that require movement of water and those that do not, known respectively as falling head and constant head permeameters.

Permian The final period of the *Palaeozoic Era, about 299–251 Ma ago. It is named after the central Russian province of Perm. The period is often noted for the widespread continental conditions that prevailed in the northern hemisphere and for the extensive nature of the southern hemisphere glaciation. Many groups of animals and plants (including the rugose corals, trilobites, and blastoid echinoderms) vanished at the end of the Permian in one of the most extensive of all mass extinctions. It was during this period that the Pteridophyta were superseded as the dominant vegetation by the gymnosperms and the first gliding flight by tetrapods occurred (about 260 Ma ago).

permineralization *See* FOSSILIZATION.

peroxyacetyl nitrates *See* PHOTOCHEMICAL SMOG.

perthophyte A plant-*parasitic fungus that derives its nutrients from dead tissues within a living host plant.

perturbation A disturbance to an *ecosystem that may be due to a natural catastrophe, such as a fire, or may be deliberately induced as part of an experimental process to test the stability (resilience, *see* STABILITY (2) or *inertia) of the ecosystem.

Peru–Chile Trench The oceanic *trench which marks the boundary between the subducting Nazca Plate and the South American Plate.

Peru Current (Humboldt Current) The oceanic water that flows northwards along the west coast of Chile and Peru, driven by the westward flow of the South *Equatorial Current (itself driven by south-east trade winds). It is essentially a 'continuity current': water flows into the low-sea-level region left by the South Equatorial current. This eastern *boundary current is slow-moving, broad, and shallow, and is noted for a prominent area of *upwelling bringing cold bottom waters to the surface. As the near-surface concentrations of nutrient elements are high there is an abundance of marine life associated with this current.

pest An animal that competes with humans by consuming or damaging food, fibre, or other materials intended for human consumption or use. Many such species are harmless or ecologically beneficial (e.g. raptors, otters, and seals); others (e.g. most insect pests) are harmless until their populations increase rapidly in response to a virtually unlimited (to them) resource (e.g. a farm crop). *See also* WEED.

peta- **(P)** A prefix used with *SI units meaning the unit × 10^{15}. The prefix is derived from the Greek *pente*, meaning five, because 10^{15} is 1 000 raised to the fifth power.

petiole **1. (plant)** The stalk attaching a leaf to the plant. *See* EPINASTIC GROWTH. **2. (animal)** *See* GASTER.

petrifaction *See* FOSSILIZATION.

petrified forest **(submerged forest)** An area of *peat containing eroded tree stumps which is exposed along a coastline at low tide. Its presence indicates a rise in sea level or a lowering in the level of the land. Post-glacial petrified forests are common in the estuaries of south-west England.

petrocalcic horizon An indurated (*see* INDURATION) *calcic horizon that is cemented by a high concentration of calcium carbonate, often comprising 40 per cent by weight of the mineral material, and which is impenetrable to plant roots or to spades used for digging.

petrogypsic horizon A surface or subsurface *soil horizon cemented by gypsum so strongly that dry fragments will not slake in water. The cementation restricts penetration by plant roots. This is a *diagnostic horizon.

petroleum **(crude oil)** Naturally occurring liquid hydrocarbons formed by the anaerobic decay of organic matter. Oil is rarely found at its original site of formation but migrates to a suitable structural or lithological trap. Petroleum is frequently associated with salt water and with gaseous hydrocarbons. *See also* NATURAL GAS and OIL SHALE.

pH A value on a scale 0–14 which gives a measure of the acidity or alkalinity of a medium. A neutral medium has a pH of 7, acidic media have pH values of less than 7, and alkaline media of more than 7. The lower the pH the more acidic is the medium, the higher the pH, the more alkaline. The pH value is the logarithm of the reciprocal of the hydrogen ion concentration, expressed in moles per litre ($pH = \log_{10} 1/H^+$). Most pH values in natural systems lie in the range 4–9. Human blood has a pH of 7.4, ocean water 8.1–8.3, water in saline environments may have a pH around 9.0 or higher, and water in acidic soils may have a pH of 4.0 or less.

phaeomelanic Red or yellow, as applied to the colour of mammalian hair. *Compare* EUMELANIC.

Phaeozems Any soil with a *mollic horizon, apart from *Chernozems and *Kastanozems. Phaeozems are a reference soil group in the *FAO *soil classification.

phage *See* BACTERIOPHAGE.

phagotroph *See* CONSUMER ORGANISM.

phagotrophy A mode of nutrition in which particulate food is ingested.

phalanx growth form The distribution that results when the *rhizomes or *stolons by which a plant spreads by *clonal dispersal are short and often long-lived. The clonal shoots are closely spaced and the clone advances along a densely packed front, like a Roman phalanx, *Compare* GUERILLA GROWTH FORM.

phallotoxin A member of a group of poisonous substances present in the fruit bodies of certain species of *Amanita* fungi (especially *A. phalloides*). Ingestion by humans causes vomiting and diarrhoea, and liver damage may occur.

phanerophyte One of *Raunkiaer's life-form categories, being a plant whose *perennating buds or shoot apices are borne on aerial shoots. Such plants are the least protected of those in Raunkiaer's scheme and therefore are most typical of environments where drought, cold, and exposure to strong winds are relatively infrequent. Several subcategories are recognized, the most universal being: (*a*) evergreen phanerophytes without bud scales (tropical trees) or with bud scales; and (*b*) deciduous phanerophytes. These groups may be subdivided further according to height as nano- (less than 2 m), micro- (2–8 m), meso- (8–30 m), and mega- (more than 30 m) phanerophytes. Other subgroups are epiphytic, stem succulent, and herbaceous phanerophytes, the last being confined to tropical environments. *Compare* CHAMAEPHYTE; CRYPTOPHYTE; HEMICRYPTOPHYTE; and THEROPHYTE.

Phanerozoic The period of geological time that comprises the *Palaeozoic, *Mesozoic, and *Cenozoic Eras. It began 542 Ma ago at the end of the *Proterozoic and is marked by the accumulation of sediments containing the remains of animals with mineralized skeletons. The name means the period of 'visible or obvious life', but is no longer used in this sense, merely defining the base of the Cambrian.

phase An individually distinct and homogeneous part of a system. For example, liquid water and water vapour are each single phases; a mixture of the two constitutes a two-phase system.

phase diagram A graphical method for examining stability in biological systems (e.g. host-parasite relationships). The system variables (the numbers of host and parasite organisms) at different times are plotted against one another, and the time taken to change is shown by a line connecting the coordinate points for successive time values. The form of the resulting curve indicates the stability of the system. A line spiralling inward indicates damped oscillations favouring ultimate stability; an outward spiral suggests ultimate instability. By testing different values for the number of organisms it is possible to establish limits for stability. The technique has considerable practical relevance in agriculture.

phenetic classification The grouping of biological organisms on the basis of observed physical similarities. *See* PHYLOGENETIC SYSTEMATICS.

phenogram A tree-like diagram that is used in analysis to show similarity or dissimilarity among specimens or groups of specimens.

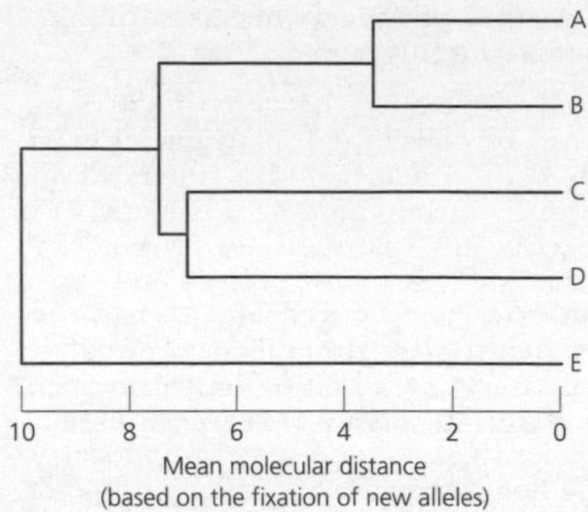

Phenogram

phenology **1.** The study of the periodicity of leafing, flowering, and fruiting in plants; these are generally triggered by periodicities in the climate. **2.** The study of the impact of climate on the seasonal occurrence of flora and fauna (dates of flowering, migration, etc.) and of the periodically changing form of an organism, especially as this affects its relationship with its environment.

phenotype The observable manifestation of a specific *genotype; those observable properties of an organism produced by the genotype in conjunction with the *environment. Organisms with the same overall genotype may have different phenotypes because of the effects of the environment and of gene interaction. Conversely, organisms may have the same phenotype but different genotypes, as a result of incomplete dominance, *penetrance, or expressivity.

phenotypic adaptation *See* ADAPTATION.

phenotypic plasticity The capacity of a *phenotype to vary, owing to environmental influences on the *genotype (e.g. in the shape of a plant or the colour of its flowers).

phenotypic variance The total variance observed in a *character. *See also* GENETIC VARIANCE and HERITABILITY.

phenylmercury compounds *See* ORGANOMERCURY COMPOUNDS.

pheromone A chemical substance, produced and released into the environment by an animal, which then elicits a physiological and/or behavioural response in another individual of the same species. For example, pheromones are released by a variety of glands on the abdomen, head, and wings of insects.

philopatry Literally, love of one's native habitat. It is the tendency of an individual not to disperse from its natal location, its home area. Most animal species show some degree of philopatry. The house sparrow (*Passer domesticus*), for example, is in decline in many parts of the world and its recovery is poor partly because of its low capacity to spread into vacated habitats. The corncrake (*Crex crex*) is a migratory species, but is also philopatric in that it settles in the locality where it originated, and is therefore slow to expand its range. One sex may be more philopatric than the other. For example, male birds tend to be more philopatric than females, and female mammals than males. *Compare* DISPERSAL.

phobotaxis A random change in the direction of locomotion of a *motile micro-organism or cell which is made in response to a given stimulus.

phoresy A method of dispersal in which an animal clings to the body of a much larger animal of another species and is carried some distance before releasing its grip and falling.

phosphate rock A rock or deposit that is composed largely of inorganic phosphate, commonly calcium phosphate. Apatite ($Ca_5(PO_4)_3(F,Cl,OH)$) is the most frequently encountered phosphate mineral and the principal source of phosphate fertilizer.

phosphorus **(P)** An element that is an essential nutrient for all living organisms. Plants require it in the oxidized form, as orthophosphate (PO_4^{2+}). The growth of phosphorus-deficient plants is usually reduced and their leaves become dark green or blue-green and a reddish pigment may develop.

phosphorylation *See* LIGHT REACTIONS.

photic zone The layer of water within which organisms are exposed to sunlight. It is divided at the *compensation level, with the *euphotic zone above and the *dysphotic zone below. *See* OCEAN DIVISIONS. The upper, well-lit layers of terrestrial *ecosystems with complex structure (e.g. rainforest canopies) are sometimes referred to as the euphotic zone. The lower, shaded regions are the oligophotic zone.

photo- From the Greek *phos, photos*, meaning 'light', a prefix meaning 'light'.

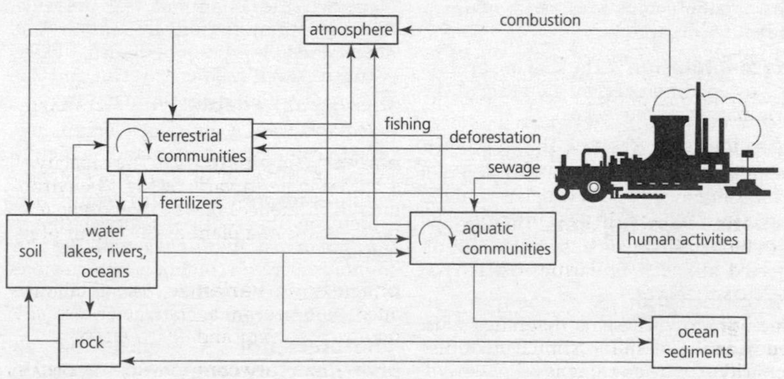

Phosphorus

photo-autotroph A *phototroph that uses carbon dioxide compounds as its main or sole source of carbon.

photoblastic Applied to seeds that use light to signal when conditions are right for germination.

photochemical reaction A chemical reaction that is induced by light (e.g. *photosynthesis).

photochemical smog A hazy condition of the atmosphere caused by the reaction of hydrocarbons with molecules of nitrogen oxide in sunlight, which produces complex organic molecules of peroxyacetyl nitrates (PAN). In humid conditions these molecules produce *smog. Such phenomena are common in large urban areas (e.g. the Los Angeles basin and Athens) where there are stable atmospheric conditions and a high level of hydrocarbon input from incomplete combustion in car engines. Natural photochemical reactions occur in the high atmosphere with the absorption of radiation by oxygen to produce ozone. *See also* OZONE LAYER.

photodisintegration The decomposition of a compound in the presence of light, particularly sunlight.

p

photodissociation The splitting of a molecule into smaller molecules or atoms as a consequence of its absorption of electromagnetic radiation.

photoheterotroph A *phototroph that uses organic compounds as its main or sole source of carbon.

photo-inhibition The slowing or stopping of a plant process by light (e.g. in the germination of some seeds).

photokinesis A change in the speed of locomotion (or frequency of turning) in a *motile organism or cell which is made in response to a change in light intensity. The response is unrelated to the direction of the light source. *Compare* PHOTOTAXIS.

photolithotroph A *phototroph in which photosynthesis is associated with the oxidation of an inorganic compound (or element, in the case of some photosynthetic bacteria).

photolysis The breakdown of water ($H_2O \rightarrow H + OH$) by electromagnetic radiation. *See* LIGHT REACTIONS.

photolytic cycle A naturally occurring sequence of chemical reactions that take place in the lower atmosphere, driven by the energy of ultraviolet radiation, in which nitrogen dioxide (NO_2) breaks down and reforms, with ozone (O_3) as an intermediate product:

$$NO_2 + UV \rightarrow NO + O$$
$$O + O_2 \rightarrow O_3$$
$$O_3 + NO \rightarrow NO_2 + O_2$$

If the air contains unburned hydrocarbons from vehicle exhausts the natural cycle is disrupted and other compounds are produced, leading to *photochemical smog.

photomorphogenesis The regulation of plant form and growth by light.

photonasty A nastic response (*see* NASTY) of a plant organ to the stimulus of light.

photo-organotroph A *phototroph (e.g. certain photosynthetic bacteria) in which *photosynthesis is associated with the oxidation of organic compounds.

photoperiod The relative periods of light and darkness associated with day and night.

photoperiodism The response of an organism to periodic, often rhythmic, changes either in the intensity of light or, more usually, to the relative length of day. Many activities of animals (e.g. breeding, feeding, and migration) are seasonal and determined by photoperiodism. Many plant activities, such as flowering and leaf-fall, are also photoperiodic in nature.

photophore A luminous organ, modified from a mucous gland, which is found in the skin of fish. Rows of photophores are present in many deepsea fish, which can apparently produce flashes of blue-green to orange light at will. The photophores secrete a compound which glows when activated, or they contain colonies of phosphorescent bacteria.

photoreceptor A pigment that absorbs light, the energy of which is used by the organism possessing it.

photorespiration A process in plants that reduces the efficiency of *photosynthesis by the *C_3 pathway. The active site of rubisco, the enzyme that catalyses the fixation of CO_2 at the start of the *dark reactions, accepts either O_2 or CO_2. The two gases therefore compete and if the CO_2 concentration in the leaf is low, as it is during photosynthesis when CO_2 is being consumed, rubisco adds O_2 rather than CO_2 to ribulose biphosphate (RuBP). This alters the sequence of reactions, resulting in the release of some of the CO_2 that was fixed at the start of the cycle. The process is called respiration because it absorbs O_2 and releases CO_2, but it does so without yielding any energy. *Glycine max* (soybean) loses up to half of the CO_2 fixed during the dark reactions due to photorespiration.

photosynthetic quotient In *photosynthesis, the volume of oxygen released as a proportion of the volume of carbon dioxide used.

photosynthesis The series of metabolic reactions occurring in certain *autotrophs, whereby the energy of sunlight, absorbed by *chlorophyll, powers the reduction of carbon dioxide (CO_2) and the synthesis of organic compounds. In green plants, where water (H_2O) acts as both a hydrogen donor and a source of released oxygen, photosynthesis may be summarized by the empirical equation:

$$CO_2 + H_2O \xrightarrow[\text{light}]{\text{chlorophyll}} [CH_2O] + H_2O = O_2\uparrow$$

See DARK REACTIONS; LIGHT REACTIONS; and PHOTO-INHIBITION.

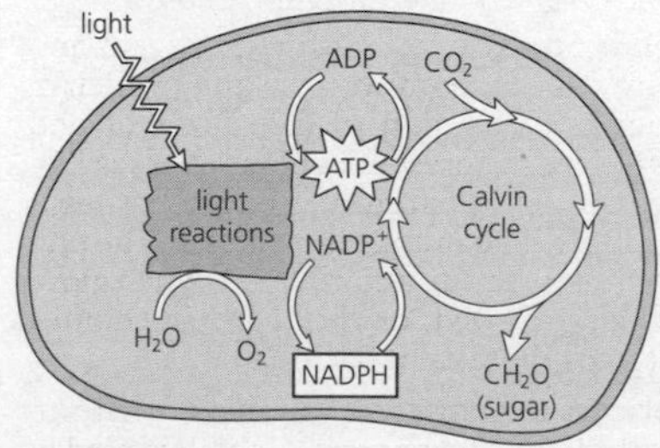

Photosynthesis

phototaxis A change in direction of locomotion in a *motile organism or cell which is made in response to a change in light intensity. The response is related to the direction of the light source. *Compare* PHOTOKINESIS.

phototroph An organism that obtains its energy from sunlight, in most cases by *photosynthesis.

phototropic *See* HELIOTROPIC.

phototropism *See* HELIOTROPISM.

phragmosis *See* SOLDIER.

phreatic From the Greek *phrear, phreatos*, meaning 'well', applied to water that is below ground level but can be reached by wells.

phreatic eruption A volcanic eruption caused by the interaction of hot *magma with surface lake water, sea water, or *groundwater. The water immediately surrounding the magma is heated and volatilized. Its expansion builds up pressure on the envelope of water surrounding it. When the pressure exceeds the confining pressure of the overlying water column the water vapour expands explosively to produce a steam-dominated eruption. Where significant amounts of magmatic material are ejected in addition to steam the activity is said to be 'phreatomagmatic'.

phreatic zone **(zone of saturation)** The soil or rock zone below the level of the *water-table, where all voids are saturated. *See also* VADOSE ZONE.

phreatomagmatic eruption *See* PHREATIC ERUPTION.

phycobiont The principal algal or cyanobacterial symbiont in a *lichen.

phycology **(algology)** The study of algae.

phycovirus A *virus that infects and replicates in algae.

phyletic Pertaining to a line of descent.

phyletic evolution Evolutionary change within a lineage, as a result of gradual adjustment to environmental stimuli.

phyletic gradualism A theory holding that *macro-evolution is merely the operation of *micro-evolution, which operates gradually and more or less continuously over relatively long periods of time. Thus gradual changes will eventually accumulate to the point at which descendants of an ancestral population diverge into separate species, genera, or higher-level taxa.

phyllody The condition in which parts of a flower are replaced by leaf-like structures; it is a symptom of certain plant diseases.

phyllosphere The micro-environment on and below the surface of a leaf.

phylogenetics The taxonomical classification of organisms on the basis of their degree of evolutionary relatedness.

phylogenetic systematics The study of biological organisms, and their grouping for purposes of classification, based on their evolutionary descent. *See also* CLADISTICS.

phylogenetic tree A variety of *dendrogram in which organisms are shown arranged on branches that link them according to their relatedness and evolutionary descent.

p

phylogenetic zone *See* LINEAGE ZONE.

phylogeny Evolutionary relationships within and between taxonomic levels, particularly the patterns of lines of descent, often branching, from one organism to another (i.e. the relationships of groups of organisms as reflected by their evolutionary history).

phylogerontism (racial senescence) The condition of an evolutionary lineage that is on the verge of extinction, according to an outmoded view of evolution which asserted that lineages proceed through a life cycle, from youth to senility. *See also* GENERIC CYCLES, THEORY OF.

phylum In animal taxonomy, one of the major groupings, coming below subkingdom and kingdom, and comprising superclasses, classes, and all lower taxa. Sometimes, confusingly, the term is used to mean 'lineage'. It is the root of 'phyletic', 'phylogeny', etc.

physical factor A factor in the *abiotic environment which influences the growth and development of organisms of biological *communities.

physiognomy The form and structure of natural communities. *Compare* MORPHOLOGY.

physiological ecology The study of the physiological functioning of organisms in relation to their environments.

phyto- A prefix meaning 'pertaining to plants', from the Greek *phuton*, meaning 'plant'.

phyto-alexin An antifungal substance produced by a plant in response to infection or damage.

phytocoenosis The primary producers (*see* PRODUCTION) that form part of the *biocoenosis in a *biogeocoenosis.

phytogeography (floristics) The study of the geography of plants, particularly their distribution at different taxonomic levels. Patterns of distribution are interpreted in terms of climatic and *anthropogenic influence, but above all in terms of earlier continental configurations and migration routes.

phytopathology The study of plant diseases.

phytophagous Feeding on plants.

phytoplankton The photosynthetic *plankton and primary producers of aquatic *ecosystems, comprising mainly diatoms in cool waters, *dinoflagellates being more important in warmer waters.

phytosociology The classification of plant *communities based on floristic rather than life-form or other considerations. The most renowned and extensively used phytosociological scheme is that developed by J. *Braun-Blanquet (1927 and later) and his associates at Zurich and Montpellier. A similar, fairly widely applied scheme was developed by G. E. *du Rietz (1921 and later) and others at Uppsala. These schemes are now being replaced by computer-based methods of quantitative phytosociology. *See* ASSOCIATION ANALYSIS and ORDINATION METHOD.

phytotoxin A substance poisonous to plants.

pico- The Spanish *pico*, meaning 'beak' or 'peak' (i.e. a point), used as a prefix (symbol p) with *SI units to denote the unit $\times 10^{-12}$.

piedmont The tract of country at the foot of a mountain range (e.g. the Po Valley, Italy, at the foot of the Alps). The word is derived from the Italian *piemonte*, meaning 'mountain foot'.

piedmont glacier A lobe of ice formed when a *valley glacier extends beyond its restraining valley walls and spreads out over the adjacent lowland or *piedmont zone. Much of the glacier surface is, therefore, at a low altitude and may show rapid *ablation. An example is the Malaspina Glacier, Alaska.

piezometer An observation well designed to measure the elevation of the *water-table or *hydraulic head of *groundwater at a particular level. The well is normally quite narrow and allows groundwater to enter only at a particular depth, rather than throughout its entire length.

piezometric surface *See* POTENTIOMETRIC SURFACE.

pileus The Latin *pileus*, meaning 'cap' used to describe an accessory cloud occurring as a small cap on or above a cumuliform cloud. The cloud is associated with *cumulus or *cumulonimbus. *See also* CLOUD CLASSIFICATION.

pillow lava Long piles of basaltic lava pods that have the general appearance of a stack, often many hundreds of metres thick, of discrete stone pillows, each 'pillow' rarely being more than 1 m in diameter. The morphology indicates that the 'pillows' continued to behave as fluid bodies after the chilled carapace had formed. This provides good evidence of submarine eruption: lava entering water acquires a glassy outer skin as heat is conducted rapidly from the surface. Because water absorbs heat more readily than air, with little increase in its own temperature, the rapid surface cooling allows the molten plastic state of the pillow interior to be maintained longer than it would be in air. Pillows have been observed forming under water from lava entering the sea off Hawaii.

pilose Covered with fine hair or down.

pilotage The steering of a course from one place to another by using familiar landmarks.

pine barren An area of pine forest in which the various species of pine usually develop as small or medium-sized trees. The barrens coincide with poor, sandy, and to a lesser extent marshy soil, and owe their ecological character in part to centuries of burning. They occur in the south-eastern USA and on the coastal plain of the Atlantic and Gulf of Mexico from New Jersey to Florida (excluding its southern tip) and into Texas.

pin-frame **(point frame)** A device for obtaining a quantitative estimate of vegetation *cover. Pin-frames are typically made from lightweight wood, aluminium, or plastic and comprise a cross-bar with pin-holes supported on legs of adjustable height. The pin-frame is set up so that the cross-bar sits above the vegetation to be sampled. Pins are lowered through the pin-holes and the plants hit by the pin-tips are recorded. Where the vegetation is of variable height, records of top-cover (the first plant encountered by the pin) and bottom-cover may be taken. The pin diameter used will affect the results and must be standardized for comparative work.

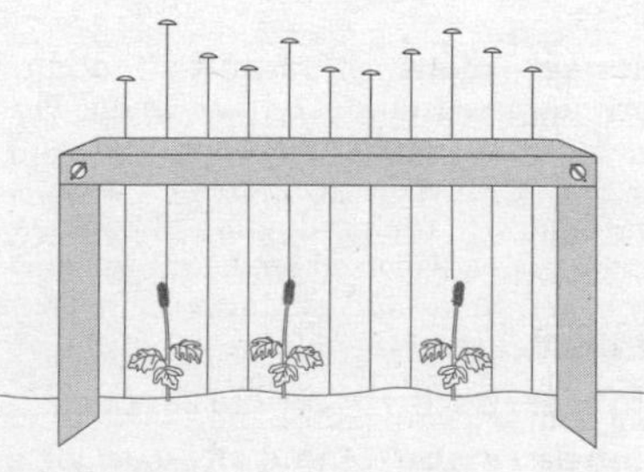

Pin-frame

pingo An ice-cored, dome-shaped hill, oval in plan, 2–50 m high and 30–600 m in diameter, developed in an area of *permafrost. The larger examples have breached crests in which ice may be exposed. They are probably due to local freezing of water that has migrated from adjacent uplands, or to the late freezing of the ground beneath a lake. Pingos can also develop along valley sides where water under pressure emerges as a spring line and contributes to the ice development from below. Fossil pingos take the form of circular or arced, raised ramparts with depressions in their centres that usually contain water bodies, and often occur in groups. Such fossil geomorphological features provide evidence of past *periglacial conditions. *See also* PALSA.

pinnacle reef *See* REEF.

pinnate Borne on either side of a central stalk; like a feather in appearance.

Pinus aristata *See* BRISTLECONE PINE.

Pinus longaeva *See* BRISTLECONE PINE.

PIOCW *See* PACIFIC and INDIAN OCEAN COMMON WATER.

pioneer phase **1.** Generally, the 'new plant' stage of any cyclical pattern of vegetation change. In forestry, it is sometimes called the *gap phase. **2.** Specifically, a stage in the cyclical pattern of *community change in *heathlands, when individual *Calluna vulgaris* plants are young (up to 10 years old) and provide little cover, so there is maximum diversity of associated species. Although *C. vulgaris* *biomass is low, productivity in the new shoots is high. *Compare* BUILDING PHASE; DEGENERATE PHASE; HOLLOW PHASE; and MATURE PHASE.

pioneer plant A plant that occurs early in a vegetational *succession. Pioneer species possess characters that suit them to their ecological *niches, notably rapid growth, the production of copious, small, easily dispersed seed, and the ability to germinate and establish themselves on open sites.

pioneer species *See* PIONEER PLANT.

pioneer stage The early stage of a plant *succession.

pipkrake (needle ice) Columnar ice found beneath individual stones or patches of earth in a *periglacial environment. It is a result of the relatively high thermal conductivity of such materials, leading to freezing under them when the temperature falls. Heaving of less than 0.1 m may occur by this process.

piston corer *See* HYDRAULIC CORER.

piston sampler *See* PEAT-BORER.

pitfall trap A device for capturing ground-dwelling invertebrates which comprises an appropriately baited vessel (e.g. a jamjar containing a small piece of rotting meat), which is buried so its mouth is at ground level. The mouth is covered by a roof, to shelter the vessel from rain and prevent larger animals from entering, raised sufficiently to allow access for target species.

pit organ *See* TEMPERATURE-SENSITIVE ORGAN.

placic horizon A subsurface *soil horizon, formed most readily in humid tropical or cold conditions, which is cemented by iron and organic matter, by iron and manganese, or by iron alone.

plaggen A *soil horizon, produced by human activity, which is more than 500 mm deep and results from long-continued manuring, often with materials enriched by phosphate. The name is from the German *Plagge*, meaning sod.

plagioclimax A term that is almost synonymous with *biotic climax, although they are sometimes given different meanings. Generally, both refer to a stable vegetation *community arising from a *succession that has been deflected or arrested directly or indirectly as a result of human activities. In a deflected succession, the resulting stable community, even when composed entirely of *native species, is one that would not have occurred in the absence of human intervention (e.g. most lowland heath communities of western Europe arose as a consequence of forest clearance, subsequent grazing pressure, and controlled burning). Associated changes in the physical environment may mean that even when these pressures are removed succession to the original *climax community is no longer possible. In an arrested succession, the stable community is a naturally occurring successional phase and it should be possible for the natural succession to continue once the disturbing factor has been removed (e.g. the cutting of reed beds arrests the natural succession to alderwood in a *hydrosere). Some authors restrict 'plagioclimax' to deflected successions; some use 'plagioclimax' where human intervention is more direct, reserving 'biotic climax' for more indirect effects (e.g. grazing by non-domesticated but introduced animals, such as the rabbit in Britain). 'Biotic climax' may also be applied to a natural, undisturbed succession in which the form of the final community is determined by a naturally occurring biological agent (e.g. grasslands in *guano-enriched coastal areas).

plagiogeotropic *See* GEOTROPISM.

planation surface **1.** A synonym for *erosion surface. **2.** The final stage produced by the erosion of folded sedimentary rocks.

plankton Aquatic organisms that drift with water movements, generally having no locomotive organs. The phytoplankton comprise mainly diatoms, which carry out photosynthesis and form the basis of the aquatic *food-chains. The zooplankton (animals) which feed on the diatoms may sometimes show weak locomotory powers. They include protozoans, small crustaceans, and, in early summer, the larval stages of many larger organisms. Plankton are sometimes divided into net plankton (more than 25 μm diameter) and nannoplankton, which are too small to be caught in a plankton net. The word is derived from the Greek *plagktos*, meaning 'wandering'.

planktonic geochronology The use of planktonic organisms (e.g. globigerinid foraminiferids or microscopic algae) to provide a relative dating of sediments deposited in marine waters. Radioactive-decay methods applied to planktonic organisms may also yield an absolute date or information on palaeoclimates.

Planosols Soils that have within 100 cm of the surface a *soil horizon that has been exposed to stagnant water for a prolonged period. Planosols are a reference soil group in the *FAO *soil classification.

plan position indicator **(PPI)** A radar display in which the echo from a target appears as a bright mark against a dark field, its position on the screen indicating the direction and distance of the target from the scanner.

plant association **(association)** **1.** In the *Zurich–Montpellier school, the basic unit of vegetation, an abstract entity that is defined floristically from field data or *relevés. Each association has a distinctive *faithful species and a group of *constant (i.e. high presence) species which give it a coherent structure. The companion species of the association often form the faithful species of the next and succeeding hierarchical levels, alliances, and orders into which similar associations are grouped. **2.** In the British and American phytosociological traditions (*see* PHYTOSOCIOLOGY), a community that is united *physiognomically as well as floristically; commonly it is a *climax community in which the species *dominants are those of the upper vegetation layer. There are usually several co-dominant species; if there is only one dominant the community is called a 'consociation'. *See also* FORMATION.

Plantae The taxonomic kingdom that includes all multicelled plants.

plantation A closely set stand of trees (other than an orchard) which has been planted by humans and usually comprises one or two species grown on land cleared for the purpose and harvested like an arable crop. Plantations do not maintain themselves.

plant available water **(PAW)** The amount of water present in the soil that is available to plants. The *field capacity marks the upper bound of PAW and the *permanent wilting point the lower bound.

plantigrade Applied to a *gait in which the entire foot makes contact with the ground (e.g. in humans).

plant sociability In the description and analysis of plant *communities, a measure of the distribution pattern and organization of a species. The *Zurich–Montpellier phytosociological scheme (*see* PHYTOSOCIOLOGY) uses a five-point sociability scale: (1) growing once in a place, singly; (2) grouped or tufted; (3) growing in troops, small patches, or cushions; (4) growing in small colonies, extensive patches, or forming carpets; and (5) growing in great crowds or pure populations. *See also* DOMIN SCALE.

plant strategies A term introduced by J. P. Grime, who defines it as groupings of similar or analogous genetic characteristics which recur widely among species or populations and cause them to exhibit similarities in ecology. Grime first proposed a threefold broad division of plant strategies into *competitors, *ruderals, and *stress-tolerators. *See also* FUNCTIONAL TYPE.

plasmid A particle found in the cytoplasm of a bacterial cell that carries one or more genes and can replicate itself autonomously. Genetic information on the plasmid is passed from each cell to its daughter cell, and sometimes to neighbouring cells. Plasmids normally remain separate from the chromosome, but some may become integrated into it temporarily and replicated with it incidentally. Plasmids are important in the genes they carry (e.g. they may confer resistance to particular antibiotics to their bacterial hosts).

plasmodesmata *See* C_4 PATHWAY.

plastachron The time that elapses between periodic events in the growth of a plant.

plastachron index A standardized set of data with which *plastachrons for a particular specimen can be compared.

platform *See* SHELF.

platyrrhine In primates, applied when the nostrils face more or less to the sides and are well separated. All New World monkeys have platyrrhine nostrils. *Compare* CATARRHINE.

play Voluntary, seemingly paradoxical behaviour (i.e. a goal usually associated with the behaviour is not attained because the activity is not pursued to its conclusion, or because it is misdirected), often occurring in bouts preceded by signals exchanged between participants, during which movements may be performed in apparently random succession, and in which certain sequences may be repeated many times.

playa **(salina)** The lowest part of an *intermontane basin or *bolson, which is frequently flooded by run-off from adjacent highlands or by local rainfall. The surface is generally flat, with *mudflats and locally small *dunes. The name (*playa* is the Spanish word for 'beach' and *salina* for 'salt-mine') was first applied to the arid basin-and-range province between the Colorado Plateau and Sierra Nevada, in the western USA, but is now used to describe such areas throughout the world.

pleiomorphism *See* PLEOMORPHISM.

pleiotropy The situation in which one *gene is responsible for a number of apparently unrelated *phenotypic effects.

Pleistocene The first of two epochs of the *Quaternary, which is held conventionally to have lasted from approximately 1.806 Ma ago until the beginning of the *Holocene, about 11 000 years ago. The earlier date has been determined by recent evidence obtained from deep-sea cores. The epoch is marked by several glacial and interglacial episodes in the northern hemisphere.

Pleistocene refuge A favourable area in which species (called *relicts) have survived periods of climatic change (glaciation in high latitudes, seasonally dry climates in the lowland humid tropics) during the *Pleistocene Epoch.

Pleistogene The most recent period of the *Cenozoic Era, comprising the *Pleistocene and *Holocene Epochs. It began 1.806 Ma and continues to the present day.

pleomorphism **(pleiomorphism)** A form of *polymorphism in which distinctly different forms occur at particular stages during the life cycle of an individual (e.g. the larval, pupal, and imago forms of many insects).

plesiomorph (adj. plesiomorphic) A primitive character state, comprising features that are shared by different groups of biological organisms and are inherited from a common ancestor. The word means 'old-featured'. *Compare* APOMORPH.

plesiomorphic *See* PLESIOMORPH.

plesion An unranked, quasi-taxonomic category to which newly discovered organisms can be assigned. The plesion is inserted at the appropriate level in order to avoid disturbing an established classification.

pleurocarpous moss A type of moss in which the female sex organs (archegonia) and *capsules are borne on short, lateral branches, and not at the tips of branches. Pleurocarpous mosses tend to form spreading carpets rather than erect tufts. *Compare* ACROCARPOUS MOSS.

pleurothetic Applied to a resting position common in members of the Bivalvia (bivalve *molluscs), in which individuals settle on to the substrate on their sides.

plexus In data analysis, a diagram in which the lengths or widths of interconnecting lines reflect the relative similarity of the samples. It provides a useful visual presentation of similarity/dissimilarity indices.

plicate Folded, as of a palm leaf.

plinthic horizon A *soil horizon that is rich in *plinthite.

plinthite A constituent of a *mineral soil, comprising a large proportion of iron and aluminium oxides, *clay, and quartz, which has developed through a combination of *leaching and gleying (*see* GLEY) in well-weathered tropical soils. On drying, plinthite changes irreversibly to an ironstone *hardpan.

Plinthosols Soils that have an iron-rich *soil horizon containing more than 25 per cent *plinthite within 50 cm of the surface. The plinthic material hardens when exposed to the air. Plinthosols are a reference soil group in the *FAO *soil classification.

Pliocene The last of the *Tertiary epochs, about 5.332–1.806 Ma ago.

plotless sampling Sampling without the use of *quadrats or *transects. Individual *pin-frame records may be regarded as plotless samples. Plotless sampling is most often used for surveys of forest vegetation, especially where a rapid inventory is required. Various approaches have been developed (e.g. the *nearest-neighbour sampling method).

plumose Plumed.

plum pox (sharka disease) A disease of plum and related trees, the symptoms of which vary according to the species affected, but which usually include the appearance of pale or dark rings or spots on leaves and fruit. Plum pox is a disease of considerable economic importance and is a notifiable disease in Britain. It is caused by a virus and transmitted by aphids.

plunging breaker A wave that breaks by plunging forward in the direction of motion, so that its crest falls into the preceding trough and encloses a pocket of air. The wave-form is then lost. Typically it occurs when a fairly low wave approaches a steep shingle beach. It is characterized by a smooth forward and under face, and a convex top.

pluvial period A prolonged phase of markedly wetter climate that occurs in a normally dry or semi-arid area. The major glaciations of North America were represented by pluvial periods in the south-west of the continent. The dry episodes between pluvials are termed interpluvials.

PM-10 *See* PARTICULATE MATTER.

PM-25 *See* PARTICULATE MATTER.

pneumatophore A specialized 'breathing' root developed in some plant species that grow in waterlogged or strongly compacted soils (e.g. mangroves). The aerial part of the root contains many pores through which gases are exchanged with the atmosphere and from which they diffuse by means of intercellular spaces throughout the submerged portion of the root.

poached soil *See* PUDDLED SOIL.

pocket rot A type of timber decay, often caused by the fungus *Heterobasidion* *annosum*, in which the decay occurs in small, distinct areas or pockets. *See also* PARTRIDGE WOOD.

podsol *See* PODZOL.

podsolization *See* PODZOLIZATION.

podzol **(podsol)** A soil profile formed at an advanced stage of *leaching by the process of *podzolization, and identified by its acid *mor *humus, *eluviated and bleached E *soil horizon, and an iron-coloured B horizon, enriched with a variety of translocated materials. Podzols are a reference soil group in the *FAO *soil classification. *See also* SPODOSOLS.

podzolization **(podsolization)** An advanced stage of *leaching, in which iron and aluminium compounds, *humus, and clay minerals are removed from the surface *soil horizons by an organic leachate solution, and some of these materials are translocated in lower B horizons.

poikilohydry The inability of an organism to compensate for fluctuations in the availability of water or evaporation, so its internal water content varies according to the humidity of its surroundings. *Compare* HOMOIOHYDRY.

poikilotherm **(exotherm)** An organism whose body temperature varies according to the temperature of its surroundings. Fish are poikilotherms. *Compare* ECTOTHERM. *See also* ENDOTHERM and HOMOIOTHERM.

point bar *See* BAR.

point frame *See* PIN-FRAME.

point mutation A *mutation that occurs at only one *nucleotide site. The term is often used synonymously with *substitution.

point quadrat A *quadrat sampled by the *pin-frame method.

Poisson distribution The basis of a method whereby the distribution of a particular attribute in a population can be calculated from its mean occurrence in a *random sample of that population, provided the population is large and there is a less than 0.1 probability that the attribute will occur. For a given mean the distribution can be calculated, giving the probabilities that a sample will contain 0, 1, 2, 3, . . . examples of the particular attribute. The distribution is named after the French mathematician S. D. Poisson (1781–1840).

polar air An *air mass, originating in latitudes 50–70°, which may have a maritime or continental source. Maritime polar air has high *relative humidity, is warmed in its passage towards the equator or over warmer seas, and becomes unstable. Continental polar air is stable in its source region and is associated with very cold surface conditions in winter (e.g. Siberian air moving across Europe).

polar-air depression A *non-frontal depression in the northern hemisphere, which typically results from the southward movement of unstable polar or Arctic maritime air along the eastern side of a large-scale meridional high-pressure ridge. *See also* COLD LOW.

polar climate The climatic type associated with regions inside the Arctic and Antarctic Circles. A gradation of climatic characteristics exists towards the poles, from *tundra conditions to those of perpetual frost. *See also* KÖPPEN CLIMATE CLASSIFICATION.

polar-desert soil A *mineral soil that lacks identifiable *soil horizons, and has almost no surface *humus. It is associated with arid, polar-desert environments where precipitation is less than 130 mm annually, plant cover less than 25 per cent, and thawed, active soil is 20–70 mm deep.

polar front The main boundary line between polar and tropical *air masses along which *depressions develop in mid latitudes, especially over the oceans. The front is in general displaced towards the equator in winter and the poles in summer, though large displacements in either direction take place over shorter periods in individual sectors of the hemisphere.

polar-front jet stream The *jet stream that is observed in different positions over middle and higher latitudes at heights of 10–13 km, and is associated with the boundary zone between polar and tropical air above the *polar front. Maximum velocity averages 60 m/s, but can be twice this in extreme cases. The jet is typically discontinuous but at times can extend almost around the globe. It is more persistent in winter in response to the stronger temperature gradient between north and south than in summer. The jet is related to the development of surface frontal depressions of middle latitudes.

polar glacier *See* GLACIER.

polarity time-scale *See* MAGNETOSTRATIGRAPHIC TIME-SCALE.

polar molecule A molecule in which, though it does not carry a net electric charge, the electrons are unequally shared between the nuclei. In the water molecule, for example, the pull of the oxygen nucleus on the shared electrons is greater than the pull of the hydrogen nuclei. As a result the oxygen end of the molecule is slightly negatively charged, and the hydrogen ends of the molecule are each slightly positively charged. The molecule is said to have a dipole moment and can attract other molecules with a dipole moment.

polar night vortex *See* POLAR STRATOSPHERIC CLOUD.

polar stratospheric cloud (PSC) Clouds that form in winter in the *stratosphere over Antarctica and less commonly over the Arctic at altitudes of 15–25 km. The clouds form in the very still air in the polar night vortex produced by the strong *geostrophic wind that circulates in winter. There are two principal types of PSCs. Type 1 consist of liquid droplets, about 0.001–0.1 mm in diameter, composed of a mixture of sulphuric (H_2SO_4) and nitric (HNO_3) acids dissolved in water. These clouds form at about 15 km where the temperature of approximately 195 K (−78°C) is above the frost point. Type 2 PSCs are made from ice crystals and form at about 25 km where the temperature is approximately 188 K (−85°C). Type 2 PSCs are seldom seen over the Arctic, where air temperatures are usually too high for them to form. The reactions that deplete stratospheric ozone (*see* OZONE 'HOLE') take place on the surface of PSC ice crystals.

polder A low-lying, flat area reclaimed from the sea and protected by embankments or dykes; especially along the Netherlands North Sea coast.

pole A single-trunked tree, usually a juvenile, which is smaller in girth and height than a mature tree or *standard. A *coppice pole is a stem arising from a coppice stool.

polje A large, flat-floored depression bounded by steep valley walls, which is found in a *karst environment. It is classically described for the Dinaric region of Bosnia Hercegovina, where it may result from faulting or from solutional processes controlled by a local base level.

pollard To behead a tree at about head height, usually about 2 m above ground level, in order to produce a crown of small branches, suitable for firewood, fencing, etc., beyond the reach of deer or farm livestock. *See* POLLARDING. *Compare* COPPICE.

pollarding A system of management in which the main stem of a (usually young) tree is severed about 2 m above ground level, favouring the development of lateral branches. Repeated pollarding leads to the formation of a slightly swollen boll in the main stem immediately below the lateral branches and frequent pollarding, common with willows (*Salix* species), produces many thin-stemmed lateral branches. Less frequent pollarding was a traditional management practice in wood-pastures, where grazing pressure made *coppicing impracticable. Pollarded wood-pastures were characteristic of slower-growing woodlands on infertile soils or in upland areas, and pollarding was also used to

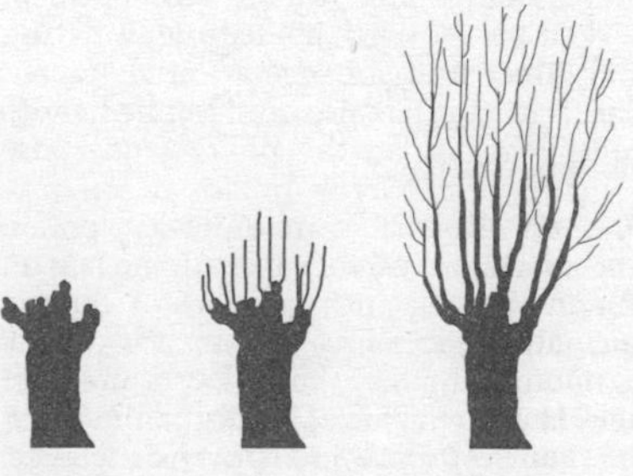

Pollarding

p

produce particular sizes and shapes of timber needed for structural uses. In Britain, the ancient and ornamental woodlands of the New Forest include classic *relict wood-pastures containing pollarded beeches and oaks.

pollen Collectively, the mass of microspores or *pollen grains produced within the anthers of a flowering plant or the male cones of a gymnosperm. Different pollen types are described according to their shapes, apertures, etc. Furrows on the surface of the pollen grain are called 'colpi' (sing. *colpus) or 'sulci' (sing. sulcus); the words are synonymous. Monosulcate pollen has a single colpus, tricolpate pollens have three, furrow-like colpi (sulci) arranged 120° apart, and there are many variants of this type.

pollen analysis The study of fossil *pollen and *spore assemblages in sediments, especially for the purpose of reconstructing the vegetational history of an area. The outer coat (exine) of a pollen grain or spore is very characteristic for a given family, genus, or sometimes even species. It is also very resistant to decay, particularly under anaerobic conditions, and virtually all spores and pollen falling on a rapidly accumulating sediment, anaerobic water, or *peat are preserved. Both pollen and spores are generally widely and easily dispersed and therefore they give a better picture of the surrounding regional vegetation at the time of deposition than do macroscopic plant remains (e.g. fruits and seeds) which tend to reflect only the vegetation of the immediate locality. With careful interpretation, pollen analysis enables examination of climatic change and human influence on vegetation, as well as sediment dating and direct study of vegetation character. The technique has also been applied, more controversially, to the pollen and spore contents of modern and fossil soil profiles. Studies of contemporary pollen and spores are useful in medicine (e.g. in allergy studies and patterns of disease spread), in commerce (e.g. for the examination and quality control of honey), in agriculture (e.g. for plant and animal disease control), and even in forensic science. *See also* PALYNOLOGY.

pollen assemblage zone *See* POLLEN ZONE.

pollen diagram A standardized pictorial summary of the *pollen record for a particular location; the vertical axis represents depth below ground-surface level and the proportions or absolute amounts of the various pollen types occurring at different levels are shown by bar histograms or by points on a continuous curve. Conventionally, similar patterns are grouped together on the diagram, with arboreal types shown first, followed in turn by shrubs, herbs, and *spores.

pollen grain A microspore in flowering plants, which germinates to form the male gametophyte.

pollen rain The *pollen grains and *spores that fall on a particular site.

pollen zone **(pollen assemblage zone)** One of the parts into which *pollen diagrams are divided on the basis of the total pollen assemblage found within them. Zone boundaries are placed at points where there is greatest evidence for change, hence the assemblage of pollen within a zone is relatively homogeneous. Pollen zones are defined by their content of pollen types and can be dated by reference to *radiocarbon or other methods, thus enabling them to be established as *chronozones. Pollen assemblage zones are thus intended to delimit periods during which the surrounding vegetation has been relatively stable. This may be due to climatic stability within that period, or to a lack of change in human land management. Various schemes have been introduced in the past, particularly by K. Jessen in Denmark and Sir Harry Godwin in Britain, which have attempted to construct regional pollen zones. In Britain, for example, a series of eight pollen zones was established by Godwin to cover the late *Devensian (last glacial) and the *Holocene (*Flandrian). Lack of synchroneity even within the British Isles, however, has largely led to the abandonment of the scheme. Pollen assemblage zones are now constructed for each individual site and regional comparisons and correlations can then be made on the basis of chronology.

pollutant A by-product of human activities which enters or becomes concentrated in the environment, where it may cause injury to humans or desirable species. In addition to chemical substances, the term also embraces noise, vibration, and alterations to the ambient temperature.

pollution The defilement of the natural environment by a *pollutant.

poly- From the Greek *polus*, meaning 'many', a prefix meaning 'many'.

polyandry The mating of a female with more than one male at one time (usually taken to be during the course of a single breeding season). Simultaneous polyandry occurs where the female mates with more than one male to produce a clutch of eggs or brood of young bearing genes from each male. Successive polyandry occurs where the female mates with one male, lays eggs or gives birth to the young, but plays little part in their parental care (usually cared for by this first male), instead moving away and mating with another male. The existence of simultaneous polyandry is usually difficult to prove, and polyandry appears to be mainly of the successive type. Polyandry is generally much rarer than *polygyny, presumably because the mother is not usually sexually receptive following successful mating, and since she bears the young she is more involved with their care and less able to seek another mate. *See also* PROMISCUITY.

polyaromatic hydrocarbon (PAH) One of a range of persistent, toxic compounds that are produced mainly by the incomplete combustion of hydrocarbon fuels. Many are carcinogens, mutagens, or both. Those of low molecular weight occur mainly as vapour in the air, those of higher molecular weight as particles in soil or water.

polychlorinated biphenyl (PCB) One of a range of compounds, first synthesized in 1881 and manufactured in 1929 and used mainly as liquid insulators in heavy-duty electrical transformers. They were detected in the environment in 1966 and associated with reduced reproduction in marine birds and mammals; they are also believed to compromise the immune system. Restrictions on their use were imposed in North America and Europe in the early 1970s, but they continue to cause contamination, because of their persistence and as a consequence of careless disposal of old equipment containing them.

polychore *See* WIDE DISTRIBUTION.

polychronic species *See* POLYTOPISM.

polyclimax theory *See* CLIMAX THEORY.

polycyclic landscape (polyphase landscape) A landscape that has been acted on by the erosional processes associated with two or more partially completed cycles of erosion (*see* DAVISIAN CYCLE). A diagnostic feature is an abrupt break of slope (*see* KNICK POINT) in the profiles of both rivers and hill-slopes.

polyembryony The production of two or more individuals from a single egg by the division of the embryo at an early stage of development. Some parasitic wasps may produce hundreds of offspring from a single fertilized egg.

polyethism Functional specialization in different members of a colony of social insects, which leads to a division of labour within the colony. The various functions may be carried out by individuals of different morphology (caste polyethism) or of different ages (age polyethism).

polygamous Applied to a plant species in which combinations of male, female, and hermaphrodite flowers occur on the same or different plants.

polygamy In animals, a pattern of mating in which an individual has more than one sexual partner. *See also* POLYANDRY; POLYGYNY; and PROMISCUITY. *Compare* MONOGAMY.

polygyny In animals, a pattern of mating in which a male has more than one female partner. *Compare* MONOGAMY and POLYANDRY; *see also* POLYGAMY and PROMISCUITY.

polyhaline **1.** Applied to a highly salt-tolerant species that is not usually found outside very saline environments (e.g.

Limonium vulgare, common sea lavender). **2.** The second most saline zone of a *salt-marsh, according to the Venice system for the classification of *brackish waters. *See also* HALINITY.

polyhedrosis A disease of insects caused by certain DNA-containing viruses. The larval stages of flies, butterflies, moths, and sawflies may be affected, and the disease is frequently fatal.

polylectic Applied to bee species which collect pollen from a wide range of flowering plants.

polymerase chain reaction An enzymatic technique for producing a large number of copies of a specified type of DNA segment. PCR may be specific to one particular DNA sequence within the *genome, or more general, as in *random amplified polymorphic DNA.

polymictic Applied to lakes (e.g. those in high altitudes in the tropics) whose waters are circulating virtually continuously. If periods of stagnation occur they are very short. *Compare* OLIGOMICTIC.

polymorphic Occurring in several different forms. *Compare* MONOMORPHIC.

polymorphism **1.** In genetics, the existence of two or more forms that are genetically distinct from one another but contained within the same interbreeding population. The polymorphism may be transient or it may persist over many generations, when it is said to be balanced. Classical examples of polymorphisms are the presence or absence of banding in *Cepaea* snails, the number of spots on the wings of ladybirds, and eye colour in humans. All these are visible polymorphisms that can readily be seen in nature. Some polymorphism, however, is cryptic and requires biochemical techniques to identify phenotypic differences. **2.** In social insects, the presence of different castes within the same sex. *See also* POLYTYPISM.

Polynesian floral region R. Good's (1974, *The Geography of the Flowering Plants*) Polynesian subkingdom within his *Palaeotropical floral kingdom, which contains a small, derived flora with only 9 or 10 endemic genera (*see* ENDEMISM), scattered through the various islands. The affinities are with Malesia and Australia, and there is no distinct 'Polynesian' flora as such. The great majority of the floral elements have been derived from adjacent floras. *See also* FLORAL PROVINCE; FLORISTIC REGION; and MELANESIA AND MICRONESIA FLORAL REGION.

polyoestrus In female mammals, the condition in which more than one *oestrus cycle occurs in a breeding season or in a year.

polyp The soft-bodied, usually sedentary form of Cnidaria, consisting of a cylindrical trunk which is fixed at one end, with the mouth surrounded by *tentacles at the other end. In Siphonophora the polyp has been modified for a *pelagic, colonial existence.

polypedon (soil individual) Two or more contiguous *pedons, which are all within the defined limits of a single *soil series.

polypeptide A linear polymer composed of 10 or more *amino acids linked by *peptide bonds.

polyphagous Applied to an organism that feeds on many types of plant or prey. *Compare* MONOPHAGOUS and OLIGOPHAGOUS.

polyphase landscape *See* POLYCYCLIC LANDSCAPE.

polyphenism (polypheny) The occurrence within a population of *phenotypic differences, commonly specializations, which lack a genetic basis.

polyphyletic *See* MONOPHYLETIC.

polyphyly The occurrence in taxa of members that have descended via different ancestral lineages. True polyphyly has traditionally been distinguished from errors of classification, especially at the higher taxonomic levels, where organisms, as a result of *convergent or *parallel evolution, have been placed wrongly in the same natural group; but modern phyletic taxonomists would hold that any *taxon found to be polyphyletic is unnatural, and so an 'error', and must be disbanded.

polyploid *See* POLYPLOIDY.

polyploidy The condition in which an individual possesses one or more sets of *homologous chromosomes in excess of the normal two sets found in *diploid organisms. It is caused by the replication within a nucleus of complete chromosome sets without subsequent nuclear division. Examples are triploidy ($3n$), tetraploidy ($4n$), hexaploidy ($6n$), and octoploidy ($8n$). Polyploids are not always *allopolyploids. For example, modern bread wheat (*Triticum aestivum*) has 42 chromosomes as a consequence of *allopolyploidy, a type of polyploidy. It arose from the interbreeding of emmer wheat (*T. turgidum*) with *T. tauschii*, a wild wheat that is found in Iran. *T. turgidum* has 28 chromosomes and *T. tauschii* has 14, so the *zygote would have had 21 chromosomes. This number doubled to 42, thus producing cells with homologous chromosome pairs and a species (*T. aestivum*) capable of producing fertile offspring.

polypoid Pertaining to or resembling a *polyp.

polythetic Applied to methods that use several or all possible criteria or attributes as the basis for each subdivision of a sample population or for agglomeration of individuals or groups. *See also* AGGLOMERATIVE METHOD; DIVISIVE METHOD; and HIERARCHICAL CLASSIFICATION METHOD; *compare* MONOTHETIC.

polytopic evolution *See* POLYTOPISM.

polytopic species *See* POLYTOPISM.

polytopism (polytopic evolution) A type of monophyletism (*see* MONOPHYLETIC) in which a new (polytopic) taxon arises in more than one place from *conspecific parents. The chances of this happening simultaneously in each instance, except perhaps in the case of subspecies, are remote; polytopic species may therefore be regarded also as polychronic species.

polytypic Of a species: divided into subspecies; varying geographically. *See* POLYTYPISM.

polytypism The occurrence of *phenotypic variations between populations or groups of a species that are geographically distinct. It is contrasted with *polymorphism, which is variation within a population or group. A species with systematic geographical variation (subspecies or *clines) is said to be polytypic.

pome *See* FRUIT.

Pomeranian One of a series of *Weichselian recessional *moraines in northern Germany, which postdates the Frankfurt series of moraines and predates the Velgart series.

ponente The local name for a westerly wind that blows in the Mediterranean.

pool-and-riffle An alternation between a deep zone (the pool) and a shallow zone (the riffle) along the sand and/or gravel bed of a stream. The pool-to-pool spacing is about 5–7 times the width of the channel. The sequence is found in both straight and *meandering channels; in the latter case the pool occurs in the meander bend on the concave side, and the riffle between bends.

pooter A device that is used to collect and transfer small animals. It consists of a jar, sealed by a cover pierced by two tubes pointing in opposite directions. The operator sucks on the end of one tube (made flexible for convenience and the end inside the jar covered with muslin), drawing the specimen up the other (open) tube and into the jar.

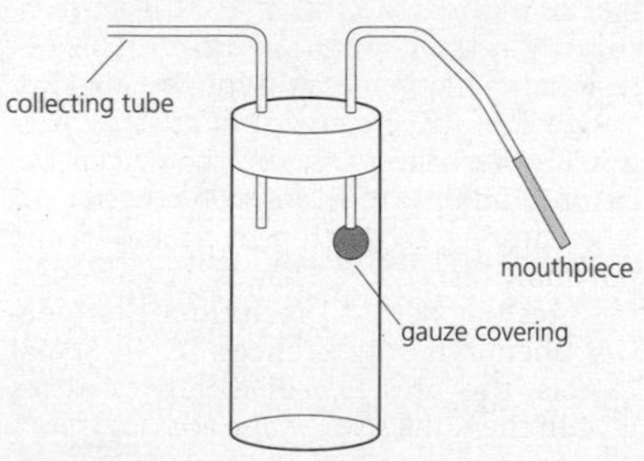

Pooter

Poperinge *See* MOERSHOOFD.

population A group of organisms, all of the same *species, which occupies a particular area. The term is used of the number of individuals of a species within an *ecosystem, or (statistically) of any group of like individuals. *See also* ECOTYPE.

population dynamics The study of factors that influence the size, form, and fluctuations of populations. Emphasis is placed on change, energy flow, and nutrient cycling, with particular reference to homoeostatic controls. Key factors for study are those influencing natality, mortality, immigration, and emigration. *See* DEMOGRAPHY.

population ecology The study of the interaction of a particular species or genus population (or sometimes of a higher taxon) with its environment. *See also* AUTECOLOGY.

population eruption *See* POPULATION EXPLOSION.

population explosion (population eruption) A sudden rapid increase in size of the population of a species or genus. The most violent explosions occur when a species is introduced into a new locality where it finds unexploited resources of suitable food, shelter, etc., and a lack of negative controls such as predators or parasites. Examples include the population explosions of the prickly pear (*Opuntia*) and of *Rhododendron ponticum* following the introduction of these species into Australia and Britain respectively. In exceptionally favourable circumstances, a species may extend its range at least temporarily as the population erupts. This occasionally happens with birds such as nutcrackers (*Nucifraga caryocatactes*), which spread westwards from their central European homelands. *See* *R*-SELECTION.

population genetics The study of inherited variation in populations of organisms, and its modulation in time and space. Population genetics relates the heritable changes in populations to the underlying individual processes of inheritance and development. Such studies generally involve the estimation of gene frequencies and the influences of *selection, *mutation, and *migration upon these frequencies in natural (and experimental) populations.

population regulation Limitations on the density of a *population which are imposed by *density-dependent factors.

population size The number of individuals in a population. *Compare* EFFECTIVE POPULATION SIZE.

pore (pedol.) A void surrounded completely by soil materials and created by the packing of mineral and organic particles. Pores can be filled by any proportion of air or water.

pore space The total continuous and interconnecting void space in the bulk volume of soil.

porosity The percentage of the total bulk volume of a body of rock or soil that is occupied by *pore space. The figure may represent: (*a*) (absolute porosity), which is the total of all pore spaces present in a rock or soil, not all of which will be interconnected and thus able to contain and transmit fluids; or (*b*) (effective porosity), which is the proportion of the rock that consists of interconnected pores.

porrect Stretched out.

positive feedback In a system, the mechanism by which a process intensifies or accelerates, as each cycle of operation establishes conditions that favour a repetition. Unless checked, positive feedback may lead to loss of control within the system and its eventual failure. *Compare* NEGATIVE FEEDBACK.

post-climax In the *mono-climax model of *climax vegetation development, *communities differing from the *climatic climax, owing to the presence of more favourable conditions for vegetation development. For example, forest may occupy river valley sites within temperate grassland *biomes because of the greater availability of water in summer in such locations. *See also* CLIMAX THEORY; *compare* PRE-CLIMAX.

postdisplacement An alteration in the *ontogeny of a descendant such that some developmental process begins later than in its ancestor, and may not have been completed by the time maturity is reached.

post-glacial *See* FLANDRIAN and HOLOCENE.

post-medieval woodland Woodland that is known to have originated on a site that was not wooded prior to about 1650.

potamodromous Applied to fish that undertake regular migrations in large freshwater systems. *See also* AMPHIDROMOUS and DIADROMOUS. *Compare* ANADROMOUS and CATADROMOUS.

potamon That part of a river in which the water is typically slow-moving, still-surfaced, deep, and relatively warm, favouring *limnophilous, *stenothermous organisms that are thrifty in their use of dissolved oxygen. *Compare* RITHRON.

potassium (K) An element that is required for healthy plant growth. Potassium *ions neutralize *anionic *macromolecules and organic acids and thus control the *water potential of cells. They pass readily through membranes and are important in the movements of leaves and guard cells. A potassium deficiency leads to reduced growth and to dark or blue-green coloration in the leaves, which may also develop a purple-brown pigment. In animals, potassium is involved in muscle and kidney function.

potassium–argon dating (K–Ar method) A dating technique based on the radioactive decay of potassium (^{40}K) to argon (^{40}Ar). This potassium isotope has a half-life of 1.3 billion (10^9) years, and the minimum age limit for this dating method is about 250 000 years.

potato blight A term that may refer to either late blight or early blight. Early blight is much the less serious of the two diseases and is caused by the fungus *Alternaria solani*; dark, concentrically zoned spots appear on the leaves of infected plants. Both early and late blights can also attack tomato plants. *See also* LATE BLIGHT OF POTATO.

potato scab A term that may refer to either common scab or powdery scab. In common scab, caused by *Streptomyces scabies*, the scabs on the tubers are corky and superficial; the disease is commonest on light soils and in dry conditions. Powdery scab is caused by *Spongospora subterranea*, and is most common on heavy soils and in wet seasons; the scabs differ from those of common scab in that they are powdery.

potential evapotranspiration (*PE*) The amount of water that would evaporate from the surface and be transpired by plants were the supply of water unlimited. It is calculated from the mean monthly temperature, with corrections for day length, and was devised by C. W. Thornthwaite as part of his system of *climate classification (*see* THORNTHWAITE CLIMATE CLASSIFICATION). From *PE* minus precipitation an approximate index can be calculated of the extent to which the water available for plants falls short of the amount they are capable of transpiring. *Compare* ACTUAL EVAPOTRANSPIRATION.

potential instability (convective instability) An atmospheric condition in which otherwise stable air would become unstable if forced to rise (e.g. over high ground) thereby reaching its saturation point. Large *cumulus with much precipitation often results from the forced uplifting of such air. *See also* INSTABILITY and STABILITY.

potential-natural *See* NATURAL.

potentiometric surface (piezometric surface) A hypothetical surface defined by the level to which water in a confined *aquifer rises in observation boreholes. In practice, the potentiometric surface is mapped by interpolation between borehole measurements. As with the *water table in an unconfined aquifer, the slope of the potentiometric surface defines the *hydraulic gradient and the horizontal direction of *groundwater flow. *See* HYDRAULIC HEAD.

potometer An instrument that is designed to measure water uptake in a plant and, indirectly, to estimate *transpiration rates.

powdery mildew Either a fungus of the order Erysiphales or a plant disease caused by such a fungus. The leaves, fruit, etc. of an infected plant bear characteristic powdery white patches of conidium-bearing mycelium; tiny black *fruit bodies containing haploid *spores (perithecia) may also be visible. Many types of plant may be affected, usually each by its own particular species or strain of fungus.

p.p.b. Parts per billion (10^9).

PPI *See* PLAN POSITION INDICATOR.

p.p.m. Parts per million.

praecipitatio The Latin *praecipitatio*, meaning 'I fall', used to describe a supplementary cloud feature in which precipitation from the cloud base falls to the ground. The feature is usually seen with *cumulus, *cumulonimbus, *stratus, *stratocumulus, *altostratus, and *nimbostratus. *See also* CLOUD CLASSIFICATION.

praemorse **(premorse)** Having the end terminated abruptly, as though bitten off.

Praghian *See* DEVONIAN.

prairie A temperate grassland of northern America, dominated by more or less *xeromorphic grasses, which fall into three groups based on stature (tall, mid, and short) with a progressive decrease in rainfall. Various herbaceous broad-leaved *annuals and *perennials are mixed in with the grasses. *See* PALOUSE PRAIRIE.

pre- From the Latin *prae*, meaning 'before', a prefix meaning 'in front of', 'earlier than', 'more important than', or 'better than'.

pre-adaptation An *adaptation evolved in one *adaptive zone which, quite by chance, proves especially advantageous in an adjacent zone and so allows the organism to radiate into it. No *selection for a future environment is implied. The concept is very similar to *exaptation, but is often thought to have teleological implications.

pre-Boreal The first *Flandrian (*Holocene, or post-glacial) stage, a time of rapid forest spread, from about 10 300–9 600 BP. Pre-Boreal refers to climatic conditions and in vegetational terms is equivalent to Pollen Zone IV of the standard British and European post-glacial pollen chronology. *See* POLLEN ZONE.

Precambrian A name that is now used only informally to describe the longest period of geological time, which began with the consolidation of the Earth's crust and ended with the beginning of the *Cambrian Period 542 Ma ago. The Precambrian lasted approximately 4 000 Ma; the rocks of this period of geological time are usually altered and few fossils with hard parts or skeletons have been found within them, although Precambrian limestone rocks in Australia, Siberia, and parts of the USA contain *stromatolites, which are believed to have been formed by cyanobacteria; frond-like impressions of a supposed plant (*Charnia*) have also been found. The *Ediacaran fauna is late Precambrian. Precambrian rocks outcrop extensively in shield areas such as northern Canada and the Baltic Sea. In modern usage the Precambrian has been replaced by the *Hadean, *Archaean, and *Proterozoic Eons.

precipitable water The quantity of rainfall that would result from condensation and precipitation of the total moisture in a column of air in the atmosphere. Most atmospheric moisture is contained in the lower atmosphere, below about 5 500 m. On average, an atmospheric column of 1 m^2 cross-section contains vapour equivalent to 5–25 mm depth of rainfall. The average residence time of moisture in the atmosphere is about nine days.

precipitation **1.** All the forms in which water falls to the ground (i.e. as rain, sleet, snow, hail, drizzle, or other more specialized forms) and also the amounts measured. Sometimes precipitation seen falling from clouds evaporates before reaching the ground. **2.** The deposition of dust or other substances (e.g. pollutants) from the air.

precipitation-efficiency index Devised in 1931 by C. W. Thornthwaite, an index based on the ratio of mean monthly rainfall and temperature values in relation to evaporation rates. Summation of monthly values gives an annual precipitation-efficiency index (P–E), which is used to define major climatic regions. *See also* THORNTHWAITE CLIMATE CLASSIFICATION.

pre-climax **1.** The *community immediately preceding the *climax, especially where this is forest. For example, on light soils in Britain, birchwood is often the pre-climax to oakwood. **2.** In the *monoclimax model of climax vegetation development, the name given to communities differing from the *climatic climax owing to environmental conditions that are less favourable for vegetation development than those of the surrounding region. For example, grassland might develop in *rain-shadow areas in the lee of a mountain range, in an area that was otherwise forested. *See also* CLIMAX THEORY; *compare* POST-CLIMAX.

precocial Applied to young mammals which are born with their eyes and ears open and are able to stand and walk, regulate their body temperature, and excrete without assistance. The young of grazing animals (e.g. cattle, sheep, and horses) are precocial. *Compare* ALTRICIAL.

predation The interaction between species populations in which one organism, the predator, obtains energy (as food) by consuming, usually killing, another, the prey. Most typically, a predator is an animal that catches, kills, and eats its prey; but predation also includes feeding by insectivorous plants and grazing interactions (e.g. the complete consumption of unicellular phytoplankton by zooplankton). Predation is analogous to *parasitism in being both a means of feeding and a cause of immediate harm, but differs in that predators are usually larger than prey organisms (but not always, e.g. piranha fish) and the prey is usually killed. The distinction is not clear-cut, especially for insect populations. In detail, a continuous gradation of interactions is found from predation to parasitism. The term predation is sometimes applied to the grazing of plants by animals. In general, grazing activity removes part of a plant but does not kill it, hence it is closer to parasitism than predation. The consumption of seeds, however, can be regarded as a form of predation if it involves the destruction of the embryo.

predation analysis The mathematical modelling of predator–prey relationships (as, e.g., in the *Lotka–Volterra equations). Using these models, computer simulations of predator and prey behaviour can be made and tested against field results. This gives a useful insight into the development of *community structure, and more particularly into the behaviour of pest organisms and other species which are important commercially or in *conservation.

predation compensation The mechanisms by which a prey population may maintain a constant size despite large fluctuations in the amount of predation. Compensation occurs because intraspecific competition changes with varying predation. When predation increases natural mortality (e.g. from lack of food) decreases and when predation decreases resources must be shared more thinly and natural mortality increases.

predator *See* PREDATION.

predictive dormancy *See* DORMANCY.

predisplacement An alteration in the *ontogeny of a descendant (a type of *heterochrony) such that some developmental process begins earlier than in its ancestor, and so has progressed further by the time maturity is reached.

preen gland *See* PREENING.

preening *Grooming behaviour that is performed by birds for the maintenance of feathers. Preening cleans feathers, but also oils them with a substance secreted by a preen gland near the tail; the bird applies this substance with the tip of its bill. The oil renders the feathers supple and water-resistant, and ingested oil supplies the bird with some vitamin D.

preferential species In *phytosociology, a species that is present with varying abundance in several *communities, but is especially abundant and vigorous in one particular community. It belongs to fidelity class 3 in the *Braun-Blanquet phytosociological scheme. *Compare* ACCIDENTAL SPECIES; EXCLUSIVE SPECIES; INDIFFERENT SPECIES; and SELECTIVE SPECIES.

preformation *See* EPIGENESIS.

prehensile Capable of grasping.

premolar In mammals, one of the cheek teeth located between the *canines

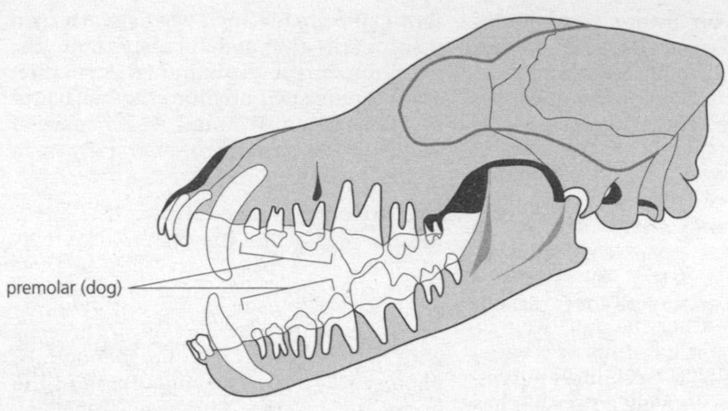

Premolar

and *molars; premolars are preceded by milk teeth.

premorse *See* PRAEMORSE.

pre-oral Relating to the region in front of the mouth.

Pressler borer *See* INCREMENT BORER.

pressure-gradient force (PGF) 1. The force acting on air that is due to pressure differences. Horizontal variations in pressure create a tendency for movement from higher to lower pressure. This is only one component of the forces acting on the actual wind, however, so air does not normally flow at right angles across the *isobars. Other forces are associated with the rotation of the Earth beneath the moving wind, and with a centrifugal force where the path of the wind is curved. In practice, the air moves nearly parallel to the isobars above the friction layer. *See also* CORIOLIS EFFECT and GEOSTROPHIC WIND. **2.** The force acting on a water mass which is due to pressure differences over distance. Horizontal variations in pressure create a tendency for movement from higher to lower pressure areas. *See also* GEOSTROPHIC CURRENT.

pressure head The potential energy possessed by a unit weight of water at any point when compared with a pressure of one *atmosphere at the same elevation. For *groundwater, it is measured by the depth of submergence between the measurement point and the water level in a *piezometer. *See also* ELEVATION POTENTIAL ENERGY; HYDRAULIC HEAD; and POTENTIOMETRIC SURFACE.

pressure melting The melting of ice in response to stress. It occurs because the freezing temperature of water falls as pressure increases, at about 1 °C for every 140 bars (140×10^5 N/m^2). The term 'pressure melting point' refers to the temperature at which ice just begins to melt under a given pressure.

pressure potential The hydrostatic pressure to which water in a liquid phase is subjected. It was formerly known as wall pressure or turgor pressure. In a turgid plant cell, pressure potential is usually positive but the pressure potential of xylem in a transpiring plant, which is under considerable tension, will be negative.

prevailing climax The most common undisturbed, stable *community that occurs in a region. In any given area various stable communities are found in response to local differences in *habitat. Gradual transitions from one stable type to another occur as environmental controls change. The steady-state community occupying the largest number of non-extreme habitats in the area is the prevailing climax. This approximates to the climatic climax communities for the area. Unlike monoclimax, prevailing climax does not involve consideration of possible long-term climatic changes, or associated habitat changes caused by weathering and *erosion. Thus it is a more practicable and readily applied concept. *See also* CLIMAX THEORY.

prevernal In the early spring. *Compare* AESTIVAL; HIBERNAL; SEROTINAL; and VERNAL.

prey *See* PREDATION.

primary forest **1.** A *climax forest comprising *primary or climax species, i.e. a forest that either has not been severely disturbed or has fully recovered from disturbance by a *secondary succession. **2.** A forest occupying a site that has been continuously forested (in Britain since the last ice advance) even though it may have been clear-felled, provided that the clear-felling does not break the forest continuity (i.e. the forest regenerated or was replanted).

primary forest species **1.** Tree species that are confined to *primary forest. **2.** Tree species that occur in primary forest but also in localities where their presence is explained by planting or by their survival from the clearance of what was once primary forest (e.g. in hedges and on stream banks).

primary productivity **(primary production)** The rate at which *biomass is produced by photosynthetic and chemosynthetic *autotrophs (mainly green plants) in the form of organic substances, some of which are used as food materials by the autotrophs themselves. Gross primary productivity (GPP) is the total rate of *photosynthesis and chemosynthesis per unit ground area, including that portion of the organic material produced which is used in *respiration during the measurement period. Net primary productivity (NPP) is the rate of production allowing for the amount lost to plant respiration during the measurement period. Net *ecosystem productivity (NEP) is the rate of accumulation of organic material, allowing for both plant respiration and heterotrophic consumption during the measurement period, i.e. NEP = GPP − (respiration by autotrophs + respiration by heterotrophs). Rates of storage at higher *trophic levels are termed secondary productivities. Strictly, the primary production of an ecosystem (as distinct from its productivity, which is a rate) is the amount of organic material accumulated per unit ground area. It is usual to indicate a period of time, however, since otherwise such data have limited value. Thus in practice the terms 'production' and 'productivity' are often used interchangeably. It is also vital to indicate whether production figures relate to net or gross, and to primary or ecosystem productivity, or to some portion of these, as with a harvested crop. In the oceans, photosynthesis by *phytoplankton in the upper 100 m (the *euphotic zone) accounts for most primary production. Waters in tropical areas are less productive than waters in temperate regions, because in tropical waters the water column undergoes no seasonal vertical mixing and so becomes *oligotrophic. Areas of *upwelling of nutrient-rich deep waters have high productivity.

primary sexual character An organ that produces *gametes. Examples are the testes of male and the ovaries of female animals and the male anther and female carpel of flowering plants.

primary succession **(prisere)** A *succession initiated on a newly produced bare area (e.g. following glaciation, major earth movements, or a newly formed riverbank). Since no living remnants of a previously existing *community are present, successional stages and interactions of the plants and the physical environment are usually fairly clear, at least initially. *Compare* SECONDARY SUCCESSION.

primary woodland Woodland occupying a site which has been continuously wooded (in Britain since the last ice advance) even though it may have been clear-felled, provided that the clear-felling does not break the woodland continuity (i.e. the woodland regenerated or was replanted).

primitive Applied to a *character (as a synonym of *'plesiomorphic') or, occasionally, to a whole organism that preserves the character state of an ancestral stage.

principal components analysis A statistical technique for determining the amount of variance in a set of data, in which the first of a series of axes explains the maximum variance, the second the maximum of the variance remaining, and so forth, each axis being at right angles to the preceding one. *Compare* RECIPROCAL AVERAGING.

Priscoan The name formerly given to the earliest of the three subdivisions of the *Precambrian. The time covered by the Priscoan is now ranked as the *Hadean eon, lasting from the formation of the Earth until 3 800 Ma ago.

prisere *See* PRIMARY SUCCESSION.

probable mutation effect The idea that, as most *mutations act to produce failures of development, a non-essential organ will eventually be reduced, simply by repeated mutations accumulating in a population. This idea has been used to explain the loss of hind limbs in whales, etc.

proboscis A tubular protrusion from the anterior of an animal.

process-response system In *geomorphology, a natural system that is formed by the combination of at least one morphological and one *cascading system. It therefore shows how form and geomorphological process are related. An example is the coastal process-response system, in which the cascading system of wave energy advances from deep water to the edge of the *swash zone, and is linked to various morphological features of the shallowing zone.

procumbent Projecting forward, more or less horizontally.

pro-delta The furthest offshore portion of a *delta, lying at the toe of the delta front, and characterized by a relatively slow rate of fine-grained deposition.

producer In an *ecosystem, an organism that is able to manufacture food from simple inorganic substances (i.e. an autotroph, most typically a green plant). *Compare* CONSUMER ORGANISM.

production **1.** The total mass of organic matter that is manufactured in an *ecosystem during a certain period of time. Net production is the yield of the producers and consumers and is the amount of living matter in the ecosystem. **2.** In energy-flow studies, that part of the assimilated food or energy which is retained and incorporated in the *biomass of the organism, but excluding the reproductive bodies released by the organism. This may be regarded as growth. In energy-flow measurements, production is expressed as energy per unit time, per unit area. **3.** *See* PRIMARY PRODUCTIVITY.

production ecology The branch of ecology dealing with energy flow and nutrient cycling within *ecosystems.

production/respiration ratio (P/R ratio) The relationship between gross *production and total *community respiration. Where P/R = 1 a steady-state community results. If P/R is persistently greater or less than 1 then organic matter either accumulates or is depleted respectively.

productivity The rate at which the *biomass increases per unit area. *See* PRIMARY PRODUCTIVITY.

profile A vertical section through all the constituent *horizons of soil, from the surface to the relatively unaltered parent material.

profile transect A *line transect from which a record is taken of the ground level at which each species is found.

profundal zone The bottom and deep-water area of freshwater ecosystems which lies beyond the depth of effective light penetration (*see* COMPENSATION LEVEL). In shallow freshwater systems, such as ponds, this zone may be missing. *Compare* BENTHIC ZONE.

progenesis **1.** A form of *paedomorphosis (more complete than *neoteny) in which development is cut short by the early onset of maturity and the adult descendant exactly resembles a juvenile stage of its ancestor. **2.** The early stages of *ontogeny (less common usage).

progenote The first organism to evolve on Earth and, therefore, the ancestor of all subsequent organisms.

proglacial Applied to the area adjacent to a *glacier. A proglacial lake is a body of water impounded in such an area and is often inferred for areas of *Pleistocene glaciation from the evidence of strand-lines, lake sediments, deltas, and *overflow channels.

prognathous Having a head that is approximately horizontal, the mouth being at the front.

progradation The outward building of a sedimentary deposit, such as the seaward advance of a *delta or shoreline, or the outbuilding of an *alluvial fan.

progressive evolution A steady, long-term improvement of evolutionary *grade, which has allowed plants and animals to become ever more independent of the aquatic environment in which they first evolved. For example, the sequence bryophyte, pteridophyte, gymnosperm, angiosperm represents a progressive evolutionary trend.

progressive feeding The provision of larval food by female wasps and bees at regular intervals throughout the period of larval growth. It is practised by all the social vespid wasps and the honey-bees (*Apis* species). It is uncommon in solitary (non-social) species, though it is found in some sphecid and nyssonid wasps.

progressive succession The normal sequential development of *communities, from simple communities with few species and low productivity to the optimum sustainable in a given environment. It is contrasted with retrogressive *successional change (*retrogression).

Prokarya In the widely used five-kingdom classification of living organisms, a superkingdom containing the kingdom *Bacteria.

prokaryote An organism, usually unicellular, in which the cells lack a true nucleus, the *DNA being present as a loop in the cytoplasm. Other prokaryotic features include the lack of chloroplasts and mitochondria and the possession of small ribosomes.

prolicide The killing of offspring.

promiscuity A form of *polygyny or *polyandry in which a member of one sex mates with more than one member of the other sex, but each relationship is an ephemeral one and terminates after mating, whereas in the other cases a bond of some kind forms.

pronking Jumping high into the air several times in succession while running, a behaviour practised by springbok (*Antidorcas marsupialis*) on the African grasslands, perhaps as a warning to a predator that its presence has been noticed and it should therefore abandon its pursuit since the springbok will escape. *See also* STOTTING.

pro-oestrus *See* OESTRUS CYCLE.

propagule Any structure that functions in propagation and dispersal (e.g. a *spore or seed). *See also* MIGRATION.

prophage The *genome of a *bacteriophage whose host bacterium responds to its presence lysogenically (*see* LYSOGENY and LYTIC RESPONSE).

propolis Resins of plant origin collected by honey-bee (*Apis* species) workers and used to seal gaps in the nest or hive, or to reduce the nest entrance to a suitable size. Sometimes it is also used to entomb large trespassers into the nest (e.g. mice) which have been killed by workers acting as guard bees. It is supposedly of medicinal value.

proprioceptor **1.** A receptor (i.e. sense organ) that detects pressure, position, or movement. **2.** More generally, a receptor which is sensitive to bodily changes that are not caused by substances taken into the respiratory tract or alimentary canal.

protalus rampart A ridge of rock debris, less than 10 m high, found near the base of a steep inland rock face. It consists of frost-shattered material that has been carried some distance from the face down the steep surface of a basal snow bank.

protandry **(adj. protandrous)** **1.** In plants, the maturation of anthers (i.e. male organs) before carpels (female organs). **2.** The production of sperm in males before females produce eggs (e.g. some roundworms). **3.** The arrival of males before females at breeding grounds (e.g. many territorial bird species). *Compare* PROTOGYNY.

protein A polymer that has a high relative molecular mass of *amino acids. Proteins occupy a central position in the architecture and functioning of living matter. Structurally they are divided into two groups, globular and fibrous proteins; and they are often found associated with a non-protein component, forming so-called conjugated proteins. Functionally they act variously as *enzymes, as structural elements (e.g. hair and collagen), as hormones, as respiratory pigments, as contractile elements, as *antibodies, and as hereditary factors.

Proterozoic The *eon of geologic time that followed the *Archaean and preceded the present *Phanerozoic Eons. It is divided into three *eras. The Palaeoproterozoic lasted from 2 500 Ma to 1 600 Ma, the Mesoproterozoic from 1 600 Ma to 1 000 Ma, and the Neoproterozoic from 1 000 Ma to 542 Ma. The Neoproterozoic culminated with the first abundantly fossiliferous period, the *Ediacaran.

protist A single-celled member of the *Protoctista.

Protista *See* PROTOCTISTA.

proto-Atlantic *See* IAPETUS OCEAN.

protocooperation **(facultative mutualism)** An interaction between organisms of different species in which both organisms benefit, but neither is dependent on the relationship. *Compare* MUTUALISM.

Protoctista In the widely used five-kingdom classification system for living organisms, one of the kingdoms within the superkingdom *Eukarya. In the three-domain classification system, a kingdom within the *domain Eukarya. Protoctists are aquatic *eukaryotes, but they are neither animals, nor fungi, nor plants. The kingdom includes naked and shelled *amoebae, foraminiferans, zooflagellates, ciliates, *dinoflagellates, *diatoms, *algae (including seaweeds), slime moulds, slime nets, and *protozoa. Single-celled organisms were formerly known as 'protists' and the kingdom containing them was the Protista. This ranking was abandoned when it became evident that multicellularity evolved many times and that multicellular organisms are closely related to single-celled forms. The name Protoctista means 'first established' from the Greek *protos* (first) and *ktistos* (established); 'Protista' is no longer used.

protogyny **(adj. protogynous)** A condition in which the female parts develop first (e.g. the females develop eggs before males produce sperm, or females arrive at breeding grounds before males). *Compare* PROTANDRY.

protohemicryptophyte *See* HEMICRYPTOPHYTE.

prototroph A strain of bacteria that has the nutritional requirements of the wild type or non-mutant species. *Compare* AUXOTROPH.

protozoa **(sing. protozoon)** Single-celled *eukaryotes that feed heterotrophically (*see* HETEROTROPH), principally on *bacteria, although some are parasites (*see* PARASITISM) or *symbionts of mammals. Taxonomically they are placed within the *Protoctista. There are four groups of protozoa: flagellates (Mastigophora); amoebae (Sarcodina); sporozoans (Sporozoa, also called Apicomplexa); and ciliates (Ciliophora). This is not a *phylogenetic classification, however, and the groups are not closely related.

protozoon An individual member of the *protozoa.

province **1.** A large region or area that may be considered as a discrete unit because of similar features which occur throughout it. **2.** In geography, an area of land or sea in one climatic belt.

provinciality The occurrence of species within well-defined biogeographic areas or *provinces. Each province contains a distinct assemblage of species, some of which are endemic (*see* ENDEMISM).

provirus A viral *genome which is incorporated into a host genome.

P/R ratio *See* PRODUCTION/RESPIRATION RATIO.

pruinose Covered with a greyish, usually waxy, bloom.

psammo- Prefix derived from the Greek *psammos*, meaning 'sand'.

psammo-littoral zone The water's-edge zone of sandy shores, both marine and freshwater, where the microscopic plant and animal *communities forming the *psammon are most prolific.

psammon The microscopic flora and fauna of the interstitial spaces between sand grains of seashore and lake-shore areas.

psammophyte A plant that grows in loose sand.

psammosere The characteristic sequence of changes associated with stages in a plant *succession developed on sand-dunes.

PSC *See* POLAR STRATOSPHERIC CLOUD.

pseudergate A member of the labouring caste in lower families of termites, one which is capable of *metamorphosis into a reproductive.

pseudofossil A naturally occurring object that resembles a *fossil, the resemblance resulting purely from chance.

pseudogene A nonfunctional member of a *multigene family.

pseudointerference Where consumers exhibit the *aggregative response, the decline in their rate of consumption as their density on foraging patches increases. The effect resembles *mutual interference, but in fact is a purely density-dependent consequence of high consumer density.

pseudopodium (pl. pseudopodia) A protrusion, usually temporary, from the cell body of an *amoeba that is used for locomotion and for engulfing food particles.

pseudorumination *See* REFECTION.

pseudospecies An assemblage of individuals of somewhat dubious taxonomic status that are regarded as a single species for the sake of numerical analysis of data.

pseudo-steppe An area with *steppe-like vegetation which occurs outside Eurasia. The term 'steppe' strictly refers to the temperate grassland of Eurasia. However, it has also been applied to vegetation on the southern fringe of the Sahara (the Sahel zone), in parts of Namibia, and in south-western Australia.

psychrometer A *hygrometer that has a *wet-bulb and a *dry-bulb thermometer. *See also* WHIRLING PSYCHROMETER.

psychrophile An *extremophile (domain *Archaea) that thrives in environments where the temperature is low, usually below 15°C.

psychrosphere In an ocean, the region of cold water that lies beneath the warmer water of the *thermosphere.

psychrotroph An organism that can grow at low temperatures (e.g. 15°C or below) but that grows better at higher temperatures.

pterochore *See* CHORE.

pterochory Dispersal of winged seeds by the wind.

pteropod ooze Deep-sea *ooze in which at least 30 per cent of the sediment consists of the shells of small planktonic (*see* PLANKTON) gastropods (known as pteropods or 'wing-footed' snails). The shells are aragonitic and, as *aragonite solubility increases rapidly with depth, pteropod ooze is restricted to water depths less than 2 500 m.

pterygote Applied to members of the Pterygota, the larger of the two subclasses of insects, which possess wings or are secondarily wingless (i.e. they are descended from winged ancestors). *Compare* APTERYGOTE.

pubescent Covered with fine, soft hair.

puddled soil (poached soil) Soil in which the structure has been destroyed by the physical impact of raindrops, by tillage when wet, or by trampling by animals.

pulse **1.** The edible seeds of any leguminous plant (Fabaceae). **2.** The rate of the heartbeat in vertebrates. **3.** See PULSE LABELLING.

pulse labelling A technique in which radioisotopes are used for the measurement of the rates of synthesis of compounds within living cells. A suspension of cells or organelles is exposed to a small quantity of an isotope for a brief period (seconds or minutes), hence the term 'pulse'. This is achieved through the addition to the suspension of a much larger quantity of the stable (unlabelled) isotope of the same compound, following the required period of exposure to the radioisotope. The effect of competition between the two isotopes is to reduce to a negligible level the further uptake of the latter. Measurement of the levels of activity in samples under various experimental conditions can yield useful information regarding the factors influencing the uptake and metabolism of compounds.

pulverulent Dusty; powdery; covered with powder.

pulvinate Cushion-shaped; swollen; convex.

puna Arid alpine vegetation on high plateaux along the western side of the Andes. It comprises sparse, tufted grasses, and large, hardy, cushion plants. The pamirs of Tibet are a similar vegetation type.

punctate Applied to any structure that is marked by pores or by very small, point-like depressions.

punctiform Dot-like in appearance.

punctuated equilibrium The theory, first proposed in 1972 by Niles Eldredge and Stephen Jay Gould, that *evolution is characterized by geologically long periods of stability during which little speciation occurs, punctuated by short periods of rapid change, species undergoing most of their morphological changes shortly after breaking from their parent species.

purga *See* BURAN.

purine A nitrogen *base composed of two adjoining ring structures, one of which has five members and the other six. The purine bases in the *nucleotides of nucleic acids are adenine and guanine.

push moraine *See* MORAINE.

puszta Hungarian grassland, similar to the North American prairie.

putrefaction The *anaerobic digestion by bacteria of proteinaceous material (e.g. meat) with the concomitant production of malodorous substances.

puy **1.** A volcanic hill in the Auvergne region of France. **2.** Any steep-sided tower of volcanic rock (e.g. the Devil's Tower, Wyoming, USA, and Shiprock, New Mexico, USA). It consists of the resistant central plug or neck of a former volcano.

PWP *See* PERMANENT WILTING POINT.

pycnocline The change of density with depth in a water column.

pyramid of biomass A diagrammatic expression of *biomass at different *trophic levels in an ecosystem, usually plotted as dry matter or *calorific value per unit area or volume. Typically this gives a gradually sloping pyramid, except where the sizes of organisms vary dramatically from one trophic level to another. In this case, the higher metabolic rate of the smaller organisms may result in a greater biomass of *consumers than of *producers, giving an inverted pyramid. Aquatic

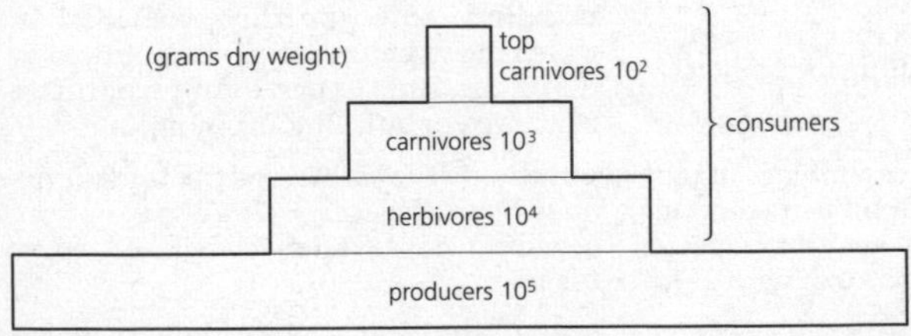

Pyramid of biomass

communities in winter typically show inverted biomass pyramids. *Compare* PYRAMID OF ENERGY and PYRAMID OF NUMBERS; *see also* ECOLOGICAL PYRAMID.

pyramid of energy A diagrammatic expression of the rates of flow of energy through the different *trophic levels of an ecosystem. It reflects the rates of photosynthesis, respiration, etc. (and not the standing crop, as in the pyramid of *biomass), and can never be inverted since energy is dissipated through the ecosystem. It is the most fundamental and most useful of the three ecological pyramids. *Compare* PYRAMID OF BIOMASS and PYRAMID OF NUMBERS; *see also* ECOLOGICAL PYRAMID.

pyramid of numbers A diagrammatic expression of the numbers of individual organisms present at each *trophic level of an ecosystem. It is the least useful of the three types of ecological pyramid since it makes no allowance for the different sizes and metabolic rates of organisms. Typically it slopes more steeply than the other pyramids and may be inverted, e.g. when based on studies of temperate woodlands in summer. *Compare* PYRAMID OF BIOMASS and PYRAMID OF ENERGY; *see also* ECOLOGICAL PYRAMID.

pyriform Pear-shaped.

pyrimidine A nitrogen *base composed of a single, six-membered ring structure. The pyrimidine bases in the *nucleotides of nucleic acids are cytosine and thymine in DNA and cytosine and uracil in RNA.

pyro- A prefix derived from the Greek *pur*, meaning 'fire', which means 'associated with fire' (e.g. *pyrophyte, literally 'fire plant').

pyroclastic Applied to rock fragments produced by a volcanic explosion; the name means 'fire-broken'.

pyroclimax *See* FIRE CLIMAX.

pyrolysis The heating of organic molecules without oxygen, to produce hydrocarbons (char) which have a high calorific value. Fuel produced by this process can be concentrated and stored. Organic waste may be used.

pyrophyte A plant adapted to withstand or to achieve a competitive advantage from fire (e.g. *Themeda triandra*, a grass widespread in Kenyan *savannahs).

p-zone A biostratigraphic zone that is distinguished by *pelagic *fossils (e.g. ammonites and graptolites). The term was proposed in 1965 by T. G. Miller. *Compare* B-ZONE.

QO_2 *See* OXYGEN QUOTIENT.

Q technique A method for analysing data, in which observations (*N*) form the columns and the variables or attributes (*n*) form the rows in a table or matrix. *Compare* R TECHNIQUE; *see also* INVERSE ANALYSIS and NORMAL ANALYSIS.

quadrat A basic sampling unit of vegetation surveys. Traditionally 1 m^2 quadrats were used to sample short, non-woody *communities such as *grasslands and *heathlands. More recently circular and rectangular *quadrats have been used, as well as squares of all sizes, depending on the purpose of the survey (e.g. 10 m^2 and 20 m^2 quadrats are commonly used in woodland studies).

qualitative inheritance An inheritance of a character that differs markedly in its expression among individuals of a species: variation in that species is discontinuous. Such characters are usually under the control of major genes. A principal example is gender. *Compare* QUANTITATIVE INHERITANCE.

quantitative character *See* QUANTITATIVE INHERITANCE.

quantitative inheritance The inheritance of a character (known as a quantitative character or quantitative trait) that depends upon the cumulative action of many genes, each of which produces only a small effect. Examples of such quantitative characters include clutch size in birds, milk production in cattle, weight and skin pigmentation in humans, spore production in ferns, height of trees, and nectar production in buttercups. Usually the character shows continuous variation (i.e. a gradation from one extreme to the other). *Compare* QUALITATIVE INHERITANCE.

quantitative trait *See* QUANTITATIVE INHERITANCE.

quantum evolution (quantum speciation) Traditionally, the rapid speciation that can occur in small populations isolated from the large, ancestral population, and which are therefore subject to the *founder effect and to *genetic drift. Typically it results when a population shifts into another *adaptive zone. Initially the number of individuals will be small and rapid genetic and *phenotypic change is feasible in such circumstances, particularly if the new *niche is unoccupied. The concept of quantum evolution has been given special significance in the theory of *punctuated equilibrium.

quantum speciation *See* QUANTUM EVOLUTION.

quarrying *See* GLACIAL PLUCKING.

quasi-sympatric speciation The separation of one species into two by *adaptation of different sub-populations into different *niches.

Quaternary The sub-era of the *Cenozoic Era of geologic time that began 1.806 Ma ago and continues to the present day. It coincides with the *Pleistogene Period. The Quaternary comprises the *Pleistocene and *Holocene Epochs, during which there have been numerous ice-sheet advances in the northern hemisphere. By the Pleistocene most of the world's faunas and floras had a modern appearance. The name Quaternary has now been abandoned and, although very commonly referred to, is used only informally.

queen In a colony of social insects, the primary female reproductive.

quickflow The part of a storm rainfall which moves quickly to a stream channel via *surface run-off or *interflow, and forms a flood wave in the channel.

quiescence A period of rest or reduced activity that follows a *consummatory act.

race An interbreeding group of individuals all of whom are genetically distinct from the members of other such groups of the same *species. Usually these groups are geographically isolated from one another, so there are barriers to *gene flow. Examples are island races of birds and mammals, such as the Skomer vole and the St Kilda wren.

racial senescence *See* PHYLOGERONTISM.

radar tracking An observational technique, used in field studies of animal behaviour, in which subjects are tracked by radar. By watching a plan-position indicator or from time-lapse photography of the screen, information can be obtained about the movements of birds and insects. Birds fitted with metal tags can be distinguished from untagged birds because of the stronger echo returned by the tags.

radial drainage A drainage pattern consisting of streams that extend radially from a central zone. It is typical of the patterns developed on freshly constructed land-forms (e.g. volcanoes) and on areas of domed uplift.

radial symmetry The arrangement of the body of an animal in which parts are arranged symmetrically around a central axis. Such an arrangement allows the animal to interact with its environment from all directions. It is most commonly associated with a *sessile way of life. *Compare* BILATERAL SYMMETRY. The term is also applied to certain types of floral structure where a bisection in any plane will result in symmetric halves, e.g. buttercups.

radiation budget (energy budget) The difference between the amount of incoming solar radiation and the amount of outgoing terrestrial radiation. The balance is in deficit (i.e. more energy leaves the Earth's surface than reaches it from the Sun) at night. Overall, the highest positive net balance is found in low latitudes.

radiation ecology The branch of ecology concerned with the effects of radioactive materials on living systems and on the pathways by which they are dispersed through ecosystems, including their dispersal through the *abiotic environment. The term is used especially with regard to those materials released through human agency.

radiation fog A condensation effect that occurs over land surfaces on clear nights with light breezes, caused by surface radiation cooling. Favoured initial conditions are very humid air, with wet and cold surfaces (e.g. marshes). The fog, most common in winter, is generally cleared by the Sun's warmth in the morning, but thick fog over wet surfaces in winter may persist much longer, particularly if an upper cloud layer screens the Sun.

radiation inversion A *temperature inversion in the lower atmosphere which is due to radiation cooling of the ground at night. *See also* RADIATION NIGHT.

radiation night A night with clear skies, when terrestrial long-wave radiation cannot be partly returned to the surface by cloud. There is rapid cooling of the air close to the ground, particularly when there is little wind, giving low minimum surface temperatures.

radiatus The Latin *radiatus*, meaning 'with rays', used to describe a variety of cloud with parallel bands which, in perspective, appear to meet at the horizon. The effect is seen in *cumulus, *stratocumulus, *altocumulus, *altostratus, and *cirrus. *See also* CLOUD CLASSIFICATION.

radical substitution The substitution of one *amino acid for another with markedly different chemical properties.

radioactive dating *See* RADIOMETRIC DATING.

radioactive tracer A radioactive isotope whose movement can be monitored, which is used to trace the pathways by which individual substances move through an organism, a living system, the abiotic environment, etc. Non-radioactive chemical analogues of certain substances may be used for the same purpose if their movement can be monitored (e.g. caesium, which can be substituted for potassium).

radioactive waste Substances from nuclear processes which are contaminated and not reusable. Low-level waste includes clothing and materials which have been used when handling radioactive sources (e.g. in hospitals and research laboratories). High-level waste is mainly from the fission process in nuclear power stations or from military waste. The high level of radioactivity is caused by short-lived isotopes; their decay leads to lower levels of radioactivity, but from longer-lived isotopes with half lives of up to one million years (although, in most cases, their emissions fall to levels approximately equal to those of natural background radiation after about 500 years and it is not necessary to safeguard them for periods longer than this). During the first phase such waste is stored in corrosion-resistant containers, but the long-term requirement is to find geologically stable repositories so that any leakage will not return to the surface.

radiocarbon dating **(^{14}C dating)** A dating method for organic material that is applicable to about the last 70 000 years. It relies on the assumed constancy over time of atmospheric ^{14}C : ^{12}C ratios (now known not to be valid), and the known rate of decay of radioactive carbon, of which half is lost in a period (the 'half-life') of every 5 730 years ± 30 years. The earlier 'Libby standard', 5 568 years, is still widely used. In principle, since plants and animals exchange carbon dioxide with the atmosphere constantly, the ^{14}C content of their bodies when alive is a function of the radiocarbon content of the atmosphere. When an organism dies, this exchange ceases and the radiocarbon fixed in the organism decays at the known half-life rate. Comparison of residual ^{14}C activity in fossil organic material with modern standards enables the calculation of the age of the samples. Since the method was first devised it has been realized that the atmospheric ^{14}C content varies, as the cosmic-ray bombardment of the outer atmosphere that generates the ^{14}C varies. Correction for these fluctuations is possible for about the last 12 000 years by reference to the ^{14}C contents of long *tree-ring series (e.g. those for *bristlecone pines).

radiogenic Applied to isotopes produced by the process of radioactive decay.

Radiolaria A subclass of *protozoa (class Actinopodea) which possess more or less elaborate skeletons of silica. Most are *pelagic marine organisms. Radiolarian fossils have been found in *Cambrian rocks and they are important stratigraphic fossils for *Mesozoic and *Cenozoic deep-sea sediments. *See* RADIOLARIAN EARTH and RADIOLARIAN OOZE.

radiolarian earth An unconsolidated or semi-consolidated bed of *radiolarian ooze. It is a relatively rare deposit, as most ancient radiolarian oozes have been compacted into hard *radiolarites.

radiolarian ooze A deep-sea *ooze in which at least 30 per cent of the sediment consists of the siliceous radiolarian tests. Radiolarian-rich oozes occur in the equatorial regions of the Pacific and Indian Oceans where the depth exceeds the *carbonate-compensation depth (around 4 500 m in the central Pacific). *See also* DIATOM OOZE.

radiolarite A compacted, siliceous, sedimentary rock composed mainly of the siliceous tests of marine zooplankton (*see* PLANKTON) belonging to protozoan subclass Radiolaria. The sediment from which radiolarites are formed is called a *radiolarian ooze.

radiometric dating **(radioactive dating)** The most precise method of dating rocks, in which the relative percentages of 'parent' and 'daughter' isotopes of a given radioactive element are estimated. Early methods relied on uranium and thorium minerals, but potassium–argon, rubidium–strontium, samarium–neodymium, and carbon-14–carbon-12 are now of considerable importance. Uranium-238 decays to lead-206 with a half-life of 4.5

billion (10^9) years, rubidium-87 decays to strontium-87 with a half-life of 50.0 billion years, and potassium-40 decays to argon-40 with a half-life of 1.5 billion years. For carbon-14 the half-life is a mere 5 730 ± 30 years (*see* RADIOCARBON DATING). It is important that the radioactive isotope be contained within the sample being dated. Carbon-14 is contained within plant material, but potassium-40, argon-40, and uranium-238 are contained satisfactorily only within crystals. *Igneous rocks are the most suitable for dating. *Fossils occur mostly in sedimentary rocks, however, so absolute dates can be calculated for them less commonly than might be supposed. The only exceptions are fossils occurring in glauconite, a *clay mineral containing potassium and argon which forms *authigenically on the bottom of shelf seas.

radiosonde An instrument package suspended beneath a helium-filled balloon that transmits meteorological data by radio as it rises through the atmosphere. When the balloon bursts the instruments fall by parachute. There are approximately 500 upper air weather stations in the world. Each station releases two radiosondes a day, one at midnight and the other at noon Greenwich time. *See also* RAWINSONDE and WIND SONDE.

radiotelemetry *See* BIOTELEMETRY.

radio tracking An observational technique used in field studies of animal behaviour, in which the animal is fitted with a battery-powered, fixed-frequency radio transmitter, usually attached to a collar, and the observer is equipped with a receiver and directional antenna.

raindrop A water droplet formed by condensation of water vapour in a cloud that is heavy enough to fall from the cloud and large enough to reach the surface of land or sea before evaporating in the unsaturated air beneath the cloud. Droplets reaching the surface range in diameter from about 100 μm in fog to 0.2 mm in drizzle and 5.0 mm in a heavy shower.

rain forest *See* TEMPERATE RAIN FOREST and TROPICAL RAIN FOREST.

rain-gauge A device for measuring the amount of rainfall. Two types are widely

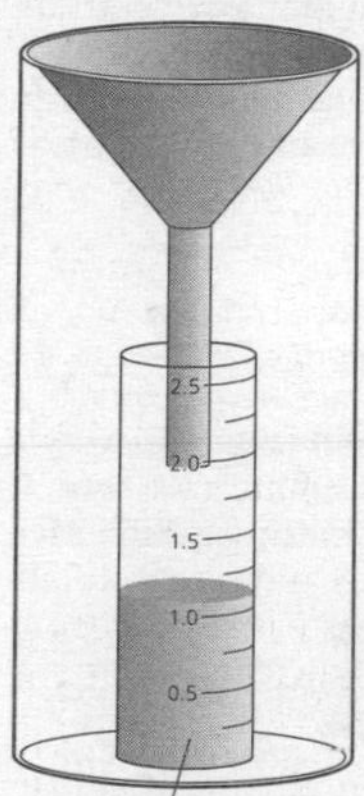

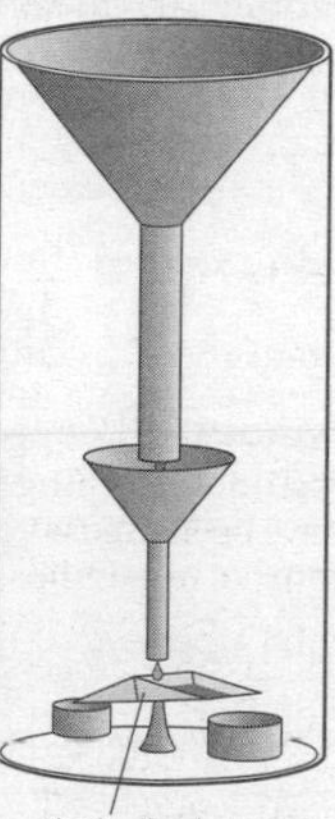

Rain-gauge

used. The standard rain gauge, approved for use at all weather stations, is designed to collect rain while minimizing losses due to evaporation. The gauge comprises a cylinder 20 cm in diameter, closed at one end, that is set on open ground with its upper rim 1 m above ground level. Rain falling on to the gauge enters a collecting funnel that directs it into a calibrated measuring tube, 6.32 cm in diameter, inside the cylinder. The cross-sectional area of the measuring tube is therefore one-tenth the area of the mouth of the collecting funnel, so the amount of rainfall is calculated by dividing the depth of water in the tube by 10. Alternative gauges include the tipping-buckets type, in which rainwater passes from a heated collecting funnel to a smaller funnel and from there to one of two buckets mounted on a rocker. Each bucket holds 0.25 mm of water. When a bucket is full it tips downward, making an electrical contact that is transmitted to a recording pen, and empties its water, while the other bucket is positioned to collect water falling from the funnel. The distance between a rain gauge and a large obstruction such as a tree or building should be equal to or greater than the height of the obstruction. The gauge should also be shielded from the wind by placing it inside a horizontal circular hoop 5–100 cm in diameter with baffles hanging from it. Other designs maintain an automatic record of the height of water in the collecting tube, or of the weight of the collected water.

rain-making An attempt to induce rainfall by 'seeding' supercooled water clouds. *See also* CLOUD SEEDING.

rainout *Condensation nuclei which deliver nutrients when they are carried to the surface in precipitation. *See* WETFALL; *compare* WASHOUT.

rain shadow The reduction of rainfall to the lee side of a mountain barrier, which results in relatively dry surface conditions (e.g. in the mountains of the south-western USA, where the wetter western slopes of the Coast Range and the Sierra Nevada contrast with the desert areas of Nevada and eastern California on the lee side of the mountains).

rain-splash *See* RAIN-WASH.

rain-wash The transfer of material across the surface and down a hill-slope as a result of rainfall. Normally it consists of two components: rain-splash, which is the detachment and subsequent down-slope transfer of small soil particles by raindrop impact; and soil-wash, which is the down-slope movement of material by surface water flow. Rainwash and *creep are the two main hill-slope processes.

raised beach A former beach that is now above the level of the present shore-line as a result of earth movement or of a general fall in sea level. Such beaches are frequently described and correlated in terms of height above present sea level.

raised bog An *ombrogenous bog community, typically dome-shaped in section, which develops on former lake sediments, estuarine sites (*see* ESTUARY), uniform *clay substrates, and sometimes on the surfaces of *valley bogs. In the British Isles, living (i.e. still growing upward) raised bogs are confined to western areas with a maritime climate, most notably central Ireland. The living bog surface shows a complex of hummocks and pools (*see* REGENERATION COMPLEX) with different **Sphagnum* species occupying different positions in relation to the degree of waterlogging. Hummocks are typically colonized by wet heath species, e.g. heathers (*Calluna vulgaris* and *Erica tetralix*) and sedges (e.g. *Eriophorium* species). The highest point of the dome may be more than 12 m above the *water-table of the surrounding land. The dome is sometimes referred to as a cupola. The sloping edges of the dome are termed the rand, and the *rheotrophic mires that surround the dome are the lagg. *See also* OMBROGENOUS BOG. *Compare* BLANKET BOG.

ramet An individual member of a *clone that is capable of separation from other members and then capable of independent life. An individual strawberry plant or clover plant developed from a *stolon are ramets.

ramiflorous Borne on the branches. *Compare* CAULIFLOROUS.

ramp *See* SHELF.

Ramsar Convention on Wetlands of International Importance Especially as Waterfowl Habitat A convention drafted at a conference held in Ramsar, Iran, in 1971 which aims to protect the international chain of *habitats used by migratory water birds. Important sites are identified and designated by signatory governments and data collected from them is stored at the Conservation Monitoring Centre, Cambridge, England. By November 2004, 141 nations had signed the convention and 1 387 'Ramsar sites' had been designated, with a combined area of 122.7 million hectares.

rand The sloping edges of the dome of a *raised bog.

random amplified polymorphic DNA (RAPD) A technique in molecular biology for the rapid assignation of DNA-based *character states for *phylogenetic analysis. The technique uses the *polymerase chain reaction to amplify any genomic region containing an arbitrary sequence. The resulting amplified DNA fragment sizes are then used as characters for phylogenetic analysis.

randomize To arrange a set of numbers in random order; in a computer program, to generate a series of *random numbers.

random number One of a series of numbers each of which has an equal probability of occupying a particular position in the series. Tables of random numbers are published and computers can generate random numbers (or pseudorandom numbers, which are selected at random from a stored sequence).

random sample A sample in which each individual measured or recorded (e.g. organism, site, or *quadrat) is independent of all other individuals and also independent of prominent features of the area or other unit being sampled. In ecological field surveying, random quadrat sampling is most easily achieved by superimposing a grid over the sample area, and identifying a series of random co-ordinates, using random-number tables, at which to locate the quadrats. Alternatively, a *random-walk technique may be used.

random-walk technique A method of *random sampling in which the number of paces between sample points is determined by random numbers, usually drawn from random-number tables, and from each sample point a right-angle turn determines the direction of the next point, a coin being tossed to decide whether to turn left or right. Where several sets of samples are taken it is best to start each set from the same point; if this is impossible it is important that the range of numbers from which a selection is made must allow each point within the study to have an equal chance of being sampled each time.

range **1.** The spread of environmental conditions within which a particular species occurs. **2.** The entire geographical area over which a species occurs. **3.** Extensive, open grazing lands. **4.** *See* HOME RANGE.

range management The use of extensive, open grasslands for cattle production or, where appropriate, for the exploitation for human use of semi-domesticated or wild grazing animals (e.g. game cropping on the African *savannah) by methods that maintain the ecosystem. These require the determination of the optimum stocking density that will permit long-term cropping of the grassland without deterioration of the pasture.

range zone A unit of strata that is defined by the presence and time range of a particular *fossil taxon. A range zone comprises the entire vertical and horizontal extent of the given organism. When used formally, the term is capitalized and the qualifying fossil name given in italics, with the generic name capitalized and the specific name in lower-case letters (e.g. the stratigraphic range of the Late *Jurassic ammonite *Cardioceras cordatum* delimits the *Cardioceras cordatum* Range zone.

rank–abundance diagram A graph in which the contribution species make to a community (in number, ground covered, or biomass) is plotted against their ranks, with the most common species ranked first.

RAPD *See* RANDOM AMPLIFIED POLYMORPHIC DNA.

raphide Needle-like crystals of silica or calcium oxalate which are found within the cells of certain plants (e.g. Aracaceae).

rapid flow **(shooting flow)** *See* CRITICAL FLOW and FROUDE NUMBER.

Rapoport's rule A rule, proposed in 1982 by E. H. Rapoport, stating that there is a positive relationship between the range of latitudes over which a species occurs and the latitude of the centre of its *range. In other words, species that inhabit the higher latitudes tend to have wider geographical ranges. It appears to work for many organisms, but is not universal in its application.

raptor From the Latin *rapere*, meaning 'to ravish', a bird of prey of the order Falconidae (formerly known as Raptores).

rarity The relative abundance of a species and, therefore, its vulnerability to extinction. The *International Union for Conservation of Nature and Natural Resources measures the vulnerability of a species according to five criteria: (*a*) the rate at which its numbers are observed, inferred, or projected to be declining; (*b*) in association with (*a*), whether the species occurs as a single, small population or a few, small, fragmented ones; (*c*) in association with (*a*), whether the species occupies a small geographic range or area; (*d*) the size of the population; (*e*) a mathematical estimate of the predicted risk of extinction within a specified time. From this assessment, species are allocated a position on a continuum of increasing threat with three categories: as 'critical'; 'endangered'; or 'vulnerable'. Species that are known to be at risk of extinction, but fail to qualify for any of the main categories, are classified as 'susceptible'.

Rassenkreis A *polytypic species, or a group of subspecies linked by *clines. The word is German, meaning 'circle of races'. *See also* RING SPECIES.

ratite A flightless, running *bird that has no keel on its sternum.

Raunkiaer, Christen (1876–1960) A Danish ecologist who suggested that the earliest flowering plants grew under tropical conditions and that as their descendants spread to less hospitable regions they evolved a range of mechanisms in order to survive. This led him to develop a system for classifying types of plant according to the positions of their perennating buds in relation to the soil surface (*see* RAUNKIAER'S LIFE-FORM CLASSIFICATION).

Raunkiaer's life-form classification A classification of plants, proposed by the Danish botanist C. *Raunkiaer, based on the position of perennating buds in relation to the soil surface. *See also* CHAMAEPHYTE; EPIPHYTE; GEOPHYTE; HEMICRYPTOPHYTE; PHANEROPHYTE; and THEROPHYTE.

ravine wind A wind that passes through an upland barrier along a narrow valley or ravine. The wind is generated by a pressure gradient between the two ends of the valley and the force of the wind is often enhanced as a result of the channelling effect caused by constriction in the valley.

rawinsonde A *radiosonde balloon that also carries a radar reflector. It is tracked to determine wind speed and direction. *See also* WIND SONDE.

Rayleigh scattering The scattering of electromagnetic radiation by spherical particles with radii that are less than 10 per cent that of the wavelength of the incident radiation. Such scattering by air molecules produces the blue colour of the sky. Particles such as dust and smoke, which are significantly smaller than 0.4 μm (the wavelength of the blue/violet or lower limit of the visible spectrum) can also scatter visible radiation. Reddish colours at sunset and sunrise result from Rayleigh scattering; these longer wavelengths pass directly through the atmosphere to the observer, while particles in the air scatter out radiation of shorter wavelengths. The phenomenon was described by the English physicist Baron Rayleigh (John William Strutt, 1842–1919). *See also* MIE SCATTERING.

R/B ratio *See* RESPIRATION–BIOMASS RATIO.

reaction 1. (pedol.) The degree of acidity or alkalinity of a soil, expressed as a value on the *pH scale. *See also* ACID SOIL and ALKALINE SOIL. **2. (ecol.)** In plant succession, the ability of an individual plant species or vegetation community to modify the physical environment and so favour further successional development. For example, the presence of the first colonizing plants in a hydrosere will tend to reduce water movement, favouring accelerated silting and so paving the way for plants typical of shallower water to arrive as new colonizers.

reaction time In *geomorphology, the time taken for a system to react to a sustained change in external conditions. Representative reaction times are difficult to define, because of variations in the resistance of systems to change and in the magnitude of the external change (e.g. a sand-bed river channel reacts more readily to change than does a rock-floored channel). *See also* RELAXATION TIME.

reafference The stimulation of an animal as a result of the movements of its own body. *Compare* EXAFFERENCE.

reafforestation *See* REFORESTATION.

realized niche The *niche a viable population of a species occupies in the presence of competitor species. *Compare* FUNDAMENTAL NICHE.

recalcitrant seed A seed that cannot survive desiccation, so germinates instantly; its storage life is usually only a few days, rarely a few weeks. Most *climax *tropical rain-forest tree species produce recalcitrant seeds and so must be stored as growing plants rather than seed. *See* GENE BANK. *Compare* ORTHODOX SEED.

recapitulation of phylogeny A theory proposed by E. F. *Haeckel, asserting that ontogeny (the development of the individual) recapitulates or reflects the phylogeny (the evolutionary history of the individual). The theory as such has been rejected as vague, von Baer's *biogenetic law providing sufficient explanation for the observations on which it was based. Apart from *postdisplacement, most types of *heterochrony can result in recapitulation.

Recent *See* HOLOCENE.

receptor A cell or group of cells that are specialized to respond to stimuli of particular types, which may be external (light, sound-waves, odours, etc.) or internal (i.e. changes within the body). When the intensity of the stimulus exceeds a threshold the receptor releases an impulse into the nerve cells connected to it.

recessional moraine *See* MORAINE.

recessive gene A gene whose phenotypic effect is expressed in the *homozygous state, but masked in the presence of a dominant *allele in organisms *heterozygous for that gene. Usually the dominant gene produces a functional product and the recessive gene does not. One or two doses per nucleus of the dominant allele lead to the expression of its *phenotype; the recessive allele is expressed only in the absence of the dominant allele (i.e. two doses of the recessive gene).

recharge **1.** The downward movement of water from the soil to the *water-table. **2.** The volume of water that is added to the total amount of *groundwater in storage in a given period of time.

recharge area **1.** The geographical area of an *aquifer in which there is a downward movement of water towards the *water-table. **2.** The area that acts as a *catchment for any particular aquifer.

recipient *See* GRAFT.

reciprocal averaging **(correspondence analysis)** An *ordination method that combines gradient analysis and a system of successive approximation. Its chief merit lies in the clarity with which strong floristic gradients are displayed. Applying the method to a series of samples each containing a number of species begins by assigning an arbitrary (e.g. random) number to each species; these numbers are known as species scores. The scores are weighted according to the abundance of each species in each sample, and the weighted average of the species scores in each sample yields a sample score. The weighted average of all the sample scores yields a new species score. The species and sample scores are then standardized by subtracting the means and dividing by the standard deviation. The procedure is repeated until successive iterations produce little or no changes in the values.

reciprocal predation A variety of *mutual antagonism in which two species each prey on the other.

recognition sequence The sequence, often a short *palindromic sequence, recognized by a *restriction endonuclease.

recombination The exchange of DNA between molecules, resulting in the formation of new sequences.

recreatability **(salvageability)** In conservation, the extent to which a *community or ecosystem could be re-established following disturbance (by motorway building, open-cast mining, etc.) without any noticeable change. Certain long-established communities (e.g. in Britain *primary woodlands) are thought not to be recreatable since they would take many hundreds of years to re-establish their present detailed structure and composition and it is virtually impossible to reconstruct the same long-term environmental influences as fashioned their present form. Research continues into ways of recreating the important features of certain communities and some success has been achieved with grassland.

recreation ecology The study of the effects on ecosystems of recreation. The ecosystems impacted upon by recreation ecology can be as diverse as marine coral reefs, freshwater lakes, sand dunes, and montane grassland. Understanding the consequences of recreation upon ecosystems is becoming increasingly urgent with the rise in *ecotourism.

rectangular drainage *See* DRAINAGE PATTERN.

recurrence surface In a peat stratigraphy, the sudden transition from a highly humified peat to a fresh unhumified peat. This change reflects the resumption of wetter conditions and peat growth following a drier period, when heath rather than bog plants colonized the older humified peat. Increasingly, these features of peat stratigraphy are coming to be regarded as important indications of past climate change. Objective identification of recurrence surfaces in a stratigraphic profile is best achieved by measuring the degree of *humification of *peat. This involves extracting humic acid and measuring its density by spectrophotometry. *See also* REGENERATION COMPLEX.

red clay **(brown clay)** A brown or red, very fine-grained, deep-sea deposit composed of finely divided clay material that is derived from the land, transported by winds and ocean currents, and deposited far from land in the deepest parts of the ocean basin, especially in mid-latitudes. Red-clay deposits cover about a quarter of the Atlantic and Indian ocean floors and almost half the Pacific ocean floor.

red core of strawberry A serious disease of strawberries caused by the oomycete *Phytophthora fragariae*. Infested plants are stunted and their roots become dark brown or black with a characteristic red core (stele).

redd A gravelly spawning bed or nest in branches of rivers, made by salmon.

redirected behaviour Behaviour that is related to a stimulus, but is misdirected (e.g. an attack upon an inanimate object by an animal that cannot or dare not attack another animal which is the true target of its *aggression). It results from *conflict.

redox potential A scale that indicates the reduction (addition of electrons) and oxidation (removal of electrons) for a given material. The position on the scale is expressed as an electric potential in millivolts, normally in the range 0–1 300 or 0–1 400 mV. The *pH of the sample must be known since this can alter the reading.

redox reaction A reaction that involves simultaneous *reduction and *oxidation.

red podzolic soil **(krasnozem)** A soil *profile formed at an advanced stage of *weathering and *leaching by the process of *podzolization; it is similar in appearance and properties to a *podzol but associated with the greater degree of chemical weathering and higher iron-oxide concentrations of a humid, tropical environment. *See also* ULTISOLS.

Red Queen effect An evolutionary principle, first proposed in 1973 by L. Van Valen, that much of the evolution of a lineage consists simply of keeping up with environmental changes (mainly tracking a deteriorating *environment), rather than occupying or adapting to new environments. The name is derived from the Red Queen in Lewis Carroll's *Through the Looking-Glass*, who said: 'Now, *here*, you see, it takes all the running *you* can do, to keep in the same place. If you want to get somewhere else, you must run at least twice as fast as that!' *See also* ROMER'S RULE.

red rust An important disease of the tea plant (*Camellia sinensis*). Orange-brown velvety areas appear on the leaves of infected plants. The disease is caused by algae of the genus *Cephaleuros*.

Red Sea An elongate basin, 2 000 km long, with shorelines 360 km apart at the widest point and only 28 km apart where it joins the Gulf of Aden, the Red Sea has an inner *median valley, associated with a positive gravity anomaly, containing basalts and hot brines. The sea is thought to be at the young stage of the *Wilson cycle of an ocean.

red tide A phenomenon in which blooms of certain *dinoflagellates colour the water of seas or estuaries red or reddish. The dinoflagellates, particularly species of *Gonyaulax*, are concentrated, along with their toxins, by filter-feeding shellfish, and when the shellfish are subsequently eaten by humans poisoning results. Fish and shellfish may also die by poisoning.

reduction A chemical reaction in which atoms or molecules either lose oxygen or gain hydrogen or electrons.

reduction potential (electrode potential) A quantitative measure of the ease with which a substance is oxidized or reduced. It represents the voltage necessary to prevent the flow of electrons to or from a half-cell containing aqueous solutions of the oxidized and reduced forms of the test substance when it is connected to a hydrogen half-cell.

reef A rigid, wave-resistant structure that is built up by carbonate organisms. Types of reef include patch reefs (small and circular in shape); pinnacle reefs (conical in form); barrier reefs (separated from the coast by a lagoon); fringing reefs (attached to a coast); and atolls (isolated reefs enclosing lagoons). Factors influencing reef growth include: (*a*) water temperature (optimum 25°C); (*b*) water depth (must be less than 10 m); (*c*) *salinity (normal marine salinity is necessary); (*d*) wave action (intense wave action favours coral growth); and (*e*) turbidity (coral growth requires clear water and an absence of terrigenous suspended sediment). The diversity of species found in a reef is related to salinity and water temperature, with stressful conditions resulting in a reduction of species present. Some reefs that formed in the distant past were made in other ways and not by corals. The earliest known reefs, formed about 2 billion years ago, consist of fossil *stromatolites.

reef flat A pavement of naturally cemented, large skeletal debris and *reef debris which forms to the protected rear (leeward) of the crest of a reef. Water depth in this zone is shallow, only a few metres at most. Sand shoals and small islands or *cays may be present on the reef flat.

reef front An irregularly sloping ramp that extends from the *surf zone to a depth of approximately 100 m on the windward, open-sea side of a *reef. Abundant skeletal growth grades downwards into the sediment of the *fore-reef zone. The robust, branching form of the coral *Acropora palmata*, common in present-day reefs, typifies this reef area.

reef trap A stratigraphic oil or gas trap produced by porous *reef limestones (the *reservoir rock) covered by impermeable strata.

re-entrant An embayment or recess in a *scarp or along a main valley, often of sufficient length to form a tributary.

refection (pseudorumination) In some mammals, a type of feeding in which food passes rapidly through the gut and is ingested for a second time as it leaves the anus. *See also* CAECOTROPHY.

reflex action An involuntary action that is made by an animal in response to a stimulus.

reflexed Bent backwards at a sharp angle.

reforestation (reafforestation) 1. The establishment of a particular type of woodland by planting into an existing, different woodland type. **2.** The replacement of a tree crop by natural or artificial means on land from which a previous wood has been removed.

refuge A site, defined in space and time, within which particular organisms are sheltered from the competitive effect of other species.

refugia *See* REFUGIUM.

refugium (pl. refugia) An isolated area where extensive changes, most typically due to changing climate, have not occurred. Plants and animals formerly characteristic of the region in general find a refuge from unfavourable conditions in these areas. An example might be a mountain summit projecting above a glaciated lowland region or a part of the Amazon rain forest that did not experience a seasonal climate during *Pleistocene glaciations. *See also* RELICT.

reg 1. Stony desert. **2.** A gravel veneer, normally consisting of small, rounded pebbles, which mantles a Saharan plain and has a gradient as low as 1:5000. The pebble layer may be underlain by a stony soil, or it may be a lag deposit. *Compare* SERIR.

regelation A process by which water that has been released by pressure melting beneath a temperate *glacier is refrozen. The process takes place in a relatively thin zone, the 'regelation layer', which may be only a few centimetres thick. Regelation is associated with the incorporation of bedrock materials in the debris-rich 'sole' of the glacier, which is restricted to the thickness of the regelation layer.

regeneration complex The now discredited idea that the cycle of hummock and hollow, typical of a growing *raised bog, resulted from a complex of developmental phases forming a mosaic of patches, each at a different stage of development from its neighbour. At one time it was believed that hollows and hummocks replaced one another in a predictable sequence, but detailed stratigraphic studies indicate that pools and hummocks have a tendency to persist in their locations once they have been formed.

regeneration niche The specific niche that must exist in order for a plant to become re-established in an area it formerly occupied. A plant that can tolerate a wide range of environmental conditions when mature may nevertheless require very specific conditions for the germination and establishment of its seeds. For example, the Californian redwoods require burned soil and many grassland plants need disturbed, bare microsites for their germination.

regolith A general term for the layer of unconsolidated (non-cemented), weathered material, including rock fragments, mineral grains and all other superficial deposits that rest on unaltered, solid bedrock. It reaches its maximum development in the humid tropics, where depths of several hundreds of metres of weathered rock are found. Its lower limit is the weathering front. Soil is regolith that is able to support rooted plants. *Compare* SAPROLITE.

Regosols A reference soil group in the *FAO *soil classification that comprises all soils that do not belong to one of the other reference soil groups, i.e. *Acrisols, *Albeluvisols, *Alisols, *Andosols, *Anthrosols, *Arenosols, *Calcisols, *Cambisols, *Chernozems, *Cryosols, *Durisols, *Ferralsols, *Fluvisols, *Gleysols, *Gypsisols, *Histosols, *Kastanozems, *Leptosols, *Lixisols, *Luvisols, *Nitisols, *Phaeozems, *Planosols, *Plinthosols, *Podzols, *Solonchaks, *Solonetz, *Umbrisols, and *Vertisols.

regression (marine) The withdrawal of water from parts of the land surface caused by a fall in sea level relative to the land. Shallow-water sediments overlie sediments characteristic of deeper water. *Compare* TRANSGRESSION.

regular distribution *See* UNDERDISPERSION.

regular sample (systematic sample) One of a number of samples taken at regular intervals (e.g. by the use of regularly spaced *quadrats along some environmental gradient such as a valley side). Though less reliable in certain circumstances than random sampling, regular sampling may be more practicable and more economical in the time it takes. The chief disadvantages are the possibility that the interval selected may resonate with some unsuspected environmental variable, so giving biased results, and that the form of the sample does not conform with the one theoretically assumed for many statistical tests.

rehabilitation ecology Where it is impossible to restore a site to its original condition, the establishment on it of a community which is similar to the original. *Compare* REPLACEMENT ECOLOGY and RESTORATION ECOLOGY; *see also* RECREATABILITY.

reindeer moss The common name for the *lichen *Cladonia rangiferina* and similar species, which form extensively branched podetia up to about 8 cm tall; in *C. rangiferina* the apices of the branches are slender and curved, typically all in the same direction. These lichens are found, for example, on *peaty ground in mountains and in

Arctic and sub-Arctic regions where, together with other lichens, they form a major component of the winter diet of reindeer, known in North America as caribou (*Rangifer tarandus*).

reinforcement The strengthening of a response to a stimulus by reward or punishment.

rejuvenation The marked increase in the rate of *erosion which takes place when a land mass is relatively elevated. Streams respond by incision, with the development of *terraces and *knick points, and finally a *polycyclic landscape emerges. Similar effects can occur after climate change.

relative age The position of an event, *fossil, etc. in relation to others of its kind within a time sequence (e.g. 'an early *Cambrian trilobite' or 'a late *Jurassic marine *transgression'). No age in years is implied. *Compare* ABSOLUTE AGE. *See also* DATING METHODS.

relative dating *See* GEOCHRONOLOGY.

relative humidity The water-vapour content of air at a given temperature, expressed as a percentage of the water-vapour content that would be required for saturation at that temperature. Generally, the relative humidity decreases during the day, with increase in temperature, and increases at night as the temperature falls.

relative-importance value In ecology, a concept similar in principle to importance value, but based on original, somewhat relative, density–frequency–dominance values. It is the actual basal area, frequency, or density of a particular species, without reference to the density, frequency, and dominance values of other species present in the area.

relative pollen frequency (RPF) An expression of *pollen data from sediments for each species, genus, or family, as a percentage of the total pollen count or the total tree pollen. It is the traditional and most widely used method for preparing pollen diagrams. *Compare* ABSOLUTE POLLEN FREQUENCY; *see also* PALYNOLOGY and POLLEN ANALYSIS.

relaxation time The time taken by a geomorphological system to adjust to a sustained change in the nature and/or intensity of an external process, usually by an alteration in the shape of the land-form or landscape. Relaxation times vary. The width of a river channel may increase within 10 years, possibly triggered by a major flood event, in response to an increased discharge; following a climate change, it may be 10^5–10^6 years before a glaciated mountain range loses the imprint of ice.

releaser *See* SIGN STIMULUS.

relevé In *phytosociology, the basic field unit recorded. The area should be uniform in floristic composition, and of uniform relief and soil type. The unit size varies with the type of *community, but should at least embrace the *minimal area. Site and soil data, as well as vegetation form, composition, and species are noted.

relic *See* RELICT.

relict (relic) Applied to organisms that have survived while other related ones have become extinct. Often the term refers to species that formerly had a much wider distribution and have survived locally through periods of unfavourable conditions (e.g. glacial periods or land submergence) by existing in regions called refugia (*see* REFUGIUM), while becoming extinct elsewhere (e.g. some Arctic-alpine plants). They may be part of a *relict community (e.g. *Dryas octopetala* (mountain avens) in Britain, which was widespread during glacial times but is now restricted to a few mountain tops). It may also refer to a surviving species of a group, the other species of which have become extinct (e.g. the coelacanth or *Ginkgo biloba*, the maidenhair tree, which survived only in Chinese monastery gardens). *See also* RELICT SEDIMENT.

relict coppice An area of abandoned *coppice woodland.

relict community A *community that formerly had a much wider distribution but now occurs only very locally. Such contraction can be caused by various factors including climatic change (e.g.

glaciation). Communities widespread during glacial times may now be restricted to mountain tops (e.g. Upper Teesdale (England) contains an isolated flora characteristic of much of lowland Britain during glacial periods).

relict sediment A sediment lying on the *continental shelf which was deposited by processes no longer active in the area where the sediment now occurs. Relict sediments are remnants from an earlier environment and are now in disequilibrium. Approximately 50 per cent of the present continental shelves are covered by relict sediments deposited during the period of lower sea levels in the *Pleistocene.

relict soil *See* PALAEOSOL.

remanent magnetization The magnetization that remains after the removal of an externally applied field; it is exhibited by ferro-magnetic materials.

removal time *See* RESIDENCE TIME.

rendzina A *brown earth soil of humid or semi-arid grassland that has developed over calcareous parent material. The term is obsolete under the *USDA Soil Taxonomy, where rendzinas may fall within the orders *Inceptisols or *Mollisols.

renewable resource A resource produced as part of the functioning of natural systems at rates comparable with its rate of consumption (e.g. food production by photosynthesis). Limits to renewable resources are determined by flow rate and such resources can provide a sustained yield. *Compare* FINITE RESOURCE.

renewal cycle A *biogeochemical cycle.

reniform Kidney-shaped.

repetitive DNA Any DNA sequence which occurs many times within a *genome; it may be in a *tandem array or dispersed repeats.

replaceability The ease with which an *ecosystem can be replaced if it is damaged or lost. Those ecosystems with a low replaceability are regarded as more worthy of selection for conservation.

replacement The substitution of one *amino acid for another as a consequence of a nonsynonymous *mutation at the DNA level.

replacement ecology Where it is impossible to restore a site to its original condition, the establishment on it of a community entirely different from the original. *Compare* REHABILITATION ECOLOGY and RESTORATION ECOLOGY.

replacement series An experiment design in which a series of plots are planted with two plant species, the proportions of the species differing in each plot, but the plant density remaining constant in all plots. The aim is to detect and measure competition between the species.

replication slippage A genetic process in which *deletions and *insertions of small contiguous repeats occur because of misalignment between DNA strands. This causes parts of template DNA to be copied more than once or to be missed during DNA replication.

representativeness A measure of how typical a particular site is of the *ecosystem it represents. Highly typical or representative sites are generally regarded highly when considered for selection as *nature reserves.

reproductive A fertile male or female member of a colony of social insects. In ants, bees, and wasps the male (drone) dies soon after insemination, leaving the queen as the colony reproductive. In termites the male (king) lives with the queen, periodically inseminating her. Supplementary reproductives, present in termite colonies, take over colony reproduction if the founding king and queen (the primary reproductives) die.

reproductive allocation The investment of resources (e.g. of food energy and time) for the purpose of reproduction (i.e. courtship, mating, and raising of young).

reproductive effort The proportion of its resources that an organism expends on reproduction.

reproductive isolating mechanism (RIM) The means by which different *species are kept reproductively isolated.

These may be: (*a*) chromosomal (if cross-mating occurs, the incompatibility of the *karyotypes makes any hybrid inviable or sterile); (*b*) mechanical (the two species cannot mate because they are of different sizes, or because the genitalia are shaped differently); (*c*) ethological (the courtship rituals of the two species diverge at some point so that an incorrect response is given and the sequence is brought to a stop); or (*d*) ecological (the two species occupy different microhabitats and normally do not meet). Other mechanisms include the breeding seasons being out of phase, or members of one species being unattractive to members of the other. Many RIMs, especially ethological ones, amount merely to mate preference, so that in the absence of a preferred partner (of the same species) a member of a different species will be accepted. In this way hybrids between different species may be bred in captivity and may even be found to be fully fertile.

reptile A member of a large and diverse class (Reptilia) of *poikilothermic vertebrates, which arose in the *Carboniferous from labyrinthodont amphibians (Labyrinthodontia). They were the dominant animals of the *Mesozoic world and gave rise to the birds and mammals. Reptiles have a body covering of ectodermal scales, sometimes supported by bony plates (scutes). There is no gilled larval phase; development is by amniote egg, but *ovovivipary is common. Reptiles are air-breathing from hatching onwards. There are about 6 000 species on all continents except Antarctica.

rescue effect A species arriving on an island may already be represented there and so may have the effect of reducing the chance of the extinction of that species from the island (i.e. of 'rescuing' it). The rescue effect will be greater on islands which are closer to the mainland source of species than more remote islands because the immigration rate will be higher. *Compare* EQUILIBRIUM THEORY.

resequent Applied to a land-form whose orientation is similar to that of the inferred original feature, but which has passed through a complex subsequent history. For example, a resequent fault-line *scarp faces the same way as the original fault scarp.

reserve **1.** Resources of coal, ore, or minerals which can be mined legally and profitably under existing conditions. The indicated reserve is the estimate of ore computed from boreholes, outcrops, and developmental data, and projected for a reasonable distance on geologic evidence. An inferred reserve is an estimate based on relationships, character of deposit, and past experience, without actual measurements or samples; it should include the limits between which the deposit may lie. A potential reserve is ore not yet discovered but whose presence is suspected; the term is sometimes used for ore not commercially viable at the present time. A proved reserve is a resource reliably established by tunnels, bore-holes, or mining. **2.** *See* NATURE RESERVE.

reserve selection The process of determining whether an area should be given the protective status of a *nature reserve. Various criteria have been suggested for evaluating a site for conservation purposes. Derek Ratcliffe has produced the most widely used set, which includes area, habitat diversity (both of which contribute to species richness), *replaceability, *fragility, *vulnerability, *representativeness, position in a geographical sequence, *rarity, naturalness, and its potential for research or education.

reservoir **1.** A surface body of water whose flow is artificially controlled by means of dams, embankments, or sluice gates in such a way that the water remains static until it is allowed to flow for a specific purpose (e.g. flood control or public water supply). **2.** An underground rock formation that has sufficient void space to act as a store for water, natural gas, or oil.

reservoir pool A large store of a nutrient at some stage in a *biogeochemical cycle. Reservoir pools are mainly abiotic, but may also be biotic, as in the case of the *biomass of a forest, which represents a considerable store of various elements, especially carbon. Exchanges between the reservoir pool and the *active pool are typically slow by comparison with exchange within the active pool. Human activity, such as the mining of mineral resources and the manufacture and application of fertilizer, may profoundly alter this exchange rate, generally releasing an excess into the active pool which can be accommodated only by establishing a new equilibrium. This may in turn produce unfavourable conditions, manifested as chemical pollution (e.g. excess phosphorus in *eutrophication, excess sulphur in *acid precipitation and lake acidification).

reservoir rock Any porous rock in which oil, gas, or water may accumulate; it is usually sandstone, limestone, or dolomite, but sometimes fractured *igneous or *metamorphic rock.

reshabar A regional south-easterly wind that affects mountain slopes in southern Kurdistan (the plateau and mountains in south-eastern Turkey, northern Iraq, northern Syria, and western Iran). The strong, swirling wind is hot and dry in summer but brings cold conditions in winter.

residence time **1. (removal time)** The time that a given substance remains in a particular compartment of a *biogeochemical cycle. **2.** The length of time water remains within an *aquifer, lake, river, or other water body before continuing around the *hydrological cycle. The time involved may vary from days for shallow gravel aquifers to millions of years for deep aquifers with very low values for hydraulic conductivity. Residence times of water in rivers are a few days, while in large lakes residence time ranges up to several decades. Residence times of continental *ice sheets are hundreds of thousands of years, of small *glaciers a few decades. **3.** The average time a particular substance spends in solution in sea water between the time it first enters and the time it is removed from the ocean. **4.** The average time that water remains in soil or rock. **5.** The average time that a water molecule or particulate pollutant spends in the atmosphere. The residence time for pollutants ranges from a few weeks in the lower *troposphere to several years in the upper *stratosphere, before it is scavenged out by *precipitation. For water molecules the overall average is believed to be 9–10 days.

residual In statistics, a data variability that is not accounted for by a particular statistical test. The residuals of individual data values (i.e. the difference between an observed and a computed value) often give ecologists insight into possible environmental influences on individual data records.

resilience *See* STABILITY 2.

resinite *see* COAL MACERAL.

resorb To re-absorb; i.e. to metabolize substances or structures that were produced metabolically by the body. For example, in some mammals, foetuses are resorbed if they are not viable.

resource allocation The way in which the products of *photosynthesis are allocated to different organs within a plant. This changes during the course of a plant's life. Initially, there is a high level of allocation to roots, followed rapidly by shoot and leaf development, and later in the strengthening of support structures. The degree of allocation to reproduction differs between species (*see* *R*-SELECTION and *K*-SELECTION).

resource partitioning *See* DIFFERENTIAL RESOURCE UTILIZATION.

respiration **1.** Oxidative reactions in cellular metabolism that involve the sequential degradation of food substances and the generation of a high-energy compound, ATP (adenosine triphosphate) in aerobic respiration with the use of molecular oxygen as a final hydrogen acceptor; ATP, carbon dioxide, and water are the products thus formed. The reactions involved in respiration are the reverse of those in *photosynthesis, i.e. $C_6H_{12}O_6 + 6O_2 \rightarrow 6CO_2 + 6H_2O$. **2.** The physico-chemical processes involved in the transportation of oxygen to and carbon dioxide from the tissues. **3.** (external respiration) The act of breathing.

respiration–biomass ratio **(R/B ratio)** The relationship between total *community *biomass (i.e. standing crop) and respiration. With larger biomass, respiration will increase but the increase will be less if the individual biomass units or organisms are large (reflecting the inverse relationship between size and metabolic rate). Natural communities tend towards larger organisms and complex structure, with low respiration rates per unit biomass.

respiration quotient **(RQ)** The ratio of the amount of carbon dioxide expired to the amount of oxygen consumed during the same period. (RQ = carbon dioxide consumed divided by oxygen utilized).

restoration ecology The establishment on a disturbed site of the plant and animal community which existed there prior to the disturbance. *Compare* REHABILITATION ECOLOGY and REPLACEMENT ECOLOGY.

restriction The cleavage of double-stranded DNA by a *restriction endonuclease.

restriction endonuclease An *endonuclease that cleaves double-stranded DNA by *hydrolysis at specific nucleotide sequences.

restriction fragment length polymorphism **(RFLP)** A *mutation occurring in a sequence of DNA that causes the *restriction fragment pattern to change discernibly by causing a change in the specific *nucleotide sequence recognized by the *restriction endonuclease.

restriction fragment pattern The characteristic number and sizes of DNA fragments obtained by *restriction, as determined by *electrophoresis. Such fragments are often used as *characters in *phylogenetic analysis.

resurgence *See* SPRING.

reticulate Marked with a network pattern (e.g. the veins in a typical leaf).

reticulate evolution The development of a network of closely related taxa within and at the species level, particularly by chromosome doubling or *polyploidy. Polyploidy is more common in plants than in animals, and so reticulate evolution is more likely in the former than in the latter.

reticulate method A term sometimes used to describe a non-hierarchical clustering technique.

retroelement Any DNA or RNA sequence that is able to produce *reverse transcriptase.

retrofection The transfer by a *retrovirus *virion of RNA from one cell to another where it is reverse transcribed and incorporated into the host *genome.

retrogene A *retrosequence which has retained a protein coding function similar or identical to that of the original gene.

retrogression **(retrogressive succession)** A *successional change, usually from an existing *climax community, leading to a less diverse and less structurally complex *community. The change frequently involves a reduction in *biomass. Retrogression is usually triggered by an environmental factor (e.g. a *pollutant) and disturbance by humans is often involved, e.g. in removing top predators or tree cover and thereby setting in motion a downward development in *ecosystem complexity.

retrogressive succession *See* RETROGRESSION.

retron A *retroelement which lacks the ability to transpose.

retroposon A *retroelement which is able to transpose, but does not produce a *virion.

retropseudogene A *retrosequence that retains no function.

retrosequence A DNA sequence which is the incorporated product of *reverse transcription, but does not itself encode for *reverse transcriptase. Such sequences are noticeably devoid of *introns.

retrotransposon The product formed when a *transposon is copied by RNA and is free within a cell. A retrotransposon can use reverse transcriptase to manufacture complementary DNA (thus copying the original transposon). *Retroviruses are thought to have evolved from retrotransposons; these occur in most cells and are considered to be unhelpful and sometimes harmful (e.g. HIV).

retrovirus A *virus whose genetic material consists of RNA and which is able, by means of the enzyme reverse transcriptase, to manufacture its DNA equivalent and insert it into the *genome of its host species. It has been suggested that this could be a way of overriding Weissmann's doctrine, and in effect causing Lamarckian inheritance. *See* NEO-LAMARCKISM.

return flow *See* INTERFLOW.

return period The frequency, based on statistical analysis of past records, with which a particular environmental hazard may be expected.

reversal time-scale *See* MAGNETOSTRATIGRAPHIC TIME-SCALE.

reverse transcriptase The *enzyme that catalyzes *reverse transcription.

reverse transcription The synthesis of single-stranded DNA from an RNA template.

revolute Curved or curled back, most commonly used of the edge of a leaf.

revolving storm *See* TROPICAL CYCLONE.

Reykjanes Ridge The part of the Mid-Atlantic Ridge to the south-west of Iceland, whose axis, marked by a *median valley, continues into the active graben across Iceland.

RFLP *See* RESTRICTION FRAGMENT LENGTH POLYMORPHISM.

rheology The study of deformation and flow in materials (e.g. ice and water), including their elasticity, viscosity, and plasticity.

rheophilous Applied to an organism or community of organisms that thrives in running water.

rheotaxis A change in the direction of locomotion in a motile organism or cell made in response to the stimulus of a current, usually a water current.

rheotrophic Applied to a *mire system that is fed by the flow of water. *Compare* OMBROTROPHIC.

rhizome A horizontally creeping underground stem which bears roots and leaves and usually persists from season to season.

rhizoplane (root surface) *See* RHIZOSPHERE.

rhizosphere The area of soil immediately surrounding plant roots, which is altered by their growth, respiration, exchange of nutrients, etc. Within this zone a further zone is sometimes distinguished, called the rhizoplane or root surface.

rhourd A large, star-shaped or pyramidal sand dune (a 'sand mountain') that may be 100–200 m high and has been described in the Algerian Sahara. It may form where two zones of sand-laden wind cross one another. *See* DRAA.

rhythm *See* CIRCADIAN RHYTHM.

ria A drowned river valley in an area of high relief. Classic examples are found in some of the peninsulas of western Europe, notably western Ireland, where they have resulted from the post-glacial rise in sea level.

ribbon lakes *See* TUNNEL VALLEY.

ribbons Straight to sinuous, long, narrow, thin bodies of sand that develop on sediment-poor, tide-swept shelves (*see* SHELF), oriented parallel to the tidal stream. More generally, the large-scale geometry of a preserved sand-body with a width-to-length ratio in excess of 1 : 100 and a thickness-to-width ratio greater than 1 : 10.

ribonucleic acid *See* RNA.

ribosomal RNA *See* RNA.

ribosome The cell organelle which is involved in the translation of *messenger-RNA into protein. The structure is composed of ribosomal *RNA and proteins.

rice stunt A virus disease which affects rice and other grasses. In addition to stunting, infected rice plants show yellowish streaks on the leaves. The virus is transmitted by sap-sucking leafhoppers.

richness *See* SPECIES RICHNESS.

Richter, Charles Francis (1900–1985) An American physicist and geologist, who is best known for his logarithmic scale of earthquake magnitudes. This was first proposed in 1927 and later revised in collaboration with Beno Gutenberg (1889–1960). *See* RICHTER SCALE.

Richter denudation slope A hillslope that develops at the foot of a cliff which is retreating fairly rapidly, chiefly because of rock fall. The slope has a uniform gradient, is cut across bedrock, and stands at the angle at which the *talus accumulates. With each unit of cliff retreat the related rock fall builds up on older talus, and so the foot of the cliff steadily rises. The Richter slope is revealed when the talus is removed, or it may remain hidden beneath a thick skin of mobile debris. E. Richter described such slopes in the Alps in 1900, and they are named after him.

Richter scale The most widely used system for reporting the intensity of an earthquake, developed by C. F. *Richter, and calculated from the amplitude of seismic waves, the period of the dominant wave, and the angular distance from the recording station to the earthquake focus. The scale is logarithmic and ranges from 0 to 10; a tremor with a value of 2 can barely be felt and structural damage is likely from earthquakes greater than 6.

ridding An area from which a woodland crop has been taken.

ridge **1. (wedge)** An extension of high pressure from an *anticyclone into a zone where generally lower pressure prevails. **2.** The poleward meanders of the flow of the upper westerly winds over midlatitudes. **3. (mid-oceanic ridge, oceanic ridge)** A long, linear, elevated, volcanic structure that often lies along the middle of the ocean floor. Such ridges tend to occupy central positions because the oceans have formed by the symmetrical spreading of two lithospheric plates from the ridge sites. Oceanic ridges occur in all the Earth's oceans, but may be offset from a central position (e.g. the east Pacific ridge, where one side of the oceanic crust is being consumed along a subduction zone. **4.** *See* RIDGE-AND-RAVINE TOPOGRAPHY and RIDGE and RUNNEL.

ridge-and-ravine topography A landscape that consists of a network of branching valleys and intervening low ridges, and which is similar to that of a maturely dissected *peneplain (*see* DAVISIAN CYCLE). It is well displayed in the central Appalachians, USA. The term has no genetic implications.

ridge and runnel A series of asymmetrical ridges that run parallel to the coast and are separated by shallow troughs (runnels) 100–200 m wide. This topography is developed on the foreshore of *mesotidal or *macrotidal beaches and is favoured by moderate wave-energy conditions acting on a flat beach with an abundant supply of sediment.

ridge crest The highest part of a mid-ocean *ridge, typically 2–3 km above the level of the *abyssal plains. The crest of a slow-spreading ridge (e.g. the Mid-Atlantic Ridge) is split by a *median valley; that of a fast-spreading ridge (e.g. the East Pacific Rise) has no median valley and the crest has more subdued topography.

riegel A rock bar that extends across the floor of a *glacial trough. It may be caused by a local reduction in the erosive ability of a *valley glacier or by a local increase in bedrock strength, perhaps owing to a reduction in joint density. It may alternate with a rock basin to give an irregular long profile.

riffle *See* POOL-AND-RIFFLE.

rift valley An elongate trough of regional extent, bounded by two or more faults. Many rifts on land are associated with volcanic activity and many contain lakes. The East African rift system is an outstanding example. Some rifts are thought to be at the embryonic stage of ocean development of the *Wilson cycle; others may become 'failed rifts' (or 'failed arms') and fill with sediment to become aulacogens. The rift valley developed along the axis of a slow-spreading oceanic *ridge is known as the *median valley (or axial rift or axial trough). 'Graben' (the German word for 'ditch') can be used synonymously for 'rift valley' and also for an infilled, fault-bounded trough of any size.

rill-wash Eroded material that is concentrated into more or less intermittent trickles and rills on inclined slopes, due to run-off of water.

RIM *See* REPRODUCTIVE ISOLATING MECHANISM.

rime The white deposit of ice that results from crystal growth on objects that are at a temperature below the freezing point. Supercooled water droplets in fog freeze on contact with such surfaces.

ring-diffuse species *See* TREE RING.

ring-porous species *See* TREE RING.

ring species A group of subspecies that are contiguous along a *cline. Members of each population are able to mate successfully with members of adjacent populations, but the group as a whole forms a ring, with sufficient morphological differentiation in some places to prevent interbreeding between overlapping populations. Gulls of the genus *Larus* comprise a circumpolar ring species in the northern hemisphere. Moving westwards from Britain, the herring gull (*L. argentatus*) occurs in North America (where one variant has developed into a distinct species, *L. glaucoides*), but is somewhat different from the British race. Between central Asia and north Europe, the races increasingly resemble the black-backed gull (*L. fuscus*) and in northern Europe the two species overlap and do not naturally interbreed.

r

ring spot **1.** A ring of pale or yellowish coloration which appears on the leaves of plants infected with certain viruses. **2.** A fungal disease of brassicas in which rounded brown spots, about 1 cm across, appear on mature leaves.

riparian Pertaining to a river-bank (from the Latin *ripa*, meaning 'bank').

rip current A strong, narrow current, usually of short duration, which flows seaward from the shore. The presence of a rip current can be detected as a visible band of agitated water flowing seawards, usually as a gap in the line of the incoming waves. Rip currents mark the swift return movement of water piled up on the shore by incoming waves and onshore winds.

ripple A small-scale ridge of sand produced by flowing water, wind motion, or wave action. The wavelength or spacing of ripple crests is usually less than 50 cm and the heights are less than 20 cm.

rip-rap A loose foundation layer of large, irregular, unscreened rock fragments used under water or in soft material for protection and to prevent the erosion of dams, sea walls, bluffs, or other structures exposed to wave action. It is used extensively in irrigation works and river improvements.

Riss The third of four glacial episodes named after Alpine rivers and established in 1909 by A. Penck and E. Bruckner. It is perhaps the equivalent of the *Saalian of northern Europe and the *Wolstonian of the East Anglian succession.

Riss/Würm interglacial An Alpine *interglacial stage that may be the equivalent of the *Eemian stage of northern Europe or the *Ipswichian of the East Anglian succession. It is the last interglacial and immediately precedes the last glaciation, the *Devensian or, in European usage, Weichselian.

rithron That part of a river in which the water is typically fast-moving, broken-surfaced, shallow, and relatively cold, favouring *rheophilous, cold-water *stenothermous organisms with a high demand for dissolved oxygen. *Compare* POTAMON.

ritualization The modification of patterns of behaviour and often (but not always) of their *motivation and function, and their subsequent use in *communication, often in stereotyped form.

rival Pertaining to a stream or small river.

river capture The process whereby a stream is able to tap and so capture the discharge of a neighbour. The capturing stream normally extends by headward *erosion along an outcrop of soft rock until it meets and diverts a second, less favoured, transverse system. A right-angled bend, the 'elbow of capture' is typical of the junction between capturing and captured streams. *See also* AVULSION.

river continuum concept A holistic view of rivers, first proposed by Robin L. Vannote and others in 1980, which permits a broad zonation of river systems based on the utilization of energy through the orderly processing of organic matter by the resident biota. Upstream, the river receives *allochthonous material from adjacent and overhanging vegetation, supplying coarse particulate organic matter. This is broken down by 'shredder' organisms in a system that is largely *heterotrophic (P/R < 1, *see* PRODUCTION/RESPIRATION RATIO), because it operates in deep shade. The shredders produce fine particulate organic matter. This is carried downstream, where it is processed by 'collectors'. As the stream widens, primary productivity increases and the shredders are replaced by grazers, living alongside collectors. Collectors predominate once more under mainly heterotrophic conditions still further downstream, where the river widens and becomes too deep for benthic plants. Predators occur throughout. The concept cannot be applied to all rivers (e.g. to those that originate above the tree-line) and breaks down when a river is blocked or passes through a lake.

river profile The slope along the bed of a river, expressed as a graph of distance-from-source against height. In detail it is typically compound, with the profiles of individual segments reflecting the local rock types. It may be broken by *knick points.

river terrace (stream terrace) A fragment of a former valley floor that now stands well above the level of the present *flood-plain and is usually covered by *fluvial deposits. It is caused by stream incision, which may be caused by uplift of the land, a fall in sea level, or a change in climate.

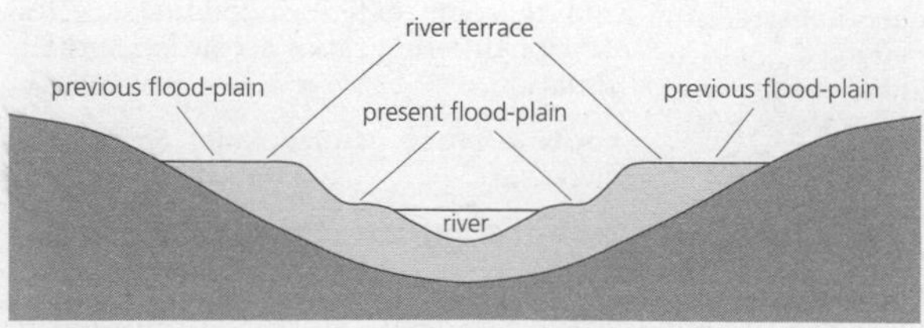

River terrace

RNA (ribonucleic acid) A *nucleic acid that is characterized by the presence of D-ribose and the *pyrimidine base uracil. It occurs in three principal forms, as *messenger-RNA, ribosomal-RNA, and *transfer-RNA, all of which participate in protein synthesis.

roaring forties A popular maritime name for the prevailing westerly winds which are commonly strong over the oceans in temperate latitudes of the southern hemisphere, particularly between about 40°S and 50°S.

robin's pincushion gall (rose bedeguar) A red and green or crimson, hairy growth on wild (dog) roses (*Rosa canina*), which is formed by the cynipid wasp *Diplolepis rosae*. Its eggs are laid in leaf buds, although the galls appear to grow from the twig or stem. The galls are multilocular, with an average of 30 progeny emerging in June. The average size of the galls is approximately that of a small pea, but the hairs may attain a length of 35 mm. Reproduction is normally *parthenogenetic (offspring arising from unfertilized ova), less than 1 per cent of reared progeny being males.

roche moutonnée (glaciated rock knob, stoss-and-lee topography) A mound-like land-form of glacial *erosion, consisting of a smoothed, streamlined, up-glacier surface and a broken, shattered, lee flank, which probably results from a combination of abrasion, frost-shattering, and the plucking out of blocks by the *glacier, although crushing has been suggested as a contributory mechanism.

rock bench *See* VALLEY-SIDE BENCH.

rock drumlin *See* DRUMLIN.

rock flour Finely ground rock debris produced chiefly by abrasion beneath a *glacier. It may be removed by meltwater streams, which consequently develop a typically milky appearance.

rock glacier A tongue-like mass of large, angular blocks, finer debris, and ice, found especially in middle-latitude Alpine regions (there are about 1 000 active examples in the Swiss Alps). It may have a core of ice, in which case it possibly originated as a debris-covered glacier, or it may be a *permafrost phenomenon.

rock pavement *See* RUWARE.

rogen moraine A field of morainic (*see* MORAINE) ridges that lie at right angles to the direction of a former ice advance. Individual ridges may vary between 10 and 30 m in height, may be more than 1 km in length, and may be 100–300 m apart. They are often linked by cross-ribs. This landscape probably formed beneath a *glacier, but the details of its origin are uncertain. It is named for its fine development around Lake Rogen, Sweden.

roguing The manual removal of infected or inferior specimens from an otherwise healthy crop of plants.

Romer's rule The proposal first made by the American palaeontologist Alfred Sherwood Romer (1894–1973), that the effect of many important evolutionary changes is to enable organisms to continue in the same way of life, rather than to adapt to a new one. For example, the evolution of bony elements that strengthened the limbs of fish enabled them to crawl over land to find new ponds when the climate started to become drier. The concept is close to that of the *Red Queen effect.

root **1.** The lower part of a plant, usually underground, by which the plant is anchored and through which water and mineral nutrients enter the plant. **2.** In a *phylogenetic tree, the common ancestor to all of the taxonomic units being studied.

rooted tree A *phylogenetic tree in which the common ancestor is identified, usually by the incorporation of a known outgroup, thus resolving the direction of evolution.

root frequency In the measurement of vegetation frequency, records based solely on the species rooted in a *quadrat, as distinct from shoot frequency, which includes those species rooted outside the quadrat but with foliage overlapping into the quadrat.

root nodule **(actinorrhiza)** A small, *gall-like growth on the roots of certain plants, especially leguminous plants (Fabaceae) but also others including alder (*Alnus* species), bog myrtle (*Myrica gale*), sea buckthorn (*Hippophaë* species), sumach (*Coriaria* species), California lilac (*Ceanothus* species), and the cycads (Cycadophyta). The nodules develop as a consequence of infection of the roots by bacteria (*Rhizobium* or *Bradyrhizobium* species in the case of legumes, *Actinobacteria in non-legumes). Bacterial colonies then inhabit the root nodules and benefit the plant by fixing atmospheric nitrogen, much of which becomes available to the plant.

root–shoot ratio The ratio of the weights of the roots to the shoots of a plant. Plants with a higher proportion of roots can compete more effectively for soil nutrients, while those with a higher proportion of shoots can collect more light energy. Large proportions of shoot production are characteristic of vegetation in early *successional phases, while high proportions of root production are characteristic of *climax vegetational phases, and to some extent individual species develop different ratios according to their situation.

root surface **(rhizoplane)** *See* RHIZOSPHERE.

rose bedeguar *See* ROBIN'S PINCUSHION GALL.

rose diagram A circular histogram plot which displays directional data and the frequency of each class. *See* WIND ROSE.

rosette plant A plant (e.g. *Bellis perennis*, daisy, *Plantago*, plantain) whose leaves are spread in a horizontal plane from a short *axis at ground level. Rosette plants are generally found in sparse or low-growing vegetation. The advantages include a reduced likelihood of being grazed or mown. *See also* HEMICRYPTOPHYTE.

Rossby waves *Troughs extending towards the equator and *ridges extending towards the poles that form long waves in the circumpolar flow of the upper air, particularly in the mid and upper *troposphere, with a typical wavelength of around 2 000 km; in the northern hemisphere the main troughs are characteristically at about 70°W and 150°E and three or four waves usually occur in the circumpolar westerly wind flow over mid-latitudes. The Rossby waves influence the formation of surface *depressions which tend to develop on a *frontal wave ahead of an upper trough. They are named after the Swedish-American meteorologist C. G. Rossby (1898–1957).

rotor cloud A cloud formed in moist air by condensation in the upper part of an eddy that has been generated beneath the wave-form in stable air on the lee side of a mountain barrier. The closed eddy system can result in local reversal of wind direction in the general airstream.

round dance *See* DANCE LANGUAGE.

RPF *See* RELATIVE POLLEN FREQUENCY.

RQ *See* RESPIRATION QUOTIENT.

***r*-selection** A selection for maximizing the intrinsic rate of increase (*r*) of an organism so that when favourable conditions occur (e.g. in a newly formed *habitat) the species concerned can rapidly colonize the area. Such species are opportunists (*see* FUGITIVE SPECIES). An opportunist strategy is advantageous in rapidly changing environments, as in the early stages of a *succession. *See also* BET-HEDGING, BIOTIC POTENTIAL, and POPULATION EXPLOSION; *compare* *K*-SELECTION.

R technique The most usual way to analyse data, in which variables or attributes (*n*) form the data, and the observations for different sample sites or individuals (*N*) form the rows in a table or matrix. *Compare* Q TECHNIQUE; *see also* INVERSE ANALYSIS and NORMAL ANALYSIS.

Rübel, Eduard (1876–1960) A Swiss phytogeographer who, in collaboration with J. *Braun-Blanquet, helped to develop the classification system of the *Zürich–Montpellier school of phytosociology. This was based originally on the system of A. F. W. *Schimper.

ruderal (noun and adj.) A plant, or applied to a plant that is associated with human dwellings or agriculture, or that colonizes waste ground. Ruderals are often weeds which have high demands for nutrients and/or are intolerant of competition. In the classification of *plant strategies proposed by J. P. Grime, a plant species found in areas of low stress and high disturbance. *Compare* COMPETITOR and STRESS-TOLERATOR.

ruderal strategy *See* GRIME'S HABITAT CLASSIFICATION.

rufous Reddish-brown in colour.

rugose Wrinkled.

rugulose Finely wrinkled.

runaway hypothesis A hypothesis proposed by R. A. *Fisher in 1930 to explain the consequences of female selection of a particular male trait (e.g. the length of the tail in a bird). Over successive generations such selection would favour increasingly extreme development of the trait (i.e. the tails of males would become longer) until the fitness of the male was reduced. (This tendency has been demonstrated experimentally by shortening or lengthening the tails of male birds.) Eventually, males would be so overspecialized as to bring the species to extinction, were it not for the restraining influence of *natural selection, which halts the development before that stage can be reached. *See also* HANDICAP PRINCIPLE.

runnel *See* RIDGE and RUNNEL.

Russian borer *See* PEAT-BORER.

rust A plant disease caused by a fungus of the class Urediniomycetes. The characteristic symptom is the development of spots or pustules bearing masses of powdery *spores which are usually rust-coloured, yellow, or brown. Infected plants may also show distortions or *gall-like swellings. *Compare* WHITE RUST.

ruware **(rock pavement)** An area of bare rock with a slightly domed profile which outcrops locally at the surface of a tropical plain. It is formed when the weathered profile is stripped from a sound rock surface, and may be seen as the first stage in the emergence of a dome or *tor.

***R* value** Usually the multiple correlation coefficient, as distinct from *r*, the simple correlation coefficient.

Ryukyu Trench The oceanic *trench which forms the boundary between the oceanic Philippine Plate and the continental crust of the Eurasian Plate. The Philippine Plate is subducting obliquely under the Eurasian Plate.

S **1.** *See* SULPHUR. **2.** *See* SVEDBERG UNIT.

s *See* SELECTION COEFFICIENT.

Saalian A northern European glacial stage dating from about 0.25 to 0.1 Ma which may be equivalent to the *Wolstonian of the East Anglian succession.

sabkha A wide area of coastal flats bordering a *lagoon, where evaporites, dominated by carbonate–sulphate deposits, are formed. It is named after such an area on the south-eastern coast of the Arabian Peninsula.

sacfry Recently hatched fish larvae that are still in possession of the yolk sac.

Saffir–Simpson Hurricane Scale A scale introduced in 1955 to describe conditions in *tropical cyclones. The scale ranks such storms in five categories according to their sustained wind speeds, surface air pressure in the eye, and height of the *storm surge each produces. The wind speed in a category 1 hurricane is between 119 km/h and 153 km/h; that in a category 5 hurricane is greater than 250 km/h.

sag and swell topography *See* KNOB AND KETTLE.

Sahel zone The semi-arid southern border of the Sahara in western Africa, which supports a very dry type of *savannah including scattered, thorny trees. In recent years the desert has advanced appreciably into the Sahel zone, partly at least as the result of poor land management. *Compare* GUINEA ZONE.

Sahulland The name often given to the tropical portion of the combined Australia–New Guinea land mass, as it existed at times of low sea level during the *Pleistocene. The faunas of the two present-day components have great similarities, the differences being due mainly to the fact that New Guinea is largely forested and Australia is largely open country. The Sahul shelf, linking New Guinea with Australia, is less than 200 m below the present sea level.

St Anthony's fire *See* ERGOT.

salic horizon A *soil horizon, usually below the surface, which contains not less than 2 per cent salt and with a figure of 60 or more for the value calculated as the thickness of the horizon in centimetres multiplied by the percentage of salt. It is a *diagnostic horizon.

salination *See* SALINIZATION.

saline **1.** Pertaining to salt. **2.** An aqueous solution of sodium chloride, with or without other salts, and approximately isotonic with body fluids, which is employed for the temporary maintenance of living cells and tissues.

saline giant A thick and extensive salt deposit, produced by the evaporation of a large hypersaline sea. In north-western Europe, for example, the Zechstein salts, which are mined in northern Germany for rock salt and potash, were formed during the *Permian by the evaporation of water from a partially barred marine basin covering more than 250 000 km^2 (*see* ZECHSTEIN SEA). Similar extensive evaporite deposits were also formed further south during the *Miocene by the evaporation of part of the Mediterranean.

saline-sodic soil Soil that contains more that 15 per cent exchangeable sodium, a saturation extract with a conductivity of more than 4 mmhos/cm (25 °C) and in the saturated soil usually has a *pH of 8.5 or less. Either high concentration of salts or high pH, or both, interfere with the growth of most plants.

saline soil Soil that contains enough soluble salt to reduce its fertility. The lower limit is usually defined as 0.4 siemens per metre (4 mmhos/cm).

salinity A measure of the total quantity of dissolved solids in water, in parts per thousand (per mille) by weight, when all organic matter has been completely oxidized, all carbonate has been converted to oxide, and bromide and iodide to chloride. The salinity of ocean water is in the range 33–38 parts per thousand, with an average of 35 parts per thousand.

salinization **(salination in US usage)** The process of accumulating soluble salts in soil, usually by an upward capillary movement from a saline *groundwater source.

SALR *See* SATURATED ADIABATIC LAPSE RATE.

saltation A major process of particle transport in air and water, which involves an initial steep lift followed by travel and then a gentle descent to the bed.

saltatorial Applied to limbs that are adapted for jumping.

saltatory Leaping movement (e.g. of crickets and grasshoppers).

salt-dome trap A salt diapir (i.e. an intrusion) which has pushed up existing sediments into a dome structure, trapping gas, oil, or water in the pores of the permeable rocks adjacent to and above the salt dome beneath a cap rock. The rocks ahead of the salt diapir are often severely faulted and may give rise to fault traps. Oil may also accumulate in the porous top of the salt diapir.

salt fingering A suggested mixing process between layers of saline and less saline ocean water. The vertical water movements between the water masses of different *salinity occur in small columns ('fingers') a few millimetres across that penetrate only a small distance, producing a mixed layer. The mixing process may then be repeated at the two interfaces that are present, and a number of layers may develop. It is thought to occur where warm, saline water overlies cooler, less saline water (e.g. where the saline Mediterranean water flows out into the less saline Atlantic).

salt flat An extensive flat surface, found in hot deserts, consisting of salts that have accumulated in a shallow saline lake or *playa; evaporation produces a crust of varying hardness.

salt lake A lake in which the concentration of mineral salts is typically about 100 parts per thousand or greater and dominated by dissolved chlorides (e.g. the Dead Sea, which contains 64 parts per thousand NaCl and 164 parts per thousand $MgCl_2$).

salt marsh Vegetation often found on mud banks formed at river mouths, showing regular zonation reflecting the length of time different areas are inundated by tides. Sea water has a high salt content, which produces problems of *osmotic pressure for the vegetation, so that only plants adapted to this environment (halophytes) can survive.

salt pan A basin in a semi-arid region where chemical precipitates (evaporites) are deposited, owing to the concentration by evaporation of natural solutions of salts. The least soluble salts (calcium and magnesium carbonates) are precipitated first, on the outside of the pan, followed by sodium and potassium sulphates. Finally, in the centre, sodium and potassium chlorides and magnesium sulphate are deposited. This pattern, slightly distorted through tilting, is seen in Death Valley, California, USA.

salt stress *Osmotic forces exerted on plants when they are growing in a *salt marsh or under other excessively saline conditions.

salt wedge An intrusion of sea water into a tidal *estuary in the form of a wedge along the bed of the estuary. The lighter fresh water from riverine sources overrides the denser salt water from marine sources unless mixing of the water masses is caused by estuarine topography. Salt wedges are found in estuaries where a river discharges through a relatively narrow channel.

salvageability *See* RECREATABILITY.

sampling *See* OVERDISPERSION; PLOTLESS SAMPLING; RANDOM SAMPLE; REGULAR SAMPLE; STRATIFIED RANDOM SAMPLE; and TRANSECT.

sand **1.** In rocks, according to the commonly used (Udden–Wentworth) scale, particles between 62.5 and 2 000 μm. **2.** In pedology, mineral particles of diameter 2–0.02 mm in the international system, or 2–0.05 mm in the *USDA system. **3.** A class of soil texture.

sand-body A finite unit of *sand (or *sandstone) usually accumulated in response to one type of depositional process (e.g. as a channel, beach-bar, or barrier system). The distribution of the sand and the three-dimensional geometry is controlled largely by the nature of the depositional regime under which it accumulated (i.e. channel sands may be sinuous, and beach or shore-bar sands may be linear and parallel to the shore, etc.).

sand pillar *See* DUST DEVIL.

sand ribbon A longitudinal strip of *sand, up to 15 km long, 200 m wide and less than 1 m thick, standing on, and surrounded by, an immobile gravel floor. Sand ribbons are developed on the sea floor of the *continental shelf where there is a paucity of sand, with water depths of 20–100 m and fast-flowing currents.

sandstone (arenite) A type of sedimentary rock, formed of a lithified *sand, and comprising grains between 63 μm and 1 000 μm in size, bound together with a mud matrix and a mineral cement.

sandstorm The lifting of sand and dust particles, often to great altitude, by turbulent winds. Visibility is greatly reduced.

sandur (pl. sandar) *See* OUTWASH PLAIN.

sand volcano A conical body of *sand, resembling a small volcano, which is rarely more than a few metres wide and less than 50 cm high. Internally the sand volcano consists of a massive central plug, surrounded by laminated sand paralleling the external form. Sand volcanoes are formed by the extrusion of highly liquefied sand through a local vent in a confining layer at the surface.

sand wave A large-scale, transverse ridge of *sand that is characteristic of *continental-shelf areas (e.g. the southern North Sea). The wavelength or spacing of sand-wave crests is 30–500 m and the height is 3–15 m. The down-current migration of sand waves leads to the formation of large-scale cross-bedding.

Sangamonian The third (0.12–0.075 Ma) of four *interglacial stages recognized in mid-continental North America. It followed the *Illinoian glacial episode and is the approximate equivalent of the *Riss/Würm interglacial of the Alps. Early warm climates and later cooling climates are represented in well-exposed pollen spectra.

sanitary landfill *See* LANDFILL.

sapling A young tree, not yet a *pole or of useful or timber size, arising from a seed or sucker.

saponin Any member of a class of glycosides that form colloidal solutions in water and foam when shaken. Saponins have a bitter taste, hydrolyse red blood cells, and are very toxic to fish. They occur in a wide variety of plants, including *Saponaria officinalis* (soapwort), which produces a lathery liquid, once widely used for washing wool and still used for delicate textiles, including antique ones.

sapro- From the Greek *sapros*, meaning 'putrid', a prefix meaning 'decayed' or 'rotten'.

saprobe *See* SAPROTROPH.

saprolite Chemically rotted rock *in situ*; often the lower portion of a weathering profile. The saprolite on granite is locally called 'grus' or 'growan', although the latter term may include material broken down by *mechanical weathering. *Compare* REGOLITH.

sapropel Organic ooze or sludge accumulated in anaerobic conditions in shallow lakes, swamps, or on the seabed. It contains more hydrocarbons than *peat. When dry it is dull, dark, and tough and it may be a source of oil and gas.

sapropelic coal *See* COAL.

sapropelite A sapropelic *coal, consisting of organic material, particularly algae, which accumulated in stagnant lake bottoms or the floors of anoxic shallow seas.

saprophage An organism that consumes other, dead, organisms. Saprophages form part of the twofold division of the *heterotrophs (organisms that feed on other organisms) and consist mainly of bacteria and fungi, but also some invertebrate animals, such as insect larvae. They break down complex compounds obtained from dead organisms, absorbing some of the simpler products, but releasing most of the products as inorganic nutrients which can then be used by other organisms. *See also* CONSUMER ORGANISM.

saprophyte *See* SAPROTROPH.

saprotroph (saprobe, saprovore) Any organism that absorbs soluble organic nutrients from inanimate sources (e.g. from dead plant or animal matter, from dung, etc.). If the organism is a plant or is plant-like it is called a saprophyte; if it is an animal or is animal-like it is called a saprozoite. *See also* CONSUMER ORGANISM.

saprovore *See* SAPROTROPH.

saprozoite *See* SAPROTROPH.

sap stain A stain (often bluish-grey) of the sapwood in freshly cut timber, caused by any of several types of fungi (e.g. *Penicillium*, *Ceratocystis*, *Cladosporium*). The mechanical strength of the timber is usually little affected.

SAR *See* SODIUM ADSORPTION RATIO.

Sarcodina *See* PROTOZOA.

Sargasso Sea The calm centre of the anticyclonic *gyre in the North Atlantic, comprising a large eddy of surface water, the boundaries of which are demarcated by major current systems such as the *Gulf Stream, *Canaries Current, and *North Atlantic Drift. The Sargasso Sea is a large, warm (18°C), saline (36.5–37.0 parts per thousand) lens of water, which is characterized by an abundance of floating brown seaweed (*Sargassum*).

sarmentose Producing long, flexuous runners or *stolons.

satellite DNA Any DNA which differs enough in its base composition to form a separate fraction from the majority of genomic DNA on centrifugation. This bias in base composition is often due to highly repetitious DNA. Satellite DNA may consist of dispersed repeats (e.g. *long interdispersed elements and *short interdispersed elements) or may be arranged in a *tandem array. In a tandem array it may have very short repeating units, 2–10 base pairs in the case of microsatellite DNA, or slightly longer ones, 10–100 base pairs in the case of minisatellite DNA. The number of repeating units within a single satellite tandem array changes rapidly in evolutionary terms, due to the processes of *replication slippage and *unequal crossing over. Consequently, the length of an array can be used as a highly informative character in *phylogenetic analysis at the intraspecific level. By using many such arrays a highly specific set of *character states can be established, by which an individual organism may be identified; this is a DNA fingerprint.

satiation A process that leads to the cessation of an activity, applied most commonly to *feeding behaviour. Satiation may be associated with physiological changes. It occurs before the point at which *appetite is satisfied completely and continuation of the activity becomes physically impossible.

saturated adiabatic lapse rate (SALR) The *adiabatic cooling rate of a rising *parcel of air which is saturated and in which condensation is taking place as it rises, so that the energy release of the latent heat of vaporization moderates the adiabatic cooling. The reduction of the rate of cooling below the *dry adiabatic lapse rate of 9.8°C/km varies with temperature, because of the greater energy release by condensation from air at higher temperatures. Thus at a given atmospheric pressure, air at 20°C may have an SALR as low as 4°C/km, and at −40°C the SALR may be close to 9°C/km. The *stability or *instability of the atmosphere at any given time for vertical motion is determined by whether the *environmental lapse rate of temperature within it is less than or greater than the adiabatic lapse rate (i.e. less than or greater than the rate of decrease of temperature of rising parcels of air).

saturated air Air that contains the maximum amount of water vapour possible at the given temperature and pressure (i.e. the *relative humidity is 100 per cent).

saturated flow The movement of water through a soil that is temporarily saturated. Most of the loosely held water moves downward, and some moves more slowly laterally.

saturation deficit At a given temperature, the difference between the actual vapour pressure of moist air and the saturation vapour pressure.

saturation moisture content **(SMC)** The maximum amount of water that can be contained in a rock when all *pore spaces are filled with water; it is expressed as the percentage of the dry weight of the rock.

saurian Of or resembling a lizard. Applied loosely to lizard-like animals, and also applied to fossils, life habits, etc. of extinct reptiles.

saurochory Dispersal of spores or seeds by snakes or lizards.

savannah An extensive tropical vegetation dominated by grasses with varying admixtures of tall bushes and/or trees in open formation. Savannah occurs in diverse tropical environments, although most experience a dry season. Much savannah is no doubt *climatic climax, although extensive tracts are anthropogenic *fire climaxes and others are *edaphically controlled; it is generally difficult to distinguish one type from the other.

savannah woodland A savannah in which trees and *shrubs form a generally light canopy. The trees and bushes are generally *deciduous, yet *evergreens are usually also well represented. Some tall trees occur, but most are stunted and gnarled. They frequently have thick, corky, fire-resistant bark.

saxicolous Growing on stones, rocks, walls, etc.

scab A general term for any plant disease in which the symptoms include the formation of dry, corky scabs. Examples include *apple scab and *potato scab.

scabrous Rough-surfaced; bearing short, stiff hairs, scales, or points.

Scandinavian ice sheet An *ice-cap which developed over Scandinavia during the *Quaternary. It has been suggested, largely on the basis of the degree of downwarp and the partial recovery of the land surface in Scandinavia and the surrounding areas, that the ice was 2 600 m thick.

scar A steep, cliff-like slope of bare rock, developed in the near-horizontally bedded *Carboniferous limestone of the Yorkshire Dales, England. The steepest and highest scars are normally associated with the outcrop of the purest and most massively bedded limestone. Often a *scree is formed at the base.

scarious Dry and membranaceous.

scarp **(abbr. of escarpment)** A steep slope or cliff found at the margin of a flat or gently sloping area, usually against the dip of the rocks; it may occur with a dip slope to give a *cuesta. The gradient of the relatively steep face is maintained by the *erosion of a relatively weak stratum which typically underlies the resistant cap rock that maintains the form of the scarp. Erosion may be achieved by *spring sapping, *sheet-wash, and *mass wasting. Many varieties are recognized and distinguished in terms of origin. A 'fault scarp' results when a fault displaces the ground surface so that one side stands high. A 'fault-line scarp' is produced by *erosion on one side of an ancient fault: obsequent and resequent varieties are recognized. A 'composite fault-line scarp' results from a combination of erosion and faulting. Erosional scarps result from vertical incision or from the headward enlargement of *pediments.

scarp-and-vale topography A landscape that consists of a roughly parallel sequence of cuestas (*scarps and dip slopes) and intervening valleys ('vales'). It dominates most of lowland Britain, where *Mesozoic sediments dip gently towards the east and south-east.

scarp-foot knick An abrupt change in gradient that often occurs in semi-arid environments between a *pediment and the adjacent *scarp. It is the boundary between the zones of *scarp retreat and pedimentation.

scarp retreat The recession of the relatively steep hill-slope that terminates a *butte, *mesa, *cuesta, or any elevated, plateau-like surface. Several geomorphological processes may be involved, and under semi-arid conditions these may give rise to a hill-slope that retreats parallel to itself.

scatter diagram A graphic representation of the relationship between two variable quantities. It can be used in taxonomic methods, e.g. plotting oak leaf *petioles against an index of elliptical or obovate lamina shapes, which allows a distinction to be made between *Quercus robur* (English oak) and *Q. petraea* (durmast oak). It can also be used to plot the distribution of organisms in relation to a particular environmental factor. When the two variables are plotted against each other it is possible to determine the degree of correlation between them.

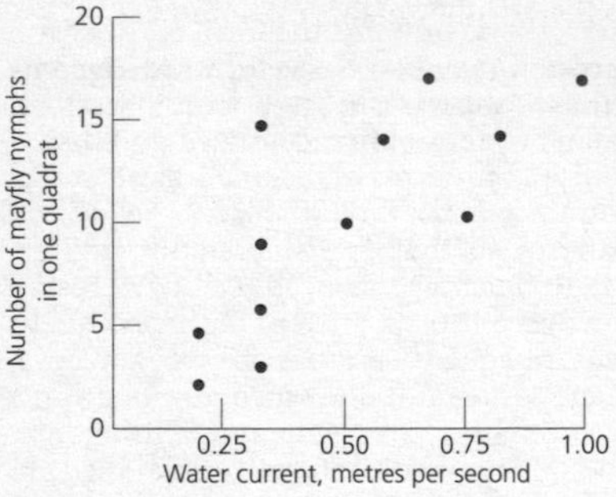

Scatter diagram

S

scavenger *See* DETRITIVORE.

scavenging The capture and removal by rain or snow of particulate matter in the atmosphere.

scent marking The use by an animal of scented secretions, faeces, urine, or saliva as a means of communication: the substance is deliberately placed on the ground, on an object, or on another animal.

Schimper, Andreas Franz Wilhelm (1856–1901) A German botanist and ecologist who was a professor at the Bonn Botanical Institute (1886–98) and at Basle (1898–1901). He showed (1881) that starch grains are not formed in cytoplasm, and (1883) that plastids originate from the division of pre-existing plastids. His major work, *Pflanzengeographie auf physiologischer Grundlage* (first edition 1898, first English edition, *Plant-Geography upon a Physiological Basis*, 1903), includes his classification and a systematic account of world vegetation based on physiological adaptation, and did much to establish a sound method for ecological investigation.

schistosomiasis (bilharzia) Any one of a group of diseases of humans and some other mammals caused by infestation by blood flukes (trematodes of the genus *Schistosoma*). The trematode requires two hosts; the first is an aquatic snail in which larvae emerging from eggs hatched in water develop through several stages, finally to fork-tailed forms (cercariae) which escape from the snail, swim freely in the water, and attach themselves to and penetrate the skin of their second host, a mammal. In the mammalian body they move through blood vessels, feeding upon glycogen in the blood plasma and growing, eventually migrating to a site that varies from species to species, where they mature, mate, and lay eggs. The eggs leave the body in urine or faeces. Symptoms vary according to the species causing the infestation, but may include swelling and tenderness of the liver, dropsy, enlargement of the spleen, watery skin eruptions, fever, cough, haemorrhage, diarrhoea, and reduced function of the affected organs. The diseases are common in low latitudes.

schizo- The Greek *schizo*, meaning 'to split', used as a prefix meaning 'split' or 'divided'.

Schmidt–Lambert net *See* EQUAL AREA NET.

school Loosely, an aggregation of fish or marine mammals which are observed swimming together, possibly in response to a threat from a predator. More strictly, a grouping of fish, drawn together by social attraction, whose members are usually of the same species, size, and age, the members of the school moving in unison along parallel paths in the same direction. Sudden changes at the leading edge of the school are followed almost instantaneously by the remainder of the group, with the fish on the flank becoming the new leaders. In general, the overall size and shape of the school, as well as the cruising depth and speed, vary from one species to another.

schooling Among fish, the formation of groups of individuals as a result of social attraction.

scia- *See* SKIA-.

scion *See* GRAFT.

sciophilous *See* SKIOPHYLLOUS.

sciophyte A shade plant (*see* SKIA-).

scirocco **(sirocco)** A regional name for one of the types of warm winds from the south which occur around the Mediterranean. It moves ahead of an eastward-travelling *depression and brings hot, dry, dusty conditions to Algeria and the Levant. To the north, where its humidity increases very rapidly as it crosses the sea, it brings moist air to the coast of Europe.

scler- From the Greek *skleros*, meaning 'hard', a prefix that means 'hard' or 'tough'.

sclerophyllous vegetation Typically scrub, but also forest, in which the leaves of the trees and *shrubs are *evergreen, hard, thick, leathery, and usually small. These adaptations allow the plants to survive the pronounced hot, dry season of the *Mediterranean-type climate in which sclerophyllous vegetation is best developed.

sclerotinite *See* COAL MACERAL.

scoria Loose, rubbly, basaltic ejecta that accumulate around eruptive volcanic vents (or similar type to Stromboli), eventually building up as a scoria cone, the height of which may range from a few tens of metres to up to 300 m, with a slope determined by the angle of repose of the loose material.

scramble competition *Competition for a resource that is inadequate for the needs of all, but is partitioned equally among contestants, so no competitor obtains the amount it needs and in extreme cases all die. *Compare* CONTEST COMPETITION.

scree An accumulation of coarse rock debris that rests against the base of an inland cliff, produced by the *weathering and release of fragments from the cliff face. Screes are widely found in upland areas affected by past or present *periglacial conditions and in hot, rocky deserts.

scree slope *See* TALUS.

scrub vegetation A general term for vegetation dominated by *shrubs, i.e. low, woody plants, which typically forms an intermediate *community between grass or heath and high forest. *Successional change is not necessarily implied, though the term is often used for the transitional stage in a succession to *climax woodland when shrubby plants predominate.

scud A popular name for stratus fractus (fractostratus), a fragmented low cloud moving quickly beneath rain clouds.

sea **1.** A large body of usually saline water which is smaller in size than an ocean. **2.** Chaotic waves generated by the action of the wind on the surface layers of the ocean. *See also* OCEAN WAVE and SWELL.

sea breeze *See* LAND AND SEA BREEZES.

sea-floor spreading The theory that the ocean floor is created at the spreading (accretionary) plate margins within the ocean basins. *Igneous rocks rise along conduits from the mantle, giving rise to volcanic activity in a narrow band along the mid-ocean ridges. The newly formed oceanic crust spreads perpendicularly away from the ridge.

sea fret A popular local name for sea fog, common in spring and summer in Cornwall and on the south, east, and north-east coasts of England. *See also* HAAR.

sea ice Ice that floats on the surface of the sea in polar regions. It exists throughout the year in the central Arctic and in some Antarctic bays, extending in winter across the entire Arctic and far out to sea around Antarctica. As ice crystals form from sea water, so salt is excluded and eventually returned to the sea. Sea ice therefore contains no salt, except where pockets of sea water become trapped in the ice.

seamount An isolated, submarine mountain, of volcanic origin, which rises more than 1 000 m above the ocean floor and has a sharp, crested summit that is, in most cases, 1 000–2 000 m below the ocean surface.

searching Behaviour by an animal which is directed towards the provision of some necessity (e.g. food, nesting material, or a mate) that has not yet been located.

searching image The mental image of an object that is apparently possessed by an animal searching for that object, the existence of which is inferred from observation of the behaviour of animals.

seatearth A clay-rich fossil soil, found immediately beneath a coal seam, which represents the soil in which the coal forming vegetation grew.

secchi disc A disc used in a simple method for measuring the transparency of water. The disc is 20 cm across and divided into alternate black and white quadrants. It is lowered into water on a line until the difference between the black and white areas just ceases to be visible, at which point the depth is recorded. The secchi disc provides a convenient method for comparing the transparency of water at different sites. The disc was invented by the Italian Jesuit and eminent astrophysicist Angelo Secchi (1818–78).

S

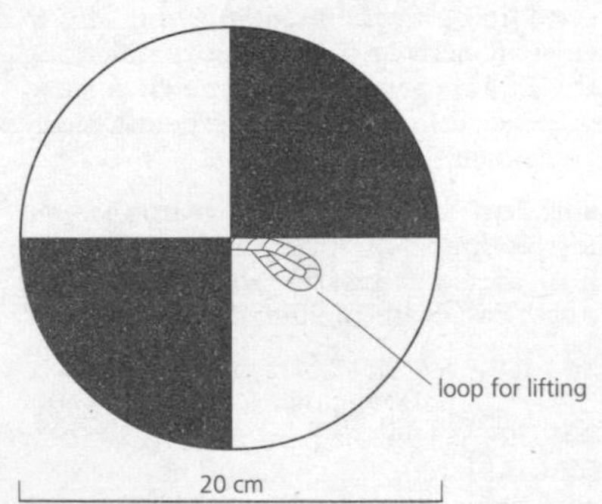

Secchi disc

secondary consumer A carnivore that preys upon herbivores.

secondary depressions *Depressions initiated as part of a 'family' or sequence of wave disturbances along a *cold front; they are in the rear of the first depression of the series. *See also* FRONTAL WAVE.

secondary dormancy *See* DORMANCY.

secondary forest A forest growing on an area cleared of a previous forest. It is composed exclusively or partially of *pioneer species. *See* PRIMARY FOREST.

secondary front A frontal development in the colder air in the rear of a *depression, in the form of a further *trough of low pressure that follows that main *cold front.

secondary plant compound A chemical compound produced by a plant that serves no primary function in plant metabolism (e.g. an *alkaloid).

secondary productivity *See* PRIMARY PRODUCTIVITY.

secondary sexual character A characteristic of animals that differs between the two sexes, but excluding the gonads and the ducts and associated glands that convey the *gametes. Examples are mammary glands, external genitalia, antlers in ungulates, and certain plumage patterns (e.g. peacock's tail) in birds.

secondary succession A *succession initiated by the disruption of a previously existing seral or *climax community by some major environmental disturbance and leading to a marked change in the stable vegetation *community. Secondary successions occur, for example, following fire, after removal of grazing pressure, when previously cultivated areas are abandoned (as in shifting cultivation), or when a forest has been cleared (*see* SECONDARY FOREST). Interactions between plants and the physical environment tend to be less clear in secondary than in *primary successions.

secondary woodland A woodland occupying a site that has not been wooded continuously throughout history (in Britain since the last ice advance). It may be the product of natural *succession or of planting on formerly unwooded land. In the tropics, secondary woodland (i.e. forest) is pure or regrowing following clear-felling; it contains fewer species than primary forest.

secondary-woodland species Tree species that are frequent and often abundant in secondary woodland. *See* PIONEER PLANT.

secular variation Any long-term variation over a period of about 10 to 20 years or longer.

secund Arranged on one side.

sedentary Applied to organisms that are attached to a substrate but are capable of limited movement. *Compare* SESSILE.

SEDEX *See* SEDIMENTARY EXHALATIVE PROCESSES.

sediment Material derived from preexisting rock, from *biogenic sources, or precipitated by chemical processes, and deposited at or near the Earth's surface.

sedimentary cycle The *weathering of an existing rock, followed by the *erosion of *minerals from it, their transport and deposition, and their burial. First-cycle sediments contain less resistant minerals and rock fragments. If this material is reworked through a second cycle, the less resistant minerals will be eliminated or altered to more stable products. The more sedimentary cycles a sediment has passed through, the more mature it will become and it will be dominated by well-rounded, resistant minerals.

sedimentary exhalative processes (SEDEX) The processes that are associated with the upwelling of mineralizing fluids into submarine sedimentary environments, whereby mineral deposits, usually of base-metal sulphides, are formed.

sedimentary rock Rock formed by the deposition and compression of mineral and rock particles, but often including material of organic origin, and exposed by various agencies of *denudation.

sedimentation factor (sedimentation value, S factor, S value) A measure of the rate at which a molecule, organelle, or particle settles under standard conditions of centrifugation. It is equal to the acceleration, measured in *Svedberg units, and allows particles to be separated and identified.

sedimentation value *See* SEDIMENTATION FACTOR.

seed **1.** In the sexual reproduction of seed plants (Spermatophyta), the discrete body from which a new plant develops. Formed from a fertilized ovule, the seed comprises an outer coat (testa) enclosing a food store and an embryo plant. **2.** Any plant or animal structure concerned with propagation.

seed bank A store in which seeds are held as a means of conserving plant species. Maintained at a constant temperature of 0°C and dried to a moisture content of about 4 per cent, seeds of many species remain viable for up to 20 years. *See* GENE BANK. *Compare* SOIL SEED BANK.

seed rain The fall to the ground of wind-dispersed seeds or spores.

seed shadow An area of ground on to which few wind-dispersed seeds fall, because its surface is well covered by vegetation or sheltered from prevailing winds.

seep *See* SPRING.

seepage The slow but often steady flow of water between one water body and another. The term is often used to describe leakage to underlying *aquifers through stream beds or the emergence of *groundwater into a stream channel, but it may also relate to flow between different aquifer units.

seepage velocity The apparent velocity with which *groundwater moves through the bulk of the porous medium. Actual velocity is higher than seepage velocity by a factor which combines the effects of porosity and the tortuosity of the actual flow path among and around the mineral grains.

seiche A stationary or standing wave in an enclosed body of water (e.g. a bay or lake). Seiches are usually the product of intense storm activity.

seif dune A linear sand dune, consisting of curved, sword-like components, which is found in hot deserts. Typically it is developed by the elongation of a *barchan arm, and built up by winds blowing from two principal directions.

Seif dune

seismonasty A nastic movement (*see* NASTY) in plants in response to sudden stimulation by touch.

seistan wind A regional wind from the north of Seistan, eastern Iran, which blows with great force and constancy for a period of months in summer.

selection A process that results from the differential reproduction of one *phenotype as compared with other phenotypes in the same population. This determines the relative share of different *genotypes which individuals possess and propagate in a population. The relative probability of survival and reproduction of a phenotype is termed 'fitness' or 'Darwinian fitness'.

selection coefficient (*s*) A measure of the relative excess or deficiency of fitness of a *genotype compared with another genotype in the population. If $s = 100$ then 1 out of a 100 individuals of a given genotype fails to reproduce.

selection differential The difference between the average value of a *quantitative character in the whole population and the average value of those selected to reproduce the next generation.

selection intensity A measure of the difference in *fitness values within a population.

selection pressure The pressure exerted by the environment, through *natural selection, on *evolution. Thus weak *selection pressures result in little evolutionary change and vice versa.

selective species In *phytosociology, a plant species found most frequently in a particular *community, but also present occasionally in others. It is fidelity class 4 of the *Braun-Blanquet scheme. *Compare* ACCIDENTAL SPECIES; EXCLUSIVE SPECIES; INDIFFERENT SPECIES; and PREFERENTIAL SPECIES.

selective value *See* ADAPTIVE VALUE.

selfish DNA A segment of DNA which reproduces itself, but conveys no advantage to the *genome in which it resides. *Transposons are thought to be examples of selfish DNA.

selfish herd A theory proposed in 1971 by W. D. Hamilton according to which the risk to an individual of predation is reduced if that individual places another individual between itself and the predator. When many individuals behave in this way an aggregation is the inevitable result and, because the risk is least near the centre and greatest at the edge, individuals of high social status will tend to occupy the centre and subordinate individuals will be pushed to the edge.

self-mulching soil A soil that mixes itself: its surface layers shrink and swell, forming deep cracks into which soil falls.

self-thinning A progressive decline in the density of a population of growing plants. During the course of the self-thinning process, individuals become larger as the population density declines. If the logarithm of density is plotted against the logarithm of individual weight, then a straight line of negative slope is obtained, and the slope is generally $-3/2$. This relationship holds for a wide range of plants, from annual weeds to forest trees, and has become known as the 'self-thinning rule'.

selva A term for tropical rain forest, applied specifically to the Amazon Basin.

semelparity (big-bang reproduction) The condition of an organism that has only one reproductive cycle during its lifetime. *Compare* ITEROPARITY.

semi-arid climate A climate with an *aridity index of 0.2–0.5.

semi-desert scrub A transitional formation type situated between true *desert and more thickly vegetated areas (e.g. between thorn forest and desert or between *savannah and desert). The vegetation is sparser than that of the thorn forest and succulents are more common, as a consequence of the drier climate. Most of the plants are shallow-rooted, and so able to exploit before it evaporates any precipitation that *percolates into the surface layer of the soil.

semi-natural community Vegetation altered by human influence or management in the past, which has taken on a natural aspect owing to the length of time over which the influences have persisted. For example, *heathland and chalk grassland in Great Britain have long been subject to management and members of each *community have adapted to it. In chalk grassland many plant species are low-growing *rosette plants which avoid being grazed. *Compare* NEAR-NATURAL COMMUNITY.

semi-natural woodland An *ancient woodland site that has been managed over a period, resulting in some change in structure and species composition, and which consists of stands of mainly *native trees that have not been obviously planted. On a recent woodland site, the term includes all stands that have originated mainly by natural regeneration.

semiparasite *See* PARASITISM.

semi-permeable Applied to a membrane whose structure allows the passage of only solvent molecules. A membrane that allows the passage of small molecules but prevents the passage of larger ones is called 'differentially permeable'. The Institute of Biology recommends that biological membranes be described as 'partially permeable'.

semi-species A group of organisms that are taxonomically intermediate between a race and a species, with reduced outbreeding and gene flow, i.e. with incomplete reproductive isolating mechanisms. Semi-species are thought to represent advanced stages of *speciation.

senescence The complex deteriorative processes that terminate naturally the functional life of an organ or organism.

sense codon A *codon that codes for an *amino acid. *Compare* NONSENSE CODON.

sensitive period A period of time during which a young animal is most impressionable and therefore most likely to acquire learned behaviour. This is the period during which *imprinting occurs.

sensitivity Of a landscape, the likelihood that a change in the controls of a *system will produce a recognizable response. Sensitivity has also been thought of as the ratio of disturbing to resisting forces; the relation of forces to a particular threshold condition; and the ability to recover from a disturbance (*see* RELAXATION TIME). **2.** *See* IRRITABILITY.

sensitivity analysis The consideration of a number of factors involved in the mathematical modelling of an *ecosystem and its components. These include feedback and control, and the stability and sensitivity of the system as a whole to changes in some parts of the system. Predictions can be made from the analysis.

sensitization The increase in the likelihood that a particular and significant stimulus will produce a response in an animal repeatedly exposed to it. *Compare* EXTINCTION and HABITUATION.

sequester In the strict sense, to bind a metal ion into a *chelate; more broadly, to take up and fix. The word is also used in describing processes within *biogeochemical cycles, as when forests or oceans sequester carbon.

serac An ice pinnacle on the surface of a *glacier which results from tensional failure in the more rigid upper crust when the glacier moves over a slope, spreads out over a plain, or passes round a bend in its valley.

seral stage A phase in the sequential development of a *climax community. *See* SUCCESSION.

sere The characteristic sequence of developmental stages occurring in plant *succession. *See* COMPETITION; ECESIS; MIGRATION; NUDATION; REACTION; STABILIZATION.

serein The fall of rain from an apparently clear sky. The phenomenon may be explained by the evaporation of cloud particles following the formation of rain droplets, or by the movement of cloud away from the overhead position as the rain approaches the ground.

serir A veneer of mixed sand and gravel mantling a Saharan plain and transported originally by *sheet-wash and *braided-stream activity. Subsequently it was weathered and modified under more arid conditions.

serotinal In late summer. *See* AESTIVAL; HIBERNAL; PREVERNAL; and VERNAL.

serotinous *See* SEROTINY.

serotiny (*adj.* **serotinous)** In certain plants, especially trees (e.g. jack pine (*Pinus banksiana*), lodgepole pine (*P. contorta*), and many species of *Eucalyptus*), the retention of seeds in pods or cones on the tree, often for many years, until a disaster, most commonly the heat of a fire, causes their release. After fire, the seeds fall on ground fertilized by ash in a site cleared of competitors.

serpentine barrens Impoverished, often *scrubby or *heathland vegetation associated with serpentine rocks. On weathering these rocks release an excess of *magnesium into the soil, and this often inhibits the development of the natural *climax in the areas concerned.

sesquioxides A general term for the hydrated oxides and hydroxides of iron and aluminium.

sessile **1.** Lacking a stalk. **2.** Attached to a substrate; non-motile. *Compare* SEDENTARY.

setation *See* CHAETOTAXY.

sewage fungus A slimy growth found in sewage and sewage-polluted waters. It consists of filamentous bacteria associated with fungi and protozoa.

sex **1.** The sum of the characteristics concerned with sexual reproduction and the raising of young, by which males, females, and hermaphrodites may be distinguished. **2.** The act of sexual intercourse.

sex ratio The relative number of males to females in a population. The measure is usually applied to a given age class. The primary sex ratio is that immediately after fertilization (normally 1 : 1); the secondary sex ratio is that at birth or hatching; and the tertiary sex ratio is that at maturity.

sex reversal A change in functioning such that a member of one sex behaves as a member of the other. Some organisms (e.g. certain molluscs) make this change as part of their normal *life cycle, with most individuals functioning as males when young and then passing through a transitional stage to a period when they function as females. Sex reversal may otherwise be experimentally induced (e.g. by hormone transplants in humans) or environmentally induced (e.g. by temperature effects on the production of male or female zygotes in some turtles).

sexual dimorphism The occurrence of morphological differences (other than *primary sexual characters) that distinguish males from females of a species of organism (e.g. male deer often have larger antlers than females, and the males of many birds have differing (often more brightly coloured) plumage). These are all *secondary sexual characters, but not all sexual dimorphism is directly sex-related. For example, female sparrowhawks (*Accipiter nisus*) are bigger than males, a difference that is possibly related to resource partitioning of prey, females taking larger birds, or to the needs of incubation.

sexual selection A theory proposed by *Darwin, that in some *species males compete for mates and that characteristics enhancing their success in mating would have value and be perpetuated irrespective of their overall value in the struggle for existence. Such *characters would be used either in male display to attract females (inter-sexual selection) or in combat between rival males (intra-sexual selection), and both could act at the same time. *See also* HANDICAP PRINCIPLE and RUNAWAY HYPOTHESIS.

S factor *See* SEDIMENTATION FACTOR.

shade temperature The temperature of the air, conventionally measured in a standard shelter or screen that protects the thermometer from rain and from direct sunshine, but allows the free passage of air.

shale Fine-grained, fissile, *sedimentary rock composed of *clay-sized and *silt-sized particles of unspecified mineral composition. The noun may be qualified by an adjective (e.g. black shale, paper shale, and *oil shale).

shamal A regional north-westerly wind which brings hot, dry conditions in summer to Iraq and the Persian Gulf. It blows with great force during the day.

Shannon–Wiener index of diversity (information index) A measure derived from information theories that were developed by Claude Elwood Shannon (1916–2001) and Norbert Wiener (1894–1964) and published in 1949 by Shannon and Warren Weaver (1894–1978). It is widely used in ecology because it is an index that combines richness with evenness. The index is usually represented by the symbol H, and is given by $H = -\Sigma_i^s p_i \log p_i$, where s is the total number of species in the community and p_i is the proportion of species i in the community. Looking at it another way, this is the probability that any random individual taken from the community belongs to species i. The higher the H value, the more diverse the community. The evenness component (J) can then be calculated by dividing H by H_{max}, i.e. the maximum possible diversity for a given number of species and individuals. H_{max} is given by $\ln s$, where s is the total number of species. The greatest value possible for J is 1.0, which represents a perfectly even distribution of the individuals among the species.

sharp sand *Sand composed of angular not rounded grains, with little foreign material; it is used in mortar and in some potting composts.

sheep-walk *See* TERRACETTE.

sheet flood *See* FLASH FLOOD.

sheet-wash A geomorphological process in which a thin, mobile sheet of water flows over the surface of a hill-slope and may transport the surface *regolith. It is important in semi-arid regions, and may also be significant in temperate zones if the vegetation cover has been removed. *See* RAIN-WASH.

shelf A gently sloping or near horizontal, shallow, marine platform. Horizontal shelf areas, particularly in carbonate-dominated areas, are called 'platforms' if they have precipitate margins; more uniformly sloping, shallow, marine, carbonate areas are known as 'ramps'.

Shelford, Victor Ernest (1877–1968) An American ecologist, who was primarily responsible for introducing animals into studies of climax communities and the successions leading to them. He was the author of *Naturalist's Guide to the Americas* (1926), *Laboratory and Field Ecology* (1926), and *The Ecology of North America* (1963).

Shelford's law of tolerance A law proposed in 1911 by V. E. *Shelford, stating that the presence and success of an organism depend upon the extent to which a complex of conditions is satisfied (e.g. the climatic, topographic, and biological requirements of plants and animals). The absence or failure of an organism can be controlled by the qualitative or quantitative deficiency or excess of any one of several factors which may approach the *limits of tolerance for that organism.

shelf zone *See* SUBLITTORAL ZONE.

shifting cultivation (slash-and-burn agriculture) The traditional agricultural system of semi-nomadic people, in which a small area of forest is cleared by burning, cultivated for 1–5 years, and then abandoned as soil fertility and crop yields fall and weeds encroach. Ideally vegetation *succession subsequently returns the plot to *climax woodland, and soil fertility is gradually restored. Shifting cultivation of this type was once practised worldwide but in modern times it has been primarily associated with tropical rain-forest areas.

The system is best suited to low population densities. With increasing population pressure, abandoned plots are often cleared again before a full climax community has been restored, leading eventually to nutrient depletion of the system and degradation of forest to open *savannah-type woodland or *scrub.

shimmer (atmospheric boil) The continuously varying distortion of objects seen at low levels caused by variations in surface atmospheric properties, as over strongly heated ground.

shingle Beach pebbles, normally well rounded as a result of abrasion, whose diameters are typically 0.75–7.5 cm. They are made of resistant materials such as flint, which is the dominant constituent of the shingle beaches of south-eastern England, and they may show lateral sorting (e.g. the shingle of the Chesil Beach, Dorset, England, steadily increases in size over 29 km from west to east).

shoal retreat massif A large *sand accumulation that is preserved on the *continental shelf during and after a marine *transgression. The massifs represent former *estuary-mouth sand bars (inlet-associated shoals) or former zones of *longshore-drift convergence (cape associated shoals). In the Middle Atlantic Bight of the east coast of the USA the massifs are up to 70 km long by 20 km wide and can be found offshore from the present Hudson, Delaware, and Chesapeake estuaries, and longshore-drift convergence zones such as Cape Hatteras.

S

shock disease In animals, a response to overcrowding in which physiological changes to the endocrine (hormone-producing) system lead to physical deterioration, reduced reproductive success, and fighting, which may be followed by a catastrophic decline in numbers. Shock disease has been observed in many small mammals (e.g. field vole, *Microtus agrestis*), but is believed to be rare among invertebrates.

shoestring sand An irregular, sinuous *sand-body resembling a shoe-lace in shape, which often represents the preserved sandy deposit of a meandering river channel (*see* MEANDER).

shoot frequency *See* ROOT FREQUENCY.

shooting flow *See* CRITICAL FLOW and FROUDE NUMBER.

shoreface The subtidal coastal zone between the low-water mark and a depth of about 10–20 m. The lower limit of the shoreface corresponds to the position at which waves begin to affect the seabed (wave base), and therefore wave action governs the processes active in this area.

shore platform (marine platform, marine terrace, marine flat, marine bench, wave-cut bench, wave-cut platform) An inter-tidal bench cut into a land mass by the action of waves and associated processes. It is terminated landward by a sea cliff, and slopes gently seaward at about 1°.

short-day plant A plant in which flowering is favoured by short days and correspondingly long nights. There are two groups of such plants, some species in which there is an absolute requirement for these conditions for the onset of flowering, and other species in which flowering is merely hastened by them. In fact, however, the term is somewhat misleading in that the critical factor is the period of darkness and short-day plants require a night of more than a minimum duration. Horticulturists exploit this, for example, by exposing chrysanthemums to a brief flash of light during the night to delay flowering until Christmas.

short interdispersed element (SINE) A DNA sequence, of less than 500 *base pairs with a *copy number over 100 000, that occurs throughout the *genome rather than in a *tandem array.

shred In forestry, to remove the branches from a *standard tree.

shrub (bush) A woody plant which branches below or near ground level into several main stems, so has no clear trunk. It may be *deciduous (e.g. hawthorn) or *evergreen (e.g. holly). At the end of each growing season there is no die-back of the axes. *Compare* HERB; SUBSHRUB; and TREE.

shrub layer In the stratified structure of a forest, the layer of woody plants shorter than the *underwood, comprising *shrubs and saplings of those tree species that form the *canopy.

SI *See* SYSTÈME INTERNATIONAL D'UNITÉS.

Si *See* SILICON.

sib Shortened form of *'sibling'.

Siberian high The region of high average pressure in the colder seasons of the year over Siberia. The intensity of the high pressure is increased by the density of the very cold air at the surface over the Siberian plains.

sibling Offspring of the same parents; a brother and/or sister.

sibling species *Species that are identical in outward appearance or very nearly so. Despite the similarity, however, they qualify as species by being reproductively isolated. *See* REPRODUCTIVE ISOLATING MECHANISMS.

sibship All the *siblings in a family.

sichelwannen Smooth, crescent-shaped features cut into a flat or sloping rock surface by the action of a *glacier. They are typically 1–10 m long and 5–6 m wide. The horns of the crescents point down-glacier. Their origin is unclear; the action of saturated *till and of subglacial meltwater have both been invoked. The name is German, its literal meaning being 'sickle tubs'.

sieve deposit A well sorted, matrix-free conglomerate, which forms where the *sediment transported and deposited comprises only particles the size of pebbles and gravel.

sievert (Sv) The *SI unit of radiation dose equivalent, being a radiation dose absorbed of 1 joule per kilogram of mass, weighted to take account of the different effects of ionizing radiation on different biological tissues.

sigma-t density (σ-t density) The density of a sea-water sample measured at atmospheric pressure (i.e. at the sea surface). It is defined as density minus 1 000, where the density is measured in kg/m^3. Sea water at 0°C and 35 parts per thousand *salinity has a density of 1028 kg/m^3, or a σ-t density, at a pressure of 1 bar, of 28.

sigmoid growth curve *See* S-SHAPED GROWTH CURVE.

significant wave height The average height of the highest third of waves in a given group of waves. This height is usually used as a standard when describing the wave characteristics of a given area.

sign stimulus (releaser) Part of a stimulus that is sufficient to evoke a behavioural response in an animal (e.g. the patch of red colour that provokes an aggressive reaction in *Erithacus rubecula*, the European robin).

silage (ensilage) A type of foodstuff for livestock, prepared from green crops (e.g. grass); the crops are stored in a pit or silo and the *bacteria present on the plants carry out *fermentation (sometimes hastened by the use of chemical additives), the products of which preserve the plant material from further decay and loss of nutritional value.

silcrete *See* DURICRUST.

silent substitution A *substitution that does not affect the *phenotype of an organism. The term embraces synonymous substititions and substititions that are not in genes.

silica Silicon dioxide (SiO_2), which occurs naturally in three main forms: (*a*) crystalline silica includes the *minerals quartz, tridymite, and cristobalite; (*b*) cryptocrystalline or very finely crystalline silica includes some chalcedony, chert, jasper, and flint; and (*c*) amorphous hydrated silica includes opal, diatomite, and some chalcedony.

siliceous Containing *silica. Some kinds of wood are siliceous, and therefore blunt cutting tools.

siliceous ooze A fine-grained *pelagic deposit of the deep-ocean floor, which tends to occur at depths in excess of 4 500 m, of which more than 30 per cent consists of siliceous material of organic origin, mainly the remains of radiolarians (*see* RADIOLARIA) and *diatoms.

siliceous sinter A *silica-rich precipitate found around the mouth of a geyser or hot spring with water carrying large amounts of dissolved minerals, which are precipitated when the water cools suddenly on exposure to the atmosphere.

silicon **(Si)** An element which is required by certain plants in small quantities. It does not seem to play a role in metabolism but some plants accumulate a large amount (as *silica) in the walls of epidermal and vascular tissues, and thereby reduce water loss and retard fungal infection. Silica forms an important part of the cell walls of diatoms and is present in many timbers.

silt **1.** In geology, according to the most widely used (Udden–Wentworth) scale, particles between 4 μm and 62.5 μm in size. **2.** In pedology, mineral soil particles that range in diameter from 0.02–0.002 mm in the international system or 0.05–0.002 mm in the *USDA system. **3.** A class of soil texture.

siltstone A lithified *silt, comprising grains between 4 μm and 62.5 μm in size.

Silurian The third of six periods of the *Palaeozoic Era, approximately 443.7–416 Ma ago. Its end is marked by the climax of the Caledonian orogeny (mountain-building episode) and the filling of several Palaeozoic basins of deposition. It is the period during which land plants first appeared.

silver leaf A disease affecting a number of ornamental and fruit trees of the Rosaceae, especially plum. The characteristic symptom is the silvery sheen that appears on the leaves of infected trees. Branches show a dark brown discoloration and die back. The causal agent is the fungus *Chondrostereum purpureum*.

silver spoon effect The life-long reproductive advantage (i.e. increased fitness) enjoyed by an individual that had access to abundant resources during the early part of its life.

silviculture The management of forests or woodlands for the benefit of the entire *ecosystem, regardless of whether the land is being exploited commercially for the production of timber and other wood products. Sometimes the term is used interchangeably with 'forestry', but is more comprehensive. *Compare* ARBORICULTURE.

similarity coefficient Any measure of the similarity of two samples. In ecological work the similarity index devised in 1913 by J. Czekanowski has been widely used. This measures similarity as $C = 2W/(a + b)$, where a and b are the quantities of all the species (or other commodity) found in the two units to be compared, and W is the sum of the lesser values for those species common to both units. Complete similarity thus scores 1, complete dissimilarity 0. *See also* AFFINITY INDEX.

simoom A regional wind in desert areas of Africa and Arabia, which blows for only a short period at a time. Its whirlwind effect carries sand and brings very hot, dry conditions.

Simpson, George Gaylord **(1902–84)** An American palaeontologist who supported the idea of 'neo-Darwinian' evolution, and initially fiercely opposed the theory of *continental drift, basing his objections on his studies of fossil mammals, especially those of Madagascar. He proposed that the dispersion of species was caused by *'sweepstake routes'. He was a leading authority on the palaeontology of South American mammals, discoverer of *Hyracotherium* (the 'dawn horse'), and collaborated with Louis and Mary Leakey in their work on hominoid evolution. From 1936 he worked at the American Museum of Natural History, New York, eventually becoming chairman of its department of geology and palaeontology, and he was professor of vertebrate palaeontology first at Columbia University and later at Harvard. In 1970–82 he was a professor at the University of Arizona.

Simpson's diversity index An index of species diversity devised in 1949 by E. H. Simpson, given by $D = 1 - \Sigma p_i^2$, where D is the diversity index, p_i is the proportion of individuals in the ith species, and Σ means 'sum of'.

SINE *See* SHORT INTERDISPERSED ELEMENT.

single To reduce the regrowth from a *coppice stool to allow a single *pole to grow on to form a *standard tree.

singularity In meteorology, a short seasonal episode of weather that lasts a few days and commonly occurs about specific dates of the year (e.g. the periods of fine, dry, warm weather commonly occurring in September and early October in Britain and central Europe, and known as 'old wives' summer' (or 'Indian summer'), which are the result of slow-moving *anticyclones).

sink A natural reservoir that can receive energy or materials without undergoing change.

sinking *See* DOWNWELLING.

Sino-Japanese floral region Part of R. Good's (1974, *The Geography of the Flowering Plants*) *Boreal Realm, which corresponds geographically with the Sino-Himalayan-Tibetan mountains, northern and central China, and Japan. The region is differentiated by more than 300 endemic genera (*see* ENDEMISM), and contains in fact the richest flora in the northern temperate zone. A vast number of garden plants are native to the region, as are a number of important economic plants. *See also* FLORAL PROVINCE and FLORISTIC REGION.

sinuate Curved; having a wavy margin.

sirocco *See* SCIROCCO.

sister taxa In *phylogenetics, two taxa connected through a single *internal node.

site of special scientific interest (SSSI) In Britain an area which, in the view of the statutory authority, is of particular interest because of its fauna, flora, or geological or physiographic features. Once designated, the owner of the site is required to notify the relevant authorities and obtain special permission before undertaking operations that would alter its characteristics.

SI units (Système International d'Unités) The international system of units of measurement which evolved over a number of years from the m.k.s. system (based on the metre, kilogram, and second) and was finally adopted in 1960 by the 11th Conférence Générale des Poids et Mesures. The system is widely used in trade and is recommended for all scientific work. There are seven basic units: metre (m, length); kilogram (kg, mass); second (s, time); ampere (A, electric current); kelvin (K, thermodynamic temperature); candela (cd, luminous intensity); and mole (mol, amount of substance). There are also two supplementary units: radian (rad, plane angle); and steradian (sr, solid angle). From these, a further 18 units are derived: hertz (Hz, frequency); joule (J, energy); *newton (N, force); watt (W, power); *pascal (Pa, pressure); coulomb (C, electric charge); volt (V, electric potential); ohm (Ω, electric resistance); siemens (S, electric conductance); farad (F, electric capacitance); weber (Wb, magnetic flux); tesla (T, magnetic induction); lumen (lm, luminous flux); lux (lx, illuminance); *gray (Gy, absorbed radiation dose); *becquerel (Bq, radioactivity); and *sievert (Sv, radiation dose equivalent). See Appendix.

size structure In a plant community, the number or proportion of individual plants falling within particular size ranges (e.g. of forest trees with girths of less than 1 m, 1.0–1.5 m, 1.6–2.0 m, 2.1–2.5 m, etc.). *See also* STAND TABLE.

skia- (scia-, skio-) The Greek *skia*, meaning 'shadow', used as a prefix meaning 'pertaining to shade or darkness'.

skiaphilic *See* SKIOPHILIC.

Skinner, Burrhus Frederic (1904–90) An American psychologist, who pioneered the study of behaviourism and made important contributions to the understanding of *operant conditioning. He was drawn to behaviourism by the work of I. P. *Pavlov. Among his many publications, the best known include *The Behavior of Organisms* (1938), *Walden Two* (1948), and *Science and Human Behavior* (1953). He was professor of psychology at Indiana University (1945–48) and professor of psychology at Harvard University from 1948 until his retirement in 1974.

Skinner box In laboratory studies of animal behaviour, a cage in which an animal may learn that the performance of a particular activity (e.g. pressing a bar) is rewarded (e.g. with food so that its behaviour may be conditioned. It was invented by B. F. *Skinner. *See also* OPERANT CONDITIONING.

skio- *See* SKIA-.

skiophilic (skiaphilic, sciophilic) Shade-loving.

skiophilous (sciophilous) Applied to the response to light shown by organisms that have adapted to living entirely in the shade.

slack *See* DUNE SLACK.

slaking The breaking up of earth materials when they are exposed to water or air.

slash-and-burn agriculture *See* SHIFTING CULTIVATION.

slate Low-grade, regionally metamorphosed rock (*see* METAMORPHISM and METAMORPHIC ROCK), which is highly fissile and fine-grained. The fissility results from the parallel alignment of numerous fine *minerals. The smooth, hard, impermeable surface produced when slate is split makes it commercially valuable for roofing, cladding buildings, and for making such items as billiard-table tops, laboratory benches, and blackboards.

sleep movements *See* NICTONASTY.

sleet *Precipitation in the form of a mixture of rain and melting snow. In North America the term is applied to ice pellets of less than 5 mm in diameter.

slick A quiescent area on the surface of a water body; the relative smoothness of such an area compared to adjacent waters is usually caused by a thin surface film of oil which changes the surface tension.

slickenside **1.** A set of linear marks on a fault or bedding plane caused by the frictional movement of one rock body against another. The plane may be coated by a *mineral (often quartz or calcite) which is also striated in the direction of movement. **2.** The polished surface left by the passage of a mud slide. **3.** In clay soils, the natural crack surfaces produced by swelling and shrinkage.

sling psychrometer *See* WHIRLING PSYCHROMETER.

slip-off slope *See* MEANDER.

slop *Poles produced by *coppice.

slope profile The two-dimensional form of a hill-slope when measured down the steepest gradient, traditionally divided into a number of units, each reflecting a distinctive geomorphological process. For example, in 1957 L. C. King identified four elements in his ideal profile: a crest (or 'waxing slope' or 'convex slope') dominated by *creep; a *scarp (or 'free face') affected by *rill activity and *mass movement; a debris (or 'constant') slope where *talus accumulated; and a *pediment (or 'waning slope') modified by *sheet-wash. Subsequently, a nine-unit model has gained some acceptance.

SLOSS principle An acronym for Single Large Over Several Small, the proposal that a single large *nature reserve will contain more species than several small reserves with the same total area. Whether this is so depends on the extent to which the species in the small reserves overlap. If there is little overlap between reserves (i.e. each small reserve contains many species that are not found in any of the other reserves), several small reserves may be preferable to a single large one. Small areas of *habitat may also afford more protection against disease or the accidental introduction of predators.

slump structure A sedimentary structure consisting of overturned folds, formed by the mass sliding down-slope of the semi-consolidated *sediment.

SMC *See* SATURATION MOISTURE CONTENT.

smog Naturally occurring fog mixed with visible (smoke) and/or invisible pollutants. *See also* PHOTOCHEMICAL SMOG.

SMOW (Standard Mean Ocean Water) A sea-water sample which comprises the international standard for D:H and $^{18}O:^{16}O$ ratios. Differences in isotopic composition are expressed as parts per mille deviations from the isotopic composition of this standard.

smolt The stage in the life of salmon and similar fish in which the sub-adult individuals acquire a silvery colour and migrate down the river to begin their adult lives in the open sea. *See also* PARR.

smudge A disease which can affect onions, shallots, and leeks. Dark patches, often in the form of concentric rings, appear on the bulbs. The causal agent is the fungus *Colletotrichum circinans*.

smudging The burning of materials (e.g. oil) to produce a smoke layer that reduces the effect of radiation cooling of the air above the ground surface. It is used as a protective measure (e.g. in fruit-growing areas, especially in frost hollows).

smut A plant disease caused by a fungus of the order Ustilaginales. Many types of plant can be affected, but smuts are particularly important in cereals and other grasses. The symptoms include the formation of masses of black soot-like *spores, and infected plants often show some degree of distortion.

snag The standing part of a tree trunk that has snapped above ground level.

snout The steep, terminal zone of a *glacier; it is usually heavily loaded with debris.

snowblitz theory A theory which proposes that following a severe winter with heavy snowfall, snow persists in lowland areas throughout the summer. This increases the *albedo and thus reduces the amount of solar warming of the ground. More snow is added during the next winter, and more snow may thus accumulate year by year. An *ice-cap may develop, and glaciation may occur after only a few hundred years.

snowflake The result of the accumulation of ice crystals in a varied array of shapes. Very low temperatures usually result in small flakes; formation at temperatures near freezing point produces numerous crystals in large flakes.

snow gauge A device for measuring the amount of snowfall. It is a modified *rain gauge, in some designs with the collecting funnel heated to melt the snow and collect it as liquid. Snowfall can also be measured by pressing an open-ended cylinder vertically through the snow to remove a core and transferring the core to a measuring beaker calibrated to correct for the difference between the diameters of the beaker and coring cylinder. A ruler will also measure the depth of snow. Knowing the air temperature at the time the snow fell allows the depth to be converted to an equivalent depth of liquid water. Snowfall is always reported as its rainfall equivalent, most simply by melting the snow, if the information is needed for meteorological purposes, rather than for other uses such as reporting road conditions. This is because snowflakes vary greatly in size, depending on temperature, and the smaller they are the more air they trap between them, increasing the bulk of the snow.

snow grain A small ice particle usually precipitated as a flattish grain.

snow line The lower limit of permanent snow cover. The height of the line varies with latitude; locally it also varies with aspect, because of the relationship to prevailing winds and the quantity of snow deposited, and to summer temperatures, etc. *See also* FIRN LINE.

snow-patch vegetation In *tundra and to a lesser extent in alpine environments, late-melting snow patches exert a marked influence on the vegetation. Vegetation that occurs beneath the small snow patches and the peripheral parts of large ones is protected from the rigours of winter, but as the snow melts comparatively early in the summer months, the vegetation is also able to capitalize on most of the *growing season: in these situations, therefore, it is rather luxuriant. Conversely, the larger snow patches, which melt well into or towards the end of the short growing season, tend greatly to restrict the development of vegetation.

soboliferous Forming clumps.

sociability scale A five-point scale used in vegetation analysis to indicate the degree of clumping or gregariousness of an individual plant species, obtained as a visual impression. 1 on the scale indicates a shoot growing singly; 5 indicates shoots growing in large mats or pure populations.

social behaviour The interactive behaviour of two or more individuals, all of which belong to the same species.

social facilitation Intensification (i.e. *facilitation) of a behaviour (e.g. feeding) that is associated with increased population density.

social wasps The common name most usually applied to wasps belonging to the subfamilies Vespinae, Rhopalidiinae, Polistinae, and Polybiinae of the family Vespidae (superfamily Vespoidea).

society A group of individuals, all of the same species, in which there is some degree of *co-operation, *communication, and division of labour.

sociobiology The integrated study of social behaviour, based on the premiss that all behaviour is adaptive. It gives special emphasis to social systems considered as ecological adaptations, and attempts a mechanistic explanation of social behaviour in terms of modern biology and, in particular, evolutionary theory.

sodication In soils, an increase in the percentage of exchangeable sodium. Sodium adsorbs on to soil *cation-exchange sites, causing soil aggregates to disperse, which closes soil *pores and renders the soil impermeable to water. *See also* SODIUM ADSORPTION RATIO.

sodic soil **1.** Soil with a sodium content sufficiently high to interfere with the growth of most crop plants. **2.** Soil with more than 15 per cent exchangeable sodium.

sodium **(Na)** An element that is found in all terrestrial plants. It is not essential except in certain C_4 salt-tolerant plants, but it has a role in crassulacean acid metabolism. In animals, it is the principal circulating *cation in body fluids and contributes to maintaining physiological excitability in tissues. Sodium is the one element that is essential for animals, which need it for nerve function, but that is not required or accumulated by most plants. Since it is abundant in nature, particularly in oceanic areas, sodium supply in water or soil is not normally a problem, but animals living in continental interiors may find difficulty in obtaining sufficient of the element. Moose (*Alces alces*) in central Canada, for example, become short of sodium, but during the summer season they obtain a supply from aquatic plants, which contain more sodium than terrestrial plants.

sodium adsorption ratio **(SAR)** The tendency for sodium *cations to be adsorbed at *cation-exchange sites in soil at the expense of other cations, calculated as the ratio of sodium to calcium and magnesium in the soil (as the amount of sodium divided by the square root of half the sum of the amounts of calcium and magnesium, where ion concentrations are given in moles per gram). A low sodium content gives a low SAR value. In practice, allowance must be made for other reactions within the soil which do not involve sodium but affect concentrations of calcium and magnesium. The SAR value is most likely to be changed by irrigation water.

SOFAR channel **(sound channel)** An acronym for the *SO*und *F*ixing *A*nd *R*anging channel, a zone in the oceanic water column at a depth of about 1 500 m, where the velocity of sound is at a minimum value. Sound passing through the zone is refracted upwards or downwards back into the zone, with little loss of energy, causing sound energy to be trapped in a zone of well-defined depth. The SOFAR channel may be used for the transmission of sound over long distances, exceeding 28 000 km, and can be used to track free-drifting, subsurface, neutrally buoyant floats.

soft rot Any of a range of plant diseases in which the tissues of the infected plant are softened or even liquefied, typically becoming slimy and malodorous. A common cause of soft rot in carrots and other vegetables is bacterial infection by *Erwinia carotovora*.

softwood The wood of coniferous trees, comprising fibres and tracheids (vessels being absent). *Compare* HARDWOOD.

soil **1.** The natural, unconsolidated, mineral and organic material occurring on the surface of the Earth; it is a medium for the growth of plants. **2.** In engineering geology, any loose, soft, and deformable material (e.g. unconsolidated *sands and *clays).

soil air The soil atmosphere, comprising the same gases as the atmosphere above ground, but in different proportions: it occupies the *pore space of the soil.

soil association **1.** A group of soils forming a pattern of soil types which are characteristic of a geographical region. **2.** A mapping unit used to denote the distribution of soil types when the scale of the map does not require or permit the identification of individual soils. *See also* SOIL COMPLEX.

soil borrow The transference of material from elsewhere for refilling excavations, etc.

soil classification A system, analogous to those used in biological *taxonomy, that arranges soil types according to their distinguishing characteristics. Russian scientists were the first to attempt to classify soils in the latter part of the 19th century (*see* DOKUCHAEV, VASILY VASILIEVICH) and many Russian soil names are still used (e.g. *chernozem, *solonetz, and *podzol). By 1975 American scientists at the US Department of Agriculture had devised a classification they called *Soil Taxonomy. This system is widely used outside the United States. There are also many national classifications. In 1961 representatives from the Food and Agriculture Organization (FAO) of the United Nations, the United Nations Educational, Scientific, and Cultural Organization (UNESCO), and the International Society of Soil Science (ISS) met to discuss preparing an international classification. This was completed in 1974, updated in 1988, and has been amended several times since. Based on *diagnostic horizons, it divides soils into 30 reference groups and 170 possible subunits.

soil complex A mapping unit that is used to denote the distribution of soils. It is more precise than a *soil association and is used where soils of different types are mixed geographically in such a way that the scale of the map makes it undesirable, or impractical, to show each one separately.

soil conservation The protection of the soil by careful management to prevent physical loss by erosion and to avoid chemical deterioration (i.e. to maintain soil fertility).

soil formation The action of the combined primary (*weathering and humification) and secondary processes to alter and to rearrange mineral and organic material to form soil, involving the differentiation of soil *profiles and the formation of loose soil from consolidated rock material. *See also* PEDOGENESIS.

soil horizon A relatively uniform layer of soil, more or less parallel to the soil surface, which is physically, chemically, and/or biologically distinguishable from the layers above and below it. Soil horizons are grouped primarily into O, A, B, and C horizons. O horizons (formerly known as Ao horizons) comprise organic material at the surface. A horizons are surface horizons of mixed organo-mineral composition. Where mineral matter has been lost the A horizon is sometimes called the E (for eluviated) horizon. Where they are present B horizons are usually located in the middle of the sequence, and they are horizons into which material (mineral and organic) is deposited, thus altering the character of the horizon. C horizons are soil *parent materials, weathered but not otherwise altered by pedogenic processes. The underlying unweathered material is sometimes called the D or R horizon. In addition to these, surface litter

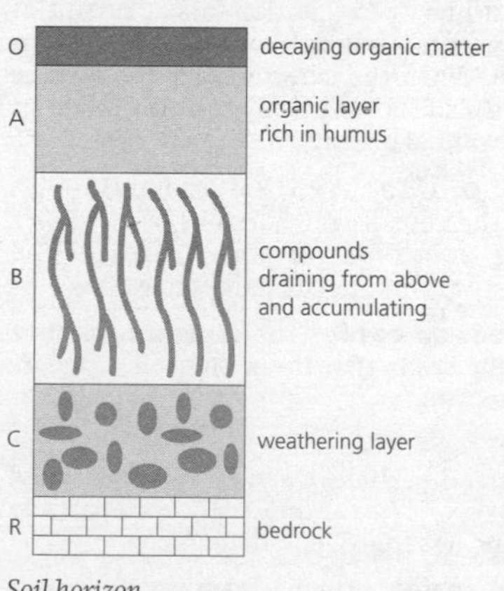

Soil horizon

S

may form an L horizon, above a layer of fermented material (F horizon) and, below that, humified material (H horizon), and a mineral crust, often cemented, is sometimes called the K horizon. *See* PEDOGENESIS.

soil individual *See* POLYPEDON.

soil management A variety of practices and operations with respect to soil which aid the production of plants; normally they are planned to allow for sustained yield in the future.

soil-moisture content The ratio of the volume of contained water in a soil compared with the entire soil volume. When a soil is fully saturated, water will drain easily into the underlying unsaturated rock. When such drainage stops, the soil still retains *capillary moisture and is said to contain its *field-capacity moisture content. Further drying of the soil (e.g. by evaporation) creates a soil-moisture deficit, which is the amount of water which must be added to the soil to restore it to field capacity, measured as a depth of *precipitation.

soil-moisture deficit *See* SOIL-MOISTURE CONTENT.

soil-moisture index *See* MOISTURE INDEX.

soil-moisture regime The changing state of soil moisture through the year, which reflects the changing balance of monthly *precipitation and *potential evapotranspiration above the ground surface. When the latter exceeds the former the period is one of soil-moisture deficit in the annual regime.

soil profile A vertical section through all the constituent horizons of a soil (*see* SOIL HORIZON), from the surface to the relatively unaltered *parent material.

soil seed bank The ungerminated but viable seeds that lie in the soil. *Compare* SEED BANK.

soil separates The size divisions of the mineral particles (*sand, *silt, and *clay) that comprise the fine earth, each particle being less than 2 mm in diameter.

soil series The basic unit of soil mapping and classification, comprising soils all of which have similar profile characteristics and developed from the same *parent material.

soil structure The grouping of individual soil particles into secondary units of *aggregates and *peds; this grouping is like an internal scaffolding of the soil.

soil survey **1.** The systematic examination and mapping of soil in the field. **2.** The chemical analysis of soil in order to detect geochemical anomalies.

soil taxonomy The system of *soil classification that was devised by workers in the US Soil Survey, within the *USDA. It divides soils into 11 orders: *Alfisols, *Andisols, *Aridisols, *Entisols, *Histosols, *Inceptisols, *Mollisols, *Oxisols, *Spodosols, *Ultisols, and *Vertisols. These orders are further subdivided into suborders, great groups, families, and soil series, defined by *diagnostic horizons.

soil variant A soil, the properties of which are sufficiently different from those of adjacent soils to justify the use of a new *soil-series name, but which occupies too small a geographical area to warrant the issuing of such a new name.

soil-wash *See* RAIN-WASH.

soil-water zone **(unsaturated zone, vadose zone)** The zone between the ground surface and the *water-table. Water is able to pass through this zone to reach the water-table, but while in the zone it is not given up readily to wells because it is held by soil or rock particles and capillary forces. *See* GROUNDWATER.

sol A substance composed of small solid particles dispersed in a liquid to form a continuous, homogeneous phase (i.e. a type of colloid, e.g. a completely fluid mud).

sola *See* SOLUM.

solarization **(heliosis)** The inhibition of *photosynthesis at very high light intensities, due mainly to the photo-oxidation of certain of the compounds involved.

soldier A sterile member, specialized for defence, of an ant or termite colony. The head is usually enlarged, and it or the tip of the abdomen may be used to block entrance holes to the nest (phragmosis).

solifluction **(solifluxion)** The downhill movement of *regolith that has been saturated with water. It was originally described in *periglacial regions (*see* GELIFLUCTION), but the term was subsequently widened to include all environments. The thick regolith of the humid tropics is particularly prone to solifluction after intense rainfall.

solifluxion *See* SOLIFLUCTION.

soligenous mire A *mire that receives water from rain and slope run-off. *See* AAPA MIRES.

solitaria *See* LOCUST.

solitary bees The majority of bees, which are not social (i.e. there is no worker caste and the bees do not live in colonies). Each nest is the work of an individual female and, although nests may be aggregated, there is no co-operation among individuals.

solitary wasps Wasps of the suborder Apocrita (order Hymenoptera) that do not live in colonies (e.g. potter wasps, mason wasps, and spider-hunting wasps). Mason wasps make nests from particles of sand and soil stuck together with saliva. The nests are found in the soil, in crevices in wood, and in stone walls, and are individually provisioned with paralysed prey for the larva.

solodic soil Leached, formerly saline soil, associated with semi-arid tropical environments, in which the A *horizon has become slightly acid, and the B horizon is enriched with sodium-saturated *clay. The term was used in soil classification systems derived from early Russian systems based on the work of V. V. *Dokuchaev, but it is not included in the *USDA Soil Taxonomy.

Solonchaks Soils with a layer enriched in soluble salts that is more than 15 cm thick and lies at the surface or only a little way below it. Solonchaks are a reference soil group in the *FAO *soil classification.

solonetz *Mineral soil at a transitional stage of *leaching or solodization (*see* SOLODIC SOIL) of saline soils, in semi-arid, tropical environments, which has a *sandy, acid A *horizon and a B horizon partially enriched with sodium *clay. The term was used in early systems of *soil classification, but in the *USDA *soil taxonomy solonetz soils are included in the order *Aridisols and the name is not used. Solonetz are a reference soil group in the *FAO *soil classification, however.

solum **(pl. sola)** The upper part of a *soil profile, above the *parent material, in which processes of soil formation occur, and within which most plant roots and soil animals are found.

solute potential *See* OSMOTIC POTENTIAL.

solution channel An elongate void within a rock, which has been enlarged by the solution of rock by moving *groundwater. Solution channels are most commonly associated with carbonate rocks, and groundwater in them can flow as fast as in a river on the ground surface. *See* KARSTIC AQUIFER.

solution pipe A cylindrical, near-vertical pipe that is developed at an intersection between rock joints in a *karst environment. It is caused by a local increase in the rate of carbonation resulting from enhanced drainage.

sombric horizon A subsurface *soil horizon of well drained, *mineral, tropical and subtropical soils into which *humus has leached downwards. Base saturation is less that 50 per cent. It is a *diagnostic horizon.

song A complex pattern of sound that is produced by means of specialized organs, which are used by many animals in communication.

sonication The emission of ultrasound in order to disrupt cells that have clumped together or to break down cell walls. The technique has many applications, including the treatment of waste water. Certain plants, e.g. potatoes (*Solanum* species), produce flowers that release pollen only when stimulated to do so by sonication from pollinating insects.

sooty mould A dark, soot-like, fungal growth which appears on plants infested with sap-sucking insects such as aphids or scale insects; these insects produce a sugary fluid ('honeydew'), and the sooty-mould fungi grow on this. Sooty moulds may belong to various genera of the Dothideales.

sorting An expression of the range of grain sizes present in a *sediment. A well sorted sediment is characterized by a narrow range of grain sizes; a poorly sorted sediment contains a wide range of grain sizes.

sound channel *See* SOFAR CHANNEL.

source region (for air masses) An extensive area of land or water over which essentially uniform surface conditions prevail and where *air masses develop their initial properties.

source rock **1.** A *sediment (usually *shale or limestone) in which hydrocarbons originate; it contains more than 5 per cent organic matter and has the potential to generate petroleum. **2.** Any parent rock from which later sediments are derived.

south Brazilian floral region Part of R. Good's (1974, *The Geography of the Flowering Plants*) *Neotropical floral kingdom, which contains a large flora with about 400 endemic genera (*see* ENDEMISM); a high proportion of its several thousand species is also endemic. Many important economic and ornamental plants derive from this region (e.g. cassava). *See also* FLORAL PROVINCE and FLORISTIC REGION.

southerly burster A regional wind in southern and south-eastern Australia, which is characterized by a rapid shift in direction from north-west to south in the rear of a *cold front. Such winds are especially prevalent between October and March. The change to a southerly wind can bring a great increase in wind speed accompanied by a rapid and marked fall in temperature. Such conditions are akin to *line squalls, and are related to the South American *pamperos.

southern bacterial wilt *See* GRANVILLE WILT.

southern beech *See* BICENTRIC DISTRIBUTION.

Southern Ocean *See* ANTARCTIC OCEAN.

southern oscillation A fluctuation of the intertropical atmospheric circulation, in particular in the Indian and Pacific Oceans, in which air moves between the south-eastern Pacific subtropical high and the Indonesian equatorial low, driven by the temperature difference between the two areas. The general effect is that when pressure is high over the Pacific Ocean it tends to be low in the Indian Ocean, and vice versa. The phenomenon is strongly lined to *El Niño.

south temperate oceanic-island floral region Part of R. Good's (1974, *The Geography of the Flowering Plants*) Antarctic floral kingdom, which contains a very small flora scattered among the islands of the oceans surrounding Antarctica. There are only two endemic genera (*see* ENDEMISM), one of which is disjunct between several islands. Despite the great distance separating the islands, and their varying latitudes, there is a notable floristic constancy. A characteristic vegetation is tussocky grassland, dominated by *vicarious species of *Poa*. *See also* FLORAL PROVINCE and FLORISTIC REGION.

south-west Australian floral region Part of R. Good's (1974, *The Geography of the Flowering Plants*) Australian floral kingdom, which is a very rich floral region with a high degree of *endemism, in many respects rivalling that of the Cape region of South Africa. The same families are prominent in both floras and they have many growth forms in common. *See also* FLORAL PROVINCE and FLORISTIC REGION.

spacer DNA The DNA that occurs between genes.

spatial summation *See* SUMMATION.

spatulate Having an end that is broad and flattened, like a spatula.

special adaptation *See* GENERAL ADAPTATION.

special creation The belief that the origin of life and the diversity of life result from acts of God whereby each species was created separately. *Evolution is implicitly rejected as the explanation of these phenomena. *See* CREATIONISM and CREATION 'SCIENCE'.

specialization A degree of *adaptation of an organism to its environment. A high degree of specialization suggests both a narrow *habitat or *niche and significant *interspecific competition.

speciation The separation of populations of plants and animals, originally able to interbreed, into independent evolutionary units which can interbreed no longer, owing to accumulated genetic differences. In *cladistics, the origin of one or more new species occurs inferentially by *cladogenesis.

species **(sing. and pl.)** Literally, a group of organisms that resemble one another closely: the term derives from the Latin *speculare*, 'to look'. In taxonomy it is applied to one or more groups (populations) of individuals that can interbreed within the group but cannot exchange genes with other groups (populations), or, in other words an interbreeding group of biological organisms which is isolated reproductively from all other organisms (*see* BIOSPECIES). A species can be made up of groups in which members do not actually exchange genes with members of other groups (though in principle they could do so), as, for example, at the two extremes of a continuous geographical range. However, if some gene flow occurs along a continuum, the formation of another species is unlikely to occur. Where barriers to gene flow arise (e.g. physical barriers, such as sea, or areas of unfavourable habitat) this reproductive isolation may lead by either local selection or random genetic drift to the formation of morphologically distinct forms termed races or subspecies. These could interbreed with other races of the same species if they were introduced to one another. Once this potential is lost, through some further evolutionary divergence, the races may be recognized as species, although this concept is not a rigid one. Most species cannot interbreed with others; a few can, but produce infertile offspring; a smaller number may actually produce fertile offspring. The term cannot be applied precisely to organisms whose breeding behaviour is unknown. *See* MORPHOSPECIES and PALAEOSPECIES.

species–area curve *See* MINIMAL AREA.

species diversity *See* DIVERSITY.

species group **(superspecies)** A complex of related species that exist *allopatrically (in different geographical areas from one another). They are grouped together because of their morphological similarities, and this grouping can often be supported by experimental crosses in which only certain pairs of species will produce *hybrids.

species longevity The persistence of species for long periods of time.

species richness The number of species present in a *community, mea-

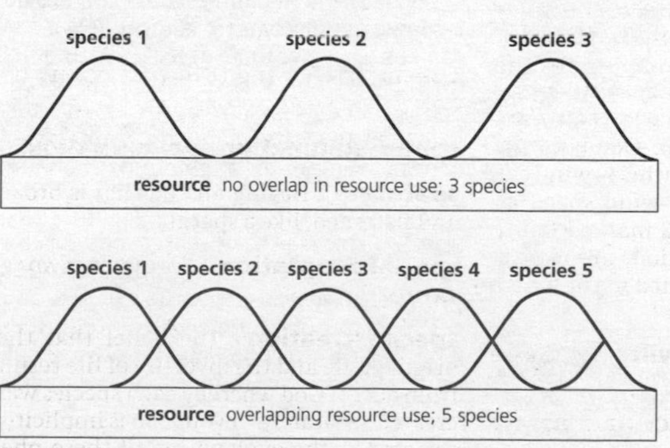

Species richness

sured as the number of species per unit of ground area. This contrasts with *diversity, which takes account of the way individuals are apportioned among the different species. Several factors influence species richness, the most important being the resources available and the extent to which species overlap in their exploitation of them. More resources will support more species, and the more that species overlap in their use of resources the more species the habitat will support. *Predation also affects richness, by preventing species from reaching their *carrying capacity and thereby allowing more species to share the resources.

species selection *See* ORTHOSELECTION.

species tree A *phylogenetic tree which depicts the evolutionary relationships of a set of species. Inferred trees, based on gene trees or other character trees, are often presented as estimates of the true species tree.

specific humidity The mass of water vapour in a unit mass of air. *See also* HUMIDITY; MIXING RATIO; and RELATIVE HUMIDITY.

specific retention The ratio of the undrained water to the total water in a rock, the undrained water being water contained in rock voids or *pore spaces, from which it cannot be recovered by drainage or pumping. It is retained against the action of gravity by molecular attraction and *capillarity.

S

specific yield **1.** The ratio of the water drained from a rock under the influence of gravity, or removed by pumping, to the total volume of the rock voids or *pore space in the drained rock. The difference is caused by the retention of water in the rock, owing to molecular attraction and *capillarity. *See* SPECIFIC RETENTION. **2.** The volume of water released by a falling *water-table from a given volume of fully saturated rock.

spelean Pertaining to caves.

spermaceti Highly vascular tissue, filled with a mixture of fatty esters (oils) which solidify at about 32°C, produced by sperm whales (Physeteridae) and stored in the upper part of an enlarged snout. A full-grown sperm whale may carry one tonne of spermaceti. The function of spermaceti is not known with certainty, but since its density changes as it liquefies and solidifies it may permit heat loss from the body and help provide neutral buoyancy to the animal, and it may contribute to the production of the sounds by which the animal echolocates.

sperm competition The competition to fertilize an ovum which occurs among sperm from different males when the female has mated with more than one partner. It is believed that certain sperms, different in appearance from ordinary sperms and incapable of fertilizing the ovum, but present in large numbers, are involved in repelling or disabling sperms from the rival male.

Sphagnum A genus of mosses, distributed worldwide, that are found, often abundantly, in wet, acidic *habitats (*bogs, *marshes, pools, *moors, wet woodland, damp grassland, etc.) There are many species, which are often difficult to distinguish. The plants are characteristically branched, with branches in fascicles of 2–8. The leaves are nerveless and composed of two main types of cell: narrow, green, living cells and inflated, colourless, dead cells. The dead cells readily fill with water, allowing the plant to hold many times its own weight of water. The capsules are roughly spherical; when ripe, the capsule wall shrinks when it dries, increasing the internal pressure until the lid is blown off, ejecting the spores. *Sphagnum* can absorb water up to at least 20 times its own dry weight. It is a *bryophyte (moss) genus that covers more of the Earth's surface than any other plant, and probably accounts for more plant *biomass than any other genus.

spheroidosis A disease of insects caused by viruses of the entomopoxvirus group. The disease may affect larval and other stages of beetles (Coleoptera), flies (Diptera), butterflies and moths (Lepidoptera), and grasshoppers, locusts, and crickets (Orthoptera).

spicate In spikes.

spicule A small needle or spine.

spilling breaker (surf wave) An over-steepened wave in which the unstable top of the wave spills down the front of the wave-form as it advances into shallower water, so that the wave gradually diminishes in height until it moves up the beach as *swash.

spillway A *glacial drainage channel cut by water during glaciation, of which there are two principal varieties: (*a*) channels cut by water escaping from a glacially impounded lake (*overflow channel); (*b*) channels cut by meltwater released from a *glacier (*meltwater channel) or cut by streams and rivers deflected by a glacier. Impressive examples were developed in central Europe (the Urstromtäler of northern Germany and Pradoliny of Poland) when the Scandinavian ice sheet diverted streams flowing north from the southern highlands.

spinney A small wood, originally meaning a wood that consists, or formerly consisted, of thorns (*Crataegus*).

spire A young timber tree, the lowest branch of which is at a considerable height.

spissatus The Latin *spissatus*, meaning 'thickened', used to describe a species of *cirrus cloud which has sufficient thickness to appear grey even when the cloud is between the Sun and the observer. *See also* CLOUD CLASSIFICATION.

spit An elongated accumulation of sand or gravel that projects from the shore into a water body. *Longshore drift of material is usually responsible for the development of a spit.

splicing The process of *intron removal from *messenger RNA before it reaches maturity.

split gene A gene that contains *introns.

spodic horizon A subsurface *soil horizon in which organic matter together with aluminium and often iron compounds have accumulated amorphously. It is a *diagnostic horizon in the *USDA Soil Taxonomy. The name is from the Greek *spodos*, meaning wood.

Spodosols An order of soils in which subsurface *horizons contain amorphous materials comprising organic matter and compounds of aluminium and often iron that have accumulated illuvially (*see* ILLUVIATION). Such soils form in acid material, mainly coarse in texture, in humid cool to temperate climates, often beneath coniferous forests.

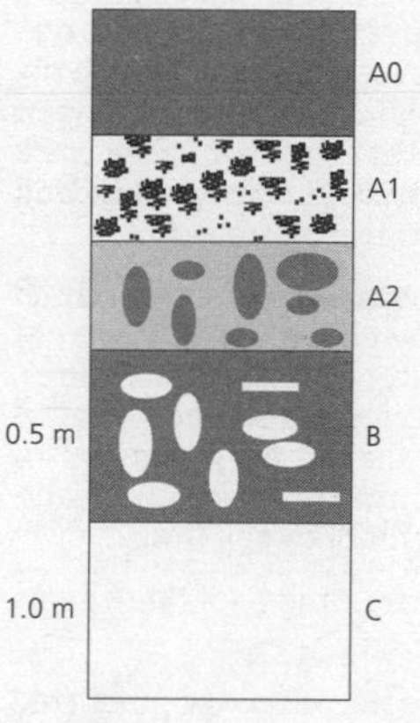

Spodosols

sporangium (pl. sporangia) A plant structure in which *spores are formed.

spore **1.** A reproductive unit, usually consisting of a single *haploid cell, that is capable of developing into a new organism without fusing with another cell. The release of spores is the main method of dispersal in fungi, algae, bryophytes, and pteridophytes. **2.** A differentiated bacterial cell which may function as a *propagule or as a resistant structure that allows the organism to survive adverse environmental conditions, often for protracted periods.

sporinite *See* COAL MACERAL.

sporophyte The *spore-producing *diploid generation in the life cycle of plants. In higher plants, such as angiosperms and gymnosperms, the sporophyte is the dominant generation, forming the conspicuous plant. In lower plants, such as mosses, liverworts, and hornworts, the *gametophyte is the dominant and conspicuous generation. *See also* ALTERNATION OF GENERATIONS.

Sporozoa *See* PROTOZOA.

sporozoans *See* PROTOZOA.

sport In plants, a sudden deviation from type; a *mutation. Many horticultural varieties of plants are sports and have been propagated vegetatively.

SPOT (Système Probatoire d'Observation de la Terre) A series of five French observation satellites, launched between 1986 and 2002, that scan a surface track 67 km wide with 20 m multispectral and 10 m monospectral resolution.

spraing A disease of potatoes, which may be caused by either of two different viruses, or by adverse growth conditions. Tubers from infected plants show characteristic crescent-shaped brown marks in the flesh when cut. The viruses are transmitted by eelworms or by soil fungi.

spray Wood trimmings, sold in faggots for kindling.

spread In the terminology preferred by some ecologists, the extension of a species's *range in response to climate change, in contrast to the more regular movements described as migration.

spring **1.** A flow of water above ground level that occurs where the *water-table intercepts the ground surface. Where the flow from a spring is not distinct (i.e. it does not give rise to obvious trickles) but tends to be somewhat dispersed, the flow is more correctly termed a 'seep'. The reappearance of surface water that had been diverted underground in a *karst region is a type of spring known as a 'resurgence'. A major variety is the 'Vauclusian spring', named after the Fontaine de Vaucluse, southern France, and descriptive of the upward emergence of an underground river from a flooded *solution channel. **2.** *See* SPRINGWOOD.

spring sapping The erosion of a hillslope around the site where a *spring emerges. The processes involved may include the collapse of saturated material, surface stream *erosion, and chemical *weathering. Spring sapping occurs towards the bases of chalk escarpments in southern England, where its effect may have been enhanced by frost activity under former *periglacial conditions.

spring tide A tide of greater than the mean range (i.e. the water level rises markedly above and falls markedly below the mean tide level). Spring tides occur about every two weeks, when the Moon is full or new, and are at their maximum when the Moon and the Sun are in the same plane as the Earth. *Compare* NEAP TIDE.

springwood (spring) **1.** High forest that has grown from shoots from tree stumps. **2.** The ground growth or shoots of new *coppice that emerge from the stools of a felled coppice. *See also* TREE RING.

SPS *See* GLOBAL POSITIONING SYSTEM.

spur A ridge that descends towards a valley floor from the higher ground above. It may be due to an outcrop of resistant rock, or it may develop on the concave side of a winding stream as a result of incision.

squall A short-lived weather condition with strong winds, which increase by at least 16 knots (30 km/hr). It may include thunder and heavy precipitation. *See also* SQUALL LINE.

squall line A series of *squalls, in which the wind speed increases by up to 50 per cent then dies away gradually, associated with very vigorous *cumulonimbus clouds that merge to form a continuous line up to 1 000 km long and advance at right angles to the line. The line begins to form ahead of a *cold front that is advancing beneath warm, moist air. Each storm cloud lasts for only an hour or two before exhausting its supply of moisture and dissipating, but the gust front produced by cold, subsiding air leaving the cloud undercuts warm air ahead and to the right of the cloud, triggering the formation of a new cloud. The squall line becomes detached from the cold front and advances ahead of it into the warm air.

squamulose Bearing or consisting of small scales (squamules).

S-shaped growth curve (sigmoid growth curve) A pattern of growth in which, in a new environment, the population density of an organism increases slowly initially, in a positive acceleration phase; then increases rapidly, approaching an exponential growth rate as in the *J-shaped curve; but then declines in a negative acceleration phase until at zero growth rate the population stabilizes. This decline reflects increasing environmental resistance which becomes proportionately more important at higher population densities. This type of population growth is termed density-dependent, since growth rate depends on the numbers present in the population. The point of stabilization, or zero growth rate, is termed the saturation value (symbolized by *K*) or *carrying capacity of the environment for that organism. *K* represents the upper asymptote of the sigmoidal or S-shaped curve produced when changing population numbers are plotted over time. It is usually summarized mathematically by the *logistic equation. *See* DENSITY-DEPENDENCE; *compare* J-SHAPED GROWTH CURVE.

SSSI *See* SITE OF SPECIAL SCIENTIFIC INTEREST.

stability **1.** An atmospheric condition in which air that is forced to rise tends to return to its pre-existing level in the absence of the uplifting force. If the *adiabatic lapse rate of uplifted air is greater than the *environmental lapse rate, the vertically displaced air will become colder than the surrounding air and as its density increases it will tend to sink back. *Compare* INSTABILITY. **2.** In an *ecosystem, either inertia (i.e. the ecosystem's resistance to perturbation) or resilience (i.e. the speed with which a perturbed ecosystem returns to its original state). There is considerable debate regarding the relationship between stability and such ecosystem features as *diversity.

stabilizing selection (maintenance evolution, normalizing selection) *Natural selection that selects against extremes at either end of the range of *phenotypes (e.g. in humans babies with low or high birth weights are less likely to survive than those with a birth weight close to 3.4 kg). It occurs in environments that change little from one location to another or over time. *Compare* DIRECTIONAL SELECTION and DISRUPTIVE SELECTION.

stable isotope Any naturally occurring, non-radiogenic isotope of an element. Many elements have several stable isotopes.

stack A pillar or block of rock, with near-vertical sides, which stands adjacent to a present or former sea cliff. Typically, it has been isolated from the main cliff by wave erosion concentrated along steeply inclined joints or faults.

stade (stadial in continental European usage) A term that is difficult to define with precision, but which refers to a single period of increased cold or advancing ice, forming a subdivision of a cold stage within the overall division of a glacial period into periods of cold interspersed with warm, or warmer, periods.

stadial *See* STADE.

staff gauge A graduated pole or board that is placed in or beside a watercourse and from which it is possible to measure directly the height of the water surface relative to a known datum elevation.

stage **1.** The elevation of the water surface of a river with reference to a fixed datum level, hence 'rising' and 'falling' stages. **2.** In palaeoclimatology, a climatic and partly geologic–climatic term usually defined by a series of sediments or a sequence of *fossil assemblages and named at a type locality. For example, the *Hoxnian (a temperate stage) is named for organic *interglacial deposits at Hoxne, Suffolk, England. *See* OXYGEN-ISOTOPE STAGE. **3.** The degree of development of a land-form or landscape over time, traditionally described as 'youthful', 'mature', and 'old age' (*see* DAVISIAN CYCLE). The recognition of such stages implies an orderly evolution and this is now seen as unlikely for many parts of the Earth's land surface. **4.** The part of a microscope on which the specimen to be examined is placed.

stage hydrograph *See* HYDROGRAPH.

stagnant ice A mass of *ice which is no longer moving and melts *in situ*. It may arise because it has been detached from the *accumulation zone.

stalactite An elongated body of dripstone descending from the roof of a cave in a *karst environment. It is produced by calcite precipitation as excess carbon dioxide diffuses into the air from water droplets entering a cave environment.

stalagmite A pinnacle of dripstone rising from the floor of a cave in a *karst environment. It is produced by the precipitation of calcite as excess carbon dioxide diffuses into the air when water droplets strike the floor.

stand (ecol.) **1.** The standing growth of plants (e.g. trees). **2.** In vegetation classification, a distinctive plant association that may be recognized elsewhere. The composition may vary slightly but the recognition of stands enables comparisons between different vegetation *communities to be made. Sometimes the suffix -etum is added to the stem of the generic name of the dominant species.

standard **1.** A tree that is allowed to grow to its full height. **2.** A single-trunked tree that is large enough to be converted to sawn timber. **3.** A cultivated plant that stands without support because it has been grafted on to a robust, upright stem (e.g. a standard rose).

standard deviation In statistics, the extent to which data depart from the mean, given by: $\sigma = \sqrt{(\Sigma(x - \bar{x})^2/N)}$, where σ is the standard deviation, $\Sigma(x - \bar{x})$ the sum of the deviations of each datum (x) from the mean ($\bar{x}$), and N is the number of samples.

S

standard error of the difference The difference between the mean values of two sets of data, where the *standard deviations of both sets (σ_1 and σ_2) are known. The standard error of the difference (*SE*) is given by: $SE = \sqrt{(\sigma_1^2/N_1 + \sigma_2^2/N_2)}$, where N_1 and N_2 are the sizes of the first and second samples respectively.

standard error of the mean In statistics, an estimate of the range by which the means of a number of data sets deviate about the mean of those means. The standard error is given by: $\sigma/\sqrt{N}$, where σ is the *standard deviation of the original distribution and N is the size of the sample.

standard mean open water *See* SMOW.

Standard Positioning Service *See* GLOBAL POSITIONING SYSTEM.

standing crop *See* BIOMASS.

standing wave A wave in which the surface of the water oscillates vertically between fixed points ('nodes'), without any forward progression, the crest at one moment becoming the trough at the next. The points of maximum vertical rise and fall are called 'antinodes'. At the nodes the water particles show no vertical motion but exhibit the maximum horizontal motion. Standing waves may be caused by the meeting of two similar wave groups travelling in opposing directions. *See* SEICHE.

stand table A histogram showing the number of individuals in a plant community within each size range.

star dune A complex *aeolian sand dune in which a series of slip faces radiate about a central point, producing a rough star shape. Such dunes are the product of highly variable wind directions.

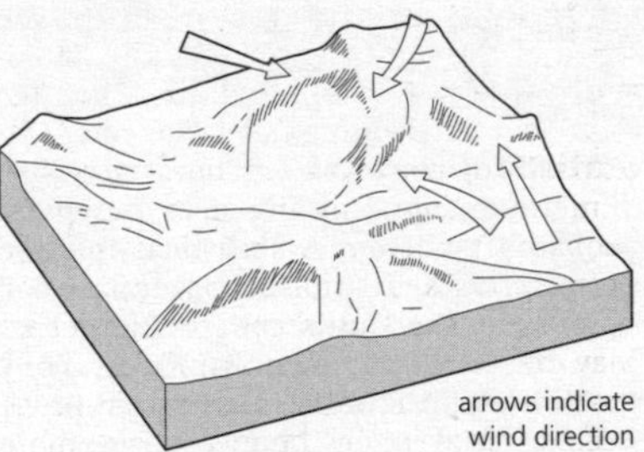
arrows indicate
wind direction

Star dune

star phylogeny In a *phylogenetic tree, the occurrence of a multifurcation with many short branches connected at the internal node. Such topologies are often inferred to represent a recent population expansion event from a common ancestor (the founder lineage). This is often seen in populations which have undergone a *founder effect.

startle response The reaction of an animal to sudden danger, by a *threat display (e.g. when certain butterflies and moths reveal eyespots on the wings) or flight (e.g. on average it takes 20 milliseconds for a puff of air from an approaching missile to set an American cockroach running).

stasigenesis The situation in which an evolutionary lineage persists through time without splitting or otherwise changing. So-called 'living fossils' provide examples of stasigenesis.

stasipatric speciation A rapid *speciation that may occur among small breeding populations that are not completely isolated genetically or spatially. Such speciation may occur either *parapatrically or *sympatrically. M. J. D. White, who proposed the concept in 1978, pointed out that most species have their own unique *karyotype, and suggested that the karyotype changes might have been the actual promoters of reproductive isolation.

stasis **(evol.)** A period of little or no evolutionary change; the 'equilibrium' that alternates with 'punctuations' in the theory of *punctuated equilibrium.

static allometry *See* ALLOMETRY.

static life table A *life table compiled from the age structure of a population at a particular time.

statistical method In modern usage, a method for analysing data based on probability theory. A statistical method permits the calculation of a value based on observations about some problem that may be tested for significance by comparison with the values that might be expected to arise by chance. Two main categories of statistical methods have been developed: classical or parametric tests, and the more recent non-parametric or distribution-free tests. Parametric statistical methods may be applied only to data on an interval scale, and typically they make assumptions about the background population from which the sample is taken, most often that it is normally distributed. Where data are in nominal or ordinal form, or where assumptions about the distribution of data on which a parametric test is based cannot be justified, then non-parametric (distribution-free) methods can be used. In general, parametric tests are more rigorous than non-parametric tests. Formerly, and more colloquially, statistical methods embraced any form of data gathering and analysis. *Compare* NUMERICAL METHOD.

statistical significance The condition in which the difference between two means is equal to or greater than 1.96 times the *standard error of the mean. Since there is an approximately 20 : 1 likelihood that such a difference occurs by chance, 95 per cent of the means will be enclosed, assuming a normal distribution of the data. The result is thus significant at the 95 per cent level or, to express it more precisely, it is significant at level $p = 0.05$. Higher levels of significance are possible and yet more convincing, e.g. $p = 0.01$ is equivalent to 1 chance in 100 (99 per cent probability) that the original (null) hypothesis of the means being acceptably the same is correct. *See also* CONFIDENCE LIMITS.

steam fog *See* ARCTIC SEA SMOKE.

steinkern *See* FOSSILIZATION.

stellate Star-shaped; radiating in arrangement.

stemflow Precipitation that is intercepted by vegetation and runs down the branches and stems of plants, arriving at the soil via the main trunk.

steno- From the Greek *stenos*, meaning 'narrow', a prefix that is used in ecology with adjectives describing environmental factors, denoting a limited tolerance by an organism of those factors. *Compare* EURY-.

stenoecious Applied to an organism that can live in only a restricted range of *habitats.

stenohalic *See* STENOHALINE.

stenohaline **(stenohalic)** Very sensitive to changes in *salinity; unable to tolerate a wide range of *osmotic pressures.

stenopaic Applied to an eye in which the pupil is narrow and slit-like (e.g. as in a domestic cat).

stenophagic Applied to organisms that have a highly specialized diet.

stenothermal *See* STENOTHERMOUS.

stenothermous **(stenothermal)** Unable to tolerate a wide temperature range.

stenotopic Tolerant of only a narrow range of environmental factors.

steppe **(Eurasian steppe)** A vast, temperate *grassland *biome that stretches from the River Danube in eastern Austria to Dunbey in China. Generally, the dominant vegetation comprises drought-resistant *perennial grasses, but the actual species composition varies from east to west and north to south, according to the rate of precipitation. Near the Black Sea, large feather grasses (*Stipa*) and sheep's fescue (*Festuca ovina*) prevail. Here the climate is warm and humid in spring, supporting *ephemeral species (e.g. *Tulipa*) which soon die as the long, hot, dry summer follows. In the central steppe region the spring is cold, supporting few ephemeral species, but in wetter years large numbers of *vegetatively reproducing plants survive (e.g. *Artemisia*). There are four discernible belts of latitude, with the highest rainfall in the north: *meadow steppe; dry herbage/turf grass steppe in which steppe herbage dominates; arid turf grass steppe which has less steppe herbage; and desert/scrub/turf grass steppe.

steppe meadow *See* MEADOW STEPPE.

stereotaxis *See* THIGMOTAXIS.

sterile **1.** Of an organism, unable to produce reproductive structures (i.e. unable to reproduce). **2.** Of land, unable to support the growth of plants, especially cultivated crops. **3.** Of an environment, object, or substance, completely free of all living organisms, including all micro-organisms of any type or form.

Stevenson screen A widely used shelter that contains meteorological instruments, arranged in such a way that they give standard readings. The screen consists of a box, with sides ventilated by louvres, a ventilated floor and upper part, and an air space between an inner and outer roof. The box contains thermometers for measuring temperature, which are protected from direct solar radiation and from radiation from surrounding objects, and a *wet-bulb thermometer for measuring humidity. The screen was invented by the Scottish civil engineer Thomas Stevenson (1818–87), the father of Robert Louis Stevenson, the author. It came into use in the late 1860s.

stillstand **1.** A period of geologic time characterized by unchanging sea levels (i.e. a state of neither *regression nor *transgression). **2.** In a *raised bog, the condition in which productivity is balanced by the decomposition of *peat, so growth ceases.

stilt root A tree root that arises from the lower trunk and runs obliquely to the ground, providing additional support for the tree. Mangroves and certain palm trees (Arecaceae) have stilt roots. *Compare* BUTTRESS ROOT.

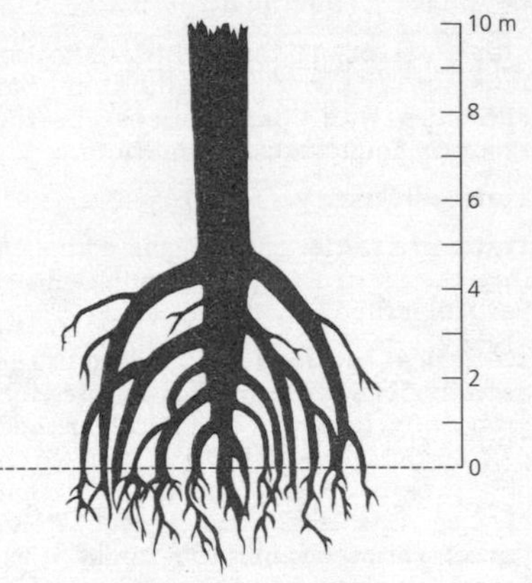

Stilt root

sting The ovipositor of ants, bees, and wasps (division Aculeata), which has lost its egg-laying function and serves as a means of injecting venom to paralyse, but not kill, the prey of hunting wasps, and is used as a means of defence by bees. The sting of honey-bees (*Apis* species) and some social wasps is barbed and remains in the skin of the victim after the wasp or bee has become detached. In honey-bees the muscular venom sac of detached stings continues to pump venom, and it also emits an alarm *pheromone to alert other workers, which may be recruited to join the attack.

stink gland In most plant and water bugs of the suborder Heteroptera, glands that produce fluids believed to be distasteful to potential predators.

stinking smut *See* BUNT.

stipitate Having a stipe or stalk.

stochastic model A mathematical representation of a system that takes account of probability, so that a given input yields a number of possible results. The outcome therefore becomes unpredictable. *Compare* DETERMINISTIC MODEL.

stock **1.** An *igneous intrusion that has steep contacts with the surrounding country rocks. It is approximately circular in plan and has an area of 20 km^2 or less. **2.** *See* GRAFT.

stolon A stem that grows horizontally and produces new plants at its tip; a runner (e.g. as in the strawberry).

stone circle *See* PATTERNED GROUND.

stone garland *See* PATTERNED GROUND.

stone net *See* PATTERNED GROUND.

stone polygon *See* PATTERNED GROUND.

stone steps *See* PATTERNED GROUND.

stone stripes *See* PATTERNED GROUND.

stool **1.** A tree stump that is capable of producing new shoots. **2.** The permanent base of a *coppiced tree.

storage coefficient **(storativity)** The volume of water given up per unit horizontal area of an *aquifer and per unit drop of the *water-table or potentiometric surface. It is a dimensionless ratio and always less than unity. In unconfined aquifers it is equal to the specific yield (*see* SPECIFIC YIELD (2)), but in confined aquifers it depends on elastic compression of the aquifer, and is usually less than 10^{-3}.

storativity *See* STORAGE COEFFICIENT.

storm beach An accumulation of coarse beach sediments built above the high-water mark by storm action. Gravel, shell debris, and other coarse materials are thrown into ridge or bank structures by waves during heavy storms.

storm bed A bed of *sediment deposited by a storm. Storm beds are usually the product of shallow marine wave activity and are often known as 'event deposits' (i.e. they are the product of a short-lived, high-energy, sedimentary environment).

storm surge A rise or piling-up of water during a storm, as a result of wind stresses acting on the surface of the sea and of atmospheric-pressure differences. If a storm surge occurs at the time of highest *spring tides, flooding of coastal areas may result, as happened in the Netherlands and East Anglia, England, in 1953.

stoss and lee The up-*glacier and down-glacier slopes respectively of a rocky obstacle that has been glaciated. The stoss slope is smoothly abraded, the lee slope roughly plucked. A landscape dominated by such features is said to have 'stoss-and-lee topography'. *See* ROCHE MOUTONÉE.

stoss-and-lee topography *See* ROCHE MOUTONNÉE and STOSS AND LEE.

stotting A series of high, stiff-legged jumps made by gazelles while they are running. This behaviour alerts the herd to the presence of a predator, but the gazelle's probable reason for stotting is to warn a predator that it has been seen and its pursuit is therefore futile, because the quarry will escape. The word may be derived from the Scots word *stot* meaning 'to bounce'. *See also* PRONKING.

Strahler climate classification A little-used system for describing climates devised in 1969 by A. N. Strahler, in which world climates are related to the main *air masses that produce them, as: (*a*) equatorial/tropical air masses, producing low-latitude climates; (*b*) tropical and polar air masses, producing mid-latitude climates; and (*c*) polar and Arctic air masses, producing high-latitude climates. Subsets of these are based on variations in temperature and *precipitation to give 14 regional types, plus upland (highland) climates which are regarded as a separate category. *See also* KÖPPEN CLIMATE

CLASSIFICATION and THORNTHWAITE CLIMATE CLASSIFICATION.

stramineous Straw-coloured.

strandflat A type of *shore platform, up to 60 km wide, which is found along the coasts of Greenland, Iceland, Norway, and Spitzbergen. It may be the result of combined glacial and marine processes.

strandline The shoreline of a marine or *lacustrine environment. The term is applied most commonly to ancient shorelines. The development of a strandline requires that the relative positions of land and water remain stable long enough for features to form. Subsequent displacement may be caused by a change in the level of the water or of the land.

strangler A plant that depends on another for physical support and ultimately suppresses the support plant by enclosing it in aerial roots (e.g. *strangling fig). Stranglers are most typical of *tropical rain forests.

strangling fig A fig tree that germinates high in a host tree and sends roots to the ground which eventually anastomose (see *anastomosis) and, as they grow, envelop and kill the supporting tree. The banyans differ in having roots descending from the limbs as well, so that a grove of closely growing, stout roots develops.

strata *See* STRATUM.

stratification **1.** The arrangement of *sediments, *sedimentary rocks, *soils, etc. in layers (strata). **2.** The placing of seeds between layers of moist peat or sand and exposing them to low temperatures (e.g. by leaving them outdoors through the winter) in order to encourage germination. *Compare* VERNALIZATION.

stratified random sample (partial random sample) In statistics, a modification of the random sample that is particularly useful when obvious heterogeneity exists in the *community, area, etc. to be investigated. In such instances a simple random sample may fail to record sufficient replicates of a particular subcategory, or may do so only very inefficiently, so preventing a proper statistical monitoring of variability. In a stratified random scheme sample points are shared among the main types present, then allocated at random within those subcategories.

stratiformis From the Latin *stratus*, meaning 'flattened' or 'spread out' and *forma*, meaning 'appearance', the name of a species of cloud consisting of an extensive level sheet or layer, found in *altocumulus, *stratocumulus, and sometimes *cirrocumulus. *See also* CLOUD CLASSIFICATION.

stratigraphy The study of stratified deposits, including sedimentary rocks, lake sediments, and peat deposits, together with any *fossils these may contain. It provides information relating to the changing conditions of the past. *See also* CHRONOSTRATIGRAPHY.

stratigraphic trap (lithologic trap) An oil or gas trap formed as a result of lithologic variations (e.g. interbedded lenses of sands and silts in a deltaic environment).

stratocoenosis The *community of a particular vegetation or physical *habitat layer (e.g. the canopy layer of a woodland or the *hypolimnion of a stratified lake).

stratocumulus From the Latin *stratus*, meaning 'flattened' or 'spread out' and *cumulus*, meaning 'heap', the name of a cloud composed of sheets or layers of grey to whitish appearance, typically with dark patches which are not fibrous. *See also* CLOUD CLASSIFICATION.

stratopause The level that marks the maximum height of the *stratosphere, at around 50 km. After high temperatures in the upper stratosphere (about 0°C at the stratopause) temperature decreases with increasing altitude in the *mesosphere above.

stratosphere The atmospheric layer above the *troposphere, which extends on average from about 10 to 50 km above the Earth's surface. The stratosphere is a major stable layer whose base is marked by the *tropopause, and where temperatures overall average approximately −60°C. Temperature in the lower stratosphere is isothermal but increases markedly in the upper part, to reach a maximum of about

0°C at the *stratopause. High stratospheric temperatures result from absorption of ultraviolet radiation (0.20–0.32 μm wavelengths) by oxygen near the stratopause; slight vertical mixing causes some atomic oxygen to move downwards to form ozone concentrated at 15–30 km (*see* OZONE LAYER). Owing to the very low air density, even the small amount of ozone concentrated in the upper stratosphere is extremely effective in absorbing radiation, thus giving high temperatures at 50 km. The isothermal condition at the base of the stratospheric inversion layer creates stability, which generally limits vertical extensions of cloud and leads to the lateral spreading of high *cumulonimbus cloud with characteristic anvil heads.

stratum **(pl. strata)** A layer of rock; unlike 'bed', 'stratum' has no connotation of thickness or extent and, although the terms are sometimes used interchangeably, they are not synonymous.

stratus The Latin *stratus*, meaning 'flattened' or 'spread out', used to describe a cloud of flat, uniform base and of grey appearance, through which the Sun may be outlined clearly when the cloud is not too dense. *See also* CLOUD CLASSIFICATION.

streak A general term for a virus disease in a (usually monocotyledonous) plant in which streaks of yellow colour or necrosis occur on the leaves.

stream drift The downstream transport by flowing water of bottom-dwelling invertebrate animals that spend part of their time in open water.

stream flood *See* FLASH FLOOD.

stream grade *See* GRADE.

streamline **1.** In a flowing fluid, a hypothetical line which indicates the local direction of flow. **2.** A shape which allows a body to offer minimum resistance to a fluid through which it moves; to impart such a shape to a body.

stream order A measure of the position of a stream (defined as the reach between successive tributaries) within the hierarchy of the *drainage network. A commonly used approach allocates order '1' to unbranched tributaries, '2' to the stream after the junction of the first tributary, and so on. It is the basis for quantitative analysis of the network.

stream power The rate at which a stream can do work, especially the transport of its load, and measured over a specific length. It is largely a function of channel slope and discharge and is expressed by $\Omega = \gamma Qs$, where Ω is the power, γ is the specific weight of water, Q is the discharge, and s is the slope. Streams tend to adjust their flow and channel geometry in order to minimize their power (*see* LEAST-WORK PRINCIPLE).

stream terrace *See* RIVER TERRACE.

stress A physiological condition produced by excessive pressures that are detrimental to the organism. The pressures may be environmental, or in animals psychological, in origin. Stress usually affects the behaviour of animals (e.g. bullying may lead to avoidance behaviour and loss of appetite) and the physiological condition of plants (e.g. stress due to lack of water causes *wilting).

stress-tolerator In the classification of *plant strategies proposed by J. P. Grime, a plant species found in areas of high stress and low disturbance. *Compare* COMPETITOR and RUDERAL.

striate Marked with fine lines, ridges, or furrows.

striation A narrow groove or scratch cut in exposed rock by the abrasive action of hard rock fragments embedded in the base of a sliding *glacier. Striation provides a useful clue to the direction of former ice movement in glaciated areas.

stridulate To produce sound by rubbing a file across a membrane. Insects have a wide variety of mechanisms for sound production. In some of the most common examples a file on one wing rubs across a roughened surface on the other wing, or a file on the leg is drawn across the edge of the wing. The volume of sound production achieved by these mechanisms is often startling. The sound may be amplified by resonation in a wing or (e.g. in mole crickets) by causing a column of air to resonate in a chamber excavated in the ground and shaped for the purpose.

strike **1. (noun)** The compass direction of a horizontal line on an inclined plane (i.e. it is at right angles to the angle of inclination, known as the dip). **(verb)** To lie in the direction of such a line. **2. (noun)** The discovery of an economically valuable source of a mineral. **(verb)** To make such a discovery.

strike ridge An elongated hill developed along the *strike of a bed which is more resistant than its adjacent strata.

strike stream *See* SUBSEQUENT STREAM.

strike valley *See* SUBSEQUENT STREAM.

strip mining *See* OPEN-CAST MINING.

stromatolite A rock-like or firmly gelatinous structure, built up over long periods of time from many layers or mats of *cyanobacteria together with trapped sedimentary material. Stromatolites are found mainly in shallow marine waters in warmer regions. Some are still in the process of being formed (e.g. in Shark Bay, western Australia). Fossil stromatolites dating from the late *Archaean are also known, although it is not certain that these were formed by cyanobacteria. Stromatolites that formed about 2 billion years ago produced the earliest known reefs, quite different from present-day coral *reefs.

structural trap A trap formed by the deformation of porous and non-porous *strata as a result of folding, faulting, etc., in which oil, gas, or water may accumulate.

structuring method Any technique for sorting data to reveal the important patterns. *Classification and *ordination methods are examples of structuring methods.

strychnine $C_{21}H_{22}N_2O_2$, an *alkaloid that is produced by *Strychnos* species, especially *S. nux-vomica*. In animals, it is a stimulant of the nervous system in general and in large doses causes convulsions. It can be lethal.

sub- The Latin *sub*, meaning 'under' or 'close to', used as a prefix meaning 'beneath' or 'lying below'.

subalpine forest A conifer-dominated forest which occurs mainly in the subalpine zone of temperate latitudes, but with a few extensions south of the Tropic of Cancer. The elevation of the zone increases with decrease in latitude. Generically this forest is closely related to the boreal forest and indeed the two types have a minority of species in common. *See also* TROPICAL SUBALPINE RAIN FOREST.

Subarctic Current *See* ALEUTIAN CURRENT.

sub-Atlantic A colder, wetter climatic phase which followed the more continental climate of *sub-Boreal times. The change from sub-Boreal to sub-Atlantic conditions in Britain began about 2 850 BP and roughly coincided with the transition from Bronze to Iron Age cultures. The sub-Atlantic marks a period of renewed *peat growth on *bog surfaces that in late sub-Boreal times were sufficiently dry and humified (*see* HUMIFICATION) to support heath vegetation (e.g. *Calluna vulgaris*, ling or heather). This renewed peat growth gives a major recurrence surface, the Grenz horizon, which at one time was considered to define the boundaries between Godwin pollen zones VIIb and VIII. *Radiocarbon dating has shown that there are many such recurrence horizons, however, and this use of Godwin pollen zones is now obsolete. *See* POLLEN ZONE.

sub-Boreal From Scandinavian evidence, a period during which the climate was cooler than during the preceding *Atlantic *climatic optimum, but not so cold and wet as during the *sub-Atlantic phase that followed. Nowadays the term sub-Boreal is used only loosely.

subboscus *See* BOSCUS.

subclimax Strictly, the penultimate stage in a *succession to a climatically controlled *climax community (as in *monoclimax theory). Typically, a subclimax community persists for a long time; for example, the forests of the *Boreal Period may be considered as subclimax to the early Atlantic forests of the post-glacial *climatic optimum.

subcritical flow **(tranquil flow)** *See* CRITICAL FLOW.

subformation In *phytosociology, a vegetation grouping used by the *Uppsala school of phytosociology, and denoting a geographically distinctive unit of a major formation.

subfossil *See* FOSSIL.

subglacial At the base of a *glacier. The term is usually applied to meltwater or *drift.

sublimation Direct evaporation from ice. In meteorology, 'deposition' is the term applied to the reverse process, in which water vapour changes directly to the solid phase. *See also* ABLATION.

sublimation nucleus *See* ICE NUCLEUS.

sublittoral zone **1. (infralittoral zone, shelf zone, subtidal zone)** The sea-shore zone lying immediately below the *littoral (intertidal) zone and extending to about 200 m depth or to the edge of the *continental shelf. Red and brown algae are characteristic of this area. Typical animals include sea anemones and corals on rocky shores, and shrimps, crabs, and flounders on sandy shores. It is approximately equivalent to the *circalittoral zone. **2.** *See* LIMNETIC ZONE.

submarine canyon A deep, steep-sided valley cut into the *continental shelf or slope, the axis of which slopes seaward at up to 80 m/km.

submerged forest *See* PETRIFIED FOREST.

submergence marsh The lower zone of a *salt marsh, from the mean high-water level of neap tides to the general level of mean high water. Typically, this zone experiences more than 360 submergences per annum, usually with more than one hour of submergence during daylight each day. Continuous exposure never exceeds nine days. *Compare* EMERGENCE MARSH.

subpolar glacier *See* GLACIER.

subsequent stream A stream that follows a line of geologic weakness; it tends to extend headwards actively, and may acquire further tributaries through the process of *river capture. It is called a 'strike stream' when its trace follows the geologic *strike, and the associated valley is called a 'strike valley'.

subsessile Nearly *sessile.

subshrub A plant, smaller than a *shrub, which produces wood only at its base and has abundant growth branching upwards from the base, the upper stems dying back at the end of each growing season. *Compare* HERB and TREE.

subsoiling The breaking up of subsoils, usually because they are compacted, without inverting them. Subsoiling is usually performed with a chisel-like device that is pulled through the soil.

subspecies Technically, a *race of a *species that is allocated a Latin name. The number of races recognized within a species and the allocation of names to them is something of an arbitrary procedure. Systematic and *phenotypic variations do occur within species, but there are no clear rules for identifying them as races or subspecies except that they must be (*a*) geographically distinct, (*b*) populations, not merely morphs, and (*c*) different to some degree from other geographic populations.

substitution The replacement of one *nucleotide in a DNA sequence by another as a result of a *mutation event. *See also* NONSYNONYMOUS SUBSTITUTION, SILENT SUBSTITUTION, and SYNONYMOUS SUBSTITUTION.

substrate **1. (biochem.)** The reactant acted upon by an enzyme. **2. (substratum)** Any object or material upon which an organism grows or to which an organism is attached; an underlying layer or substance.

substratum *See* SUBSTRATE.

subsurface flow The flow of water at a shallow depth beneath the ground surface; it may be influenced by relatively impermeable layers which enlarge lateral flow. A temporary rise in the *water-table can produce saturated *interflow (saturated throughflow) analogous to saturated *overland flow. The subsurface flow can re-emerge at the surface at or near the base of hill-slopes. Throughflow is flow through the soil, whereas interflow is flow through the rock above the water-table.

subtidal Applied to that portion of a *tidal-flat environment which lies below the level of mean low water for *spring tides. Normally it is covered by water at all states of the tide. The word is often used as a general descriptive term for a sub-aqueous but shallow marine depositional environment.

subtidal zone *See* SUBLITTORAL ZONE.

subtropical high Surface high-pressure cells that are especially prominent and persistent over oceans at around 30° latitude. The *anticyclones develop below the *subtropical jet stream from subsiding air. The development tends to shift towards the equator in winter and towards the pole in summer. The high pressure is weaker over continents in summer. *See also* AZORES HIGH and BERMUDA HIGH.

subtropical jet stream The *jet stream of subtropical latitudes. The change from a westerly direction in winter to an easterly one in summer influences the surface wind changes that mark the summer *monsoon. The jet is related to a marked temperature gradient in the upper *troposphere. The westerly jet moves towards the equator in winter and is associated with subsiding air and settled surface weather. In summer the jet moves towards the pole. At the seasonal extremes it tends at times to merge with the *polar-front jet.

S

succession The sequential change in vegetation and the animals associated with it, either in response to an environmental change or induced by the intrinsic properties of the organisms themselves. Classically, the term refers to the colonization of a new physical environment by a series of vegetation communities until a final equilibrium state, the *climax, is achieved. The presence of the colonizers, the pioneer plant species, modifies the environment so that new species can join or replace the initial colonizers. Changes are rapid at first but slow to a more or less imperceptible rate at the climax stage, composed of climax plant species. The term applies to animals (especially to *sessile animals in aquatic *ecosystems) as well as to plants. The characteristic sequence of developmental stages (i.e. nudation, migration, ecesis, competition, reaction, and stabilization) is termed a sere. During succession, the ecosystem grows in *biomass, reaching the maximum biomass at climax. This means that during the course of succession the gross primary production exceeds the total ecosystem respiration, the excess accumulating as biomass. When equilibrium is eventually attained, the ecosystem (community) respiration has risen to a point where it equals gross primary production. *See* ALLOGENIC; AUTOGENIC; and DEGRADATIVE.

succulent Fleshy.

sucker **1.** An underground shoot arising adventitiously from the roots or lower stem of a tree or shrub and emerging from the soil to form a new plant, initially nourished by the parent plant. In cultivated species where grafting (*see* GRAFT) is practised (e.g. roses and fruit trees), production of suckers from the stock may seriously detract from the vigour of the grafted scion. The term may also be applied to the modified root of a parasite which enables it to extract nutrients from the host. **2.** An organ with which an animal attaches itself to a surface.

Sudanese park-steppe floral region Part of R. Good's (1974, *The Geography of the Flowering Plants*) African floral subkingdom within his *Palaeotropical floral kingdom, which has a not very rich flora in which species of *Acacia*, grasses, and palms tend to dominate the vegetation. *See also* FLORAL PROVINCE and FLORISTIC REGION.

sudden oak death A disease caused by the fungus-like organism *Phytophthora ramorum* that has resulted in extensive tree death in the United States and has been reported in a few localities in Britain in 2003. *P. ramorum* infects a wide range of trees, including oaks, but is also found in some shrubs, such as *Rhododendron* and *Viburnum*.

suffocation disease A disease of rice in which reddish- brown patches spread from the tips of the leaves downwards; the roots of the plants blacken and rot. The condition is encouraged by poorly drained soils and appears to be associated with the production of toxic levels of hydrogen sulphide by bacteria in the soil.

suffruticose chamaephyte *See* CHAMAEPHYTE.

sulcate Marked with ridges, grooves, or furrows.

sulcus A furrow or groove. On the surface of a *pollen grain, a *colpus.

sulphur **(S)** An element that is needed for plant life. It is found covalently bound, especially in proteins, where it stabilizes their structures. It is also involved in *oxidation and *reduction reactions. Sulphur-deficient plants become chlorotic (*see* CHLOROSIS) and etiolated (*see* ETIOLATION).

sulphur dioxide **(SO_2)** The gaseous product of the oxidation of sulphur, which is released into the atmosphere when fuels containing sulphur compounds are burned. In the atmosphere, a series of reactions oxidize SO_2 to particles of sulphate (SO_4). Sulphate is also produced in the atmosphere from dimethyl sulphide (DMS, $(CH_3)_2S$), released by some species of *phytoplankton (especially *coccolithophorids). Sulphate crystals act as *condensation nuclei, but lower the *pH of the water into which they dissolve and SO_2 is implicated in *acid precipitation.

sulphur-reducing organism **(domains *Archaea and *Bacteria)** The *phenotype which utilizes sulphur-containing compounds, rather than oxygen, in its respiratory pathways, often releasing hydrogen sulphide. Both *archaebacteria and bacteria include sulphur-reducing phenotypes.

sumatra A regional *squall, usually occurring at night, in the Malacca Strait, accompanied by high winds which veer (*see* VEERING) from southerly to south-westerly and north-westerly. Extensive *cumulonimbus cloud brings heavy rain, with thunder and lightning.

summation The addition of stimuli of different kinds, which may be perceived at more than one time (temporal summation) and in more than one place (spatial summation), to produce a co-ordinated response in an animal. Summation is one of the integrating properties of the central nervous system.

sunbathing The exposure of its body to sunshine by an animal as a means of thermoregulation (mainly in *poikilotherms) or to stimulate the production of vitamin D in the skin. Many animals prefer to rest in warm (but not hot) sunshine, thereby conserving metabolic energy that otherwise would be needed to maintain body temperature.

suncracks *See* DESICCATION CRACKS.

Sundaland A geographical unit composed of Malaya, Sumatra, Java, Borneo, and Palawan, with the intervening small islands, which all lie on the shallow-water (less than 200 m) Sunda shelf, which was

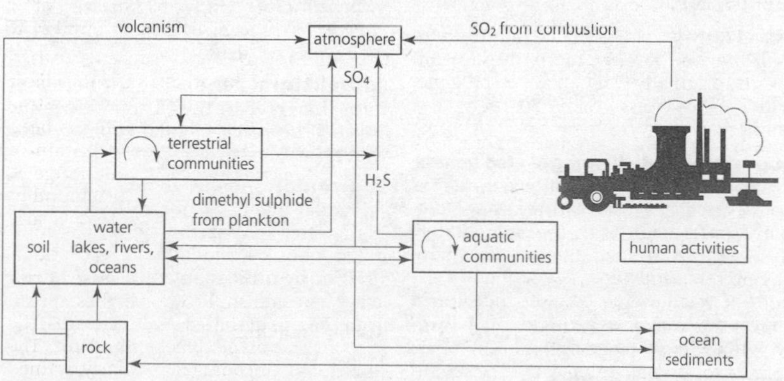

Sulphur

exposed during periods of low sea level in the *Pleistocene. The fauna of this region is fundamentally homogeneous and differs slightly from that of areas further north-west, which are more seasonal, despite the fact that today Malaya is linked by an isthmus to the Asian mainland.

Sundance Sea A shallow marine embayment that extended over what are now the states of Wyoming and South Dakota during the Middle and Late *Jurassic. The southern edge of this sea (in modern Colorado) was bordered by *tidal flats, the marine connection being northwards through the present-day mid-west and Canada.

sunshine recorder *See* CAMPBELL–STOKES SUNSHINE RECORDER.

super- The Latin *super*, meaning 'on top of', used as a prefix meaning 'directly over', 'over', or 'above.'

super-adiabatic lapse rate A fall of temperature with increasing altitude which is greater than the usual *dry adiabatic lapse rate; it occurs in conditions of intense heating over land or sea.

supercooled cloud Cloud containing pure water droplets at temperatures considerably below the nominal freezing temperature. With very pure water (i.e. free from pollutants), supercooling of liquid drops can occur down to around –40°C; *altocumulus cloud, for example, is usually composed of water droplets at temperatures well below 0°C. *See also* CLOUD SEEDING.

S

superfamily In genetics, a group of genes derived by one or more *duplication events, originally from an ancestral gene, which show less than 50 per cent similarity.

superimposed drainage (epigenetic drainage) A *drainage pattern that has been established on an earlier surface (perhaps conformable with the immediately underlying strata, and standing well above the present landscape). Subsequently the pattern was lowered by river incision so that it now lies across geologic structures to which it bears no relation.

supernormal stimulus An artificial stimulus that produces in an animal a response that is stronger than would be evoked by the natural stimulus it resembles. For example, in some birds incubation behaviour is stimulated by the presence of an egg, and the larger the egg the stronger the stimulus; in such birds a very large artificial egg may be incubated in preference to a much smaller real egg.

superorganism *See* ORGANISMIC.

superparasite *See* HYPERPARASITE.

supersaturation **1.** The condition of air in which the *humidity is above the level required for saturation at a given temperature (i.e. the *relative humidity is greater than 100 per cent). Supersaturation results when the temperature of air containing no *condensation nuclei falls below its *dew point. **2.** The condition of a chemical solution, for example a salt solution, which on losing solvent through evaporation comes to contain an excess of the solute. The solute then begins to crystallize, as, for example, in soda lakes or the Dead Sea.

superspecies *See* SPECIES GROUP.

supertramp A species that disperses efficiently, colonizes readily, and has a wide range, but is excluded by competition from habitats supporting a rich variety of species.

supra- The Latin *supra*, meaning 'above', 'beyond', or 'earlier in time', used as a prefix meaning 'above' or 'in a superior position to'.

supraglacial On the surface of a *glacier. The term is usually applied to melt-water or *drift.

supralittoral zone The seashore zone immediately above the *littoral fringe and beyond the reach of tidal submergence, though affected by sea spray.

supratidal Applied to that portion of a *tidal flat that lies above the level of mean high water for *spring tides. It is inundated only occasionally by exceptional tides or by tides augmented by a *storm surge.

surf Breaking waves in the area between the shoreline and the outermost limit of breaking waves.

surface inversion A *temperature inversion in the lower atmospheric layers, extending upwards from the Earth's surface. The condition results, for example, from radiation cooling of the ground and the air above, or from *advection of warm air over cold surfaces.

surface run-off (overland flow, Hortonian flow) The flow across the land surface of water which accumulates on the surface when the rainfall rate exceeds the infiltration capacity of the soil. The rate of infiltration, and therefore the possibility of surface run-off, is determined by such factors as soil type, vegetation, and the presence of shallow, relatively impermeable, *soil horizons. Saturated overland flow can occur when a temporary rise of the *water-table inhibits infiltration and causes flow over the surface.

surface wind The wind close to the Earth's surface, the velocity of which is usually measured at a standard height of 10 m. Surface-wind velocity is reduced by the frictional effect of the underlying surface. The actual wind is a balance of *pressure-gradient force, *Coriolis effect, and frictional effects.

surf wave *See* SPILLING BREAKER.

'survival of the fittest' *See* NATURAL SELECTION.

survivorship curve A graphical description of the survival of individuals in a population from birth to the maximum age attained by any one member. Usually it is plotted as the logarithm of the number of survivors as a function of age. If a population has a constant mortality rate the graph will be a straight line. Following a classification of survivorship curves proposed by R. Pearl in 1928 and considered further by E. S. Deevey in 1947, convex curves are commonly labelled Type I, straight curves as Type II, and concave curves as Type III.

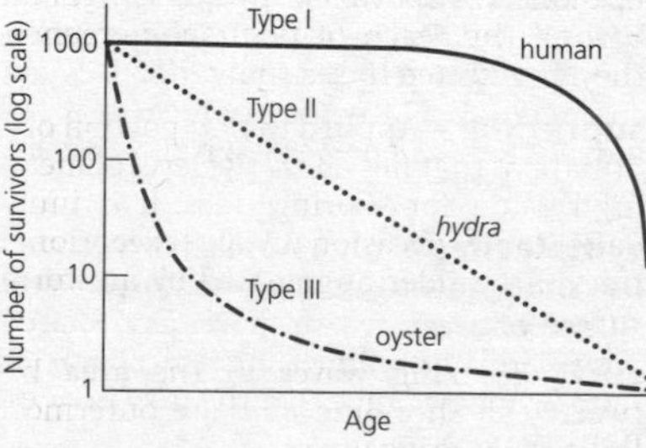

Survivorship curve

susceptible species *See* RARITY.

suspended load The part of the total load of a stream that is carried in suspension. It is made up of relatively fine particles that settle at a lower rate than the upward velocity of water eddies and its highest concentration is in the zone of greatest turbulence, near the bed. It reaches a maximum in shallow streams of high velocity.

sustainable development *See* SUSTAINABILITY.

sustainability (sustainable development) Economic development that takes full account of the environmental consequences of economic activity and is based on the use of resources that can be replaced or renewed and therefore are not depleted. The concept was introduced in the late 1970s and was emphasized strongly in the *World Conservation Strategy*, published in 1980 by the *IUCN in collaboration with the UN Environment Programme and the World Wildlife Fund (now the World Wide Fund for Nature). *Our Common Future*, published in 1983 by the World Commission on Environment and Development (the Brundtland Commission), defined it as development that 'seeks to meet the needs and aspirations of the present without compromising the ability to meet those of the future'.

Sv **1.** *See* SIEVERT. **2.** *See* SVERDRUP.

S value *See* SEDIMENTATION FACTOR.

Svedberg unit (S) The unit of measurement in which sedimentation coefficients are expressed. It is equal to 10^{-13} seconds and is usually given for the solvent water at 20°C. It is named after the Swedish physical chemist Theodor Svedberg (1884–1971) and is written with no space between the number and the symbol (e.g. 70S).

sverdrup (Sv) The unit of fluid flow. 1 Sv = $10^6 m^3/s$. It is named after the

Norwegian geophysicist Harald Ulrik Sverdrup (1888–1957).

swamp A wet area, dominated by emergent aquatic vegetation, that is normally covered by water all year and is not subject to drying out during the summer. In European usage, the term is usually applied to herbaceous wetland ecosystems, such as reed beds, but in American usage 'swamp' is used only of forested wetlands. The American equivalent of the European swamp is a 'marsh'. *Compare* FEN.

swash The turbulent uprush of water that occurs when a wave breaks on a beach. It is an important mechanism for transporting sand and shingle landwards.

sweat bees Small, stingless bees belonging to the families Halictidae and Apidae, which are attracted to human sweat, especially in warm-temperate and tropical climates. Although stingless, some (e.g. *Melipona*) can bite.

sweep net A finely meshed net, usually with a round mouth, used for sampling insects from vegetation. Its advantages lie in its simplicity and speed, and its ability to collect relatively dispersed insect species on top of the vegetation. It cannot be used on very short vegetation.

sweepstake dispersal route A term coined by G. G. *Simpson in 1940 to describe a possible route of faunal interchange which is unlikely to be used by most animals, but which will, by chance, be used by some. It requires a major barrier that is occasionally crossed. Which groups cross and when they cross are determined virtually at random.

swell Long-period waves that have built up sufficient energy to move away from the area where wind stresses created them. The waves assume a uniform pattern and move even through areas where winds are weak or absent. The longer-period waves move faster than shorter-period waves, so the waves spread out as they move. Swell waves generated south of New Zealand have been recorded arriving on the coast of Alaska. The size of the waves that become swell is determined by the *fetch.

sym- *See* SYN-.

symbiont A symbiotic organism. *See also* SYMBIOSIS.

symbiosis A general term describing the situation in which dissimilar organisms live together in close association. As originally defined, the term embraces all types of mutualistic and parasitic relationships. In modern use it is often restricted to mutually beneficial species interactions (i.e. *mutualism). *Compare* COMMENSALISM and PARASITISM.

sympatric Applied to species or other taxa with ranges that overlap. *Compare* ALLOPATRIC and PARAPATRIC.

sympatric speciation The development of new taxa from the ancestral taxon, within the same geographic range (it is geographically possible for interbreeding to occur between the potential new taxa, but for some reason this does not happen). Because of the difficulty of envisaging what the reasons might be, few authorities until recently accepted the reality of sympatric evolution, except for certain special kinds of organisms; but recent studies have shown that chromosomal mutation and behavioural differences (e.g. in birdsong) can set up a partial barrier to interbreeding, sufficient to permit sympatric (*stasipatric) speciation; and the restricted interbreeding between ecological races of a species could promote separation. *Centrifugal speciation appears to be so common in some groups of animals (e.g. mammals) that some sympatric mechanism must be occurring. *Compare* ALLOPATRIC SPECIATION.

sympatry The occurrence of *species or other taxa together in the same area. The differences between closely related species usually increase (diverge) when they occur together, in a process called character displacement, which may be morphological or ecological. *Compare* ALLOPATRY.

symplesiomorph Applied to a character state, primitive (*plesiomorph) and shared between two or more taxa. Shared possession of a symplesiomorph character state is not evidence that the taxa in question are related.

sympodial Applied to a type of branching in which the apex itself divides and extension growth occurs from these lateral axes, not from the original tip. *See* MONOPODIAL.

syn- From the Greek *sun*, meaning 'with', a prefix meaning 'together', 'with', or 'resembling'. Before the letters b, m, and p the prefix is written as 'sym-'.

synaeresis The process of subaqueous shrinkage of clays by the loss of pore water.

synaeresis cracks Irregular, radiating, lenticular-shaped cracks, found on bedding surfaces and often resembling the form of a bird's foot. They form by subaqueous shrinkage rather than desiccation, and are therefore not an indication of subaerial exposure. *Compare* DESICCATION CRACKS.

synanthrope An animal that benefits from environmental modifications made by humans to such an extent that it becomes closely associated with humans. Examples include the feral pigeon, house sparrow, and house mouse.

synapomorphic Applied to *apomorphic features possessed by two or more taxa in common. If the two groups share a character state that is not the *primitive one, it is plausible that they are related in an evolutionary sense, and only synapomorph character states can be used as evidence that taxa are related. Phylogenetic trees are built up by discovering groups united by synapomorphies.

syncline A basin or trough-shaped fold in which the upper component *strata are younger than those below.

synecology The study of whole plant and animal *communities, including the study of terrestrial *ecosystems, biological aspects of oceanography, and applied problems of human management and alteration of ecosystems. *Compare* AUTECOLOGY.

synergism The result of combined factors, each of which influences a process in the same direction but which, when combined, give a greater effect than they would acting separately.

syngeneic *See* ISOGENEIC.

synonym In taxonomy, a plant name that differs from the official name. Usually, it is an older name that does not conform to the rules governing priority in the application of names.

synonymous substitution In a DNA sequence which codes for a protein, a *substitution that does not result in an *amino acid change in the protein.

synoptic meteorology The presentation of the current weather elements of an extensive area at a particular time. Sea-level and upper-level synoptic charts display weather conditions by symbols at selected synoptic stations. 'Synoptic' is from the Greek *sunoptikos*, meaning 'seen together'.

synthetic theory A modern theory of *evolution, incorporating Darwinian thinking, Mendelian genetics, and an understanding of genes and genetic change at the molecular level. *Compare* MODERN SYNTHESIS.

syntype In taxonomy, all specimens in a type series in which no *type specimen was designated.

synusia **(pl. synusiae)** A life-form vegetation community (e.g. the shrub layer of temperate forest or the photophotic (crown) *epiphytes of *tropical rain forest).

synzoochory *See* ZOOCHORY.

syrinx The vocal apparatus of a bird.

system A distinct entity that consists of a number of interacting parts such that the removal or failure of one part may incapacitate the entity as a whole.

systematic sample *See* REGULAR SAMPLE.

Système International d'Unités *See* SI UNITS.

Système Probatoire d'Observation de la Terre *See* SPOT.

tachytely A rate of *evolution within a group which is much faster than the average or *horotelic rate. Such accelerated evolution typically occurs when an organism enters a new *adaptive zone and initiates an *adaptive radiation to fill the available *niches. *Compare* BRADYTELY.

tactile Pertaining to the sense of touch.

tafoni A hollow, produced by localized *weathering on an exposed rock face. The rock usually breaks down by granular disintegration or by flaking and the hollow tends to grow upwards and backwards.

taiga The name applied by many authorities to the whole of the *boreal forest, but by some only to the more open, park-like tracts along the northern fringe of the boreal forest, otherwise known as *lichen woodland.

tailings Fine waste material from a mineral-processing plant, which is too poor for further treatment. It is often stored in a *tailings dam.

tailings dam A dam made from material that is sufficiently permeable for moisture to drain through it over a regulated period. The dam is constructed to retain the water-sodden, fine-grained materials (*tailings).

take-all A disease of wheat, barley, and other grasses, caused by the fungus *Gaeumannomyces graminis*. The *pathogen infects the roots of the plant, causing root rot and blackening of the stem base; the ears are bleached and empty (whiteheads).

talik *See* PERMAFROST.

talus (scree slope) A sloping mass of coarse rock fragments accumulated at the foot of a cliff or slope.

taluvium A hill-slope deposit that consists of a mixture of coarse, rocky rubble and finer particles. The term is a hybrid, derived from *talus and colluvium (relatively fine debris that has been washed downhill).

Tame Valley An *interglacial in the *Upton Warren interstadial, within the Würm (*Devensian) glacial.

tandem array Repetitive DNA, where the repeating units are contiguous. Some genes of high *copy number occur in tandem arrays e.g. ribosomal DNA in *eukaryotes). *Satellite DNA is also often arranged in tandem arrays. Such DNA regions are prone to the genetic processes of *unequal crossing over and *replication slippage.

Tangled Bank hypotheses A set of hypotheses, proposed by G. Bell in 1982, which emphasize the importance of the spatial heterogeneity found in a complex environment (i.e. a Tangled Bank) in the maintenance of sexual reproduction. The fitness of a *genotype that is successful in one location but less so in another may not be inherited by its offspring. It may be advantageous, therefore, for siblings to disperse at random into microsites nearby, regardless of the local success of the genotype. It is argued that this allows greater penetration of microsites than would be possible for *clones of a parthenogenetically reproducing organism and reduces competition among siblings. The 'Tangled Bank' is taken from the opening sentence of the final paragraph of *Darwin's *Origin of Species*: 'It is interesting to contemplate an entangled bank, clothed with many plants of many kinds, with birds singing on the bushes, with various insects flitting about, and with worms crawling through the damp earth, and to reflect that these elaborately constructed forms, so different from each other, and dependent on each other in so complex a manner, have all been produced by laws acting around us.'

tannin A generic term for complex, non-nitrogenous compounds containing phenols, glycosides, or hydroxy acids, which occur widely in plants. They are toxic substances with astringent properties, whose principal function appears to be to render plant tissues unpalatable to herbivores. Tannins often build up during the individual development of leaves, rendering them less palatable to herbivores as they age.

Tansley, Sir Arthur George (1871–1955) A British ecologist and conservationist who emphasized *ecology as an 'approach to botany through the direct study of plants in their natural conditions' (*Practical Plant Ecology*, 1923). He also pointed out the fact that since plants exist in communities the ecologist should be concerned with the structure of communities, or 'plant sociology'. This view became central to most British and American ecological theory. Tansley coined the term *ecosystem in 1935 (in 'The use and abuse of vegetational terms and concepts', *Ecology* **16**: 284–307), although it may have been used earlier by his colleague Roy Clapham. Tansley's book *Types of British Vegetation* (1911) paved the way for vegetation description in Britain. Tansley was a lecturer at the University of Cambridge (1907–23), where much of his ecological work was done, and professor of botany at the University of Oxford (1927–37). He was instrumental in founding the British Ecological Society (1913) and was its first president. His many books include *The British Islands and their Vegetation* (1939) and *Britain's Green Mantle* (1949). *See also* CLEMENTS, FREDERIC EDWARD.

tapetum **1.** In many nocturnal mammals, a layer of cells either in the retina or outside it, in the choroid (e.g. in cats), which contain crystals of zinc and a protein (often riboflavin, which fluoresces). The tapetum reflects light back through the retina and so increases the sensitivity of the eye to dim light. At night, when the pupil is fully dilated, reflected light from the tapetum will cause 'night shine' if the animal turns to look in the direction of a sudden bright light (e.g. car headlights). **2.** A layer of nutritive cells within the *sporangium of a fern.

taphonomy The study of the processes that affect an organism after its death, including those of *fossilization.

tar spot A disease of sycamore (*Acer pseudoplatanus*) in which black spots appear on the leaves, following infection by *Rhytisma acerinum*. The name may also refer to the fungus itself.

taungya A Burmese word that is now widely used to describe the practice, in many tropical countries, of establishing tree plantations by planting and tending tree seedlings together with food crops. Food cropping is ended after 1–2 years as the trees grow up.

tautonym In taxonomy, a name in which the genus and species are identical (e.g. *Gulo gulo*, the wolverine). Such names are permitted by the *ICZN (i.e. for animals), but not by the ICBN (for plants).

taxa *See* TAXON.

taxis A change in direction of locomotion in a motile micro-organism or cell, made in response to certain types of external stimulus (e.g. the presence of particular chemicals (chemotaxis), changes in light intensity (phototaxis), or changes in temperature (thermotaxis), etc.).

taxon **(pl. taxa)** A group of organisms of any taxonomic rank (e.g. family, genus, or species).

taxonomy The scientific classification of organisms. By extension, the term is also given a wider use (e.g. soil taxonomy). The word is derived from the Greek *tasso*, meaning 'arrange'.

TCDD *See* DIOXIN.

***t*-distribution** An estimate of the extent by which values in a small set of data deviate from the mean; it is used to determine the variation within a set of data and to compare two sets of data. As the size of the sample increases the distribution of data approaches normal and the method becomes redundant.

tegulicolous Growing on tiles (e.g. on a roof).

teleology The belief that observed phenomena may be explained in terms of a predetermined purpose or design they appear to serve.

teleutospore *See* TELIOSPORE.

telinite *See* COAL MACERAL.

teliospore (teleutospore) A resting *spore formed by *fungi of the classes Urediniomycetes (*rust fungi) and Ustilaginomycetes, which includes many parasitic species.

telotaxis The movement of an animal in response to a stimulus directly towards or away from the source of stimulation, guided by the use of one sense organ which, by virtue of its structure, provides information regarding the direction of the stimulus (e.g. the eye provides information regarding the direction of a light source).

temperate climate A mid-latitude climate influenced from time to time by both tropical and polar *air masses. Temperature criteria provide subdivisions into warm, cool, or cold-temperate climates. *See also* CLIMATE CLASSIFICATION; KÖPPEN CLIMATE CLASSIFICATION; STRAHLER CLIMATE CLASSIFICATION; and THORNTHWAITE CLIMATE CLASSIFICATION.

temperate deciduous forest *Deciduous summer forest dominated by broad-leaved hardwoods, which occurs over large tracts in the mid-latitudes of Europe, North America, and eastern Asia, but which is restricted in the southern hemisphere to parts of Chilean Patagonia. Most of the original forest has been cleared for cultivation and pasture.

temperate glacier *See* GLACIER.

temperate grassland A type of vegetation which includes in the northern hemisphere the North American *prairies and the Eurasian *steppe, and in the southern hemisphere the *veld of South Africa, the *pampas of Argentina, and the Canterbury Plains of New Zealand. Much of this grassland has been ploughed up or converted into pasture for domesticated animals. Although most of this grassland occurs in parts of the temperate zone where there is a seasonal deficit of moisture, there is often enough to permit tree growth. Presumably fires, both natural and anthropogenic, have been important in the development of this grassland.

temperate rain forest Forest that develops in temperate regions where rainfall is high (typically 1 500–3 000 mm) or fog is frequent (e.g. on the Pacific coast of North America). Broad-leaved evergreen trees are common, conifers are often present, and there are many *epiphytes and climbers. Such forests occur in coastal areas of the south-eastern United States and north-western North America, in southern Chile, in parts of Australia and New Zealand, and in southern Japan and China.

temperature inversion An atmospheric condition in which the typical *lapse rate is reversed and temperature increases vertically through a given layer. In the *troposphere an inversion layer marks conditions of great stability (i.e. a region in which vertical motion is strongly damped, with an absence of turbulence). An inversion acts as a ceiling, preventing further upward convection, and is generally the limit for cloud development. Marked and persistent inversions occur at lower levels, with subsiding air in major anticyclonic cells (e.g. the *Azores high-pressure zone and cold *anticyclones over continents).

temperature-sensitive organ (pit organ) An apparatus, developed in some snakes, which can detect heat radiated from prey and air motion. Paired organs lie in front of the eye in pit vipers (Crotalidae). The lip pits of some boids (Boidae) perform the same function.

temporal summation *See* SUMMATION.

temporary wilting *Wilting that occurs in hot weather when the rate of *transpiration exceeds the rate at which a plant can absorb moisture from the soil. The plant recovers when the temperature falls. *Compare* PERMANENT WILTING POINT.

tendency The degree of behavioural *motivation that may be inferred from the observed behaviour of an animal.

tepee A fold-like structure, shaped like a tepee tent, developed in a *calcrete *soil profile, tidal area, or the margin of a *salt lake, caused by fluctuations in water levels and changes in chemical precipitation.

tephigram A diagram showing the vertical variation in properties of the atmosphere (i.e. temperature and humidity are plotted as a function of pressure, with lines indicating *dry and *saturated adiabatic lapse rates). The changes in temperature and humidity of lifted *air parcels can be compared against the environment (surrounding air) curve, revealing the *stability or *instability of the *air mass and the level at which condensation will occur in the uplifted air. *See also* ENVIRONMENTAL LAPSE RATE.

tephra All the particles or fragments ejected from a volcano, irrespective of size, shape, or composition. The term is usually restricted to material that falls from the air. Tephra composed of particles of dust-like size is dispersed extremely widely following a volcanic eruption and layers of it may be preserved in ice and peat deposits far from the site of the eruption. Analysis of the chemical composition of this fine tephra permits the identification of the precise source and the identity (by means of dating methods) of the particular eruption, and therefore can provide excellent time datum horizons in the study of ice or peat *stratigraphy and *palaeoecology.

tera- From the Greek *teras*, meaning 'monster', a prefix denoting malformation (e.g. *teratogenic). Used in conjunction with *SI units, the prefix (symbol T) means the unit $\times 10^{12}$.

teratogenic Applied to any substance that interferes with the normal development of a foetus or embryo.

terminal node In a *phylogenetic tree, the point at the end of a branch representing a progenic taxon.

terminating stimulus An external or internal stimulus that causes a behaviour response to end because it is completed (e.g. the visual stimulus of a completed nest will terminate nest-building).

termitophile Applied to a species that must spend part of its life closely associated with termites. It may scavenge or steal food, or may prey on the termites.

terpene *See* ISOPRENE.

terrace A nearly flat portion of a landscape which is terminated by a steep edge. It may be produced by any one of a range of processes, so the following varieties are recognized: *altiplanation terrace, *kame terrace, *river terrace, *shore platform, and *solifluction terrace.

terracette (sheep-walk) A small-scale land-form consisting of a long, narrow, stepped feature developed in unconsolidated material mantling a steep hillslope. It may result from slight slippage in material standing at an angle too great for stability.

terra firme *See* AMAZON FLORAL REGION.

terrain (N. American terrane) An area of ground with a particular physical character; an area or region with a characteristic geology (e.g. metamorphic terrain). Many British authors now use the 'terrane' spelling, except for 'terrain correction' and when the word is used in the sense of such expressions as 'rough terrain'.

terrane *See* TERRAIN.

terra rossa European soils, red in colour, which developed over limestone on residual material rich in iron oxide. They are deep and ancient, some of them being pre-*Pleistocene. The *USDA Soil Taxonomy classifies them as *Inceptisols or *Mollisols.

terrestrial radiation Long-wave electromagnetic radiation (wavelengths 4–100 μm, with a peak at 10 μm) from the Earth's surface and atmosphere.

territoriality The establishment, demarcation, and defence of an area by animals, normally during mating ritual. Once *territory has been established the animals can exist without disturbance and with sufficient food for the offspring. Evidence shows that among territorial species individuals without a territory rarely breed.

territory The area occupied by an animal, or by a pair or group of animals, which it or they will defend against intruders. *Compare* RANGE.

Tertiary The first sub-era of the *Cenozoic Era, which began about 65.5 Ma ago and ended 1.806 Ma ago. The Tertiary followed the *Mesozoic and comprises two periods, *Palaeogene and *Neogene. During the Tertiary *angiosperms superseded the *gymnosperms as the dominant plants. Although the name Tertiary is widely used, in years to come it is likely to become obsolete in formal use.

tertiary consumer A carnivore that preys upon other carnivores; a member of the topmost *trophic level of a *food web.

testa *See* SEED.

Tethyan realm The *faunal realm based in the region of the *Tethys Sea. Characteristically *Jurassic to *Cretaceous in age, it comprised a warm-water, tropical to subtropical fauna and flora. The name may also be applied to warm-water fauna and flora of *Mesozoic age outside the area of Tethys, especially when used in contrast to the *Boreal (northern) and Austral (southern) Realms.

Tethys Sea **(Neotethys)** The sea that more or less separated the two great *Mesozoic supercontinents of *Laurasia (in the north) and *Gondwana (in the south). That *land bridges between the two supercontinents existed for much of the Mesozoic is attested by the cosmopolitan character of dinosaur fauna and some plants and by the existence of two distinct floral assemblages in Laurasia and Gondwana.

tetra-odotoxin *See* TETRODOTOXIN.

tetrapod A vertebrate animal that possesses four limbs or is descended from an animal with four limbs. Amphibians, reptiles, birds, and mammals are tetrapods, although snakes, amphisbaenans, and many lizards have lost their limbs in the course of later evolution.

tetrapodomorph A fossil animal that is *morphologically intermediate between fishes and *tetrapods.

tetrodotoxin **(tetra-odotoxin)** A powerful poison which can cause general paralysis in humans after ingestion. It is found in the body, particularly the intestines, liver, and gonads, of pufferfish (Tetraodontidae) and their relatives. Only after careful filleting should these fish be eaten.

texture In pedology, the proportions of *sand, *silt, and *clay in the fine earth of a soil sample, which give a distinctive feel to the soil when handled and are defined by classes of soil texture. The particles that give soil its texture are defined by their size according to UK, US, and international systems of classification.

thallophyte A plant that is not differentiated into root, stem, and leaves; the plant body is known as a thallus.

thalweg A line joining the lowest points of successive cross-sections, either along a river channel or, more generally, along the valley that it occupies. The word is German and means 'valley course'.

thanatocoenosis *See* DEATH ASSEMBLAGE.

thanatosis **(death feigning)** Defensive behaviour in which a prey animal (e.g. an opossum or certain snakes) feigns death. It is usually employed only when escape is impossible. The technique is often effective against predators that kill only living prey. The simulated death posture is maintained for only a short time, after which the animal 'recovers'.

thelytoky Obligatory *parthenogenesis, such that populations consist entirely of females, with occasional functionless males. It is the only genetic system in which fertilization (the union of egg and sperm) is eliminated completely. For example, in some aphids the sexual stage of the life cycle has disappeared and popula-

Fraction	UK	US	International
Stones and gravel	>2.0 mm	>2.0 mm	>2.0 mm
Coarse sand	2.0–0.2 mm	2.0–0.2 mm	2.0–0.2 mm
Fine sand	0.2–0.06 mm	0.2–0.05 mm	0.2–0.02 mm
Silt	60–2 μm	50–2 μm	20–2 μm
Clay	<2 μm	<2 μm	<2 μm

t

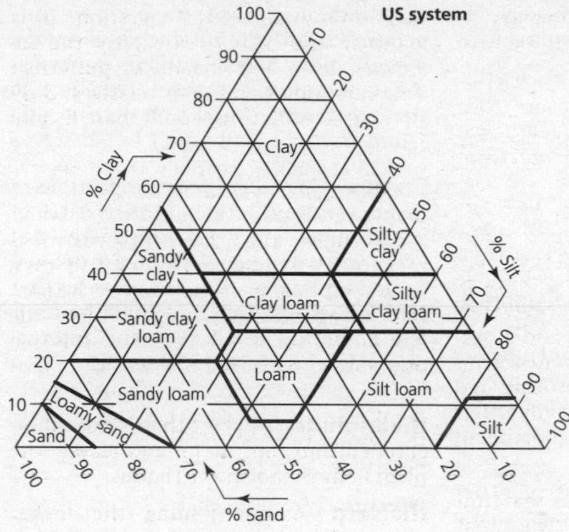

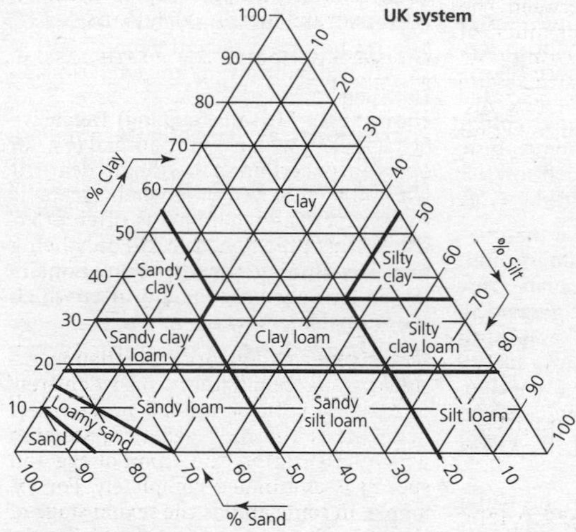

Texture

tions consist exclusively of females; *Eriosoma lanigerum* (woolly apple aphid), a North American species, was introduced to Europe, where it became thelytokous on apples because it has no access to its primary host, the American elm, which presumably supplies a chemical stimulus necessary for sexual reproduction. Some species of parasitic wasps are also thelytokous. Evolution in thelytokous species can therefore take place only by favourable mutations occurring in a single individual, and persisting in the line descending from that individual.

theory of differentiation *See* DIFFERENTIATION, THEORY OF.

theory of generic cycles *See* GENERIC CYCLES, THEORY OF.

thermal (noun) Air that rises in a column over some localized heat source, as a result of its reduced density. When thermals rise to the *condensation level, short-lived clouds (e.g. fair-weather *cumulus) may form. In continued strong thermals, the cloud may develop into large cumulus or *cumulonimbus. When the upward growth reaches a stable layer, the cloud top may spread laterally.

thermal equator The zone of highest mean temperature over the Earth, either in the annual or long-term average or at a given moment. On the long-term average, it is located around 5°N latitude. This position north of the geographical equator results from the generally rather higher temperature of the northern hemisphere as compared with the southern hemisphere; this is because the glaciated Antarctic continent maintains colder summers in the southern hemisphere than does the Arctic, with a much smaller land area, in the northern hemisphere. *See also* COLD POLE.

thermal low Basically a *non-frontal depression associated with strong daytime heating of land surfaces, mainly in summer, which occurs, for example, in Arizona, Spain, and in northern Indian low-pressure cells. The occurrence of thermal lows is also influenced by imbalances in the main wind-streams in the upper *troposphere.

thermal stratification In the water column of a lake, a condition that may develop during the summer in which the *thermocline and *pycnocline change over a short vertical distance, at the *metalimnion. This prevents mixing between the waters of the *epilimnion and *hypolimnion. It occurs because of the limited depth to which solar radiation penetrates and wind movement mixes the water. There is almost no temperature gradient within the epilimnion, but in the metalimnion it may be up to 1°C/m; temperature is fairly constant throughout the hypolimnion. *Photosynthesis occurs in the epilimnion; organisms residing in the hypolimnion subsist on detrital rain falling from the epilimnion, but dissolved oxygen is rapidly used and is not replenished. As solar warming decreases, thermal stratification breaks down, water mixes fully, and the hypolimnion is recharged with oxygen. This is called 'overturn' and occurs in autumn or winter. Inverse stratification occurs where the epilimnion surface freezes. Dense water, at 4°C, sinks to the bottom, and water that is less dense but cooler (about 3.94°C) floats above it. Fish can, therefore, survive the winter in hypolimnion water at temperatures above freezing.

thermic *See* PERGELIC.

thermistor An instrument for measuring temperature, based on a semiconductor, the electrical resistance of which decreases as its temperature increases. Thermistors are sufficiently robust for use in soil, water, or for prolonged periods in the open.

thermoclastis (insolation weathering) Physical *weathering in which the stresses caused by the alternate heating and cooling of a rock become sufficient to cause failure.

thermocline Generally, a gradient of temperature change, but applied more particularly to the zone of rapid temperature change between the warm surface waters (*epilimnion) and cooler deep

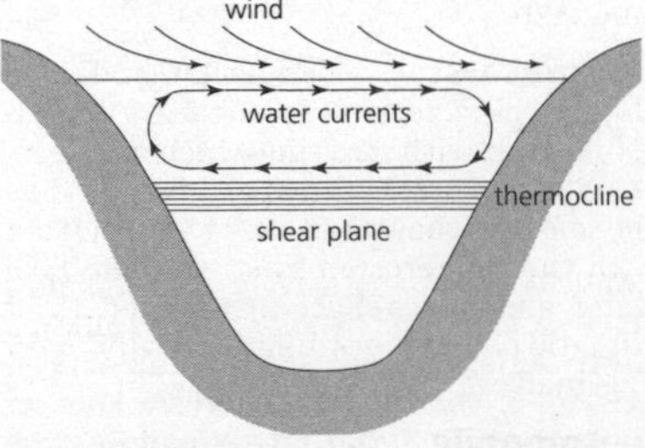

Thermocline

waters (*hypolimnion) in a thermally stratified lake in summer. In the oceans this zone of rapid temperature change starts 10–500 m below the surface and can extend down to more than 1 500 m. In polar regions the thermocline is generally absent, because the ocean surface is covered with ice in winter and solar radiation is small in summer.

thermoduric Applied to an organism (usually a micro-organism) that can tolerate relatively high temperatures; often used specifically for micro-organisms that survive the process of pasteurization.

thermograph An instrument that gives a continuous record of temperature for a day or for a week. The device uses a helical strip of two metals with differing coefficients of expansion. The resulting opening and closing of the coil operates a pen which produces a line over a calibrated chart on a round clock drum.

thermohaline circulation **1.** Vertical circulation induced by the cooling of surface waters in a large water body. This cooling causes convective overturning and consequent mixing of waters. **2.** A system of ocean currents that is driven by the increase in density that occurs close to the edge of the North Atlantic sea ice. Water adjacent to the ice is chilled to 0°C, the temperature at which sea water reaches its maximum density, and accumulates salt that is expelled as water freezes, increasing the density of adjacent water. This dense surface water sinks to the ocean floor, becoming the *North Atlantic deep water, and continues to move southward as the *great conveyor.

thermo-hygrograph *See* HYGROTHERMOGRAPH.

thermokarst A *periglacial landscape that is characterized by enclosed depressions (some with standing water) and so resembles a *karst landscape. It is caused by the selective thaw of ground ice associated with thermal erosion by stream and lake water and may reflect the influences of climatic changes or human activity upon *permafrost.

thermophile An *extremophile (domain *Archaea) that thrives in environments where the temperature is high, typically up to 60°C. *Compare* HYPERTHERMOPHILE.

thermophilous species A warmth-loving species. In pollen analysis (*see* PALYNOLOGY) the term refers in particular to a species, genus, or family characteristic of warmer environments than those otherwise indicated by the pollen record.

thermoremanent magnetism (TRM) *See* ARCHAEOMAGNETISM.

thermosphere The upper zone of the atmosphere, above about 80 km, where solar radiation of the shortest wavelengths is absorbed. In this zone, which includes the *ionosphere, temperature increases with height, but because of the very low atmospheric density there, the heat capacity is minute.

therophyte One of *Raunkiaer's life-form categories, being a plant that completes its life cycle rapidly during periods when conditions are favourable and survives unfavourable conditions (e.g. heat or competition) as seed; it is thus an annual or ephemeral plant. Therophytes are very typical of desert environments and cultivated land. *Compare* CHAMAEPHYTE; CRYPTOPHYTE; HEMICRYPTOPHYTE; and PHANEROPHYTE.

Thiessen polygons A diagram used in studies of plant competition in which the position of each plant is plotted and joined by a straight line to each of its neighbours. The joining lines are then bisected by lines drawn at right angles to them, forming polygons. The size of polygons indicates the density of plants.

thigmotaxis (stereotaxis) A change in direction of locomotion in a *motile organism or cell which is made in response to a tactile stimulus (touch), and where the direction of movement is determined by the direction from which the stimulus is received. It may inhibit movement, causing the organism to come into close contact with a surface. Thigmotaxis is commonly observed in insects.

thigmotropism (haptotropism) The tropic response of a plant organ to the stimulus of touch.

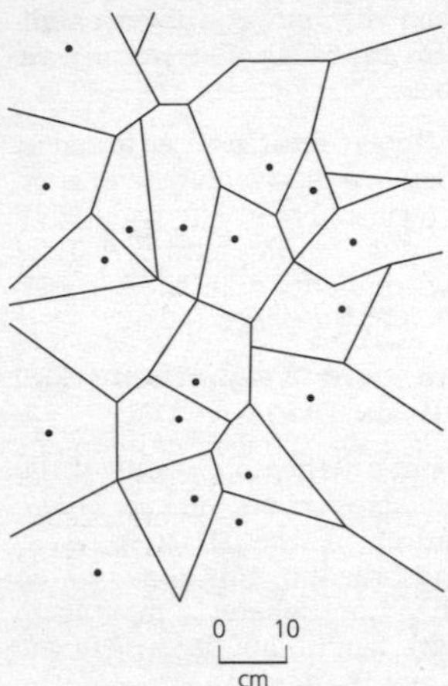

Thiessen polygons

thin In a cultivated crop, to remove some plants in order to increase the area available to others. In forestry, to remove some of the trees in an area to enable those that remain to grow larger, especially in girth.

thiobiont A member of the *thiobios.

thiobios The organisms that inhabit anoxic sediments rich in sulphides.

thixotropy The property possessed by some materials of changing from gel to liquid under shearing stress (e.g. when shaken) and returning to the original state when at rest. Some clay-mineral muds (e.g. montmorillonite and bentonite) have this property. The change is completely reversible and no change in water content or composition occurs.

Thomson, Sir Charles Wyville (1830–82) An Irish biologist and oceanographer, who was appointed professor of natural history at Edinburgh University in 1870. He was especially interested in life in the deep oceans and proposed and then led the five-year **Challenger* expedition (1827–6), in the course of which 4 717 new marine species were discovered. Following the expedition, Thomson began the arrangement of the collections and publication of the results. His studies of temperature differences led him to postulate the circulation of ocean waters.

thorium–lead dating A radiometric dating method based on the radioactive decay of ^{232}Th, to yield $^{208}Pb + 6He^4$, with a half-life of 13 900 million years. The method is not totally reliable, in most cases because of lead loss, and it is usually employed in conjunction with other methods.

thorn forest (thorn scrub, thorn woodland) A tropical vegetation formation with thorny shrubs and bushy trees, perhaps with a few taller trees, set in a sparse ground flora in which grasses are often lacking. It is in this last respect that thorn forest differs principally from *savannah woodland, the paucity of grasses reflecting the increased aridity of the climate. Thorn forest merges in even drier regions with semi-desert scrub, and in wetter climates with *tropical seasonal forest.

thorn scrub *See* THORN FOREST.

Thornthwaite climate classification A system for describing climates devised in 1931 and revised in 1948 by the American climatologist Charles Warren Thornthwaite (1889–1963) which divides climates into groups according to the vegetation characteristic of them, the vegetation being determined by precipitation effectiveness (P/E, where P is the total monthly precipitation and E is the total monthly evaporation). The sum of the monthly P/E values gives the P/E index, which is used to define five humidity provinces, with associated vegetation. A P/E index of more than 127 (wet) indicates rain forest; 64–127 (humid) indicates forest; 32–63 (subhumid) indicates grassland; 16–31 (semi-arid) indicates steppe; less than 16 (arid) indicates desert. In 1948 the system was modified to incorporate a moisture index, which relates the water demand by plants to the available precipitation, by means of an index of *potential evapotranspiration (PE), calculated from measurements of air temperature and day length. In arid regions the moisture index is negative because precipitation is less than the PE. The system also uses an index of thermal efficiency, with accumulated monthly temperatures ranging from 0, giving a frost climate, to more than 127, giving a tropical climate. *Compare* KÖPPEN CLIMATE CLASSIFICATION and STRAHLER CLIMATE CLASSIFICATION.

thorn woodland *See* THORN FOREST.

threat A form of *communication by which an animal may keep rivals or potentially dangerous animals of other species at bay without fighting.

three-cell model An approximate representation of the general circulation of the atmosphere. Air rises over the Equator and subsides over the sub-tropics; these are *Hadley cells. Air subsides over polar regions and spreads into lower latitudes; these are polar cells. The Hadley and polar cells drive a third set of Ferrel cells in middle latitudes. *See also* ROSSBY WAVES.

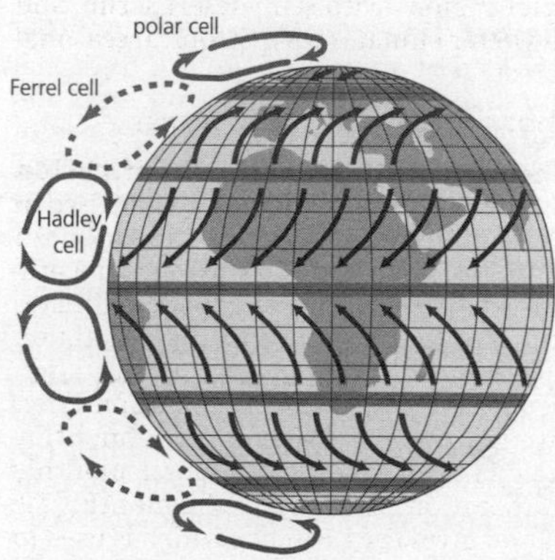

Three-cell model

throughfall The part of *precipitation which, having been intercepted by vegetation, then falls on to the ground surface. *See* INTERFLOW, STEMFLOW, and SUBSURFACE FLOW.

throughflow *See* INTERFLOW and SUBSURFACE FLOW.

thufur An earth hummock some 0.5 m high and 1–2 m wide, which is found in contemporary and past *periglacial environments. Normally it has a core of sediment, suggesting that a form of differential heaving by ground ice was responsible for its development.

thymine A *pyrimidine *base found in nucleic acids.

tidal barrage A dam or barrier built across a tidal channel to allow a power-generating plant to operate. The siting of the barrage requires there to be a *tidal range in excess of 5 m and a large water body open to the sea, with a narrow entrance. One barrage and generating plant in operation since 1960 is on the Rance estuary in Brittany, France. It generates electricity by the use of two-way turbines, as water flows in and out of the *estuary through the barrage. The Severn estuary and Morecambe Bay are potential sites for tidal barrages in the British Isles.

tidal current An alternating, horizontal movement of water, associated with the rise and fall of the tide. Offshore tidal currents tend to exhibit rotary patterns; those in areas near coasts follow rectilinear paths and reverse periodically (ebb and flow currents). Tidal currents can often reach 2.5 m/sec near shores.

tidal flat An area of intertidal sand flat, mud flat, and *marsh developed in some *lagoons in *mesotidal areas, and in protected bays and estuarine areas along *macrotidal coasts. Extensive tidal flats occur in the Wash (eastern England), Waddenzee (Netherlands), and along the German coast of the North Sea. Tidal flats also occur in warmer climates, as in the Persian Gulf, where carbonate and evaporite deposits develop. In tropical areas tidal flats tend to be colonized by mangrove swamps.

tidal inlet A narrow channel that connects the open sea with a lagoon. Tidal inlets often occur in barrier island systems and are typified by small-scale deltas at each end of the inlets, resulting from the high-velocity *tidal currents that flow through the channels.

tidal range The difference in height between consecutive high and low waters. The tidal range varies from a maximum during *spring tides to a minimum during *neap tides. In tide tables daily high- and low-water heights are given for each geographical locality mentioned.

tidal stream The flow of water in and out of estuaries, bays, and other restricted coastal openings associated with the rise and fall of the tide. Landward (flood) and seaward (ebb) streams often follow different paths in shallow-water areas, so forming ebb/flood avoidance cells and a braided pattern of sandbanks.

tide The periodic rise and fall of the Earth's oceans, caused by the relative gravitational attraction of the Sun, Moon, and Earth. The effect of the Moon is about twice that of the Sun, giving rise to the *spring–*neap cycle of tides. Variation in tides is caused by: (*a*) changes in the relative positions of the Sun, Moon, and Earth; (*b*) uneven distribution of water on the Earth's surface; and (*c*) variation in the seabed topography. Semidiurnal tides are those with two high and two low waters (period 12 hours and 25 minutes) during a tidal day (24 hours and 50 minutes). Diurnal tides have one high and one low water during a tidal day.

tile drain Short lengths of concrete or ceramic pipe placed end to end to make a drain, which is laid at any appropriate depth and spacing to remove water from the soil by allowing water to enter at the joints. Nowadays continuous, slotted, plastic piping is often used.

till A collective term for the group of sediments laid down by the direct action of glacial ice without the intervention of water. The sediments may be classified in terms of particle size or grouped according to the basic process of debris release. Subglacial melt gives rise to lodgement till; surface *ablation gives ablation till, followed by *flow till after further movement; and the general thaw of static ice produces melt-out till. Although the material is unsorted by water action, the alignment of stones within till can indicate the direction of ice movement. *See also* BOULDER CLAY.

tiller In grasses, a lateral shoot arising at ground level.

till fabric analysis The measurement of the orientation and direction of slope of the fragments contained within a *till, and the subsequent plotting and analysis of the data. These may indicate the direction of a former glacial advance (particles tend to lie parallel to the direction of ice movement).

tillite A lithified deposit of *boulder clay or *till.

till plain A smooth plain underlain by *till. It may be well preserved, as in the mid-western USA, or dissected by later *erosion, as in the lowlands of central and eastern England.

tilth The physical condition of the soil that determines its suitability for plant cultivation.

timber **1.** Wood in the form of unsquared logs. **2.** Tree trunks that are suitable for beams or for sawing into planks.

timber line **(waldgrenze)** A line that marks that altitudinal limit of trees that are in a close canopy and grow erect and tall. It occurs below the baumgrenze or *tree line proper, and below the *kampfzone (in which the trees often show the *krummholz condition). It is commonly depressed to below its natural level by fires caused by human activities.

time series A data set in which the intervals are of equal time and arranged in order of occurrence. The series may be for individual or averaged values which can be analysed by statistical techniques, including spectrum or harmonic analyses.

Tinbergen, Nikolaas (1907–88) A Dutch-born, but later British, zoologist who shared (with Konrad *Lorenz and Karl von *Frisch) the 1973 Nobel Prize for Physiology or Medicine for studies of animal behaviour under natural conditions. Tinbergen was especially noted for his studies of social organization among gulls. He became professor of animal behaviour at the University of Oxford in 1966, and emeritus professor in 1974.

tissue culture *See* CELL CULTURE.

tjaele **(frost table)** A frozen surface at the base of the *active layer, which moves downwards as thaw occurs. The tjaele should not be confused with the upper limit of *permafrost, the permafrost table.

toadstool A loose term for any umbrella-shaped fungal fruit body, or for any such fruit body that is inedible or poisonous. *Compare* MUSHROOM.

tobacco mosaic A disease of tobacco, tomato, and related plants, in which the leaves become distorted and are marked with a characteristic mosaic of light and dark green. It is caused by the tobacco mosaic virus, which contains RNA.

tobacco necrosis virus A virus containing RNA, which can infect a wide range of plants; symptoms of infection range from small leaf lesions to *necrosis of the entire plant. It can be transmitted by the soil fungus *Olpidium brassicae*.

tobacco rattle A disease which occurs in tobacco, tomato, potato, and other plants. Symptoms vary with the nature of the plant infected. The causal agent is a virus which consists of two entirely separate particles, both of which must be present in the plant for replication of the complete virus to occur.

toeset The base of a *Gilbert-type delta, usually characterized by fine-grained sediment deposited at the toe of the prograding (*see* PROGRADATION) *foreset. *See also* TOPSET.

tolerance, limits of *See* LIMITS OF TOLERANCE.

tolerant strategy *See* GRIME'S HABITAT CLASSIFICATION.

tombolo A *spit that links an island to the mainland or to another island, formed by deposition when waves are refracted round the island.

tomentose Woolly; covered with a fine mesh of hairs.

Tonga–Kermadec Trench The oceanic *trench in the western Pacific Ocean which forms part of the boundary between the Indo-Australian and Pacific Plates.

tonhäutchens *See* CUTAN.

tool use The use by an animal of an object external to itself and not attached to any substrate, which is held, carried, or otherwise manipulated in order to achieve an objective.

topogenous mire **(topogenous peat)** A type of bog that forms under climatic conditions of reduced rainfall, with consequent lower humidity and summer drought, which restrict the growth of wetland vegetation to areas where precipitation is concentrated (e.g. valley bottoms).

topogenous peat *See* TOPOGENOUS MIRE.

topology In *phylogenetics, the branching pattern of a *phylogenetic tree.

toposequence A sequence of soils in which distinctive soil characteristics are related to topographic situation.

topset The upper, nearly horizontal layers of sediment deposited on a *Gilbert-type delta, generally characterized by the coarsest sediment found on the prograding (*see* PROGRADATION) delta. *See also* FORESET and TOESET.

topsoil **1.** The superficial layer of soil moved in cultivation. **2.** The A horizon of a *soil profile. **3.** Any surface layer of soil.

tor A mass of exposed bedrock, standing abruptly above its surroundings, and typically but not exclusively developed on granitic rocks. It may be formed by selective subsurface *weathering followed by the removal of the weathered debris, by differential frost-shattering, or as an end-product of *scarp retreat under semi-arid conditions.

tornado A relatively small-scale (about 100 m diameter) 'twisting' or rotating column of air, like a funnel, with high wind speeds and great destructive force over the narrow path of its movement. Such systems are especially frequent in unstable air conditions in the central parts of the USA.

Torrey, John **(1796–1873)** An American botanist and chemist, who held professorships in chemistry, mineralogy, geology, and natural history before being appointed New York State botanist in 1836. While still a student he was one of the founders in 1817 of the Lyceum of Natural

History, which later became the New York Academy of Sciences. He originated the *Flora of North America* and between 1838 and 1843 collaborated on its early sections with his former pupil, Asa *Gray (who completed it).

torrid zone General climatic name for the region between the northern and southern tropics, broadly the equatorial zone.

Tournefort, Joseph Pitton de (1656–1708) A French botanist who became a professor at the Jardin du Roi in Paris and is remembered for producing a system of plant classification and nomenclature in the 1690s. His *Institutiones Rei Herbariae* (1700) might be described as the most important precursor of *Linnaeus's work. He also wrote *A Voyage into the Levant* (1718) in which he was the first to describe the common azalea and rhododendron, as well as other plants he collected during this journey. The genus *Tournefortia* of the Boraginaceae was named for him.

tower karst A form of karstic morphology (*see* KARST) that is developed mainly in low latitudes, characterized by residual hills of limestone rising from a flat plain. The hills have near-vertical sides, resembling towers.

toxin **1.** Any poisonous substance of plant or animal origin. **2.** A microbial product which is poisonous to animals or plants. The symptoms of many types of human disease are the result of the production of one or more toxins by the pathogen. Toxins usually act at specific sites in the body (e.g. neurotoxins affect nerves, enterotoxins affect the gut).

t

T-peg A device for measuring the rate of surface *creep which consists of a metal rod to which a cross-piece is attached. The apparatus is inserted into the *regolith and the cross-piece is levelled. Soil creep causes the cross-piece to tilt and the degree of inclination is measured with a graduated spirit-level. The relative movement between the ground surface and the depth of insertion can then be inferred.

trace element **1.** An element that occurs in minute but detectable quantities in minerals and rocks, much less than 1 per cent. All elements except the most common rock-forming ones (O, Si, Al, Fe, Ca, Na, K, Mg, and Ti) generally occur as trace elements, except where they are locally concentrated in their *ores. **2.** *See* ESSENTIAL ELEMENT.

trace fossil **(ichnofossil)** A structure formed in a sediment by the action of a living organism (e.g. a tube, burrow, footprint, or groove made by crawling over a surface) and preserved when the sediment becomes a sedimentary rock. Traces are most commonly found at interfaces between different rock types (e.g. between sandstone and shale) and are classified in various ways, including their forms and the places of their occurrence. *See also* FOSSILIZATION.

tracer A substance that is used to follow the passage of *groundwater in places where it cannot be observed directly. Typical tracers include fluorescent dyes and salt. The presence of radioactive isotopes (e.g. tritium and carbon-14) may also be used as tracers in that they allow the age of groundwater to be determined. The presence of small amounts of other substances may also be used to make deductions about the origin and flow path of groundwater.

traction carpet *See* BED LOAD.

traction load *See* BED LOAD.

trade-wind inversion The inversion of temperature *lapse rate with height over a major zone of the *trade-wind belt, which is very significant in tropical meteorology. Moist tropical air, extending up to 2–3 km above the surface, is 'boxed in' or 'trapped' by dry, clear, warmer air above, resulting from subsidence in subtropical *anticyclones. The inversion forms where the subsiding air meets a surface flow of cooler maritime air. The top of a cloud layer marks the base of the inversion.

trade winds The prevailing winds that blow towards the equator from the *subtropical highs, from the north-east in the northern hemisphere and from the south-east in the southern hemisphere. They form part of the *Hadley cell circulation and are the most dependable of all winds. Their name is derived from the old meaning of the word 'trade' as 'established track'. The winds blow at an average 18 km/h in the northern hemisphere and 22 km/h in the southern hemisphere. They are most reliable on the eastern sides of the Atlantic, Pacific, and Indian Oceans, and are stronger in winter than in summer.

trait Any detectable *phenotypic property of an organism; a *character.

tramontana A local wind in the Mediterranean region, which brings dry, cold conditions from the north across the mountains.

tramp species Species that have been spread around the world inadvertently by human commerce. For example, the brown weed *Sargassum muticum* (japweed or strangleweed) was inadvertently introduced into British coastal waters, probably with the importation of the Japanese oyster (*Crassostrea gigas*); its common name refers to its effects on pipes and outboard engines. *Eichhornia crassipes* (water hyacinth), taken to the USA from Venezuela by Japanese exhibitors at the 1884 Cotton Exposition in New Orleans, escaped and is now a serious weed of many rivers in low latitudes throughout the world. The brown rat (*Rattus norvegicus*), originally from south-east Asia, and the house mouse (*Mus musculus*), originally from the Russian–Turkish border region, are well known animal examples.

tranquil flow (subcritical flow) *See* CRITICAL FLOW.

transduction The transfer of host DNA from one cell to another by a *virus.

transect (isonome) A line used in ecological surveys to provide a means of measuring and representing graphically the distribution of organisms, especially when they are arranged in a linear sequence (e.g. up a seashore, or across a woodland margin) or to investigate an environmental gradient (e.g. of salinity across a *salt marsh). Recordings are made at intervals along the line. A transect is particularly useful for detecting transitions or distribution patterns. *See* BELT TRANSECT and LINE TRANSECT.

transfer ribonucleic acid *See* TRANSFER-RNA.

transfer-RNA (transfer ribonucleic acid, t-RNA) A generic term for a group of small *RNA molecules, each composed of 70–80 *nucleotides arranged in a cloverleaf pattern stabilized by hydrogen bonding. They are responsible for binding *amino acids and transferring these to the *ribosomes during the synthesis of a polypeptide (i.e. during translation). At the *ribosomes, which are attached to the *messenger-RNA (m-RNA), the 'reading frame' indicates the three m-RNA nucleotides that form the next triplet codon in the sequence: whichever t-RNA molecule carries the complementary anticodon can associate with the ribosome such that the amino acid that it bears can be joined on to the end of the growing polypeptide.

transfluence *See* GLACIAL BREACH.

transgenic Applied to an organism which contains genetic material from another organism, usually supplied by molecular biological techniques.

transgression, marine An advance of the sea to cover land areas, caused by a rise in the sea level relative to the land. *Compare* REGRESSION.

transition The substitution of a *purine base containing nucleotide for another with a purine base, or a *pyrimidine base containing nucleotide for another with a pyrimidine base. Transitions generally occur more frequently in evolution than *transversions.

translation The production of a polypeptide sequence of *amino acids from the information encoded in the *nucleotide sequence of *messenger-RNA. The process takes place at the *ribosome.

translocation **1.** The movement of dissolved substances within a plant, usually from the site of synthesis or uptake to centres of growth or storage. **2.** The movement of soil materials in solution or in suspension from one *horizon to another. **3.** The movement of a DNA segment from one point in the *genome to another, resulting in no change in *copy number.

translucidus The Latin *translucidus*, meaning 'transparent', used to describe a variety of cloud, occurring in extensive layers or sheet form, which is translucent, allowing the Sun or Moon, and occasionally some stars, to be visible. This cloud itself may be *stratus, *stratocumulus, *altostratus, or *altocumulus. *See also* CLOUD CLASSIFICATION.

transmissivity The rate at which *groundwater is transmitted through a unit width of an *aquifer under a unit hydraulic gradient.

transmutation **1.** The transformation of one element into another by radioactive decay. **2.** The change of one species or type to another.

transpiration The loss of water vapour from a plant to the outside atmosphere. It takes place mainly through the stomata of leaves and the lenticels of stems. Its function is disputed. It may reduce leaf temperature, but its absence from some tropical plants would suggest that this is not essential. It may also be important in mineral absorption and *translocation. However, it may be merely an inevitable concomitant of gaseous exchange which, to be efficient, requires open stomata; as gases are exchanged, water is lost.

transposition The copying of genetic information from one point in the *genome, followed by *insertion into another, resulting in an increase in *copy number of the DNA segment involved.

transposon A DNA element that can insert itself at random into a plasmid or bacterial *chromosome, independently of the host cell-recombination system. In addition to other genes, transposons carry genes that confer new phenotypic properties on the host cell (e.g. resistance to some antibiotics).

trans-Saharan seaway The marine seaway that extended from *Tethys in the north through what are now Libya, Chad, Niger, and Nigeria, to the newly developing South Atlantic Ocean during two intervals in the Late *Cretaceous. On both occasions ammonite and ostracod faunas are known to have migrated through this seaway. At the southern end (in present-day Nigeria) there was a structural control in the form of the Benue Trough, but the remainder of the seaway appears to have been controlled only by global sea-level change.

transverse dune *See* DUNE.

transversion The substitution of a *purine containing *nucleotide for a *pyrimidine containing nucleotide, or vice versa. *Compare* TRANSITION.

trapdoor spiders Spiders of the families Antrodiaetidae, Actinopodidae, and Ctenizidae (order Araneae), most species of which are 3 cm long and live in silk-lined tubes dug into the ground and covered by silk lids. Passing insects are attacked and pulled into the tubes with great rapidity, and the tubes are enlarged as the spiders grow. Trapdoor spiders are found in the Americas, Africa, and Australia.

tree A woody plant with a single main stem (the trunk), that is unbranched near the ground; some trees (e.g. oak and ash) have multi-trunked forms. At the end of each growing season there is no die-back of aerial parts, apart from the loss of foliage. *Compare* HERB; SHRUB; and SUBSHRUB.

tree-borer *See* INCREMENT BORER.

tree line **(baumgrenze)** A line through the last of the stunted trees, forming the latitudinal or altitudinal limit beyond which the climate is too cold for trees to grow. This is normally the region where the mean summer temperature is lower than 10°C. The tree line marks the boundary between *tundra vegetation and the bare rock, snow, and ice of the high Arctic and Antarctic. On mountainsides the height of the tree line varies with latitude (i.e. it is higher in the tropics than in higher latitudes) and climate type. Trees below the tree line are erect and grow in dense stands. Trees close to the tree line are sparsely scattered and prostrate or nearly so. *See* KRUMMHOLZ.

tree ring **(annual ring, growth ring)** A sheath of cells appearing as one of a series of concentric rings in the cross-section of a woody stem. Each ring is usually the result of a single yearly growth flush starting in spring and ceasing in the late summer. The new wood (xylem) cells arise from renewed activity of the vascular cambium. A sharp boundary usually occurs between the rings since cells formed in spring are typically large, thin-walled, and appear pale in colour when compared with the small, thick-walled cells of late summer. In some dicotylodenous species vessel distribution also varies, with most vessels occurring in the early (spring) wood. These ring-porous species contrast with ring-diffuse species in which the vessels are distributed evenly throughout the growth ring. The growth check that causes variable cambial activity may be temperature or water stress or both. In some extreme environments growth may not always be renewed on an annual basis, leading to absent rings ('missing years') and false series (multiple rings in the same year). Dating by tree rings is difficult in these circumstances. *See also* DENDROCHRONOLOGY; DENDROECOLOGY; and FIRE SCAR.

tree-ring analysis *See* DENDROCHRONOLOGY; DENDROCLIMATOLOGY; and DENDROECOLOGY.

tree-ring index An annual *tree-ring width that has been standardized for age. Most dendrochronological work (*see* DENDROCHRONOLOGY) is based on tree-ring indices or standardized series, rather than on the original ring widths.

tree veld In South Africa, grassland with an open or light cover of trees which gives it a parkland aspect. It is thought that in the absence of fire much, if not all, of the veld would develop into forest or *scrub.

trellis drainage pattern *See* DRAINAGE PATTERN.

trench **(oceanic trench)** An elongate depression of the ocean floor which runs parallel to the adjacent volcanic islands or continent. Oceanic trenches are up to 11 km deep, typically 50–100 km wide, and may be thousands of kilometres long. In cross-section the trench slopes are usually asymmetric, with a steeper slope on the landward side. Most trenches are associated with subduction zones.

trend surface analysis A special case of multiple regression analysis, in which the independent variables are spatial co-ordinates. This enables assessment of spatial trend in data values. The technique is used, in particular, in geographical studies.

trial-and-error learning *Learning in which an animal comes to associate particular behaviours with the consequences they produce. This tends to reinforce the behaviour (i.e. the behaviour is likely to be repeated if the consequences are pleasant, but not if they are unpleasant). Such learning is believed to involve a process of classical *conditioning followed by *operant conditioning.

Triassic The earliest of the three periods of the *Mesozoic Era, which lasted from 251 Ma ago to 199.6 Ma ago. As a result of the mass extinctions of the late *Palaeozoic, Triassic communities contained many new faunal and floral elements. Among these were the ammonoids, modern corals, various molluscs, the dinosaurs, and certain gymnosperms.

tribe In plant *taxonomy, a rank between family and genus, comprising genera whose shared features serve to distinguish them from other genera within the family. The names of tribes bear the suffix -eae. Tribes may be grouped to form subfamilies and divided to form subtribes.

trichiation *See* CHAETOTAXY.

trigonous Triangular in cross-section.

trillion One thousand billion (i.e. 10^{12}).

trim line A boundary (or zone) between frost-shattered and glacially scoured bedrock in an upland region. It marks the junction between glacial and *periglacial activity and so indicates the position of a former glacial limit.

triquetrous Triangular in cross-section with a sharp angle at each corner.

tritium clock The decay of the radioactive isotope of tritium (T), which is used to measure the age of water up to about 30 years and to monitor the movement of *groundwater. The tritium is produced naturally in the upper atmosphere by the action of fast cosmic-ray neutrons on ^{14}N and combines with hydrogen and oxygen to form HTO, which is then dispersed throughout the hydrosphere. The tritium has a half-life of 12.26 years.

TRM (thermoremanent magnetism) *See* ARCHAEOMAGNETISM.

t-RNA *See* TRANSFER-RNA.

troph- A prefix, or part of a compound word (e.g. oligotrophic), derived from the Greek *trophe*, meaning 'nourishment', and associating the word in which it occurs with food or nutrition.

trophallaxis Food sharing.

trophic Pertaining to nutrition, food, or feeding.

trophic cascade The movement through three or more *trophic levels of a *food-web of nutrients released by organisms high in the web (i.e. secondary or tertiary consumers).

trophic fountain The movement through three or more *trophic levels of a *food-web of nutrients suddenly released in large amounts near the base of the web. For example, the death after mating of periodical cicadas deposits large numbers of cicada corpses on the forest floor, each corpse containing about 10 per cent nitrogen. As the bodies decompose this nitrogen enters the *detrital pathway, passing from there into the soil solution and returning to green plants as a major nutrient boost to the primary *producers.

trophic level A step in the transfer of food or energy within a chain. There may be several trophic levels within a system, for example, *producers (autotrophs), primary *consumers (herbivores), and secondary consumers (carnivores); further carnivores may form fourth and fifth levels. There are rarely more than five levels since usually by this stage the amount of food or energy is greatly reduced. *See also* ECOSYSTEM.

trophic level assimilation efficiency *See* ECOLOGICAL EFFICIENCY.

tropical air An *air mass that acquires its characteristics in the tropics, as (*a*) maritime tropical air, originating over oceans; or (*b*) continental tropical air, originating over land masses (especially over North Africa and the Middle East). Movements of these air masses poleward, with modification of their original characteristics, can have marked influence on weather in mid-latitudes.

tropical cyclone (revolving storm) A local area of low pressure (cyclone) that generates extreme winds and rainfall, and produces *storm surges that often cause coastal flooding. Known as hurricanes in the Atlantic and Caribbean, typhoons in the Pacific and Indian Oceans and China Seas, and cyclones in the Bay of Bengal, with many other more regional names, tropical cyclones form only where the sea-surface temperature is greater than 27°C over a large area. They cannot form closer to the equator than latitude 5°, because they need the *Coriolis effect to set the system turning. The storms develop from tropical disturbances that intensify into tropical storms before becoming tropical cyclones. A tropical cyclone has a central eye surrounded by an eyewall of towering *cumulonimbus clouds. Winds and rainfall are most intense in the eyewall. Inside the eye the air is calm, warmer than the air in the eyewall, and the sky is largely clear. Tropical cyclones move at 16–24 km/h in a westerly direction in either hemisphere, then turn away from the equator and accelerate. The storms are classified by the *Saffir–Simpson hurricane scale according to their sustained wind speeds, surface atmospheric pressure in the eye, and height of storm surge. The weakest hurricane has sustained winds greater than 121 km/h; the most violent has winds of more than 250 km/h.

tropical forest A category of vegetation comprising a variety of formations including rain forest, seasonal forest or *monsoon forest, and *thorn forest.

tropical moist forest A convenient term that describes *tropical rain forest plus *tropical seasonal forest (i.e. the forests of the wetter tropics).

tropical montane forest Two distinct formations, lower and upper montane forest, each of which has a characteristic structure, *physiognomy, and flora; the two are usually sharply bounded. Lower tropical montane forest is tall. Upper tropical montane forest is often a forest of short trees 10–12 m high in a single stratum. When their trunks and boughs are misshapen and covered in mosses, *lichens, and liverworts this is *mossy forest, which coincides with belts of mist and cloud at heights of 2 000–3 000 m. Above the cloud zone (e.g. in New Guinea), the trees are taller and the 'mosses' less in evidence. On the highest tropical mountains, subalpine forest occurs as a further zone.

tropical rain forest A term invented in 1898 by the botanist A. F. W. *Schimper (*tropische Regenwald*) to describe the forests of the permanently wet tropics. His definition still stands: 'evergreen, at least 30 m tall, rich in thick-stemmed lianes, and in woody as well as herbaceous epiphytes'. Nearly all independent plants have the form of trees and at maturity most attain 10–30 m tall. By convention, the trees are divided into four strata (but these are mere abstractions) of which the topmost is the emergent stratum of usually single, huge trees (up to 60 m or more) standing head and shoulders above the continuous canopy.

tropical seasonal forest A tropical forest that grows in regions with a marked dry season. It is a distinct formation type, dominated by both *evergreen and *deciduous broad-leaved trees, flanking the rain forest in areas which have a marked dry season. The deciduous species are mostly in the canopy and become commoner as the climate gets more seasonal. There is some defoliation, especially of the biggest trees, during the dry season, the degree depending on the severity of the moisture deficit. The structure of the forest is also simpler than that of the rain forest, with fewer tree strata, and less luxuriant growths of climbing and herbaceous plants. Examples are found on all the continents with tropical territory, but are especially extensive in Central and South America, continental south-eastern Asia, and northern Australia. In Africa *savannah and related types of vegetation replace much of this kind of forest. *See* MONSOON FOREST.

tropical subalpine rain forest A formation of low, stunted trees with tiny leaves which is found just below the tree line on the very tallest mountains in the humid tropics.

tropic movement *See* TROPISM.

tropism (tropic movement) A directional response by a plant to a stimulus. It may be positive or negative (i.e. towards or away from the source of the stimulus). The plant may respond by growth or *turgor changes so that parts of it bend towards, away from, or at right angles to the direction of the stimulus. The suffix -tropism is used in relation to responses to particular stimuli (e.g. phototropism is a response to light, geotropism to gravity, and chemotropism to a chemical substance).

tropopause The boundary separating the lowest layer of the atmosphere (*troposphere), in which air temperature generally decreases with height, from the layer above (*stratosphere), in which temperature remains constant or increases with height. The altitude of the tropopause varies according to sea-surface temperature and season, but also over shorter periods, from an average of 10–12 km over the poles (occasionally descending to 8 km or below) to 17 km over the equator.

troposphere The layer of the atmosphere between the Earth's surface and the *tropopause, within which the air temperature on average decreases with height at a rate of about 6.5 °C/km, though variations that sometimes occur include inversions (temperature increase with height within some limited layer). Most of the atmospheric turbulence and weather features occur in this layer, which contains almost all the atmospheric water vapour and most of the *aerosols in suspension in the atmosphere (although there is also an important aerosol layer at about 22 km).

tropotaxis The movement of an animal, typically in a straight line, in response to a stimulus directly towards or away from the source of the stimulus. It is made possible by the possession by the animal of more than one receptor, so that the strength of a stimulus to either side of its body may be detected simultaneously and compared.

trough **1.** An extension of low atmospheric pressure from the central regions of a low-pressure system into a zone where generally higher pressure prevails. **2.** A meander towards the equator in the flow of the upper westerly winds over middle latitudes. (The 'equatorial trough', where *trade winds meet, is synonymous with the *intertropical convergence zone'.)

trowal A Canadian meteorological term for the line of the upper front of an *occlusion and the region about it where the warm air, having been lifted off the surface, is still relatively low. It is a *trough or valley of warm air that is still undergoing lifting, and is normally marked by clouds and *precipitation.

true age *See* ABSOLUTE AGE.

truncated spur A blunt-ended, sloping ridge which descends the flank of a valley. Its abrupt termination is usually due to *erosion by a *glacier which tends to follow a straighter course than the former river.

tsunami A seismic sea wave of long period, produced by a submarine *earthquake, underwater volcanic explosion, or massive gravity slide of seabed sediment. In the open ocean, such waves are barely noticeable even though they may be travelling at 700 km/h, but on reaching shallow water they build up to heights of more than 30 m and can cause severe damage in coastal areas.

tuba The Latin *tuba*, meaning 'trumpet', used to describe a supplementary cloud feature of *cumulonimbus or sometimes *cumulus, characterized by a column or cone of cloud projecting from the cloud base. *See also* CLOUD CLASSIFICATION.

tuber A swollen stem or root that functions as an underground storage organ. Stem tubers (e.g. in potatoes) often produce buds from which aerial stems arise the following season. Root tubers produce no buds, or produce buds only at the point where the tuber is attached to the stem of the plant.

tufa **(calc-tufa)** A sedimentary rock formed by the deposition or precipitation of calcium carbonate, or more rarely silica, as a thin layer around saline springs, or by the encrustations on *stalactites and *stalagmites.

tuff A rock formed by the compaction of particles ejected from a volcano, in which individual particles are less than 2 mm in size.

tulip tree *See* BICENTRIC DISTRIBUTION.

Tullgren funnel A device used to extract small invertebrate animals from a dry soil sample. The sample is placed in a container with a base made from gauze, with a mesh designed to hold soil particles but permit the animals to pass. The container is arranged over a funnel, with a light above. The heat causes the animals to move away from the top of the sample, through the gauze sheet and into the funnel from which they can be collected. Most species are collected after two hours, but complete extraction takes 2–3 days.

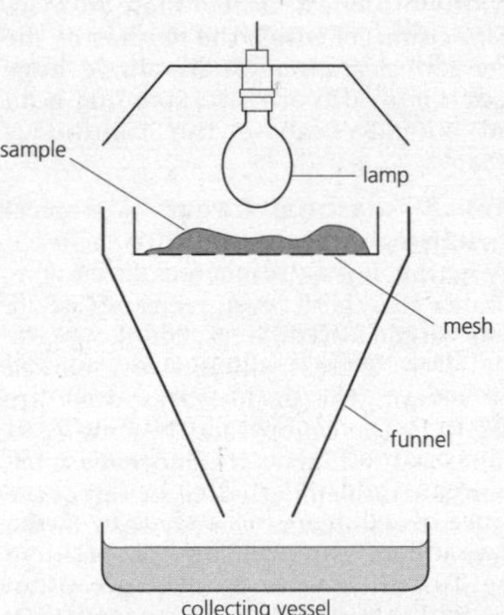

Tullgren funnel

tundra A treeless plain of the Arctic and Antarctic characterized by a low, 'grassy' sward. Actually, although grasses are rarely absent, sedges (*Carex* species), rushes (*Juncus* species), and wood rushes (*Luzula* species) are the dominant plants, together with *perennial herbs, dwarf woody plants, and various bryophytes and *lichens.

Tundra Soil One of the Great Soil Groups, within suborder 1 of the order Zonal Soils of the 1949 *USDA system of soil classification, based originally on the work of V. V. Dokuchaev, but now superseded by the USDA Soil Taxonomy in which Tundra Soils are classified as *Inceptisols. They occur on ground that drains poorly (mainly because of *permafrost), and are acid, 30–60 cm deep, have a high content of organic matter at the surface, and a microrelief formed by freezing and thawing. Their formation, and the decomposition of organic matter, is inhibited by the low temperature.

tunnel valley A valley cut by a subglacial stream escaping from beneath an *ice sheet. It is well developed in Denmark (where such valleys are called *tunneldale*) and in Germany (*Rinnentaler*). Individual examples may be up to 75 km long and 100 m deep, with steep sides and flat floors. The long profile may be irregular, as a result of water under pressure being locally forced uphill, and so the valley may now be occupied by a string of lakes ('ribbon lakes').

turbidite A sedimentary deposit laid down by a *turbidity current.

turbidity current A water current that flows as a result of a density difference created by dispersed sediment within the body of the current. Such currents occur off *delta fronts, in lakes, and in oceans, and are initiated by the disturbance of sediments on a slope by strong wave action, *earthquake shock, or slumping. Turbidity currents in the oceans are thought to move rapidly (at speeds of up to 7 m/s) down the *continental slope or submarine canyons along the seabed, and to deposit originally shallow-water sediments at the foot of the slope or on the *abyssal plain.

turgor The rigidity of a plant and its cells and organs, resulting from hydrostatic pressure exerted on the cell walls.

turgor pressure *See* PRESSURE POTENTIAL.

turnover **1.** The proportion of a *population that is lost (e.g. by death or emigration) or gained (e.g. by reproduction or immigration) in a given period. **2.** The replacement of species by extinction within an area and their replacement by newly evolved or immigrant species. **3.** The ratio of the energy entering a *community or *ecosystem by primary production (*see* PRIMARY PRODUCTIVITY) to the *biomass.

turnover rate A measure of the movement of an element in a *biogeochemical cycle. Turnover rate is calculated as the rate of flow into or out of a particular nutrient pool, divided by the quantity of nutrient in that pool. Thus it measures the importance of a particular nutrient flux in relation to the pool size. *Compare* TURNOVER TIME.

turnover time The measure of the movement of an element in a *biogeochemical cycle; the reciprocal of *turnover rate. Turnover time is calculated by dividing the quantity of nutrient present in a particular nutrient pool or reservoir by the flux rate for that nutrient element into or out of the pool. Turnover time thus describes the time it takes to fill or empty that particular nutrient reservoir.

two-way table *See* CONTINGENCY TABLE.

type specimen **(holotype)** An individual plant or animal chosen by taxonomists to serve as the basis for naming and describing a new species or variety. *Compare* LECTOTYPE; NEOTYPE; PARATYPE; and SYNTYPE.

typhoon The name given to a *tropical cyclone that forms over the Pacific and China Seas.

ubac Applied to the north-facing or shaded slopes of Alpine valleys, which often remain forested, because their climate makes them less suitable for cultivation and settlement than the sunnier *adret slopes.

ULR *See* UNIT LEAF RATE.

Ultisols An order of *mineral soils, identified by an *argillic B *soil horizon with a base saturation of less than 35 per cent, and red in colour from iron oxide concentration. Ultisols are leached, acid soils, associated with humid subtropical forest environments. *See also* RED PODZOLIC SOILS.

ultraviolet radiation (UV) Electromagnetic radiation at wavelengths between 100 and 400 nanometres (nm), lying just beyond the high-energy (violet) end of the visible-light band of the solar spectrum. Radiation at UV wavelengths comprises about 5 per cent of the total energy the Earth receives from the Sun (55 per cent is in the infrared and 40 per cent in the visible light wavebands). The UV waveband is conventionally divided into three: UVA, 315–400 nm; UVB, 280–315 nm; and UVC, 100–280 nm. All the UVC radiation and some of the UVB is absorbed by the atmosphere (mainly in the *ozone layer). Prolonged exposure to UVB causes sunburn in fair-skinned humans and can lead to non-melanoma skin cancers; a link between UVB exposure and melanomas is probable, but other factors are also implicated. Certain plant cells are also damaged by UVB, but threats to *phytoplankton are limited, because UV is absorbed strongly by sea water (25 per cent is absorbed in the uppermost metre of water and none penetrates below 20 m).

umbric epipedon A surface *soil horizon similar to a *mollic epipedon but with a base saturation of less than 50 per cent. It is a *diagnostic horizon. The name is from the Latin *umbra*, meaning shade.

Umbrisols Soils with an *umbric epipedon. Umbrisols are a reference soil group in the *FAO *soil classification.

unavailable water Water that is present in the soil but cannot be absorbed rapidly enough by plants for their needs because it is held so strongly to the surface of soil particles.

uncinus The Latin *uncinus*, meaning 'hooked', used to describe a species of *cirrus cloud that is hooked at the end of its filaments and on the upper parts. *See also* CLOUD CLASSIFICATION.

unconditional stimulus *See* CONDITIONING.

unconfined aquifer *See* AQUIFER.

unconformity trap A *stratigraphic trap formed by the folding, uplift, and *erosion of porous *strata, followed by the deposition of later beds which can act as a seal for oil, gas or water. Although common structures, these traps contain only 4 per cent of the world's oil, perhaps because of losses that occur during uplift and erosion.

unconsolidated sediments Sedimentary materials that have not become compacted and fused into hard rock. They include lake muds and peats.

undercliff A stretch of land that lies parallel to and below a major cliff and is often covered by *mass-movement features.

underdispersion (regular distribution) In plant ecology, the situation in which the pattern of individuals of a given plant species within a *community is not random but regular, with similar numbers recorded in all *quadrats. *See* PATTERN ANALYSIS.

underfit stream *See* MISFIT STREAM.

underflow The flow of *groundwater in *alluvial sediments, parallel to and beneath a river channel. It forms a significant fraction of the total river flow in coarse gravel alluvium.

understorey species The trees of the lower canopy levels in a forest ecosystem, as distinct from emergent, crown, or upper-storey species. Some species (e.g. hazel and banana) are characteristic of the understorey, but others may be younger specimens of emergent species.

undertow The general seaward flow of water beneath individual breaking waves, in contrast to the more localized *rip-current return flow.

underwood **1.** A wood, either growing or cut, which consists of *coppice or *pollard regrowth, or small shrubs and saplings grown from seed or *suckers. **2.** The lower storey of a forest crop.

undulatus The Latin *undulatus*, meaning 'waved', used to describe a variety of cloud whose layers undulate. *See also* CLOUD CLASSIFICATION.

unequal crossing over The *crossing over of genetic information between two *homologous chromosomes at different chromosomal positions, resulting in a disproportional exchange. In regions of repetitive DNA this results in changes in the *copy number of repeats on both chromosomes and successive events will ultimately result in the fixation of one type of repeat.

ungulate **1.** Any hoofed, grazing mammal, which is usually also adapted for running. Hoofed mammals occur in several mammalian groups, and the term 'ungulate' no longer has any formal taxonomic use. **2.** Hoof-shaped.

unguligrade Applied to a *gait in which only the tips of the digits, covered with hoofs, touch the ground (e.g. in artiodactyls and perissodactyls). The limbs are moved as a whole by the action of shoulder and hip muscles.

unicentric distribution The occurrence of endemic species (*see* ENDEMISM) in very local areas. In Scandinavia, for example, about 40 endemics are found in just one or two areas; these areas are thought by some authorities to have been refuges from the glaciation of the late *Pleistocene (i.e. they were *nunataks above the ice sheets). *Compare* BICENTRIC DISTRIBUTION.

uniform flow **1.** A flow of water, the velocity and discharge of which do not vary along the length of its channel. **2.** In *groundwater, a flow, the velocity and direction of which are the same at all points in the field of flow.

uniformitarianism The principle proposed by James Hutton (1726–97) and paraphrased as 'the present is the key to the past', that the surface of the Earth has been formed and shaped by processes similar to those which can be observed today. This is a considerable oversimplification, since processes that occurred in historical times may not be occurring now, or may not be observable now, and vice versa. *See also* ACTUALISM.

uniramous Unbranched.

uniseriate Arranged in a single row.

unisexual flower A flower that possesses either stamens or carpels but not both. A plant my be unisexual (*dioecious), possessing only male flowers or female flowers; or it may be *monoecious with male and female reproductive organs borne in the same flower or in different unisexual flowers but on the same plant.

United States Department of Agriculture *See* USDA.

unit hydrograph *See* HYDROGRAPH.

unit leaf rate **(ULR)** The rate of *photosynthesis per unit area of leaf. *Primary productivity is thus the unit leaf rate multiplied by the *leaf-area index.

univoltine Applied to species in which one generation reaches maturity each year.

unsaturated zone *See* PHREATIC ZONE and SOIL-WATER ZONE.

Uppsala school of phytosociology A set of floristic methods intended mainly for the classification of vegetation *communities, developed by G. E. *du Rietz and others in Uppsala, Sweden, starting in 1921, at first quite independently of the work of the *Zurich–Montpellier (Z–M) team. More recently the approaches have converged as many ideas have been modified in the light of Z–M work in an attempt to link the two. The main distinction between them lies in the emphasis placed by the Uppsala scheme on *exclusive or *preferential species of high *constancy or dominance, rather than *fidelity, to define the basic community types.

Upton Warren A warm *interstadial during the Middle *Devensian, 43 000–40 000 years BP, when faunal and floral evidence suggests that the landscape was devoid of trees.

upwelling In oceans or large lakes, a water current or movement of surface water produced by wind, which brings colder water, loaded with nutrient, to the surface from a lower depth. Ocean upwellings occur off Peru, California, West Africa, and Namibia, and increase the nutrient content of the surface waters, leading to an abundance of marine and bird life. Upwelling also occurs in the open oceans where surface currents diverge, as deep waters rise to the surface to replace the departing waters, and all along the equator as a result of the effects of the *trade winds.

Ural Sea (Obik Sea) A *Palaeocene-*Eocene seaway that extended from the present-day Caspian Sea in the south to the Arctic Ocean in the north, covering the area immediately east of the Ural Mountains. The area is now covered by the extensive plain across which flows the River Ob.

uranium–lead dating A radiometric dating technique that uses the decay of ^{238}U and ^{235}U, which are present in all naturally occurring uranium in the ratio 137.7 : 1 and both of which decay to lead in stages that involve 14 steps, but are different for each isotope. ^{238}U (half-life 4 510 Ma) decays to ^{206}Pb; ^{235}U (half-life 713 Ma) decays to ^{207}Pb. Thorium–lead dating, in which ^{232}Th (half-life 13 900 Ma) decays to ^{207}Pb, is also included in this range of dating methods.

urban climate The modified surface-layer atmospheric conditions that are caused by the influence of large 'built-up' areas. Changes include pollution, reduction in strong wind speeds towards a city centre, turbulence of air around buildings, warming of air by the heat output from city structures, and increased evaporation and removal (drainage and run-off) of water.

urceolate Flask-shaped.

USDA (United States Department of Agriculture) The department of the United States federal government that exists to serve the needs of those engaged in agriculture and rural communities. It comprises a number of agencies, each with its own functions, one of which is the US Soil Survey. In 1960, the USDA published *Soil Classification; A Comprehensive System*, prepared by the Soil Survey. This method of classifying soils hierarchically was renamed the US Soil Taxonomy in 1970. This system of *soil classification is widely used, but since 1974 it has been largely superseded by the international classification developed by scientists at the Food and Agriculture Organization (FAO) of the United Nations, the United Nations Educational, Scientific, and Cultural Organization (UNESCO), and the International Society of Soil Science (ISS). *See* SOIL TAXONOMY.

UV *See* ULTRAVIOLET RADIATION.

uvala An irregularly shaped hollow in a *karst terrain, which is generally 500–1 000 m in diameter and may be 100–200 m deep. It is the result of the coalescence of a number of swallow-holes (dolines).

vacuum activity Patterns of behaviour that occur in the absence of the external stimuli that normally elicit them.

vadose zone *See* PHREATIC ZONE; SOIL-WATER ZONE; and WATER-TABLE.

vagile Applied to a plant or animal that is free to move about. *Compare* SESSILE.

vagility **1.** The inherent power of movement possessed by individuals or diaspores. Vagility in plants is often greater than commonly realized; *spores and seeds may float in the air for several miles and are very efficient means of *dispersal. **2.** The tolerance of an organism of a wide range of environmental conditions; vagility may be qualified as 'high' or 'low'.

valency A measure of the number of other ions of a chemical element that can be combined with a particular atom.

valley bog A *mire community that develops in wet valley bottoms, valleys with some downstream impedance, or badly drained hollows. Many European valley mires have layers of charcoal in their *stratigraphy below the *peat, suggesting that fire may originally have initiated peat formation by creating a charcoal deposit that effectively seals off basal soils from water penetration and causes waterlogging. The *groundwater is base-poor and conditions are acidic. This type of mire is flow-fed (rheotrophic), so technically it is a poor *fen type of community rather than a true *bog. The supply of nutrients is determined by the concentration of elements in the drainage water (which is usually low) and the rate of water flow through the system. The central part of the mire often has the fastest water flow and is hence less *oligotrophic than the lateral parts of the mire. In extensive valley mires the acidic, lateral mire expanses may become elevated and *ombrotrophic if rainfall is adequate. *Compare* BLANKET BOG; OMBROGENOUS PEAT; and RAISED BOG.

valley bulging An upward arching of the bedrock along the axis of a valley. It may not be visible at the ground surface, owing to subsequent *erosion, but it is revealed by the distortion of the geologic structure. It may be the result of *frost heave or the compressive forces set up when two opposing valley sides approach each other.

valley glacier A long, relatively narrow ribbon of ice that is confined between valley walls.

valley-side bench **(rock bench)** A terrace-like land-form that stands on the flank of a valley, but lacks a veneer of *alluvium. It may have originated as a true *river terrace from which the alluvium has been stripped by subsequent *erosion or it may result from the exposure of a nearly level *stratum of resistant rock.

valley train An accumulation of fluvioglacial deposits laid down in a valley by meltwater escaping from a decaying *glacier. The surface slopes quite steeply down-valley, and is incised by shifting *braided streams.

valley wind An *anabatic wind that blows up-valley during the day in otherwise calm conditions, or a *katabatic, down-valley, night wind. *See also* MOUNTAIN WIND and RAVINE WIND.

value *See* MUNSELL COLOUR.

Van Valen's 'law' When plotted as cumulative curves on a logarithmic scale, the duration frequencies of many species tend to show a more or less straightline relationship with time. In effect, taxa generally become extinct regardless of their age. Duration frequencies vary greatly among different groups, e.g. on average, bivalve genera last ten times longer than mammalian genera. The 'law' is named after L. Van Valen, who proposed it in 1973.

vardar **(vardarac)** A type of *ravine wind, which blows in the Moravia–Vardar valley, bringing cold conditions from the north to the Thessaloniki area of Greece.

variance *See* MEAN SQUARE.

variation Differences displayed by individuals within a species, and which may be favoured or eliminated by *natural selection. In sexual reproduction, reshuffling of genes in each generation ensures the maintenance of variation. The ultimate source of the variation is *mutation, which produces fresh genetic material.

variegation The phenomenon in some plants in which patches of two or more different colours occur on the leaves or flowers; variegation may be an inherited characteristic, often originally a *sport, or may be due to *virus infection. The persistence of variegation in some plants suggests that it has adaptive significance. In plants such as *Sansevieria*, variegation may serve a camouflage function, especially in dappled light. A similar avoidance of *predation may be served by the 'V'-shaped variegation pattern on clover leaves that makes them appear smaller than they actually are.

varve A banded layer of *silt and *sand deposited annually in lakes, especially those close to *ice sheets. A coarse-grained layer, which is often lighter in colour, results from spring and summer ice-melt, eroding materials into the lake. The finer particles remain longer in suspension and settle more slowly through the winter, forming a darker, fine-grained layer. One varve consists of one light band and one dark band. Varves can be counted to calculate the age of glacial deposits (a technique called varve analysis or varve count). Since the pattern of thicknesses of successive varves is often distinctive, correlations can be made between widely separated deposits, using the same principle as that of *dendrochronology.

V

varve analysis *See* VARVE.

varve chronology *See* VARVE.

varve count *See* VARVE.

varzea *See* AMAZON FLORAL REGION.

Vavilov, Nikolai Ivanovich (1887–1943) A Russian plant geneticist whose extensive field studies, in Iran, Afghanistan, Ethiopia, China, and Central and South America, led him to the view that the greatest variation in species occurs in certain restricted areas (centres of diversity) which he believed identified the regions in which those species originated (*centres of origin). He returned from his travels with a large collection of specimens he intended to use for study and to breed new varieties. He was professor of botany at the University of Saratov (1917–21) and later became head of the Lenin All-Union Academy of Agricultural Sciences. He was elected a member of the Academy of Sciences of the USSR in 1929, and a foreign member of the Royal Society of London in 1942. He was opposed by T. D. *Lysenko, who had him removed from his positions in 1940, and Vavilov is believed to have spent the years from 1940 in prison and to have died at Magadan, Siberia.

vector An organism that carries a disease-causing organism from an infected individual to a healthy one; the vector may transfer the pathogen passively or may itself be infected by it.

veering A clockwise shift in the direction of the wind. The reverse change is called *backing.

vegetation index In remote sensing, a classification of the ranges of brightness in different wavelength bands corresponding to the reflectance of electromagnetic radiation by vegetation cover, and sometimes specific vegetation types, as opposed to rock, soil, or water cover.

vegetation mosaic The pattern of different plant *communities, or stages of the same community. The term is applied particularly to communities that show cyclical change (e.g. the *Calluna* cycle on heathlands), with examples of all stages being present together in a typically extensive and well-developed community.

vegetative Applied to a stage or structure that is concerned with feeding and growth rather than with sexual reproduction. Vegetative reproduction is asexual reproduction.

vegetative propagation (vegetative reproduction) A reproductive process that is asexual and so does not involve a recombination of genetic material (i.e. a form of *apomixis). It involves unspecialized plant parts which may become reproductive structures (e.g. roots, stems, or leaves). Compared with sexual reproduction, it represents a saving of material and energy for the plant. It is especially common among grasses.

vegetative reproduction *See* VEGETATIVE PROPAGATION.

vegetative state **1.** A stage in the *life cycle of a plant when reproduction proceeds asexually by detachment of some part of the plant body and its subsequent development into a complete plant. **2.** The non-infective state in a phage during which the *genome multiplies actively and controls synthesis by the host (a bacterial cell) of the materials necessary for the production of infective particles (i.e. more phages and phage DNA). These are released by lysis of the host cell.

vein-banding A symptom of some virus diseases of plants in which bands of lighter or darker colour occur along the main veins of a leaf.

vein-clearing A symptom of some virus diseases of plants in which the veins become unnaturally clear or translucent.

veld Extensive grasslands in the east of the interior of South Africa, often with a scattering of trees or bushes. In these instances the names *tree veld and bush veld respectively are applied. True grass veld is confined to the high terrain.

velocity The speed at which a body is moving in a specified direction. Speed is a scalar quantity that can be described only by its magnitude. Velocity is a vector quantity that can be described only by its magnitude and direction of application. For example, wind speed can be reported as *x* km/h; wind velocity must be reported as *x* km/h from *y*°, usually written as km/°.

velocity profile The variation of water velocity with vertical distance from the bed of a river, or of wind velocity with distance from the ground. *See* VON KARMANN–PRANDTL EQUATION.

velum An accessory cloud feature of *cumulus or *cumulonimbus, characterized by a widespread veil on or above the upper surface of the cloud. *See also* CLOUD CLASSIFICATION.

vendavale A strong, local, south-westerly wind affecting the Straits of Gibraltar, associated with a *depression, and bringing *squalls and heavy rainfall.

Venezuela and Guiana floral region Part of R. Good's (1974, *The Geography of the Flowering Plants*) *Neotropical floral kingdom, a region which has affinities with the adjacent Andean and Amazon floral region, but nevertheless constitutes an independent unit with about 100 endemic genera (*see* ENDEMISM). The flora is poorly known and has yielded no plants of any great value. *See also* FLORAL PROVINCE and FLORISTIC REGION.

Venice system A system for the classification of *brackish water based on the percentage of chloride contained in the water. *See* HALINITY.

ventral **1.** Of a plant or animal, the surface closest to the ground or *substrate. **2.** Of a plant organ, the *adaxial surface. **3.** Of a chordate animal, the surface or structure furthest from the notochord (in vertebrates, the spine).

veranillo In South America, the short period of finer weather that occurs during the summer wet season.

verano In tropical America, the drought season that occurs in winter.

vernal In the late spring. *See* AESTIVAL; HIBERNAL; PREVERNAL; and SEROTINAL.

vernalization The treatment of germinating seeds with low temperatures to induce flowering at a particular preferred time. For example, winter varieties of wheat can be sown in the spring and then be exposed to a temperature just above 0°C for a few weeks. The result of this is that they behave like spring varieties and flower in the same year (otherwise they would continue to grow vegetatively and would not flower until the following year). The stimulus is perceived by the *apical meristem (either in the embryo or as an apical bud), and some plant hormones such as gibberellin can be used to achieve the same effect. *Compare* STRATIFICATION.

verrucose Warty in appearance.

vertebratus A variety of cloud, usually with *cirrus, in which cloud elements have a skeletal arrangement, in a form resembling vertebrae. *See also* CLOUD CLASSIFICATION.

vertic horizon A *soil horizon composed of clayey material with polished *ped surfaces, often with *slickensides, or wedge-shaped soil *aggregates. The name is from the Latin *vertere*, to turn.

verticillate Arranged in a whorl or whorls.

verticillium wilt A plant disease caused by a fungus of the genus *Verticillium*. The infected plant may show *wilting, with bending downwards and yellowing of the lower leaves. When the stem is cut well above ground level, characteristic brown streaks can be seen inside the stem, following the conducting elements.

Vertisols An order of *mineral soils which contain more than 30 per cent by weight of swelling clay (e.g. montmorillonite), and expand when wet and contract when dry to produce a self-inverting soil and an undulating (*gilgai) microrelief. Vertisols are associated with seasonally wet and dry environments, and are extensive in the tropics, forming beneath grassland. Vertisols are a reference soil group in the *FAO *soil classification.

vestigial organs Atrophied or nonfunctional organs that are well developed and fully functional in other members of the group. They result from the adoption of a way of life in which the organs are no longer required (e.g. rudimentary pelvic bones are found in some whales, but are no longer required, since whales have no hind limbs as such). Such features are difficult to reconcile with the concept of *special creation. *See* EXCESS BAGGAGE HYPOTHESIS.

viability The probability that a fertilized egg will survive and develop into an adult organism.

vicariad *See* VICARIANCE.

vicariance (vicariad, vicarious species) The geographical separation of a *species so that two closely related species or a species pair result, one species being the geographical counterpart of the other.

vicariance biogeography A school of biogeographical thought, derived from *panbiogeography, whose supporters maintain that the distribution of organisms depends on their normal means of dispersal (e.g. disjunctions are explicable in terms of new barriers (rivers, rises in sea level, etc.) having split formerly continuous ranges, rather than in terms of the organisms hopping over already existing barriers). Thus they reject *sweepstake routes and similar concepts, postulating instead former *land bridges and even vanished continents where there is sufficient coincident plant and animal distribution.

vicarious distribution The distribution that results from the replacement of one member of a species pair (i.e. two closely related species derived from a common ancestor) by the other, geographically (as opposed to ecologically). There are many animals and herbaceous plants with a vicarious distribution between North America and Europe. In zoology, subspecies are conspecific vicariants.

vicarious species *See* VICARIANCE.

vigilance The readiness of an animal to detect certain specified events that occur unpredictably in its environment.

Villafranchian The land-mammal stage that spans the Upper *Pliocene and Lower *Pleistocene (3–1 Ma ago).

virga (fall-stripes) The Latin *virga*, meaning 'rod', used to describe a supplementary cloud feature caused by trails of *precipitation falling from the under surface of cloud but not reaching the ground. This phenomenon is often seen with *cumulus, *cumulonimbus, *altocumulus, *stratocumulus, *cirrocumulus, *nimbostratus, and *altostratus. *See also* CLOUD CLASSIFICATION.

virion An individual *virus particle.

virus A type of non-cellular 'organism' which has no metabolism of its own. It consists mainly or solely of a nucleic acid genome (RNA or DNA) enclosed by protein; in some cases there is also a lipoprotein envelope. In order to replicate (multiply), a virus must infect a cell of a suitable host organism, where it redirects the host-cell metabolism to manufacture more virus particles. The progeny viruses are released, with or without concomitant destruction of the host cell, and then can infect other cells. All types of organism are susceptible to infection by viruses; virus infections may be asymptomatic or may lead to more or less severe disease.

viscotaxis A change in direction of locomotion in a *motile organism or cell, which is made in response to a change in the viscosity of the surrounding medium.

vitric horizon A *soil horizon that contains more than 10 per cent volcanic glass or other volcanic material. The name is from the Latin *vitrum*, meaning glass.

vitrinite *See* COAL MACERAL.

vivipary The method of reproduction in which: (*a*) a plant (e.g. mangroves) produces seeds that germinate within and obtain nourishment from the fruit; (*b*) a plant (e.g. some grasses) reproduces vegetatively from shoots rather than an inflorescence; (*c*) an animal produces young at a stage of development in which they are active, the growth of the embryo occurring within the mother's body, which nourishes it. *Compare* OVIPARY and OVOVIVIPARY.

vocalization The production of sound by the passage of air across vocal cords.

void ratio A measure of the porosity of rocks and soils; the void ratio $(e) = Vv/Vs$, where Vv is the volume of air in voids and Vs is the volume of solid particles.

volatile A substance with a high vapour pressure, which passes readily into a gaseous phase. Common volatiles include water vapour, carbon dioxide, sulphur dioxide, hydrochloric acid, and there are many more. In coal, volatiles comprise the mixture of combustible gases (hydrogen, carbon dioxide, and methane) with other substances, which is given off when the coal is heated without air being present. Peat contains more than 50 per cent volatiles, lignites about 45 per cent, anthracite 10 per cent, and graphite less than 5 per cent.

volcanic dust Dust, ash, or other particulate matter that is commonly suspended in the atmosphere after volcanic eruptions. After explosive eruptions the dust may be thrown to heights of 20–30 km or more. The fall-out times of dust particles are quite short, a matter of days or weeks, depending on altitude and precipitation. Volcanogenic *aerosols, usually sulphates, may linger for months, spreading as a long-lived veil in the *stratosphere over much of the Earth.

voluntary behaviour Non-habitual behaviour that is associated with the somatic nervous system or (possibly in some species) with operant control of the autonomic nervous system.

volunteer plant A plant that has resulted from natural propagation, as opposed to having been deliberately planted by humans.

von Karman–Prandtl equation An equation that describes the logarithmic variation of water velocity within a channel from zero flow at the stream bed to a maximum velocity at the water surface. Originally developed in aerodynamics, the equation also describes the profile of wind velocity above the ground. *See* EDDY VISCOSITY.

vulnerability The extent to which an *ecosystem is under threat. For example, an area of heathland close to a housing development may be particularly vulnerable because of the high likelihood that further expansion of housing would encroach upon it. The criteria for selecting *nature reserves include this consideration. *Compare* FRAGILITY.

vulnerable species *See* RARITY.

Waalian An *interglacial that occurred in northern Europe from about 1.3 to 0.9 Ma. It may be equivalent to the *Donau/Günz *interglacial of the Alpine areas and the *Aftonian of North America.

wadi **(ouadi)** The Arabic name for an *ephemeral river channel in a *desert area. Flow may occur very occasionally.

waggle dance *See* DANCE LANGUAGE.

waldgrenze *See* TIMBER-LINE.

Wallace, Alfred Russel **(1823–1913)** An English naturalist who was a contemporary of Charles *Darwin. He worked in the East Indies and as a result of his observations there came independently to the theory of evolution by natural selection. His ideas were presented together with those of Darwin at a meeting of the Linnean Society of London on 1 July 1858. *See also* WALLACE'S LINE.

Wallacea *See* WALLACE'S LINE.

Wallace's line The important zoogeographical division which separates the *Oriental and *Notogean zoogeographical regions. Alfred *Wallace, a zoogeographer and contemporary of Charles *Darwin, first demarcated the boundary between the Oriental faunal region and the Australasian region with its distinctive marsupials and birds. The boundary, known to this day as Wallace's line, passes east of Java and Bali, northward through the Strait of Makassar (separating Borneo and Sulawesi), then extends eastward, south of Mindanao in the Philippines. There is a zone of mixing called 'Wallacea', and strictly the line defines the extreme western limit of Australasian mammals and the eastern limit of the main oriental fauna.

Walter climate diagram A graphic representation of climatic conditions at a particular place, which shows seasonal variations and extremes as well as mean values and therefore provides a succinct and easily accessible summary of a local climate. The diagram shows the height above sea level, the number of years during which temperature and precipitation observations continued, mean annual temperature and precipitation, mean daily minimum temperature in the coldest month and maximum in the warmest month, the lowest and highest temperatures recorded, mean daily variation in

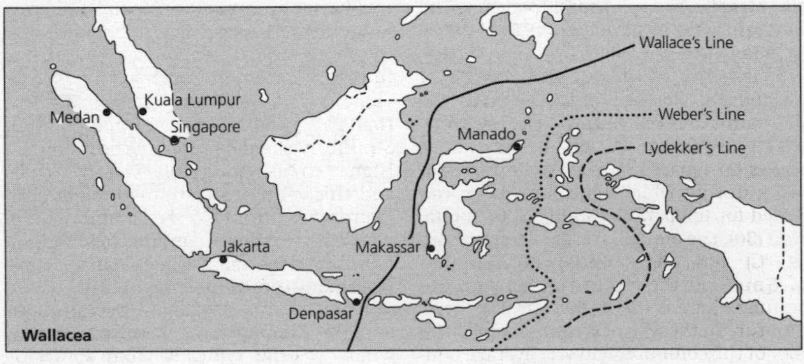

Wallace's line

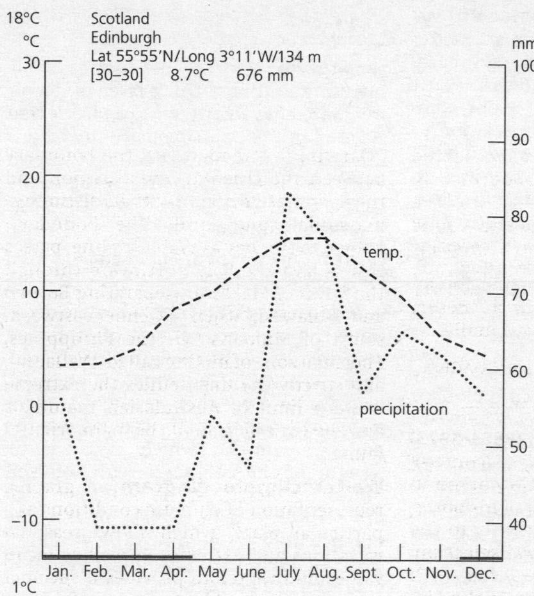

Walter climate diagram

temperature, mean monthly temperature and precipitation, relative period of drought and relative humid season, mean monthly rainfall in excess of and below a given value, reduced supplementary precipitation curve, months in which the mean daily minimum temperature was below freezing, months with frosts, and the mean duration of frost-free days. The diagram was devised in 1979 by the German biogeographer Heinrich Walter. The example illustrates the climate of Edinburgh, Scotland. In the heading, the diagram states the country, city, latitude, longitude, and elevation. Below that it shows (in square brackets) the number of years during which data have been collected for temperature (30) and precipitation (30), the annual average temperature (8.7°C) and total annual precipitation (676 mm). At the top of the temperature column there is the highest average temperature in the warmest month and at the foot of the column the lowest average temperature in the coldest month. The graph shows the average monthly temperature and precipitation. The heavy line at the bottom indicates the months during which frosts are likely (January–April and December).

warm front A surface where advancing warm air displaces colder air (e.g. in mid-latitude *depressions) where, owing to the convergence of the *air masses and the difference of density between them, the warm air tends to rise over the cold air. Slopes of warm fronts are typically less than 1 : 100 and the ascent of air is gradual. Stratiform cloud develops in the rising air. High *cirrus cloud followed by lower and thickening *altostratus indicate the approaching front. As the frontal contact with the ground approaches, heavy *nimbostratus and much rain may occur. Passage of the front is marked by a rise of temperature, clearing of precipitation, and (in the northern hemisphere) the wind *veering typically from south or south-easterly to south-westerly.

Warming, Johannes Eugenius Bülow (1841–1924) A Danish botanist whose work laid the foundations for and greatly stimulated the study of modern plant ecology. He maintained that plant communities should be studied in relation to their surroundings and developed a basis for their classification. This view was expounded in his book *Plantesamfund* (1895, published in English in 1925 as *Oecology of Plants*). From 1882 to 1885 he was professor of botany at the Royal Institute of Technology in Stockholm and from 1885 to 1911 professor of botany and director of the botanical gardens at Copenhagen.

warm sector A tongue of relatively warm air of tropical or old polar or maritime origin, which appears between colder *air masses. Such a tongue occupies the area between the *warm and *cold fronts in a developing mid-latitude *depression. Within the warm sector, pressure, wind, and temperature remain fairly steady. Cloud and *precipitation depend on the precise condition of the generally stable air; cloud may be produced from *orographic uplift or *fog from passage over a cool sea surface.

warning coloration *See* APOSEMATIC COLORATION.

warping A traditional farming practice in which a river is permitted to flood low-lying ground temporarily in order to supply silt in which crops can be grown without irrigation.

warp soil Soil consisting principally of *silt deposited on land by deliberate flooding (*warping).

washboard moraine *See* MORAINE.

washout Nutrients that dissolve in atmospheric water droplets and are brought to the surface in precipitation. *See* WETFALL; *compare* RAINOUT.

washover delta *See* WASHOVER FAN.

washover fan (washover delta) A fan-shaped body of sediment that is transported landward by marine waters flowing through or across a coastal barrier (e.g. a *barrier bar or island). Such bodies are formed especially during storms, when the barriers are likely to be breached.

washplain A nearly flat surface of *alluvium mantling a thick layer of deeply weathered (*see* WEATHERING) bedrock that is found in a *savannah environment. It is washed by seasonally flooding streams which are unable to incise, owing to the lack of abrasive *bed load and the large volume of sediment.

waste An area of unenclosed land used for common pasture.

water-absorption test A test to determine the moisture content of *soil as a percentage of its dry weight (British Standard 1377, 1967). The sample is weighed, dried in an oven, then reweighed under standard conditions. It is calculated as the moisture content, which is equal to: (weight of the container with wet soil minus the weight of the container with dry soil) divided by (weight of the container with dry soil minus the weight of the container), then multiplied by 100 to express it as a percentage.

water inventory **1.** An inventory to show how water is consumed, used, or otherwise involved in a particular process or place (e.g. within a steel plant or a house). **2.** Globally, approximately 97 per cent of the Earth's water occurs in the oceans. Of the fresh water, 75 per cent is locked up in *ice sheets and *glaciers and almost 25 per cent is *groundwater. Lakes, reservoirs, swamps, river channels, biospheric water, atmospheric water, and soil moisture together account for the remainder. The amount of water at each stage of the *hydrological cycle is called the water storage.

water potential **1. (chemical potential, osmotic potential)** The difference between the energy of water in the system being considered and of pure, free water at the same temperature. The water potential of pure water is zero, so that of a solution will be negative. If there is a gradient of water potential between two plant cells, water will diffuse down the gradient until equilibrium is reached. **2.** *See* CAPILLARY MOISTURE.

watershed **1.** *See* DIVIDE. **2.** *See* CATCHMENT.

waterspout A *tornado that forms over water or that forms over land and then moves over water. The waterspout consists of a vortex of air spiralling inward to a central region of low pressure and made visible by the condensation of atmospheric water vapour as it enters the low-pressure region. A spray ring around the base of the waterspout consists of surface water whipped into spray by the air being drawn into the vortex.

water storage *See* WATER INVENTORY.

water-table The upper surface of *groundwater or the level below which the material is permanently saturated with water. The region below the water table is the phreatic or vadose zone.

Watt, Alexander Stuart (1892–1985) The Scottish plant ecologist who discovered the cyclical pattern in changes in plant *communities (*see* BUILDING PHASE; DEGENERATE PHASE; HOLLOW PHASE; MATURE PHASE; and PIONEER PHASE) developed from his extensive studies of the Chiltern beechwoods and the brecklands of East Anglia. Watt was born at Monquhittar, Aberdeenshire and graduated from Aberdeen University in 1913 with a degree in agriculture. He worked with Sir Arthur *Tansley at Cambridge in 1914, returning to Aberdeen University in 1915 as a lecturer in forest botany and forest zoology. After war service he returned to Cambridge in 1919 to complete his studies for the BA degree and resumed his lectureship at Aberdeen. He was elected a Fellow of the Royal Society of London in 1957 and a Fellow of the Linnean Society in 1975, and held visiting professorships at the Universities of Colorado (1963) and Khartoum (1965).

wave base The depth beneath a water mass below which wave action ceases to disturb the sediments. Wave-base depth is approximately equal to half the wavelength of the surface waves.

wave clouds Clouds occurring in an airstream affected by wave motion, usually after it has passed over a mountain barrier. Lenticular (lens-shaped) clouds and cloudlets appearing stationary in the crests of the waves are a characteristic sight in climates experienced on the lee side of mountain barriers.

wave-cut bench *See* SHORE PLATFORM.

wave-cut platform *See* SHORE PLATFORM.

wave depression A low-pressure area developed at the apex of a developing wave distortion along a *front. A series of such systems is typical of mid-latitude *depression sequences.

wave diffraction An effect seen as waves pass through an opening in a breakwater into protected waters. The waves fan out from the opening into the region beyond, but as they do so their height is diminished.

wave refraction The process by which the direction of waves moving in shallow water at an angle to the submarine contours is altered. The part of the wave train travelling in shallower water moves more slowly than that still advancing in deeper water. The lines of the wave crests therefore become more parallel with the submarine contours closer to the coast.

wave ripple mark *See* OSCILLATION RIPPLE.

weakening In *synoptic meteorology, a decrease in pressure gradient around a pressure system over time. Changes of this kind bring about a weakening of the winds. *See also* INTENSIFICATION.

weathering The breakdown of rocks and *minerals at and below the Earth's surface by the action of physical and chemical processes. Essentially it is the response of Earth materials to the low pressures, low temperatures, and presence of air and water that characterize the near-surface environment, but which were not typical of the environment of formation. There are several varieties of rock breakdown. Simple disintegration may occur, resulting in the production of coarse, angular blocks, of peels or skins (the process of 'desquamation'), of *sands, and of *silts. Minerals may be removed in solution, and *chemical weathering may form new, often easily eroded substances. Biological processes contribute to weathering. For example, *lichens growing on exposed rock faces release acids that etch into the rocks.

weathering front The junction between chemically weathered (*see* CHEMICAL WEATHERING) rock or *regolith and sound rock. Where the front lies between regolith and unweathered bedrock it may be exposed by subsequent *erosion to form an *etchplain.

weathering profile A vertical section, from the ground surface to unaltered bedrock, which passes through *weathering zones. It is usually best developed in the humid tropics, where depths of 100 m have been recorded but where 30 m is more common. The nature of the profile is a complex response to climatic and geologic controls, and to long-term changes in external conditions.

weathering zone A distinctive layer of weathered material that extends roughly parallel to the ground surface. It differs physically, chemically, and mineralogically from the layers above and/or below.

Weber's line A line of supposed 'faunal balance' between the *Oriental and the Australasian faunal regions within Wallacea. *See* WALLACE'S LINE.

weed A plant in the wrong place, being one that occurs opportunistically on land or in water that has been disturbed by human activity (*see* RUDERAL) or on cultivated land, where it competes for nutrients, water, sunlight, or other resources with cultivated plants. Under different circumstances the plant may itself be cultivated (e.g. it may grow from seed or propagate vegetatively from the residue of a previous crop). The presence of weeds among crop plants can result in reduced yield because of competition; in gardens weeds may be unsightly; in waterways they can become a hindrance to navigation (e.g. the water hyacinth, *Eichhornia crassipes*).

Weichselian *See* DEVENSIAN.

weighted average The value assessed from a number of samples, where each sample is given a different value of importance according to its reliability.

West African rain-forest floral region Part of R. Good's (1974, *The Geography of the Flowering Plants*) African floral subkingdom, which contains a rich flora that has provided some valuable economic plants (e.g. *Coffea liberica*) and others of more local importance, including timber trees. *See also* FLORAL PROVINCE and FLORISTIC REGION.

West and Central Asiatic floral region Part of R. Good's (1974, *The Geography of the Flowering Plants*) *Boreal Realm, which contains a limited, specialized flora, because much of the region comprises dry mountains and deserts. There are about 150 endemic genera (*see* ENDEMISM), virtually all small. Barley and some types of wheat probably originated in this region. *See also* FLORAL PROVINCE and FLORISTIC REGION.

West Australia Current The oceanic current that flows north along the western Australian coast. The flow is strong and steady in summer, but is much reduced during the winter months. The *salinity (34.5 parts per thousand) and temperature (3–7°C) are both relatively low.

Western Boundary Undercurrent *See* CONTOUR CURRENT.

West Indian mammal subregion An *island subregion which, according to the analysis by Charles H. Smith, is distinct from both the *Holarctic and *Latin American regions by reason of its high insular *endemism.

western intensification The tendency of currents along the western margins of all oceans to be particularly strong, swift, and narrow, flowing northwards in the northern hemisphere and southwards in the southern hemisphere. Currents at the eastern margins of all oceans tend to be slower and more diffuse. *See* GYRE.

West Wind Drift *See* ANTARCTIC CIRCUMPOLAR CURRENT.

wet-bulb depression The extent to which the temperature recorded by a ventilated *wet-bulb thermometer falls below the dry-bulb air temperature.

wet-bulb thermometer A thermometer, the bulb of which is kept moist by a thin cloth (e.g. muslin) bag connected by a wick to a bath of clean (preferably distilled) water. As long as the air is not saturated, evaporation from the muslin keeps the wet-bulb thermometer at a lower temperature than the dry-bulb thermometer beside it, with which its readings are compared. The depression of the wet-bulb temperature gives a measure of the saturation deficit, from which the *relative humidity and *dew-point can be calculated.

wetfall The delivery to surface communities of nutrients carried in precipitation, in solution (e.g. sulphur and nitrogen oxides) or as insoluble particles (e.g. dust and ash particles). *Compare* DRYFALL.

wetland A term that was defined by the *Ramsar Convention on Wetlands of International Importance in 1971 as: 'all areas of marsh, fen, peatland, or water, whether natural or artificial, permanent or temporary with water that is static or flowing, fresh, brackish, or salt, including areas of marine water the depth of which at low tide does not exceed six meters.' This definition embraces openwater *habitats and seasonally or permanently waterlogged land areas, including lakes, rivers, and estuarine and freshwater marshes. Wetland habitats, especially *marsh and *bog areas, are among the most vulnerable to destruction since they can be drained and reclaimed for agriculture or forestry, drained for pest control (e.g. to eliminate breeding grounds for malaria-carrying mosquitoes), or modified for water supply, flood control, hydroelectric power schemes, waste disposal, etc.

wet rot A type of timber decay found only in wood with a high moisture content. It is caused by the cellar fungus, *Coniophora puteana*.

Wheelerian The first of two stages in the *Pleistocene of the west coast of North America, underlain by the Venturian (*Pliocene), overlain by the *Hallian, and roughly contemporaneous with the Lower Pleistocene Series of southern Europe.

whirling psychrometer (sling psychrometer) A *psychrometer with a handle, which allows rapid rotation of mounted *wet- and dry-bulb thermometers to ensure air flow around the bulbs. See illustration below.

whirlwind A spiral wind storm around a low-pressure centre. In arid areas dust may be carried upwards several hundred metres.

white blister *See* WHITE RUST.

white rot **1.** A disease of onions, shallots, and leeks, caused by the fungus *Sclerotium cepivorum*. Leaves of infected plants turn yellow, and a fluffy, white *mycelium appears on the bulb. Small, black sclerotia

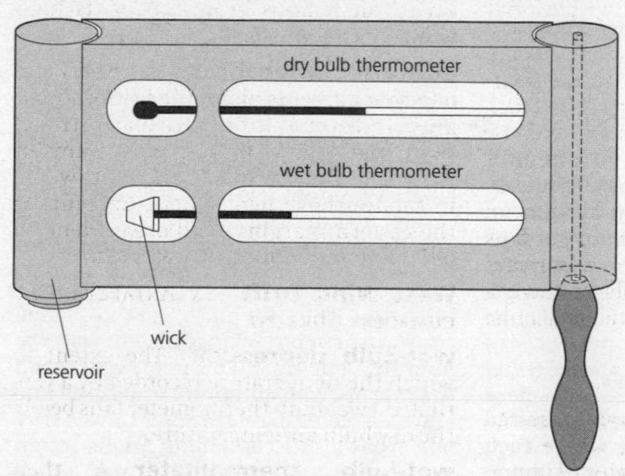

Whirling psychrometer

can be seen in the mycelium and these can survive in the soil for many years. **2.** A type of timber decay in which the cellulose, hemicellulose, and lignin of the wood are decomposed, leaving the wood soft, white, and fibrous.

white rust **1. (white blister)** A disease affecting crucifers and other plants, caused by fungi of the genus *Albugo*. White, blister-like spots are formed on the leaves, and infected plants may show some distortion. **2.** A disease of chrysanthemums caused by the fungus *Puccinia horiana*. Buff or white pustules appear on the undersides of the leaves and may become brown and *necrotic. This disease first appeared in Britain in 1963 and is now notifiable.

white smoker *See* HYDROTHERMAL VENT.

whitings Extensive patches of white, turbid water consisting of a dense suspension of *aragonite mud. These mud suspensions are seen from time to time in carbonate-dominated shallow seas, and are due mainly to the disturbance of the muddy sea floor by schools of fish or by turbulence. It was suspected that some whitings were the result of direct precipitation of aragonite from hypersaline, carbonate-saturated waters, but this has been shown not to be the case.

wide distribution **(polychore distribution)** The situation in which taxonomic groups of plants (and to a lesser extent animals) have a very extensive distributional range, spanning several floral kingdoms or regions (e.g. 'wides' may be cosmopolitan, subcosmopolitan, tropical, or temperate).

wilderness An extensive area of land which has never been permanently occupied by humans or subjected to their intensive use (e.g. for mineral extraction or cultivation) and which exists in a *natural or nearly natural state. Wilderness areas are selected for their ecological wholeness, rather than for the presence of any particular *biota, landscape, or recreational attraction. In the USA, where wilderness areas have been formally designated, no economic use is allowed except by presidential decree in extreme emergency. The areas are free from traffic, and the number and activities of visitors are carefully controlled. Elsewhere, the concept merges with that of national parks, wilderness areas often being zones of more restricted public access within the park areas.

wildlife Any undomesticated organisms, although the term is sometimes restricted to wild animals, excluding plants.

willy-willy The name occasionally given to a *tropical cyclone that forms near western Australia. Although the name is no longer in current use it appears in many older textbooks.

Wilson cycle The hypothesis proposed by the geophysicist John Tuzo Wilson (1908–93) that an ocean develops through six distinct stages driven by the movement of crustal plates. The cycle begins with the lifting and extension of the crust to form a *rift valley. Further crustal subsidence and sea-floor spreading produce a narrow sea with approximately parallel sides (e.g. the *Red Sea). Further plate movement causes the sea to widen into an ocean flanked by continents (e.g. the *Atlantic Ocean). The system becomes unstable and part of the cooled crust, away from the mid-oceanic ridge (*see* RIDGE (3)) sinks, forming a *trench with an associated island arc (e.g. the *Pacific Ocean). The ocean is then shrinking and continues to do so, with wedges of rock being lifted to form young mountain ranges (e.g. the *Mediterranean). Finally, when all the crust between the continents has been subducted, the continents converge and join along a suture (e.g. in the Himalayas).

wilt A type of plant disease in which wilting (i.e. loss of plant turgidity) is a principal symptom. Wilts are frequently caused by infection of the plant by a fungus belonging to the Deuteromycotina. *See, for example*, VERTICILLIUM WILT.

wilting The limpness found when plant tissues contain insufficient water to hold the cells rigid. This may occur when the rate of *transpiration exceeds the rate at which water is able to enter the root system from a soil containing ample water, causing temporary wilting from which the plant recovers when the transpiration rate falls. It may also be the result of a deficiency of water in the soil. When water shortage is prolonged or acute, the plant may reach a point from which recovery is impossible even if abundant water is subsequently supplied. *See* PERMANENT WILTING POINT.

wilting coefficient *See* PERMANENT WILTING POINT.

wilting point *See* PERMANENT WILTING POINT.

Windermere interstadial (Late-Devensian interstadial) A relatively warm period that occurred towards the end of the last (*Devensian) glaciation in Britain. The event took place about 13 000–11 000 radiocarbon (*see* RADIOCARBON DATING) years BP. It includes the *Bølling, Older *Dryas, and *Allerød chronozones of Scandinavia (Britain did not experience the Older Dryas). The pollen sequence shows a sharp rise followed by a marked fall in birch and juniper, marking the beginning and end respectively of the interstadial.

wind rose A quantitative diagram (*rose diagram) that represents the relative frequencies of different wind directions and wind speeds at a climatic station over a period of time. The direction of the lines indicates the wind direction and their length indicates the frequency with which wind blows from that direction.

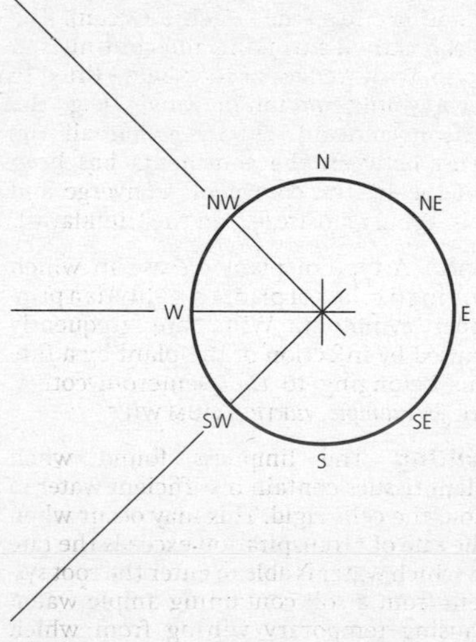

Wind rose

windrow A streak of foam or row of floating debris, aligned in the prevailing wind direction, formed on the surface of a lake or ocean. Where winds blow across a water surface, vertical circulation cells are set up in near-surface waters. These circulation cells are alternately right- and left-handed vortices, and windrows form along the lines of convergence between adjacent cells at the water surface.

wind shear A change in the speed or direction of the wind with horizontal or vertical distance.

wind sonde A weather balloon that carries only a radar reflector and is tracked to determine wind speed and direction. There are about 500 upper air weather stations in the world as a whole. Each station releases two wind sondes each day, at 0600 and 1800 Greenwich time, midway between the launches of *radiosondes. *See also* RAWINSONDE.

Winkler method The standard chemical procedure for measuring oxygen in water.

Wisconsinian The last (75 000–10 000 years ago) of four glacial episodes recognized in North America. As with previous glacials there were several advances of the ice, and these glacial deposits include sequences of *tills laid down by ice in the western North American Cordillera. More evidence from areas peripheral to the *ice sheets suggests that temperatures were perhaps 6 °C cooler than they are now. This glacial is perhaps equivalent to the *Würm glacial of the Alpine areas.

Wisconsin School A group of ecologists, led by J. T. Curtis and his associates, who developed a range of simple *ordination methods in 1950 and later, while studying the vegetation of Wisconsin. Their scheme has largely been replaced by more modern classification and ordination methods.

Wolstonian The glaciation that followed the *Hoxnian Interglacial; a pale, chalky *till (the Gipping Till) occurs in East Anglia, England, and the stone orientation and the included erratics suggest a movement from the north. The general stratigraphy needs clarification since it is not certain that the till rests on Hoxnian deposits and there are a number of morainic deposits attributed to this glacial which occur in Norfolk. During this ice advance a large lake, Lake Harrison, was ponded up in the Midlands. The Wolstonian perhaps correlates with the *Saalian.

wood **1.** The secondary xylem of dicotyledons and conifers, which forms a dense growth during secondary thickening, providing the mechanical support which allows *trees to grow to a considerable height. **2.** An area of trees, often associated with a particular name (e.g. Hayley Wood) that denotes a district area. **3.** The produce of *coppice or underwood that is not of timber size.

woodland **1.** A vegetation *community that includes widely spaced, mature trees. The tree crowns are typically more spreading in form than those of *forest trees. Crowns do not touch and do not form a closed canopy. Woodland is often defined as having 40 per cent canopy closure or less. Between the trees, grass, *heathland, or *scrub communities typically develop, giving a park-like landscape. **2.** A general term for a wooded landscape, often used generally (e.g. broad leaved woodland) or to describe a number of separate wooded areas (e.g. the Estate Woodlands). Colloquially, the terms forest and woodland are often used interchangeably in Britain.

wood-pasture Land on which trees grow and where farm livestock or deer are grazed systematically. *See also* POLLARDING.

worker In social insects, the caste that rears the brood, maintains the structure of the nest, and forages for food. Usually workers are sterile. They are female in ants, bees, and wasps, and either male or female in termites. In termites, a true worker caste (incapable of metamorphosis) occurs only in Termitidae. In the lower families colony labour is performed by nymphs and *pseudergates.

Würm *See* DEVENSIAN.

xanthophyll *See* CAROTENOID.

xenobiotic A chemical substance that enters the environment, where no organism has previously been exposed to it. Many pesticides, cosmetics, cleaning materials, and preservatives are xenobiotics.

xenology In phylogenetics, the study of genes which have been separated by a *horizontal gene transfer event. *Compare* ORTHOLOGY and PARALOGY.

xeric A dry, as opposed to a wet (hydric) or intermediate (mesic) environment.

xeromorphic Applied to organisms which show morphological adaptations that appear to be able to withstand drought.

xerophile A plant of warm, dry *desert environments in Alphonse de *Candolle's (1874) classic temperature-based scheme of world vegetation zones. *Compare* MEGATHERM.

xerophyte A plant (usually a *xeromorph) that can grow in very dry conditions and is able to withstand periods of drought. The adaptations include an ability to store water, waxy leaves and leaves reduced to spines to avoid water loss through *transpiration, and short life cycles (ephemeral) that can be completed when sufficient water is available.

xerosere The characteristic sequence of *communities reflecting the developmental stages of a plant *succession that begins in an arid environment.

xylophagous Wood-eating.

xylophilous Preferring to grow on wood.

yardang A streamlined, wind-sculptured hill ranging in length from metres to kilometres and developed in any bedrock that is at least weakly consolidated. Yardangs are restricted to deserts that have high aridity with minimal plant cover and soil development, and are dominated by strong, unidirectional winds for most of the year.

Yarmouthian The second (0.23–0.17 Ma) of four *interglacial stages recognized in mid-continental North America. It follows the *Kansan glacial episode and is equivalent to the upper part of the *Günz/Mindel Interglacial of the Alps. At various times during the Yarmouthian climates were both warmer and cooler than they are now.

year Not an *SI unit, but a measurement of time that is commonly used in reporting ages. It is usually abbreviated as 'a' (from the Latin *annus*, meaning 'year'), with a prefix indicating a number, e.g. ka (kilo-years, 1 000 years), Ma (mega-years, 10^6 years), and Ga (giga-years, 10^9) years.

yellowcake Concentrated, precipitated, and dried uranium oxide.

yellows A general term for any plant disease in which yellowing or *chlorosis is a characteristic symptom. Yellows diseases are often caused by *viruses but may be caused by bacteria or fungi.

yield–depression curve A graph on which the *drawdown is plotted against the yield of a pumped well or borehole. The resulting plot is invariably curved, and is used to determine the optimum pumping rate for a water supply.

Yoda's power law A numerical description of the process of self-thinning among plant seedlings. Beyond a certain density of sowing, the number of surviving plants is not related to the initial seed density; instead a constant relationship is evident between the density of survivors and their total *biomass. In 1963, K. Yoda and others summarized this relationship as $W = C\varrho^{-3/2}$ where W is the dry weight of surviving plants, ϱ is the density of the surviving plants, and C is a constant reflecting the growth characteristic of the species concerned.

Younger Dryas *See* DRYAS.

Zechstein Sea An Upper *Permian, shallow, gulf sea or depositional sequence of rocks that developed in northern Germany and the North Sea basin. The depositional sequence laid down in this sea consisted of carbonates and evaporites. *See also* SALINE GIANT.

Zelzate *See* DENEKAMP.

zeolites A group of naturally occurring hydrated alumina silicates of sodium, potassium, calcium, and barium, in which the water molecules are weakly held. The hydration–dehydration reaction has some useful applications, as do the ionexchange properties of zeolites.

zephyr A prevailing light and warming breeze from the west at the time of the summer solstice (in the northern hemisphere).

zinc **(Zn)** An element that is required by plants. It is found bound to a variety of enzymes, stabilizing them and also being involved in catalysis. Deficiency in plants prevents the expansion of leaves and internodes, giving a rosette style of plant. It is a growth factor in some rodents and a constituent of certain mammalian enzymes.

Zn *See* ZINC.

zonal Applied to features (e.g. soils and vegetation) characteristic of a particular region that is approximately bounded by lines of latitude (i.e. a region lying parallel to the equator).

zonal flow The winds that blow in a mainly west-to-east or east-to-west direction, and particularly to the main, broad airstreams of the general or large-scale atmospheric circulation (e.g. the zonal westerly winds of middle latitudes). The zonal (or circulation) index is a conventional measurement indicating the strength of the west-to-east airflow over middle latitudes.

zonal fossil *See* INDEX FOSSIL.

zonal index *See* ZONAL FLOW.

zonation **1.** The broad distribution of vegetation according to latitude and altitude. The control is primarily climatic, and similar vegetation zones are encountered on the flanks of high tropical mountains to those found at sea level between the tropics and the poles. **2.** The division of an *ecosystem into distinct vertical layers that experience particular *abiotic conditions. This is particularly clear in the distribution of plants and animals on a rocky seashore, where different species inhabit a series of horizontal strips or belts of the shore, approximately parallel to the water's edge. In many places the strips (zones) are sharply bounded by the differently coloured seaweeds that populate them. **3.** The division of vegetation in relation to a successional sequence (e.g. in sand-dunes), implying that spatial zonation may correspond to temporal processes. This assumption, which may be flawed, is often used as the basis for succession studies.

zone In stratigraphy, a unit of rock characterized by a clearly defined fossil content. *See* INDEX FOSSIL.

zone fossil *See* INDEX FOSSIL.

zone of aeration *See* SOIL-WATER ZONE.

zone of saturation *See* PHREATIC ZONE.

zoochory **(synzoochory)** Dispersal of spores or seeds by animals.

zoocoenosis The secondary producers (consumers) that form part of the *biocoenosis in a *biogeocoenosis.

zoogeographical region *See* FAUNAL REGION.

zoogeographical zone A geographical unit that is distinguished by a fauna which is more or less distinctive at the *species, genus, family, or order level, by virtue of its habitat, or past or present isolation.

zoogeography The study of the geographical distribution of animals at different taxonomic levels, particularly of mammals from the order down to *species level. Emphasis is given to the explanation of distinctive patterns in terms of past and/or present factors, particularly *migration routes.

zooneuston Those animals associated with the surface of water, mainly in freshwater habitats, and influenced by surface tension. The group is divided into those associated with the upper surface of the film, to which the term 'epineustic' is applied, and those associated with the lower surface, described as 'hyponeustic'. Many organisms show adaptation to this environment (e.g. some bugs of the order Hemiptera have pads at the ends of their legs that enable them to walk across the water surface). *See also* NEUSTON.

zoonosis A disease which can be transmitted from animals to humans.

zooplankton *See* PLANKTON.

zooxanthellae Unicellular dinoflagellates that live symbiotically (*see* SYMBIOSIS) with certain corals.

Zurich–Montpellier school of phytosociology (Montpellier school of phytosociology) A group led by J. *Braun-Blanquet and his associates, who developed a set of floristic methods for vegetation classification (in 1927 and later) at Zurich and Montpellier. These have been widely adopted in continental Europe although they are less accepted elsewhere. The aim was to provide a framework for the classification of the vegetation of the world, but in practice the scheme is most useful in regional and national surveys. The approach depends on detailed field surveying to identify vegetation associations, which can then be grouped hierarchically into alliances, orders, classes, etc., with the vegetation circle (global scale) being the most complex hierarchical level. Suffixes added to the genitive stem of the generic names of the plants label the *communities so identified and indicate the hierarchical status of the community:

RANK	ENDING
class	-etea
order	-etalia
alliance	-ion
association	-etum
sub-association	-etosum
variant	(specific name used)

An extensive ecological literature discusses the system and introduces many modifications. The most often quoted objections relate to the use of *homogeneous stands only in the description of vegetation; the concept of *minimal area as it is used to define homogeneity; and in particular to the use of fidelity, and the associated problem of defining *faithful species in order to characterize the associations. *Compare* UPPSALA SCHOOL.

zygote The fertilized ovum of a plant or animal.

Estimates of Population Parameters

1. Population size
   Capture–recapture (for small population sizes)

$$N = s/p \quad \text{where} \quad p = r/s$$

   where
   s is the sample size captured and marked, and the sample size of the second capture;
   r is the number of marked individuals in the second capture;
   N is the population size
2. Measures of diversity
   (*a*) Species diversity:
   (i) Simpson's index

$$D = \frac{1}{\sum_{i=1}^{s} P_i^2}$$

   where
   P_i is the proportion of individuals/biomass to the total sample of the *i*th species;
   s is the total number of species
   (ii) Shannon–Wiener index

$$H = -\sum_{i=1}^{s} P_i \ln P_i$$

   (*b*) Population diversity:
   (i) Nucleotide diversity (π)

$$\pi = \Pi/L$$

   where

$$\Pi = \frac{1}{\left[n(n-1)/2\right]} \Sigma\Pi_{ij}$$

   and

$$E(\Pi) = \theta$$

$$\theta = 4Ne\mathit{u}$$

   where
   L is the length of DNA sequence;
   n is the number of sequences examined;
   Π_{ij} is the number of nucleotide differences between the *i*th and *j*th sequences;
   Ne is the effective population size;
   u is the mutation rate
   (ii) Allele frequency data
   homozygosity of population *j* (H_j)

$$Hj = \sum_{i=1}^{L} p_{ij}^{\ 2}$$

   heterozygosity (allele diversity) of population *j* (h_j)

$$h_j = 1 - H_j$$

   average heterozygosity of s populations (h_s)

$$h_s = \sum_{j=1}^{s} h_j / s$$

gene frequency variations between populations (F_{ST})

$$F_{ST} = (h - h_s)/h \quad \text{where} \quad h = 1 - \sum_{\iota=1}^{L} (\overline{p}_\iota)^2$$

where

L is the number of alleles examined of a given locus;

p_{ij} is the gene frequency of allele *i* in population *j*;

$\overline{p}_i$ is the average gene frequency of allele *i* in s populations

(*c*) Genetic distance:

(i) Genetic distance between two DNA sequences (K) using the Kimura 2 parameter model

$$K = {}^1/_2 \ln a + {}^1/_4 \ln b$$

where $$a = \frac{1}{1 - 2P - Q}$$

and $$b = \frac{1}{1 - 2Q}$$

where

P is the proportion of transitions;

Q is the proportion of transversions;

K is the number of substitutions per base site

(ii) Estimate of genetic distance based on allele frequencies (D)

$$b = 2\sum_{j=1}^{s} \frac{n_j h_j}{s(2\overline{n} - 1)}$$

$$a + b = \frac{\sum_{j=1}^{s} n_j \sum_{i=1}^{L} (p_{ij} - \overline{p}_i)^2}{n(s-1)} + \frac{(2n-1)b}{2n}$$

where $$\overline{n} = \sum_{j=1}^{s} n_j / s$$

and $$n = \frac{\overline{n}s}{s-1} - \frac{\sum_{\varphi=1}^{s} n_\varphi^{\,2}}{\overline{n}s(s-1)}$$

$$\Theta = a/(a+b)$$

$$D = -\ln(1 - \Theta)$$

where

n_j is the sample size of population *j*

Geologic Time-Scale

Eon/Eonothem	Era/Erathem	Sub-era	Period/System	Epoch/Series	Began Ma
PHANEROZIC		*Quaternary*	Pleistogene	Holocene	0.115
				Pleistocene	1.806
	Cenozoic	*Tertiary*	Neogene	Pliocene	5.332
				Miocene	23.03
			Palaeogene	Oligocene	33.9
				Eocene	55.8
				Palaeocene	65.5
	Mesozoic		Cretaceous	Upper	99.6
				Lower	145.5
			Jurassic	Upper	161.2
				Middle	175.6
				Lower	199.6
			Triassic	Upper	228
				Middle	245
				Lower	251
	Palaeozoic	Upper	Permian	Lopingian	260.4
				Guadalupian	270.6
				Cisuralian	299
			Carboniferous	Pennsylvanian	318.1
				Mississippian	359.2
			Devonian	Upper	385.3
				Middle	397.5
				Lower	416
		Lower	Silurian	Pridoli	418.7
				Ludlow	422.9
				Wenlock	428.2
				Llandovery	443.7

Geologic Time-Scale (*Continued*)

Eon/Eonothem	Era/Erathem	Sub-era	Period/System	Epoch/Series	Began Ma
			Ordovician	Upper	460.9
				Middle	471.8
				Lower	488.3
			Cambrian	Furongian	501
				Middle	513
				Lower	542
PROTEROZOIC	Neoproterozoic	Ediacaran			600
		Cryogenian			850
		Tonian			1000
	Mesoproterozoic	Stenian			1200
		Ectasian			1400
		Calymmian			1600
	Palaeoproterozoic	Statherian			1800
		Orosirian			2050
		Rhyacian			2300
		Siderian			2500
ARCHAEAN	Neoarchaean				2800
	Mesoarchaean				3200
	Palaeoarchaean				3600
	Eoarchaean				3800
HADEAN	Swazian				3900
	Basin Groups				4000
	Cryptic				4567.17

Source: International Union of Geological Sciences, 2004.

Note: Hadean is an informal name. The Hadean, Archaean, and Proterozoic Eons cover the time formerly known as the Precambrian. Quaternary is now an informal name and Tertiary is likely to become informal in the future, although both continue to be widely used.

SI Units (Système international d'unités)

Quantity	Name of unit	Symbol	Equivalent	Reciprocal
length	metre	m	3.281 feet	1 ft = 0.3048 m
mass	kilogram	kg	2.2 pounds	1 lb = 0.454 kg
time	second	s		
electric current	ampere	A		
thermodynamic temperature	kelvin	K	1°C = 1.8°F	1°C = 1 K
luminous intensity	candela	cd		
amount of substance	mole	mol		

Supplementary units

Quantity	Unit	Symbol
plane angle	radian	rad
solid angle	steradian	sr

Derived SI units

Quantity	Name of unit	Symbol	Equivalent	Reciprocal
frequency	hertz	Hz		
energy	joule	J	0.2388 calories	1 cal = 4.1868 J
force	newton	N	0.225 pounds force	1 lbf = 4.448 N
power	watt	W	0.00134 horse power	1 hp = 745.7 W
pressure	pascal	Pa	0.00689 pounds force/sq. inch	1 lbf/sq. in = 145 Pa
electric charge	coulomb	C		
electric potential difference	volt	V		
electric resistance	ohm	Ω		
electric conductance	siemens	S		
electric capacitance	farad	F		
magnetic flux	weber	Wb		
inductance	henry	H		
magnetic flux density	tesla	T		
luminous flux	lumen	lm		
illuminance	lux	lx		
absorbed dose	gray	Gy		
activity	becquerel	Bq		
dose equivalent	sievert	Sv		

Multiples used with SI units

Name of multiple	Symbol	Value (multiply by)
atto	a	10^{-18}
femto	f	10^{-15}
pico	p	10^{-12}
nano	n	10^{-9}
micro	μ	10^{-6}
milli	m	10^{-3}
centi	c	10^{-2}
deci	d	10^{-1}
deca	da	10
hecto	h	10^{2}
kilo	k	10^{3}
mega	M	10^{6}
giga	G	10^{9}
tera	T	10^{12}
peta	P	10^{15}
exa	E	10^{18}